Benchmark Papers in Analytical Chemistry

PUBLISHED VOLUMES AND VOLUMES IN PREPARATION

ION-EXCHANGE CHROMATOGRAPHY
Harold F. Walton
THERMAL ANALYSIS
Wesley W. Wendlandt and L. W. Collins
DIGITAL COMPUTERS, PARTS I and II
Thomas L. Isenhour and Joseph B. Justice, Jr.
SOLVENT EXTRACTION
Henry Freiser
FLUORESCENCE AND PHOSPHORESCENCE SPECTROSCOPY
Thomas C. O'Haver

Benchmark Papers in Analytical Chemistry / 1

A BENCHMARK® Books Series

ION-EXCHANGE CHROMATOGRAPHY

Edited by

Harold F. Walton
University of Colorado,
Boulder

Dowden, Hutchinson & Ross, Inc.
STROUDSBURG, PENNSYLVANIA

Distributed by
HALSTED PRESS
A Division of
John Wiley & Sons, Inc.

Benchmark Papers in Analytical Chemistry, Volume 1
Library of Congress Catalog Card Number: 75-31610
ISBN: 0-470-91936-1

78 77 76 1 2 3 4 5
Manufactured in the United States of America

LIBRARY OF CONGRESS CATALOGING IN PUBLICATION DATA
Main entry under title:
Ion-exchange chromatography
(Benchmark papers in analytical chemistry / 1)

Bibliography: p. 427
Includes indexes.
1. Ion exchange chromatography—Addresses, essays, lectures.
I. Walton, Harold Frederic, 1912-
QD79.C453166 544'.923 75-31610
ISBN: 0-470-91936-1

Exclusive Distributor: **Halsted Press**
A Division of John Wiley & Sons, Inc.

ACKNOWLEDGMENTS AND PERMISSIONS

ACKNOWLEDGMENTS

AMERICAN INSTITUTE OF PHYSICS—*Journal of Chemical Physics*
Fractionation of the Lithium and Potassium Isotopes by Chemical Exchange with Zeolites

AMERICAN PHYSICAL SOCIETY—*Physical Review*
New Element Mendelevium, Atomic Number 101

PERMISSIONS

The following papers have been reprinted or translated with the permission of the authors and copyright holders.

AMERICAN ASSOCIATION FOR THE ADVANCEMENT OF SCIENCE—*Science*
The Separation of Purine and Pyrimidine Bases and of Nucleotides by Ion Exchange

AMERICAN ASSOCIATION OF CLINICAL CHEMISTS—*Clinical Chemistry*
Analysis of Urine for Its Ultraviolet-Absorbing Constituents by High-Pressure Anion-Exchange Chromatography

AMERICAN CHEMICAL SOCIETY
Analytical Chemistry
Automatic Recording Apparatus for Use in the Chromatography of Amino Acids
Fast Liquid Chromatography: An Investigation of Operating Parameters and the Separation of Nucleotides on Pellicular Ion Exchangers
Ion Exchange Chromatography of Amino Acids. Effect of Resin Particle Size on Column Performance
Ion Exchange Selectivity Scales for Cations in Nitric Acid and Sulfuric Acid Media with a Sulfonated Polystyrene Resin
Ligand Exchange Chromatography of Amphetamine Drugs
Rapid Analysis of Ribonucleosides and Bases at the Picomole Level Using Pellicular Cation Exchange Resin in Narrow Bore Columns
Journal of the American Chemical Society
The Anion-Exchange Separation of Ribonucleotides
The Anion-Exchange Separation of Zirconium and Hafnium
Anion Exchange Studies: I. Separation of Zirconium and Niobium in HCl–HF Mixtures
Anion Exchange Studies: IV. Cobalt and Nickel in Hydrochloric Acid Solutions
Anion Exchange Studies: VI. The Divalent Transition Elements Manganese to Zinc in Hydrochloric Acid
The Chemical Identification of Radioisotopes of Neodymium and of Element 61
The Exchange Adsorption of Ions from Aqueous Solutions by Organic Zeolites: IV. The Separation of the Yttrium Group Rare Earths

Ion Exchange as a Separations Method: IV. A Theoretical Analysis of the Column Separations Process
An Ion-Exchange Study of Possible Hybridized 5f Bonding in the Actinides
A Laboratory Method for Separating Nitrogen Isotopes by Ion Exchange
Production of Bromite and Perbromate Ions in Crystalline $CsBrO_3$ Irradiated with Cobalt-60 γ Rays and the Vibrational Spectrum of the BrO_4^- Ion
The Separation of Monosaccharides by Ion Exchange
The Separation of Rare Earths by Ion Exchange: I. Cerium and Yttrium
The Separation of Rare Earths by Ion Exchange: VI. Conditions for Effecting Separations with Nalcite HCR and One-Tenth Per Cent Citric Acid–Ammonium Citrate Solutions

Journal of Chemical Education
Travel of Molecules and Disturbances in Chromatographic Columns: A Paradox and Its Resolution

The Journal of Physical Chemistry
An Improved Method for Operating Ion-Exchange Resin Columns in Separating the Rare-Earth Elements
Salting-Out Chromatography: I. Alcohols

AMERICAN SOCIETY OF BIOLOGICAL CHEMISTS, INC.—*The Journal of Biological Chemistry*
Chromatography of Amino Acids on Sulfonated Polystyrene Resins

THE CHEMICAL SOCIETIES IN DENMARK, FINLAND, NORWAY, AND SWEDEN—*Acta Chemica Scandinavica*
An Automated Procedure for Separation of Monosaccharides on Ion Exchange Resins
Die Verteilung von Äthylalkohol und Wasser bei Ionenaustauschern auf Harzbasis
Die Verteilung von Glukose bei Ionenaustauschern auf Harzbasis in Äthylalkohol-Wassergemischen / The Distribution of Glucose Between Ion-Exchange Resins and Alcohol–Water Mixtures

ELSEVIER SCIENTIFIC PUBLISHING COMPANY

Analytica Chimica Acta
Accurate Determination of Traces of Sodium and Potassium in Rocks by Ion Exchange Followed by Atomic Absorption Spectroscopy
Anion-Exchange Separation of Iron, Cobalt, and Nickel
Distribution Coefficients and the Cation-Exchange Behaviour of Elements in Hydrochloric Acid—Ethanol Mixtures
Separation of Noble Metals from Base Metals by Means of a New Chelating Resin
Solubilization Chromatography: III. Ethers, Carboxylic Acids, and Hydrocarbons

Journal of Chromatography
Ligand-Exchange Chromatography of Racemates: Influence of the Degree of Saturation of the Asymmetric Resin by Metal Ions on Ligand Exchange
Separation of Some Lower Oxy-Anions of Phosphorus by Anion-Exchange Chromatography

THE FARADAY SOCIETY—*Transactions of the Faraday Society*
Theory of Chromatography: Part 9. The "Theoretical Plate" Concept in Column Separations

MACMILLAN JOURNALS LTD.—*Nature*
"Ligand Exchange": A Novel Separation Technique

MASSON ET CIE—*Analusis*
Théorie du développement par déplacement en chromatographie d'échange d'ions: application à la séparation de corps très voisins / Theory of Development by Displacement in Ion-Exchange Chromatography: Application to the Separation of Very Similar Substances

MICROFORMS INTERNATIONAL MARKETING CORPORATION—*Talanta*
Non-aqueous Solvents in Anion-Exchange Separations

PRESTON TECHNICAL ABSTRACTS COMPANY—*Journal of Chromatographic Science*
High Speed Separations of Nucleotides and Nucleic Acid Bases by Column Chromatography Using Controlled Surface Porosity Ion Exchangers

SPRINGER-VERLAG
Mikrochimica Acta
Anionenaustauschtrennungen der mit Tributylphosphat extrahierbaren Elemente: I. Trennung des Urans vom Eisen in Gegenwart von Tributylphosphat / Anion-Exchange Separations of Elements Extractable by Tributyl Phosphate: I. Separation of Uranium from Iron in the Prescence of Tributyl Phosphate
Die Naturwissenschaften
Kontinuierliche Aminosäuretrennung an einer Austauschersäule durch Gradientenelution mit einem Puffer / Continuous Amino Acid Analysis of an Ion-Exchanging Column by Gradient Elution with a Buffer

SWEDISH CHEMICAL SOCIETY—*Svensk Kemisk Tidskrift*
Über die Verwendung von basenaustauschenden Stoffen in der analytischen Chemie II / On the Application of Base-Exchanging Substances in Analytical Chemistry II

UNITED NATIONS INTERNATIONAL ATOMIC ENERGY AGENCY
Proceedings of the International Conference on the Peaceful Uses of Atomic Energy Held in Geneva 1955
Anion Exchange Studies of the Fission Products
Proceedings of the Second United Nations International Conference on the Peaceful Uses of Atomic Energy
Ion Exchange Properties of Hydrous Oxides

PREFACE

Separations are an important part of chemical analysis, and ion-exchange chromatography is a powerful tool for chemical separations. Columns of ion-exchanging materials are used to separate all manner of inorganic ions from one another, as well as organic ions and organic compounds that are not ionized at all. The wide scope of ion-exchange chromatography is the product of the labors of many, many people. It has grown in response to practical needs, and it is hard to identify publications that mark novel or radical advances. Certain trends may be noted, however. I have divided the papers of this collection into twelve groups, each group describing a particular kind of activity or trend, plus one introductory paper that may mark the earliest use of an ion-exchange resin in chemical analysis.

Certain names stand out—the names of pioneers and innovators and the names of persons who have made many contributions to the development of ion-exchange chromatography. To illustrate the work of a particular researcher I have generally chosen the first paper of a series, because the first paper usually sets out the aims of the investigation and is the easiest for the general reader to understand; besides, it is usually quite brief. There is much to be said for the historical approach to a new subject, and I hope that this collection may serve as a guidebook for newcomers to this highly practical field.

I have restricted myself to chromatography in columns; thin-layer and paper chromatography are not included. The theory of chromatographic columns is developed in the first group of papers. The theory of the ion-exchange process itself is not discussed. The specialist who wants to know more about particular separations is advised to consult the reviews of ion-exchange chromatography that have appeared in even-numbered years in the "Fundamental Reviews" issues of *Analytical Chemistry*.

It is unfortunate that so few of the papers reproduced here are in languages other than English. This lack may bespeak a narrowly parochial point of view, but one must recognize that ion-exchange chromatography has developed in response to practical needs, particularly in the area of atomic energy, and that these developments have occurred principally in English-speaking countries. (One must not overlook Scandinavia, but the Scandinavians usually publish in English!) A few papers in German and French are included, and with each I have provided a partial translation. The translations are not intended to be literal but are meant to tell as truthfully as possible what the author has to say.

I want to thank the authors who so generously sent reprints and figures, and in particular F. H. Spedding, G. E. Boyd, and O. Samuelson, who sent much additional material and wrote long letters about the development of ion-exchange chromatography as they saw it. It would have been easy to make this volume twice as long.

HAROLD F. WALTON

CONTENTS

PART V: SEPARATION OF METALS BY ANION EXCHANGE OF CHLORIDE AND FLUORIDE COMPLEXES

PART VI: MIXED SOLVENTS IN ANION AND CATION EXCHANGE

PART VII: CHROMATOGRAPHY OF INORGANIC IONS

PART XII: CHROMATOGRAPHY OF NONIONIC COMPOUNDS

CONTENTS BY AUTHOR

ION-EXCHANGE CHROMATOGRAPHY

INTRODUCTION

Ion exchange was discovered in 1850, nearly 40 years before Arrhenius proposed his theory of ionization. Its discoverers, two English agricultural chemists, J. T. Way (Way, 1850) and H. S. Thompson (Thompson, 1850) called it "base exchange." Way's paper, "On the Power of Soils to Absorb Manure," described how a solution of potassium chloride, percolated through a bed of soil, gave up its potassium to the soil and received in its place calcium and magnesium in amounts that taken together were chemically equivalent to the potassium that had been absorbed. A similar experiment was made with ammonium sulfate.

"Chromatography" was so named by Mikhail Tswett (Tswett, 1906), who in 1905 used a column of precipitated calcium carbonate to separate the pigments of green leaves. He extracted the dried leaves with petroleum ether and poured the extract into the top of the column. He saw that the pigments were absorbed, forming a thin, dark-green disc. He poured more petroleum ether into the column. The pigments were gradually washed down the column, and as they moved, they separated into different colored zones. The lowest, fastest moving zone was red. Above it was a yellow zone, and above this were two green zones, one bluish green, the other a yellowish green. They contained, respectively, carotin, xanthophyll, chlorophyll A, and chlorophyll B.

As more solvent was passed, the carotin flowed out of the column and was collected in a receiver. The other constituents could be collected in turn. Movement could be accelerated by mixing a little alcohol, a *polar solvent,* with the petroleum.

Since Tswett's work, the name "chromatography" has been extended to apply to the separation of any dissolved constituents, whether colored or not, by partition between an absorbent and a flowing solvent. It is not necessary to see colors in the column, but only to detect and identify the dissolved substances as they emerge from the column. It probably occurred to several people to use ion-exchanging materials as absorbents in chromatography, but the only such absorbents that were available before 1935 were the *synthetic zeolites,* dried aluminosilicate gels that were used to soften water. These materials were attacked by dilute acids and alkalies, which seriously limited their use in chromatography. Nevertheless, some analytical uses were devised, notably Folin's method for determining ammonium ions in urine (Folin and Bell, 1917). The urine was shaken with powdered zeolite, which absorbed ammonium ions, releasing sodium ions in exchange. The zeolite was separated, washed, and treated with dilute sodium hydroxide to release the ammonia, which was measured colorimetrically with Nessler's reagent.

In 1938, T. I. Taylor and H. C. Urey (Paper 12 in this volume) achieved a partial separation of the isotopes of lithium and potassium on a 100-ft-long, column of aluminosilicate exchanger mounted in the stairwell of the chemistry building of Columbia University. Their paper is included in this collection. It is noteworthy that this, the first real attempt at ion-exchange chromatography, should have been for a task so difficult as the separation of isotopes.

A new era in ion exchange began in 1935, when the first organic ion exchangers, the *ion-exchange resins,* were made by B. A. Adams and E. L. Holmes (Adams and Holmes, 1935) at the National Chemical Laboratory, Teddington, England. They were condensation products of formaldehyde with phenol and with phenolsulfonic acid. They were stable in acid solutions, and could exchange hydrogen ions just as easily as any other ions. Other acid-resistant cation exchangers were made by treating bituminous coal with fuming sulfuric acid. These materials, as well as the phenolic resins, had the drawback of being unstable in alkaline solutions. Nevertheless, some early attempts at ion-exchange chromatography were made with these materials, such as the partial separation of yttrium, lanthanum, and neodymium by R. G. Russell and D. W. Pearce (Russell and Pearce, 1943) and of cadmium, copper, and nickel by Kozak and Walton (Kozak and Walton, 1945).

The first person to investigate systematically the use of ion exchange in chemical analysis was Olof Samuelson. In 1939, he published a paper (Samuelson, 1939) on the separation of interfering cations and anions, for example iron and phosphate, by absorbing the cation (iron) on a resinous cation exchanger and replacing it in the solution by an ion that would not interfere in the analysis, such as the hydrogen ion. He also showed that one could measure the total electrolyte concentration of a solution by passing the solution through a cation-exchange resin in the hydrogen form, converting the dissolved salts to their corresponding acids, and then titrating these acids. In a second paper, which is reproduced here (Paper 1), he described the use of an ion-exchanging column for measuring sulfate in a solution of aluminum sulfate and sodium in a vanadium catalyst (after converting the vanadium to the vanadate anion).

Samuelson's doctoral thesis (Samuelson, 1944) is a classic document in ion-exchange chromatography. It was followed by two books, *Ion Exchangers in Analytical Chemistry* (Samuelson, 1953) and *Ion Exchange Separations in Analytical Chemistry* (Samuelson, 1963). More recently, Samuelson and his associates have developed methods for the chromatography of carbohydrates on ion-exchange resin columns, and three of his papers on this topic are included in our collection (Papers 41–43).

In 1939, with the outbreak of World War II, came the development of atomic energy for military purposes, primarily under the Manhattan Project in the United States. The fission of uranium gave rise to most of the elements in the periodic table, and, in particular, elements of the rare-earth or lanthanide series. These had to be separated, and researchers turned to ion-exchange chromatography for this purpose. Two groups worked on this project, one at Ames, Iowa, under the leadership of F. H. Spedding (see Spedding, 1947) the other at Oak Ridge, Tennessee, under G. E. Boyd. Both groups worked independently for reasons of secrecy, and it is difficult and pointless to establish priorities, but is appears the Oak Ridge group was the first to separate tracer amounts of the rare earths, whereas the Ames group was the first to separate relatively large amounts. After the war the work was declassified and presented in a series of papers at a symposium at the national meeting of the American Chemical Society in New York, September 1947. The papers were published in the November 1947 issue of *Journal of the American Chemical Society.* Four of these papers are included in collection (Papers 2 and 6–8).

One of these papers, that by Ketelle and Boyd (Paper 7) describes the use of a new resin called Dowex-50, which, with minor modifications, is the standard resin used today for cation exchange. It is made by polymerization, not condensation. The method of production yields spherical beads, unlike the brittle gels formed by condensation. The beads can be made in any desired diameter, from a few microns up to 1 to 2mm, and they are chemically stable over a very wide range of conditions, even in strong alkali. They are made by copolymerizing styrene and divinylbenzene; the more divinylvenzene used, the denser (more highly crosslinked) is the polymer network. After the polymer is produced, ionic functional groups of various kinds are introduced by chemical treatment. Cation exchangers of the type of Dowex-50 are made by sulfonation. By introducing quaternary ammonium groups, a strongly basic anion exchanger is produced.

Ion-exchange resins based on crosslinked polystyrene were first produced by G. F. D'Alelio in 1944 (D'Alelio, 1945) and were the subject of several important patents.

The separation of the rare-earth elements by ion exchange did not rely on differences in the affinities of the ions for the resin. Rather, the resin was used as an undiscriminating support to hold the ions in place while another process discriminated among the ions and produced the separation. This process was the formation of complex ions in the solution. Citrate was added to the eluent. Citrate ions form uncharged complexes with the trivalent lanthanide ions; the most stable complex is formed with the smallest cation, that of lutecium. Lutecium therefore came out of the column first; lanthanum, with the largest cation and the least stable citrate complex, came out last.

Most inorganic ion-exchange separations, in fact, rely on the formation of complex ions in the solution. Formation of uncharged or negatively charged complex ions works to pull metal ions off a cation-exchange resin; the same process acts to bind metal ions to an anion-exchange resin. The anion exchange of metal chloride complexes in hydrochloric acid solutions is one of the most potent and versatile ways of separating metal ions in solution. It was developed primarily by K. A. Kraus, F. Nelson, and others at Oak Ridge National Laboratory (Paper 19).

It is impressive to note how closely the developments in ion-exchange chromatography have followed developments in atomic energy. Fully half the papers chosen for this Benchmark volume have come from laboratories concerned with atomic energy research. In the Manhattan Project, Boyd, Coryell, Marinsky, Schubert, Spedding, and others separated the rare earths (see Part II), while Cohn (Papers 31 and 32) and Khym (Paper 40) used ion exchange to attack biological problems related to atomic energy research. The first theoretical studies of ion-exchange columns were done in atomic energy laboratories in the United States and Britain by Mayer, Tompkins, and Glueckauf (Papers 2 and 3). Nuclear fission led to the discovery of transuranium elements; these were separated by ion exchange; in one spectacular case, the discovery of the element mendelevium, it was predicted where this element would emerge from an ion-exchange column, and the element was identified by the appearance of only five atoms. This was the work of Ghiorso, et al. (Paper 11). R. M. Diamond (Paper 10) has also made fundamental studies of ion-exchange selectivity and the behavior of complex ions in ion-exchange resins.

Continuing the "roll call" of contributions from atomic energy laboratories, we note the work of K. A. Kraus and his collaborators (Paper 28) on inorganic ion exchangers based on the hydrous oxides of transition metals. Zirconium oxide, zirconium phosphate, and zirconium tungstate have been intensively studied and widely used. Their advantages are (1) resistance to radiation, and (2) high selectivity for the heavier alkali and alkaline earth metal ions. They separate rubidium and cesium from one another and from potassium and sodium, and are used to remove radioactive cesium and strontium from fission products. The selectivity is a property of the exchangers themselves, and does not depend on complex ions in solution. In fact, the alkali and alkaline earth ions form very few complexes in solution.

Important pioneer work on inorganic exchangers was performed by C. B. Amphlett (Amphlett, 1958) at the Atomic Energy Research Establishment, Harwell, England. A great number of hydrous oxide combinations with ion-exchanging properties have been prepared over the last 15 years, many of them by M. Qureshi and his group in India (Qureshi et al., 1972).

From the French atomic energy laboratories comes the work of R. Rosset on the separation of the boron isotopes (Paper 14), as well as Rosset's important studies of the selectivities of chelating resins and other resins having special functional groups.

Of the biomedical implications of atomic energy we have mentioned the continuing work of Cohn on the chromatography of nucleic acid derivatives by ion exchange. The last group of papers in this collection, Part XII, includes a contribution from Oak Ridge National Laboratory by C. D. Scott (Paper 46) on the chromatography of ultraviolet-absorbing constituents of urine. This paper is a monumental contribution to clinical chemical analysis and illustrates a very important current trend, the use of ion-exchange resins as stationary phases for liquid chromatography of nonionic organic compounds.

Related to nuclear energy through the analysis of ores and intermediates in the production of nuclear fuel is the extensive work of J. Korkisch and his school at the University of Vienna, Austria. They have exploited the use of mixed solvents in ion-exchange separations. Korkisch was not the first to demonstrate the profound effects of the solvent on anion and cation exchange; this honor probably belongs to J. S. Fritz (see Paper 20) of the Ames Laboratory for Atomic Research at Iowa State University. Korkisch, however, has published a vast amount of data on distribution ratios in many solvent systems for both anion and cation exchange, and he has described many separation schemes, some of which, but by no means all, are collected in his book *Modern Methods for the Separation of Rarer Metal Ions* (Korkisch, 1969). He calls his approach "combined ion exchange and solvent extraction." It is described in more detail in Papers 21 and 22.

The use of ion exchange to analyze rocks and minerals has been brought to high precision by F. W. E. Strelow and his associates at the National Chemical Laboratory in Pretoria, South Africa (Strelow, 1973).

They believe that if one needs a really reliable and accurate analysis of the chemical constituents of a complex mixture, such as a rock, there is no substitute for separating the constituents, isolating each in a pure state, and then measuring its amount. Most metals are best measured by titration with EDTA unless the amount is small, when a method such as atomic absorption spectrometry may be better. As everyone knows who has worked in rock analysis, the results of "round robin" analyses of the same material by different laboratories are sometimes discouraging and even chaotic, particularly where minor constituents are concerned. Strelow showed that ion-exchange separations can bring some order out of chaos, particularly if reagent blanks are controlled.

Most of Strelow's separations use stepwise changes of eluent; that is, one reagent, or one particular concentration of reagent, is used to bring one element out of the column, then another reagent or a different concentration of the same reagent is used to bring the next element out, and so on. In his earlier work Strelow used organic complexing agents, such as oxalate and malonate; more recently he has relied on mixed aqueous-nonaqueous solvents and simple inorganic acids, like hydrochloric and nitric, that are easily purified. Before the column separations were developed, exhaustive studies were made of distribution ratios.

Ion-exchange chromatography is an eminently practical matter, and it has developed to meet the needs of industry and medicine. The Swedish paper industry provided the encouragement for the work of Olof Samuelson and his colleagues on the chromatography of carbohydrates, using ion-exchange resins and water–alcohol mixtures as solvents. This work is described in Papers 41–43 and illustrates the use of ion-exchange resins as selective absorbents for nonionized compounds. Probably the most common example of ion-exchange chromatography as a routine analytical tool is the analysis of amino acid mixtures, which is so important in biochemical and medical studies. It was developed to a high degree of perfection by D. H. Spackman, W. H. Stein, and S. Moore (Paper 37) at the Rockefeller Institute for Medical Research in New York, and by P. B. Hamilton (Paper 38) at the A. I. DuPont Research Institute at Wilmington, Delaware.

We have already mentioned great activity today in the chromatography of organic compounds on ion-exchange resins. Resins are good stationary phases for this purpose because the ionic groups become hydrated (or solvated), causing the resin to swell and become permeable; the nonionic portion, the hydrocarbon polymer network, acts as an absorbent or solvent for organic compounds, even if these compounds are not themselves ionized. Weak acids may be bound by cation-exchange resins at low pH values but not at higher pH values, because even a small degree of ionization brings about the exclusion of their anions from the resin on account of the Donnan effect. Singhal and Cohn (1972) pointed out that this exclusion can differentiate between acids of different strengths. Analogous pH effects are found in the absorption of weak bases on cation- and anion-exchange resins.

The affinities of nonionized and even ionized compounds on ion-exchange resins depend on several factors, including quite subtle differences in the resin polymer structure, and are thus very hard to predict. Nevertheless, differences exist, making possible a wide range of chromatographic separations.

Owing to present-day liquid chromatography techniques, large affinity differences are not needed to make effective separations possible. The important thing is to get narrow bands, and this is possible by using spherical resins of small and uniform particle size. Sensitive detection devices are also available for organic compounds. Gradient elution through solvent programming is, of course, routine.

William Rieman III, of Rutgers University, was one of the first to explore the absorption and chromatography of nonionic organic compounds on ion-exchange resins. He showed the importance of the solvent and that the difference in behavior between cation- and anion-exchange resins was relatively minor.

Rieman also spent much time trying to separate optical isomers by ion-exchange chromatography. In recent years, some success has been achieved in this area, notably by Davankov, who employed a form of ligand-exchange chromatography to separate *d*- and *l*-amino acids. Ligand exchange is one of the newer ways of using ion-exchange resins to separate organic compounds. A cation-exchange resin is loaded with ions of a metal, such as copper, that forms complexes with electron-donor molecules. The donor molecules, or ligands, are generally eluted with ammonia. Examples of this technique are cited in Part XII, along with Rieman's pioneer papers on "solubilization" and "salting-out" chromatography (Papers 44 and 45).

As a final comment on current directions in ion-exchange chromatography, we must note that the distinctions between ion-exchange and non-ion-exchange are becoming blurred. If the only function of the fixed ions and their counter-ions is to promote swelling and penetration by organic molecules, it may be that nonionic polymers with a macroporous (or "macroreticular") structure would serve just as well. Such polymers are, in fact, very effective as collectors to absorb trace concentrations of drugs from urine, or traces of phenolic contaminants from potable water. Polystyrene-type resins are used, and these have an affinity for aromatic solutes. If their structure is sufficiently homogeneous, they could be used as column packings for liquid chromatography.

The worker in liquid chromatography today has many types of packings at his disposal, and he will chose the one that serves his purpose best. This may be a gel-type ion-exchange resin, a macroporous resin, or a pellicular resin; it may be a surface-bonded packing, with carbon chains chemically attached to a silica support; and packings may be effective that have no ionic groups at all. It is out of place in a Benchmark volume to speculate about future developments.

It has been hard to select individual papers for inclusion in this volume, because ion-exchange chromatography today is the result of many contributions, most of them rather trivial in themselves, but each of them advancing the state of the art in response to a specific need. I have tried to identify the scientists who have contributed most to the field, then selected, perhaps arbitrarily, one or more of their papers to illustrate their work. Even the choice of investigators was to some extent arbitrary, and I particularly regret not citing more Russian, Japanese, and Eastern European sources. The preponderance of English-language references is natural, because, for historical reasons, the ground-breaking developments in ion-exchange chromatography accompanied those in atomic energy, which occurred in English-speaking countries. Nevertheless, I think that a Russian, Japanese, or Hungarian reviewer might have chosen a different set of Benchmark papers.

REFERENCES

Adams, B. A., and E. L. Holmes. 1935. *J. Soc. Chem. Ind. (Lond.), 54T,* 1

Amphlett, C. B. 1958. *Proc. 2nd Intern. Conf. Peaceful Uses Atomic Energy, 28,* 17

D'Alelio, G. F. 1945. U.S. Patents 2,366,007 and 2,366,008 (Dec. 26, 1945).

Folin, O., and R. D. Bell. 1917. *J. Biol. Chem., 29,* 327.

Korkisch, J., 1969. *Modern Methods for the Separation of Rarer Metal Ions,* Elmsford, N.Y., Peragamon Press.

Kozak, R., and H. F. Walton. 1945. *J. Phys. Chem., 49,* 471.

Qureshi, M., S. A. Qureshi, J. P. Gupta, and H. S. Rathore. 1972. *Separation Sci., 7,* 615.

Russell, R. G., and D. W. Pearce. 1943. *J. Amer. Chem. Soc., 65,* 595.

Samuelson, O. 1939. *Z. Anal. Chem., 116,* 328.

——. 1944. *Studier rörande Jonbytande fasta Ämnen.* Esselte Aktiebolag, Stockholm.

——. 1953. *Ion Exchangers in Analytical Chemistry,* New York, Wiley.

——. 1963. *Ion Exchange Separations in Analytical Chemistry,* New York, Wiley.

Singhal, R. P., and W. E. Cohn. 1972. *Anal. Biochem., 45,* 585.

Spedding, F. H. 1947. *Discussions of the Faraday Society, 7,* 214.

Strelow, F. W. E. 1973. *Ion Exchange and Solvent Extraction,* Vol. 5, Chap. 2, New York, Marcel Dekker.

Thompson, H. S., 1850. *J. Roy. Agric. Soc. Engl., 11,* 68.

Tswett, M., 1906. *Ber. Deut. Botan. Ges., 24,* 316, 384 (A translation of part of this paper, by L. S. Ettre, appears in *Anal. Chem., 43,* 20A, 1971.)

Way, J. T. 1850. *J. Roy. Agric. Soc. Engl., 11,* 313.

Editor's Comments on Paper 1

1 SAMUELSON
Über die Verwendung von basenaustauschenden Stoffen in der analytischen Chemie II
English translation: *On the Application of Base-Exchanging Substances in Analytical Chemistry II*

Any account of ion exchange in chemical analysis must include the name of Olof Samuelson. The paper reproduced here is his second paper, which is more detailed than the first, and appeared in a less accessible journal than the journal that carried his first paper (Samuelson, 1939).

One might question whether the subject matter is really ion-exchange chromatography. Paper 1 does, however, concern the use of ion-exchange columns in chemical separations. It describes the measurement of total electrolyte concentration by a method that is now routine, and that is especially useful in analyzing natural waters. It also describes the removal of interfering ions, in this case the separation of sodium ions from vanadate ions in order to measure the sodium content of a vanadium catalyst. It is noteworthy that today, when selective spectroscopic techniques such as atomic absorption are available, ion-exchange separations are still used where trace amounts of alkali-metal cations are to be measured in complex mixtures (see Part VII).

Ion-exchange equilibria and the *salzfreie Wasserhaut* (salt-free water layer) are discussed. Samuelson saw that the Donnan equilibrium applied to the distribution of dissolved ions between the surrounding solution and the solution within the exchanger, making the solution in the exchanger contain a relatively small proportion of co-ions, or ions having the same charge sign as the fixed ions of the exchanger. Hydration of the fixed ions likewise excludes neutral salt from the fluid in the exchanger.

REFERENCE

Samuelson, O. 1939. *Z. Anal. Chem.*, *116*, 328.

1

Reprinted from *Svensk Kem. Tidskr.*, **51**, 195–206 (1939)

Über die Verwendung von basenaustauschenden Stoffen in der analytischen Chemie II.*

Von *Olof Samuelson*.

Einleitung.

In der ersten Mitteilung dieser Serie (1) wurde gezeigt, dass wir in den organischen Basenaustauschern ein analytisches Hilfsmittel haben, das auf verschiedenen Gebieten verwendbar ist. Die Analysenmethodik gründet sich darauf, dass man mit Hilfe eines geeigneten Organolites[1] einfach und schnell sämtliche Kationen einer Probelösung gegen z. B. Wasserstoffionen austauschen kann. Aus dem Organolit können, wenn die Kationen bestimmt werden sollen, diese durch Behandeln mit Säure oder einer geeigneten Salzlösung verdrängt und quantitativ bestimmt werden.

Von den früheren Veröffentlichungen über die Verwendung von basenaustauschenden Stoffen für analytische Zwecke wurde nur die Arbeit

[1] Nach dem Vorschlag von *H. Burell* (Ind. Eng. Chem. Industr. Edition *30* [1938], 358) wird das Wort Organolit als Bezeichung für die technischen organischen Basenaustauscher verwendet.

**Editor's Note:* An English translation follows this article.

von *C. Kullgren* (2) zitiert. Jedoch haben auch einige andere Autoren anorganische Basenaustauscher für spezielle Analysen angewandt. Schon im Jahre 1917 wurde eine Arbeit von *Folin* (3) veröffentlicht, der einen Basenaustauscher zur Bestimmung des Ammoniaks im Harn verwandte. *A. Bahrdt* (4) hat ferner in seiner Arbeit über eine volumetrische Bestimmungsmethode von Schwefelsäure im Trink- und Gebrauchswasser die Calcium- und Magnesiumionen des Wassers mit Hilfe eines anorganischen, mit Natriumionen gesättigten Basenaustauschers gegen Natrium ausgetauscht. Die Sulfationen wurden mit einer abgemessenen Menge einer Bariumchloridlösung ausgefällt und der Überschuss an Bariumchlorid nach der bekannten Palmitatmethode von *Blacher* zurücktitriert. Mit Hilfe der jetzt bekannten organischen Basenaustauscher würde sich dieser Umsatz ohne Zweifel viel bequemer und mit grösserer Sicherheit durchführen lassen.

In dieser Mitteilung sollen einige spezielle Methoden beschrieben werden, welche seit einiger Zeit im hiesigen Laboratorium im Gebrauch sind. Zunächst soll aber eine allgemeine Frage erörtert werden, welche für die Analysenmethodik von grosser Bedeutung ist.

Über die Ionenkonzentrationen der festen Phase und ihre Bedeutung für die Analysenmethodik.

Wenn man einen trockenen Basenaustauscher mit einer Salzlösung schüttelt, wird die Konzentration der Lösung grösser als die der ursprünglichen. Die sauren Gruppen des Basenaustauschers werden dabei hydratisiert und nehmen einen Teil des Wassers in Anspruch, eine Erscheinung, die u. a. von *F. Kurmies* und *B. Alten* (5), *C. Du Rietz* (6), *U. Hofmann* und *K. Giese* (7) an verschiedenen basenaustauschenden Stoffen studiert worden ist. Über die Natur dieser Hydratation gehen die Ansichten noch auseinander, und die meisten Autoren haben sich deshalb damit begnügt, die von dem Basenaustauscher aufgenommene Wassermenge als eine salzfreie Wasserhaut zu rechnen. Die salzfreie Wasserhaut ist eine Fiktion weil in dem Wasservolumen, das den austauschfähig gebundenen Kationen zur Verfügung steht, sämtliche Ionen der freien Lösung — zwar in anderen Konzentrationen als in der Aussenlösung — vorhanden sein dürften. *U. Hofmann* und *K. Giese,* die in dieser Beziehung verschiedene Tonmineralien untersucht haben, sprachen die Vermutung aus, dass die Kationen des Tons nahe der Tonoberfläche häufiger wären als in grösserem Abstand. In der Nähe der Tonoberfläche würde die Salzkonzentration geringer sein als z. B. in der Aussenlösung. Mit steigendem Abstand von der Tonoberfläche würde die Salzlösung *allmählich* konzentrierter werden, in dem Masse, als die Zahl der Kationen des Tons abnimmt, bis die Konzentration der freien Lösung erreicht ist.

Andere Autoren, u. a. *C. Du Rietz,* nehmen an, dass die Ionen des Basenaustauschers sich in einer homogenen, von der Aussenlösung ziemlich scharf abgetrennten Lösung befinden. Diese Lösung wollen wir als die f-Lösung bezeichnen. In den Grenzschichten zwischen der f-Lösung und der Aussenlösung haben wir selbstverständlich einen Konzentrationsausgleich. Eine rechnerische Ermittelung von den Konzentrationen der f-Lösung unter Berücksichtigung dieses Ausgleichs muss jedenfalls äusserst kompliziert sein, und deshalb muss darauf verzichtet werden. Unter Annahme von Homogenität der f-Lösung und der von dieser scharf abgetrennten Aussenlösung können wir auf die Ionenverteilung zwischen diesen beiden Lösungen die *Donnan*-Gleichung anwenden. Nach den Untersuchungen von *C. Du Rietz* ist diese Berechnungsweise als wohlbegründet anzusehen und gibt eine Möglichkeit, die Ionen-Konzentrationen der f-Lösung zu berechnen.

$$\left[\frac{(A_f)}{(A_w)}\right]^{\frac{1}{a}} = \left[\frac{(B_f)}{(B_w)}\right]^{\frac{1}{b}} = ----- = \left[\frac{(P_w)}{(P_f)}\right]^{\frac{1}{p}} = ----$$

wo (A), (B), — — — die Aktivitäten der Kationen mit den Valenzen a, b, — — — und (P), — — — die der Anionen mit den Valenzen p — — — bezeichnen. Die Indizes f und w bezeichnen die Zugehörigkeit zur f-Lösung bzw. zur Aussenlösung.

Aus der Konzentrationserhöhung der Aussenlösung beim Schütteln einer Salzlösung bekannter Konzentration mit einem trockenen Basenaustauscher kann, wie oben erwähnt worden ist, die salzfreie Wasserhaut bestimmt werden. Da die Konzentrationsunterschiede oft recht klein sind, ist es besser, die Probe abzunutschen und die Ionenmenge der am Basenaustauscher haftenden Lösung zu bestimmen und mit der Behandlungsflüssigkeit zu vergleichen. Nach der Methode, die von *C. Du Rietz* verwendet wurde, schlämmt man die abgenutschte Probe in einer recht konzentrierten Lösung eines indifferenten Elektrolyten auf, wodurch die ungleichmässige Ionenverteilung grössenteils und berechenbar aufgehoben werden soll. Von dieser Aufschlämmung wird eine Probe genommen und die Lösung analysiert. Nach dieser Bestimmung kann die anhaftende Ionenmenge berechnet und somit die salzfreie Wasserhaut ermittelt werden. Die ungleichmässige Ionenverteilung lässt sich durch diesen Zusatz nicht vollständig aufheben und deshalb müssen Korrektionen eingeführt werden. Eine Schätzung der Korrektion in dieser konzentrierten Lösung dürfte aber sehr unsicher sein.

Bei den technischen Organoliten sind, soweit bekannt, Untersuchungen von der Hydratation nicht veröffentlicht worden. Wegen deren Bedeutung für die Analysenmethodik ist es aber von grossem Interesse, dieses Problem zu studieren. Ein neues Verfahren zur Bestimmung der salzfreien Wasserhaut wurde deshalb ausgearbeitet. Dieses hat gegenüber der Methode von *C. Du Rietz* den Vorteil, dass die unsicheren Korrektionen fortfallen. Die Bestimmung lässt sich am besten durch ein Beispiel erklären.

4,228 g eines mit Wasserstoffionen gesättigten Sulfonsäureorganolites wurde in einem Filtertiegel (Gewicht 23,305 g) eingewogen und 0,05-n Salzsäure so lange durchfiltriert, bis die Säure unverändert ablief. Dann wurde die Säure abgesaugt und der Tiegel sofort gewogen. (Gefunden: 35,238 g.) Die dem Organolit anhaftende Lösung wurde darauf mit destilliertem Wasser verdrängt. Die erhaltene Lösung wurde mit 0,01-n Natronlauge gegen Methylrot titriert. Laugeverbrauch: 20,60 ccm.

Mit dem spez. Gewicht der Säure = 1 erhält man den totalen Wassergehalt der abgenutschten Probe zu 7,705 g. Diese Wassermenge stammt aus der anhaftenden 0,05-n Salzsäure und der f-Lösung. Die f-Lösung enthält ausser den an den Sulfonsäuregruppen gebundenen Wasserstoffionen auch eine kleine Menge Chloridionen nebst der damit äquivalenten Menge Wasserstoffionen. Aus der Donnan-Gleichung ist unmittelbar ersichtlich, dass die Chloridionenkonzentration der f-Lösung bei Verdünnung der Aussenlösung zunehmend kleiner wird. Es ist deshalb möglich, alle Chloridionen der abgenutschten Probe mit destilliertem Wasser auszuwaschen. Die Menge lässt sich z. B. durch Bestimmung der entsprechenden Wasserstoffionen festlegen. Wenn wir zunächst bei der Berechnung von den Chloridionen der f-Lösung absehen und so rechnen, als ob diese der anhaftenden Säure angehörten, erhalten wir für die fiktive salzfreie Wasserhaut δ ccm Wasser/g Organolit

$$4{,}228\,\delta = 7{,}705 - \frac{20{,}60 \cdot 0{,}01}{0{,}05}$$

$$\delta = 0{,}848.$$

Wenn die Austauschkapazität mit Σ Milliäquiv./g Organolit bezeichnet wird und die Ionen der festen Phase als vollständig dissoziiert anzusehen sind, was bei Sulfonsäureorganoliten auch für die Wasserstoffionen der Fall ist, kann die Ionenkonzentration der f-Lösung in erster Annäherung durch folgende Gleichung berechnet werden:

$$n_{sf} = \frac{\Sigma}{\delta} \text{ Äquiv./Liter.}$$

Bei dem hier erwähnten Beispiel war $\Sigma = 2{,}5$ und somit

$$n_{sf} = \frac{2{,}5}{0{,}848} = 2{,}95 \text{ Äquiv./Liter.}$$

Wenn wir die wirkliche Normalität n_f der f-Lösung in unserem Falle als Sulfonatäquiv./Liter definieren, erhalten wir:

$$n_f = [H_f] - [Cl_f].$$

Aus der Substanzbilanz in Bezug auf die Chloridionen der abgenutschten Probe erhält man für das wirkliche Volumen — V ccm/g Organolit — der f-Lösung den folgenden Ausdruck:

$$(7{,}705 - \delta) \cdot 0{,}05 = (7{,}705 - V)\, 0{,}05 + V \cdot [Cl_f]$$

und demnach

$$V = \frac{\delta\, 0{,}05}{0{,}05 - [Cl_f]}$$

$[Cl_f]$ kann aus der *Donnan*-Gleichung geschätzt werden:

$$[Cl_f] = \frac{[Cl_w]\,[H_w]}{[H_f]} \cdot \frac{\gamma_{Cl_w} \cdot \gamma_{H_w}}{\gamma_{H_f} \cdot \gamma_{Cl_f}},$$

wo γ die Aktivitätsfaktoren der betreffenden Ionen bedeutet. Bei der numerischen Auswertung haben wir:

$$[Cl_w] = [H_w] = 0{,}05.$$

Für $[H_f]$ setzen wir $n_{sf} = 2{,}95$ und erhalten

$$[Cl_f] = \frac{\overline{0{,}05}^2}{2{,}95} \cdot \frac{\gamma_{Cl_w} \cdot \gamma_{H_w}}{\gamma_{H_f} \cdot \gamma_{Cl_f}} = 0{,}00085 \frac{\gamma_{Cl_w} \cdot \gamma_{H_w}}{\gamma_{H_f} \cdot \gamma_{Cl_f}}$$

Da eine numerische Schätzung der Aktivitätsfaktoren der f-Lösung jedenfalls recht unsicher wird, kann darauf verzichtet werden, um so mehr, als es jetzt unmittelbar ersichtlich ist, dass $[Cl_f]$ in Vergleich mit 0,05 klein ist. Deshalb dürfte V nur um einige Prozente grösser als δ sein. Wir erhalten demnach

$$n_f = \frac{\Sigma}{V} = 2{,}9 \text{ Äquiv./Liter.}$$

Zum Vergleich sei hier erwähnt, dass *C. Du Rietz* bei starkem Sulfitzellstoff die Normalität 0,2—0,4 gefunden hat. In dieser Untersuchung wurden einige Sulfonsäureorganolite recht verschiedener Charakter studiert und immer Normalitäten von der Grössenordnung 2—3 gefunden. Ich halte es deshalb für recht wahrscheinlich, dass die von *Du Rietz* gefundene salzfreie Wasserhaut zum grossen Teil der hydrophilen Natur der Zellulose zuzuschreiben ist.[1] Aus den Versuchen von *Hofmann* und *Giese* geht hervor, dass die Normalität der f-Lösung bei Bentonit ebenfalls von derselben Grössenordnung ist wie bei den hier erwähnten Sulfonsäureorganoliten.

Die technisch wichtige Frage nach den Konzentrationsverhältnissen an verschiedenen Basenaustauschern muss weiter untersucht werden. Für die Analysenmethodik dürfte aber das oben angeführte genügen. Es ist leicht einzusehen, dass eine Arbeitsmethodik wie die von *A. Bahrdt* (8), wo eine Lösung auf das Austauschfilter gegossen wird und, nachdem die erst durchgelaufenen Anteile verworfen worden sind, ein aliquoter Teil abgemessen und analysiert wird, theoretisch unrichtig ist, weil die salzfreie Wasserhaut des Basenaustauschers, wie aus der *Donnan*-Gleichung hervorgeht, von der Natur — besonders der Valenz — der betreffenden Ionen abhängt. Wenn ein hartes Wasser zwecks Sulfatbestimmung nach *Bahrdt* durch einen anfangs trockenen Natriumzeolit filtriert wird, muss demnach wegen der Bildung der salzfreien Wasserhaut die Sulfationenkonzentration der ersten Anteile des Filtrates vergrössert werden. Dieser Teil wird verworfen und die nächsten Anteile werden analysiert. Die Sulfatkonzentration dieses Teiles ist aber kleiner als die der Probe, denn in dem Masse, als die Natriumionen des Basenaustauschers allmählich durch Calcium- und Magnesiumionen ersetzt

[1] Diese Theorie könnte vielleicht die von *E. Hägglund* und *G. Sävö* (Svensk Papperstidning n:r 2/1937) gefundenen Ergebnisse über die Auslösegeschwindigkeit der Lignosulfonsäure bei dem Sulfitkochprozess erklären. Vgl. *Du Rietz* (6) Seite 106.

werden, wird die salzfreie Wasserhaut verkleinert und die Probelösung deshalb verdünnter. Die Grösse dieses Fehlers ist von der Natur des Basenaustauschers abhängig, und der Fehler wird demnach umso kleiner, je grösser die Normalität der f-Lösung des betreffenden Basenaustauschers ist. Bei sehr verdünnten Lösungen, wie bei der Methode von *Bahrdt,* wird der Fehler prozentuell klein und ist deshalb ohne praktische Bedeutung. Bei Austausch von Ionen hoher Wertigkeit, wie z. B. Aluminium, gegen Wasserstoff nach der unten beschriebenen Methode, wo man praktisch mit einer verhältnismässig konzentrierten Lösung arbeitet, konnte diese Arbeitsmethodik einen Fehler bis 1 % und bei einem Basenaustauscher von der Natur des Sulfitzellstoffes mehrere Prozente bedeuten. Es ist deshalb immer ratsam, das ganze Filtrat zu verwenden und den Basenaustauscher nachher auszuwaschen, was übrigens keine Schwierigkeiten bereitet. Wenn man aus irgendwelchen Gründen die andere Arbeitsweise anwenden will, darf man es niemals unterlassen, die Bedeutung dieser Fehlerquelle an den betreffenden Basenaustauschern zu bestimmen.

Über die Sulfatbestimmung in Aluminiumsulfat.

Wegen der grossen technischen Bedeutung des Aluminiumsulfates, z. B. für die Wasserreinigung und in der Papierindustrie, ist von verschiedenen Autoren vorgeschlagen worden, die für die Fabrikspraxis langwierige gewichtsanalytische Sulfatbestimmung als Bariumsulfat durch eine alkalimetrische Titration unter Verwendung von geeigneten Indikatoren zu ersetzen. Bei dem direkten Titrieren findet aber wegen der Adsorptionsfähigkeit des entstehenden Aluminiumhydroxydniederschlages ein Mitreissen von Sulfationen im Niederschlag statt, und die Resultate fallen zu niedrig aus. Es ist auch versucht worden, das Sulfat durch Ausfällung mit Bariumchlorid und nachfolgende Titration des entstehenden Aluminiumchlorides mit Lauge zu bestimmen. Nach unserer Erfahrung werden hier ebenfalls die Resultate zu niedrig. Eine Kompensationsmethode, die recht gute Übereinstimmung gibt, ist von *S. A. Tolkatscheff* und *Julja Grigorjewna Titowa* (9) ausgearbeitet worden, aber das Verfahren ist anderseits recht kompliziert.

Mit Hilfe eines organischen Basenaustauschers lässt sich indessen die Aufgabe einfach und genau lösen. Die Aluminiumionen der Probelösung lassen sich nämlich durch eine Filtration durch eine mit Wasserstoffionen gesättigten Organolitschicht quantitativ gegen Wasserstoffionen austauschen. Die dabei entstehende Schwefelsäure kann dann durch eine alkalimetrische Titration einwandfrei bestimmt werden.

Durch Auflösen verschiedener Handelsprodukte in destilliertem Wasser und nachfolgendes Filtrieren wurden die Probelösungen dargestellt. Bei den Versuchen wurden genau abgemessene Mengen mit einem Sulfatgehalt von etwa zwei Milliäquivalenten verwendet. Wie in der vorigen Arbeit wurde ein Sulfonsäureorganolit und ein Trichter vom Typus eines Cadmiumreduktors verwendet. Die Austauschkapaziät des Basenaustauschers betrug etwa 3,5 Milliäquivalente/Gramm. Durch einige Vorversuche wurde gezeigt, dass eine Organolitmenge von 3 g erforderlich war, um praktisch einen vollständigen Umsatz zu geben. Bei den Versuchen wurden gewöhnlich der Sicherheit wegen 3,2—4 g verwendet. Nach der Filtration wurde mit ausgekochtem destilliertem Wasser gewaschen, bis das Waschwasser neutral ablief und die gesamte Lösung mit 0,1-n Natronlauge gegen Methylrot titriert. Die Lauge war gegen Kaliumbijodat nach den Vorschriften von Kolthoff (10) eingestellt. Die ganze Analyse kann in zehn Minuten durchgeführt werden.

Tabelle I.

Präparat	Gefunden: Milliäquivalente Schwefelsäure	
	Gewichtsanalytisch	Alkalimetrisch nach der Organolitmethode
E. Merck, purum...	1,996; 2,001	2,003; 1,998 je 50 ccm Probelösung
Handelsprod. I.	2,040; 2,036	2,040; 2,040; 2,041; 2,039 je 50 ccm Probelösung
Handelsprod. II....	1,895; 1,898	1,909; 1,909; 1,908; 1,911; 1,907; 1,909; 1,912 je 50 ccm Probelösung
Handelsprod. III... »Eisenfrei«	1,930	1,930; 1,928 je 50 ccm Probelösung
Handelsprod. IV ... »Eisenfrei«	1,995; 1,985	1,997; 1,997; 1,999; 1,999; 1,998; 1,997; 1,995 je 50 ccm Probelösung
Handelsprod. V	2,426	2,429; 2,429; 2,432; 2,430; 2,430 je 25 ccm Probelösung
Dieselbe Lösung vor der Analyse mit 25 ccm Wasser verdünnt	—	2,432; 2,429; 2,431

Der Organolit wurde nach jedem Versuch mit 2-n Salzsäure regeneriert. Bei einer Organolitmenge von 3,2 g wurden 150 ccm Salzsäure gebraucht. Wenn eine Organolitmenge von 4 g verwendet wurde, konnten mit diesem Säurevolumen die Aluminiumionen mit befriedigender Sicherheit nicht entfernt werden, weshalb hier 200 ccm 2-n Salzsäure verwendet wurde. Nach dem Auswaschen mit destilliertem Wasser war der Organolit wieder gebrauchsfertig. Die mit dieser Methode erhaltenen Resultate wurden mit den gewichtsanalytisch bestimmten verglichen. Bei der gravimetrischen Sulfatbestimmung wurde mit Bariumchlorid nach der Arbeitsweise von *Moser, Cohn* (11) gefällt. Die Versuchsresultate sind in Tabelle I zusammengestellt.

Die Übereinstimmung zwischen den Bestimmungen nach der Organolitmethode untereinander ist sehr gut, und die Unterschiede liegen innerhalb der Fehlergrenzen bei der alkalimetrischen Titration. Die Übereinstimmung mit den gewichtsanalytisch bestimmten Werten ist ebenfalls befriedigend. Wenn das Aluminiumsulfat neben Sulfat andere Anionen enthält, werden diese selbstverständlich mittitriert, und für diese muss, wenn man hier ebenfalls die Organolitmethode verwenden will, eine Korrektion eingeführt werden. In den gewöhnlichen Handelsprodukten dürften aber solche Verunreinigungen kaum vorkommen, jedenfalls nicht in den hier verwendeten Produkten.

Über die Bestimmung von Alkalimetallen in Anwesenheit von Vanadin.

Bei Untersuchungen über Vanadinkatalysatoren, die seit einigen Jahren im hiesigen Forschungslaboratorium durchgeführt worden sind, ist mehrmals das Problem aufgetaucht, Alkalimetalle in Anwesenheit von Vanadin quantitativ zu bestimmen. Früher wurde dafür die von *Berl-Lunge* (12) empfohlene Methode nach *Roscoe* verwendet. Nach dieser wird das Vanadin in neutraler Lösung als Bleivanadat ausgefällt und abfiltriert. Der Überschuss an Blei wird als Bleisulfat ausgefällt, das nach dem üblichen Verfahren entfernt wird. Die Alkalimetalle werden dann nach dem Eindampfen als Sulfate gewogen. Diese Methode ist sehr zeitraubend und gibt nach unserer Erfahrung befriedigende Resultate nur, wenn die grösste Sorgfalt verwendet wird.

Nach der Organolitmethode kann die Aufgabe in folgender Weise recht einfach gelöst werden. Die Probelösung, in der das Vanadin als Vanadat vorliegen soll, wird durch eine mit Ammoniumionen gesättigte Organolitschicht filtriert und mit Wasser nachgewaschen. Im Filtrat befinden sich sämtliche Anionen der Probelösung als Ammoniumsalze. Die Alkalimetalle werden vom Organolit quantitativ aufgenommen. Aus der Organolitschicht werden die Alkalimetalle mit einem Überschuss an Salzsäure verdrängt und die erhaltene vanadinfreie Lösung, welche die Alkalimetalle der Probelösung und den Überschuss an Ammoniumionen als Chlorid nebst Salzsäure enthält, wird eingedampft und die Alkalimetalle wie gewöhnlich bestimmt.

Wegen der recht geringen Löslichkeit von Ammoniummetavanadat in Wasser muss man mit einer verhältnismässig verdünnten Lösung arbeiten. Dies bereitet aber keine Schwierigkeiten, weil die vom Organolitfilter ablaufende Lösung nicht weiter verarbeitet wird.

Etwa 5 g Sulfonsäureorganolit mit einer Austauschkapazität von etwa 3,5 Milliäquivalenten/Gramm wurden bei dieser Untersuchung für jede Analyse verwendet. Die Kationenmenge der Probelösungen war etwa 2 Milliäquivalente und der verwendete Organolitüberschuss somit etwas grösser als bei den früheren Versuchen. Die Auswaschung der Filterschicht geht aber recht schnell, und es ist der Sicherheit wegen besser, einen grossen Überschuss zu verwenden, weil es sehr schwierig ist, den Umsatz auf Vollständigkeit zu prüfen. Ein Versuch, Spuren von Alkalimetallen in dem Filtrate nachzuweisen, muss selbstverständlich komplizierter sein als die ganze Analyse.

Die technischen Sulfonsäureorganolite enthalten ausser den Sulfonsäuregruppen meistens auch andere saure Gruppen, namentlich Phenolgruppen. Bei Wasserstoffionenaustausch sind diese praktisch für die Analysenmethodik ohne Bedeutung. Wenn man aber in neutralen oder alkalischen Lösungen arbeitet, können sie, auch wenn sie nur in kleinen Mengen vorhanden sind, störend einwirken. Beim Nachwaschen mit destilliertem Wasser können nämlich die an den Phenolgruppen gebundenen Metalle hydrolytisch abgespaltet werden. Die Bedeutung dieser Fehlerquelle wurde durch den folgenden Versuch studiert.

I. Das Organolitfilter wurde mit n-Natriumchloridlösung gerade so lange behandelt, bis die ablaufende Lösung gegen Methylorange neutral war und mit 100 ccm ausgekochtem destilliertem Wasser chloridfrei gewaschen.

I a. Das Filter wurde darauf mit weiteren 200 ccm ausgekochtem, destilliertem Wasser gewaschen und die ausgelöste Menge Natriumionen nach Eindampfen als Clorid bestimmt.

I b. Dasselbe Filter wurde mit 200 ccm gewöhnlichem, destilliertem Wasser behandelt und die Natriummenge im Filtrat bestimmt.

II. Das Filter wurde mit einem Überschuss von 0,1 molarer Natriumbicarbonatlösung behandelt und mit 200 ccm ausgekochtem, destilliertem Wasser nachgewaschen.

II a. Weitere 200 ccm ausgekochten, destillierten Wassers wurden durchfiltriert und die Natriummenge im Filtrat bestimmt.

II b. Das Filter wurde wie oben behandelt, aber die letzte Auswaschung mit gewöhnlichem, destilliertem Wasser ausgeführt.

Zusammenstellung.

Vorbehandlung	Waschflüssigkeit	Ausgelöst. Milliäquiv./200 ccm Filtrat
Ia Natriumchlorid	Aq. dest. ausgekocht	0,000
Ib »	Aq. dest. nicht ausgekocht	0,000
IIa Natriumbicarbonat	Aq. dest. ausgekocht	0,029
IIb »	Aq. dest. nicht ausgekocht	0,059

Bei den Analysen ist die Gefahr der Auswaschungsstörungen selbstverständlich kleiner, weil hier, nachdem die Lösung durchgelaufen ist, die letzten Teile der Filterschicht aus reinem Ammoniumorganolit bestehen sollen. Nach unserer Erfahrung ist es aber ratsam, die Probelösung mit Säure zu neutralisieren, wenn die Lösung anfangs alkalisch ist. Der Organolit wird ausserdem durch Behandeln mit Ammoniumsulfat- oder Ammoniumchloridlösung mit Ammoniumionen gesättigt, gerade bis die durchgelaufene Lösung gegen Methylorange neutral ist, nicht aber mit beispielsweise Ammoniumcarbonatlösung, damit wenigstens in den letzten Teilen der Filterschicht hauptsächlich nur die Sulfonsäuregruppen mit

Ammoniumionen abgesättigt werden. Die anhaftende Lösung wird mit destilliertem Wasser verdrängt.

Die Probelösungen wurden durch Auflösen von Vanadinpentoxyd in einer genau abgemessenen Menge Alkalilauge und nachherige Verdünnung auf ein bekanntes Volumen bereitet. In besonderen Proben wurden die eingeführten Alkalimengen durch Versetzten einer bekannten Menge der ursprünglichen Lauge mit Salzsäure und Eindampfen als Chlorid bestimmt. Durch Blindproben wurde untersucht, dass die verwendeten Chemikalien keine schädlichen Verunreinigungen enthielten. Die abgemessenen Proben wurden auf etwa 200 ccm verdünnt und durch die Organolitschicht filtriert. Darauf wurde mit ausgekochtem, destilliertem Wasser gewaschen, bis das Waschwasser vanadinfrei ablief, was durch Prüfung mit Wasserstoffperoxyd nach Zusatz von Schwefelsäure untersucht wurde. Die totale erforderliche Waschwassermenge war 100 bis 150 ccm. Durch Behandeln mit 2-n Salzsäure wurden die Alkali- und Ammoniumionen des Basenaustauschers verdrängt. Bei den hier verwendeten Ionen- und Organolit-Mengen zeigte es sich, dass bei vorsichtigem Arbeiten eine Säuremenge von 150 ccm genügte, um die quantitative Verdrängung zu erzielen. Der Sicherheit wegen wurden aber 200—250 ccm bei den Analysen verwendet. Eine Prüfung auf Vollständigkeit dieser Umsetzung ist recht einfach durchzuführen und muss wenigstens bei den ersten Proben ausgeführt werden. Bei Serienbestimmungen unter genau denselben Umständen kann aber darauf verzichtet werden. Das Filtrat wird in einer Platinschale eingedampft. Die Alkalimetalle können als Chloride gewogen werden. Das Entfernen des von den Kristallen festgehaltenen Wassers bereitet bekanntlich grosse Schwierigkeiten, weil bei Zerspringen der Kristalle leicht Verluste eintreten können. Die Vorsichtsmassnahmen, die beispielsweise in *Treadwells* (13) bekanntem Lehrbuch vorgeschrieben werden, sind für Serienanalysen sehr langwierig. Nach *Hillebrand-Lundell* (14) können die Schwierigkeiten beim Trocknen bisweilen vermieden werden, wenn man die auf ein kleines Volumen eingedampfte Lösung mit konzentrierter Salzsäure versetzt. Dabei wird das Chlorid sehr feinkörnig ausgefällt und kann ohne Verknistern getrocknet werden. Diese Fällungsmethode ist recht primitiv und aus der Destillationskurve für Chlorwasserstoff-Wassergemische (15) ist unmittelbar ersichtlich, weshalb die Methode nur bisweilen mit Erfolg verwendet werden kann. Der Chlorwasserstoff ist in konzentrierter Lösung leichter flüchtig als das Wasser. Wenn die Lösung vom Anfang kein oder nur wenig Chlorwasserstoff enthält, kann das beim Zusatz von Salzsäure ausgefällte Alkalichlorid beim fortgesetztem Eindampfen, wenn die Konzentrationsverhältnisse richtig gewählt worden sind, wenigstens zum Teil unverändert bleiben, aber enthält die Lösung schon vom Anfang grosse Mengen Salzsäure, wie es bei der hier beschriebenen Methode der Fall ist, wird beim fortgesetztem Eindampfen das ausgefällte Natriumchlorid wieder aufgelöst, weil hier der Chlorwasserstoff schneller als das Wasser verdampft. Dieses Verfahren führt demnach nicht zum Ziel, und es wurden deshalb andere Zusätze geprüft. Die Löslichkeitsverhältnisse und Destillationskurven dieser recht komplizierten Systeme waren aber nicht bekannt und deshalb musste rein empirisch gearbeitet werden. Es zeigte sich jedoch bald, dass mit Propylalkohol (sec.) gute Resultate erhalten werden konnten. Die Probelösung wurde wie bei der oben erwähnten Methode auf ein möglichst kleines Volumen eingedampft und mit etwa dem doppelten Volumen sec. Propylalkohol versetzt. Das Alkalichlorid fiel sehr feinkörnig aus und beim fortgesetztem Eindampfen wurden wenigstens keine die Bestimmung störende Mengen wieder aufgelöst. Wurde der Propylalkohol zu früh zugesetzt, musste beim fortgesetzten Eindampfen noch etwas mehr Propylalkohol zugefügt werden, um das Alkalichlorid in solcher Form erhalten zu können, dass beim Trocknen kein Zerspringen der Kristalle stattfand.

Nach dem Trocknen wurde das Ammoniumchlorid abgeraucht und schwach geglüht. Das Alkalichlorid wurde, um die Wägung der Platinschale zu umgehen, mit 0,1-n Silbernitratlösung potentiometrisch titriert. Sollen mehrere Alkalimetalle getrennt werden, ist selbstverständlich eine gravimetrische Summenbestimmung vorzuziehen. Die Versuchsergebnisse sind in Tabelle II zusammengestellt.

Tabelle II.

Probelösung	% Alkali als Vanadat	Alkaligehalt: Milliäquivalente zugesetzt	Alkaligehalt: Milliäquivalente Gefunden
Natriummetavanadat........	100	2,163; 2,160	2,164; 2,160
» + Sulfat	70	2,163; 2,160	2,166; 2,160
» + »	50	2,163; 2,160	2,159; 2,163
» + Chlorid	50	2,163; 2,160	2,159; 2,165
» + »	30	2,163; 2,160	2,164; 2,164
Kaliummetavanadat	100	2,271; 2,275	2,270; 2,268
» + Sulfat	50	2,271; 2,275	2,273; 2,270

Die Übereinstimmung zwischen den zugesetzten und gefundenen Alkalimengen ist sehr befriedigend, und im Vergleich mit den früher bekannten Methoden für diese Bestimmung gibt dieses Verfahren nicht nur sichere Resultate, sondern bedeutet gleichzeitig eine bedeutende Zeitersparnis. Das Verfahren eignet sich sehr gut für Serienbestimmungen, was übrigens allgemein für die Analysen nach der Organolitmethode gilt, weil mehrere Filtrationen, Waschungen u. s. w. nebeneinander ohne Mühe ausgeführt werden können.

Dem Vorstand des Laboratoriums, Herrn Dozenten Dr. *O. Stelling*, möchte ich für wertvolle Ratschläge verbindlichst danken.

Hälsingborg, Forschungslaboratorium von Reymersholms Gamla Industri Aktiebolag, September 1939.

Zitierte Literatur:

(1) *O. Samuelson:* Z. anal. Chem. *116* (1939), 328.
(2) *C. Kullgren:* Svensk Kemisk Tidskr. *43* (1931), 99.
(3) *Folin.* Vgl. *A. Kolb:* Chem. Ztg *48* (1924), 557.

(4) *A. Bahrdt:* Z. anal. Chem. *70* (1927), 109; Vgl. auch ***M. Kehren*** und ***H. Stommel:*** Chem. Ztg *51* (1927), 913 und Chem. Ztg *52* (1928), 163.
(5) *F. Alten* und *B. Kurmies:* Ztschr. angew. Chem. Beih. *21* (1935).
(6) *C. Du Rietz:* Dissertation, Stockholm Techn. Hochschule 1938.
(7) *U. Hofmann* und *K. Giese:* Kolloid-Z. *87* (1939), 21.
(8) *A. Bahrdt* l. c.
(9) *S. A. Tolkatscheff* und *Julja Grigorjewna Titowa:* Z. anal. Chem. ***105*** (1936), 96. Weitere Literaturangaben ebenda.
(10) *Kolthoff-Menzel:* Die Massanalyse II, Berlin 1931.
(11) *A. Kurtenacker:* Analytische Chemie der Sauerstoffsäuren des Schwefels, Stuttgart 1938.
(12) *Lunge-Berl:* Chemisch-technische Untersuchungsmethoden II,2 S. 1663 (Berlin 1932). Weitere Literaturangaben in: *F. Ephraim:* Ahrens Sammlung Chemischer und chemisch-technischer Vorträge IX, Stuttgart 1904, S. 182 und *A. Rüdisule:* Nachweis, Bestimmung und Trennung der chemischen Elemente VI, S. 286 (Bern 1923).
(13) *Treadwell:* Analytische Chemie II, Leipzig 1935.
(14) *Hillebrand-Lundell:* Applied Inorganic Analysis, New York 1929.
(15) *E. Berl-H. Staudinger:* Ztschr. angew. Chem. *43* (1930), 1019.

1

ON THE APPLICATION OF BASE-EXCHANGING SUBSTANCES IN ANALYTICAL CHEMISTRY II

Olof Samuelson

This condensed translation was prepared expressly for this Benchmark volume by Harold F. Walton, University of Colorado, Boulder, Colorado, from "Über die Verwendung von basenaustauschenden Stoffen in der analytischen Chemie II," Svensk Kem Tdiskr., **51,** *195–206 (1939)*

INTRODUCTION

In the first paper of this series (1) it was shown that in the organic base exchangers we have an analytical tool that can be applied in various ways. The procedures are based upon the fact that, with the help of a suitable "organolite," one can quickly and simply exchange all the cations of a test solution for hydrogen ions. If these cations are to be determined later, they can be displaced from the organolite by passing an acid or a suitable salt solution and then determined quantitatively.

Of earlier publications on the use of base-exchanging materials for chemical analysis we have cited only the work of C. Kullgren (2). Nevertheless, other authors have used inorganic base exchangers for special kinds of analysis. The work of Folin (3) on the use of a base exchanger to determine ammonia in urine appeared as long ago as 1917. A. Bahrdt (4), studying the volumetric determination of sulfate ions in drinking water, used an inorganic exchanger saturated with sodium ions to remove calcium and magnesium ions from the water; the sulfate ions were then precipitated with a measured excess of barium chloride and the uncombined barium ions were titrated by the well-known palmitate method of Blacher. Undoubtedly, this exchange of ions could be performed more easily and with more certainty by using today's organic base exchangers.

In this report we shall describe some special methods that have been in use in our laboratory for some time. First, however, we shall discuss a general matter that is of great importance in analytical methodology.

ON THE IONIC CONCENTRATION OF THE STATIONARY PHASE AND ITS SIGNIFICANCE IN THE ANALYTICAL METHODOLOGY

When a dry cation exchanger is shaken with a salt solution, the concentration of the solution becomes greater than it was originally. The acidic groups of the cation exchanger become hydrated and take away part of the solvent, a phenomenon that has been studied by several workers . . . .

Other authors, including C. Du Rietz, believe that the ions of the exchanger are situated in a homogeneous solution, sharply separated from the outer solution . . . . Assuming the interior solution to be homogeneous, we can apply the Donnan equilibrium equation to the distribution of ions between the two solutions.

DETERMINATION OF SULFATE IN ALUMINUM SULFATE

Because of the great technical importance of aluminum sulfate in, for example, water purification and the paper industry, various authors have proposed replacing the tedious gravimetric sulfate determination by an alkalimetric titration, with the help of a suitable indicator. In direct titration with alkali, however, some sulfate ions are carried down by the highly absorptive aluminum hydroxide precipitate, and the results are too low. Another approach has been to precipitate the sulfate with barium chloride and then titrate the liberated aluminum chloride with alkali. In our experience this method, too, gives low results. A compensative method has been worked out by Tolkatscheff and Titowa (9). This gives good results, but the procedure is quite complicated.

With the help of an organic base exchanger the problem may be resolved easily and precisely. The aluminum ions of the test solution can be exchanged quantitatively for hydrogen ions by filtering through a layer of organolite that is saturated with hydrogen ions. The sulfuric acid formed can now be determined accurately by titration with alkali.

Test solutions were prepared by dissolving various commerical products in distilled water and filtering. Measured portions with a sulfate content of about 2 mEq were taken for each analysis. An organolite of the sulfonic acid type was used, as previously described, in a filter tube of the form of a cadmium reductor. The exchange capacity of the material was about 3.5 mEq/g. Preliminary experiments showed that 3 g of organolite gave complete exchange, but 3.2 to 4 g were used in most of the experiments to be on the safe side. After passing the test solution, the exchanger was washed with boiled distilled water until the washings were neutral, and then the entire solution was titrated with 0.1 *N* sodium hydroxide, using methyl red . . . . The entire analysis can be done in 10 minutes.

After each experiment the organolite was regenerated with 2 *N* hydrochloric acid. For 3.2 g of organolite, 150 cm^3 of acid were used. If 4 g of organolite were used, this volume of acid could not be relied upon to remove all the aluminum ions, and so 200 cm^3 of 2 *N* hydrochloric acid were used in this case. After washing with distilled water, the organolite was again ready for use. The results obtained by this method were compared with those obtained by gravimetric analysis performed according to the method of Moser and Cohn (11). The experimental data are collected in Table I.

The agreement between the determinations made by the organolite method is very good; the differences lie within the experimental error of the alkalimetric titration . . . . If the aluminum sulfate contains other anions besides sulfate, these anions are of course titrated too; if the organolite method is to be used in such cases, a correction must be made. In most commercial products, however, such impurities are not present. At any rate, they were not present in the products used here.

ON THE DETERMINATION OF ALKALI METALS IN THE PRESENCE OF VANADIUM

In investigations of vanadium catalysts made in this laboratory a few years ago, the problem often arose of determining alkali metals in the presence of vanadium. We used to use the method of Roscoe as recommended by Lunge-Berl (12). In this method the vanadium was precipitated in neutral solution as lead vanadate. The excess of lead was precipitated as lead sulfate, which was removed in the usual way. After evaporation, the alkali metals were weighed as sulfates. This method is very time consuming and in our experience gives satisfactory results only with the greatest care.

The problem can be solved very easily by using an organolite, in the following way. The test solution, containing vanadium as vanadate, is filtered through a layer of organolite saturated with ammonium ions, which is then washed with water. In the filtrate are all the anions of the test solution in the form of their ammonium salts. The alkali metals are quantitatively retained by the organolite. They are displaced from the organolite layer by an excess of hydrochloric acid, and the resulting solution, which is free from vanadium and contains the alkali metals of the test solution together with an excess of ammonium chloride and hydrochloric acid, is evaporated to dryness. The alkali metals are determined in the usual way.

Because of the low solubility of ammonium metavanadate, one must use relatively dilute solutions. This causes no difficulty, however, because the solution flowing through the organolite is discarded.

[*Editor's Note:* References and tables for this translation will be found in the preceding original article.]

Part I

THEORY OF ION-EXCHANGE COLUMNS

Editor's Comments on Papers 2 Through 4

2 MAYER and TOMPKINS
Ion Exchange as a Separations Method: IV. A Theoretical Analysis of the Column Separations Process

3 GLUECKAUF
Theory of Chromatography: Part 9. The "Theoretical Plate" Concept in Column Separations

4 HELFFERICH
Travel of Molecules and Disturbances in Chromatographic Columns: A Paradox and Its Resolution

Paper 2 is one of the symposium papers published in the November 1947 issue of the *Journal of the American Chemical Society* that reported work on ion exchange performed under the Manhattan Project. The authors used the *theoretical-plate concept,* a concept originally used to describe fractional distillation and adapted to chromatographic columns by Martin and Synge (1941). Mayer and Tompkins applied the concept to ion exchange by using the mathematics of countercurrent extraction and considering the partition ratio of the exchanged ions between the solution and the resin to be constant. Band spreading was related to the number of theoretical plates in the column.

Glueckauf (Paper 3) noted that the spreading of bands observed in practical ion-exchange chromatography was considerably greater than the Mayer–Tompkins treatment would predict, at least if the theoretical plate had a thickness approximating the diameter of a resin bead. He developed a modified plate theory in which the particles of absorbent were considered to be spheres of uniform diameter, within which the ionic concentrations were uniform; the flow of solvent, however, was considered to be continuous, with the concentration of the solution changing continuously as it flowed out of the plate. In other words, the stationary phase was discontinuous, whereas the moving phase was continuous. A differential equation was derived and solved by Laplace transforms. The height of the theoretical plate emerged as the coefficient of a second differential whose magnitude depended not only on the diameter of the resin beads, but also on the rate of mass transfer between the stationary and moving phases. Another feature of the theory was that the solute entering the column did not have to occupy just one plate, as in the Mayer–Tompkins treatment, but could extend over a number of plates.

In a second major paper ("Principles of Operation of Ion-Exchange Columns") Glueckauf (1955) quantitatively related plate heights to mass-transfer rates and flow rates, considering the rates of diffusion within the particles of resin and across the concentration gradients surrounding the resin spheres *(particle diffusion* and *film diffusion).* Particle radius, diffusion coefficients, and the distribution ratio all affect plate height. Graphs and equations from this paper have been reproduced in a number of books, such as *Chromatography* (Heftmann, 1975).

Paper 4 is a short article by Helfferich in which he makes the point that the rate of travel of a disturbance, or *concentration pulse,* along a column is not the same as the rate of travel of the molecules themselves. This is a very simple idea that is often overlooked, yet the practical importance is great. If one is performing liquid chromatography with a mixed solvent, and a sample is introduced in a solvent whose composition is different from that of the main stream, a "front" will travel along the column at a rate that normally will be different from the rate of solvent flow. If a sample is introduced at a concentration high enough that the distribution ratio is not constant, but increases or decreases according to the concentration, its elution profile will not be symmetrical but lopsided, because the peak of the band travels at a different rate from the base. This behavior can be understood by reference to this short paper by Helfferich.

The idea of the rate of travel of a concentration gradient was elaborated and applied to mixtures of several solutes by Helfferich and Klein (1970).

REFERENCES

Glueckauf, E. 1955. *Ion Exchange and Its Application,* London, Society of Chemical Industry.

Heftmann, E. (ed.). 1975. *Chromatography,* 3rd ed., New York, Van Nostrand Reinhold.

Helfferich, F., and G. Klein. 1970. *Multicomponent Chromatography,* New York, Marcel Dekker.

Martin, A. J. P., and R. L. M. Synge. 1941. *Biochem. J., 35,* 1385.

2

Reprinted from *J. Amer. Chem. Soc.*, **69**, 2866–2874 (Nov. 1947)

Ion Exchange as a Separations Method. IV. A Theoretical Analysis of the Column Separations Process[1]

By Stanley W. Mayer and Edward R. Tompkins

Introduction

Considerable success has been achieved in the separation of elements of very similar characteristics by the use of a method developed for the separation of the products of uranium fission. This method, which is similar to chromatographic analysis, consists of adsorbing a mixture of cations in a narrow band at the top of an ion exchange column and subsequently eluting these cations stepwise with solutions of complexing agents. The separation depends upon differences in the affinity of the exchanger for the cations and differences in the dissociation constants of their complexes. The two competing reactions may be represented by the equations:

$$(1)\quad M^{+n} + nR^- \rightleftarrows MR_n;\ \text{and}$$

$$(2)\quad M^{+n} + yH_xA^{-s} \rightleftarrows M(H_xA)_y^{-s}$$

in which M^{+n} is a cation, R^- is the ion exchanger anion, H_xA^{-s} some complex forming anion and s is the final charge on the complex.

Although this method of separation of the fission products and later of the rare earths was developed empirically it gave very satisfactory results. The empirical approach was, however, very time consuming and, at best, resulted in the selection of a set of conditions for the column operation which were acceptable, since the true optimum conditions required too much time to evaluate.

This difficulty was partially resolved by a series of equilibrium studies of the column reactions.[2] As a result of that study, it was possible to evaluate the effects of various factors on the distribution of a cation between the resin and solution phases at equilibrium. In order that the equilibrium data could be applied to column conditions, a satisfactory theory of column operation was necessary.

Wilson[3] had attempted a theoretical approach to chromatographic analysis in 1940. He based his development on the assumption that the band width of an adsorbed substance does not change during the chromatographic development. When he noted that the band did actually widen and the boundaries did become diffuse, he ascribed this to diffusion and lack of equilibration of the solute between the two phases. DeVault[4] extended Wilson's treatment but still failed to predict fully the actual behavior of the solutes during the chromatographic development process.

(1) This document is based on work done under Contract W-35-059-eng-71 for the Atomic Energy Project, and the information contained therein will appear in Division IV of the National Nuclear Energy Series (Manhattan Project Technical Section) as part of the contribution of the Clinton Laboratories.

(2) E. R. Tompkins and S. W. Mayer, This Journal, **69**, 2859 (1947).

(3) J. N. Wilson, *ibid.*, **62**, 1583 (1940).

(4) D. DeVault, *ibid.*, **65**, 532 (1943).

A theory of mass transfer, first proposed by Beaton and Furnas[5a] has been quantitatively extended to adsorbent beds by G. E. Boyd, *et al.*,[5b,c] but not to computation of the degree of separation.

About a year after Wilson's publication, Martin and Synge[6] described a treatment based on the assumption that the chromatographic columns are similar in operation to distillation and extraction fractionating columns. They considered the column as consisting of a number of theoretical plates, within each of which equilibrium between the two phases occurred. They found that under the limited conditions in which the distribution coefficient did not change with the change in concentration of a solute, the width of the band, as predicted by their theory, was very similar to that observed experimentally.

The theoretical approach described below is also based on the analogy of the ion exchange separations column to the fractionating distillation or extraction column. The notations used to express the various factors (particularly the use of solution volume and resin mass rather than area and height for expressing the elements in each plate) allow a simpler representation of several of the expressions than those used by Martin and Synge. Also, this approach is directly applicable to the determination of the composition of the eluate as well as predicting the distribution of the various substances on the column.

Notation of Symbols to be Used.—The column is considered as being made up of a number of theoretical segments or plates, p, each plate having an equivalent mass of resin, m, and volume of solution, v. As flow is started through the column, these v's of solution progress down the column from plate to plate. Other notations which will be used are:

- V = (pv) = volume of solution in the column.
- n = number of v's that have entered into a plate up to any given time.
- F = (n/p) = number of V's that have passed through the column.
- L = the fraction of a solute in any v of solution.
- S = the fraction of a solute in the solid phase of the column in any plate.
- $L_{n,p}$ = the fraction of solute in solution which is in contact with plate p when the n'th v has equilibrated in that plate.
- $S_{n,p}$ = the fraction of solute in the solid phase in plate p when the n'th v has equilibrated in the plate.

(5) (a) R. H. Beaton and C. C. Furnas, *Ind. Eng. Chem.*, **33**, 1501 (1941); (b) G. E. Boyd and A. W. Adamson, This Journal, **69**, 2836 (1947); (c) G. E. Boyd, L. S. Myers and A. W. Adamson, *ibid.*, **69**, 2849 (1947).

(6) A. J. P. Martin and R. L. M. Synge, *Biochem. J.*, **35**, 1385 (1941).

Development of the Theory.—The distribution of solute in any plate may be expressed as

$$S_{n,p}/L_{n,p} = C \qquad (1)$$

where C represents the distribution of the solute between the solution and the resin in a plate, p. To maintain material balance

$$L_{n,p} + S_{n,p} = L_{n,p-1} + S_{n-1,p} \qquad (2)$$

As mentioned above, the separation procedure developed in this laboratory consists of adsorbing a mixture of cations in a narrow band at the top of the column and then eluting them differentially by the use of a complexing agent in aqueous solution. As the first increment of volume v enters the column it will come in contact with the portion of exchanger on which the cations to be separated are adsorbed. This part of the column will be designated as plate zero and the quantity of the cation on it will be considered as unity. Then, from (1) and (2), when the first volume v has equilibrated in plate zero

$$L_{1,0} + S_{1,0} = 0 + 1$$
$$S_{1,0}/L_{1,0} = C$$
$$\therefore S_{1,0} = C/(1 + C);\ L_{1,0} = 1/(1 + C)$$

Similarly

$$L_{2,0} + S_{2,0} = 0 + \frac{C}{1 + C};\ \frac{S_{2,0}}{L_{2,0}} = C$$

and

$$S_{2,0} = \frac{C^2}{(1 + C)^2};\ L_{2,0} = \frac{C}{(1 + C)^2}$$

In general

$$S_{n,0} = \frac{C^n}{(1 + C)^n};\ L_{n,0} = \frac{C^{n-1}}{(1 + C)^n}$$

For the first theoretical plate

$$L_{1,1} + S_{1,1} = L_{1,0} + S_{0,1} = \frac{C^0}{(1 + C)} + 0;\ \frac{S_{1,1}}{L_{1,1}} = C$$

$$\therefore L_{1,1} = \frac{C^0}{(1 + C)^2};\ S_{1,1} = \frac{C}{(1 + C)^2}$$

and

$$L_{2,1} + S_{2,1} = L_{2,0} + S_{1,1} = \frac{C}{(1 + C)^2} + \frac{C}{(1 + C)^2};\ \frac{S_{2,1}}{L_{2,1}} = C$$

$$\therefore L_{2,1} = \frac{2C}{(1 + C)^3};\ S_{2,1} = \frac{2C^2}{(1 + C)^3}$$

In general

$$L_{n,1} = \frac{nC^{n-1}}{(1 + C)^{n+p}};\ S_{n,1} = \frac{nC^n}{(1 + C)^{n+p}}$$

$L_{n,p}$ can be similarly calculated for $p = 2,3,4,\cdots$, to give the general equations

$$L_{n,p} = \frac{(n + p - 1)!}{(n - 1)!\,p!} \frac{C^{n-1}}{(1 + C)^{n+p}} \qquad (3)$$

and

$$S_{n,p} = \frac{(n + p - 1)!}{(n - 1)!\,p!} \frac{C^n}{(1 + C)^{n+p}} \qquad (4)$$

For large values of n and p not given in tables of factorials, Stirling's approximation may be used. (*i. e.*, $n! = e^{-n}n^n\sqrt{2\pi n}$)

$$L_{n,p} = \frac{1}{\sqrt{2\pi}} \frac{(p + n - 1)^{p+n-1/2}}{(n - 1)^{n-1/2}\,p^{p+1/2}} \frac{C^{n-1}}{(1 + C)^{p+n}} \qquad (5)$$

(When $p = 75$ and $n = 26$, Stirling's approximation gives a value of 2.434×10^{23} for the above coefficient whereas the value of the coefficient in equation (3) equals 2.425×10^{23}.)

It can be shown that $L_{n,p}$ in equation (5) has its maximum at $n = pC$

$$L_{\text{max.}} = L_{p,p} = [2\pi pC(1 + C)]^{-1/2} \qquad (6)$$

Furthermore

$$F_{\text{max.}} = n_{\text{max.}}/p$$

But

$$C = n_{\text{max.}}/p$$

Consequently

$$F_{\text{max.}} = C$$

Application of the Theory to Previous Separations.—The usual elution curve for these separations is a plot of the concentration of one of the elements being separated in the effluent (as ordinate) against the volume of eluting solvent that has passed through the column (as abscissa). Therefore, the plot of L, for the last plate of the column versus n as abcissa (as n varies from zero to its value at the termination of the separation) is a theoretical elution curve.

In Fig. 1, a typical experimental elution curve for rare earth separations[7] is compared with the calculated elution curve. The data for the peak of this experimental curve were used to calculate p and C employing equation (6) and the simple relationship derived above, showing that $C = F$ at the peak. The degree of correlation between the experimental and calculated elution curves indicates the degree of applicability of the developed theory to this experimental column run. It will be noted that both the experimental and theoretical elution curves are non-symmetrical, and in all cases the peak occurs before one-half

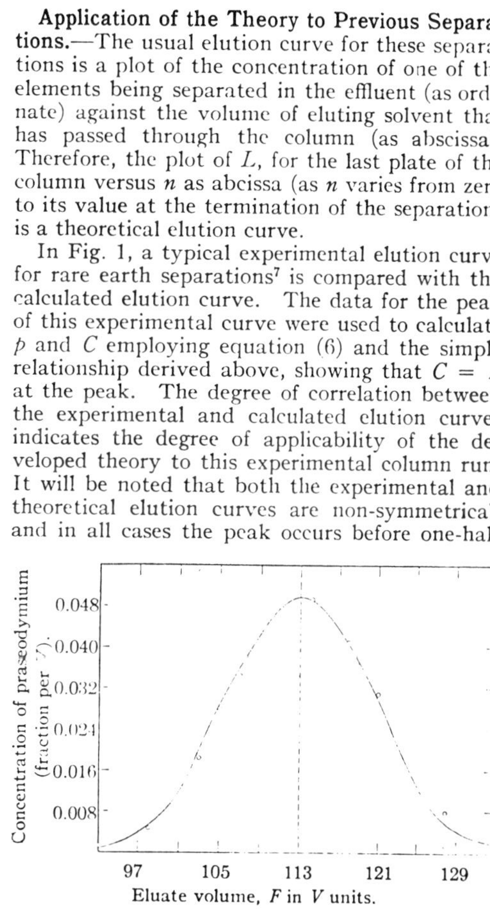

Fig. 1.—Correlation of experimental and theoretical elution curves: The full curve is calculated from (3) for $p = 240$ and $C = 113$. The experimental points are for the Pr-Ce-2 separation.[7] From the location and magnitude of the maximum Pr concentration found in the eluate, it was calculated (from equation (6)) that $p = 240$ and $C = 113$.

(7) D. H. Harris and E. R. Tompkins, THIS JOURNAL, **69**, 2792 (1947).

the material has been eluted. It is obvious, however, that the asymmetry is greater in this experimental curve than in the curve predicted by the theory.

Analysis of the Separations.—The degree of separation of two or more solutes during their elution depends upon the differences in their distribution coefficients. Determination of F at the peaks gives the corresponding C's of the materials being separated, since F and C are numerically equal at any peak. Therefore, the intervals between the peaks of the elution curves will be a measure of the differences between their C's in units of F. No matter how complete the separations may be, there will be some overlapping between the elution curves since the column must be infinitely long or the C's must differ by infinity to secure complete separation. A convenient graphical method has been developed for determining the fraction of one material present in the other (*i. e.*, the degree of separation) at any point on the elution curves. Using primes to distinguish between solute 1 and solute 2, we can write from (3)

$$\frac{L'_{n,p}}{L''_{n,p}} = \frac{\dfrac{(n+p-1)!}{(n-1)!\,p!}\,\dfrac{C'^{n-1}}{(1+C')^{n+p}}}{\dfrac{(n+p-1)!}{(n-1)!\,p!}\,\dfrac{C''^{n-1}}{(1+C'')^{n+p}}}$$

or

$$\frac{L'_{n,p}}{L''_{n,p}} = \text{antilog}\left[\log\frac{C'^{n-1}}{(1+C')^{n+p}} - \log\frac{C''^{n-1}}{(1+C'')^{n+p}}\right] \quad (7)$$

Inasmuch as p is constant in any given separation, each of the logarithmic expressions is a linear function of n (as is also their difference). Therefore, the complete curve for each of the logarithmic expressions (or the difference between them) can be drawn as a straight line when any two points have been calculated.

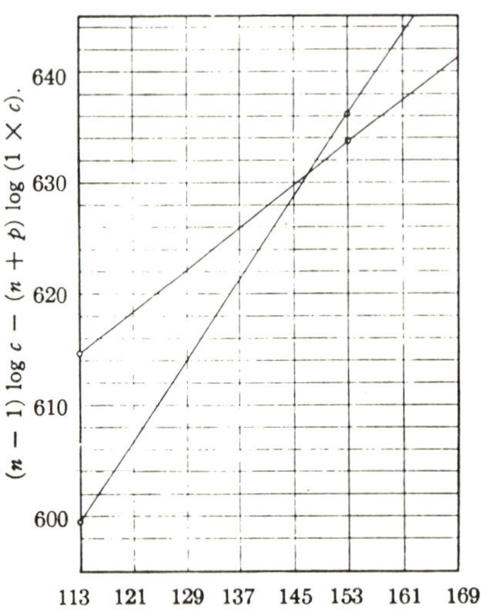

Fig. 2.—Calculated degree of separation for Pr-Ce-2.[7]

In Fig. 2, plots of the pair of lines have been made for the separation of praseodymium and cerium, by Harris and Tompkins.[6] Here C_{Pr} = 113 and the estimated C_{Ce} = 193. (The pH was changed at F = 165, so that C_{Ce} had to be estimated.) p was calculated to be 240. The two values of n plotted in this figure are F = 113 and 153. The ordinate axis is logarithmic, and the ratio $L'_{n,p}/L''_{n,p}$ is equal to the anti-logarithm of the ordinate distance between the pair of lines at points of equal F. (To calculate the quantity ratio, the ratio of fractions L'/L'' must be multiplied by the ratio of the initial quantities of the two substances put on the column.)

From Fig. 2, it is apparent that the maximum fraction of impurity occurs at F = 147, where $L'_{n,p} = L''_{n,p}$ Analytical data on the eluate showed that overlapping occurred in the range from F = 146 to 151. If it is assumed that the spectrographic method used will just detect 1% Pr in the presence of Ce (or *vice versa*) in the milligram quantities of rare earth obtained in that center portion of the elution curve, there will be detectable impurity when the pair of lines in Fig. 2 are less than two units apart at equal F. It can be seen that the lines are within two ordinate units of each other over a range of five column free-volumes of solution. This calculation is in agreement with the above analytical data. On the other hand, if 0.1% purity can be detected in this eluate, the calculated range will be 7.5.

The graphical method for calculating separation can be applied to conditions which approach those for the separation of isotopes. A calculation was made for the case where C' = 50, C'' = 51; p = 180, 900, and the criterion was a tenfold enrichment of one isotope in the other. For 180 plates the maximum enrichment (at the very beginning of the effluent) is 14 and the 90% isotope is obtained for 0.34 of the distance to the first peak. For 900 plates, however, the maximum enrichment is 10^7, and the 90% isotope is obtained for 0.77 of the distance to the first peak.

The expression for the volume of eluate that is cross-contaminated may be derived as follows. Any given ratio $L'_{n,p}/L''_{n,p}$ may be selected as the criterion for cross-contamination, and $\log L'_{n,p}/L''_{n,p}$ will be designated by the symbol I. The limits of the cross-contamination flow are determined by two values of n: n_1 (when solute one begins to be contaminated with solute two); n_2 (when cross-contamination of solute two by one stops). From (7)

$$\log\frac{L'_{n,p}}{L''_{n,p}} = (n-1)\log C' + (n+p)\log(1+C'') - (n-1)\log C'' - (n+p)\log(1+C') \quad (8A)$$

$$I = n_1[\log C' + \log(1+C'') - \log C'' - \log(1+C')] + \{-\log C' + p\log(1+C'') - \log C'' + p\log(1+C')\}$$

or $+I = n_1K + J$, where K and J are constants equal to the quantities in the brackets and braces, respectively. Similarly, $-I = n_2K + J$. The cross contaminated flow, $n_{cc.} = n_2 - n_1 = -2I/K$. But $n = pF$. Therefore

$$F_{cc.} = \frac{-2I}{pK} \quad (8B)$$

where $F_{cc.}$ denotes the cross contaminated flow.

The above derivation shows that the fraction of eluate in which insufficient separation has occurred varies inversely as the number of theoretical plates, or (other conditions being equal) inversely as the column length. Therefore, increasing the number of theoretical plates increases separation. But the relationship showing that the maximum concentration varies as $[\sqrt{2\pi}Cp(1+C)]^{-1/2}$ leads to the conclusion that the maximum concentration obtainable varies as $p^{-1/2}$. Accordingly, the longer the column, the smaller the maximum concentration. In addition, increasing the length of the column increases the time required for separation inasmuch as the rate of eluate flow is maintained constant.

Calculation of Material Eluted.—Since an elution curve is a plot of concentration *vs.* volume, the total material eluted is represented by the area under the elution curve. As has been pointed out by Martin and Synge[6] the normal curve of error can be used to approximate the theoretical elution curve, the approximation becoming closer as p and C become larger and as the column operating conditions approach equilibrium. Tables of the area under the normal curve of error are readily available. These tables give the area corresponding to values of the sigma unit t (t is defined as n/σ, with the peak at the origin). The ordinate of the normal curve at the peak $= [\sigma\sqrt{2\pi}]^{-1}$. The ordinate of the theoretical elution curve at the peak $= [2\pi pC(1 + C)]^{-1/2}$. Equating, $\sigma = [pC(1 + C)]^{1/2}$, and therefore

$$t = \frac{n - n_{max.}}{[pC(1 + C)]^{1/2}} \quad (9)$$

where n max is the value of n at the peak of the elution curve. Since $n_{max.} = pC$, this equation may be written in the form

$$t = \frac{p(F - C)}{[pC(1 + C)]^{1/2}} = \frac{p^{1/2}(F - C)}{[C(1 + C)]^{1/2}} \quad (10)$$

When C is large, C and $1 + C$ are almost equal, and

$$t = p^{1/2}(F - C)/C \quad (11)$$

Since problems involving the use of the column very frequently revolve about the questions of the degree of separation and the fractions of materials recovered, the application of equation (9) in conjunction with equation (7) is useful. As an illustration, we may refer to the previous example of calculations for isotope separation where $p = 900$, and the 90% isotope is obtained through 0.77 of the distance to the first peak. If we wished to know the quantity of 90% isotope recoverable in such a column, we can proceed as follows: $n_{max.} = pC = 900 \times 50 = 45{,}000$. $n = 0.77 \times 45{,}000$. Substituting in (9), we have $t = -6.9$. From a published table, it can be seen that this value of t corresponds to an area of less than 10^{-10}, so that a negligible portion of the isotope will be eluted 90% pure.

To calculate the enrichment factor of one isotope in the eluate when one per cent. of that isotope has been eluted, the procedure is as follows. From the table we see that t is -2.33 when the area under the curve is 0.01. Substituting in (9), we find that $n - n_{max.} = 3535$. Therefore $n = 41{,}465$. Substituting this value in (7), gives a value of 3.4 for the enrichment factor. By a comparable calculation, it is possible to ascertain the length of a column necessary to produce any fraction of each of the cations of the original mixture in any desired purity.

In the praseodymium and cerium separation of Harris and Tompkins[7] referred to above, F at the point where impurities are less than 0.01% is equal to 141. Therefore, from (11), $t = 240(141 - 113)/113 = 3.8$. Thus, the recovery of Pr in a purity greater than 99.99% is calculated to be 99.99%. In another separation, Nd-Pr-Ce-1, there was a relatively long flow of eluate between the praseodymium and cerium peaks in which the eluate contained less than 0.001% of the rare earth per ml.[7] To see whether the calculated recoveries agree with this experimental observation, one proceeds as follows (after making corrections for pH changes in the elutrient). The value for p, as calculated from the heights of the neodymium and praseodymium peaks (the Ce fractions obscured its peak), is 230 and 235, respectively. The flow between $F = 164$ and $F = 197$ contained less than 0.001% rare earth per ml. At $F = 164$, $t_{Pr} = 7$. Therefore, less than 0.01% of the Pr remained to be eluted. At $F = 197$, $t_{Ce} = -3.0$, and therefore only 0.13% of the Ce had been eluted when $F = 197$. Thus, the calculations show that only 0.14% of the two rare earths was eluted in the 1070 ml. of eluate between $F = 164$ and $F = 197$.

Table I was calculated from (7) and (9) for the conditions of a typical rare earth separation where $C' = 100$ and $C'' = 150$. It gives the per cent. of the rare earth that has been eluted before an eluate of less than the tabulated purity is obtained.

TABLE I

PER CENT. OF RARE EARTH ELUTED IN GIVEN PURITY

p	>99.9% purity	>99.99% purity
100	33	21
250	99.77	97.6
500	99.99999	99.9999
1000	99.99.....(to 15 places)	99.99...(to 10 places)

From these calculations it appears that in the rare earth separations described by Harris and Tompkins[7] (where $p \cong 250$), 97.6% of one rare earth was eluted before 0.01% impurity of the second appeared in the eluate.

Distribution on the Column.—The quantity of material in any plate may be calculated from the following expression

$$S_{n,p} + L_{n,p} = \frac{(1 + C)\,(p + n - 1)^{p+n-1/2}}{\sqrt{2\pi}\,(n - 1)^{n-1/2}\,p^{p+1/2}}\,\frac{C^{n-1}}{(1 + C)^{n+p}} \quad (12)$$

From equation (12) the band width of a substance on the column may be obtained by determining the range of p's over which not less than any given fraction (0.1% for example) is present.

The theory leads to the following picture of material movement through the column during elution. The material with the lowest C will travel through most rapidly. The fraction of the column traversed by the band of any of the materials = F/C. The rate of movement of the band down the column is Dr/C where D = length of the column, and r = rate of flow in V's per unit of time. The distance between bands equals $DF(1/C' - 1/C'')$. The fraction of that distance in which any selected degree of overlapping occurs is proportional to p^{-1}(8B). Thus the bands separate as the flow of eluting solution moves them down the column. However, each band broadens out as it moves through the column, the maximum concentration varying as $p^{-1/2}$(6). The maximum concentration also varies as C^{-1} so that when several materials are eluted from a column, the heights of the successive peaks represent successively smaller fractions of the initial quantities of the corresponding materials. From the initial data on band movements near the top of the column, the values of C can be estimated and predictions can be made as to separation of two or more substances.

Test of the Theory for Near Equilibrium Column Operation

Although the fit of the experimental curve shown in Fig. 1 is fairly good, it is somewhat more asymmetrical than would be predicted by the theory. Most of the elution curves obtained in the rare earth separations are even less symmetrical than the one shown here. Some previous work by Harris and Tompkins on the Amberlite resins had shown the relationship of flow rate to the shape of the elution curve of a solute. As the rate was decreased the curves became more and more symmetrical, approaching a limit at rates of 0.05 to 0.10 ml. per sq. cm. per minute. For this reason, it seemed probable that the failure of the experimental curves to fit the one predicted from the theory might be due chiefly to non-equilibrium conditions within the column. Further substantiation of this hypothesis was found in the general observation of several investigators here and elsewhere that decreasing the resin particle size would increase the efficiency of separation.

From the above considerations, it seemed that column runs at much slower rates, using resin of very fine mesh size, should be made to test the theory under conditions approximating equilibrium. Operating at very slow rates has the disadvantage of requiring a long time for each run. An alternative to this is to operate the columns at higher temperatures which is possible with Dowex 50.[8] The advantages of operating columns at higher temperatures was demonstrated by Ketelle[9] who was working on the separation and identification of the radioisotopes of the heavy group of rare earths. The separation of these elements, which was obtained by operating the columns at 100°, was much better than any previous separations of this group of elements. Furthermore, the elution curves were very nearly symmetrical. The chief disadvantage of using the higher temperature is the difficulty of keeping air from condensing in the column. It is necessary to remove almost completely the dissolved air from the influent solutions or the column will soon become "air bound" and the flow will stop.

Recently some additional samples of Dowex 50, consisting of agglomerates of colloidal particles, were obtained.[10] Examination of this material by the electron microscope showed that the size of the individual particles was about 0.05 micron. Since the material was obtained in agglomerates varying widely in size, it was possible to separate a fraction of the agglomerates of any mesh size desired. Such a fraction, in the range of 250 to 300 mesh, was prepared and used in some preliminary experiments designed to test its effectiveness as an exchanger for the column separations. These experiments showed that the separations which could be made by using the agglomerates of colloidal resin at room temperature were comparable, or somewhat superior, to those obtained by the use of ordinary Dowex 50 resin (with particle size about equal to that of the agglomerates) at 100°

Apparatus and Method.—The column separations were made in the usual manner[7] with the exception that conditions which would give considerable overlap of the elution curves of the cations being studied were chosen. The columns used in these studies were about 10 cm. and 15 cm. in length. The rare earth pairs Ce, Pr, and Eu, 61 were used in the two series reported here. To facilitate the study, radioisotopes of the rare earth elements were employed. The effluent from the column was passed through a device which continuously monitored and recorded the concentration of activity in this solution. The recorded curve (which was a plot of solute concentration in the eluate versus time) was the usual elution curve, since the flow rate was constant. In the study of cross contamination, samples were collected frequently, the time of their collection being marked on the recorded elution curve. This permitted a good comparison of the purity of

(8) L. A. Matheson, private communication.

(9) B. H. Ketelle and G. E. Boyd, THIS JOURNAL, **69**, 2800 (1947).

(10) Through the courtesy of W. C. Bauman, The Dow Chemical Co.

each sample, as determined experimentally, with that predicted from theory.

The degree of cross-contamination, was determined by a differential counting technique similar to that described earlier.[7] When Eu^{154} and 61^{147} were used a considerable degree of precision was possible since the former has a gamma radiation while the latter has only a weak beta radiation.[11]

Effect of Flow Rate and Particle Size on Separations.—From the theory, the relative positions of the peaks of two rare earths being separated by elution from a column would not be expected to vary appreciably with flow rate, unless the rates of the reactions of the two elements were considerably different. On the other hand, it was expected that the actual position and height of each peak and the degree of asymmetry of the elution curves would be very dependent on the flow rate. To test this and to determine the conditions necessary for near-equilibrium column operation, a series of preliminary column runs were made.

In these preliminary experiments, a column 0.855 sq. cm. × 15 cm. of colloidal agglomerates and another of equivalent size, utilizing the regular Dowex 50 resin of similar mesh size, were employed. A mixture of Pr^{143} and Ce^{141} were eluted from these columns under identical conditions and the elution curves from the two columns compared. The results of this study are shown in Fig. 3. The elution was made under conditions usually considered very unfavorable for the separation[12] (*i. e.*, 5% citrate at *p*H 3.5) and yet considerable fractionation of the two elements occurred. A careful study of the elution curves 1, 2 and 3 shows their close approximation to symmetry. Increase in the flow rate or resin particle size resulted in asymmetry and poorer separations as evidenced by curves 4, 5 and 6.

After the conditions for near-equilibrium column operation had been ascertained (Fig. 3), another separation, designed to test the theory, was made. For this experiment radioisotopes of the rare earth elements europium and 61 (in trace concentrations) were used, since they are easily determined in the presence of each other by differential counting, due to the difference in the characteristics of their radiations. They were first adsorbed at the top of the bed and then eluted with citrate of 0.25 *M* concentration at a *p*H of 2.90. The solution was 0.294 *M* in NH_4^+.

These studies were made with a column 1.0 sq. cm. × 10.4 cm. of colloidal agglomerates of Dowex 50. The weight of the resin (oven-dried basis) was 2.83 g., its density 1.45 g./cc. Thus the volume of solution, *V*, in the bed was 81.9% of the total bed volume or 8.84 ml. It has been shown[2] that

$$K_d = \frac{M_s}{M_l} \times \frac{\text{solution volume (ml.)}}{\text{resin mass (g.)}} = C \times \frac{\text{ml. of solution}}{\text{grams of resin}} \quad (13)$$

(11) Plutonium Project, THIS JOURNAL, **68**, 2411 (1946).

(12) E. R. Tompkins, J. X. Khym, W. E. Cohn, *ibid.*, **69**, 2769 (1947).

Hence

$$C \text{ will be equal to } K_d \times \frac{\text{grams of resin}}{\text{ml. of solution}} = K_d \frac{2.83}{8.84} \quad (14)$$

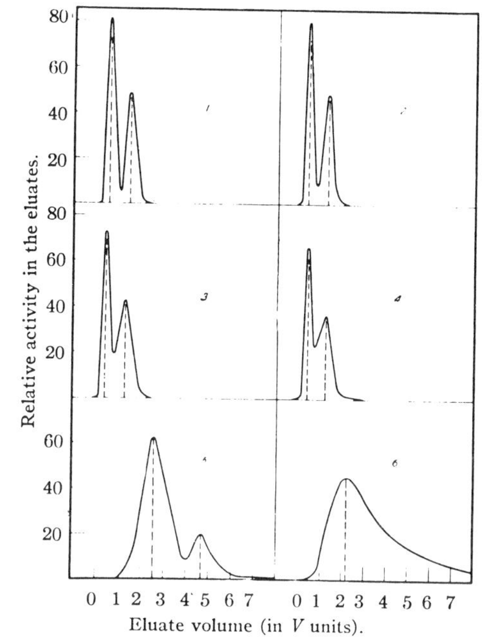

Fig. 3.—The effect of resin particle size and flow rate on the separation of Pr and Ce[a]: elutrient solution, 0.25 *M* citrate, *p*H 3.50; columns, 0.867 sq. cm. × 15 cm., Dowex 50 ammonium form, (1) colloidal agglomerates of Dowex 50, (2) Dowex 50, 170–200 mesh. In the following designations the first number represents the column used and the second the number of the run on that column.

DESIGNATIONS OF CURVES

		Flow rate, ml./sq.cm./minute
1.	Pr-Ce-1-2	0.053
2	Pr-Ce-1-5	0.207
3	Pr-Ce-1-3	0.627
4	Pr-Ce-1-4	1.152
5	Pr-Ce-2-2	0.026
6	Pr-Ce-2-1	0.230

[a] In runs Ce-Pr-2 to 5, a 50-mg. Al absorber was used during the Pr elution. This reduced the heights of the Pr curves to 43% of the value which would have been obtained if the absorber had not been used.

To determine K_d at equilibrium, 0.275 g. and 0.278 g., respectively, of this resin was shaken with 1.0 ml. of the elutrient solution containing Eu and 61 tracers. The values obtained for the K_d's were 29.9 for 61 and 19.4 for Eu. Thus the predicted values for the *C*'s were 9.56 for 61 and 6.20 for Eu. Since $C = F_{max.}$, it was possible to test the theory by these elutions at near-equilibrium conditions. Figure 4 shows the elution curves for Eu and 61 from this column, when they were eluted separately. The values for the *C*'s determined by the elution were 9.41 for 61 and

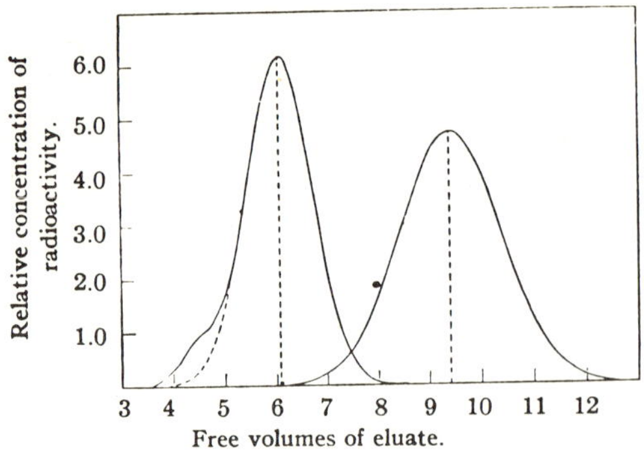

Fig. 4.—Elution curves of europium and element 61.

6.08 for Eu. The ratio of the distribution coefficients (K_d 61/K_d Eu) as determined at equilibrium was 1.542, while that determined by the column run was 1.548. A further comparison was made between the values for K_d, found here, and those in the equilibrium experiments in which Dowex 50 of 40–60 mesh was used.[2] The results of these comparisons (when corrections for differences in NH_4^+ concentration have been made) check within 3%.

To test the applicability of the theory for predicting the cross-contamination of one rare earth with the other, a further column run, in which both elements were eluted simultaneously, was made. Figure 5 shows a reproduction of the continuously recorded curves in this elution, the dotted line extensions being calculated from the theory. Conditions which would give a sufficient overlap to test the theory were chosen. It can be seen that the shapes of these curves closely approximate those for the two elements eluted separately (Fig. 4).

It is necessary to calculate the number of theoretical plates in the column to determine how nearly the degree of cross-contamination observed in this experiment conforms with that predicted from the theory. From equation (6), at the maximum

$$L_{C_{p,p}} = \frac{1}{p^{1/2}\sqrt{2\pi C(1+C)}} \text{ and } pL_{C_{p,p}} = \frac{p^{1/2}}{\sqrt{2\pi C(1+C)}}$$

From Fig. 4, for element 61, $C = 9.41$ and

$$pL_{C_{p,p}} = \frac{4.75 \times 1.00}{4602 \times 0.0025}$$

Thus $p = 103$. Calculation of p for the europium elutions gave results agreeing to within 2%.

Table II shows the comparison between the composition of each of the fractions in the over-lap region of the curves, as determined experimentally, with the values calculated from equation (7). As may be seen, there is good agreement.

The number of V's of solution in which detectable over-lap occurs may now be calculated as shown above, (8B). The criterion selected is that degree of impurity which is just detectable by the differential counting technique used here. This is 1%.

$$F_{co.} = \frac{2 \times 2}{103[\log 6.08 + \log 10.41 - \log 9.41 - \log 7.08]}$$

$$= \frac{4}{103 \times 0.0223} = 1.74$$

Thus 1.74 V's of solution contain cross-contaminated rare earths. Figure 5 shows 1.71 V's of cross-contaminated solution. The yields of each of the two elements in a purity of 99% may be calculated from equations (7) and (10) and show that 97.3% of element 61, and 95.8% of the europium were recovered in this purity.

Table III shows the comparison between the experimental values for the fraction of solute in succeeding eluate volumes and those values calculated from (10).

Discussion

The resin bed of an exchange column consists of a highly cross-linked organic matrix with attached

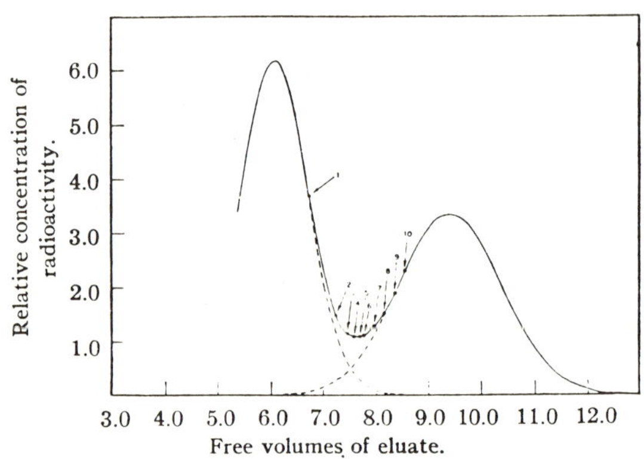

Fig. 5.—Cross-contamination in the eluates of europium and element 61.

TABLE II

CROSS-CONTAMINATION—COMPARISON OF EXPERIMENTAL DATA WITH THEORY

	Volume (V) of eluate, free-vols.	Observed ratio of Eu to 61	Fractional ratios of Eu/61 Experimental[a]	Theoretical
1	6.74	>100	>80	110
2	7.23	8.18	6.7	8.0
3	7.47	2.45	2.0	2.14
4	7.57	1.50	1.22	1.3
5	7.67	.762	.62	.74
6	7.73	.423	.34	.48
7	7.98	.164	.133	.135
8	8.18	.064	.052	.050
9	8.36	.024	.020	.021
10	8.57	.006	.005	.007

[a] Corrected for original ratio of Eu to 61.

TABLE III

AREA UNDER THE ELUTION CURVE—COMPARISON OF EXPERIMENTAL DATA WITH THEORY

(Element 61)

Volume of eluate, free-volumes From	To	Area units under experimental curve	Fraction of total area under curve Experimental	Theory
6.0	6.5	4	0.0009	0.0011
6.5	7.0	25	.0054	.0054
7.0	7.5	85	.0185	.0182
7.5	8.0	215	.0467	.0485
8.0	8.5	470	.1021	.1027
8.5	9.0	748	.1625	.1610
9.0	9.5	927	.2014	.1987
9.5	10.0	880	.1912	.1899
10.0	10.5	652	.1417	.1418
10.5	11.0	370	.0804	.0809
11.0	11.5	152	.0330	.0352
11.5	12.0	58	.0126	.0122
12.0	12.5	13	.0028	.0032
12.5	13.0	3	.0007	.0007
		Total 4602	1.00	1.00

sulfonic acid groups which provide points of combination for cations. The number of these exchangeable positions in any column is the product of the number of equivalents of exchanger times Avogadro's number. If a cation, M^+, is first adsorbed on an exchange position at the top of a resin bed of NR and then a solution of cations, N^+ is passed through the column, one of these N^+ ions will replace the M^+ from the MR; M^+ will then be carried down the column by the flowing solution. As it passes other exchange positions, NR, there will be a tendency for the reverse reaction to take place and so for it to be readsorbed. This ion, M^+, will therefore proceed down the column, being successively adsorbed and desorbed, until it finally appears in the eluate from the bottom of the column bed.

In describing the behavior of the M^+ ion in its route through the column, it is necessary to determine the number of times the exchange reactions take place. This will be governed by such factors as the rates of the exchange reactions themselves, the rates of diffusion of the ions and the rate of the movement of the solution through the column. Thus, as the ion passes any possible point of attachment there will be some finite probability that it will exchange. When large numbers of ions are eluted simultaneously, each will behave individually according to the same probability, which is dependent upon the rate factors named. Thus, it is to be expected that in the usual elution, in which more than 10^{10} ions are eluted by chromatographic development through the column, the shape of the elution curve will approach that of the probability curve (or the normal curve of error) as the flow rate approaches zero and, therefore, as equilibrium conditions are approximated. This expectation has been amply demonstrated experimentally in the series of elutions of trace quantities of several rare earths from colloidal resin columns, in which near-equilibrium conditions prevailed throughout the column.

When the quantity of solute is very small as compared to the possible exchange positions in the resin bed, it may be adsorbed originally in a very narrow band at the top of the column. As the elution is started and these ions move down the column there is an upper limit to the number of exchanges that can take place, *viz.*, the average number of exchange positions passed by each ion as it moves down the column. In the usual column separation, the number of transfers of an ion is only a very small fraction of this limiting value. As equilibrium in the column is approached, this number of transfer units, or plates approaches a limit. When this number of plates and the equilibrium distribution coefficient are known for an elution under any given set of conditions, the elution curve may be calculated as demonstrated above.

When the quantity of solute is so small as to saturate a very small fraction (*e. g.*, less than 10^{-6}) of the column area. On the other hand, when the quantity of solute is relatively large, an increase in the column area should benefit the separation, as has been previously demonstrated by a number of investigators.

In those cases where the quantity of solute is sufficient to saturate an appreciable fraction of the resin bed, each ion will undergo fewer transfers per unit length of the column, during its elution. This is a consequence of the increased competition for the available exchange positions in any horizontal plane of the resin bed. Thus, the separation between two solutes (which depends upon the number of such transfers of the solute ions) will be poorer than in the case where the solutes are present in trace concentrations. In addition, if the concentration of the solute ions in the solution exceeds a certain critical value (3×10^{-4} M for Dowex 50 and 4×10^{-7} M for IR-1), the equilibrium distribution coefficient is affected,[2] still further decreasing the degree of separation of the two solutes.

The importance of the use of a complexing agent in increasing the separation of relatively large quantities of solutes has been amply demonstrated. In some of the rare earth separations, described in an earlier paper,[7] as much as 15 to 20% of the column bed was saturated with the solute. The separation in these cases was still very good as long as the concentration of the solute ions in the liquid phase did not exceed the critical value given in the preceeding paragraph. Because the dissociation constants of the complexes of the rare earth citrates are small, the total rare earth concentration in the eluates from these columns was several orders of magnitude larger than their combined ionic concentration.

For laboratory experiments and other small scale separations, it is possible to operate columns under conditions approaching equilibrium. In such cases, the theory described above predicts the degree of separation which may be expected. Its present limitations do, however, decrease its applicability to large scale industrial processes in which columns must be operated in such a way that the distribution coefficient may not remain constant throughout the bed and equilibrium conditions are not approached. To be generally useful for the development of such large scale, industrial separations processes, it will be necessary to extend this theory by a further study of the kinetics of column reactions and the rates of diffusion of the various substances entering into these reactions.

Summary

A simple theoretical analysis of the ion exchange column separations process has been developed for the case in which the distribution coefficient of the solute between the solid and liquid phases remains constant throughout the column. Application of this theory to the rare earth separations, reported earlier by one of the authors, and for the prediction of separation in another column separation, demonstrates its usefulness. The elution curves derived from the theory fit the observed data within the limits of experimental error when near-equilibrium conditions prevail in the column. The theory is also applied to determine the purities of the rare earths separated by this process. The general applicability of the theory to processes of this type is discussed.

P. O. Box W, Oak Ridge, Tenn. Received July 14, 1947

3

Reprinted from *Trans. Faraday Soc.*, **51**(385), P. 1, 34–44 (1955)

THEORY OF CHROMATOGRAPHY

PART 9. THE "THEORETICAL PLATE" CONCEPT IN COLUMN SEPARATIONS

BY E. GLUECKAUF

Atomic Energy Res. Establishment, Harwell, Didcot, Berks

Received 10*th March*, 1954

The "theoretical plate" treatment for chromatographic separations has been re-examined, and the "discontinuous flow treatment" of Mayer and Tompkins has been replaced by a "continuous flow" model. The effect of finite width of the feed band, the purity of separated products, and the effects of non-equilibrium in the particles have been discussed for the purpose of predicting column performance.

GENERAL CONSIDERATIONS

Martin and Synge [1] were the first to realize the similarity of the chromatographic elution technique with the process occurring in distillation columns, and they developed on this basis a theory which has been very fruitful for the understanding of separations of substances following linear adsorption or exchange isotherms. Their theory has been expanded by Mayer and Tompkins [2] so as to render possible the prediction of the number of "theoretical plates" needed to obtain a required purity of the separated products. However, the calculations of the latter suffer from the fact that the physical picture on which they are based is inadequate. The shortcoming is not a minor effect. In a column of 1000 theoretical plates with a separation factor of $K_{d2}/K_{d1} = 1{\cdot}2$ between two substances, and a ratio of (amount in pore space)/(amount in adsorbent) = 1/5, the impurity ratio of the products calculated according to Mayer and Tompkins is 3 times smaller than it actually is. This discrepancy is not confined to low separation factors, and is the same, e.g., for $K_{d2}/K_{d1} = 1{\cdot}5$ and a correspondingly shorter column of 160 theoretical plates. The error introduced is particularly large, when the theory is applied to cases with $K_d < 1$ (see, e.g., ref. (6)) where the number of theoretical plates calculated errs by more than 100 %.

This result is due to the Mayer and Tompkins model, where the operation is visualized as a discontinuous process, in which a finite solution volume, namely the free volume of a theoretical plate, is equilibrated step by step with one theoretical plate full of adsorbent after another. In reality the flow of solution is continuous, even if the column itself is divided up into finite sections. The difference between a continuous and a discontinuous operation of the solution flow can be easily assessed mathematically.

SYMBOLS

A = cross-sectional area of column,
$A_\varepsilon\{t\}$ = area of the normal curve of error for the argument t,
$a = f_{(c)}/c$ = independent of c for linear isotherm,
α = void fraction of column,
β = width of band at the concentration $c = c_{\max}/e$,
c = concentration of solute (mequiv. cm^{-3}),
η = impurity ratio of separated band, e.g. $\eta_2 = \Delta m_1/(m_2 - \Delta m_2) \simeq \Delta m_1/m_2$,
$f_{(c)}$ = amount of solute contained in 1 cm^3 of column volume when adsorbent and solution is in equilibrium; $f_{(c)} = f^*_{(c)} + \alpha c$ (in mequiv. cm^{-3} of column),

K_d = distribution coefficient of solute between sorbent and solution $= f^*_{(c)}/c$,
l = distance from top of column (cm),
Δl = height of theoretical plate (cm),
m_1, m_2 = number of mequiv. of solute loaded on the column,
M = solution parameter $= v/a\Delta x$ = number of "theoretical plate elution volumes",
N = column parameter $= l/\Delta l = x/\Delta x$ = number of theoretical plates,
$N' = N - \frac{1}{2}N_0$,
q = amount of solute contained in 1 cm³ of column volume including both adsorbed and solution phase; $q = q^* + \alpha c$ (mequiv. cm^{-3} of column),
v = volume of eluting solution,
$\bar{v}$ = peak elution volume $= aX' \simeq aX$,
$\tilde{v} = \sqrt{\bar{v}_1\bar{v}_2}$,
X = total column volume $= AL$,
$X' = X - \frac{1}{2}x_0$,
x = column volume of length $l = Al$,
Δx = volume of a theoretical plate $= A\Delta l$,
c° = concentration of inflowing solute (at $x = 0$),
c_0, q_0, x_0, N_0 = parameters referring to concentration and width of original load band.

Let us picture our column as a discontinuous medium divided into units each of which represents an effective theoretical plate. Hereby we define an "effective theoretical plate" as a unit of length, within which we shall treat the concentrations of solute as uniform, both in the sorbent phase and in the solution phase, the two concentrations being assumed to be in equilibrium. Such a model is strictly applicable only at very low flow-rates where equilibrium is attained between solution and sorbent during the time of passage of the restrictions, but it can cope also with non-equilibrium operation, with certain adaptations which will be discussed elsewhere.[5] Let us consider the material balance in such a unit of thickness Δl. If the column has a cross-section of A cm², its volume will be $\Delta x = A\Delta l$. The amount adsorbed per unit volume of column (not necessarily at equilibrium) we may signify q. The distance from the top of the column may be measured either in cm (l) or in column volume (x), or by the number of theoretical plates ($N = x/\Delta x = l/\Delta l$) above and including the theoretical plate concerned.

After a volume v of solution has already passed through the column, a further amount Δv will enter the Nth section with the concentration $c_{(x-\Delta x)}$ and leave it with the concentration c_x. The amount in the section has thus changed by $v(c_{(x-\Delta x)} - c_x)$. When the volume v had passed, the section has the content $\Delta x q_{(v)}$, which after the further amount Δv of solution has changed to $\Delta x q_{(v+\Delta v)}$. It follows that

$$\Delta x(q_{(v+\Delta v)} - q_v) = \Delta v(c_{(x-\Delta x)} - c_x). \tag{1}$$

Small differences of this type can generally be expressed by means of differentials, according to Taylor's theorem,

$$c_{(x-\Delta x)} = c_x - \left(\frac{\partial c}{\partial x}\right)_v \Delta x + \left(\frac{\partial^2 c}{\partial x^2}\right)_v \frac{\Delta x^2}{2} -, \text{ etc.,} \tag{2}$$

and

$$q_{(v+\Delta v)} = q_v + \left(\frac{\partial q}{\partial v}\right)_x \Delta v + \left(\frac{\partial^2 q}{\partial v^2}\right)_x \frac{\Delta v^2}{2} +, \text{ etc.,} \tag{2a}$$

third differentials being quite negligible if

$$\Delta x \ll x.$$

We obtain therefore the mass balance equation:

$$\left(\frac{\partial c}{\partial x}\right)_v + \left(\frac{\partial q}{\partial v}\right)_x - \frac{\Delta x}{2}\left(\frac{\partial^2 c}{\partial x^2}\right)_v + \frac{\Delta v}{2}\left(\frac{\partial^2 q}{\partial v^2}\right)_x = 0. \tag{3}$$

If the solution flow is continuous, then this is equivalent to the volume units Δv being infinitesimally small, and eqn. (3) simplifies to

$$\left(\frac{\partial c}{\partial x}\right)_v + \left(\frac{\partial q}{\partial v}\right)_x - \frac{\Delta x}{2}\left(\frac{\partial^2 c}{\partial^2 x}\right)_v = 0 \tag{4}$$

and herein lies the difference between continuous and discontinuous column operation. In order to make the difference clearer we may rearrange eqn. (3). If the fractional void space is α, then

$$\Delta v = \alpha \Delta x. \tag{5}$$

If we assume local equilibrium and a linear isotherm, the amount q of solute contained in one unit volume of the column is proportional to the concentration

$$q = ac. \tag{6}$$

Then,

$$\left(\frac{\partial^2 q}{\partial v^2}\right)_x = a\left(\frac{\partial^2 c}{\partial v^2}\right)_x = a\left(\frac{\partial^2 c}{\partial x^2}\right)_v\left(\frac{\partial x}{\partial v}\right)_c^2, \tag{7}$$

where $(\partial x/\partial v)_c$ is the rate of movement of a point of constant concentration c for unit volume.

If the number of theoretical plates is reasonably large then we can neglect, as an approximation, both second differentials and this leads to

$$0 \simeq \left(\frac{\partial c}{\partial x}\right)_v + \left(\frac{\partial q}{\partial v}\right)_x = \left(\frac{\partial c}{\partial x}\right)_v - \left(\frac{\partial q}{\partial x}\right)_v\left(\frac{\partial x}{\partial v}\right)_c \tag{8}$$

leading to

$$\left(\frac{\partial x}{\partial v}\right)_c \simeq \frac{\mathrm{d}c}{\mathrm{d}q} = \frac{1}{a}. \tag{9}$$

Utilizing eqn. (7) and (9), eqn. (3) for discontinuous flow can be written with a very good approximation

$$\left(\frac{\partial c}{\partial x}\right)_v + a\left(\frac{\partial c}{\partial v}\right)_x - \frac{\Delta x}{2}\left(1 - \frac{\alpha}{a}\right)\frac{\partial^2 c}{\partial x^2} = 0. \tag{10}$$

This may be compared with eqn. (4) valid for continuous flow, which for a linear isotherm takes on the form

$$\left(\frac{\partial c}{\partial x}\right)_v + a\left(\frac{\partial c}{\partial v}\right)_x - \frac{\Delta x}{2}\left(\frac{\partial^2 c}{\partial x^2}\right)_v = 0. \tag{4a}$$

The coefficient of the second differential is responsible for the spreading of the chromatographic band, as it passes through the column. Without going into details, we can therefore say, from a comparison of the coefficients of eqn. (10) and eqn. (4) that discontinuous operation would lead to sharper bands and hence to better separations.

The reason for this can be easily illustrated. Suppose we have a large vessel which we fill repeatedly from a cold water tap and then empty down the drains. If we suddenly changed to hot water, then a thermometer in the drain pipe will register a sharp jump in temperature just after the change has taken place. This is discontinuous flow. On the other hand, we may run the water into the vessel continuously and, through an overflow, into the drainpipe. A sudden switch from cold to hot will then result only in a gradual increase of the temperature which will reach its final value only after many vessel volumes have been flowing through.

As pointed out in the beginning, the two models can give very different results as regards separation efficiency. The outcome is very similar, if $a \gg \alpha$, but this condition is usually avoided in practice, as it increases the volume of solution required for the elution of the band and thus the duration of separation runs. It is therefore necessary to develop the theory of the " theoretical plate " concept on the basis of continuous solution flow, and the opportunity is used to extend the quantitative discussion at the same time to the influence of some experimenta factors, such as the length of the original band, i.e. of column loading, and to the efficiency of separation of two bands with unequal amounts of solute.

BREAK-THROUGH CURVE OF A SOLUTION OF CONCENTRATION c° FLOWING THROUGH A PREVIOUSLY EMPTY COLUMN

We apply the boundary conditions:

(i) for $v = 0$ and $x > 0$, $c = 0$,
(ii) for $x = 0$ and $v > 0$, $c = c^{\circ}$,
(iii) $q = ac$ (local equilibrium),

to eqn. (4):

$$\frac{\partial c}{\partial x} + \frac{a\partial c}{\partial v} - \frac{\Delta x}{2}\frac{\partial^2 c}{\partial x^2} = 0. \qquad (11)$$

For simplicity we transform to the dimensionless parameters,

$x/\Delta x = N =$ the number of " theoretical plates " up to the point x (12*a*)

and $v/a\Delta x = M =$ the number of " theoretical plate elution volumes " contained in the volume v. (12*b*)

Then eqn. (11) becomes

$$\left(\frac{\partial c}{\partial N}\right)_M + \left(\frac{\partial c}{\partial M}\right)_N - \frac{1}{2}\left(\frac{\partial^2 c}{\partial N^2}\right)_M = 0. \qquad (13)$$

with the boundary conditions:

$M = 0$, $N > 0$, $c = 0$,

and $N = 0$, $M > 0$, $c = c^{\circ}$.

The solution of this problem is

$$\frac{c}{c^{\circ}} = \tfrac{1}{2} - A_{\varepsilon}\left\{\frac{N-M}{\sqrt{M}}\right\} + \exp(2N)\left[\tfrac{1}{2} - A_{\varepsilon}\left\{\frac{N+M}{\sqrt{M}}\right\}\right]. \qquad (14)$$

Here $A_{\varepsilon}(t)$ is the area of the normal curve of error defined by

$$A_{\varepsilon}(t) = (2\pi)^{-\frac{1}{2}}\int_0^t \exp\left(-\frac{t^2}{2}\right)\mathrm{d}t \qquad (15)$$

which is tabulated as function of t in many places, the one most accessible to chemists being the *Handbook of Chemistry and Physics.*

For values of $N > 3$, which are the only ones of practical interest, the second term of eqn. (14) becomes completely negligible. Eqn. (14) for application to the data of the break-through curve may be written in the form

$$c/c^{\circ} = \tfrac{1}{2} - A_{\varepsilon}\left\{\sqrt{N}\frac{(\bar{v}-v)}{\sqrt{\bar{v}v}}\right\}, \qquad (16)$$

where

$$\bar{v} = aX \qquad (17)$$

is the elution volume of the centre of the S-shaped break-through curve ($c = c^{\circ}/2$), X being the column volume. Fig. 1 shows the eluted break-through curves for various values of N as function of $\log_{10}(v/\bar{v})$. Break-through curves of this type are a convenient way of determining the value of N of the column from experimental data. We read from the observed break-through curve the value of $\bar{v}$ and also the elution volume v' of the concentration given by

$$c'/c^{\circ} = \tfrac{1}{2} - A_{\varepsilon}\{1\} = 0{\cdot}1587.$$

Then we obtain

$$N = \frac{\bar{v}v'}{(\bar{v}-v')^2}.$$

ELUTION OF A BAND OF SOLUTE OF FINITE WIDTH

Let us consider a narrow band of solute of width l_0 (or x_0, or N_0) at the top of an otherwise empty column. An eluting solution is then added at the top which moves the band gradually down the column, until it appears in the eluate.

Again we apply eqn. (13), this time to the boundary conditions for

$$\begin{aligned} M &= 0, \quad 0 < N < N_0, \quad c = c_0 \\ M &= 0, \quad N > N_0, \quad c = 0 \\ M &> 0, \quad N = 0, \quad c = 0. \end{aligned}$$

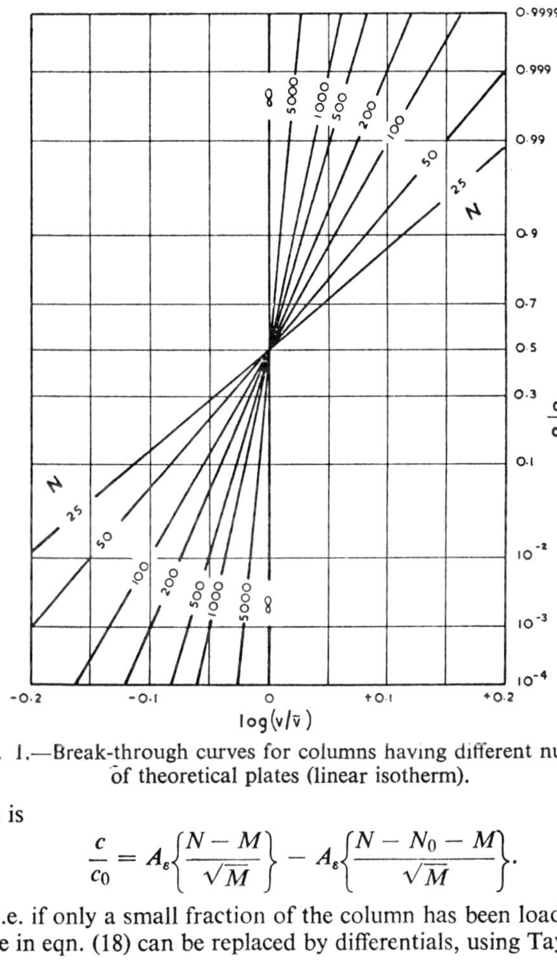

FIG. 1.—Break-through curves for columns having different numbers of theoretical plates (linear isotherm).

The solution is

$$\frac{c}{c_0} = A_\varepsilon\left\{\frac{N-M}{\sqrt{M}}\right\} - A_\varepsilon\left\{\frac{N-N_0-M}{\sqrt{M}}\right\}. \tag{18}$$

If $N_0 \ll N$, i.e. if only a small fraction of the column has been loaded with solute, the difference in eqn. (18) can be replaced by differentials, using Taylor's theorem:

$$\frac{c}{c_0} = N_0\frac{\partial}{\partial N}\left(A_\varepsilon\left\{\frac{N-M}{\sqrt{M}}\right\}\right) - \frac{N_0{}^2}{2}\frac{\partial^2}{\partial N^2}\left(A_\varepsilon\left\{\frac{N-M}{\sqrt{M}}\right\}\right) + \text{etc.,} \tag{19}$$

further terms being negligible.

According to eqn (1 ·) we have

$$\frac{\partial}{\partial N}A_\varepsilon\{t\} = \frac{1}{\sqrt{2\pi}}\exp\left(-t^2/2\right)\frac{\partial t}{\partial N} = \frac{1}{\sqrt{2\pi M}}\exp\left\{-\frac{(N-M)^2}{2M}\right\}, \tag{20}$$

$$\frac{\partial^2}{\partial N^2}A_\varepsilon(t) = \frac{1}{\sqrt{2\pi}}\exp\left(-t^2/2\right)\left[\frac{\partial^2 t}{\partial N^2} - t\left(\frac{\partial t}{\partial N}\right)^2\right] = \frac{M-N}{M\sqrt{2\pi M}}\exp\left\{\frac{-(N-M)^2}{2M}\right\} \tag{21}$$

Thus

$$\frac{c}{c_0} = \frac{N_0}{\sqrt{2\pi M}} \exp\left\{-\frac{(N-M)^2}{2M}\right\}\left(1 + N_0\frac{(N-M)}{2M}\right). \tag{22}$$

To obtain reasonably efficient separations we require narrow bands, which implies that we are interested only in conditions where $| N - M | \ll M$. Under these conditions we can replace the bracket by

$$\exp\left\{N_0\frac{(N-M)}{2M}\right\}, \text{ and } \sqrt{\frac{N}{M}} \text{ by } \exp\left\{\frac{N-M}{2M}\right\},$$

which gives

$$\frac{c}{c_0} = \frac{N_0}{\sqrt{2\pi N}} \exp\left(-\frac{(N-M)}{2M}(N - N_0 - 1 - M)\right)$$

$$\simeq \frac{N_0}{\sqrt{2\pi N}} \exp\left(-\frac{\left(N - \frac{N_0}{2} - \frac{1}{2} - M\right)^2}{2M}\right). \tag{23}$$

For applying this result to the elution curve we replace N and M again from eqn. (12*a*, *b*) by parameters more suitable for the interpretation of experimental data. Eqn. (22) or (23) then takes the form

$$c = \sqrt{\frac{N'}{2\pi}} \frac{m}{\sqrt{aX'v}} \exp\left(-\frac{N'}{2}(aX' - v)^2/aX'v\right), \tag{24}$$

where $m = ac_0x_0$ = amount of solute in the original band $= ac_0N_0\Delta x$,

$N' = N - \frac{N_0}{2} = \frac{X'}{\Delta x} = \frac{X - (x_0/2)}{\Delta x}$ = number of theoretical plates from the centre of the original band to the bottom of the column.

$aX' = v$ is the elution volume of the mass centre of the band which, for practical purposes coincides with the peak of the well-known bell-shaped elution band (see fig. 2). The form most suitable for a discussion of the elution curve is therefore

$$c = c_{\max} \exp\left\{-\frac{N'}{2}\frac{(\bar{v} - v)^2}{\bar{v}v}\right\}, \tag{25}$$

where the maximum peak height is given by

$$c_{\max} = \frac{m}{\bar{v}}\sqrt{\frac{N'}{2\pi}}. \tag{26}$$

There are two convenient ways to determine N' from the experimental data. Eqn. (26) gives

$$N' = 2\pi\left(\frac{c_{\max}\bar{v}}{m}\right)^2, \tag{27}$$

where $m = \int_0^\infty c\,dv$ is conveniently obtainable from the area under the elution peak. The other utilizes the elution volume v_e, where the concentration is $c = c_{\max}/e = 0{\cdot}368\ c_{\max}$ leading to

$$N' = \frac{2\bar{v}v_e}{(\bar{v} - v_e)^2} = 8\left(\frac{\bar{v}}{\beta}\right)^2, \tag{28}$$

where β is the band width at the concentration $c = c_{\max}/e$ (see fig. 2). These equations differ from those derived previously[2] (or derivable from their equations) in that they do not contain the "free column volume" of the earlier authors, which was entirely the result of their " discontinuous " treatment. It is interesting to note that eqn. (25), apart from negligible second-order terms discarded by the approximations made in eqn. (23), and apart from the effective shortening of the column by half the width of the original load band, shows no effect whatsoever of the width of the original band. This of course is only the case as long as we are permitted to replace the difference of eqn. (18) by the differentials of eqn. (19). We are entitled to do this as long as the original band is less than approximately $\frac{1}{4}$ of the band width when leaving the column, as otherwise the differentials of higher order can no longer be neglected. The mean width of the band when leaving the column follows from eqn. (28) as $\beta = \bar{v}\sqrt{8/N}$ so that, inside the column, just before leaving it, it had a length $x_\beta = X\sqrt{8/N}$. Clearly the original band width x_0 ought to be much smaller than this, say $\frac{1}{4}$, so that its value should be limited to

$$x_0/X < 1/\sqrt{2N}. \tag{29}$$

This means that the better the column performance (large N), the smaller must be the original band, so that the good performance is not upset on account of the starting width. For a 5000 plate column, x° should be less than 1 %, while for a 200 plate column, 5 % loading will not interfere with the separation performance, and even 10 % may still be tolerated, provided of course that a 200 plate column can satisfactorily perform the particular separation on a tracer scale.

PURITY OF SEPARATED PRODUCTS

The amount of solute in a given part of the band in the column, say from x to infinity, is

$$\Delta m = \int_x^\infty ac\mathrm{d}x = ac_0x_0\left[A_\varepsilon\left\{\frac{N-M}{\sqrt{M}}\right\}\right]_N^\infty = m\left[0{\cdot}5 - A_\varepsilon\left\{\frac{N-M}{\sqrt{M}}\right\}\right]. \tag{30}$$

This follows directly from eqn. (19) for bands, where the original width x_0 is sufficiently small to neglect the second term of eqn. (19), which as we have shown, has practically no effect on the shape of the band.

Let us now consider two bands which have been eluted from the same column (i.e. the column parameter N is identical). Even after they have been well separated there is always some overlap as all bands reach from $v = 0$ to $v = \infty$, if we have continuous solution flow. (With the " discontinuous " operation no solute arrives with the first free column volume!) If we cut the eluate into two fractions (see fig. 2), at an elution volume v between the two peaks then the minor cut-off fractions are

$$\Delta m_2 = m_2\left[0{\cdot}5 - A_\varepsilon\left\{\frac{N-M_2}{\sqrt{M_2}}\right\}\right], \tag{31}$$

and

$$\Delta m_1 = m_1\left[0{\cdot}5 - A_\varepsilon\left\{\frac{M_1-N}{\sqrt{M_1}}\right\}\right]. \tag{32}$$

Let us first consider the case where $m_1 = m_2$. Then if one requires equal purity in both bands the best elution volume for making the cut is given by $\Delta m_1 = \Delta m_2$. The equality leads to

$$\sqrt{M_1M_2} = N, \tag{33}$$

which in terms of our experimental parameters means a cut at

$$v = X\sqrt{a_1a_2} = \sqrt{\bar{v}_1\bar{v}_2}, \tag{34}$$

i.e., the best point for the cut is the geometric mean of the two peak elution volumes.

We may define the impurity content of the band 2 as

$$\eta_2 = \frac{\Delta m_1}{m_2 - \Delta m_2} \simeq \frac{\Delta m_1}{m_2}, \text{ if } \Delta m_2 \ll m_2,$$

and vice versa for band 1, and obtain from eqn. (31) and (32) and from the condition $\Delta m_1 = \Delta m_2$,

$$\eta_{1:1} = [\eta]_{m/m_2=1} = 0{\cdot}5 - A_\varepsilon\left\{\sqrt{N}\frac{(\sqrt{\bar{v}_2} - \sqrt{\bar{v}_1})}{\sqrt[4]{\bar{v}_1\bar{v}_2}}\right\}. \quad (36)$$

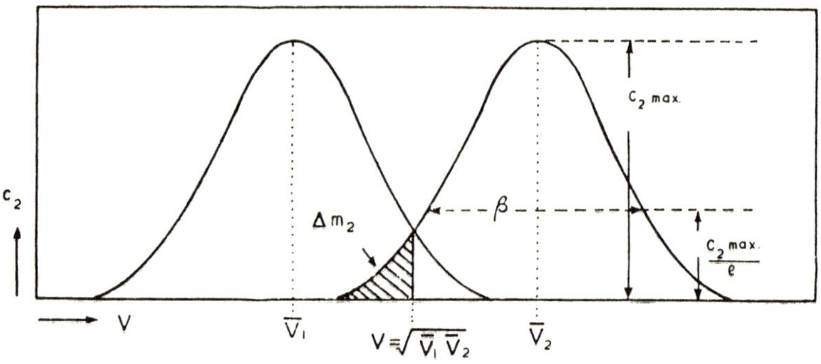

FIG. 2.—Width and overlap of eluted bands.

If $m_1 \neq m_2$, and we wish to derive the conditions required to obtain both components in equal purity, we must solve the equation

$$\Delta m_1/m_2 = \Delta m_2/m_1$$

using eqn. (31), (32). Here it is more convenient to use the experimental parameters, v, x, Δx or N, etc.:

$$\frac{m_2}{m_1}\left(0{\cdot}5 - A_\varepsilon\left\{\sqrt{N}\frac{(\bar{v}_2 - v)}{\sqrt{\bar{v}_2 v}}\right\}\right) = \frac{m_1}{m_2}\left(0{\cdot}5 - A_\varepsilon\left\{\sqrt{N}\frac{(v - \bar{v}_1)}{\sqrt{\bar{v}_1 v}}\right\}\right). \quad (37)$$

The solution for this is (see appendix)

$$v = \sqrt{\bar{v}_1\bar{v}_2} + \frac{2\bar{v}_1\bar{v}_2}{N(\bar{v}_2 - \bar{v}_1)}\frac{(m_1^2 - m_2^2)}{(m_1^2 + m_2^2)}, \quad (38)$$

which expresses quantitatively the plausible need that the cut has to be made nearer the smaller peak. We also obtain for the impurity fraction

$$\eta = \frac{\Delta m_1}{m_2} = \frac{\Delta m_2}{m_1} = \frac{2m_1m_2}{m_1^2 + m_2^2}\left(0{\cdot}5 - A_\varepsilon\left\{\frac{\sqrt{N}(\sqrt{\bar{v}_2} - \sqrt{\bar{v}_1})}{\sqrt[4]{\bar{v}_1\bar{v}_2}}\right)\right\} \quad (39)$$

Comparison with eqn. (36) shows the unexpected feature that bands of unequal amounts of solute can be separated with the same column into products of a purity higher than obtainable with equal amounts, as the factor $2m_1m_2/(m_1^2 + m_2^2)$ has its maximum at $m_1/m_2 = 1$.

With the help of eqn. (39) we can conveniently construct a family of curves showing the number of theoretical plates required to separate two species of given separation factor a_2/a_1 ($= \bar{v}_2/\bar{v}_1$) and a given mass ratio m_1/m_2 with a desired product purity η (see fig. 3). Fig. 3 is very instructive in showing that it is comparatively easy to obtain sensational purities by ion exchange chromatography provided the separation factors are above 1·2. On the other hand, it requires impractically long columns to separate substances of very low separation factors such as e.g., isotopes, even if only a moderate degree of separation is required.

APPLICABILITY OF THE THEORETICAL PLATE THEORY

Strictly speaking, the " theoretical plate " theory can be applied only in such cases where the spreading of the bands is not dominated by lack of equilibrium between the sorbent and solution phase. This situation exists

(i) when equilibrium is practically attained by using very small particle sizes or adsorbents with surface action only, and suitably low flow rates;

(ii) in gas chromatography, where the height of the theoretical plate is entirely determined by gas-phase diffusion along the direction of the column;

(iii) when the theoretical plate height is determined by channelling effects in the column, and this appears to be the case more often than one might think.

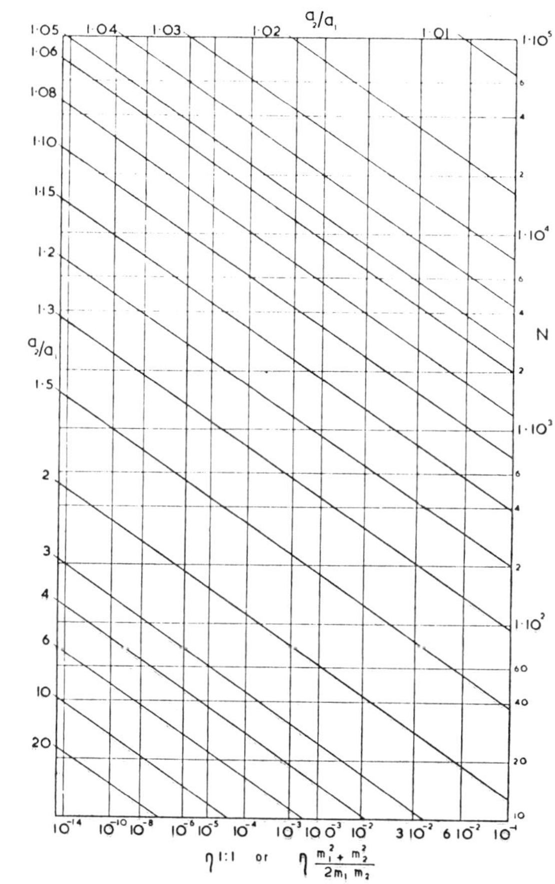

FIG. 3.—Purity of separated products as function of separation factor (a_2/a_1), number of theoretical plates N and mass ratio m_1/m_2.

If the band spreading is dominated by non-equilibrium phenomena, as it frequently is in separations employing ion exchangers, the mass balance equations contain, instead of term $-\partial^2 c/\partial x^2$, a term of the type $+\partial^2 c/\partial x \partial v$ (see ref. (3), (4)). Though the rate of movement of the centre of the band remains unaffected, the eluted bands are less symmetrical than those caused by pure theoretical plate effects. However, the term $\partial^2 c/\partial x \partial v$ can be written $-\left(\frac{\partial x}{\partial v}\right)_c \frac{\partial^2 c}{\partial x^2}$

and as the rate of movement $\partial x/\partial v$ becomes fairly constant after a sufficiently large number of theoretical plates (or transfer units) have been passed, the difference in the shape of the bands ceases eventually to be significant.

In general, one can say that the difference is no longer significant after the passage of about 200 theoretical plates (or transfer units). The efficiency of separation of two species is little affected even for smaller columns, as the greater diffuseness of the front of the following band is compensated by a greater sharpness of the rear of the preceding one. There are few chromatographic separations which can be effectively carried out with less than 100 plates in the column, and one can conclude therefore that the theoretical plate theory can deal very successfully in practice with column operations dominated by non-equilibrium phenomena. The numerical values to be used for the height of the "effective" theoretical plate have been discussed in a recent paper [5] (which also contains a more adequate list of references) and these need not be repeated here.

There is, however, one point which is of importance in connection with this paper. With "particle diffusion" dominating the disturbance term,[5] the height of theoretical plate (Δl or Δx) is not a constant property of column and operating conditions alone, but is roughly inversely proportional to the distribution coefficient a of the species. This means that different values of N (proportional to a or $\bar{v}$) apply to different species in the same separation experiment. Allowance has to be made for this, when trying to extrapolate the results obtained with one eluting solution to another solution with different eluting power. This feature affects also the discussion on purity of separated products. Most important, the volume v where optimum separation is obtained is now not represented by eqn. (36) and (38). Instead, for "particle diffusion" the optimum cut is at

$$v = \frac{\bar{v}_1 + \bar{v}_2}{2}\left[1 + \frac{N_2}{2v_2}\left(\frac{\bar{v}_2 - \bar{v}_1}{\bar{v}_2 + \bar{v}_1}\right)^2 \frac{(m_1{}^2 - m_2{}^2)}{(m_1{}^2 + m_2{}^2)}\right], \tag{40}$$

and the purity of the products is given by

$$\eta = \frac{\Delta m_1}{m_2} = \frac{\Delta m_2}{m_1} = \frac{2m_1m_2}{m_1{}^2 + m_2{}^2}\left(0{\cdot}5 - A_\varepsilon\left\{\sqrt{\frac{N_2}{2\bar{v}_2}}\,\frac{(\bar{v}_2 - \bar{v}_1)}{\sqrt{\bar{v}_1 + \bar{v}_2}}\right\}\right), \tag{41}$$

or

$$\eta = \frac{2m_1m_2}{m_1{}^2 + m_2{}^2}\left(0{\cdot}5 - A_\varepsilon\left\{\sqrt{\tilde{N}}\,\frac{(\bar{v}_2 - \bar{v}_1)}{(\bar{v}_2 + \bar{v}_1)}\right\}\right), \tag{42}$$

where $\tilde{N}$ is the arithmetic mean of the N's for the two species. While the optimum cut is considerably displaced, there is little difference in the numerical values from eqn. (39) and from (42). This means that fig. 3 can serve equally well for the prediction of separation efficiency for particle diffusion, provided that the mean value of N is employed in the use of these diagrams.

APPENDIX

Solution of the equation:

$$\frac{\Delta m_1}{m_2} = \frac{\Delta m_2}{m_1} = \frac{m_2}{m_1}\left(0{\cdot}5 - A_\varepsilon\left\{\sqrt{N}\frac{(\bar{v}_2 - v)}{\sqrt{\bar{v}_2 v}}\right\}\right) = \frac{m_1}{m_2}\left(0{\cdot}5 - A_\varepsilon\left\{\sqrt{N}\frac{(v - \bar{v}_1)}{\sqrt{\bar{v}_1 v_1}}\right\}\right). \tag{A1}$$

The value of v satisfying (A1) is not far from $\tilde{v} = \sqrt{\bar{v}_1\bar{v}_2}$. We can therefore substitute

$$v = \tilde{v} + \Delta v \tag{A2}$$

and expand the error function, which we shall write for short, $A\{F(\tilde{v} + \Delta v)\}$:

$$A\{F(\tilde{v} + \Delta v)\} = A\{F_{(\tilde{v})}\} + \Delta v \frac{\partial A}{\partial v}\{F_{(v)}\} + \ldots . \tag{A3}$$

$$\frac{\partial A}{\partial v}\{F\} = \frac{\exp(-F^2/2)}{\sqrt{2\pi}}\frac{\partial F}{\partial v} \text{ at } v = \tilde{v}. \tag{A4}$$

For large values of the argument F such as apply when $\Delta m \ll m$, the error function can be expressed by

$$0{\cdot}5 - A\{F\} = \frac{\exp(-F^2/2)}{F\sqrt{2\pi}}. \tag{A5}$$

Thus we can write (A3) in the form

$$0{\cdot}5 - A\{F(\tilde{v} + \Delta v)\} = (0{\cdot}5 - A\{F_{\tilde{v}}\})(1 - \Delta v\, F_{\tilde{v}}\, \partial F/\partial v. \tag{A6}$$

The arguments of the two error functions in (A1) are identical for $v = \tilde{v} = \sqrt{\bar{v}_1\bar{v}_2}$ and so is

$$F_{(\tilde{v})}\frac{\partial F}{\partial v}(\text{at } v = \tilde{v}) = \pm\frac{N}{2}\frac{(\bar{v}_2 - \bar{v}_1)}{v_1 v_2}, \tag{A7}$$

the left-hand function giving a negative, the other a positive value. With these substitutions we obtain

$$\Delta v = \left|\frac{1}{F_{(\tilde{v})}\dfrac{\partial F}{\partial v}}\right| \cdot \frac{m_1^2 - m_2^2}{m_1^2 + m_2^2} = \frac{2\bar{v}_1\bar{v}_2}{N(\bar{v}_2 - \bar{v}_1)}\frac{m_1^2 - m_2^2}{m_1^2 + m_2^2}. \tag{A8}$$

Substituting Δv back into (A6) and (A1) gives

$$\frac{\Delta m_1}{m_2} = (0{\cdot}5 - A\{F_{(\tilde{v})}\})\frac{2m_1m_2}{m_1^2 + m_2^2}, \tag{A9}$$

and finally substituting $\tilde{v}$,

$$\eta = \frac{\Delta m_1}{m_2} = \frac{\Delta m_2}{m_1} = \frac{2m_1m_2}{m_1^2 + m_2^2}\left(0{\cdot}5 - A\left\{\sqrt{N}\frac{(\sqrt{\bar{v}_2} - \sqrt{\bar{v}_1})}{\sqrt[4]{\bar{v}_1\bar{v}_2}}\right\}\right). \tag{A10}$$

[1] Martin and Synge, *Biochem. J.*, 1941, **35**, 1358.
[2] Mayer and Tompkins, *J. Amer. Chem. Soc.*, 1947, **69**, 2866.
[3] Boyd, Myers and Adamson, *J. Amer. Chem. Soc.*, 1947, **69**, 2849.
[4] Glueckauf, Barker and Kitt, *Faraday Soc. Discussions*, 1949, **7**, 199.
[5] Glueckauf, *J. Appl. Chem.* (Ion Exchange Conf., 1954), in press (see eqn. (14)).
[6] Simpson and Wheaton, *Chem. Eng. Progr.*, 1954, **50**, 45.

4

Reprinted from *J. Chem. Educ.*, **41**(8), 410–413 (1964)

F. Helfferich
Shell Development Company
Emeryville, California

Travel of Molecules and Disturbances in Chromatographic Columns

A paradox and its resolution

Today, there is hardly a chemist or chemical engineer who has not, at one or the other occasion, praised chromatography as a god-send. Analytical gas chromatography in particular has been developed into a very convenient and rapid technique which is all but indispensable in a modern laboratory. Awareness of the relevant physical phenomena has become common knowledge, and mathematical theories have been brought to a high degree of perfection.

Most of the work on chromatography, however, has been confined to so-called "linear" systems. In common analytical gas chromatography, for example, samples are carried along through the column by a large excess of an inert carrier gas. At such great dilutions of the samples, the partition coefficients of the sample components between the gas and the stationary liquid phase can be taken as independent of concentration, i.e., the sorption isotherms (concentration in the stationary phase as a function of concentration in the gas) of the components can be regarded as linear. More recently, interest in chromatographic behavior under "nonlinear" equilibrium conditions has increased. Here, the substances of interest are present at higher concentrations, and the concentration dependence of the partition coefficients, resulting in nonlinear sorption isotherms, cannot be disregarded.

The extention to nonlinear behavior is not without pitfalls. To be sure, adequate theories of nonlinear chromatography have been worked out long ago,[1] and this communication does not add novel features. Certain physical implications of these theories, however, have never been stressed. Newcomers with experience in linear chromatography only, if they argue on the basis of intuitive or plausible deductions, may easily go astray and have, indeed, often done so.

The purpose of the present communication is to contribute toward a better understanding of some fundamental, but not obvious, aspects of nonlinear chromatography by presenting and then resolving what appears to be, at first sight, a surprising paradox.

Case 1: Rate of Travel of Individual Molecules

Let us suppose that a binary feed gas consisting of a sorbable substance and an inert (i.e., nonsorbable) carrier gas is passed continuously from a reservoir through a chromatographic column. The sorbable substance is present at a substantial concentration, and the feed-gas composition and flow rate are kept constant. We are interested in the rate of travel of the molecules of the sorbable substance through the column.

The molecules of the substance under investigation spend part of their time in the stationary sorbent and are therefore retarded compared to the carrier gas, which remains entirely in the moving gas phase. Since the molecules make progress, at the rate of carrier-gas flow, only in the fraction α of time in which they are in the gas, the relation between the rates of travel of the molecules, v_m, and of carrier gas, v_c, is

$$v_m = \alpha v_c \tag{1}$$

The time fraction α is related to the distribution of the substance between the gas and the stationary phase by

$$\alpha = \frac{Q_g}{Q_{st} + Q_g} = \frac{1}{Q_{st}/Q_g + 1} \tag{2}$$

where Q_g and Q_{st} are the amounts of the substance in the gas and the stationary phase, respectively. Equation (1) can be rewritten in terms of the partition coefficient K, which is defined by

$$K \equiv \frac{C_{st}}{C_g} \equiv \frac{Q_{st}V_g}{Q_gV_{st}} \tag{3}$$

Here, C is the concentration per unit volume of the respective phase, V is the volume fraction of the phase, and indices g and st refer as before to the gas and the stationary phase, respectively. Equation (1) then becomes

$$v_m = \frac{V_g}{KV_{st} + V_g} v_c \tag{4}$$

The rate of travel thus is governed by the partition coefficient $K = C_{st}/C_g$.

One can observe the rate of travel experimentally by injecting into the feed a small pulse of tagged (for example, radioactive) molecules of the substance and measuring their retention time[2] (see Fig. 1, top).

Case 2: Rate of Travel of a Concentration Pulse

Let us now carry out an experiment under identical conditions except that, instead of injecting a tracer into the feed, we create a concentration disturbance in the feed such that, for a short period of time, the concentration is $C_g + \Delta C_g$ rather than C_g (see Fig. 1, middle).

[1] DeVault, D., *J. Am. Chem. Soc.*, **65**, 532 (1943); for references to more recent work, see Vermeulen, T., *Advan. Chem. Eng.*, **2**, 147 (1958); Helfferich, F., *Angew. Chem., Intern. Ed.*, **1**, 440 (1962).

[2] Helfferich, F., and Peterson, D. L., *Science*, **142**, 661 (1963).

The concentration pulse travels through the column and can be recorded by a detector in the effluent. Some flattening-out of the pulse will, of course, occur in the column, but such effects are not relevant for the present purpose and will be disregarded for the sake of clarity.

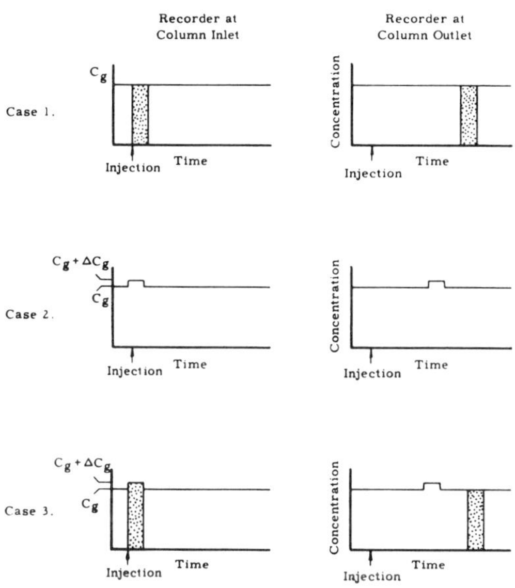

Figure 1. Concentration recordings at column inlet and outlet for Cases 1 2, and 3. Band of tagged molecules is shown as shaded area.

The rate of travel of the concentration pulse is readily calculated from a material balance around a differential layer of thickness dz of the column. (For simplicity, it is assumed in the derivation that the nonsorbable component in the gas phase is in great excess. Otherwise the rates v_c above and below the front differ from one another. The result, however, is the same except for a correction factor in the first term in the denominator of equation (6).) At the moment the front of the pulse (assumed for simplicity to remain sharp) enters the layer at the inlet side, the layer contains the amount $(C_{st}V_{st} + C_gV_g)\,q\,dz$, where q is the column cross section. At the moment the front leaves the layer at the outlet side, the layer contains $[(C_{st} + \Delta C_{st})\,V_{st} + (C_g + \Delta C_g)\,V_g]\,q\,dz$. In the time dt in which the front has moved across the layer, the latter has thus gained the amount $(\Delta C_{st}V_{st} + \Delta C_gV_g)\,q\,dz$. During this time, the amount leaving the layer at the outlet side is $V_gC_gv_cq\,dt$, and that entering at the inlet side is $V_g(C_g + \Delta C_g)\,v_c\,q\,dt$. The material balance, inflow minus outflow equals net gain, then is

$$(\Delta C_{st}V_{st} + \Delta C_gV_g)q\,dz = V_g(C_g + \Delta C_g)v_c\,q\,dt - V_gC_gv_cq\,dt \quad (5)$$

This equation can be rearranged to give the rate of travel, v_f, of the front

$$v_f = \frac{dz}{dt} = \frac{V_g}{(\Delta C_{st}/\Delta C_g)V_{st} + V_g}v_c \quad (6)$$

The same result is obtained when the calculation is carried out for the rear end of the concentration pulse. According to equation (6) the rate of travel is governed by the isotherm slope $\Delta C_{st}/\Delta C_g$ at the concentration C_g (see also Fig. 2).

Case 3: Simultaneous Travel of Molecules and Disturbances

Comparison of equation (4) for the travel rate of molecules with equation (6) for the travel rate of a concentration pulse reveals a disturbing and seemingly paradoxical fact. The two expressions are identical except that the former contains the partition coefficient $K = C_{st}/C_g$ where the latter contains the isotherm slope $\Delta C_{st}/\Delta C_g$. With a nonlinear isotherm, the two quantities are in general quite different (see Fig. 2) and, accordingly, the travel rates should be different.

To show that this difference is real rather than caused by an error or inadmissible simplification in one of the derivations, let us introduce into the feed a tracer pulse and a concentration pulse simultaneously. With a nonlinear isotherm we find that, indeed, the two pulses emerge from the column at different times (see Fig. 1, bottom).

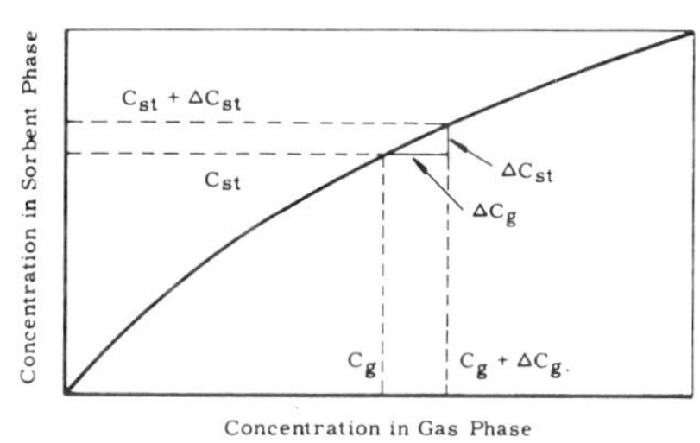

Figure 2. Typical sorption isotherm.

Discussion

The paradox is easily resolved when one takes into consideration that travel of individual molecules, observed in Case 1, and propagation of a disturbance, observed in Case 2, are physically different phenomena. Crude, but more readily perceived analogies are the differences between the propagation of a flood wave racing down a river and the flow of the river water, and between the propagation of a traffic jam on a freeway and the rate of progress of the individual cars. The flood wave reaches the river estuary way ahead of the rain-water molecules that caused the flood high upstream. A traffic jam, created by a car stopping briefly and thus forcing the next car to stop behind, is even propagated *against* the direction of traffic flow (see Fig. 3). (In chromatography, propagation of a concentration disturbance against the direction of flow could arise only with a negative isotherm slope, which is thermodynamically impossible.)

Such parables notwithstanding, intuition may still be reluctant to accept that significant differences in the rates of travel in a chromotographic column exist. An argument this auther has often heard is as follows. The molecules travel at the rate given by equation (4), governed by the partition coefficient. Now if, in Case 3, the concentration disturbance is made infinitesimal (i.e., $\Delta C_g \ll C_g$), then the tagged molecules introduced with the pulse should travel at practically the same rate as the untagged ones preceeding and following them, because the difference between the partition coefficients at $C_g + \Delta C_g$ and at C_g then is infinitesimal. Consequently, there is no obvious reason why the concentration disturbance should separate from the molecules that caused it, and why the tagged molecules should not, instead, travel with unchanged concentration $C_g + \Delta C_g$ through the column. On the basis of such seemingly logical arguments, authors have proposed to determine partition coefficients from retention times of very small concentration disturbances (Case 2) by use of equation (4). Others, aware of the fact that, with a nonlinear isotherm, the time of emergence of a particular concentration C_g is the effluent is related to the isotherm slope dC_{st}/dC_g at C_g much more generally than discussed here, have contended that the retention time of a tracer pulse (Case 1) is related to the isotherm slope by equation (6) rather than to the partition coefficient as in equation (4).

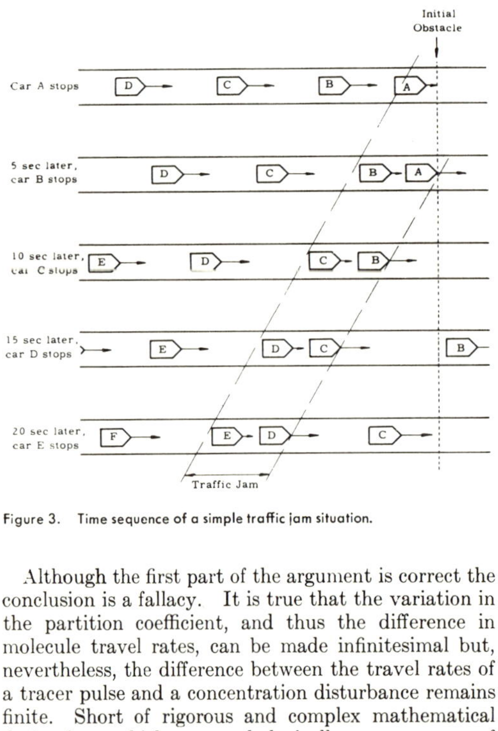

Figure 3. Time sequence of a simple traffic jam situation.

Although the first part of the argument is correct the conclusion is a fallacy. It is true that the variation in the partition coefficient, and thus the difference in molecule travel rates, can be made infinitesimal but, nevertheless, the difference between the travel rates of a tracer pulse and a concentration disturbance remains finite. Short of rigorous and complex mathematical derivations, which are psychologically a poor means of overcoming intuitive objections, only a more detailed study of the physical phenomena in the column can refute the fallacy. For this purpose, let us reconsider Case 3.

As described before, we insert into the feed of concentration C_g a pulse of tagged molecules of significantly higher concentration $C_g + \Delta C_g$ (see Fig. 4*a*). Let the sorption isotherm be negatively curved, as in Figure 2, so that the partition coefficient at the concentration $C_g + \Delta C_g$ is smaller than at the concentration C_g. If the tagged and untagged molecules would move independently of one another, the whole band of tagged molecules would then travel somewhat faster than the untagged ones, and, after passage through a few layers of the column, a situation as in Figure 4*b* should arise. The "overlap" (darkly shaded) of tagged and untagged molecules at the front of the tagged band is obviously impossible because the total concentration (tagged plus untagged) must remain within the limits C_g and $(C_g + \Delta C_g)$. In systems with only one sorbable component, as considered here, a pulse can only flatten out. Any sharpening, resulting in increased pulse height, would constitute a reduction in entropy and would thus violate thermodynamics. Therefore the front of the tagged band must act as a bulldozer, throwing up a wave of untagged molecules in front of it as shown in Figure 4*c*.

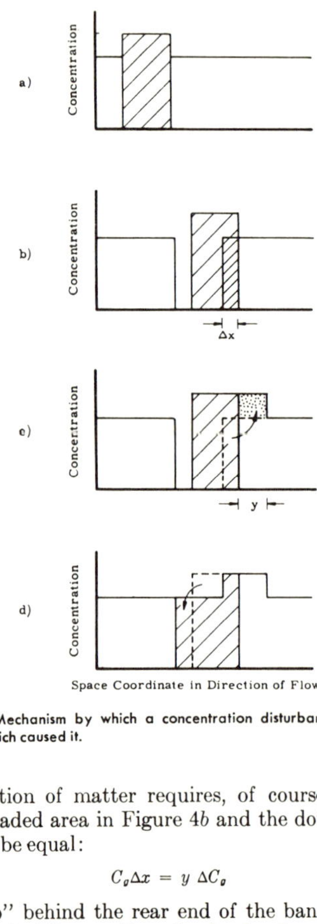

Figure 4. Mechanism by which a concentration disturbance outruns the molecules which caused it.

Conservation of matter requires, of course, that the darkly shaded area in Figure 4*b* and the dotted area in Figure 4*c* be equal:

$$C_g \Delta x = y\, \Delta C_g \tag{7}$$

The "gap" behind the rear end of the band of tagged molecules in Figures 4*b* and *c* is also impossible, and some tagged molecules must stay behind to keep this gap filled, as shown in Figure 4*d*. The net result (Fig. 4*d*) is that the concentration pulse has gained on the tagged molecules through the bulldozer effect at the front and the gap filling at the rear. Since the process will repeat itself in subsequent column layers farther downstream, the concentration pulse may eventually separate completely from the band of tagged molecules.

With a positively curved isotherm the molecules at the concentration $(C + \Delta C_g)$ will move more slowly than those at the concentration C_g. Thus, here, the bulldozer effect occurs at the rear and the gap filling at the front, so that the concentration pulse falls more and more behind the band of tagged molecules.

We can now refute, without recourse to mathematics, the earlier fallacious argument that the difference in travel rates of tracers and concentration pulses should disappear when the concentration disturbance is made infinitesimal. It is true that, in this way, the difference between the travel rates of tagged molecules at the concentration $(C_g + \Delta C_g)$ and untagged ones at the concentration C_g can be made infinitesimal, so that the quantity Δx in Figure 4*b* becomes infinitesimal. Now, however, the quantity ΔC_g is also infinitesimal and, therefore, the displacement y of the concentration pulse relative to the tagged band remains finite:

$$y = C_g \Delta x / \Delta C_g \tag{8}$$

[see equation (7)]. In other words, the bulldozer moves forward an only very short distance but, since the height of the wave thrown up is also very small, the length of that wave is nevertheless considerable.

Several additional remarks are called for. Throughout the mathematical derivations and nonmathematical deductions in this communication all effects resulting in band broadening in the column have been disregarded. The only justification for this is that inclusion of such effects, while rendering the treatment of the problem exceedingly complex, does not alter the final result. It should also be stressed that the fundamental difference between travel of individual molecules and propagation of disturbances, analyzed here in one particular system only, is a general and rather frequently encountered feature of chromatographic behavior. For example, phenomena of the same general nature are found in liquid-solid chromatography, in systems with arbitrarily complex rather than binary feed mixtures, in sorption or ion exchange operations in incompletely regenerated columns, etc. The absence of the described effect in conventional analytical gas chromatography under "linear" equilibrium conditions, where the isotherm slope and the partition coefficient are equal, should not obscure the importance of the physical factors involved.

Part II

ION-EXCHANGE CHROMATOGRAPHY OF RARE-EARTH MIXTURES

Editor's Comments on Papers 5 Through 9

5 SPEDDING et al.
The Separation of Rare Earths by Ion Exchange: I. Cerium and Yttrium

6 MARINSKY et al.
The Chemical Identification of Radioisotopes of Neodymium and of Element 61

7 KETELLE and BOYD
The Exchange Adsorption of Ions from Aqueous Solutions by Organic Zeolites: IV. The Separation of the Yttrium Group Rare Earths

8 SPEDDING et al.
The Separation of Rare Earths by Ion Exchange: VI. Conditions for Effecting Separations with Nalcite HCR and One-Tenth Per Cent. Citric Acid-Ammonium Citrate Solutions

9 NERVIK
An Improved Method for Operating Ion-Exchange Resin Columns in Separating the Rare-Earth Elements

The first successful use of ion-exchange chromatography to separate metals from one another by selective elution from a column was made during the Manhattan Project in the period 1942 to 1945. Plutonium and uranium were separated from each other and from the fission products, using columns of Amberlite IR-1, a resin prepared by condensing tannin with formaldehyde and sulfonating the product. Uranium (VI) was eluted by dilute sulfuric acid, then plutonium (IV) with oxalic acid, and finally the fission products were eluted with sodium hydrogen sulfate solution. This work remained classified for several years and was later described by Boyd and Schubert (1969). Boyd writes: "To the best of my knowledge these (separations) are the earliest of any ever made with organic ion exchangers."

The next triumph of ion-exchange chromatography, and the one that received the most publicity, was the separation of the rare-earth elements from one another. These elements constituted a large part of the fission products, and their separation by fractional crystallization of their salts was one of the most tedious and difficult operations in chemistry. Boyd and others suggested that they might be separated by ion exchange. Two groups—one at Oak Ridge, Tennessee, under G. E. Boyd, W. E. Cohn, E. R. Tompkins, and others (see Tompkins et al., 1947), the other at Ames, Iowa, under F. H. Spedding—worked on this project, and because of wartime secrecy they worked almost independently. It appears that the Oak Ridge group was the first to separate tracer quantities of the rare earths and the first to isolate element 61, promethium. The Ames group worked with macro amounts and succeeded in separating gram quantities of adjacent rare earths in purities approaching 100 percent. A patent for this process was awarded to Spedding and Voigt (1951).

The separations methods for the rare earths were declassified in 1947, and a preliminary report was published by Johnson et al. (1947). The first group of papers appeared in the *Journal of the American Chemical Society,* November 1947, in the form of a symposium. Eight papers in this issue dealt specifically with rare-earth separations, and three of them are reproduced here.

Paper 5 is the first of a series by the authors. It describes the separation of cerium from yttrium, which is not a specially difficult task, but it showed the way to more exacting separations. Paper II in Spedding's series (Spedding et al., 1947) describes the separation of praseodymium and neodymium, two adjacent rare-earth elements, which was a much more ambitious project. The Pr–Nd separation used columns 8.5 m long, whereas 2 m sufficed for separating cesium and yttrium. Citrate buffer of pH 2.65 was used in both cases; the resin was the sulfonated phenolic Amberlite IR-1.

Paper 6 is historic because it describes the isolation and identification of a new element, radioactive element 61, later named promethium. It was produced by irradiating element 60, neodymium, with neutrons. Elements 60 and 61 were separated by ion exchange.

Paper 7 looks like a modern paper in liquid chromatography. The successive elution of several radioactive elements is shown, using continuous beta-ray counting. Jacketed columns at high temperature are used, and the advantage of using fine-particle-sized resin is demonstrated. This paper reports the use of a new resin, Dowex-50, which was to become the standard cation-exchange resin for chromatographic use. This resin was produced in spherical particles, unlike the older gel-type sulfonated phenolic resins, whose particles were irregular. It is clearly seen that the spherical resin gives better chromatographic bands. Dowex-50, as everyone in the field now knows, is a styrene-divinylbenzene copolymer prepared by bead polymerization and later sulfonated, according to the patents of D'Alelio (1945). It is chemically more uniform and stable than the phenolic resins, and opened up new possibilities in ion-exchange chromatography.

Figure 7 in Paper 7 is specially noteworthy for showing the power of ion-exchange chromatography to reveal impurities in supposedly pure materials, in this case spectroscopically pure erbium oxide.

Paper 8 shows the development of a new way of separating the rare earths in bulk, called by Spedding the "displacement band technique." A quantity of mixed rare-earth salts is introduced into the column under conditions such that the ions are strongly absorbed, with a sharp front at the lower end of the band. Now a complexing eluent is passed into the column, one that in its turn displaces the rare-earth ions very strongly. The band of rare-earth ions now travels down the column as a zone or "plug" of constant length, without spreading. As it travels, however, the different rare-earth elements arrange themselves along the zone in the order of their binding strengths, with the most weakly bound at the head or lower end. After the zone has traveled a few times its own length, the separation reaches a limit of efficiency that is not improved by more travel. The different elements emerge from the column as a sequence of essentially pure fractions, with short intermediate regions of cross-contamination.

This method is used today to produce pure rare-earth elements on an industrial scale. The same technique has been used by Spedding (Paper 13) and by Rosset (Paper 14) to separate isotopes.

Paper 9 introduces *gradient elution.* When a number of ions or molecules are eluted from a column, the first two or three may emerge in rapid succession; it then becomes harder and harder to remove the rest. (This happens because free energies, rather than equilibrium constants, tend to increase linearly in a set of similar reactions.) The different ions will emerge at more or less regular intervals if the eluting strength of the flowing solution is increased continuously in the proper manner. Gradient elution is now standard practice in the ion-exchange chromatography of multicomponent mixtures, like amino acids. Its theory was treated in a paper by Freiling (1955) that appeared almost simultaneously with the paper of Nervik that is reproduced here.

REFERENCES

Boyd, G. E., and J. Schubert. 1969. *Progress in Nuclear Energy, Series III—Process Chemistry,* Vol. 4, Elmsford, N.Y., Peragamon Press.

D'Alelio, G. F. 1945. U.S. Patent 2,366,007 (Dec. 26, 1945).

Freiling, E. C. 1955. *J. Amer. Chem. Soc., 77,* 2067.
Johnson, C., L. L. Quill, and F. Daniels. 1947. *Chem. Eng. News, 25,* 2494.
Spedding, F. H., and A. F. Voigt. 1951. U.S. Patent 2,539,282.
——, A. F. Voigt, E. M. Gladrow, and N. R. Sleight. 1947. *J. Amer. Chem. Soc., 69,* 2786.
Tompkins, E. R., J. X. Khym, and W. E. Cohn. 1947. *J. Amer. Chem. Soc., 69,* 2769.

5

Reprinted from *J. Amer. Chem. Soc.*, **69**, 2777–2781 (Nov. 1947)

The Separation of Rare Earths by Ion Exchange.[1,2] I. Cerium and Yttrium

By F. H. Spedding, A. F. Voigt, E. M. Gladrow and N. R. Sleight

I. Introduction

For many years one of the most difficult processes in the field of chemistry has been the separation of the rare earths from each other into their pure states. Their chemical and physical properties are so similar that in general a single operation leads only to a partial separation or enrichment.

Ever since the beginning of the Manhattan Project there has been a constant demand for samples of rare earths of exceptional purity in gram amounts or greater. This demand arose for numerous reasons, but mainly because some of the rare earths are formed as fission fragments during fission of the heavy elements. It was highly desirable, therefore, to have a means of preparing pure rare earths so that their nuclear properties could be studied and also to allow a more thorough consideration of their chemical behavior. Their radioisotopes are less well understood than those of any other group of elements.

In general, the best means of separating these elements has been the well known but laborious method of fractional crystallization as used by James and further developed in many laboratories. Exceptions are cerium with its quadrivalent state, and samarium, europium and ytterbium with their di-valent states which do permit a means of separation from the normal trivalent rare earth ions.

A number of workers have reported studies on the application of chromatographic and ion exchange methods to the separation of the rare earths[3,4,5,6] While they obtained considerable enrichment their results were not sufficiently promising to lead to further intensive investigation or to the quantity production of pure rare earths. The history within the Manhattan District, of the use of columns of Amberlite type resins for the separation of fission products, both with and without the use of citric acid–ammonium citrate eluants at controlled *p*H has been described elsewhere and will not be discussed here.[7]

The present paper is the first of a series, from this laboratory dealing with the successful separation of macro quantities of rare earths of spectrographic purity, by adsorption on Amberlite type resins and subsequent elution with complexing agents such as citric acid–ammonium citrate solutions at controlled *p*H. This paper establishes that cerium and yttrium can be separated relatively rapidly by these methods on any desired scale.

The marked success of the process described depends on the fact that the rare earths form complexes with the citrate ions. If the *p*H is suitably adjusted, competition is set up for the rare earth ions between the citrate complexes and the active centers of the resin. Therefore, as the citrate solution washes the rare earths down the column, each rare earth ion is adsorbed and desorbed many times. Since the equilibrium constants for the rare earth citrate complexes vary slightly among the different rare earths, their rates of travel down the column differ sufficiently to lead to their separation. The repeated cycles in the columns effectively replace the thousands of individual operations required by the older methods for separating the rare earths and lead to a highly effective process analogous to the use of distillation columns.

II. Materials, Apparatus and General Procedure

1. **Materials.**—The cerium stock solution was prepared from pure $(NH_4)_2Ce(NO_3)_6$ which was dissolved in dilute nitric acid, reduced with sulfuric dioxide and precipitated as $Ce(OH)_3$. The hydroxide was filtered off, dissolved in sulfuric acid and stored. The yttrium stock solution was commercial yttrium chloride having a purity greater than 80%.

The citric acid was of commercial grade but

(1) This work was supported, in part, by a grant from The Manhattan District, U. S. Corps of Engineers.

(2) This work was done during the period of January to March, 1945, and was presented before the Manhattan Project Council, Chicago, Illinois, on April 15, 1945, and before the Chicago Section of the American Chemical Society in November, 1945.

(3) E. Lange and K. Nagel, *Z. Elektrochem.*, **42**, 210 (1936).

(4) G. Sahama and V. Kamula, *Ann. Acad. Sci. Fennicae*, **A57**, no. 3, 5 (1941).

(5) U. Croatto, *Ricerca sci.*, **12**, 157 (1941).

(6) R. G. Russell and D. W. Pearce, This Journal, **65**, 595 (1943).

(7) F. Daniels, W. C. Johnson and L. L. Quill, *Chem. Eng. News*, **25** 2494 (1947).

spectroscopic analysis showed only traces of metal ions. The term "5% citric acid solution," used throughout this paper, refers to 50 g. of the citric acid monohydrate per liter of solution. The oxalic acid used as precipitating agent was commercial grade and contained no salts capable of introducing metallic ions into the precipitate.

Cerium and/or yttrium tracers were used in all the experiments; their use in analysis will be discussed later in this paper. The tracers were obtained by chemically separating the true rare earth group and yttrium from the other fission products in neutron irradiated uranium. The cerium was then oxidized to the quadrivalent state and separated, as the insoluble iodate, from the other active rare earths. After several months the activities, other than Y, had all decayed leaving radio-pure Y^{91} with its 57 day period and beta activity of 1.4 Mev. maximum energy. The Ce isotope was Ce^{144} with a 275 day half-life. The other cerium isotope, 28 day Ce^{141}, had decayed when the measurements were made; measurements were always made after the Ce^{144} had established equilibrium with its 17.5 min. Pr daughter. This pair of isotopes emits beta rays of 0.31 Mev. and 3.1 Mev., respectively. While the tracer solutions were not entirely carrier free the amount of inactive carrier was negligible. Individual stock solutions of these tracers were employed in the subsequent experiments.

2. Calculation and Presentation of Data.—Two methods have been employed in presenting the data, each one depending upon a measurement of the activity in a 1-ml. sample of eluate as the means of measuring the amount of material eluted.

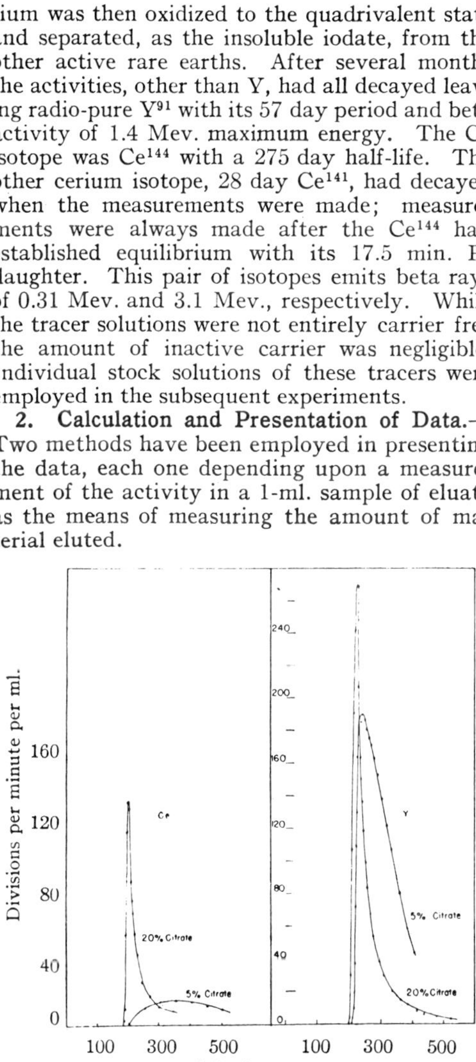

Fig. 1.—The elution of tracer amounts of Ce and Y using 5 and 20% citric acid: *p*H 2.75, flow rate 3 cm./min., column dimensions 1 cm. × 60 cm. The curves for Ce are on the left, Y on the right.

In the experiments designed to determine the nature of the elution curves for each ionic species, when only tracer amounts were employed (see Figs. 1 and 2), a very small aliquot of the original solution was evaporated on a watch glass and measured on the electroscope. This permitted a calculation of the total amount of activity used in the experiment. During the course of the elution 1-ml. samples of eluate were collected at various volume intervals and measured similarly. A graph showing activity/ml. as a function of volume was plotted. A simple integration under this curve permitted the calculation of the per cent. of the ionic species eluted at any desired point (see Fig. 3).

A somewhat similar method was employed in those experiments in which macro amounts of material were used. A known amount of inactive material was weighed out and dissolved. Tracer activity was added and the solution made up to known volume; a small aliquot was dried and measured on the electroscope. This gave the activity per unit weight of carrier or, conversely, the weight of carrier corresponding to unit activity. Samples were collected as above during the elution and their radioactivity measured. Using the conversion factor, the weight of material eluted was plotted as a function of volume of eluate. Integration under the curve to any point

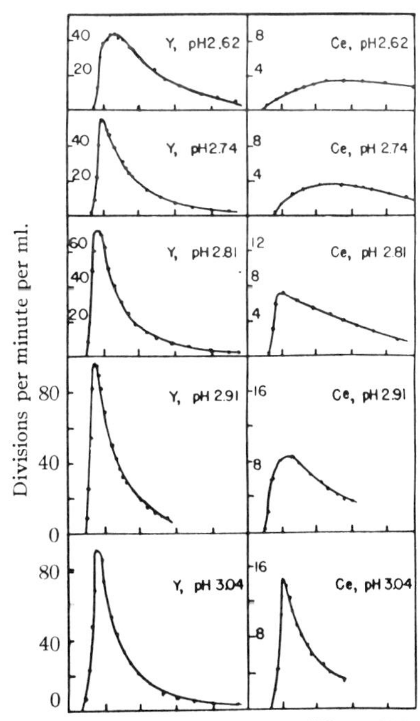

Fig. 2.—The elution of Ce and Y tracers in 5% citrate at various *p*H values: flow rate 3 cm./min.; column dimensions, 1 cm. × 60 cm.

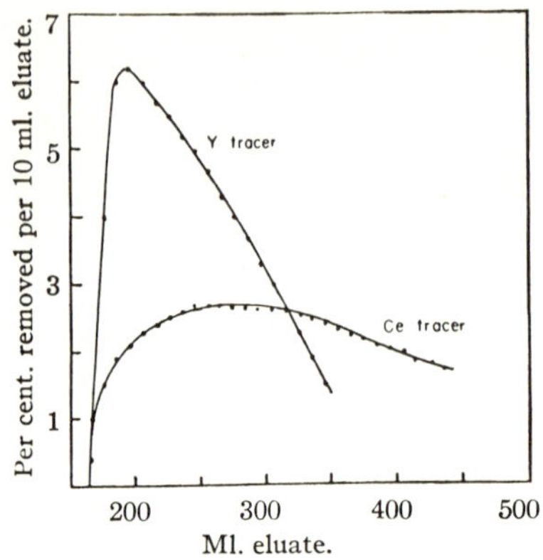

Fig. 3.—Individual elution curves for Ce and Y tracers: *p*H 2.74; flow rate, 3 cm./min.; column dimensions, 1 cm. × 160 cm.

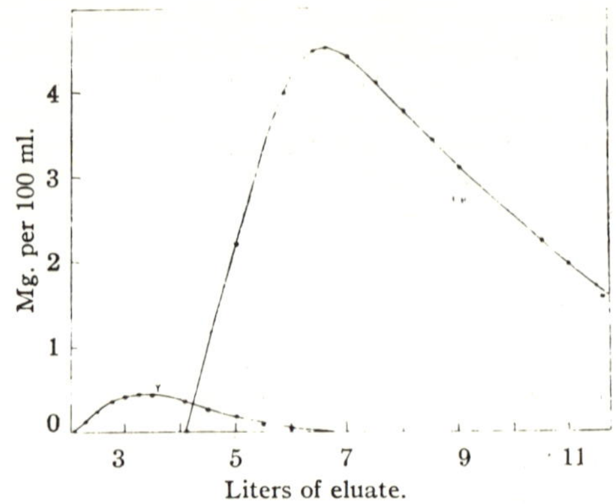

Fig. 4.—The elution of a small amount of Y in the presence of a large amount of Ce: *p*H 2.77; flow rate, 5 cm./min.; column dimensions, 1.6 × 190 cm.; composition of starting material, 10 mg. Y and 250 mg. Ce.

along the abscissa gave the weight of material eluted to that point (see Fig. 4).

In general, in studies of mixtures of Ce and Y, only one species of tracer was added. This permitted a better determination of the intermediate region between the two peaks. The total elution curve was obtained by precipitating the mixture in any volume interval as oxalate, igniting to oxide and weighing. The weight per unit volume was plotted as a function of the volume of eluate. The difference between the total weight curve and the one obtained by using activity measurements is due to the species without the tracer.

Not only does this method give a means of assaying the samples which is faster and more accurate than simple weighing but it also allows the detection of small amounts of material at or near the point of break-through. The break-through represents the total volume of eluant at which the presence of the given element is first evident. At low concentrations of rare earths the oxalates may have sufficient solubility or tendency to supersaturate so that no precipitate is formed. Simple weighing of the ignited precipitate thus has a practical lower limit of detection of 5 mg. per liter. However, tracer measurements can be made much more sensitive by increasing the specific activity of the original sample.

3. Description of Apparatus.—The principal pieces of apparatus were long vertical glass tubes filled with the cation exchanger, Amberlite IR-1 resin. A cylindrical, silvered glass Geiger–Mueller tube of the type made by the Eck & Krebs Company was mounted in a copper housing having a small window opening. This unit was connected to a counting circuit and scaler by a long lead which made the G–M tube fairly portable. By moving the counter tube up and down the column, the position of the material could be determined at any time during those experiments in which radioactive tracers were used.

The samples of eluted Ce and Y were measured on a Lauritsen quartz fiber electroscope having a multiple stage mounting in which conversion factors had been worked out between the stages to facilitate the measurement of extremes of activity. The samples containing but little activity were read from a stage close to the ionization chamber and the highly active samples were read from a stage somewhat removed from the ionization chamber.

4. Preparation of Columns.—The columns were constructed by drawing out one end of the glass tube and sealing on a small outlet tube, finally resembling a buret. A plug of glass wool was inserted at the bottom of the column to prevent washing out of the resin; the tube was dented directly above the glass wool to prevent it from pushing up the resin bed during the process of backwashing.

The resin, Amberlite IR-1, was first screened to remove the large particles (>20 mesh) and then soaked in water. The resin slurry was added to the column and backwashed with water to remove fine resin particles which tended to plug the column and thus cut down the flow rate. The resin was conditioned as follows. A solution of 5% hydrochloric acid was passed through the column converting the resin to the acid cycle. The contact time was at least thirty minutes to insure complete conversion. This process was followed with a 2% sodium chloride solution to convert the resin to the sodium cycle. Treatment with acid and salt was repeated two or more times, the resin being finally left in the acid cycle.

III. Experimental

1. Elution of Ce and Y Singly

a. Effect of Concentration of Citrate in Eluant.—Preliminary experiments were undertaken to determine the effect of variation in original concentration of citric acid in the eluant. The concentrations of citric acid tested were 1, 5 and 20% adjusted to *p*H 2.74 by the addition of concentrated ammonium hydroxide; the elution was followed with both Ce and Y tracers. The 1% solution was ineffective at the *p*H employed. The change in citrate concentration had a large effect on the rate of elution of both Ce and Y tracers. At the higher concentration they, especially Ce, were eluted in a much smaller volume of eluant. The data for the higher concentrations of citrate are presented in Fig. 1. Since the 5% solution gave satisfactory results, it was chosen as standard for the subsequent experiments.

b. Effect of *p*H Using Tracer Amounts.—The resin bed was 60 cm. in height and 1 cm. in diameter. The 5% citrate solution was used as eluant at *p*H values of 3.04, 2.91, 2.81, 2.74 and 2.62. The flow rate was maintained at 3 ml. per minute per square cm. of cross-section area. This unit (ml./min./sq. cm.) is equivalent to cm. of column length per minute and is so designated in subsequent discussion. Samples for analysis consisted of 1 ml. of eluate on watch glasses. They were dried under infrared lamps and the activity read on the electroscope. The cerium samples were read through a 70 mg./sq. cm.

aluminum absorber to remove the soft beta rays; the yttrium samples were determined without an absorber.

The data obtained are shown in Fig. 2. The elution of both Ce and Y increased with increase in *p*H. Yttrium was satisfactorily eluted at all the *p*H values tested while the elution of Ce was relatively poor at *p*H values less than 2.81; the change in shape of the curve with varying *p*H is especially striking with this element. However, at lower *p*H values yttrium would undoubtedly have shown similar changes in shape of curve.

c. Effect of Column Length Using Tracer Amounts.—The preceding experiment showed that the Ce tracer moved down the column at a slower rate than did the Y tracer. These results indicated that a longer column should improve the separation. Experiments were performed by the same procedure as in the previous experiment except that the bed length was increased to 162 cm. and the *p*H was 2.75. The superimposed elution curves are shown in Fig. 3; it is evident that the Y is eluted much more rapidly, under these conditions, than is the Ce. The Y/Ce ratios, as a function of the per cent. of Y eluted are given in the second column of Table I.

d. Elution of Macro Amounts of Ce and Y.—Since the results with tracers, on a 162 cm. column, showed a separation of Y and Ce, the same procedure was applied to macro amounts of the materials. The Y/Ce ratios, as a function of the amounts of Y eluted, are given in the third column of Table I. It is evident that a much better separation was obtained with the macro amounts than with the tracer amounts.

2. Elution of Mixtures of Ce and Y in Macro Amounts

The previous discussion has dealt with elution in which the primary purpose was to determine the nature of the elution curve for each ionic species, under identical conditions, and in the absence of interfering ions. Data are now presented on the elution of macro amounts of mixtures of Ce and Y.

TABLE I

Y/Ce RATIO AS A FUNCTION OF THE PER CENT. OF Y ELUTED

Y eluted, %	Y/Ce (tracer)	Y/Ce (macro)
10	5.0	..
20	4.0	42
30	3.7	30
40	2.9	13
50	2.6	8.3
60	2.3	5.2
70	1.9	3.7
80	...	2.7

a. Effect of *p*H.—A series of experiments was conducted with the *p*H values of the eluant varying in the range 2.50 to 2.87. The column beds were 190 cm. in length and the diameter 16 mm.; each sample adsorbed was equivalent to 250 mg. of Ce_2O_3 and 250 mg. of Y_2O_3. In one set of experiments Ce tracer and, in another set, Y tracer was added to the inert material. At *p*H values higher than 2.87 both Ce and Y were eluted so rapidly that there was little separation. The best *p*H value for separation is in the neighborhood of 2.60. The summary of the results of these experiments is given in Table II.

TABLE II

EFFECT OF *p*H ON PER CENT. Y ELUTED BEFORE Ce BREAK-THROUGH

*p*H	Per cent. Y eluted before Ce break-through
2.87	70.0
2.77	82.0
2.66	96.0
2.61	98.5
2.50	91.0

b. Effect of Original Y/Ce Ratio.—In the previous experiment the Y/Ce ratio was unity. Since, in actual practice, the ratio may be far from unity, data were obtained for a mixture equivalent to 10 mg. of Y_2O_3 and 250 mg. of Ce_2O_3, a ratio of 1 to 25. The bed length of the resin was 190 cm., the diameter 16 mm. and the *p*H 2.77. The elution curves are shown in Fig. 4. When all of the Y had been eluted, about 60% of the Ce was still in the column and was subsequently eluted pure. Even when the Y was present in such a small relative amount, more than 60% of it was obtained in a pure form before the Ce break-through. It can be concluded that the procedure described can yield the products of spectroscopic purity.

c. Effect of Length of Column.—It has been shown that a separation of Ce and Y of greater than 96% can be effected, from a 1:1 mixture of these two components, on a column 190 cm. long and 16 mm. in diameter at a *p*H range of 2.55–2.65. It was also shown that when Ce and Y are eluted independently, they move at different rates, indicating that the band fronts might move farther apart as the length of the column is increased. It, therefore, seemed of interest to see whether the separation of the bands of the two elements continually increases or whether they move apart a certain distance and then proceed at much the same rate.

A column was prepared which was 610 cm. in bed length and 25 mm. in diameter. The mixture contained 250 mg. of each component employing tracers of Ce and Y. The *p*H of the eluant was 2.66 and the flow rate 5 cm. of column length per minute. The band fronts were determined by use of the portable counting apparatus. The data obtained are given in Table III.

TABLE III

EFFECT OF VOLUME OF ELUATE ON SEPARATION OF BAND FRONTS

Eluate, liters	Separation of band fronts, cm.
16	173 ± 15
20	193 ± 15
24	218 ± 15
28	230 ± 20
32	297 ± 20

Analysis of the eluates, after the break-through, showed that only 90–95% of the Y was obtained free from Ce even though the band fronts were moving further apart. This phenomenon is probably associated with channeling and possibly somewhat with back diffusion. This experiment indicates that there is a critical bed length of column beyond which no improvement in separation can be gained by increasing the length of the column.

IV. Summary

Conditions have been described for the separation of cerium and yttrium by adsorption on Amberlite IR-1 columns and elution with 5% citric acid–ammonium citrate solutions. The separations were studied both for tracer and macro amounts, using radioactive isotopes of the elements for the determination of the degree of separation. The factors studied included concentration of eluant, *p*H of eluant, the ratio of cerium and yttrium in the starting sample and length of columns. Conditions were established by which the elements, of spectrographic purity, were separated in macro quantities. The principles established for the separation of cerium and yttrium are being applied to the development of methods for the large scale separation of adjacent rare earths; the details will be published in subsequent papers.

AMES, IOWA RECEIVED JULY 7, 1947

6

Reprinted from *J. Amer. Chem. Soc.*, **69**, 2781–2785 (Nov. 1947)

The Chemical Identification of Radioisotopes of Neodymium and of Element 61[1]

By J. A. Marinsky,[2] L. E. Glendenin[2] and C. D. Coryell[2]

Introduction

In the comprehensive studies of the radioactive species produced in the fission of uranium[3] it has been found that over thirty are members of the rare earth family (isotopes of yttrium and the group lanthanum through europium). The chemical and physical identification of these was an important part of the research program of the Manhattan Project. Standard oxidative separations and fractional precipitations[4] and the use of radiochemical methods based on chain relations served to distinguish the activities of yttrium, lanthanum, cerium, and some of praseodymium, and those[5] of samarium and europium. The characterization of the sequence praseodymium, neodymium, and element 61 presented very difficult problems[6] that were solved only with the intensification of ion exchange methods originally developed by Boyd and co-workers[7] and applied to the rare earth field by Cohn and co-workers.[8] In this paper is reported the successful separation of these three elements, the first to have been achieved with radioisotopes of neodymium and of element 61.

Earlier work on the Manhattan Project had revealed the presence of two unidentified fission products in the rare earth region. A soft beta emitting activity of ~4y half-life was discovered by Ballou[9] and independently by Goldschmidt and Morgan.[10] This activity was later studied by Seiler and Winsberg[6] who set the half-life at ~3.7y. These investigators considered the activity to be an isotope of praseodymium, neodymium, or element 61 on the basis of detailed qualitative chemical separations between lanthanum and praseodymium. Another unidentified rare earth activity was later discovered by Davies[11] through its characteristic gamma radiations. This was confirmed by Hume and Martens[12] who determined its half-life as ~11d and showed that the activity was not cerium. We have been able to identify these two activities as isotopes of element 61 and of neodymium, respectively, both of mass number 147. In addition the 47h 61^{149} has been identified among the fission products. All three of these activities, together with a previously unidentified 1.7h isotope of neodymium, are also produced as the result of the activation of neodymium with slow neutrons.

Chemical Identification

The 11d activity was first subjected to standard qualitative serpations[4,6] in order to limit the identification problem to a definite group of rare earth elements. Separations by carbonate digestion definitely eliminated yttrium and all rare earth elements of atomic number greater than 61. Oxidative fusion with sodium nitrate[4,6] demonstrated that the activity was not an isotope of lanthanum and must therefore be identified with praseodymium, neodymium or element 61. Finally, the activity was limited to an isotope of neodymium or element 61 by demonstration of separation from praseodymium with potassium hydroxide fusions.

Positive identification of the 3.7y and the 11d radioisotopes was then achieved by the recently developed ion-exchange method using a synthetic organic cation exchanger (Amberlite IR-1) of the sulfonated phenol-formaldehyde type. This new

(1) This document is based on work performed at the Clinton Laboratories operated under contract No. W-35-058-eng-71 for the Manhattan Project, classified reports of which are referred to here as "Manh. Proj. Rep." The information covered herein will appear in the Plutonium Project Record (Plut. Proj. Rec., Vol. 9B, papers 7.54.3, 7.54.4, 7.54.5, 7.54.6, and 11.1), in Division IV of the Manhattan Project Technical Series.

(2) Present address: Department of Chemistry, Massachusetts Institute of Technology, Cambridge, Mass.

(3) The Plutonium Project, Nuclei Formed in Fission: Decay Characteristics, Fission Yields and Chain Relationships, This Journal, **68**, 2411–2442 (1946). Reprints are obtainable from the offices of The American Chemical Society.

(4) A. A. Noyes and W. C. Bray, "A System of Qualitative Analysis for the Rare Elements," The Macmillan Co., New York, N. Y., 1927.

(5) L. Winsberg, Manh. Proj. Rep. CC-2310, pp. 231–244 (Jan. 1945) and CC-2966 (Apr. 1945); Plut. Proj. Rec., Vol. 9B, 7.55.1, 7.55.2, 7.56.1, 7.56.2, and 7.56.3 (1946).

(6) (a) N. E. Ballou, Manh. Proj. Rep. CC-389B (Dec. 1942), CC-920 (Sept. 1943), CC-1204 (Jan. 1944), CN-1312 (May 1945), Plut. Proj. Rec., Vol. 9B, 8.24.7 and 8.24.12 (1946); N. E. Ballou and J. A. Marinsky, *ibid.*, 8.24.11; (b) J. A. Seiler and L. Winsberg, Manh. Proj. Rep. CC-2310, pp. 227–230 (Jan. 1945); Plut. Proj. Rec., Vol. 9B, 7.54.2 (1946).

(7) (a) G. E. Boyd, J. A. Swartout, *et al.*, Manh. Proj. Rep. CN-3346 (Dec. 1945); J. A. Schubert, Manh. Proj. Rep. CN-1873 (Jan. 1945); G. E. Boyd, J. Schubert, and A. W. Adamson, This Journal, **69**, 2818 (1947); (b) A. W. Adamson, Manh. Proj. Rep. CN-1859 (Apr. 1944); G. E. Boyd, A. W. Adamson and L. S. Myers, This Journal, **69**, 2836 (1947).

(8) E. R. Tompkins, and J. X. Khym, W. E. Cohn, Manh. Proj. Rep. CL-WEC-10 (Dec. 1946); This Journal, **69**, 2769 (1947).

(9) N. E. Ballou, Manh. Proj. Rep. CC-680, p. 22 (May 1943) and CC-3418 (Feb. 1946); Plut. Proj. Rec., Vol. 9B, 7.54.1 (1946).

(10) B. Goldschmidt and F. Morgan, Canadian Proj. Rep. MC-11 (Aug. 1943).

(11) T. H. Davies, Manh. Proj, Rep. M-CN-1424 (Apr. 1944).

(12) D. N. Hume and R. I. Martens, *ibid.*, CN-1311 (June 1944).

method provides very effective separations of the rare earths,[8] even between adjacent elements. The procedure involves the adsorption of the rare earth ions on a column of the resin from dilute acid solution followed by elution with 5% citric acid adjusted to a *p*H of about 3 with concentrated ammonia.

The order of elution of the rare earths (from lanthanum through europium) and yttrium was established in a series of experiments[8,13] using radioactive cerium, lanthanum, europium and yttrium produced in fission and radioactive praseodymium and europium produced by slow neutron activation of pure oxides. Macroscopic quantities of samarium, europium, and neodymium with spectrographic analyses were also used. The results of these experiments show clearly that the order of elution of the rare earths is the reverse of that of atomic number, and that consequently the atomic number of an unknown radioisotope can be determined from the elution curves of a group of successive rare earths. Yttrium is eluted in the region of gadolinium as might be expected from its well known behavior in precipitation work. A typical set of elution curves is presented in Fig. 1.

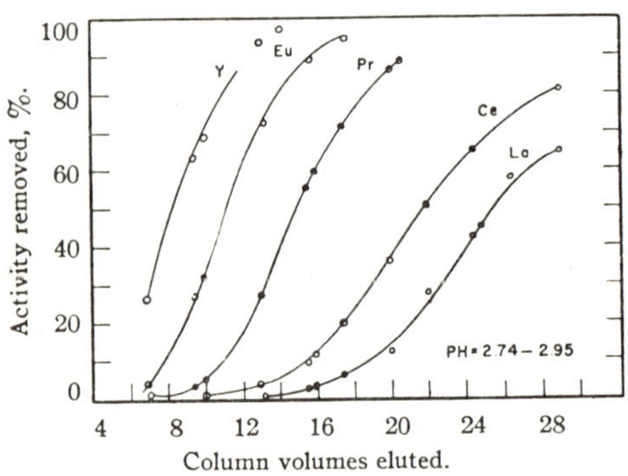

Fig. 1.—A typical set of curves showing the order of elution of the rare earths.

The 11d and the 3.7y activities were identified in the following experiment. A sample which contained essentially the praseodymium group (praseodymium, neodymium and element 61) with very little cerium, samarium, europium, and yttrium activities (by virtue of thorough chemical removal of cerium by ceric iodate precipitations[4] and of yttrium, samarium, and europium by repeated digestions with potassium carbonate[4,6]), was isolated with about 15 mg. of lanthanum carrier from a rare earth fission product elution fraction intermediate between cerium and yttrium that was supplied by Cohn and co-workers.[8] The rare earth activities in this sample were adsorbed on an Amberlite IR-1 column and eluted with a 5% solution of ammonium citrate at a *p*H value of 2.75. Fractions of the eluate were collected, and the observed beta and gamma activity in each fraction was plotted against the volume of eluate that had passed through the column. Four beta activity peaks were observed, as shown in Fig. 2. The gamma activity (11d half-life) was found to be associated with the third peak.

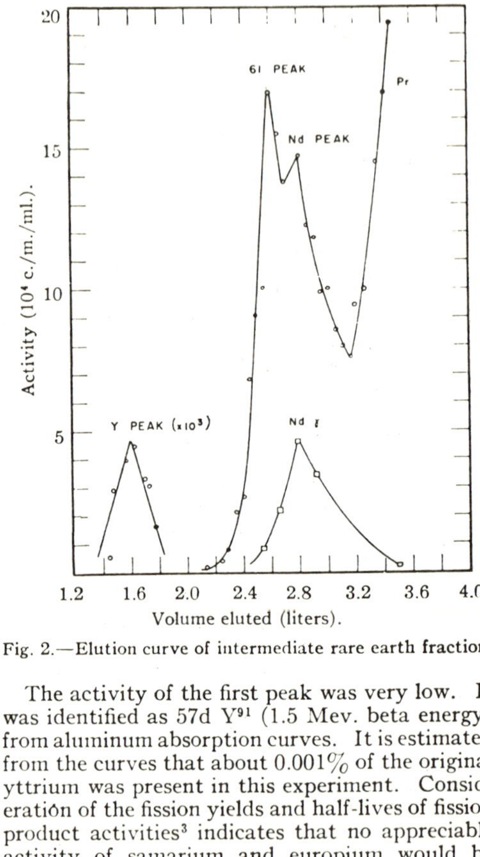

Fig. 2.—Elution curve of intermediate rare earth fraction

The activity of the first peak was very low. It was identified as 57d Y^{91} (1.5 Mev. beta energy) from aluminum absorption curves. It is estimated from the curves that about 0.001% of the original yttrium was present in this experiment. Consideration of the fission yields and half-lives of fission product activities[3] indicates that no appreciable activity of samarium and europium would be found even though contamination from these elements should exceed that from yttrium by a factor of about one thousand. The remaining peaks must thus be ascribed to element 61, neodymium, and praseodymium in that order.

Decay and absorption data on a fraction corresponding to the top of the fourth peak showed that the activity consisted only of 13.8d Pr^{143} (1.0 Mev. beta energy), as expected. Similar studies of activities elsewhere along the distribution curve showed that the element 61 peak consisted of the activity of the long-lived soft beta emitter (0.2 Mev.), and that the neodymium peak consisted of the 11d activity with moderately hard beta rays (0.4 and 0.9 Mev.) and gamma rays.

(13) J. A. Marinsky and L. E. Glendenin, Manh. Proj. Rep. CC-2829 (June 1945); J. A. Marinsky, Plut. Proj. Rec., Vol. 9B, 11.1 (1946).

TABLE I

CHARACTERISTICS OF SOME RADIOISOTOPES OF Nd AND ELEMENT 61

Nucleus	Half-life	Decay	Class[a]	Fission yield, %	Energy of radiations in Mev. Particles	γ Radiations
$_{60}Nd^{147}$	11.0d	β^-, γ, e^-, X-ray	A	2.6	~0.4 (40%) 0.9 (60%) e^-: ~0.03	0.58 (40%) X-ray ~0.04
$_{60}Nd^{(149)}$	1.7h	β^-, γ or X-ray	B	...	1.5	γ or X-rays present
61^{147}	3.7y	β^-	A	(2.6)	0.20	no γ
61^{149}	47h	β^-, γ, e^- (?) X-ray (?)	A	1.4	1.1	0.25 (low intensity) X-ray (?)

[a] The classification corresponds to that of ref. 3, namely: A, element certain, isotope certain; B, element certain, isotope probable.

This experiment[14] was the first carried out to separate radioisotopes of element 61 from those of neighboring elements. Independent work[15] and subsequent amplification[15,16] has led to great improvements in the techniques for separating a variety of individual rare earth elements.

Similar elution experiments on activities produced by slow neutron irradiation of neodymium revealed not only the occurrence of the two radioisotopes discussed above, but also led to the identification of a 47h isotope of element 61 and a 1.7h isotope of neodymium. The 47h element 61 activity was subsequently isolated from fission products by using suitable timing and chemical techniques to eliminate interference from the 40.0h La^{140}, the 33h Ce^{143}, the 47h Sm^{153}, and the 56h Np^{239}. The 1.7h Nd, of probable mass 149, undoubtedly occurs among the fission products, but has not yet been identified there.

Physical Characteristics

Decay Energies.—Radiochemical studies were made of the four activities mentioned above in order to establish their decay characteristics, genetic relationships, fission yields, and mass assignments. These properties are presented in Table I. The decay characteristics were determined from decay curves and the absorption of the radiations in aluminum and lead. Maximum energies of the beta rays were estimated from ranges in aluminum evaluated by the method of Feather.[17] Beta-gamma coincidence counting was used to establish the coincidence of the gamma ray of the 11d activity with the softer beta component and the association of X-radiation with the emission of the harder beta component.

Genetic Relationships.—Since the two isotopes of element 61 are obtained in the slow neutron activation of neodymium as well as in the fission of uranium, it was of interest to see how these are related to the radioisotopes of neodymium. A study was made of the change in the ratio of the 3.7y element 61 activity to that of the 47h element 61 activity as a result of decay of the 11d Nd activity, under conditions for which the 1.7h activity had been removed by decay.

Very pure neodymium oxide (Hilger) was irradiated in the Clinton pile for seven days. This preparation was dissolved in 6 *N* hydrochloric acid and diluted to a definite volume. Several hours after the end of irradiation an aliquot portion was adsorbed on a resin column and eluted with a 5% solution of ammonium citrate at a *p*H of 2.75. Neodymium carrier was added to the first eluted fraction, which contained 47h element 61 with very little neodymium activity. The activities of the fraction were precipitated on neodymium fluoride and finally on neodymium oxalate. The sample was mounted as fraction I and its decay was studied. Ten days after the end of irradiation another equal aliquot was treated in the manner described above to give a similar fraction II rich in element 61.

Decay curves were followed for several months on both the fractions to determine the content in the 47h, 11d, and 3.7y activities, taking the point of reference as that of the end of the irradiation. Fractions I and II were found to contain the same amount of 47h 61 activity but the second contained less activity of 11d Nd, undoubtedly because of a small difference in *p*H of the lutriant. The observed activities, without correction for external absorption (20 mg./sq. cm.) and self-scattering (from ~25 mg./sq. cm.) are given in counts per minute (c./m.) in the first three lines of Table II. It is seen that the 3.7y activity is increased with reference to the 47h activity, which can only happen if the former is produced as a result of decay of the 11d Nd.

Correlation of the exact amounts of 3.7y activ-

TABLE II

ACTIVITIES OBSERVED IN TWO ELEMENT 61 FRACTIONS

Species	I, Separation at end of irradiation c./m.	II, Separation after 10 days of decay c./m.
11d Nd	9.0×10^4	5.0×10^4
47h 61	2.0×10^6	2.0×10^6
3.7y 61	3.5×10^2	7.0×10^2
11d Nd/740	1.2×10^2	0.7×10^2
Corr. 3.7y 61	2.3×10^2	6.3×10^2

(14) L. A. Marinsky and L. E. Glendenin, Manh. Proj. Rep. CN-2809 (Apr. 1945) and CC-2829 (June 1945).

(15) F. H. Spedding, A. F. Voigt, E. M. Gladrow, N. R. Sleight, J. E. Powell, J. M. Wright, T. A. Butler, and P. Figard, Manh. Proj. Rep. CC-2980 (Apr. 1945) and CC-2720 (May, 1945); THIS JOURNAL, **69**, 2786 (1947).

(16) (a) B. H. Ketelle and G. E. Boyd, THIS JOURNAL, **69**, 2800 (1947); (b) D. H. Harris and E. R. Tompkins, *ibid.*, **69**, 2792 (1947).

(17) N. Feather, *Proc. Camb. Phil. Soc.*, **34**, 599 (1938); (b) L. E. Glendenin, *Nucleonics*, in press (1947).

ity formed during the seven-day period of irradiation and during the ten-day period of decay requires correction for the amount of this activity coming from the total decay of the 11d Nd contaminating the fractions. Taking the reasonable value of one-sixth for the ratio of counting efficiencies of the two activities and the factor 1/123 for the ratio of the half-lives, there are obtained the values given in the fourth line for the contamination correction in equivalents of 3.7y activity. The difference between these values and the observed activities in the third line gives the estimate of the true activities present in the two fractions. The activity in fraction II exceeds that in fraction I by the factor 2.7, in good agreement with the predicted value 2.96. Correspondingly the activity in fraction I is found to be 2.3×10^2 in comparison with the predicted value 2.1×10^2 computed from the activity of 11d Nd (6.65×10^5 c./m.) observed to be present before fractionation with the reference amount of 47h activity (2.0×10^6 c./m.).

It was not possible to study the occurrence of genetic relationship between 1.7h Nd and 47h element 61 by this method since their radiations are similar and the half-life of 1.7h Nd is too short to permit the required separations. It was therefore necessary to establish the relationship indirectly by varying the time of slow neutron irradiation of Nd and comparing the ratio of the two activities at the end of each bombardment with the ratio that should result from a genetic relationship. Such experiments indicated a parent-daughter relationship; however, the same neutron capture cross-section per normal atom of Nd could produce similar results.

Fission Yield.—The fission yield is defined as the percentage of the fissions (in this case of U^{235} with thermal neutrons) leading to the nucleus in question by direct formation and decay of precursors. The fission yields of the activities were determined from analysis of the decay curves of activities of the two elements isolated together from uranium that had been exposed in the pile, using neodymium carrier. The timing of the chemical operations was adjusted so that other rare earth activities were eliminated. The observed activities were corrected for decay, absorption, and degree of saturation during bombardment, and compared with the 12.8d Ba^{140} activity isolated from the same source. From the absolute fission yield value[18] of 6.1% for Ba^{140} the fission yields of 11d Nd and 47h element 61 were determined as 2.6 and 1.4%, respectively.

Mass Assignments.—The mass number of a fission product nucleus may be estimated from its fission yield since the fission yield varies rapidly with the mass over much of the fission product range. According to the yield-mass curve,[3] in which the yields of U^{235} fission product chains are plotted as a function of their masses, the fission yield values of 2.6 and 1.4% correspond to mass numbers of 147 and 149, respectively. An intense source of pure 3.7y element 61 from fission was prepared and submitted to Hayden and Lewis[19] who verified the mass assignment of 147 with the mass spectrograph. More recently the assignment of mass 149 to the 47h element 61 has been verified in the mass spectrograph by Inghram, Hess, Hayden, and Parker[20].

(18) M. S. Freedman and D. W. Engelkemeir, Manh. Proj. Rep. CC-1331 (Feb. 1944).

Discussion

Numerous studies have been made of activities from bombardments of praseodymium and neodymium in which isotopes of element 61 might be expected together with isotopes of neighboring elements from competitive nuclear reactions. Following exploratory investigations[21] the major contribution has been from Pool, Kurbatov, Quill, and co-workers[22] with important specialized studies by Wu and Segrè[23] and by Bothe.[24] The activities observed, together with the assumed nuclear reactions and the proposed assignment of the activities by element and mass number, are listed in Table III. It is apparent that considerable uncertainty persists in the assignment of the activities as can be determined by checking the reported identifications of the 1.7h Nd^{149}, the 47h 61^{149}, and the 11d Nd^{147}, whose identities have been established in this paper. This is to be expected because all assignments were made without attempts at chemical separations, which are virtually indispensable for element identification in this difficult section of the periodic system. The ion exchange column separations are certainly superior to the best standard fractionation available to these workers.

The identification of radioisotopes produced in cross bombardments (where the nuclear charge changes in the primary nuclear reaction) is in principle easier than for the (n,γ) or $(n,2n)$ reactions. The possibility of rare earth impurities in target materials cannot, however, be dismissed. Seaborg[25] in his Table of Isotopes lists all the activities in Table III only as Class E, element probable. Other reviews of the identification of radioisotopes of element 61 have been given by Seaborg[26] and Paneth.[27]

(19) R. J. Hayden and L. G. Lewis, *ibid.*, CP-3383 (Dec. 1945) *Phys. Rev.*, **70**, 11 (1946).

(20) M. G. Inghram, D. C. Hess, Jr., R. J. Hayden, and G. W. Parker, *ibid.*, **71**, 743 (1947).

(21) (a) E. Amaldi, C. D'Agostino, E. Fermi, B. Pontecorvo, and E. Segrè, *Proc. Roy. Soc. (London)*, **A149**, 522 (1935); (b) J. C. McClennan and W. H. Rann, *Nature*, **136**, 831 (1935).

(22) (a) M. L. Pool and L. L. Quill, *Phys. Rev.*, **53**, 437 (1938); (b) H. B. Law, M. L. Pool, J. D. Kurbatov and L. L. Quill, *ibid.*, **59**, 936 (1941); (c) J. D. Kurbatov, D. C. MacDonald, M. L. Pool, and L. L. Quill, *ibid.*, **61**, 106 (1942); (d) J. D. Kurbatov and M. L. Pool, *ibid.*, **63**, 463 (1943).

(23) C. S. Wu and E. Segrè, *ibid.*, **61**, 203 (1942).

(24) W. Bothe, *Z. f. Naturforsch.*, **1**, 179 (1946).

(25) G. T. Seaborg, *Rev. Mod. Phys.*, **16**, 1 (1944).

(26) G. T. Seaborg, *Science*, **105**, 349 (1947).

(27) F. A. Paneth, *Nature*, **159**, 8 (1947).

TABLE III

PREVIOUSLY REPORTED RADIOISOTOPES OF NEODYMIUM AND OF ELEMENT 61

Assumed reaction	Target element	Half-life	Mode of decay	Proposed assignment	Reference
n,γ	Nd	1h^a			21
n,γ	Nd	35m			21
n,γ	Nd	21m	β^-	Nd^{151}	22
n,γ	Nd	2h^a	β^-	Nd^{149}	22
n,γ	Nd	84h^b	β^-	Nd^{147}	22
n,γ	Nd	2.0h^a	β^-	$Nd^{147,149,151}$,	
n,γ	Nd	47.5h^b	β^-	$61^{147,149,151}$, or	24
n,γ	Nd	11.1d^c	β^-	Sm^{151}	
$n,2n$	Nd	2h^a	β^-	Nd^{149}	22
$n,2n$	Nd	84h^b	β	Nd^{147}	22
$n,2n$	Nd	2.3h^a	β^-	61^{147} or ^{149}	22
$n,2n$	Nd	47h^b	β^-	Nd^{149}	22
$n,2n$	Nd	10d^c	β^-	61^{147}	22
$n,?$	Nd	108d			22
$n,2n$	Nd	2.5h	β^+	Nd^{141}	22
$n,2n$	Nd	41h^b	β^-	Nd	22
$n,2n$	Nd	~2h^a		Nd	22
$n,2n$	Nd	short		Nd	22
$n,2n$	Nd	47.5h^b	β^-	$Nd^{147,149,151}$ or	24
$n,2n$	Nd	11.1d^c	β^-	$61^{147,149,151}$	
d,p	Nd	47h^b	β^-	Nd^{149}	22, 23
d,n	Nd	12.5h	β^-	61	22
α,n	Pr	~108d		61^{144}	22
d,n	Nd	~108d		61^{144}	22
α,n	Pr	~200d	K or I. T., γ	61^{144}	23
d,n; p,n; α,p	Nd	2.7h	β^-,γ	61	22
d,n; p,n; α,p	Nd	5.3d	β^-,γ	61	22
d,n	Nd	16d^c	β^-,γ	61	22
$d?$	Nd	~1y	γ	..	22

[a] Presumably 1.7h Nd^{149}. [b] Presumably 47h 61^{149}. [c] Presumably 11d Nd^{147}.

Added interest in the identification of isotopes of element 61 derives from the fact that significant quantities of the natural element have not been separated, and that stable isotopes may not occur. The problem of the natural existence of element 61 has been reviewed by Yost, Russell and Garner,[28] who state, "the evidence for the existence in nature of element 61 is somewhat circumstantial rather than conclusive, and the isolation of this element cannot be regarded as having been effected."

(28) D. M. Yost, H. Russell, Jr., and C. S. Garner, "The Rare Earth Elements and Their Compounds," Chapter 4, John Wiley and Sons, New York, N. Y., 1947.

Considerable quantities of 61^{147} can be prepared, however, as the result of the industrial application of nuclear power. A uranium pile operating at the power level of 1,000 kw. produces $\sim 3 \times 10^{16}$ fissions per second. From the data in Table I, it is seen that 2.6% of these will lead to chain 147. There is produced, therefore, ~16 mg. per day of Nd^{147}, which, after suitable decay, is transformed into 61^{147} of sufficient half-life for isolation and chemical study. The activity associated with 1 mg. of the species is 0.66 curie, but the radiations consist solely of soft β rays that are easily shielded and should not interfere too greatly with chemical operations.

We wish to express our gratitude to Dr. D. N. Hume for his generous advice and encouragement in this research. The collaboration of Mr. J. X. Khym on the order of elution experiments and the coöperation of Drs. W. E. Cohn and E. R. Tompkins in furnishing rare earth concentrates are gratefully acknowledged.

Summary

Studies of the elution characteristics of rare earth ions from Amberlite IR-1 absorption columns, using buffered citrate solutions, have shown that the order of elution is the reverse of atomic number, with yttrium falling near gadolinium. The position in the elution sequence can be used to identify individual rare earth elements, and useful separations can be achieved.

Column separations on fission product mixtures containing activities of praseodymium, neodymium, and element 61 have led to the first positive chemical identification of isotopes of element 61 of 47h and 3.7y half-life.

Radiochemical studies and fractionations have established the decay characteristics, genetic relation, and mass assignments of 11d Nd^{147}, 3.7y 61^{147}, 1.7h $Nd^{(149)}$, and 47h 61^{149}.

Other radiochemical work on neodymium and element 61 has been reviewed.

OAK RIDGE, TENNESSEE RECEIVED JULY 16, 1947

7

Reprinted from *J. Amer. Chem. Soc.*, **69**, 2800–2812 (Nov. 1947)

The Exchange Adsorption of Ions from Aqueous Solutions by Organic Zeolites. IV. The Separation of the Yttrium Group Rare Earths[1]

BY B. H. KETELLE AND G. E. BOYD[2]

Introduction

The current state of development of the quantitative understanding of the physics and chemistry of the rare earth elements has been summarized in an excellent recent monograph by Yost, Russell and Garner.[3] In this work the extraordinary difficulty thus far met with in the separation of the rare earths has been remarked upon. Further, the possible general utility of radio-isotopic tracer techniques was recognized as a means in the overcoming of this classical chemical problem. Since investigation of this unique group of elements evidently has been impeded by the scarcity of pure compounds, it was considered important that any promising method of separation and purification be utilized to its full potentialities. Researches on ion-exchange adsorption techniques during the recent war years[4] had indicated that an efficient method of chromatographic separation was possible, and therefore deserved examination at a later time. In fact, the separation of the following fission product element groups had been achieved[5]: Zr + Cb; La + Ce + Pr + Sr + Ba; and Te + I.

The researches described below had as their objective the production, identification and/or characterization of the radio-isotopes of the yttrium group rare earths. Since separation procedures based on ion-exchange adsorption appeared to offer almost unique possibilities for employment as an analytical tool in this problem, the initial studies were directed toward their development. Prior to the initiation of the present program, a series of empirical studies on the separation and recovery of the uranium fission products had led to the development of a "citrate elution" technique by means of which the possibility of the rapid separation of the alkaline and rare earth ions from each other on a milligram scale was demonstrated.[6] Later this procedure was adapted to the separation of cerium and yttrium,[7] and to the production of gram amounts of pure neodymium from didymium earth mixtures.[8] Shortly thereafter, the method was applied to the separation and chemical identification of two radio-isotopes of element 61 formed by the fission of U-235.[9] Quite recently, decidedly superior fractionations of mixtures of approximately equivalent amounts of Ce, Pr and Nd have been achieved by using the new synthetic organic exchanger, Dowex-50.[10]

In spite of the gratifying success met with by the groups lead by Cohn and Tompkins, and by Spedding, there was good reason, on the basis of

(1) This work was performed under the auspices of the Manhattan District at Clinton Laboratories, Monsanto Chemical Co., Oak Ridge, Tennessee, during the period May, 1945, to May, 1947.

(2) On leave from the Department of Chemistry, University of Chicago. Present address: Clinton Laboratories, Monsanto Chemical Co., Oak Ridge, Tennessee.

(3) D. M. Yost, H. Russell and C. S. Garner, "The Rare Earth Elements and Their Compounds," John Wiley and Sons, New York, N. Y., 1947.

(4) H. D. Smyth, "Atomic Energy for Military Purposes," Princeton University Press, Princeton, N. J., 1945, p. 137.

(5) J. A. Swartout, E. R. Russell, D. N. Hume and B. H. Ketelle, July, 1944, to be published.

(6) E. R. Tompkins, J. X. Khym and W. E. Cohn, THIS JOURNAL, **69**, 2769 (1947).

(7) F. H. Spedding, A. F. Voigt, E. M. Gladrow and N. R. Sleight, *ibid.*, **69**, 2777 (1947).

(8) F. H. Spedding, A. F. Voigt, E. M. Gladrow, N. R. Sleight, J. E. Powell, J. M. Wright, T. A. Butler and P. Figard, *ibid.*, **69**, 2786 (1947).

(9) J. A. Marinsky, L. E. Glendenin and C. D. Coryell, *ibid.*, **69**, 2781 (1947).

(10) D. H. Harris and E. R. Tompkins, *ibid.*, **69**, 2792 (1947).

the variation of the ionic radius within the lanthanide contraction, to believe that the separation of the yttrium group would, in fact, be considerably more difficult than the separation of the early members of the cerium group. The attainment of optimum conditions for a maximum fractionation seemed to depend upon a balancing and adjusting of several chemical and physical variables. An entirely empirical study of these over a sufficient range for the heavier rare earths would have required a prohibitive effort. Fortunately, in the light of the principles established in earlier researches,[11] the lines along which a rational approach was to be conducted seemed clear. For example, it was possible to assume in the beginning that the mass law would be applicable to the heterogeneous equilibria between the trivalent rare earth ion in solution and in the ion-exchanger. The essential nature of the mechanism governing the adsorption rate was felt to be understood. A satisfactory quantitative theory of rate-dependent adsorption processes as they occur in deep, cylindrical adsorbent beds was at hand. Finally, the classical, or rate-independent, theory of chromatography[12,13,14,15] proved to be of considerable aid in the prediction and interpretation of column separations.

As a result of this study, new information on the radio-isotopes of the heavy rare earths was obtained which will be published elsewhere. This communication will describe the results from the examination of various factors governing adsorption column separations. The unusual power of ion-exchange adsorption column techniques has been demonstrated by the separation of all of the fourteen rare earths. It is possible that the results from this work will lead in diverse ways to an improved understanding of the systematic inorganic chemistry of this most interesting family of elements.

Theoretical Considerations

In order to form a basis for the presentation of the experimental findings in the subsequent sections of this paper, it is desirable to outline the elementary principles governing the separation of cations in deep cylindrical beds of base-exchange substances. First, a consideration of the homogeneous and heterogeneous ionic equilibria existing in the aqueous solution and between this solution and the solid exchanger will make explicit those factors which determine the equilibrium distribution. If the partition coefficient can be made independent of the rare earth concentration, a prediction of the fractionating action of a deep adsorbent bed based on the theory of chromatography becomes possible. However, owing to the finite rate of flow of solution through the bed, the shape of the predicted chromatogram must be regarded as a limiting form, perturbed in actuality by the effect of a finite adsorption velocity. Accordingly, an analysis of the influence of rate-dependent adsorption on the non-steady-state conditions occurring in adsorption columns will be required. The object of these considerations will be to give a qualitative explanation of the observed chromatograms.

The Equilibrium Theory.—Chemical equations for the heterogeneous adsorption reaction may be written as follows, arbitrarily assuming the behavior of a trivalent cerium ion to be typical for any trivalent rare earth

$$Ce^{+++} + 3NH_4R \rightleftarrows CeR + 3NH_4^+ \qquad (1)$$

In addition, the homogeneous equilibrium in the aqueous solution between "free" rare earth ion, and rare earth bound in a complex ion may be written[16]

$$Ce(H_2Cit)_3 \rightleftarrows Ce^{+++} + 3H_2Cit^- \qquad (2)$$

Assuming the mass law may be applied to each of these equilibria,[17] and combining the expressions so obtained, the relation for the equilibrium distribution coefficient, K_d, is[18]

$$K_d = \frac{CeR}{Ce(H_2Cit)_3} = K_{exchange}\,K_{complex}\left(\frac{NH_4R}{NH_4^+}\right)^3 \frac{1}{(H_2Cit^-)^3} \qquad (3a)$$

$$K_d = \frac{n_{CeR}}{Ce(H_2Cit)_3} = K_{exchange}\,K_{complex}\,\frac{n_{NH_4R}}{(NH_4^+)^3}\,\frac{1}{(H_2Cit^-)^3} \qquad (3b)$$

Utilizing the appropriate expression for first dissociation constant of the citric acid, Equation (3) may be written

$$K_d = \frac{K_{exchange}\,K_{complex}}{K_1^3}\left(\frac{NH_4R}{NH_4^+}\right)^3\left(\frac{H^+}{H_3Cit}\right)^3 \qquad (4)$$

Assuming the exchange capacity of the adsorbent remains constant, the dependence of the distribution ratio upon pH is revealed by Equation (4). Since the usual procedure for adjusting this variable has been to add ammonia to the citrate

(11) (a) G. E. Boyd, J. Schubert and A. W. Adamson, THIS JOURNAL, **69**, 2818 (1947); (b) G. E. Boyd, A. W. Adamson and L. S. Myers, *ibid.*, **69**, 2836 (1947); (c) G. E. Boyd, L. S. Myers and A. W. Adamson, *ibid.*, **69**, 2849 (1947).

(12) J. N. Wilson, *ibid.*, **62**, 1583 (1940).

(13) D. DeVault, *ibid.*, **65**, 532 (1943).

(14) J. Weiss, *J. Chem. Soc.*, 297 (1943).

(15) J. E. Walter, *J. Chem. Phys.*, **13**, 229 (1945).

(16) At the present writing the stoichiometric composition of the most important rare earth citrate complex remains indefinite, although it seems probable that an uncharged structure is formed. Preliminary results indicate that either the H_2Cit^- or the HCit ions participate, and not the tribasic citrate ion as is the case with the alkaline earth complexes which carry a negative charge (J. Schubert and J. W. Richter, February, 1945, to be published). Preliminary measurements carried out in September, 1945, by B. H. Ketelle and J. W. Richter gave values of 4.3×10^{-8} and 4.5×10^{-9} for the instability constants of the neutral compounds, $La(HCit)_{1.5}$ and $Y(HCit)_{1.5}$, respectively.

(17) An investigation of heterogeneous equilibria of the type indicated by Equation (1) has been carried out by Mr. L. S. Myers, September, 1946, to be published.

(18) In the experimental measurement of the distribution ratio a determination of only the total cation in the solution (*i. e.*, total Ce(III)) is made. The assumption that this is identical with the complex ion concentration may be justified by a simple computation based on approximate values for the complex ion dissociation constants. Concentrations are expressed as thermodynamic activities. Equation (3b) will apply when ideal solid solutions are formed.

buffer,[4] a decrease in hydrogen ion will be accompanied by an increase in NH_4^+ ion concentration so that the magnitude of the distribution coefficient would be expected to decrease sharply with increase in pH. The decrease in the concentration of the undissociated citric acid will tend to offset these changes somewhat, of course. It is evident also that the lowering in the magnitude of K_d which will be produced by an increase in the initial citric acid concentration, keeping the pH constant, will be caused in part by an increased ammonium ion concentration.

The conditions under which the distribution coefficient changes are of interest, since, if this varies with the concentration in solution much poorer separations are observed. As may be seen from Equation (4), as long as the quantity of rare earth adsorbed is sufficiently small that the activity of the ammonium ion in the exchanger remains constant, the distribution ratio will not vary. A mole fraction of NH_4R constant within experimental error may be realized either with small values of K_d, with low rare earth concentrations, or with a high capacity exchanger.

Although the exact effect of temperature changes upon K_d cannot be predicted, general considerations suggest a direct proportionality will be observed. Thus, the exchange constant will decrease, whereas the complex ion instability constant will increase with the temperature so as to compensate. The variation of K_d, therefore, will be determined largely by the change of the first dissociation constant of citric acid, K_1, with temperature. At room temperatures the third dissociation constant, K_3, for citric acid has been found to decrease with temperature, whereas the first and second constants, L_1 and K_2, increase very slowly.[19] By analogy to the behavior of other organic acids, oxalic acid in particular,[20] at temperatures above 50° both K_1 and K_2 would be expected to decrease; hence, above this latter temperature, K_d should tend to increase.

In as much as the interest in this research was in the separation of the heavy rare earths, a separation factor, α, defined by the ratio of the distribution coefficients for two neighboring rare earths, has been employed. It then follows from Equation (4) that the separability of a pair of rare earths will be independent of the ammonium and citrate ion concentrations in solution, and of the pH so long as the composition of the complex ion does not vary. The separation factor would be expected to be virtually unchanged by temperature variations.

By way of illustration, it will be of interest to give a comparison of the partition coefficients for a pair of rare earth ions, say, Ce^{+++} and Y^{+++}. Thus,

$$\alpha = K_d^{Ce}/K_d^{Y} = (K_{exchange}^{Ce}/K_{exchange}^{Y})(K_{complex}^{Ce}/K_{complex}^{Y}) \quad (5)$$

It was predicted[9] and has been shown experimentally[17] that cerium is more strongly adsorbed than yttrium. A value of 1.55 has been observed for the first ratio on the right-hand side of Equation (5). The second ratio is known also, from preliminary measurements,[16] to possess a value of approximately 2.9. The role of the citrate becomes apparent: by its use, the normal difference in adsorbabilities, by means of which a separation of the two ions could be effected, is enhanced by the difference in the homogeneous equilibria in the aqueous phase.

It is desirable to stress a number of additional factors in the adaptation of the foregoing principles to the practical problem of the separation of the rare earths by an adsorption column. If the separation is conducted in such a manner that equilibrium is realized, the behavior of the adsorption column may be interpreted in terms of the theory of chromatography. The prediction made for linear isotherms is that a band with an abrupt leading and trailing edge will be formed. Further, this band will move through the adsorbent bed with a velocity given by $v/(1 + \delta/f)$ where v is the linear flow rate of solution in the bed, δ is the equilibrium distribution coefficient, and f is the fractional bed void space.[21] Correspondingly, that rare earth with the larger δ will move slower and will lag in its relative movement down the adsorption column. Using the example given above, where $K_d^{Ce} > K_d^{Y}$, yttrium would be expected in the effluent from the adsorbent before cerium.

It may be shown that the general expression for the velocity of movement of a band, $v(c)$, depends upon the adsorbate concentration in solution and will be a minimum for the case of linear adsorption. From the dependence of δ on the concentrations in solution, it can be deduced that an increase in pH or in citric acid concentration at constant pH will increase the migration rate of the band. An increase in temperature would be expected to cause a retardation. Further, the larger the exchange capacity of the adsorbent, the slower the band velocity, all other things being equal. Conversely, the larger the ammonium ion concentration, the greater the band velocity.

A number of additional useful guiding rules are derivable from the elementary theory. Values of δ may be computed from the time, t_x, required for the band to transverse the length of the column, choosing the center or the peak as the point of reference. Thus

$$\delta = (Wt_x/x) - f \quad (6)$$

where W is the volume flow rate (ml./sq. cm./min.) and x is the bed depth. The bed void may

(19) Bjerrum and Unmack, *Danske Videns Kabenes Selskabs Math. Fys. Med.*, IX, 1 (1929).

(20) H. S. Harned and L. D. Fallon, THIS JOURNAL, **61**, 3111 (1939).

(21) The bed distribution coefficient, δ, is related to K_d and to the bed density, ρ, by $\delta = K_d\rho$. If K_d is expressed as (moles/g. oven dry NH_4R)/(moles/cc. solution) then ρ will be given in the units: (g. oven dry NH_4R/cc. bed).

be measured with high reproducibility by determining the time, t_h, for a non-adsorbed (anionic) ion to pass through the bed. Then

$$f = t_h W/x \quad (7)$$

whereupon

$$\delta = (t_x/t_h - 1)(Wt_h/x) = (W/x)(t_x - t_h) \quad (8)$$

It is evident, if the correct bed densities are known, that values of K_d may be computed from observations made with adsorption columns. The validity of Equation (8) may then be checked by an independent determination of K_d in an equilibrium experiment.

Useful information concerning the minimum bed depth, x_c, or volume of pure solvent,[22] V_c, required to just effect a separation of two substances whose adsorption coefficients are δ_1 and δ_2, where $\delta_2 > \delta_1 > 1$, may be obtained also. If V_0 is taken as the volume from which the initial adsorption onto the top of a bed of cross-section A occurred, then

$$x_c = (V_0/fA)(\delta_1/\delta_2 - \delta_1) = (V_0/fA)(\alpha - 1) \quad (9)$$

The Influence of Rate Dependent Processes.—Although adsorption bands with sharp front and rear edges are predicted from the rate-independent theory of chromatography for conditions in which the equilibrium adsorption increases linearly with concentration, in actuality such a behavior is virtually never observed. The usual finding is an S-shaped leading and trailing edge in which, generally, a long near-linear middle portion may be found. Frequently these bands are symmetrical in shape, although occasionally asymmetric forms with more abrupt forward than rear edges which resemble those predicted for a non-linear adsorption isotherm are observed. As was indicated in the preceding paper in this series,[9] for the linear isotherm, the sigmoid character of the adsorption edges may be understood on the basis of a theory of rate-dependent chromatography. In the terms of this theory, the edges of the bands are characterized by a dimensionless parameter, y, related to the rate and equilibrium constant, k and δ; to the bed depth, x; to the linear flow velocity, v; and to the fractional bed void space, f, by the expression

$$y = (k\delta/f)(x/v) \quad (10)$$

When the value of y becomes infinitely large, bands with abrupt edges are approached, so that under limiting conditions the equilibrium theory and the rate dependent theory of chromatography become identical. Since relatively small values of y produce broad bands, in general, it is essential to employ conditions leading to as high a value as possible if the separation of several substances by chromatographic techniques is desired.

The factors which determine the rate constant, k, must be examined to predict the variations in y resulting from changes in the variables of the adsorption system. Studies of the mechanism of the rate processes[8] have led to a film diffusion hypothesis by means of which k was shown to depend upon the diffusion coefficient of the ion being adsorbed, D^l, the adsorbent particle radius, r_0, the fluid film thickness, Δr_0, and the equilibrium distribution coefficient, δ, by the equation

$$k = 3D^l/r_0\Delta r_0\delta \quad (11)$$

which, upon substitution into Equation (10), yields

$$y = (3D^l/r_0\Delta r_0 f)(x/v) \quad (12)$$

Now, if diffusion through a thin liquid film is rate controlling in adsorption, the same process would be expected to govern the *desorption rate* with an unchanged rate constant. The same y value would describe the forward and rear edges, and therefore the band would possess a symmetrical shape.

Equation (12) reveals that particle size will play an important role; consequently, smaller particles should produce sharper band edges. Furthermore, adsorbent granules of a rather highly uniform size are essential to reduce the classification caused by the back-washing of deep beds. On the other hand, since the film thickness, Δr_0, appears to vary approximately inversely as the linear flow rate, v, the character of the band edge will either be independent of the flow or change quite slowly with it. Deeper adsorbent beds, it is seen, should lead to sharper band edges. The magnitude of the partition coefficient, δ, will be without an effect on the edges of the band, although of course the band width will be determined by this.

From Equation (12) it is seen for maximum values of y that the magnitude of the bed void space, f, should be reduced to a minimum. This factor will be independent of the particle diameter, in the absence of wall effects, although it is determined in part by the particle shape. It is well-known that f will be a minimum for a bed composed of spherical granules of a uniform diameter.

The only temperature dependent quantity in Equation (12) is the diffusion coefficient, D^l, of the exchanging ion. Since this will increase with temperature with an activation energy of 4–5 kcal. mole^{-1}, appreciably higher y values, and hence sharper band edges, will be obtained by conducting the adsorption separations at elevated temperatures. In order to off-set the slowing down in the movement of the band, owing to the increase in δ with temperature, indicated by Equation (4), either the flow rate may be increased, or the magnitude of δ may be lowered by an increase in pH.

An increase in temperature will also bring about several additional advantages. Thus, owing to the reduced fluid viscosity, larger flow rates are

(22) In the rare earth separations, the "solvent" was the ammonium citrate buffer solution.

possible with the same hydrostatic head. This gain may be utilized by employing finer adsorbent particles thereby additionally sharpening the band edges, or, it may be used to obtain the same degree of separation in a correspondingly shorter time, since the character of the band edge will be nearly independent of flow.

The presence of a rate factor in the adsorption system is seen, therefore, to cause a broadening of the band over that predicted from the classical theory of chromatography. As a result, the separations observed will be less than was anticipated. Although the bands will move with the same velocity as before, the actual value of x_c observed will be greater than indicated by Equation (9). In principle, these observed values may be computed from the rate theory, although the criterion for band separation must be revised.

Experimental Methods and Procedures

The experimental arrangements are illustrated by Fig. 1. In a typical experiment, the ion-exchanger was contained in the Pyrex glass tube, A, 120 cm. long by 6 mm. diameter, which was surrounded by a jacket so that the temperature in the bed could be controlled either by circulation of a liquid from a thermostat, or by the passage of steam through the jacket. The ion-exchanger was retained by a small plug of Pyrex glass wool which rested upon a constriction at the lower end of the column.

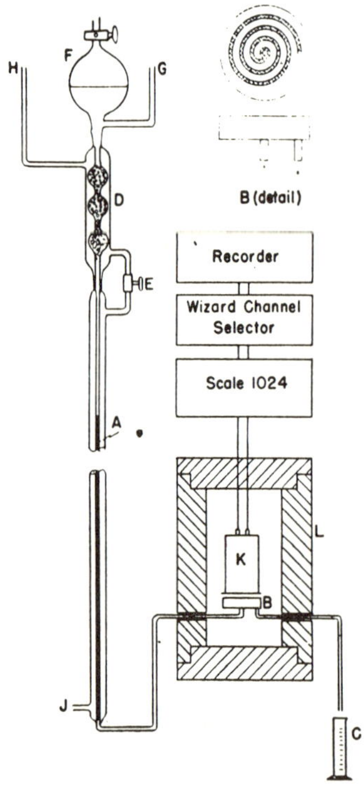

Fig. 1.—Experimental arrangement employed in adsorption column separations: A, adsorbent bed, Amberlite IR-1 or Dowex-50; B, counting cell; C, receiver; D, Allihn condenser; E, throttle valve; F, gas entrainment bulb; G, elutriant inlet; H, thermostat fluid inlet; K, mica end-window Geiger–Müller counting tube; L, two-inch lead radiation shield.

The effluent solution from the column was conducted through a transflex tube of $^1/_{16}$" inside diameter to the counting cell, B, and finally to the receiver, C, where the various samples were collected. The cell, B, is shown enlarged in the inset diagram of Fig. 1. It consisted of a cylindrical block of Lucite, $1^1/_4$" in diameter by $^1/_2$" thick. A spiral groove $^1/_{32}$" broad and $^3/_{32}$" deep was cut in one face. This channel was covered by a "window" of 3.2 mg./sq. cm. thick polystyrene which was sealed to the Lucite block with Amphenol cement. Access to each end of the spiral channel was attained through holes and threaded nipples. The volume of liquid held by this cell was 0.6 ml.

In a particular experiment the solution containing the radioactive rare earths was placed on the uppermost portions of the adsorbent bed by means of a small diameter tube inserted into the top end of the column. A separatory funnel containing water was connected by means of an air-filled rubber tube to the top of the column to inject the solution into the exchanger. No water was permitted to flow into the column. The object of this technique was to cause the mixture to be adsorbed as a narrow band with sharp lower edge at the topmost part of the bed. The volume from which adsorption took place was either 1 or else 2 ml. of 1 *M* hydrochloric acid or 5% citrate adjusted to the same *p*H as was employed subsequently to effect a separation. Usually not more than 1 to 3 mg. of rare earth sesquioxide mixture was adsorbed; frequently, only 0.1 mg. quantities were studied.

Following the adsorption step, the air above the resin bed was completely displaced by solution to develop the chromatogram, and an Allihn condenser, D, was attached to the upper end of the column. This condenser was maintained at the same temperature as the column, and thereby served to "pre-heat" the influent elutriant solution. When steam was being passed through the jacket, the temperature of the condenser was controlled by means of the valve, E, in such a manner that the influent solution was caused to boil gently. This was desirable in order that dissolved gas be removed from the solution before it reached the bed. The expelled gas was collected in the bulb, F. Thus, the eluant or wash solution from a constant head tank entered the top of the condenser through inlet, G, under the force of gravity, and then flowed through the packing of glass beads and glass wool, through the exchanger bed, A, through the counting cell, and finally to a receiver. The fluid used to control the temperature entered the condenser jacket at H, passed through valve E before entering the column jacket, and then escaped through outlet J at the lower end of the column.

The radioactivity in the effluent solutions was detected by a thin mica window bell-type G-.M. counter tube, K, which was surrounded by a two inch thick lead shield, L. The tube was connected to a scale of 1024 which in turn transmitted impulses to a Wizard recorder. The latter, through electrical contacts and relays, could be adjusted to close any of four channels on a Simplex time stamper. These channels corresponded to the digits, tens, hundreds, and thousands wheels of the Wizard recorder. By means of a selector switch, it was possible to record the time at which any channel carried a pulse, or, to record only every tenth or one-hundredth scaler when the level of radioactivity was sufficiently high that such a record sufficed. In this way, a continuous record of the radioactivities emerging from the adsorbent bed was made possible. The time between pulses was proportional to the reciprocal of the activity in the solution contained in the counting cell. It was possible, therefore, either to plot the concentration history of the effluent solution in terms of the reciprocal activity, expressed in terms of minutes per 640 counts, or else to convert the data to counts per minute and to plot this as a function of the time from the beginning of the elution. With the former type of plot inverse peaks are obtained when the activity of the emergent solution passes through a maximum.

A peak in the effluent activity was recorded for each rare earth element if a separation had been effected, and if a suitable radio-isotope indicator were present. The magnitude of the activity level reached in these peaks depended upon (a) the concentration of rare earth in solution; (b) the neutron capture cross-section and isotopic abundance of the stable isotope from which the tracer radio-isotope was formed; (c) the neutron flux during the bombardment; (d) the duration of bombardment; (e) the ability of the radiation to penetrate water, polystyrene and mica and, finally, (f) the half-life of the active isotope formed. No attempt has been made to obtain a record of beta activities having maximum energies less than 0.4 Mev. It was felt, however, that under proper conditions somewhat softer radiation might be detected.

A summary of the half-lives and energies of the rare earth isotopes which were measured in the foregoing experimental arrangement is given in Table I. Several of these activities, it will be noted, are new.

TABLE I

DECAY PERIODS AND BETA-RAY ENERGIES FOR RARE EARTH ISOTOPES

Isotope	Half-life[a]	Maximum energy, Mev.
Lu^{177}	3.4 h[b]	1.2
Yb^{177}	2.6 h[b]	1.3
Tm^{170}	125 d	1.1
Er^{171}	7.5 h[b]	1.6
Ho^{166}	27 h	1.9
Y^{90}	65 h	2.2
Dy^{165}	2.5 h	1.2
Tb^{160}	72 d	0.7
$Gd^{159,161}$	8.0 d[b]	0.7
Eu^{152}	9.2 h[b]	1.9
Sm^{153}	47 h	0.8
61^{149}	47 h[b]	1.1
Nd^{147}	11 d	0.9
Pr^{142}	19.3 h	2.1
Ce^{141}	28 d	0.6
La^{140}	40 h	2.1

[a] h = hours; d = days. [b] Longer periods present in addition.

Experimental Results

The continuous beta ray recording equipment described above was employed to determine the optimum conditions for separation, and subsequently to demonstrate a variety of rare earth fractionations. The effects on adsorption column performance of *p*H, citrate and ammonium ion concentration had been studied previously[6,7,8] and are in accord with the deductions presented in an earlier section. Preliminary investigations of the influence of eluant flow rate on the separation of the rare earths at room temperatures served to confirm the earlier prediction based on Equation (12) of the relative unimportance of this variable at rates below 2 ml./sq. cm./min. The limitation imposed by the mass of the adsorbed ions on the separations was found to be in agreement with earlier findings of Khym and Tompkins[23] and of the Ames group.[7]

Effect of Adsorbent Particle Size on Band Width.—The importance of the mesh size of the exchanger in determining the character of the desorption band had been demonstrated in several early studies.[24] Also, adsorbent particle size had been shown to determine the velocity of exchange regardless of whether diffusion in and through the solid, or through a film surrounding the particle, controls.[10] In either case, the rate increased as the particles became smaller. In terms of column separations, if the speed of exchange is increased, a more rapid flow may be employed, or, two elements can be separated to the same extent in less time. Irrespective of the flow rate, a smaller granule should give an improved separation. Several experiments were conducted to ascertain the influence of particle size in rare earth separations by ion-exchange columns.

A quantity of the synthetic organic ion-exchanger, Amberlite IR-1, was placed in a Pyrex tube and was converted to the hydrogen form (HR) by washing with ten bed volumes of 3 *M* hydrochloric acid. After rinsing with distilled water, the exchanger was air-dried and screened to give samples of known ranges of mesh size.

Two-tenths gram of adsorbent of a pre-determined particle size was shaken for ten minutes with 5 ml. of hydrochloric acid solution of *p*H 1.0 which contained 0.025 ml. of carrier-free Y^{91} stock solution after which the suspension was slurried into a glass column to form a shallow bed approximately 4 mm. long by 1 sq. cm. cross-section. The yttrium was displaced by washing with a buffer solution prepared by dissolving 50 g. of citric acid monohydrate in 1000 g. of water and adjusting the *p*H to 2.64 by the addition of concentrated ammonium hydroxide solution. The flow rate was 0.35 ml./sq. cm./min. The effluent solution was passed through a counting cell to measure its activity continuously. The plots of the desorbed activity as a function of time for each of four particle size ranges are shown in Fig. 2.

The "tailing-out" of the desorption band produced by the large particles would cause unnecessary overlapping of the successive bands and poor resolution of the earths. Consequently, a better separation would be expected with small particles of exchanger if hydrostatic pressure sufficient to maintain satisfactory flow rates were available. Although 270/325 mesh adsorbent was clearly superior, the labor involved in preparing sufficient quantities for deep beds forced a compromise on the use of 170/200 mesh size particles in the subsequent work with Amberlite IR-1.

Heavy Rare Earth Separations with Amberlite IR-1 at Room Temperatures.—The initial degree of success met with in the separation of the yttrium group rare earths may be illustrated by the results from a series of experiments whose primary objective was the isolation and characterization of the radioisotopes of erbium. Samples of Hilger's spectrographically pure erbium oxide (Lab. No. 10,374) were activated

(23) J. X. Khym and E. R. Tompkins, July, 1945, to be published.

(24) E. R. Russell, private communication, April, 1944.

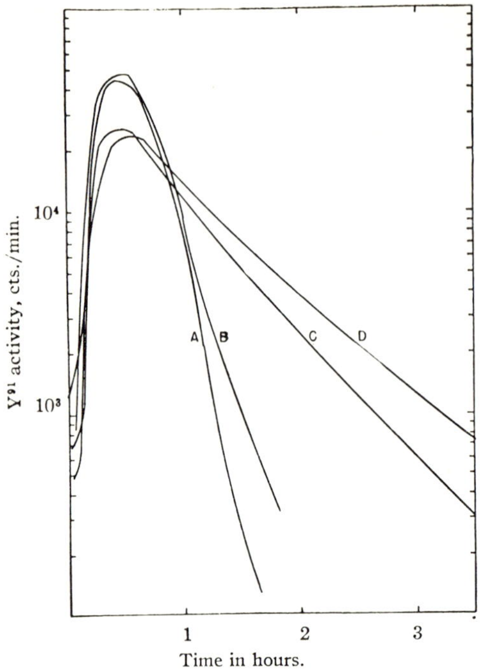

Fig. 2.—Effect of mesh size of Amberlite IR-1 on desorption band width: A, 270/325 mesh; B, 170/200 mesh; C, 50/60 mesh; D, 30/40 mesh.

with slow neutrons and the decay of the activity was measured. The slope of the decay curves changed more or less continuously, although the data could be represented by assuming several activities with different half-lives to be present. This behavior might be expected since there are four possible radio-isotopes of erbium which could have resulted from slow neutron bombardment. Since the relative isotopic abundance of stable Er^{162} is but 0.1%, it would probably be difficult to detect Er^{163} activity. There was some reason to believe that active isotopes other than those of erbium were produced, for it was impossible to obtain the same decay periods from bombardments of varying duration. Since three (at the most, four) activities with quite different half-lives would be expected, it should be possible to determine these from the decay curves observed following differing periods of irradiation. Since such an analysis proved impossible, it was suspected that the erbium oxide contained appreciable quantities of impurities. Therefore, a column purification was undertaken.

About 0.1 mg. of neutron activated Er_2O_3 was adsorbed on a 125 cm. by 0.79 sq. cm. bed of 170/200 mesh Amberlite IR-1, and a separation was obtained at room temperature in 3300 minutes by elution with 5% citrate buffer of *p*H 2.64 at 1.2 ml./sq. cm./min. As may be seen in Fig. 3, five partially overlapping activity peaks were obtained. In spite of the mediocre separation, it was possible by means of the radiation characteristics of samples of the eluate to establish the presence of thulium, holmium and yttrium impurities. The lower *p*H of 2.64 used in this fractionation of the yttrium group is to be compared with the higher values of 2.74 to 2.95 employed to separate the cerium group.[7] The almost symmetrical shape of the activity peaks in Fig. 3 is to be noted. This behavior is to be expected if a linear isotherm is obeyed. Clearly, it would be impossible to utilize the Hilger Er_2O_3 oxide preparation without further purification or separation of the activities produced in it.

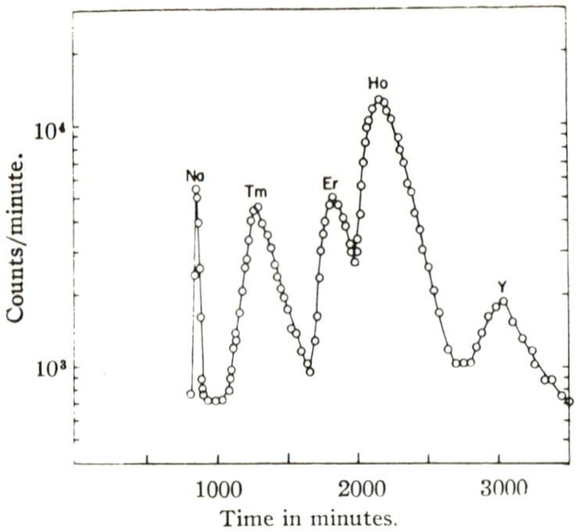

Fig. 3.—Yttrium group separation with a 170/200 mesh Amberlite IR-1 column at room temperature: bed dimensions, 125 cm. × 0.79 sq. cm.; flow rate, 1.2 ml./sq. cm./min.; 5% citrate buffer at *p*H 2.64.

The Effect of Temperature on the Separation of Thulium and Ytterbium with Dowex-50.—Further study of the separation exhibited in Fig. 3 showed that a better fractionation could be effected if the leading and trailing edges of the activity peaks could be made more abrupt. The slope of these edges, as was pointed out in the section on Theoretical Considerations, is determined by the dimensionless parameter *y*, which, by Equation (10), is dependent upon the rate constant, *k*. The rate of exchange, as shown earlier in both shallow and deep bed studies was known to increase with temperature and with decreasing particle size. These earlier investigations also disclosed, however, that work with the Amberlite IR-1 exchanger must be limited to temperatures less than 60° owing to an increasingly objectionable decomposition of this material at higher values. Fortunately, during the course of these considerations, Dr. L. A. Matheson of the Dow Chemical Company informed us of the demonstrated stability of a new ion-exchanger,[25] Dowex-50, at temperatures as high as 100°. Accordingly, we were encouraged to take up the use of this syn-

(25) See W. C. Bauman, THIS JOURNAL, 69, 2830 (1947).

thetic exchanger for studies at elevated temperatures even though preliminary rate measurements by L. S. Myers had indicated an exchange velocity at 25° of only one-third of that for Amberlite IR-1.

A comparative study of the degree of separation effected at 25° and 100° was conducted using the heavy rare earths, Yb and Tm, which had been tagged by neutron irradiation of 0.1 mg. of each element. The same 91 cm. by 0.26 sq. cm., 270/325 mesh Dowex-50 column and a flow rate of approximately 0.5 ml./min. was used at both temperatures. Since the exchange capacity of the Dowex-50 is roughly twice that of Amberlite IR-1, it was necessary to increase the *p*H of the 5% citrate eluant so as to decrease the value of the distribution ratio, δ, in order to effect the differential desorption in approximately the same time as demonstrated in Fig. 3. An initial *p*H value of 3.12 caused the appearance of Yb activity in the column effluent in only 280 minutes, and therefore was considered undesirably high. The degree of separation achieved at 25° with a *p*H of 2.98 is shown in Fig. 4 together with the fractionation observed at 100°. The predicted increase in δ with temperature was confirmed apparently by the finding that a higher *p*H (3.20) was required at 100° to accomplish the separation in the same time, although the reliability of the *p*H measurement (glass electrode) at 100° was not established.

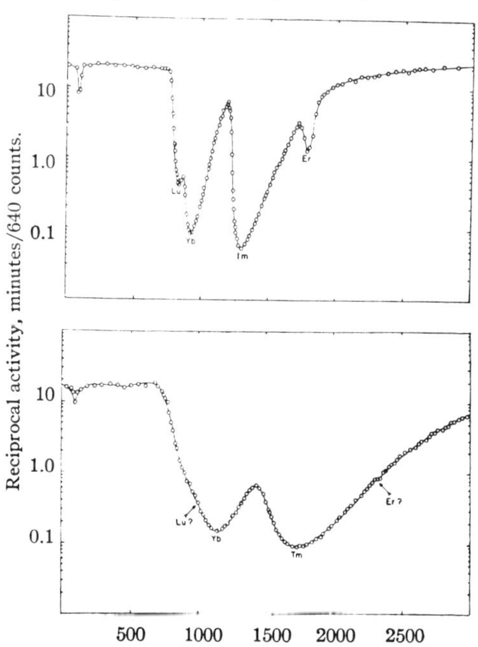

Fig. 4.—Effect of temperature on the separation of the yttrium group earths with a 270/325 mesh Dowex-50 column: bed dimensions, 91 cm. × 0.26 sq. cm.; upper curve for 100°, flow rate 0.35–0.45 ml./sq. cm./min., and 5% citrate buffer at *p*H 3.2; lower curve for 20°, flow rate 0.55 ml./sq. cm./min. and *p*H 2.98.

An examination of Fig. 4 will reveal the superiority of the rare earth fractionation at the higher temperature. Not only are the peak edges sharper at 100° than at 25°, but the presence of unsuspected Lu and Er impurities was also clearly indicated in the higher temperature experiment. Only an inconclusive suggestion of the occurrence of these elements may be determined from a careful examination of the 25° curve. Subsequent to these studies, many additional separations have been carried out at 100°. Thus far, there has been no indication of the decomposition of the Dowex-50 exchanger even after several months of operation of the same bed.

The Effect of Particle Size in the Separation of the Heavy Rare Earths at 100° with Dowex-50.—The initial shipments of Dowex-50 exchanger contained sizeable quantities of 120/140 and 270/325 mesh granules so that it was possible to determine the role of particle size in the yttrium group separations by preparing beds of the same depth from these two fractions. Separations of an identical rare earth mixture at 100° were then conducted. The fractionation obtained with the 120/140 mesh size range contained in a 98 cm. by 0.75 sq. cm. bed when a flow rate of 2.2 ml./sq. cm./min. was used is shown in Fig. 5. The eluant was a 4.75% citrate solution adjusted to a *p*H of 3.2. The mixture studied was the same Hilger Er_2O_3 used in the separation illustrated by Fig. 3. A quantitative intercomparison of the data in these two figures is not possible owing to the differing temperatures, bed dimensions and particle diameters. Qualitatively, however, it may be seen that a vastly improved separation is demonstrated in Fig. 5.

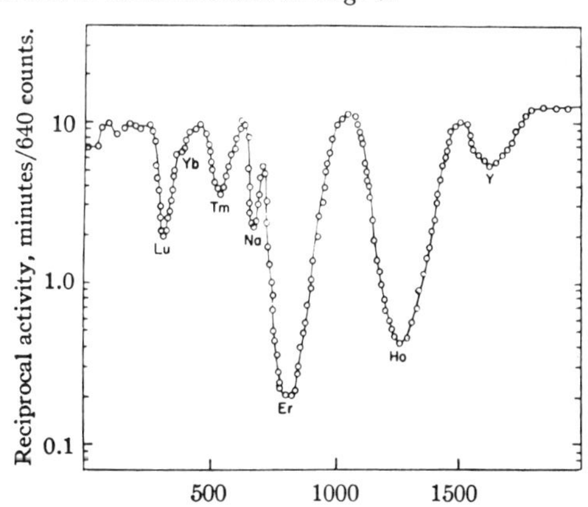

Fig. 5.—Effect of adsorbent particle size on the separation of the yttrium group rare earths with Dowex-50 columns at 100°: bed dimensions, 98 cm. × 0.75 sq. cm.; flow rate, 2.2 ml./sq. cm./min.; 5% citrate buffer at *p*H 3.2; adsorbent particle size, 120/140 mesh; compare with curve for 270/325 mesh exchanger shown in Fig. 6a.

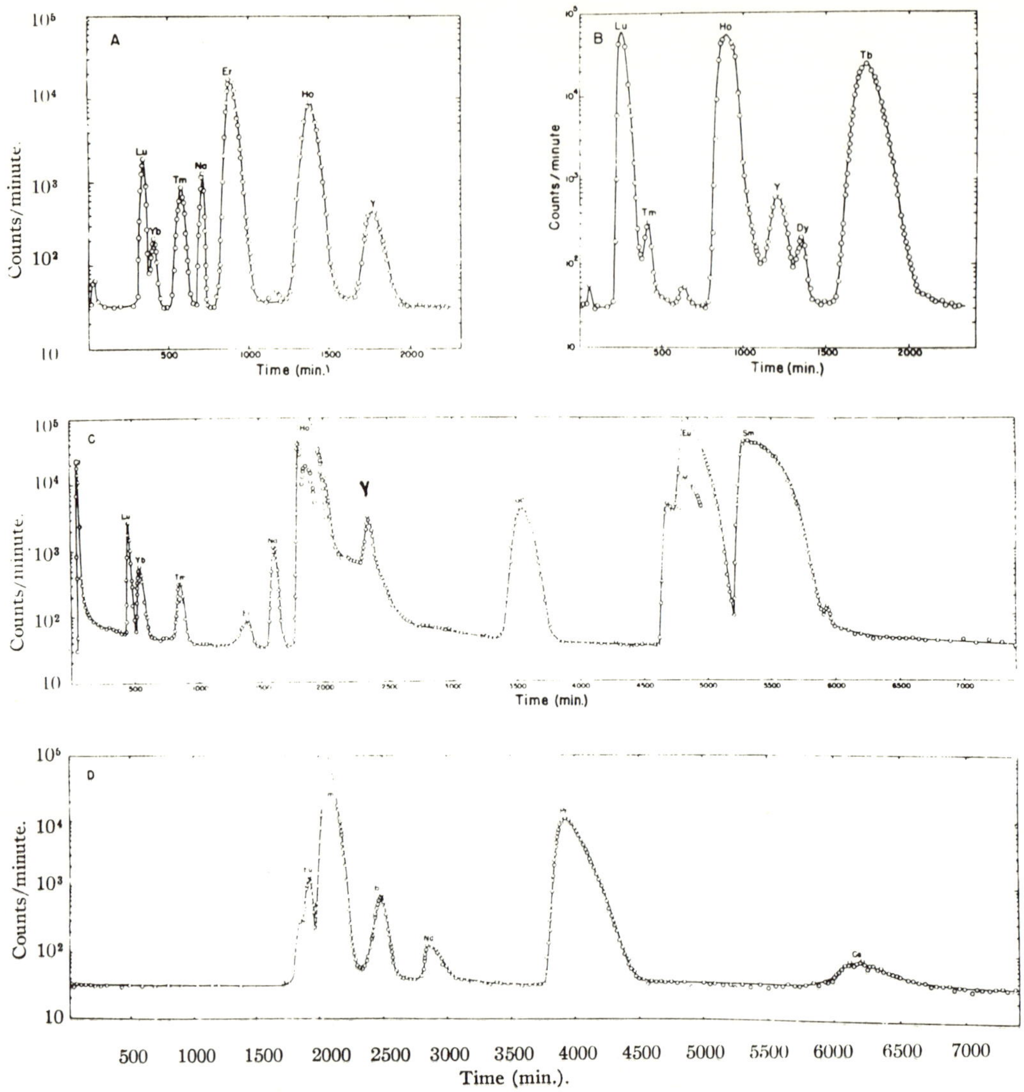

Fig. 6.—Demonstrations of rare earth separations effected with a 270/325 mesh Dowex-50 column at 100°: bed dimensions, 97 cm. by 0.26 sq. cm.; flow rate, 1.0 ml./sq. cm./min. except in *a* where 2.0 ml./sq. cm./min. was used: (a) fractionation of activities produced by neutron irradiation, 0.8 mg. spectrographic grade Er_2O_3 (Hilger) (*p*H 3.20); (b) fractionation of heavy rare earth mixture consisting of 0.1 mg. each of Lu_2O_3, Yb_2O_3, Ho_2O_3 and Tb_2O_3 [Tm, (Er), Y and Dy present as impurities] (*p*H 3.20); (c) fractionation of intermediate rare earth mixture consisting of 0.1 mg. Ho_2O_3 and 1.0 mg. each of Dy_2O_3, Gd_2O_3, Eu_2O_3 and Sm_2O_3 (Cl, Lu, Yb, Tm, Er and Na present as impurities, *p*H 3.25 for 4550 minutes, then *p*H 3.33); (d) fractionation of light rare earth mixture consisting of 0.1 mg. each of Sm_2O_3 and Nd_2O_3 plus 0.01 mg. each of Pr_2O_3, Ce_2O_3 and La_2O_3 (Eu present as impurity, 61 produced by 1.7 h $Nd^{149} \longrightarrow$ 47 h 61^{149}, *p*H 3.33 for 1610 minutes, then *p*H 3.40).

The degree of separation achieved with the bed of 270/325 mesh Dowex-50 particles may be seen in Fig. 6a. Here, except for the particle size and the bed cross section which was 0.26 sq. cm., the conditions were the same as those for the fractionation exhibited in Fig. 5. As in the previous experiments, samples of the eluate were taken and the activity was reclaimed on a small column. The identification of the activity peaks was made by a characterization of the decay periods and energies of these samples.

A comparison of Figs. 5 and 6a serves to establish the improved separation made with 270/325 mesh adsorbent. Whereas about the same time was required for a peak due to a particular element to appear in the effluent, these peaks possessed less abrupt edges with the 120/140 mesh bed and hence were broader. Also, the activity between the peaks was never less than *twice* the instrument background. Furthermore, with the larger particles, the resolution of the Lu and Yb peaks (Fig. 5) was very much less than was found with the 270/325 mesh preparation. The relatively high flow rate of 2 ml./sq. cm./min. used in both of these experiments is to be noted. In fact, because of the decrease in viscosity of the buffer solution as the temperature increased, it was possible to operate the column at flow rates well in excess of this value using a hydrostatic head of only twelve feet.

A comparison of Figs. 3 and 6a shows that in the latter the separation of the Lu, Yb, Tm, Er, Ho and Y activities was accomplished in about two-thirds the time required for the fractionation made at 25° with 170/200 mesh Amberlite IR-1. The activity peaks were well separated in the 100°, 270/325 mesh Dowex-50 experiment, for it may be seen that the activity decreased to background between the bands.

The position of the peak of sodium activity observed in Figs. 3, 6a, and 6c was of interest. It is apparent that the rare earths were desorbed in the order of increasing crystal radius. However, not only is the ionic charge of sodium less than those of the trivalent rare earths, but its ionic radius is smaller even than that of lutecium (r_{Na^+} = 0.98 Å.; $r_{Lu^{+++}}$ = 0.99 Å.). In most base-exchange reactions univalent ions are much less tightly bound and therefore more easily desorbed than are more highly charged ions. The position of sodium among the rare earths in these experiments is to be explained on the basis that the value of δ for the former comes in the same range as those for the rare earths. Alternatively, the movement of sodium through the bed was slow because of the low cation content of the buffer, ($c_{NH_4^+} = 0.2\ M$ at pH = 3.2), whereas the movement of the rare earths was increased relative to sodium because they formed complex ions with the anions of the citrate buffer.

In order to confirm the assignment of the sodium and yttrium peaks in Fig. 6a, about a milligram each of sodium, holmium and yttrium were irradiated, adsorbed, and separated at 100° under the same conditions and on the same 270/325 mesh column as was used for the erbium study. The relative positions of the activities were the same as in the experiment in which these activities were found in Hilger erbium, thus verifying the assignments made in that case.

Separation of both the Yttrium and Cerium Group Rare Earths at 100° with Dowex-50.—The same 97 cm. 270/325 mesh Dowex-50 column employed to separate the heavy rare earth impurities in the Hilger erbium oxide was used next in three consecutive experiments in which the rare earths from lutecium through lanthanum were separated. In each of these, the column was operated at 100°, and at flow rate of 1 ml./sq. cm./min.

In the first experiment 0.1 mg. each of the neutron irradiated oxides of lutecium, ytterbium, holmium and terbium were adsorbed. The elutriant solution had a pH of 3.2. The activity history of the column is plotted in Fig. 6b. The lutecium and ytterbium bands were not completely resolved, probably because the elutriant pH was somewhat higher than the optimum of 3.1. The elements thulium, erbium, yttrium and dysprosium were present as impurities in the lutecium and holmium.

The second experiment consisted of the bombardment and separation of 0.1 mg. of holmium oxide and 1.0 mg. each of the oxides of dysprosium, gadolinium, europium and samarium. The results are presented in Fig. 6c.

The desorbing buffer used during the first 4550 minutes had a pH of 3.25, and thereafter another buffer with a pH of 3.33 was used. Before the change of buffers a portable Eck and Krebs G.-M. counter tube was used to scan the column. There were two well separated activity maxima. It is believed that the upper band of activity, the last peak to be eluted, (*i.e.*, Sm) was moved appreciably closer to the lower band (*i.e.*, Eu) by the advancing boundary between the two solutions. This resulted in the europium and samarium peaks being less well separated than would ordinarily have been the case, and also in the anomalous double peak in the former.

During the early part of this separation the minima between peaks was much above the instrument background of 32 counts per minute. It is believed that the high level of these minima was due to the radiation coming from the unshielded column and not to activity in the effluent solution. Although the counter was protected by a two-inch thick lead shield, the γ radiation of the 2.5 h dysprosium and the 9.2 h europium could penetrate this easily. In fact, the level of radiation from the column was quite high until sufficient time had lapsed for these activities to decay.

With several elements the activity in the eluate increased to such a high level that it exceeded the maximum the counting equipment could accept. Therefore, aluminum absorbers were placed between the counting cell and the counter tube while the peak activity of holmium and europium passed through the counting cell. In this way it was demonstrated that the peaks, which would have been too high to measure without absorbers, were not multiple peaks due to poorly separated adjacent earths.

In the third experiment 0.1 mg. each of samarium and neodymium oxides and 0.01 mg. each of praseodymium, cerium and lanthanum were bombarded and separated. The elutriant solution had a pH of 3.33 for the first 1610 minutes, and a value of 3.4 thereafter. The results are plotted in Fig. 6d.

In this experiment again it was necessary to increase the *p*H of the elutriant to increase the rate of movement of the bands. This compressed the first three peaks which came from the column so that they appeared to have been less well separated than would have been the case had the same buffer solution been used throughout the entire experiment. The relative smallness of the neodymium activity peak was caused partially by its weak beta radiations and likewise partially by its long half-life. Actually, as much Nd as Sm was present. The clearly defined position of element 61 between Nd and Sm (see Fig. 6d) is of interest. This finding is in agreement with earlier observations,[9] and in fact serves to establish the place of this radio-isotope in the rare earth adsorption sequence beyond a reasonable doubt.

At the time the cerium had been displaced completely from the column, another peak could be detected well up in the column; this presumably was the lanthanum band.

A higher *p*H value was required to differentially desorb the cerium earths (*ca.* 3.4) compared with that for the yttrium group (*ca.* 3.2) at 100° also. A similar difference was found with the room temperature separations wherein Amberlite IR-1 was employed, and is to be interpreted as a consequence of the greater stability of the citrate complex ions of the heavy rare earths.

The Application of Ion-exchange Separations Techniques to Rare Earth Analysis.—The above described study of the impurities in Hilger erbium oxide constituted a qualitative analysis for the most difficultly recognizable impurities, namely, the other rare earths. One experiment has been made which might be classed as semi-quantitative analysis. The material used was

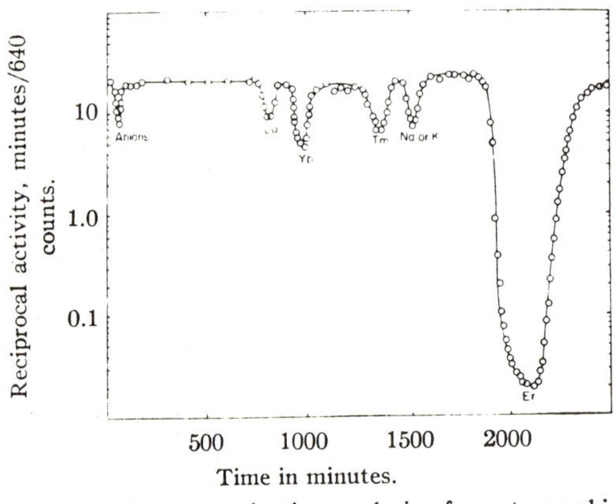

Fig. 7.—Neutron activation analysis of spectrographically pure erbium oxide: bed dimensions, 97 cm. × 0.26 sq. cm.; flow rate, 0.5 ml./sq. cm./min.; 5% citrate buffer at *p*H 3.2; temperature, 100°; 5.0 mg. of Er_2O_3 irradiated in Clinton neutron chain reactor and then fractionated by Dowex-50 ion exchange column.

erbium which had been purified to a high degree on an ion-exchange column.[26] Spectrographers skilled in rare earth analyses had failed to detect the spectral lines of any other rare earth except erbium in this sample. A 5-mg. aliquot was bombarded in the pile before processing it on the 97 cm. Dowex-50 column. An extremely good separation was obtained of traces of lutecium, ytterbium, thulium, and sodium from the erbium (see Fig. 7). No rare earths lighter than erbium were detected although the column was operated for a sufficiently long time that holmium and yttrium, if present, should have been desorbed.

The entire sample of eluant which contained the thulium was processed on a small column to reclaim the activity. This activity was assayed on apparatus at known counting geometry so that the number of disintegrations, and, therefore, the total number of atoms of Tm^{170}, could be computed from the known half-life. The activation cross-section of Tm^{169} is about 10^{-22} sq. cm. per atom. From the neutron flux of the pile, the cross-section, the duration of the bombardment, and the number of Tm^{170} atoms produced, it was possible to compute the number of Tm^{169} atoms present as impurity in the original erbium. Since the mass of erbium which was irradiated was known, it was possible to express the thulium impurity in per cent. of the total. This erbium contained not more than 0.001% (*i. e.*, 10 p.p.m.) thulium.

A considerably smaller amount of Tm impurity could have been detected if longer bombardments were made to activate the thulium more nearly to saturation.

Discussion

Taken all together, the data of Fig. 6 reveal that the sequence of relative adsorbability among the rare earths was as follows: La > Ce > Pr > Nd > 61 > Sm > Eu > Gd > Tb > Dy > Y > Ho > Er > Tm > Yb > Lu. This progression is in the same order as that of the crystal ionic radii.[27] The variation of the basicity predicted by von Hevesy[28] from theoretical considerations is likewise the same. In these terms, then, adsorbability follows basicity in the sense that strong bases will replace weaker bases from the adsorbent.

The relative position of yttrium, which fell between dysprosium and holmium, was in excellent agreement with the predictions based on the ionic radius. Owing to the very close similarity of Y with these two elements, the triad: Dy, Y,

(26) The authors wish to thank Mr. Darwin Harris for this material. Efforts to produce purified heavy rare earths on a gram scale were initiated in December, 1946, by D. H. Harris and E. R. Tompkins. By June, 1947, 10 milligram quantities of pure Lu, Tm and 100 mg. amounts of spectrographic grade Yb and Er had been separated from crude yttrium salts by successive chemical and exchange column refining procedures. This work will be published soon; private communication, E. R. Tompkins, May 28, 1947.

(27) V. M. Goldschmidt, T. Barth and G. Lunde, *Skrifter Norske Videnskaps Akad. Oslo*, 1 Math. Nat. Kl. no. 7 (1925).

(28) G. von Hevesy, "Die selten Erden vom Standpunkte des Atombaues," Julius Springer, Berlin, 1927, pp. 16–21.

Ho, in fact, constitutes what may be taken as the most difficultly separable of all of the rare earths. Although the similarities were not so great, the adsorbabilities in two other positions within the rare earths were found to be comparable, namely, the triads: Tm, Yb and Lu; and, Sm, Eu and Gd. It was of interest to find that the most recent and exact X-ray diffraction measurements on the rare earth sesquioxides[29] have indicated a dependence of the lattice spacing in these compounds on the number of electrons in the 4f shell of the atom. As may be seen from Fig. 8 where the relative radii ($r_{La^{+++}}$ = 1.000) are plotted, the rare earths are conveniently divided into two groups. The first of these, which corresponds to the filling in of the first seven 4f electrons, begins with cerium and ends with gadolinium. Likewise, the relative adsorbability decreases steadily as the atomic number increases so that the maximum similarity between pairs is reached with Eu and Gd, in excellent conformity with the known parallelism between ionic radii and adsorbability. The second subgroup, which begins with terbium and ends with lutecium, corresponds to the filling in of the second seven 4f electrons, and a maximum similarity in both the relative radii and adsorbability is found with Yb and Lu. However, the differences from member to member in the second sub-group are smaller than those between corresponding members of the first. Consequently, taken as a whole, cerium and praseodymium are the easiest of the rare earths to separate from one another, and ytterbium and lutecium are the most difficult, excepting for the separation of the quasi rare earth yttrium from dysprosium. The separability of Eu and Gd corresponds roughly to that of Er and Tm in difficulty.

Quantitative comparisons of the relative adsorbability within the yttrium group earths were made possible by utilizing the data from a much larger number of chromatograms than those shown in Figure 6, together with Equation (8) above. The ratios of the distribution coefficients of neighboring members of the sequence, presented in Table II, clearly bears out the foregoing statements. The average deviation found for these ratios from experiment to experiment, in spite of differing conditions of flow rate, pH and temperature, is seen to have been less than 2%, emphasizing the highly reproducible character of adsorption column performance, as well as confirming some of the predictions made in an earlier section. Although the absolute values for the distribution coefficients derived from column data have been computed, they will not be reported until a comparison with values of K_d determined by simple equilibrium measurements becomes possible.

TABLE II

RELATIVE RARE EARTH ADSORBABILITIES

Element	$\delta_{n+1} : \delta_n$
Lu	
	1.21 ± 0.01
Yb	
	1.44 ± .02
Tm	
	1.55 ± .02
Er	
	1.57 ± .02
Ho	
	1.30 ± .02
Y	
	1.20
Dy	

(29) Bommer, *Z. anorg. allgem. Chem.*, **241**, 273 (1939). We wish to thank Dr. Max Bredig for his helpful discussion of this point.

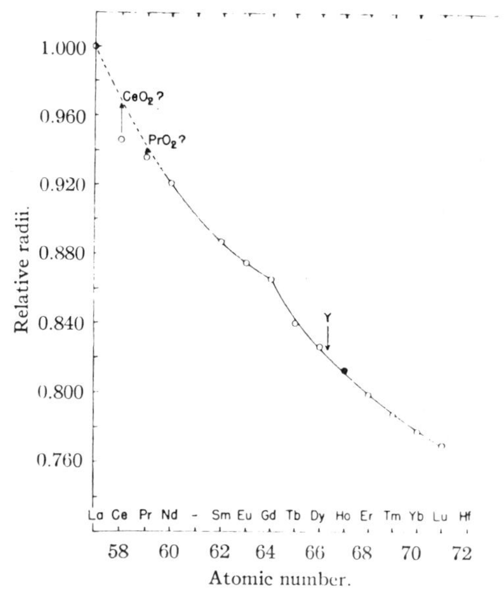

Fig. 8.—Relative (cubic) radii of trivalent rare earth ions: $R_{La^{+++}}$1.291 Å. for cubic sesquioxide; O, H. Bommer, *Z. anorg. allgem. Chem.*, **241**, 273 (1939); ●, V. M. Goldschmidt, *Ber.*, **60**, 1263 (1927); these data were computed by Dr. M. Bredig.

Acknowledgments.—The authors wish especially to thank Dr. W. C. Bauman of the Dow Chemical Company for his invaluable assistance in supplying the needs of our research with special laboratory scale preparations of spherical Dowex-50. The help given by Dr. L. A. Matheson in this and in other ways is gratefully acknowledged. The initial stages of this work were facilitated by the receipt of small quantities of spectrographically analyzed lutecium, thulium holmium and terbium sesquioxides through the generosity of Dr. C. J. Rodden of the National Bureau of Standards. Thanks are also due to Dr. E. M. Gladrow of Iowa State College for his donation of gram quantities of neodymium oxide which had been purified by him to a high degree using ion-exchange column techniques.

Summary

1. A demonstration of the separation of all the rare earths by chromatographic techniques based on the use of synthetic organic ion-exchange adsorbents was made. With a 97 cm. by 0.26 sq. cm. bed of 270/325 mesh Dowex-50 adsorbent maintained at 100°, the optimum *p*H of the 4.75% citric acid eluant was 3.2 for the separation of the yttrium earths and 3.4 for the cerium earths.

2. The employment of a temperature of 100° permitted the use of elutriant flow rates of as high as 2 ml./sq. cm./min., thereby greatly reducing the time required for a separation. In a typical experiment, milligram quantities of lutecium, ytterbium, thulium, erbium, holmium, and yttrium were separated in thirty hours.

3. The fractionation of the heavy rare earths could be made sufficiently complete that the method was established as a promising procedure for the quantitative analysis of rare earths for cationic impurities. A previously exhaustively purified sample of erbium oxide, in which all other rare earths were reported as absent by spectrographic analysis, was shown to contain 10 p.p.m. of Tm by irradiation of the Er_2O_3 with neutrons and subsequent chromatographic separation.

4. The adsorbability of yttrium was shown to lie intermediate between holmium and dysprosium in quantitative agreement with the expectations based on considerations of the ionic radii. As a result, the separation of these elements was even more difficult to accomplish than the separation of the heavier rare earths.

5. The adsorbability of ytterbium and lutecium were shown to be more nearly the same than any other pair of rare earths in conformity with the trend in ionic radii. Predictions based upon crystal radii suggested that samarium, europium and gadolinium would be the most difficult of the cerium earths to fractionate.

RECEIVED JULY 16, 1947

8

Reprinted from *J. Amer. Chem. Soc.*, **73**, 4840–4847 (Oct. 1951)

The Separation of Rare Earths by Ion Exchange. VI. Conditions for Effecting Separations with Nalcite HCR and One-tenth Per Cent. Citric Acid–Ammonium Citrate Solutions

BY F. H. SPEDDING, E. I. FULMER, J. E. POWELL, T. A. BUTLER AND I. S. YAFFE

Procedures have been developed for the successful separation of the rare earths using a high capacity resin, Nalcite HCR, and 0.1% citrate solution. The high capacity resin proved to be definitely superior to a low capacity resin, Amberlite IR-100, for the separation of the heavy rare earths. Gram quantities of Er, Ho and Dy have been obtained in purities greater than 99%. The process proved to be equally efficient when stepped up 50-fold to a pilot plant scale separation of the more abundant rare earths.

I. Introduction

Previous communications from this Laboratory[1-7] have dealt only with the use of the low capacity resins, Amberlite IR-1 and IR-100, for the separation of rare earths by ion exchange on a macro scale. Although Amberlite IR-1 proved to be undesirable due to its relative high solubility, IR-100 was found to give satisfactory separations over a rather wide range of conditions and has been successfully employed for the separation of certain rare earths from their mixtures. Harris and Tompkins[8] have stated that Dowex 50 resin is superior to the Amberlite resins IR-1 and IR-100 for rare earth separations; their comparison appears to be based on experiments performed on yttrium and cerium mixtures, using 5% citrate solutions in the *p*H range 2.75 to 3.3. Recent investigations,[4,6] using 0.1% citrate solutions, resulted in elution curves which differed significantly from those previously reported. The present communication deals with the use of the high capacity resin, Nalcite HCR, with 0.1% citrate solutions.

(1) F. H. Spedding, A. F. Voigt, E. M. Gladrow and N. R. Sleight, THIS JOURNAL, **69**, 2777 (1947).

(2) F. H. Spedding, A. F. Voigt, E. M. Gladrow, N. R. Sleight, J. E. Powell, J. M. Wright, T. A. Butler and P. Figard, *ibid.*, **69**, 2786 (1947).

(3) F. H. Spedding, E. I. Fulmer, T. A. Butler, E. M. Gladrow, M. Gobush, P. E. Porter, J. E. Powell and J. M. Wright, *ibid.*, **69**, 2812 (1947).

(4) F. H. Spedding, E. I. Fulmer, B. O. Ayers, T. A. Butler, J. E. Powell, A. D. Tevebaugh and R. Q. Thompson, *ibid.*, **70**, 1671 (1948).

(5) F. H. Spedding, E. I. Fulmer, T. A. Butler and J. E. Powell, *ibid.*, **72**, 2349 (1950).

(6) F. H. Spedding, E. I. Fulmer, J. E. Powell and T. A. Butler, *ibid.*, **72**, 2354 (1950).

(7) F. H. Spedding, *Discussions Faraday Soc.*, No. 7, 214 (1949).

(8) D. H. Harris and E. R. Tompkins, THIS JOURNAL, **69**, 2792 (1947).

II. Materials, Apparatus and General Procedures

Unless otherwise noted, the materials, apparatus and procedures are the same as those previously given.[6]

1. **Resin.**—Nalcite High Capacity Resin, a sulfonated copolymer of styrene and divinylbenzene, was used in all experiments described in this paper; according to a private communication from the Dow Chemical Co., the National Aluminate Corporation is licensed to distribute Dowex 50 resin under the trade name Nalcite HCR. Resin particles of −40 + 50 mesh size were used in the 22-mm. and the 40-mm. i.d. columns, and −30 + 40 mesh size in the 6-inch i.d. columns.

2. **Resin Beds.**—Each of the 40-mm. columns contained 450 g. of air-dried resin which gave a bed height of 60 cm. in the NH_4^+ cycle and 63 cm. in the H^+ cycle. Each 6-inch column was loaded with 9 kg. of resin, which gave a bed height of 30 in. in the H^+ cycle. The conditioning of the resin has been previously described.[6]

3. **Eluant.**—The eluting solution was 0.1% citrate, based on anhydrous citric acid; the concentration of the eluant was, therefore, 9.4% greater in citric acid than that used in previous investigations[6] in which the concentration was based on the monohydrate.

4. **Recovery of the Heavy Rare Earths.**—Certain properties of the heavy rare earths make them more difficult to recover than the light rare earths; for example, the complexing action with citrate ions is increased; the oxalates are more soluble; and the rate of precipitation[9] is slower.

(9) D. S. Martin and C. Crouthamel, *ibid.*, **72**, 1382 (1950).

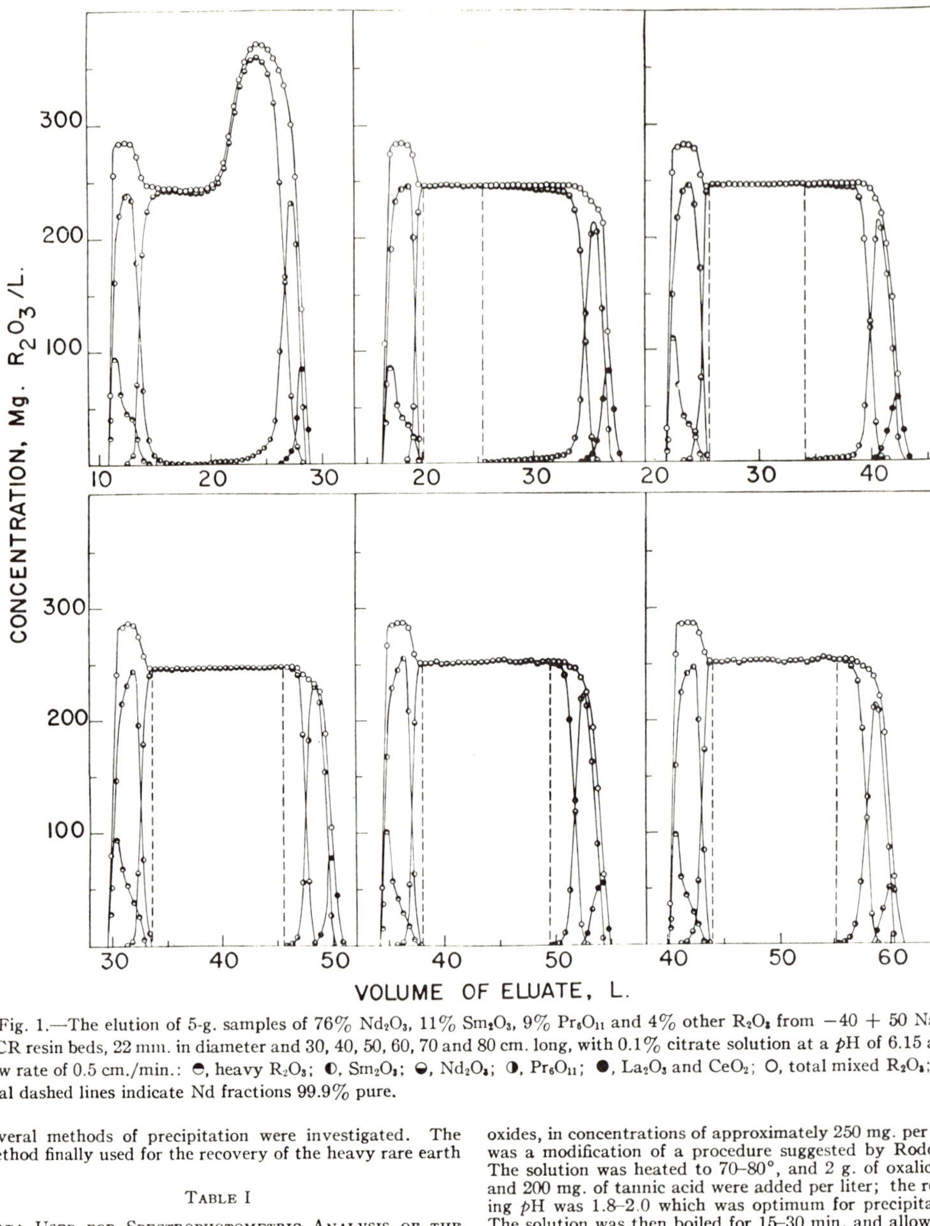

Fig. 1.—The elution of 5-g. samples of 76% Nd_2O_3, 11% Sm_2O_3, 9% Pr_6O_{11} and 4% other R_2O_3 from −40 + 50 Nalcite HCR resin beds, 22 mm. in diameter and 30, 40, 50, 60, 70 and 80 cm. long, with 0.1% citrate solution at a *p*H of 6.15 and a flow rate of 0.5 cm./min.: ◓, heavy R_2O_3; ◐, Sm_2O_3; ◒, Nd_2O_3; ◑, Pr_6O_{11}; ●, La_2O_3 and CeO_2; O, total mixed R_2O_3; vertical dashed lines indicate Nd fractions 99.9% pure.

Several methods of precipitation were investigated. The method finally used for the recovery of the heavy rare earth oxides, in concentrations of approximately 250 mg. per liter, was a modification of a procedure suggested by Rodden.[10] The solution was heated to 70–80°, and 2 g. of oxalic acid and 200 mg. of tannic acid were added per liter; the resulting *p*H was 1.8–2.0 which was optimum for precipitation. The solution was then boiled for 15–30 min. and allowed to stand 48 hr. at room temperature before filtering; over 95% of the rare earth oxalate precipitated. The rare earths remaining in solution may be recovered by readsorption, stripping and reprecipitation.

TABLE I

DATA USED FOR SPECTROPHOTOMETRIC ANALYSIS OF THE HEAVY RARE EARTHS

Element	Wave length, mμ	Slit width, mm.	Abs. coef. (liters mole⁻¹ cm.⁻¹)	Interfering elements	Corrections (based on absorbancies)
Er	522	0.02	3.13	Nd[a]	....
Ho	536	.02	4.54	Er	0.02 of Er peak at 522 mμ
Dy	908	.02	2.33	Ho, Yb[a]	0.03 of Ho peak at 536 mμ

[a] As no Yb or Nd was present in the samples analyzed, no corrections were necessary.

5. **Methods of Analysis.**—The spectrophotometric data, used in this Laboratory, for the light rare earths have been previously reported.[5] The data employed for the analyses of the heavy rare earths are listed in Table I. Spectrographic analyses were made according to the methods of Fassel and Wilhelm.[11,12]

(10) C. J. Rodden, private communication.

(11) V. A. Fassel and H. A. Wilhelm, *J. Opt. Soc. Am.*, **38**, 518 (1948).

(12) V. A. Fassel, *ibid.*, **39**, 187 (1949).

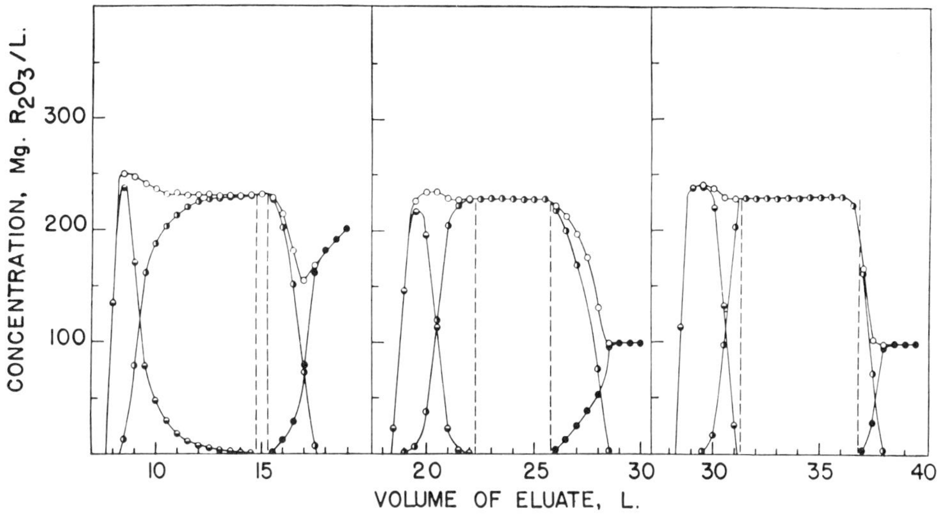

Fig. 2.—The elution of 3.5-g. samples of 43.5% Pr_6O_{11}, 10.2% Nd_2O_3, 46% La_2O_3 and traces of Sm_2O_3 and CeO_2 from −40 + 50 Nalcite HCR resin beds, 22 mm. in diameter and 20, 40 and 60 cm. long with 0.1% citrate solution at a *p*H of 6.15 and a flow rate of 0.5 cm./min.: ⊖, Nd_2O_3; ◑, Pr_6O_{11}; ●, La_2O_3; O, total mixed R_2O_3; vertical dashed lines indicate Pr fractions >99.9% pure.

III. Experimental

1. The Effect of Increased Column Length with a Constant Weight of Sample.—Six beds of −40 + 50 mesh size resin, 22 mm. in diameter, were prepared in the acid cycle, rinsed with distilled water and leveled to heights of 30, 40, 50, 60, 70 and 80 cm., respectively. The initial load, per column, was 5 g. of Lindsay Light & Chemical Co. code #630 oxide, consisting of 76% Nd_2O_3, 11% Sm_2O_3, 9% Pr_6O_{11} and 4% other R_2O_3. Elution curves and conditions for this experiment are given in Fig. 1.

In each case the total heavy rare earths was determined by difference and tended to separate and precede the Sm as expected, except for Eu and Gd. The Ce and La concentrated at the rear of the band and were also determined by difference. No Sm or Pr was obtained pure from this mixture in any case due to their low initial concentrations and the overlapping of neighboring bands. Data for the Nd separation are given in Table II. The yield of > 99.9% Nd increased until a column length of 60 cm. was reached and then remained constant within the limits of experimental error. The shapes of the elution curves were identical for all lengths greater than 30 cm. It can readily be seen from Fig. 1 that the lengths required for optimum separation and for attainment of the equilibrium type elution curve for this mixture do not coincide. The same effect was also observed in the separation of a crude Pr oxide which contained 43.5% Pr_6O_{11}, 10.2% Nd_2O_3, 46% La_2O_3 and traces of Sm_2O_3 and CeO_2; the material was obtained from Lindsay Light & Chemical Co. as "praseodymium ammonium nitrate." The elution curves for this mixture and the experimental conditions are given in Fig. 2. Data from this experiment are recorded in Table II.

Table II

The Effect of Column Length on the Separation of Nd and Pr from Crude Concentrates

Sample weight Nd crude, g.	Sample weight Pr crude, g.	Column length, cm.	Amt. obt. > 99.9% pure Nd_2O_3, g.	Amt. obt. > 99.9% pure Pr_6O_{11}, g.	Yield of Nd_2O_3 obt. >99.9% pure, %
5.0		30	0.0000		0.0
5.0		40	1.2269		32.2
5.0		50	1.8397		48.3
5.0		60	2.9654		78.1
5.0		70	2.9041		76.5
5.0		80	2.7822		73.2
	3.5	20		0.1159	7.6
	3.5	40		0.8000	52.6
	3.5	60		1.2604	82.9

2. The Effect of Proportionate Increase in Column Load and Length.—In an earlier paper concerning Amberlite IR-100 and 0.1% citrate,[6] it was reported that the separation could be made more efficient by a proportionate increase in load and length. Figure 3 shows that the same effect occurred when Nalcite HCR was used in place of Amberlite IR-100 as the exchange resin. However, the series of separations shown was not based on the optimum load per unit length. The data from these experiments are presented in Table III. The percentage yield dropped off in the case of the 100-cm. column, but this effect was due to poor packing in the bed.

Table III

The Effect of Increasing Length and Load Proportionately on the Separation of Nd from a Concentrate

Bed length, cm.	Column load as R_2O_3, g.	Amt. Nd_2O_3 obt. >99.9% pure, g.	Yield of Nd_2O_3 obt. >99.9% pure, %	Volume of eluant req. l.
20	2.50	0.0000	0.0	18.6
40	5.00	1.1042	29.1	37.8
60	7.50	2.2283	39.1	55.5
80	10.00	4.1316	54.4	74.5
100	12.50	4.9079	51.7	93.0

3. The Effect of *p*H on the Separation of Various Rare Earth Mixtures.—The effect of *p*H of 0.1% citrate, using Nalcite HCR, was in good agreement with data reported previously, using Amberlite IR-100.[6] The maximum concentration obtained for an element at a given *p*H was lower with Nalcite than with Amberlite; this effect was probably due to the more acidic nature of Nalcite HCR. The increase in separation with decreased *p*H, previously observed with Amberlite, was not as pronounced when Nalcite was used; in fact, as shown in Fig. 4. the separation of Nd from neighboring rare earths was practically independent of *p*H in the range studied. Figures 5, 7 and 8, and Tables VI and VII illustrate the effect of *p*H on the separation of other rare earth mixtures.

4. The Effect of Increased Load on Columns of Equal Length.—Data from a number of pilot plant scale runs illustrate the effect of column load on the total and percent-

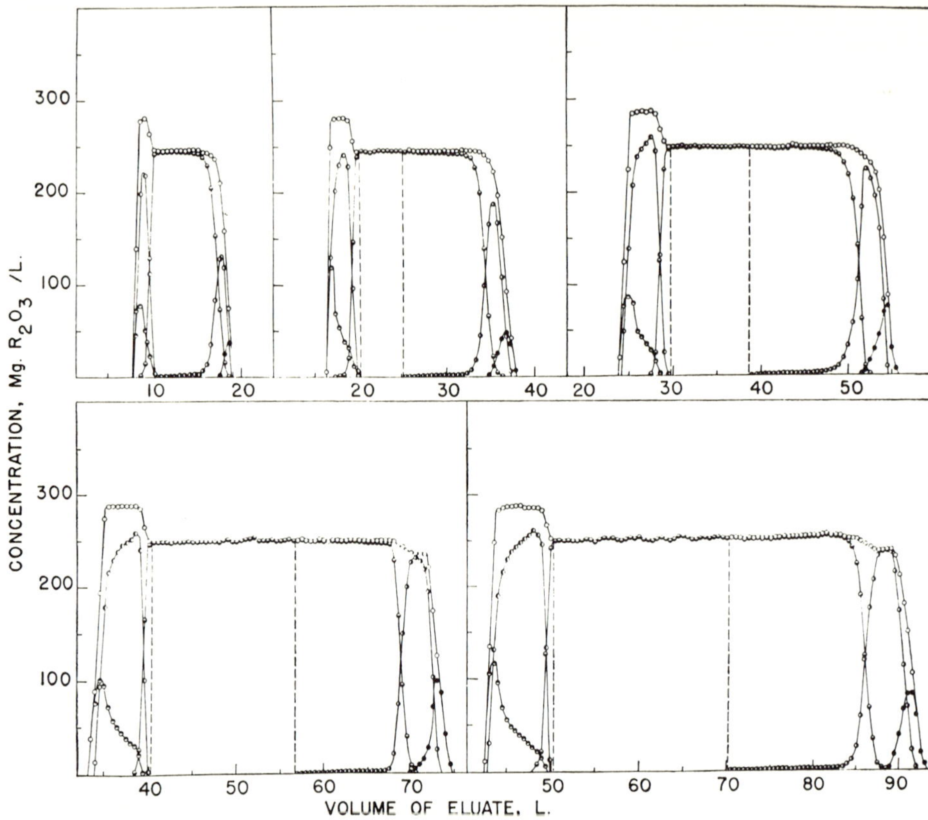

Fig. 3.—The elution 2.5, 5.0, 7.5, 10.0 and 12.5 g. samples of 76% Nd_2O_3 from −40 + 50 Nalcite HCR resin beds, 22 mm. in diameter and 20, 40, 60, 80 and 100 cm. long, with 0.1% citrate solution at a *p*H of 6.15 and a flow rate of 0.5 cm./min.: ◓, heavy R_2O_3; ◐, Sm_2O_3; ◒, Nd_2O_3; ◑, Pr_6O_{11}; ●, La_2O_3 and CeO_2; O total mixed R_2O_3; vertical dashed lines indicate Nd fractions >99.9% pure.

age yields of spectrographically pure Nd obtained from code #630 oxide. The resin beds were 6 in. in diameter and 30 in. in length. The samples were all eluted with 0.1% citrate at a *p*H of 6.1 and a flow rate of 0.5 cm./min. The average yields per column for different column loads are given in Table IV.

TABLE IV

THE EFFECT OF INCREASED LOAD ON COLUMNS OF EQUAL LENGTH USING CODE #630 OXIDE

Column load as R_2O_3, g.	No. of columns operated	Av. wt. of >99.9% Nd_2O_3 obtained per column, g.	Percentage of available Nd_2O_3 obtained >99.9% pure, %	Total volume of eluant req., l.
250	10	140.2	73.8	3050
300	10	162.0	71.0	3180
325	5	166.2	67.3	3300
375	8	162.2	57.0	3400
450	2	156.4	45.7	3550

For a column of the specified dimensions, the optimum load for this particular crude mixture appears to be approximately 325 g. when the above operating conditions are used. A slight precipitate formed in the resin bed during the elution of the 450-g. loads, but redissolved as elution progressed.

A similar study of the effect of column load was made using the Pr concentrate described in Section 1 and under the conditions given above. The average yields per column for different column loads are given in Table V. The columns were stripped immediately after the Pr was eluted since, under these conditions, La elutes much more slowly than does the Pr and has relatively little value. In every case an insoluble rare earth citrate compound precipitated in the resin bed almost immediately after the elution was begun. This precipitate was found to contain every rare earth component present in the original crude. Although this precipitate dissolved as elution progressed and the last traces completely disappeared before break-through occurred, sufficient bed length had to be provided in order to separate the rare earths in the precipitated portion. The amount of precipitate was observed to increase markedly as the column load increased. Since excessive precipitation blocked the resin bed pores and increased channeling, the optimum load is between 250 and 300 g. for a column of the above dimensions.

TABLE V

THE EFFECT OF INCREASED LOAD ON COLUMNS OF FIXED LENGTH USING A PR CONCENTRATE

Column load as oxide, g.	No. of columns operated	Av. wt. of >99.9% Pr_6O_{11} obtained per column, g.	Percentage of available Pr_6O_{11} obtained >99.9% pure, %	Total volume of eluant to elute the Pr, l.
225	10	67.6	69.1	2560
250	10	76.5	70.2	2600
300	1	79.6	61.0	2640

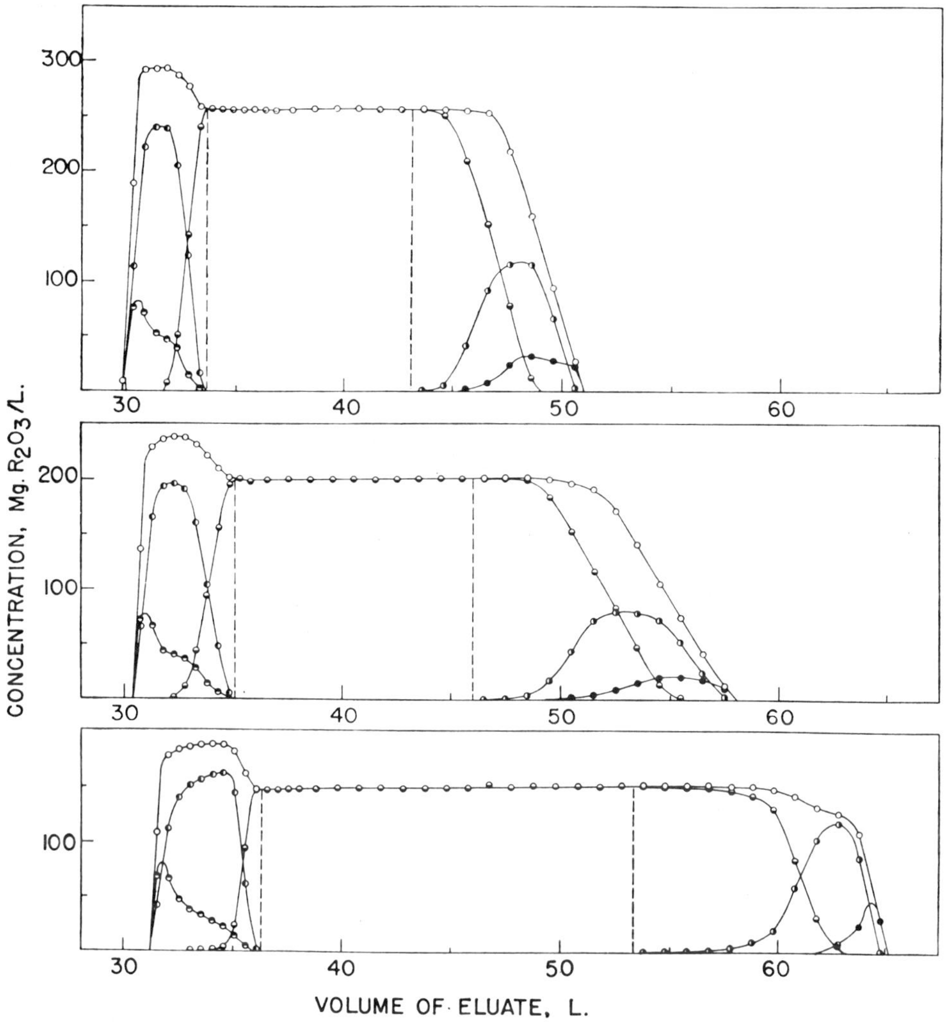

Fig. 4.—The elution of 5-g. samples of 76% Nd_2O_3 from −40 + 50 Nalcite HCR resin beds, 22 mm. in diameter and 60 cm. long, with 0.1% citrate solutions at *p*H values of 6.2, 6.0 and 5.8 and a flow rate of 0.5 cm./min.: ◒, heavy R_2O_3; ◐, Sm_2O_3; ⊖, Nd_2O_3; ◑, Pr_6O_{11}; ●, La_2O_3 and CeO_2; O, total mixed R_2O_3; vertical dashed lines indicate Nd fractions >99.9% pure.

5. Purification of Er.—The analysis of the starting material was 90% Er_2O_3, 8% Ho_2O_3 and 1% Tm_2O_3. Ten grams of the oxide were used for each column. Duplicate columns were eluted with 0.1% citrate solutions at *p*H values of 5.40, 5.70 and 6.00; the duplicate columns checked each other closely. At all three *p*H values, the Ho concentrated toward the rear of the band, and the Tm toward the front, yielding pure Er in between. Figure 5 shows that the latter portion of the Er band contained traces of Ho which could be detected only by spectrographic analysis. Table VI lists the yields of Er_2O_3 obtained at the specified purities.

TABLE VI

YIELDS OF Er_2O_3 OBTAINED AT DIFFERENT *p*H VALUES

*p*H of eluant	Percentage of available Er_2O_3 obtained >99.0% pure	>99.9% pure
5.40	73	36
	76	41
5.70	74	42
	78	52
6.00	66	35
	67	41

TABLE VII

PERCENTAGES OF PURE RARE EARTHS OBTAINED FROM MIXTURES ELUTED AT DIFFERENT *p*H VALUES

*p*H of eluant	Wt. R_2O_3 adsorbed, g.	Percentages of available oxides obtained, >99% pure Er_2O_3	Ho_2O_3	Dy_2O_3
5.40	10	65	75	33
5.70	10	73	50	26
6.00	10	50	60	19
5.70	15	86[a]	70	28

[a] Includes the yield of Er_2O_3 (3.5%) obtained > 99.9% pure.

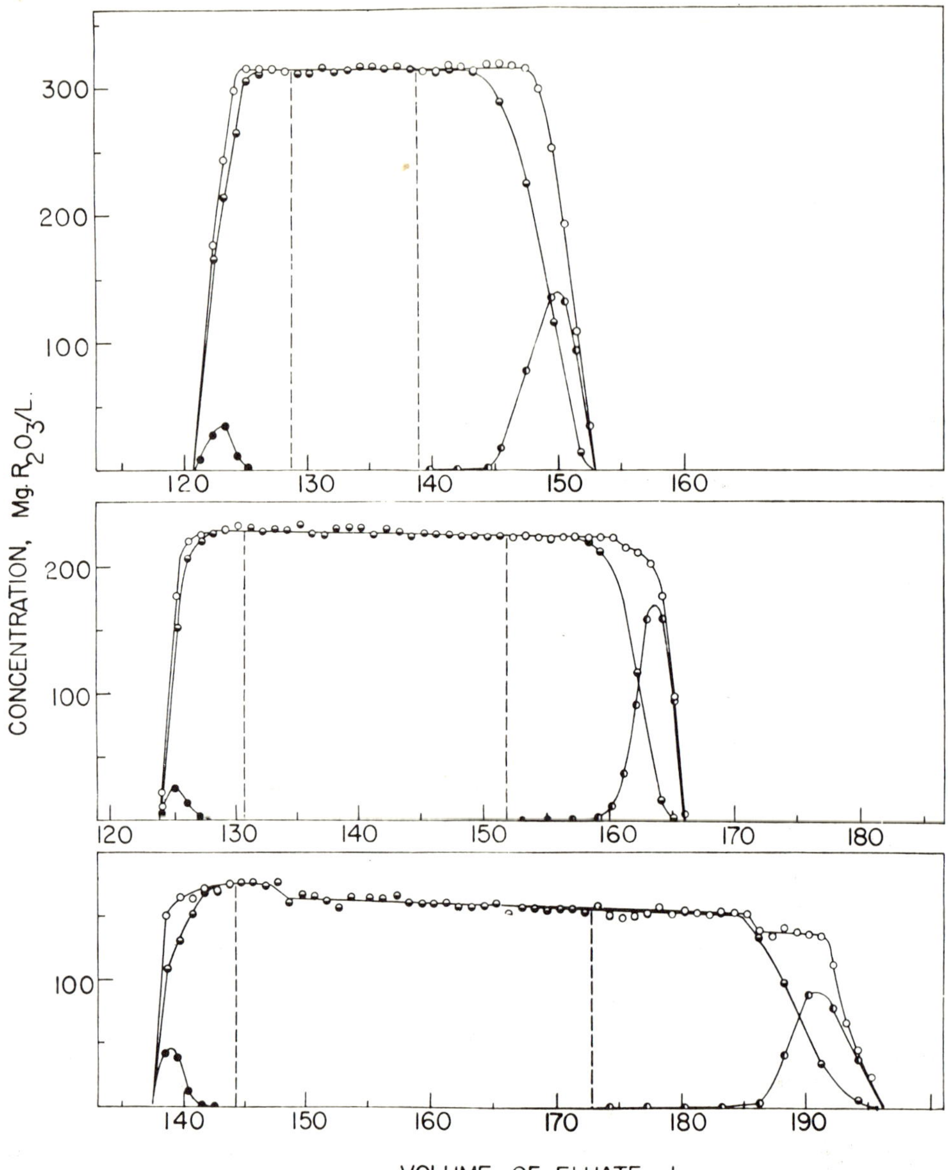

Fig. 5.—The elution of 10-g. samples of 90% Er_2O_3, 8% Ho_2O_3 and 1% Tm_2O_3 from −40 + 50 Nalcite HCR resin beds, 40 mm. in diameter and 80 cm. long, with 0.1% citrate solution at *p*H values of 6.0, 5.7 and 5.4 and a flow rate of 0.5 cm./min.: ●, Tm_2O_3; ◒, Er_2O_3; ◐, Ho_2O_3; O, total mixed R_2O_3; vertical dashed lines indicate Er fractions >99.9% pure.

6. **Separation of Er, Ho and Dy.**—The composition of the starting material was 56% Ho_2O_3, 27% Er_2O_3, 17% Dy_2O_3 and 1% Y_2O_3. Ten grams of the oxide mixture was used for each of three columns and 15.0 g. on a fourth. The 10-g. columns were eluted at *p*H values of 5.40, 5.70 and 6.00, and the 15-g. column at a *p*H of 5.70. A typical elution curve is shown in Fig. 6. Spectrophotometrically pure Er, Ho and Dy were obtained from all columns, as indicated in Table VII, but no spectrographically pure rare earths were obtained, except for a small amount of Er_2O_3 from the 15-g. column. A better Er separation was obtained at a *p*H of 5.70 than at other *p*H values; however, better separations of Ho and Dy were obtained at a *p*H of 5.40. Data in Section 4 indicate that still better separations could be obtained for these column dimensions by using a larger load, although more time would be required for elution.

7. **The Separation of Sm, Eu and Gd Mixtures.**—Ketelle and Boyd[13] have pointed out that the triad Gd, Eu and Sm constitutes what should be a difficultly separable group of rare earths; in this Laboratory a marked separation of Gd from Sm or Eu from Sm by ion exchange has never been

(13) B. H. Ketelle and G. E. Boyd, THIS JOURNAL, 69, 2800 (1947).

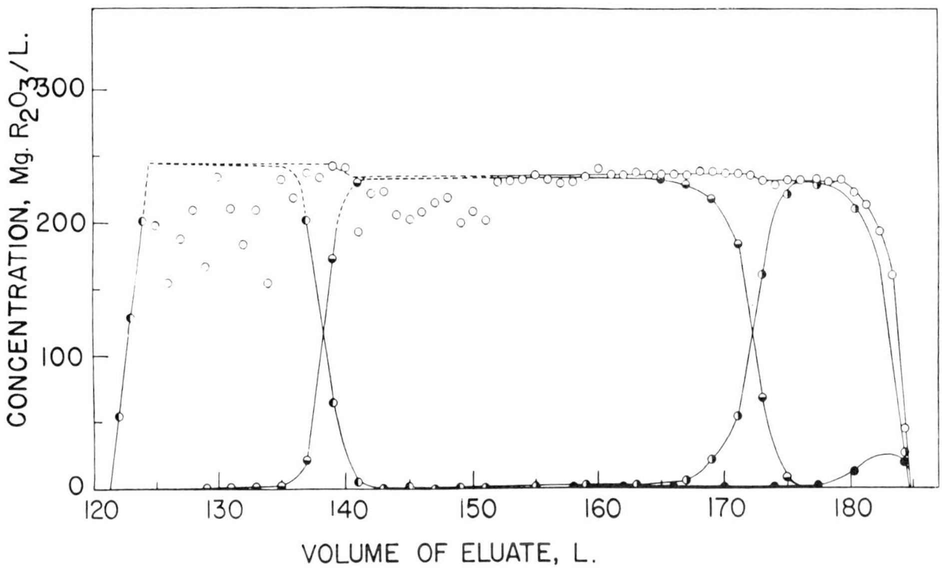

Fig. 6.—The elution of a 15-g. sample of 56% Ho_2O_3, 27% Er_2O_3, 17% Dy_2O_3 and 1% Y_2O_3 from −40 + 50 Nalcite HCR beds, 40 mm. in diameter and 60 cm. long, with 0.1% citrate solution at a *p*H value of 5.7 and a flow rate of 0.5 cm./min.: ◐, Er_2O_3; ◒, Ho_2O_3; ◑, Dy_2O_3; ●, Y_2O_3; O, total mixed R_2O_3; erratic total R_2O_3 points at the front of the curve were due to incomplete precipitation of the rate earth oxalates.

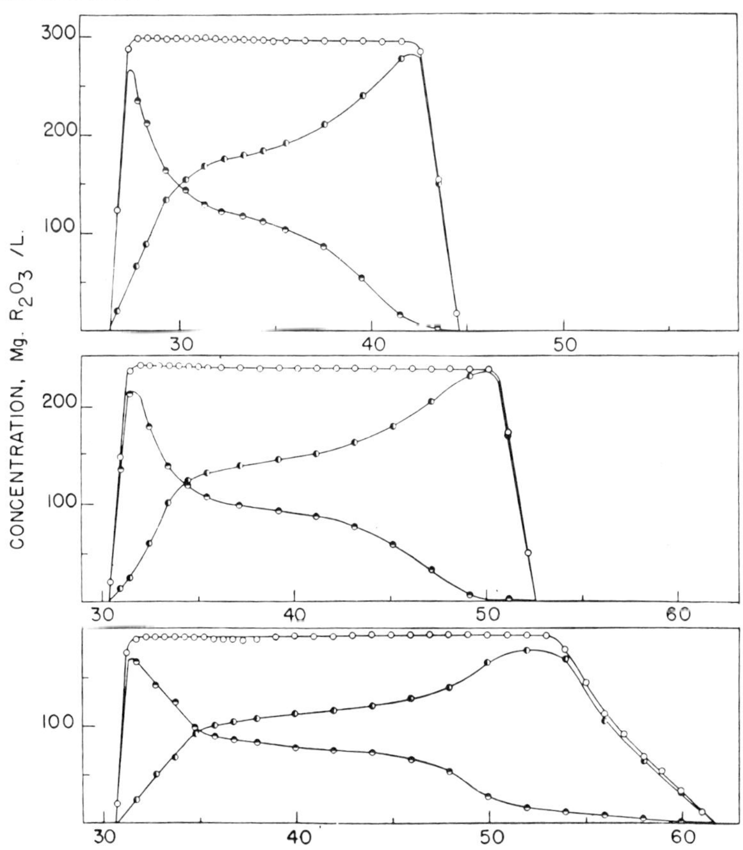

Fig. 7.—The elution of 5-g. samples of 34.7% Eu_2O_3 and 65.3% Sm_2O_3 from −40 + 50 Nalcite HCR resin beds, 22 mm. in diameter and 60 cm. long, with 0.1% citrate at *p*H values of 6.2, 6.0 and 5.8 and a flow rate of 0.5 cm./min.: ◓, Eu_2O_3; ◐, Sm_2O_3; O, total mixed R_2O_3.

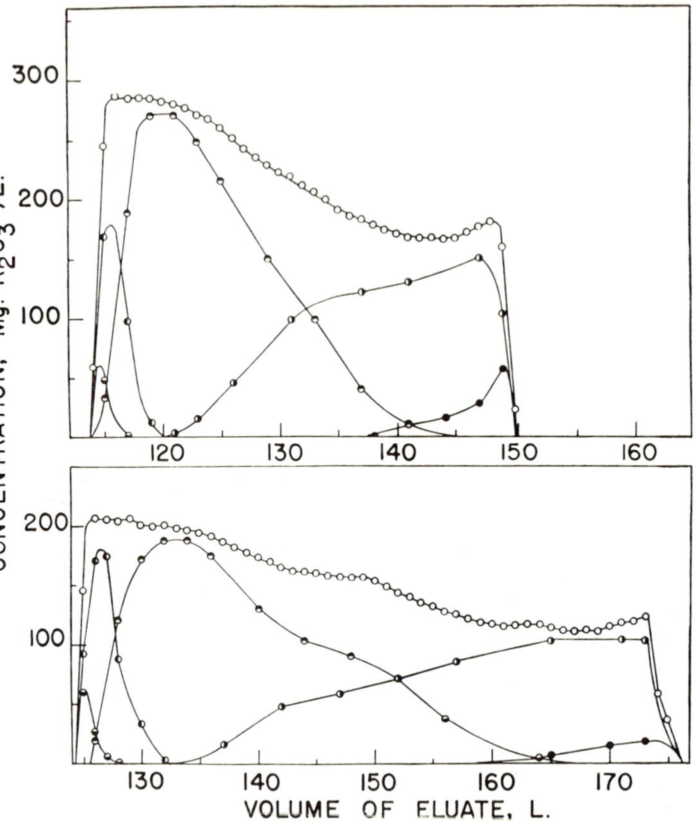

Fig. 8.—The elution of 8-g. samples of 49% Dy_2O_3, 41% Y_2O_3, 7% Ho_2O_3, 2% Tb_2O_3 and 1% Er_2O_3 from −40 + 50 Nalcite HCR beds, 40 mm. in diameter and 60 cm. long, with 0.1% citrate solution at *p*H values of 6.0 and 5.7 and a flow rate of 0.5 cm./min.: ◒, Er_2O_3; ◐, Ho_2O_3; ◓, Dy_2O_3; ◑, Y_2O_3; ●, Tb_2O_3; ○ total mixed R_2O_3.

noted under any conditions studied. This difficulty has been illustrated in previous papers concerned with the use of Amberlite IR-100 resin with 0.5[3] and 0.1%[7] citrate solutions, and it is further demonstrated by the elution curves in Fig. 7. It should be noted that conditions used in these studies gave good separations when applied to other rare earth mixtures, and that the elements Gd, Eu and Sm can be separated readily, as a group, from other rare earths. These individual elements can be obtained in a pure state by using ion exchange as a concentrating step and then applying the methods of Marsh[14] and McCoy.[15]

8. **Separation of Ho, Dy and Y**.—Ketelle and Boyd[13] reported that this triad would also be a difficultly separable group, and that, when using 5% ammonium citrate on Dowex resin at 90°, the order of elution was Ho–Y–Dy. However, data obtained in this Laboratory using Amberlite IR-100 at room temperature, indicate that, while Dy and Y are difficult to separate, the order of elution is Ho–Dy–Y. Figure 6 shows that the Y was eluted after the Dy, and that a good separation of Ho from Dy was obtained. The position of Y in the elution of Ho, Dy, Y and Tb from Nalcite HCR at room temperature is shown in Fig. 8. Although no spectrophotometrically pure Ho_2O_3 was obtained, the Ho separated from the bulk of the Dy quite sharply. While a small amount of spectrophotometrically pure Dy was obtained, there is a great deal of overlap between the Dy and Y. Long before the Dy was completely eluted, Tb began to elute. Thus it appears that, when 0.1% citrate solution is employed at room temperature, Dy–Y–Tb is a difficultly separable triad and that Dy and Ho are fairly easy to separate from one another.[16]

AMES, IOWA

RECEIVED JANUARY 12, 1951

(14) J. K. Marsh, *J. Chem. Soc.*, 523 (1942); 531 (1943).

(15) H. N. McCoy, THIS JOURNAL, **58**, 2279 (1936); **63**, 3432 (1941).

(16) NOTE ADDED IN PROOF: Recent work [F. H. Spedding and J. L. Dye, *ibid.*, **72**, 5350 (1950)] has shown that the order of elution using 5% ammonium citrate at *p*H values of 2.8-3.0 is Dy-Tb-Y at room temperature.

9

Reprinted from *J. Phys. Chem.*, **59**, 690–695 (Aug. 1955)

AN IMPROVED METHOD FOR OPERATING ION-EXCHANGE RESIN COLUMNS IN SEPARATING THE RARE-EARTH ELEMENTS[1]

By Walter E. Nervik

University of California, Radiation Laboratory Livermore Site, Livermore, Cal.

Received January 14, 1955

A "gradient elution" method has been adapted for operating an ion-exchange resin column using an eluting agent in which the *p*H is changed continuously. The method shortens the column operating time required for separation of the rare earth elements without adversely affecting their purity.

I. Introduction

Procedures for separating individual members of the rare-earth group of elements through the application of synthetic cation exchange resins have been described by many investigators.[2–5] The exact experimental conditions in each case depended on the number of elements to be separated and the degree of separation desired. In general, the rare earth elements were first adsorbed on the top portion of a long narrow column of cation-exchange resin, then selectively eluted by passing a solution containing a suitable complexing agent through the resin column and collecting portions of the eluent. The degree of separation for a given experimental arrangement depends on variables such as temperature, type of cation-exchange resin, size of the resin particles, dimensions of the column, eluting agent used, concentration and *p*H of the eluting agent, flow rate of the eluting agent, etc., but for a given eluting agent the relative position of the various rare earths remains approximately constant.[3] An experimental consequence of this phenomenon is the fact that if operating conditions are adjusted to effect separation of the heavy rare earths, the light rare earths require excessively long column operating times before being eluted. If conditions are adjusted to give separation of the light rare earths in reasonable operating times, the heavy rare earths are usually eluted too quickly to give high purity. Ordinarily the chemist has to make some compromise between purity of individual rare earths and column operating time. This series of experiments has been undertaken to determine whether introduction of an additional variable, *i.e.*, continuous change in *p*H of the eluting solution, could be used to effect complete separation of all of the individual rare earths in relatively short column operating times.

Although the author was not aware of it at the time, the elution technique which was finally developed was quite similar to the "gradient elution" chromatographic method originally described by Alm, Williams and Tiselius.[6] In addition, the physical arrangement of the various reservoirs and column equipment components was similar to that used by Busch, Hurlbert and Potter.[7]

Experimental

Materials.—Commercial 12% cross-linked Dowex-50 cation-exchange resin of "minus 400" mesh size was graded to obtain that portion which settled between 1.0 and 1.5 cm./min. in distilled water. The resin was washed with 6 *M* ammonium thiocyanate until the red ferric thiocyanate color was no longer visible, then washed in turn with distilled water, 6 *N* hydrochloric acid and distilled water again. Finally, the resin was converted to the ammonium form with 1 *M* ammonium lactate and stored in distilled water until loaded on the column.

The eluting agents were prepared from conductivity water, Baker and Adamson reagent grade 85% lactic acid, and concentrated ammonium hydroxide. All solutions were 1 *M* in total lactate concentration with the *p*H adjusted with concentrated ammonium hydroxide and measured on a Beckman Model G *p*H meter. The solutions were also made approximately 0.01 *M* in phenol to prevent deterioration of the lactate.

Rare earth activities were obtained by bombarding a target consisting of alternate layers of pure uranium and tantalum metal foil with 340 Mev. protons in the 184-inch Berkeley synchrocyclotron. The target foils were dissolved in a solution of concentrated hydrofluoric acid and nitric acid to which 5 mg. of lanthanum carrier and 1 mg. each of strontium, zirconium, niobium, molybdenum, ruthenium, rhodium, palladium, tin, antimony, tellurium and barium carriers had been added. The mixed rare earth activities were then purified as a group by fairly well-known chemical steps which included a fluoride precipitation in the presence of dichromate, a barium sulfate "scavenge" in dilute acid solution, precipitation of the rare earth hydroxides with ammonium hydroxide, passage of the rare earth group through a small Dowex A-1 anion-exchange resin column in concentrated hydrochloric acid (to remove those elements

(1) This work was performed under the auspices of the U. S. Atomic Energy Commission.

(2) B. H. Ketelle and G. E. Boyd, *J. Am. Chem. Soc.*, **69**, 2800 (1947).

(3) S. W. Mayer and E. C. Freiling, *ibid.*, **75**, 5647 (1953).

(4) F. H. Spedding, J. E. Powell and E. J. Wheelwright, *ibid.*, **76**, 612 (1954).

(5) E. C. Freiling and L. R. Bunney, *ibid.*, **76**, 1021 (1954).

(6) R. S. Alm, R. J. P. Williams and A. Tiselius, *Acta Chem. Scand.*, **6**, 826 (1952).

(7) H. Busch, R. B. Hurlbert and V. R. Potter, *J. Biol. Chem.*, **196**, 717 (1952).

which form anionic chloride complexes under these conditions), and a zirconium phosphate "scavenge" in 4 *N* hydrochloric acid. Each of these steps was performed at least twice before the rare earth group was considered pure enough to be used on the column. Previous experience had shown that activities of all of the rare earths were formed in this type of bombardment in sufficient quantities for easy detection and that no measurable amounts of fission product activities other than the rare earths were present after the initial chemical purification.

Apparatus.—Preliminary experiments indicated that "gravity feed" of the eluting agent, *i.e.*, controlling the flow rate of the eluting agent through the resin by adjusting the height of the reservoir containing the eluting agent, did not give sufficiently high flow rates for the long column dimensions and fine resin particles that were to be used in the current experiments. Therefore, an apparatus (a schematic diagram of which is shown in Fig. 1) was designed which would permit a wide range of accurate external pressures to be applied to the solution passing through the resin bed and which, in addition, would be relatively simple in operation and maintenance. The ion-exchange resin bed dimensions were 7 mm. i.d. × 60 cm.; this column was surrounded by a 7 cm. o.d. reservoir filled with water, and around the reservoir was wound a 3 inch by 60 inch "Electrothermal" heating tape. The temperature in this entire apparatus was controlled by a "Variac" variable transformer connected to the heating tape and measured by means of a thermometer in a well immersed in the water reservoir. Since normal operating temperatures were above 90°, a small water-cooled condenser was installed above the water level in the reservoir to minimize losses due to evaporation. Elution characteristics of rare earths on columns using this type of heating arrangement were found not to differ in any way from those which used the conventional heating systems with either trichloroethylene or steam vapor jacketing.

The eluting agent reservoir system consisted of two 2,000-ml. flasks arranged in such a manner that the solution in the upper flask could be made to drop into the lower flask at any desired rate by means of a stopcock control. Both of these units were connected to a "Nullmatic" pressure regulator which was found to maintain the desired pressures to ±1 inch of water. This regulator was then connected to the laboratory air pressure system through a small air filtering unit.

Procedure.—When preparing for a run, the equipment was assembled and the column unit brought up to a stable temperature of approximately 90°. The resin was boiled in distilled water to remove dissolved gases and transferred to the column unit while hot. The new resin bed was preconditioned by passing through at least 100 ml. of the eluting agent which was to be used in the run. The rare earth activities were adsorbed on a small amount of resin by equilibrating in a hot solution of low ionic strength for several minutes. The resin was then washed twice with boiling distilled water and transferred while still hot to the top of the resin bed. Since the purpose of the experiments was to determine the effect of various continuous rates of change of *p*H on the elution characteristics of the rare earths, an attempt was made to keep all experimental conditions other than the rate of change of *p*H constant. A single resin bed was used throughout, the resin being reconditioned between runs by passing suitable quantities of *p*H 7, 1 *M* lactate and *p*H 3.19, 1 *M* lactate solutions through the column. A pressure of 5 p.s.i., which gave a flow rate of approximately 0.4 ml./cm.^{2}/min., was used on all runs. The initial *p*H of the eluting agent in the lower reservoir was 3.19. The upper reservoir contained a solution of the same total lactate concentration at a *p*H of 7.0, and the flow rate of this solution into the lower reservoir was adjusted during each run to give the desired rate of change of *p*H. In most cases the initial volume of *p*H 3.19 solution in the lower reservoir was adjusted so that the flow rate of the *p*H 7 solution was approximately the same as the flow rate of the solution through the column. In order to ensure complete and continuous mixing of the solution in the lower reservoir, a small magnetic stirring device was kept in operation during each run.

After a run had begun, samples of the eluent were collected in the collecting tubes over three-minute intervals by means of an automatic sampling turntable. For the activity assay, a drop of the eluent was obtained on an aluminum plate, evaporated to dryness under a heat lamp, and the activity in each sample counted in a proportional counter. Unequivocal identification of the individual peaks could be made through the known decay characteristics of the active nuclides present in each element. Determination of the *p*H of the eluting agent was made by direct measurement with the Beckman *p*H meter on the solution in the collecting tubes.

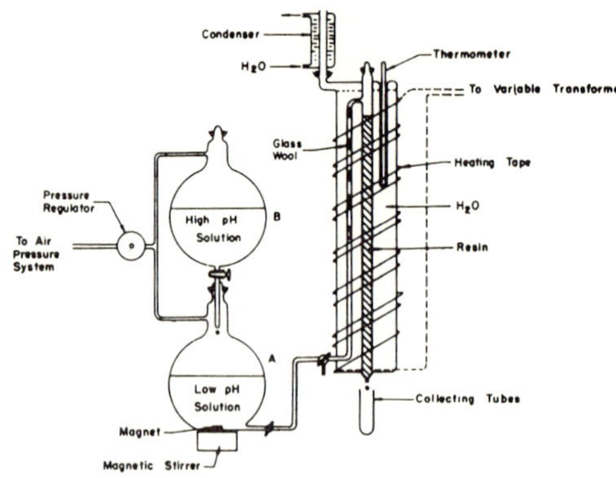

Fig. 1.—A schematic diagram of an ion-exchange column apparatus in which the *p*H of the eluting agent may be changed continuously.

III. Results and Discussion

An initial *p*H of 3.19 was used since it was found to effect satisfactory separation of the three heaviest rare earths in a reasonable column operating time under these experimental conditions. A "standard" elution curve which was obtained by passage of *p*H 3.19 solution through the column without any change in *p*H is shown in Fig. 2. While the curve shows some evidence of excessive "tailing" on several of the peaks, it has the general characteristics of a "normal" elution curve. The heavy elements (lutetium, ytterbium and thulium) are quite closely spaced but reasonably pure with a column operating time of two hours and thirty-five minutes for lutetium. The lighter elements are also separated but the column operation time rapidly becomes excessive, samarium requiring more than 38 hours for elution.

Figure 3 shows the elution curve which was obtained with an initial *p*H of 3.19 and an average rate of change of *p*H (d*p*H/d*t*) of 0.107 *p*H unit/hour. Actually, this run was made before that shown in Fig. 2, so that the short-lived activities of erbium and holmium could still be seen. Comparison of the curves of Fig. 2 and Fig. 3 shows several striking differences. The heavy elements are eluted slightly sooner with the changing *p*H (1 hour and 55 minutes *versus* 2 hours and 35 minutes for lutetium) but the separation between adjacent elements is still satisfactory. However, lighter rare earths are eluted much more rapidly in the second case than in the first without any adverse effect on their purity. Thus, with this rate of change of *p*H, samarium was eluted in 4 hours and 58 minutes as against 38 hours and 25 minutes for the constant *p*H elution. Lanthanum was eluted in 8 hours and 20 minutes while, with no change in *p*H, it may reasonably have been expected to take more than 100 hours to elute.

Additional elution curves similar to that shown in Fig. 3 have been obtained with average d*p*H/d*t*

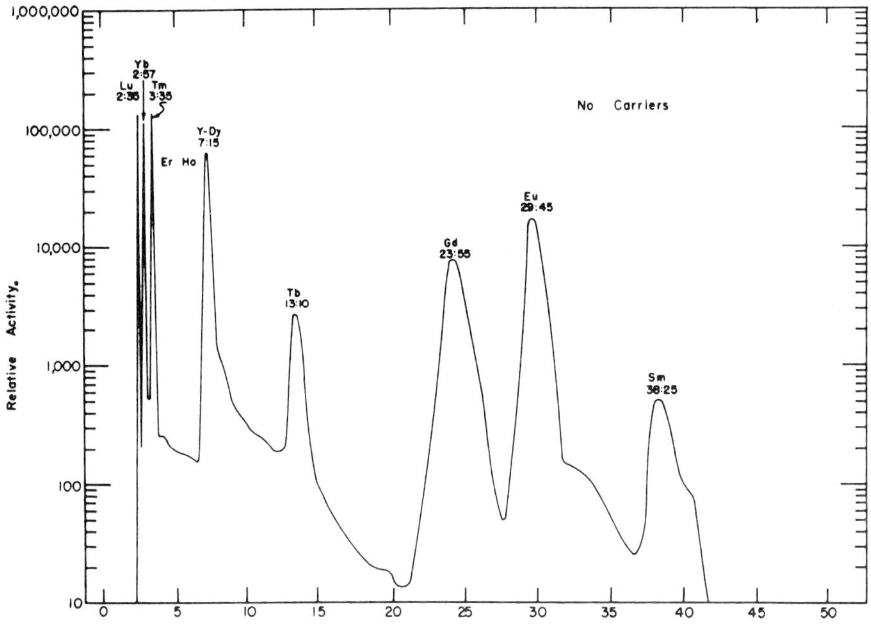

Fig. 2.—Elution curve of carrier-free rare earth tracer activities at a constant pH of 3.19.

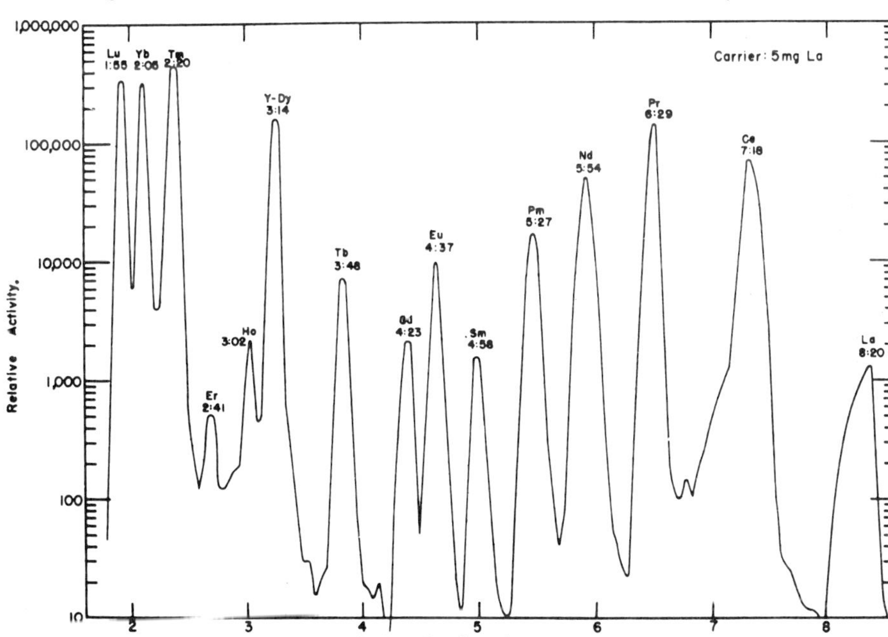

Fig. 3.—Elution curve of rare earth tracer activities plus 5 mg. of lanthanum; initial pH 3.19; pH changed continuously at 0.107 pH unit/hour.

rates of 0.007, 0.015, 0.031 and 0.067 pH unit/hour. Since the elution times for the light rare earths under constant pH condition are extremely sensitive to the experimental conditions, several elution curves were run at a constant pH of 3.19. Data for all runs are summarized in Fig. 4, where the column operation time required for elution of a given element is plotted *versus* (dpH/dt). From Fig. 4 it may be seen that the elution time for heavier rare earths, the so called "yttrium" group, is not affected greatly by any of the dpH/dt values used. However, elution of the lighter rare earths is affected markedly even by a slowly changing pH. Thus the elution time for neodymium is halved for a dpH/dt of 0.01 pH unit/hour. It may be noted also that the speed of elution is not increased very much for any of the rare earths when the pH is changed faster than approximately 0.07 pH unit/hour.

The elution curve shown in Fig. 3 represents separation of a mixture of rare earth activities in which the only carrier present was 5 mg. of lanthanum. It is well known that the presence of weighable amounts of carrier may easily distort the shape of an elution peak and affect the separation of adjacent elements. In order to determine the kind and extent of such distortions with a changing pH elution, a series of runs were made in which known amounts of various rare earths were added to the mixed tracer activities. Two elution curves which show the typical effects of massive amounts of carrier are reproduced in Figs. 5 and 6. Figure 5 shows the curve which was obtained when a mixture of 9 mg. of yttrium, 3 mg. of neodymium, 2 mg. of praseodymium and 5 mg. of lanthanum, with tracers of all of the rare earths, was run with an average dpH/dt of approximately 0.1 pH unit/hour. Peaks for those elements of which only tracers were present (lutetium, ytterbium, thulium, gadolinium, europium, samarium, promethium and cerium) show the same symmetrical shape as those in Fig. 3. However, the peaks for neodymium, praseodymium and lanthanum are definitely wider than in Fig. 3, although there is still satisfactory separation between the neodymium and praseodymium peaks. The yttrium peak, with 9 mg. of carrier present, shows approximately a fivefold increase in width over the carrier-free yttrium peak in Fig. 3. The shape of the yttrium peak is also quite different from that of a carrier-free element, the activity rising slowly on the leading edge and falling abruptly after all of the yttrium had been eluted. This effect could be noticed most strikingly on the plates which were counted. As the yttrium came off the column, deposits of yttrium carrier could be seen on the plates after the lactate solution had been evaporated. These deposits became increasingly large up to the point where the column had been operating for 6 hours and 19 minutes, after which no additional yttrium deposit was seen. The plate for 6:19 had the heaviest of the visible carrier deposits and a considerable amount of activity while the plate for 6:22 had no visible deposit but approximately four times the activity of the 6:19 plate. From the curve of Fig. 5, it may be seen that the yttrium peak was so wide that it completely displaced the terbium activity. The terbium was probably eluted in a very narrow band after all of the yttrium had come off the column. If there had been only a small amount of terbium activity present, it would have been completely hidden under the yttrium peak instead of appearing as the sharp "spike" shown in Fig. 5.

Figure 6 shows the elution curve which was obtained when a mixture of 8 mg. of yttrium, 10 mg. of europium, 8 mg. of neodymium, 2 mg. of praseodymium, and rare earth tracer activities were eluted with an average dpH/dt of 0.11 pH unit/hour. Here lutetium, ytterbium and thulium have the normal carrier-free peak shape. The yttrium peak is very wide and has the terbium "spike" as in Fig. 5. Europium also shows this saturation peak shape. The presence of relatively large amounts of europium carrier does not seem to affect the gadolinium–europium separation, but samarium is completely hidden under the back edge of the europium peak. Similarly, promethium is not affected by the presence of large amounts of neodymium carrier, but the neodymium-praseodymium peak separation is poor.

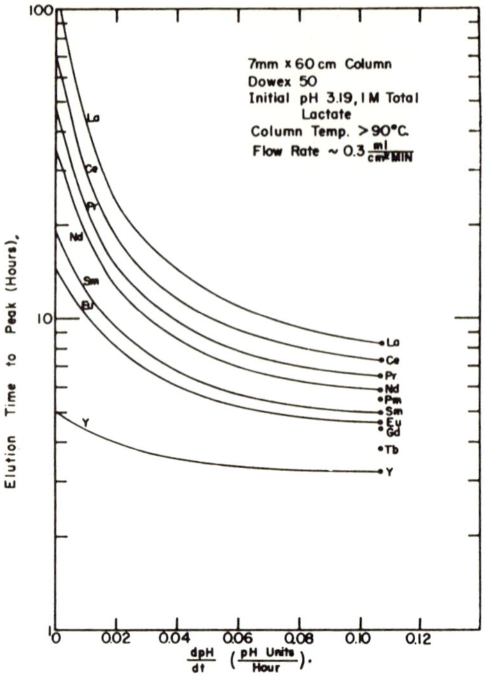

Fig. 4.—"Column operation time required for elution" *versus* "Average rate of change of pH" for several of the rare earths.

IV. Conclusion

The elution curve of Fig. 3 indicates that an eluting agent with a continuously changing pH may be used most effectively with a mixture of essentially carrier-free activities. The method is also of great aid when relatively large amounts of carrier are present, but greater caution must be taken to effect the desired separations. The curves shown here, however, suggest several general rules for operating an ion-exchange column with any given mixture of rare earth activities and carriers.

1. Any mixture of carrier-free rare-earth activities may be separated completely and quickly without difficulty.

2. Any essentially carrier-free rare earth may be easily separated from massive amounts of a rare earth of lower atomic number so long as there are no excessive massive amounts of a rare earth of

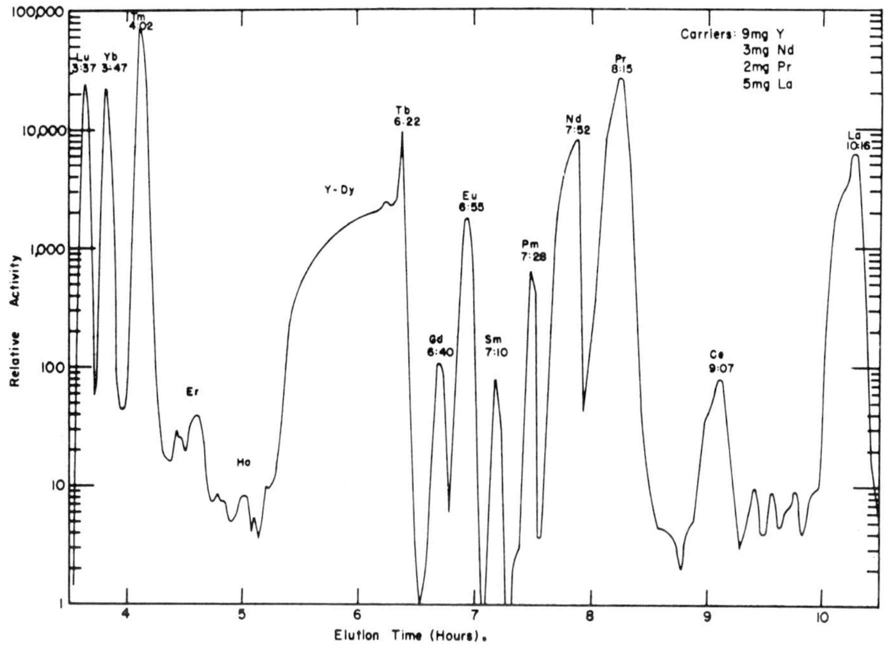

Fig. 5.—Elution curve of rare earth tracer activities plus 9 mg. of yttrium, 3 mg. of neodymium, 2 mg. of praseodymium and 5 mg. of lanthanum carriers, initial pH 3.19; pH changed continuously at 0.1 pH unit/hour.

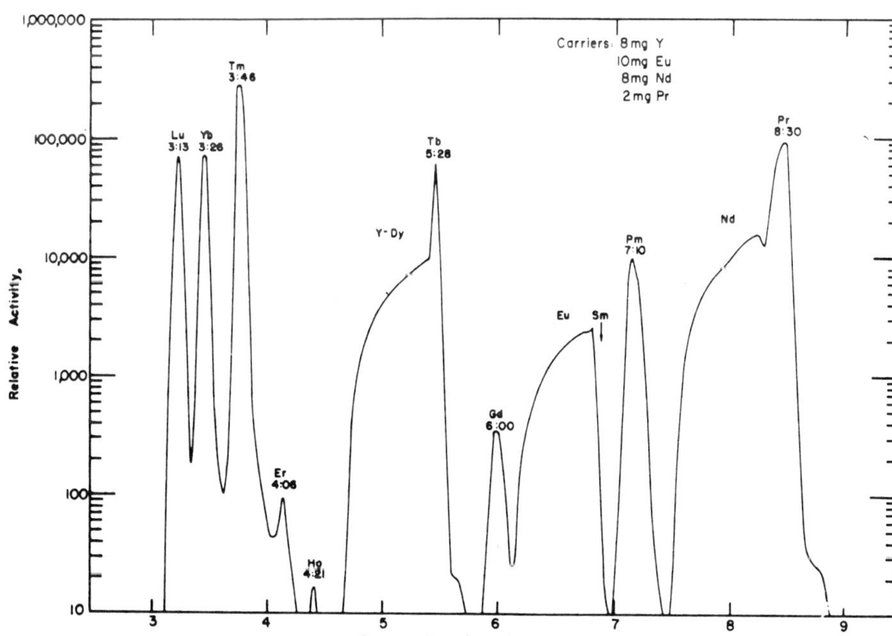

Fig. 6.—Elution curve of rare earth tracer activities plus 8 mg. of yttrium, 10 mg. of europium, 8 mg. of neodymium and 2 mg. of praseodymium carriers; initial pH 3.19; pH changed continuously at 0.1 pH unit/hour.

heavier atomic number present, *i.e.*, in Fig. 6 lutetium, ytterbium, thulium, erbium and holmium were separated from relatively large amounts of yttrium. Gadolinium was separated from europium but would not have been pure if larger amounts of yttrium carrier had been present.

3. If a small amount of one rare earth is to be separated from massive amounts of a rare earth of higher atomic number, the most effective method seems to be one in which the column is run under saturation conditions; *i.e.*, the yttrium–terbium separation in Fig. 6. The terbium fraction was not completely free of yttrium after the first run, but the amount of yttrium carrier had been greatly reduced. A second column run on the terbium fraction should then give complete separation of the yttrium and terbium peaks without any significant loss in the total terbium activity.

4. When massive amounts of adjacent rare earth elements are present, the chemist has two choices if well-defined peaks are desired. He may either use a wider column to give an "unsaturated" rare earth carrier load per unit area of resin, or he may use a lower rate of change of *p*H (and a longer column operating time) to give total separation of the two peaks in question.

These general rules have been used effectively in this Laboratory to determine the experimental conditions under which a mixture of 10 mg. of each of seven of the rare earths (yttrium, europium, samarium, neodymium, praseodymium, cerium and lanthanum) containing uranium fission product activities could be separated completely on a routine basis. As an indication of how the curves of Figs. 3 to 6 may be used to establish the operating conditions for optimum separation of a mixture of this type, the following line of reasoning was followed.

(a) From the curves of Fig. 5 and Fig. 6 an estimate of the saturation peak width *versus* carrier mass could be made. Thus, for 10 mg. of carrier, the peak would be approximately 60 minutes wide.

(b) Choosing the elements in the given mixture which were most difficult to separate, and allowing for the effects of carriers of these elements, the curves of Fig. 4 were used to estimate the highest d*p*H/d*t* which would give the desired separation of the most difficult pair. In the given mixtures, the europium–samarium pair are hardest to separate because of the presence of carriers. Allowing for a generous error in the estimated carrier peak width (*i.e.*, using a 120-minute europium–samarium peak separation instead of 60 minutes), a value of 0.01 *p*H unit per hour was chosen from Fig. 4. This rate of change of *p*H has proven to give excellent separation of this mixture of carriers with no detectable cross contamination of the peaks.

V. Acknowledgments.—The author wishes to thank Mr. D. Nethaway for his considerable assistance in assembling and operating the ion-exchange columns and Drs. P. C. Stevenson, H. G. Hicks and H. B. Levy for their aid in assaying the samples.

Part III

THE TRANSURANIUM ELEMENTS

Editor's Comments on Papers 10 and 11

10 DIAMOND et al.
An Ion-Exchange Study of Possible Hybridized 5f Bonding in the Actinides

11 GHIORSO et al.
New Element Mendelevium, Atomic Number 101

After the successful separation of the lanthanides or rare earths, it was only natural to use ion exchange to separate the actinide elements. Paper 10 gives elution data for the elements of atomic number 88 (actinium) through 98 (californium) and shows that there is a close similarity between the elution orders of trivalent actinide ions and of the corresponding trivalent lanthanide ions. It also illustrates the way in which ion-exchange distributions can give information about complex ions in solution.

Paper 11 must be one of the most remarkable papers in the history of chemistry. It describes the identification of a new and short-lived element, element 101 or mendelevium, from the spontaneous fission of only five atoms, eluted in the expected place from a column of cation-exchange resin.

The eluent used to isolate mendelevium was α-hydroxyisobutyric acid. This reagent had been shown by the same authors (Choppin et al., 1955) to be very effective in separating the actinides. It is now used routinely to separate the lanthanides, where it shows better discrimination than citric or lactic acid, especially at the low-atomic-weight end of the series.

REFERENCE

Choppin, G. R., B. G. Harvey, and S. G. Thompson. 1955. *J. Inorg. Nucl. Chem., 2,* 66.

10

Reprinted from *J. Amer. Chem. Soc.*, **76**(6), 1461-1469 (1954)

An Ion-exchange Study of Possible Hybridized 5f Bonding in the Actinides[1]

By R. M. Diamond,[2] K. Street, Jr., and G. T. Seaborg

Received October 1, 1953

A study has been made of the elution behavior of curium(III), americium(III), plutonium(III), actinium(III), plutonium(IV), neptunium(IV), uranium(IV), thorium(IV), neptunium(V), plutonium(VI), uranium(VI), lanthanum(III), cerium(III), europium(III), ytterbium(III), yttrium(III), strontium(II), barium(II), radium(II), cesium(I) with 3.2, 6.2, 9.3 and 12.2 M HCl solutions from Dowex-50 cation-exchange resin columns. These elutions indicate that in high concentrations of hydrochloric acid the actinides may form complex ions with chloride ion to a greater extent than the lanthanides. In order to explain these results, a partial covalent character may be ascribed to the bonding in the transuranium complex ions. It is shown that a reasonable structure for such covalent bonding involves hybridization of the 5f-orbitals in the actinide elements.

Study of the physical and chemical properties of the transuranium elements has revealed similarities and regularities in their behavior analogous to those in the properties of the rare earth elements. Much evidence, such as chemical, crystallographic, spectroscopic and magnetic susceptibility measurements, now leads to the view that the 5f-electron shell is being filled in the transuranium elements, similarly to the filling of the 4f-shell in the rare earths.[3]

The ion-exchange resin elution of the tripositive ions furnishes a very interesting and striking piece of chemical evidence for the similarity of the lanthanides and actinides. The two series on elution with 0.25 M ammonium citrate solution, pH 3.5 at 87°, are very similar, even to the spacings between the ions, for the elements with 6, 7, 8 and 9f-electrons: europium, gadolinium, terbium, dysprosium and the corresponding heavy elements americium, curium, berkelium and californium[4] (Fig. 1).

However, some previous work[5] in this Laboratory brought to light certain important differences in the behavior of the actinide and lanthanide elements on elution from Dowex-50 cation-exchange resin columns with hydrochloric acid solutions of various concentrations. For example, it was observed that in varying the concentration of the hydrochloric acid solutions from 3 to 6 M HCl, the actinides and lanthanides moved up in their positions of elution, but that in going from 6 to 9 to 12 M HCl solutions, the lanthanides lutetium, promethium and cerium gradually moved back, that is, they showed slower and slower elution rates. However, the elution rates of the actinides americium and curium continued to increase as a monotonic function of the hydrochloric acid concentration. Also, although lutetium, promethium and cerium kept the same relative order at the various hydrochloric acid concentrations, and while curium came off ahead of americium with 3 and 6 M HCl, as was also observed in the citrate elutions (and as would be expected from a comparison of their ionic radii), in 12 M HCl the order of elution of americium and curium was reversed, with americium moving up just ahead of curium. An explanation was suggested involving complex ion formation of the actinides with chloride ion in the higher concentrations of hydrochloric acid.

In order to confirm these results and to obtain additional evidence on the nature of the complex ion formation in the transuranium elements, the present study of the properties of the actinides and lanthanides on elution with various concentrations of hydrochloric acid was undertaken.

Experimental

First, in order to confirm that the previously observed[5] maximum in the elution rates of the lanthanide ions cerium(III), promethium(III) and lutetium(III) from Dowex-50 cation columns with hydrochloric acid solutions of varying

(1) This work was performed under the auspices of the U. S. Atomic Energy Commission.

(2) Based on the Ph.D. Thesis submitted to University of California, Berkeley, September, 1951, by R. M. D.

(3) G. T. Seaborg, National Nuclear Energy Series, Plutonium Project Record, Vol. 14B, "The Transuranium Elements: Research Papers," Paper No. 21.1, McGraw Hill Book Co., Inc., New York, N. Y., 1949.

(4) K. Street, Jr., S. G. Thompson and G. T. Seaborg, This Journal, **72**, 4832 (1950).

(5) K. Street, Jr., and G. T. Seaborg, *ibid.*, **72**, 2790 (1950).

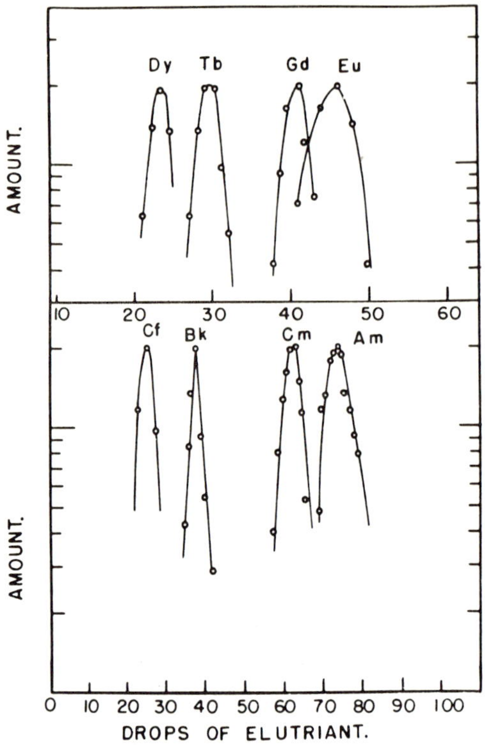

Fig. 1.—Comparison of the elution rates of californium, berkelium, curium and americium with those of their homologs dysprosium, terbium, gadolinium and europium.

concentrations is due to an actual minimum in the equilibrium distribution coefficients of the ions and not to some unknown column effect, and similarly, that in the same range of concentration of hydrochloric acid the apparent non-existence of such a maximum in the column elution rates of the actinide ions americium(III) and curium(III) is a real equilibrium effect, a series of equilibrium distribution experiments was performed at room temperature on tracer amounts of alpha particle emitting Am^{241} and β-particle emitting Pm^{147}. The resin used was colloidal Dowex-50, hydrogen form, wet graded to settle at *ca.* 0.5 cm./min., and then oven dried at 95–105° overnight. Known amounts of the radioactive ions in 5.00-ml. samples of solutions of hydrochloric acid varying in concentration from 1.12 to 13.1 *M* were shaken for about 30 minutes with small amounts (0.0076 to 1.8013 g.) of resin. The 5-ml. volumetric flasks containing the resin and solution were then centrifuged, and 0.50-ml. aliquots of the supernatant solutions were withdrawn for counting of the tracers and for titration to determine the acid concentration.

From the experimental data given in Table I, values of K_d, the equilibrium distribution coefficient, were calculated for promethium and americium at various hydrochloric acid concentrations

$$K_d = \frac{1 - F_S}{F_S} \times \frac{V_S}{W_R} = \frac{\text{(counts/g. of resin)}}{\text{(counts/ml. soln.)}}$$

where F_S is the fraction of activity in the solution phase, W_R is the weight of resin taken and V_S is the volume of solution.

The values of K_d are plotted against the corresponding hydrochloric acid concentration in Fig. 2, and the results confirm the previous observations on column elutions. The plot of K_d *vs.* hydrochloric acid concentration for americium(III) shows a steady decrease, while that for promethium(III), after decreasing at first, passes through a minimum around 7 *M* HCl, and then increases.

Next, a series of column elutions of various actinide and lanthanide ions using 3.2, 6.2, 9.3 and 12.2 *M* HCl solutions as eluting agents were performed. The elution behavior of tracer amounts of tripositive Cm^{242}, Am^{241}, Pu^{239}, $Eu^{152,154}$, Ce^{144}, Y^{90}, La^{140}, Yb^{169}, of tetrapositive Pu^{239}, Np^{237}, Th^{230}, U^{233}, of pentapositive Np^{237}, of hexapositive U^{233}, Np^{237}, Pu^{239}, of dipositive Sr^{90}, Ba^{140}, Ra^{226} and of monopositive Cs^{134} was studied.

The resin used was again Dowex-50, hydrogen form, 250–500 mesh, wet graded to settle at *ca.* 0.5 cm./min. Two columns were used, one 10 cm. long by 1 mm. in diameter, and the other 10 cm. long by 1.5 mm. in diameter. The linear flow rate was *ca.* 0.1 cm./min. Before an elution the resin column was rinsed overnight with the hydrochloric acid solution to be used as the eluting agent. To preserve reducing conditions in the 10 cm. by 1 mm. column, a drop of hydriodic acid was also added to this rinse solution, and a drop or two of nitric acid was added in the case of the 10 cm. by 1.5 mm. column to maintain oxidizing conditions. In this manner different oxidation states of the transuranium elements could be studied, and the low concentration of hydriodic or nitric acid used apparently did not affect the relative elution rates of other ions. Whenever the concentration of the eluting solution was changed, going from one experiment to another, the resin bed was backwashed with water, and then allowed to repack in the new hydrochloric acid solution.

Since high precision was not required the elutions were performed at room temperature without any thermostatting

TABLE I

EQUILIBRIUM DISTRIBUTION DATA FOR AMERICIUM AND PROMETHIUM

Expt.	Resin wt., g.	Soln. vol., ml.	Soln. normality	α-counts/min./ml. × 10^{-2}	K_d of Am	β-counts/min./ml. × 10^{-2}	K_d of Pm
1	0.000	5.0	1.12	44.7	....	40.6	....
2	.0076	4.0	1.12	10.0	1.8 × 10^3	10.3	1.6 × 10^3
8	.0061	10.0	1.12	20.4	1.9 × 10^3	20.2	1.7 × 10^3
3	.0468	5.0	3.02	28.9	58.5	28.0	48.3
4	.0825	4.5	3.02	22.2	60.0	21.4	52.8
5	.4771	5.0	5.30	20.4	12.5	19.15	11.7
6	.6697	5.0	7.61	21.1	8.36	16.8	10.6
9	.6863	5.0	9.04	25.8	5.33	15.5	11.4
7	1.1800	5.0	10.44	2.8	4.06	9.02	14.6
10	1.8013	5.0	13.06	24.1	2.37	4.25	23.7

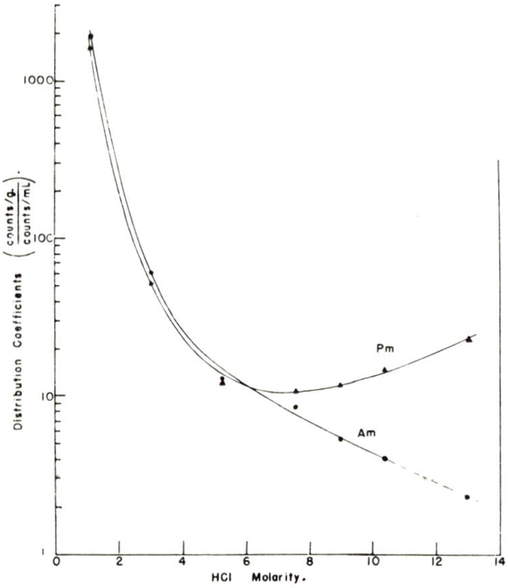

Fig. 2.—Plot of the equilibrium distribution coefficients of promethium and americium *vs.* hydrochloric acid molarity.

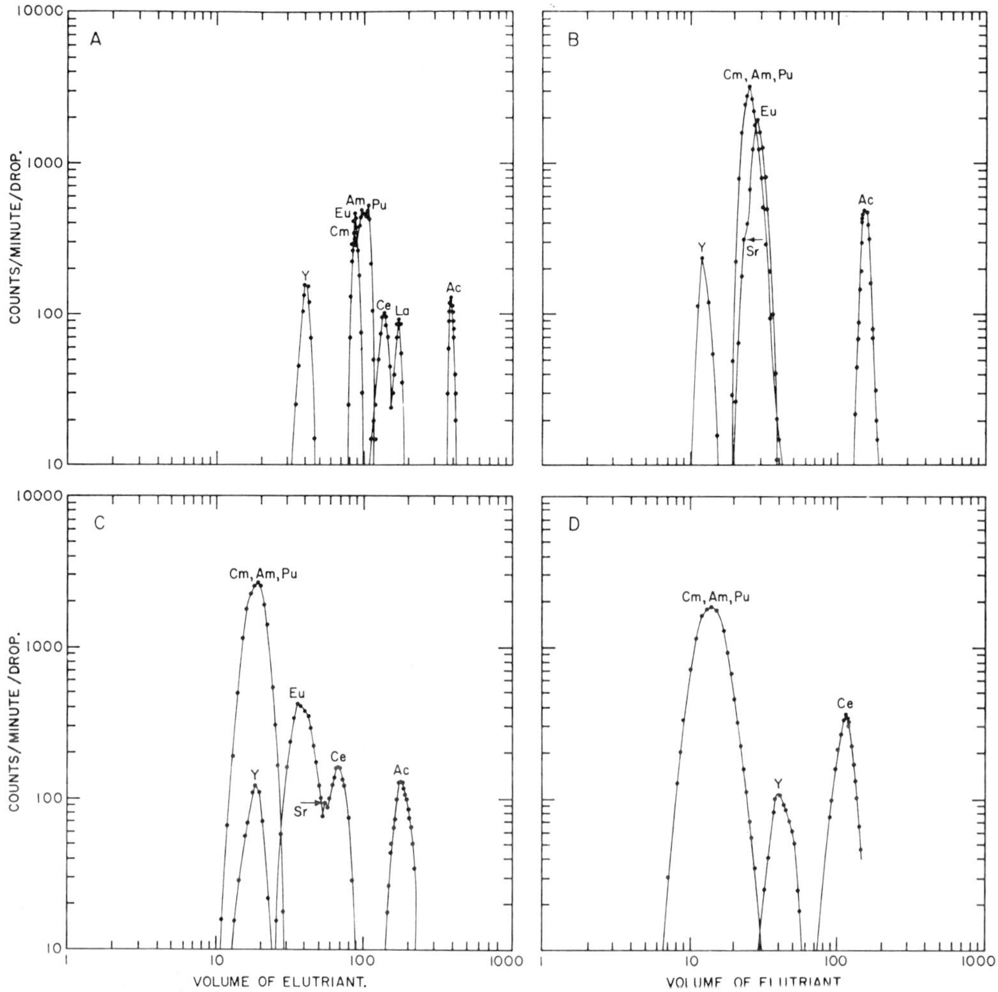

Fig. 3.—Typical elution curves in: A, 3 *M* HCl; B, 6 *M* HCl; C, 9 *M* HCl; D, 12 *M* HCl.

of the system. Preliminary equilibrium distribution coefficient determinations over a range of temperature from 15 to 80° showed only as mall temperature coefficient for hydrochloric acid elutions in contrast, for example, to the situation with citrate solutions. Nevertheless, due to the slow flow rate used, and the fact that times up to many days were required for some of the elutions, variations in the temperature up to 15–20° during a long elution were possible and probably contribute to the scattering of the results. Another possible contributing factor to such scattering of the experimental data is the gradual decrease in concentration of the 9.3 and 12.2 *M* HCl solutions by volatilization loss of HCl during a long elution.

The elution of the tracer ions was assayed by counting aliquots of evaporated solutions on plates on standard α-particle and end-window G.M. counters. Whenever necessary, the various nuclides were identified by their radiation characteristics and half-lives. Typical runs are illustrated in Fig. 3. In the case of the close curium, americium, plutonium triad, resolution of the total α-particle peak was made using the 48-channel pulse analyzer[6] with the results shown in Fig. 4. The ordinate is given as counts/minute/drop since in this type of experiment the actual number of atoms is unimportant. The abscissa, given in drops, is an arbitrary unit, as the volume of solution necessary to elute a given element, even under the near equilibrium conditions employed in this work, will vary somewhat with the particular "batch" of resin used and with the packing and previous history of the resin in the column.

Sodium dichromate was used to oxidize neptunium and plutonium to the hexapositive state, but in the 9 and 12 *M* HCl elutions the dichromate was reduced during the experiment and eluted in the first few drops as the green chromic chloride complex. However, enough plutonium remained in the (VI) state to give results in agreement with predictions based on the behavior of uranium(VI). The results for neptunium(VI) in 9 and 12 *M* HCl are very ambiguous, as the (V) and (VI) states should elute in about the same place, namely in the first column volume, and furthermore, it has been observed[7] that neptunium(VI) is unstable in strong hydrochloric acid solutions. Evidently no neptunium(VI) remained in these elutions at 9 and 12 *M* HCl, although the presence of neptunium(VI) is quite definite at 3 *M* HCl.

In order to determine the elution peak positions of ura-

(6) A. Ghiorso, A. H. Jaffey, H. P. Robinson and B. B. Weissbourd, National Nuclear Energy Series, Plutonium Project Record, Vol. 14B, "The Transuranium Elements: Research Papers," Paper No. 16.8, McGraw-Hill Book Co., Inc., New York, N. Y., 1949.

(7) J. C. Hindman, L. B. Magnusson and T. J. LaChapelle, *ibid.*, Papers 15.2 and 15.11.

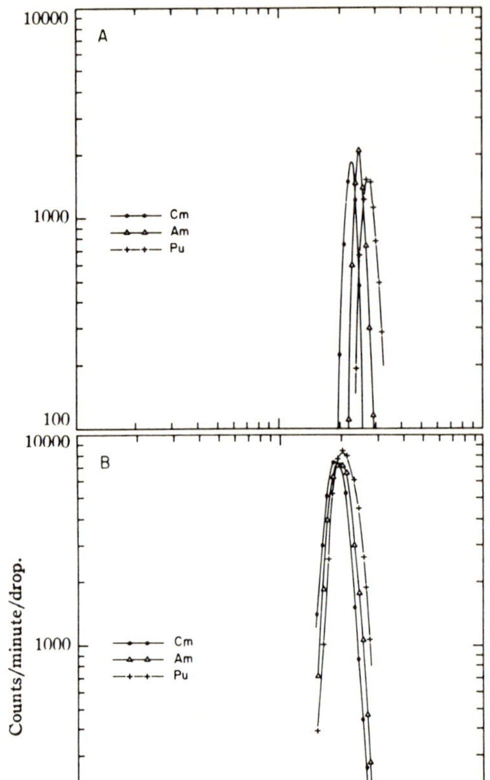

Fig. 4.—α-Particle pulse analyses of elution peaks: A, 6 M HCl; B, 9 M HCl; C, 12 M HCl.

nium(IV), a tracer solution was prepared by reducing a solution of U^{233} in 0.5 M HCl or $HClO_4$ by stirring for 15–60 minutes with a small piece of metallic U^{238}. At the hydrogen ion concentration employed, the metallic uranium apparently does not react rapidly with hydrogen ion but does reduce the uranium(VI) ions, providing a solution of tetrapositive U^{233} which does not contain too much U^{238} to load the column seriously.

A summary of what were considered successful elutions, *i.e.*, those that were not disturbed mechanically by factors beyond control, bubbles in the resin bed, overly great temperature fluctuations, etc., is given in Fig. 5 in which the average values of the elution peak positions from various runs have been plotted. The drop is again used as the elution unit volume. Although two columns of different diameter were used, the relative positions of the ions on elution should be the same, so the results have been normalized to the 10 cm. by 1 mm. column by adjusting the relative positions of the ions to that of one of the ions chosen as a standard.

Results and Discussion

The results of the equilibrium distribution experiments on americium and promethium in hydrochloric acid solutions of various concentrations, Fig. 2, illustrate very well the minimum in the K_d for the lanthanides around 7 M HCl, and the continuous decrease in K_d with increasing hydrochloric acid concentration for the actinide ions. There are at least four main factors involved in such equilibria. These are: (1) the mass action effect of hydrogen ion in the exchange reaction

$$M(H_2O)_n^{+++}\text{ (aq.)} + 3H(H_2O)_{m'}^{+}\text{(resin)} \rightleftarrows M(H_2O)_{n'}^{+++}\text{(resin)} + 3H(H_2O)_m^{+}\text{ (aq.)}$$

(2) complex ion formation of the metal ion with chloride ion

$$M(H_2O)_n^{+++} + Cl^- \rightleftarrows MCl(H_2O)_{n-1}^{++} + H_2O,$$

$$MCl(H_2O)_{n-1}^{++} + Cl^- \rightleftarrows MCl_2(H_2O)_{n-2}^{+} + H_2O\text{, etc.}$$

(3) changes in the solvation shell of each metal ion other than those due to complex ion formation, *i.e.*, dehydration

$$M(H_2O)_n^{+++} \rightleftarrows M(H_2O)_{n-1}^{+++} + H_2O$$

(4) shrinkage of the resin.

The first factor should cause the equilibrium distribution constants of both americium and promethium to decrease continuously with increasing hydrogen ion concentration, but would not allow for any differentiation in the behavior of the two classes of ions, or for an increase in K_d with hydrochloric acid concentration as is observed with the lanthanide promethium. This latter behavior, *i.e.*, the increased attraction of the resin for a rare earth ion at high hydrochloric acid concentrations, is very likely due to factors 3 and 4. Increasing the hydrochloric acid concentration in the contacting solution causes the resin to shrink, and this shrinkage of the resin causes an increase in the concentration of the functional groups inside the resin, hence shifting the ion equilibrium toward the resin and increasing the time required for elution. Also, the environment of an ion in hydrochloric acid of 12 M concentration is quite different from that which it is in 3 or 6 M; there are fewer water molecules to solvate the ion so that its hydration shell is smaller, giving the ion a smaller *effective* radius. Since it is this *effective* radius which is of importance in determining a resin's affinity for the ion,[8] a smaller effective radius should also cause the K_d to increase with increasing hydrochloric acid concentration.

(8) F. C. Nachod and W. Wood, THIS JOURNAL, **67**, 629 (1945); G. E. Boyd, J. Schubert and A. W. Adamson, *ibid.*, **69**, 2818 (1947).

In dilute solutions the predominating factor in the exchange process is the first one mentioned above so that in this concentration region all the ions studied showed a decrease in K_d with increasing hydrochloric acid concentration. Above about 1 M, however, the shrinkage of the resin begins to become appreciable and so tends to slow down the elution rate. Above 6 M HCl the result of this phenomenon and the smaller effective radius of the ion, due to partial dehydration, may overcome the mass-action effect of increasing hydrogen ion concentration and result in an increase in K_d with increasing hydrochloric acid concentration. It is proposed that the behavior of promethium is the "normal" type of elution, the americium being unusual, since the latter does not show this minimum in the value of K_d around 7 M HCl.

However, the second factor mentioned above, association and complex ion formation with chloride ion, must also be considered for these tri-positive ions. Data in the literature indicate that the rare earth and transuranium cations associate to a varying degree with chloride ion in hydrochloric acid solutions of the concentrations used in this work.[9] The differentiation of americium from promethium in hydrochloric acid solutions greater than 7 M could then be due to a greater tendency toward complex ion formation of the former, or toward desolvation of the latter.

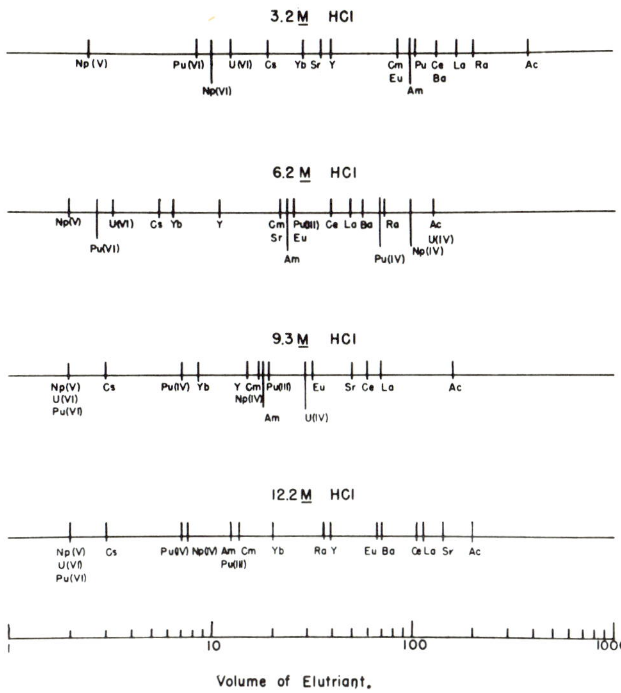

Fig. 5.—Summary of the elution peak positions of the ions studied with 3.2, 6.2, 9.3 and 12.2 M HCl.

Since as a group the actinides have ionic radii comparable to the lanthanides (Table II), in fact,

TABLE II

IONIC RADII OF ACTINIDE AND LANTHANIDE ELEMENTS[10]

	Lanthanide series Element	Radius, Å.	Actinide series Element	Radius, Å.
1	La(III)	1.04	Ac(III)	1.11
2	Ce(III)	1.02	.....	..
3	Pr(III)	1.00	.....	..
4	Nd(III)	0.99	U(III)	1.03
5	Pm(III)	(0.98)	Np(III)	1.01
6	Sm(III)	0.97	Pu(III)	1.00
7	Eu(III)	0.96	Am(III)	0.99

slightly larger values, differences in solvation of the ions are not likely. While all of the factors affecting complex ion formation are not clearly understood, *ionic* complex ion formation does not seem likely to be the answer to the differences in behavior for such similar elements of nearly the same radii. Therefore, chloride complex ion formation of a *covalent* nature must be considered for the transuranium ions. Such a concept does explain the results, and is, in fact, suggested by the differences in stability and size of the 5f-eigenfunctions in the actinides and the 4f-functions in the lanthanides. M. G. Mayer has shown on theoretical grounds that the f-eigenfunctions have a unique behavior compared to the s-, p-, or d-functions,[11] that is, they undergo a sudden change in spatial extent and energy at certain atomic numbers. The effective potential energy of an f-electron in the field of the residual atom has two negative regions after a minimum Z (about 47) and at first the outer potential minimum is dominating so the 4f-eigenfunctions have a maximum there and are "outside" functions, that is, they extend out spatially as far as, or beyond, the 5d- and 6s-orbitals. The 4f-orbitals, however, are at a higher energy level than the 5d or 6s, so these latter fill in first. With increasing atomic number, the inner potential minimum gets deeper very rapidly so that it comes to dominate the nature of the 4f-functions and in a small range of Z the shape of the function changes abruptly to that of an "inner" orbit. The rapidly decreasing potential also brings the energy down so that the 4f-level begins to fill at cerium, Z = 58, and at this point the 4f-functions are already "inside" the 5s- and 5p-electron shells. This can be seen by the negligible effect of the 4f-electrons on the valence properties of the atoms, the rare earths all having a common oxidation state, and by the very sharp absorption lines shown by the lanthanides. These lines are due to (forbidden)

(9) R. E. Connick and S. W. Mayer, THIS JOURNAL, **73**, 1176 (1951); S. W. Mayer and S. D. Schwartz, *ibid.*, **73**, 222 (1951).

(10) W. H. Zachariasen, Chapter 18, National Nuclear Energy Series, Plutonium Project Record, Vol. 14A, "The Actinide Elements," G. T. Seaborg and J. J. Katz, Editors, McGraw-Hill Book Co., Inc., New York, N. Y., 1954.

(11) M. Goeppert-Mayer, *Phys. Rev.*, **60**, 184 (1941).

f- to f transitions, and so are shielded from external influences.

Because the 4f-eigenfunctions have already shrunk at cerium, they are not very useful for covalent bonding in the lanthanide elements, since, for bonding, orbitals must not only be energetically favorable, but also must be available spatially at normal bond distances, *i.e.*, they must extend out far enough to have sufficient overlap with the bonding orbitals of the other atom or atoms. However, at a lower atomic number than that of cerium, the 4f-eigenfunctions, although at a higher energy, are appreciable beyond the 5s- and 5p-shell maxima, and so, if more energy can be gained by using them in hybridized covalent bonding than is lost in promoting or placing electrons in these higher levels, the 4f-orbitals may become bonding orbitals. It has been noted that TeF_6 hydrolyzes readily in water in contrast to the inert behavior of SeF_6 and SF_6, and the explanation offered involves such a use of 4f-orbitals.[12] Such 4f-hybridization has also been given as the explanation for the enhanced stability of the higher oxidation states of antimony, tellurium and iodine over those of arsenic, selenium and bromine, and of the ability of the former to bond a larger number of atoms.[13]

The 5f-eigenfunctions go through a transformation similar to that of the 4f after the completion of the rare earth series. At first they are "outside" functions, but again, as the atomic number increases, a critical region of Z is reached at which the eigenfunctions shrink and become "inner" functions and their energy level falls so that the 5f-shell starts to fill. In this actinide series, however, the 5f-orbitals either do not shrink as rapidly as in the 4f-series, or they have a greater spatial extension relative to the 6s- and 6p-orbitals than the 4f-orbitals have relative to the 5s- and the 5p-functions. In either case the result is a lower shielding of the 5f-electrons.

The fact that the first members of the actinide series have higher oxidation states than the characteristic tripositive state, *i.e.*, protactinium(V), uranium(IV), uranium(VI), neptunium(IV), neptunium(V), neptunium(VI), plutonium(IV), plutonium(V), plutonium(VI), in contrast to the beginning of the lanthanide series, shows that the first 5f-electrons are not bound as tightly as the 4f-electrons and that they are more easily affected by external influences, although the gradual uniform stabilization of the tripositive state indicates the presence of a regular f series, analogous to the rare earths.

The nearness of the 5f- and 6d-levels over a region of several atomic numbers makes it difficult to determine where they cross, *i.e.*, at which value of Z the first 5f-electron enters, and the lower shielding of the 5f-orbitals compared to the 4f, makes it possible that, at the beginning of the 5f-series, the levels change positions depending on the chemical binding of the element or on its degree of ionization. Thus, the gaseous atom may have a different electronic configuration than the metallic atom, and the electronic configuration of the ion may differ from compound to compound in solids and in solution. These small energy differences are evident also in the spectroscopic data; for example, spectroscopic evidence seems to show that the ground state of the neutral thorium atom has a $7s^26d^2$-configuration[14] (although X-ray data indicate the presence of a 5f-electron)[15] but the ground state of the triply ionized atom has a single 5f electron.[16]

The comparable binding energy, at the beginning of the actinide series, of the 5f-, 6d- and 7s-levels for a range of several atomic numbers, is a favorable condition for their use in hybridized bonds. Furthermore the greater spatial extension of the 5f-orbitals relative to the 4f and their slower shrinkage, might make them more available for such hybridization closer to the atomic number where the levels start to fill, rather than several numbers earlier as is probably the case with the 4f-orbitals. Thus such hybridization involving 5f-orbitals seems reasonable to account for the results reported in the present paper, *i.e.*, the greater tendency for covalent complex ion formation of the trans-uranium elements as compared to the rare earth elements.

The use of 5f-orbitals in complexes formed by uranyl nitrate has been postulated by Glueckauf and McKay[17] to explain a series of facts about its chemistry, including its unusual solubility in oxygenated organic solvents, although Katzin[18] does not believe that such coördination is unique or different in kind from that found lower in the periodic table. Still another view is presented by Connick and Hugus, who, from consideration of the X-ray diffraction evidence on the crystal structure and a thorough analysis of the hydrolytic behavior of the uranyl ion, come to the conclusion that f orbitals participate in the metal–oxygen bonding in the uranyl, neptunyl and plutonyl ions.[19]

With these considerations in mind, the experimental results of the ion-exchange studies may be interpreted. The elutions with the tripositive ions of lanthanum, cerium, europium, ytterbium, actinium, plutonium, americium, curium are summarized in Fig. 6. Yttrium(III) is also included because it acts as a pseudolanthanide, with no possibility of f-orbital bonding.

The molarity of the eluting hydrochloric acid solution is plotted as abscissa, and the elution peak position in drops, normalized to the 10 cm. by 1 mm. column, is plotted as ordinate. The elution peak position is directly related to the equilibrium distribution coefficient of the ion and is directly comparable to Fig. 2, although the curves are distorted toward too high values at high hydrochloric acid concentration, since somewhat larger amounts of resin were used to maintain the column constant to a 10 cm. length.

Certain features in Fig. 6 are of interest. The difference in behavior of curium, americium and

(12) G. E. Kimball, *J. Chem. Phys.*, **8**, 188 (1940).

(13) Z. Z. Hugus, Jr., THIS JOURNAL, **74**, 1076 (1952).

(14) P. Schuurmans, Thesis, Amsterdam, 1946.

(15) H. Russell, Jr., Los Alamos Scientific Laboratory Report LA-145 (Sept. 24, 1944); Manhattan District Declassified Document MDDC-406 (Oct. 22, 1946).

(16) P. F. A. Klinkenberg and R. J. Lang, *Physica*, **15**, 774 (1949).

(17) E. Glueckauf and H. A. C. McKay, *Nature*, **165**, 594 (1950).

(18) L. I. Katzin, *ibid.*, **166**, 605 (1950).

(19) R. E. Connick and Z. Z. Hugus, Jr., THIS JOURNAL, **74**, 6012 (1952).

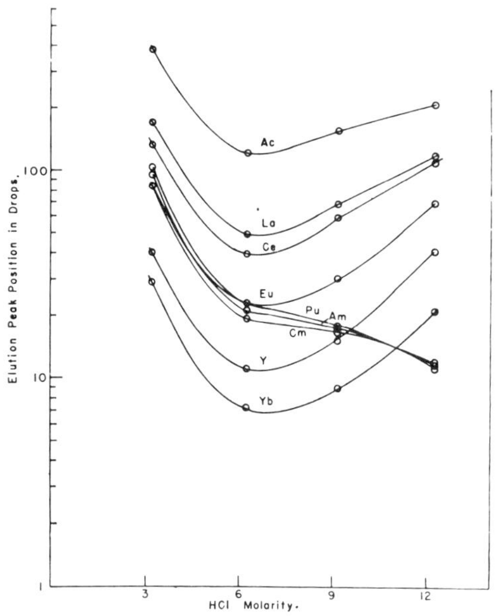

Fig. 6.—Elution peak positions of the tripositive ions *vs.* hydrochloric acid molarity.

plutonium, on one hand, and ytterbium, yttrium, europium, cerium, lanthanum and actinium on the other, is clearly evident. In hydrochloric acid above 6 *M* in concentration the actinide ions show a much larger elution rate than the rare earth ions. The latter are delayed due to the resin shrinkage and consequent increase in concentration of exchange groups, and also to the decrease of the effective radii caused by the increasing disruption of their hydration shells. While the actinides also experience a greater attraction by the resin at higher hydrochloric acid concentration, the greater extent of their complex ion formation with chloride apparently more than offsets this and is responsible for their earlier elution. This complex formation with chloride ion is borne out by other types of data. For example, absorption spectra measurements on plutonium(III) show that changes in the spectrum indicative of complex ion formation occur in hydrochloric acid solutions above 6 *M* in concentration.[20] Further, transference experiments on plutonium(III) in hydrochloric acid solutions indicate that a small but appreciable fraction of the plutonium migrates to the anode in 1 *M* HCl, and that a somewhat larger amount[21] does so in 10 *M* HCl. These facts are compatible with the actinides forming negative complex ions with chloride ion, and more detailed anion-exchange work with europium and americium in 12 *M* HCl does indeed show a small delay in the elution peak position of americium(III) compared to europium(III).[22] Similar experiments in 5 *M* HCl show no such differentiation indicating that the tendency of americium to form complex anions with chloride, although pronounced relative to the rare earths, is still quite weak requiring concentrated hydrochloric acid to become observable.

Examination of Fig. 6 shows that yttrium(III) which has a comparable radius to the actinides and lanthanides (and has d-orbitals available) but which also has such a low atomic number that the 4f-orbitals are still at too high an energy to be useful for bonding, elutes with the lanthanides, and, in fact, shows a curve rising more steeply from 6 to 12 *M* HCl than that of any of the lanthanide ions. This behavior is indicative of even less complex ion formation in this range of hydrochloric acid concentration than that displayed by the lanthanide ions Examination of Fig. 6 also shows that actinium(III) resembles the lanthanides more closely in its elution behavior than it does plutonium(III), americium(III) and curium(III), *i.e.*, its tendency toward complex ion formation is smaller than in the case of the latter ions, though greater than that of any of the lanthanides. This may indicate that the 5f-orbitals in actinium are still at too high an energy to be readily available for bonding.

In order to illustrate further the point that the actinides complex more chloride ions than do the lanthanides, the data of Table I are shown in Fig. 7,

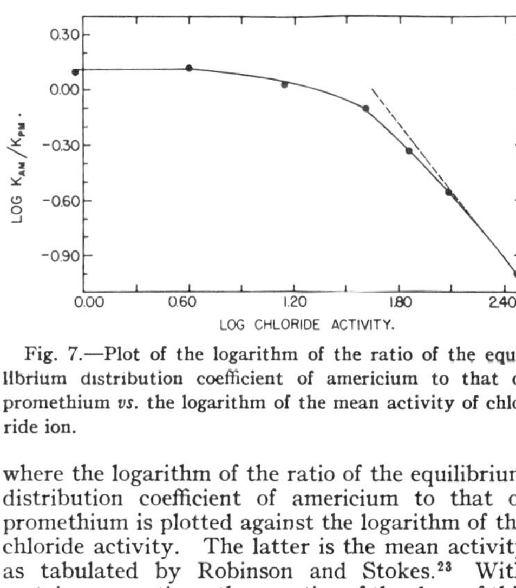

Fig. 7.—Plot of the logarithm of the ratio of the equilibrium distribution coefficient of americium to that of promethium *vs.* the logarithm of the mean activity of chloride ion.

where the logarithm of the ratio of the equilibrium distribution coefficient of americium to that of promethium is plotted against the logarithm of the chloride activity. The latter is the mean activity as tabulated by Robinson and Stokes.[23] With certain assumptions, the negative of the slope of this curve gives a measure of the number of excess chloride ions bound up with americium over those tied up with promethium. Actually, the assumptions make the slope a lower limit, so that the observed slope of about -1.2 at 12 *M* HCl may well indicate that there are, on the average, two more chlorides bound to an americium ion than to a promethium ion in strong hydrochloric acid solutions. For the reasons already discussed, this difference in complex ion formation ability is very likely due to a greater contribution to hybridization by the

(20) J. C. Hindman and D. P. Ames, National Nuclear Energy Series, Plutonium Project Record, Vol. 14B, "The Transuranium Elements: Research Papers," Paper No. 4.2, McGraw-Hill Book Co., Inc., New York, N. Y., 1949; J. C. Hindman, *ibid.*, Paper No. 4.4.

(21) C. R. McLane, J. S. Dixon and J. C. Hindman, *ibid.*, Paper No. 4.3.

(22) D. A. Orth, unpublished data, this Laboratory (1949).

(23) R. A. Robinson and R. H. Stokes, *Trans. Faraday Soc.*, **45**, 612 (1949).

5f-orbitals in the actinides than by the 4f-orbitals in the rare earths.

A smaller effect, but a very striking one, is the reversal of the elution order of the ions plutonium(III), americium(III) and curium(III) in 12 *M* HCl, from that in the less concentrated hydrochloric acid solutions, where the elution sequence is curium, americium and plutonium, as might be expected from their hydrated radii. This anomaly may be due to the already mentioned spatial contraction of the f-orbitals with increasing atomic number. For the 4f-orbitals this shrinkage has occurred already at cerium, where they commence to fill, but for the 5f-orbitals it may occur most prominently around neptunium or plutonium. (It is interesting to note that a maximum value for the force constant, as determined by infrared absorption, occurs for NpO_2^{++} in the sequence UO_2^{++}—NpO_2^{++}—PuO_2^{++}—AmO_2^{++} which might also be due to the changing availability of 5f-orbitals in this region.)[24] The 5f-orbitals in curium would then be smaller than those in americium, which would in turn be smaller than those in plutonium, and so their ability to overlap the orbitals of the chloride ions would consequently be less and the resulting bonding would be weaker. This would cause the strength of the complex ions to vary in the order plutonium > americium > curium and would give the observed reversal of the elution order. However, this explanation must be reconciled with the fact that even in 12 *M* HCl berkelium and californium elute ahead of plutonium and that they have even stronger tendencies to form complex ions with chloride ion.[25] Actually, this latter behavior is not incompatible with the idea of the spatial contraction of the 5f-orbitals, since, by analogy with the rare earths, the berkelium(III) ion is probably considerably smaller than the plutonium(III), americium(III) and curium(III) ions, and the californium(III) ion is probably even smaller. This decrease in size of the ion, that is, of the filled 6s- and 6p-orbitals, may be comparable to the shrinkage of the 5f-orbitals and thus allow the latter to be of greater effective use again in hybridized bonding. It should be mentioned that a study of the elution behavior of europium and of gadolinium, the homologs of americium and curium, showed no reversal of their relative elution peak positions in 12 *M* HCl.

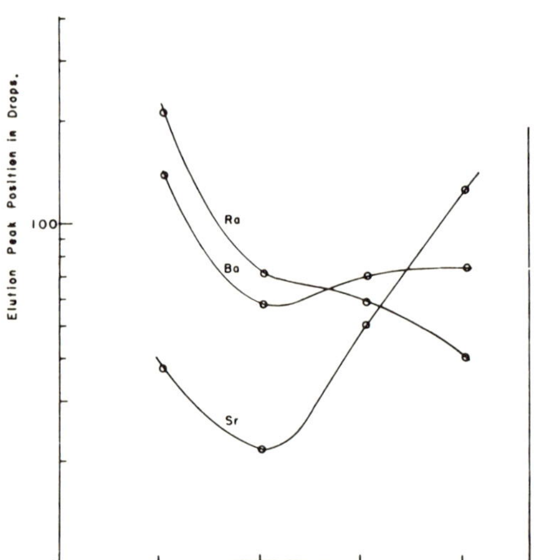

Fig. 8.—Elution peak positions of the dipositive alkaline earth ions *vs.* hydrochloric acid molarity.

(24) L. H. Jones and R. A. Penneman, *J. Chem. Phys.*, **21**, 542 (1953).

(25) S. G. Thompson and K. Street, Jr., unpublished work (1951).

Another possibility, which is perhaps not separate or incompatible with the foregoing discussion, is that the tendency of plutonium(III), americium(III) and curium(III) to form positively charged chloride complex ions is about the same, but in the process the hydration shells of the ions are necessarily disrupted. In 12 *M* HCl these complex ions are then partially desolvated so that their effective radii go in the order plutonium > americium > curium, so that their elution peak positions are reversed. The behavior of berkelium and californium, eluting even faster than plutonium in 12 *M* HCl, is again explainable as caused by a much stronger complex ion forming tendency, due to their smaller size.

Although a definite answer as to why plutonium, americium and curium show a reversal in their elution order in hydrochloric acid solutions of greater than 9 *M* concentration requires further study, a similar, but much more pronounced reversal of elution order with the alkaline earth ions, strontium(II), barium(II), radium(II), Fig. 8, indicates the importance of considering the effect of ion dehydration in concentrated hydrochloric acid solutions. With dilute hydrochloric acid these ions elute in the order to be expected from their effective (hydrated) radii, namely, in the sequence strontium, barium and radium, but at 12 *M* HCl they have reversed this order to radium, barium, strontium, and with considerable spacing between the elution peak positions. Chloride complex ion formation with these alkaline earth ions is probably not very important, certainly less so than in the case of the rare earth and transuranium ions. The larger Ra(II) ion is probably only partially hydrated even in 3 *M* HCl so it shows a continuous decrease in K_d with increase in hydrochloric acid concentration due mainly to the mass-action effect of increasing the hydrogen ion concentration. In contrast, the smaller strontium ion is certainly more hydrated in dilute hydrochloric acid solutions, but with increasing concentration of hydrochloric acid the free water left available for this purpose decreases and so the ion's effective radius becomes smaller. This, plus the effect of resin shrinkage, appears more than sufficient to overcome the mass action effect of the hydrogen ion in solutions above a concentration of 6 *M* HCl, and so Sr(II) goes back on the resin, *i.e.*, it elutes slower and slower. The behavior of the barium ion is intermediate between that of the strontium and radium ions, as would be expected.

The results for the hexapositive states of uranium, neptunium and plutonium are given in Fig. 5.

These are oxygenated to form the UO_2^{++}, NpO_2^{++} and PuO_2^{++} ions, which are often considered as large dipositive cations. At concentrations as low as 6 *M* HCl these ions come off in the first column volume, indicating strong complex ion formation with chloride ion, with little attraction by the resin. Evidence for the formation of such complex ions with chloride ion for this oxidation state can be found in the literature from transference experiments[21] and spectroscopic studies.[20] Also similar to these ions is neptunium(V) which exists in the form NpO_2^+, and washes right through the resin column even in 3 *M* HCl. Since cesium(I), a typical monopositive ion, is held up somewhat on the resin column in hydrochloric acid solutions of such concentration, the rapid elution of NpO_2^+ indicates that appreciable complex ion formation occurs with the neptunium(V) in 3 *M* HCl.

Still another series of ions can be considered in the actinides: the tetrapositive states of thorium, uranium, neptunium and plutonium. The results for these are summarized in Fig. 9 where again elution peak position is the ordinate and hydrochloric acid molarity is the abscissa. All these ions are presumably "bare" ions, *i.e.*, are not oxygenated, and so due to their high charge should be strongly adsorbed on the resin in the absence of complexing agents. However, the same high charges should also cause these ions to be strongly bound in complex ions. No data are given for 3 *M* HCl solutions, as the ions are all bound so tightly to the resin at this hydrochloric acid concentration that the elution could not be conveniently observed. With the 6, 9 and 12 *M* HCl solutions, however, the tetrapositive ions elute in the sequence plutonium, neptunium, uranium and then thorium, which is the order expected from their radii if there was no complex ion formation. As the hydrochloric acid concentration is increased, these ions elute more rapidly, showing a large variation in the value of K_d, and indicating strong complex ion formation, particularly for plutonium, neptunium and uranium.

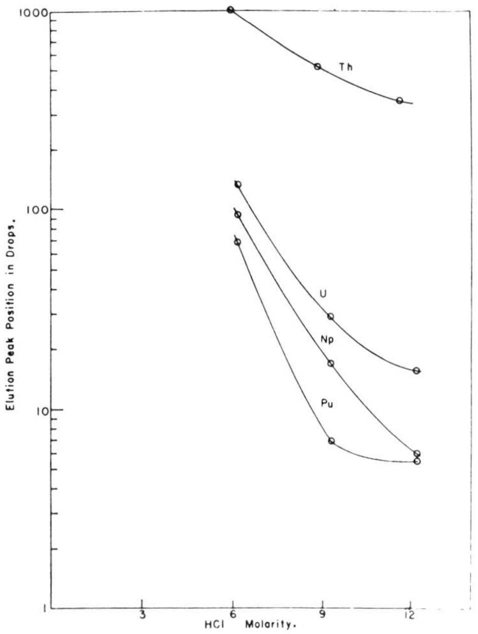

Fig. 9.—Elution peak positions of the tetrapositive ions *vs.* hydrochloric acid molarity.

Acknowledgment.—It is a pleasure to acknowledge the many helpful discussions which we have had with Professor B. B. Cunningham and Dr. J. W. Cobble concerning the possible interpretations of the experimental results reported here.

BERKELEY, CALIFORNIA

11

Reprinted from *Phys. Rev.*, **98**, 1518-1519 (1955)

NEW ELEMENT MENDELVIUM, ATOMIC NUMBER 101

A. Ghiorso, B. G. Harvey, G. R. Choppin, S. G. Thompson, and G. T. Seaborg

Radiation of Laboratory and Department of Chemistry University of California, Berkeley, California

WE have produced and chemically identified for the first time a few atoms of the element with atomic number 101. Very intense helium ion bombardments of tiny targets of 99^{253} have produced a few spontaneously fissionable atoms which elute in the *eka*-thulium position on a cation resin column.

The method of production utilized the following techniques. In a special position in the Crocker Laboratory 60-inch cyclotron a very concentrated collimated beam of 48-Mev helium ions (as much as 10 microamperes in an area $\frac{1}{32}\times\frac{1}{4}$ inch) was allowed to pass through a degrading absorber and then through a 2-mil gold foil (yielding 41-Mev helium ions). On the back side of the gold foil, approximately 10^9 atoms[1] of the 20-day 99^{253} were electroplated in the beam area. From this target the nuclear transmutation recoils were ejected in a narrow spray and caught on 0.1-mil gold foil adjacent to the target. The gold foil was quickly dissolved in aqua regia, the gold extracted with ethyl acetate, and the aqueous phase eluted through a Dowex-1 anion resin column with 6*M* HCl to complete the removal of gold and other impurities. The drops containing the actinide fraction were evaporated and the activity was then eluted through a Dowex-50 resin cation column with ammonium α-hydroxy-isobutyrate[2] to separate the various actinide elements from each other. The radiations from the various fractions were then examined with various types of counters.

The earliest experiments were confined to looking for short-lived alpha-emitting isotopes of element 101 such as 101^{255} and 101^{253} that would be expected from $(\alpha,2n)$ and $(\alpha,4n)$ reactions. For this purpose it was sufficient to look quickly in the gross actinide fraction as separated by the anion resin column alone. No alpha activity was observed that could be attributed to element 101 even when the time between the end of bombardment and the beginning of alpha pulse analysis had been reduced to 5 minutes. To ascertain that a proper recoil and chemical yield was being obtained, a small amount of Cm^{244} was included in the target so that the Cf^{246} and Cf^{244} produced by the helium ion beam would serve as a monitor.

Longer bombardments were then made, and a few spontaneous fission events were observed over a period of several hours in the actinide fractions. At this time the cation resin column procedure was then added to the chemistry so that we could distinguish between elements 101 and 100. The initial experiments were confined to fission counting of the *trans*-100 and the 100 fractions. Several experiments were consistent in their demonstration of spontaneous fissions occurring in both fractions. To determine more precisely the elution position of the element responsible for the *trans*-100 spontaneous fission activity a more elaborate experiment was undertaken. Three successive three-hour bombardments were made and in turn their transmutation products were separated quickly on a cation resin column. The isotope 99^{253} was added in each case so that together with the Cf^{246} produced it would help to calibrate the column run. Five spontaneous fission counters were then used to count simultaneously the corresponding drops of eluent from the three runs. The combined and normalized elution curve is shown in Fig. 1. That the drops did not correspond exactly in each run is indicated on the spontaneous fission histogram by the overlapping of the relative drop number limits of the activities that were placed in each counter. By combining the data from all experiments, the half-life of the spontaneous fission activity in the 101 and

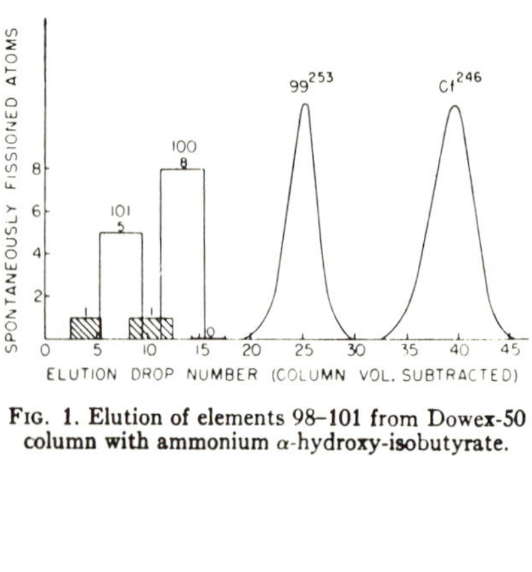

FIG. 1. Elution of elements 98–101 from Dowex-50 column with ammonium α-hydroxy-isobutyrate.

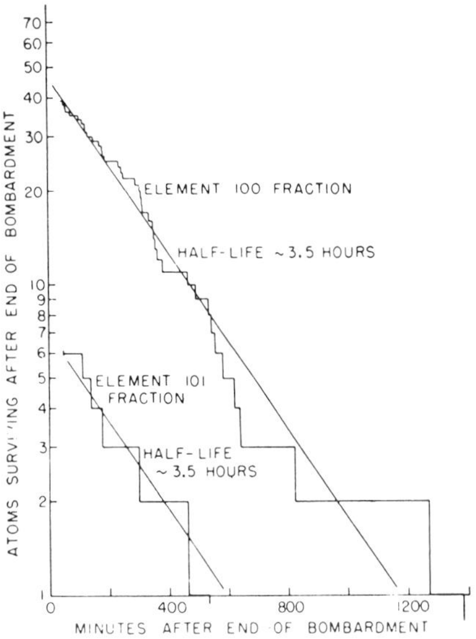

FIG. 2. Spontaneous fission decay curve of element 100 and element 101 fractions.

100 fractions was found to be approximately the same, perhaps 3 to 4 hours as shown in Fig. 2.

Obviously there is insufficient evidence to define the mass numbers concerned in this experiment. However, recent experimental observations of the spontaneous fission decay rates[3] of odd-nucleon types indicate that they are hindered by factors of thousands. If we then apply this principle to this experiment it is tempting to draw the following conclusions. By an (α,n) reaction on 99^{253} we have produced the isotope 101^{256} which decays by electron capture with a half-life of the order of a half hour to 100^{256}; this isotope then decays by spontaneous fission with a half-life of the order of 3 to 4 hours. (Since no alpha activity is observed we can rule out the 3.2-hour isotope 100^{254} as being responsible for the spontaneous fission activity on the basis of its known alpha-to-fission ratio of 1550.) In a total of eight separate experiments we have isolated only 17 atoms of 101 so it has not yet been possible to demonstrate this proposed genetic relationship between 101^{256} and 100^{256}. However, another experiment[4] has shown shown that 100^{256} does have a spontaneous fission half-life of about 3 hours.

3. By this method of production, it would not be possible to produce an element above 101.

We would like to suggest the name mendelevium, symbol Mv, for the new element in recognition of the pioneering role of the great Russian chemist, Dmitri Mendeleev, who was the first to use the periodic system of the elements to predict the chemical properties of undiscovered elements, a principle which has been the key to the discovery of the last seven transuranium (actinide) elements.

We should like to acknowledge the fact that this work was greatly facilitated by the reshaping of the magnetic field of the Crocker Laboratory 60-inch cyclotron so as to secure a more concentrated ion beam. This modification was successfully carried out under the expert supervision of G. B. Rossi. We should like to express our appreciation to C. A. Corum for his expert collaboration in the design of the special deflector channel probe used in these bombardments. To G. B. Rossi, W. B. Jones, and the crew of the 60-inch cyclotron, we extend our thank for their patience and careful manipulation of the machine which made the bombardments successful. We should like to thank T. O. Parsons, R. Silva, and A. Chetham-Strode for their considerable assistance. We should also like to express our appreciation to Professor J. G. Hamilton, director of the Crocker Laboratory, for his help and and cooperation and to Professor E. O. Lawrence for his continued interest in this program of research.

The proof that these experiments resulted in the identification of element 101 is as follows:

1. Only the very heaviest elements decay by spontaneous fission with such short half-lives.

2. The elution position immediately ahead of element 100 shows that the chemical properties are those of an element heavier than 100, and probably 101 rather than 102.

* This work was performed under the auspices of the U. S. Atomic Energy Commission.

[1] The 99^{253} target atoms were made by intense neutron bombardment of Cf^{252} and the subsequent beta decay of Cf^{253}. We would like to thank the personnel of the Materials Testing Reactor in Idaho for their very expeditious handling of this bombardment.

[2] Choppin, Harvey, and Thompson (unpublished information).

[3] Unpublished data in this laboratory.

[4] Choppin, Harvey, Thompson, and Ghiorso, following Letter [Phys. Rev. 98, 1519 (1955)].

Part IV

ISOTOPE SEPARATIONS

Editor's Comments on Papers 12 Through 14

12 TAYLOR and UREY
Fractionation of the Lithium and Potassium Isotopes by Chemical Exchange with Zeolites

13 SPEDDING et al.
A Laboratory Method for Separating Nitrogen Isotopes by Ion Exchange

14 CONRARD and ROSSET
Théorie de développement par déplacement en chromatographie d'échange d'ions: application à la séparation de corps très voisins
English translation: *Theory of Development by Displacement in Ion-Exchange Chromatography: Application to the Separation of Very Similar Substances*

It is interesting that one of the first serious attempts to use ion-exchange chromatography should have been for separating isotopes. Taylor and Urey used the gel-type aluminosilicate exchanger that was commonly used in water softeners, with no special attention to grain size, and achieved a partial separation of the isotopes of lithium and potassium. In the light of later work with better analytical tools (Lindner and Bergdahl, 1960) it is doubtful whether they obtained as good a separation as they reported, but they were pioneers.

Spedding, Powell, and Svec separated the isotopes of nitrogen by displacement chromatography, using the same principles that they had used to separate rare-earth elements. A long column of cation-exchange resin in the hydrogen form was passed into it a solution of ammonia. Ammonia combined with the hydrogen ions of the resin to give a band of ammonium ions in the column with a sharp boundary, and a sharp boundary is, of course, a necessity in the displacement-band method. Above the band of ammonium ions they introduced a solution of sodium hydroxide, which produced a second sharp boundary as the hydroxyl ions reacted with ammonium ions to form ammonia, and sodium ions took the place of ammonium ions in the resin.

As the band of ammonium ions moved down the column, the isotope ^{14}N moved ahead while the heavier isotope ^{15}N lagged behind. The isotopic discrimination was not due to ion exchange as such, for $^{14}NH_4^+$ was bound just as strongly by the resin as was $^{15}NH_4^+$, but rather because the equilibrium between ammonium ions, ammonia, and hydrogen ions favors the heavy isotope in the ammonium ion, NH_4^+, and the light isotope in the ammonia, NH_3 (see equation 8 of the paper). The profile of isotopic ratio along the length of the band changed as the band moved, in accordance with Figure 1. In the early stages of the band's travel there was a region in the middle where the isotopic ratio was constant and equal to the original ratio; but after the band had moved a certain distance, in this case about 50 times its own length, the flat "plateau" region disappeared, and the graph of isotopic ratio versus distance became a smoothly rising curve without inflection. There was a rapid rise in the fraction of the rarer isotope near the tail, or upper end, of the band.

A pilot plant for concentrating ^{15}N is described by Spedding (1963). Multiple columns and band recycling are used, and the theory is reviewed.

In connection with the French atomic energy program, Caude, Conrard, and Rosset (Rosset, 1964; Conrard and Caude, 1971) separated the isotopes of boron. Their method, similar to that of Spedding, did not depend on an ion-exchange equilibrium as such, but on the equilibrium between boric acid and borate ions. This equilibrium favors the isotope ^{10}B in the ionized form. This isotope is therefore held more strongly by the resin and it accumulates at the rear of the band. A column of anion-exchange resin in the hydroxyl form was used. Boric acid solution was poured into the column, producing a band of borate ions, and the band was driven along the column by passing hydrochloric acid. Sharp boundaries formed at the front and rear, and as the band moved, ^{11}B moved ahead while the rarer isotope ^{10}B became more concentrated toward the rear.

The boron isotope separation has been described by Conrard et al. (1972). The paper is the first of a series appearing in this journal on the practice and theory of isotope separation by displacement. The second paper (Conrard and Rosset, 1972) discusses the separation of nitrogen isotopes and points out that Spedding, in the work described in Paper 13, did not actually reach the steady-state distribution. The third and fourth papers of this series (Conrard, 1973) are theoretical.

We have chosen the paper by P. Conrard and R. Rosset from the French journal *Analusis* for reproduction here. It is primarily theoretical, and summarizes and extends the discussions of the four papers in *Separation Science.* The scope of the paper will be seen from the summary and the opening paragraphs, which we have translated. The discussion is based on the discontinuous plate theory used by Mayer and Tompkins (Paper 2) and is restricted to cases where the equilibrium constant for the exchange is close to unity. The most important conclusions, perhaps, are the equations for the steady-state distribution along the moving band. The equation relating the mole fraction of the rarer isotope, x, with its average or original mole fraction, x_0, the exchange equilibrium constant, $1 + \epsilon$, and the distance along the band expressed in numbers of theoretical plates, p, is equation (9). Equation (10) gives the coordinate of the center of the band as measured from the point at which $x = x_0$; P is the total number of plates along the length of the band. The printers did not show clearly where the parentheses come in equation (10), and we therefore reproduce equations (10) here:

$$M = \frac{1}{2\epsilon} \ln \left\{ \frac{1-x_0}{x_0} \cdot \frac{e^{2\epsilon P x_0} - 1}{e^{\epsilon P} - e^{\epsilon P(2x_0 - 1)}} \right\}$$

or

$$M = \frac{1}{2\epsilon} \ln \left\{ \frac{1-x_0}{x_0} \cdot \frac{\sinh \epsilon P x_0}{\sinh \epsilon P(1 - x_0)} \right\}$$

Figures 1 and 2 show calculated concentration profiles for isotopes of boron and uranium. Obviously, for a band of a given length, the smaller the plate height, the better is the separation.

The remainder of the paper discusses the rate at which enrichment proceeds as the band moves, and gives some guidelines for the production of separated isotopes.

REFERENCES

Conrard, P. 1973. *Separation Sci., 8,* 1, 269.
——, and M. Caude. 1971. *Compt. Rend.,* Ser. C, *272,* 764
——, and R. Rosset. 1972. *Separation Sci., 7,* 487.
——, M. Caude, and R. Rosset. 1972. *Separation Sci., 7,* 465.
Lindner, R., and T. Bergdahl. 1960. *Z. Elektrochem., 64,* 919.
Rosset, R. 1964. *Bull. Soc. Chim. France, 1964,* 607.
Spedding, F. H. 1963. *J. Chim. Phys., 1963,* 89. (The paper is in English.)

12

Reprinted from *J. Chem. Phys.*, **6**, 429–438 (Aug. 1938)

Fractionation of the Lithium and Potassium Isotopes by Chemical Exchange with Zeolites*

THOMAS IVAN TAYLOR AND HAROLD C. UREY
Columbia University, New York, N. Y.

(Received June 14, 1938)

A study has been made of the fractionation of the lithium isotopes by electrolysis and of the lithium, potassium, and nitrogen isotopes by chemical exchange with zeolites. The fractionation factor obtained by a "batch" electrolysis of lithium chloride solution was 1.039. The normal abundance ratio of the lithium and potassium isotopes was determined and found to be 11.71±0.14 for $Li^7 : Li^6$ and 14.10±0.09 for $K^{39} : K^{41}$. The fractionation factor for the lithium isotopes in exchange with sodium zeolite was found to be 1.022. A partial separation of the lithium, potassium and nitrogen isotopes has been affected using long columns filled with zeolites. Both an increase and a decrease of about 25 percent from the normal abundance ratio has been obtained for the lithium isotopes and a change of about 10 percent for the potassium and nitrogen isotopes. Some of the factors involved in the mechanism of the process are discussed.

INTRODUCTION

RECOURSE may be had to a number of different methods in attempts to separate the isotopes of an element. One may choose a method more or less directly dependent on physical properties related to the masses of the isotopes such as diffusion, distillation, or deflection in an electric and magnetic field using large capacity mass spectrographs. Electrolysis or reaction velocity may also lead to some fraction-

* Publication assisted by Ernest Kempton Adams Fund for Physical Research of Columbia University.

ation. Chemical exchange methods[1] have been used with success by Urey and his co-workers for the fractionation of a number of isotopes as, for example, oxygen,[2] carbon,[3] nitrogen,[4,5] and more recently sulfur,[6] lithium and potassium.

The use of large capacity mass spectographs has led to the separation of only minute quantities of the lithium[7] and potassium[8] isotopes. Lewis and Macdonald[9] have obtained a marked increase in the concentration of Li^6 by countercurrent extraction of a lithium amalgam with an ethyl alcohol solution of lithium chloride. They estimated a lower limit of 1.025 for the fractionation factor of the process.

Encouraged by the successful separation of the hydrogen isotopes by electrolysis, a number of investigators started work on the separation of the lithium isotopes using a flowing mercury cathode. Of these may be cited the work of Eucken and Bratzler[10] on lithium sulphate; Taylor and Urey[11] on lithium hydroxide; and Holleck[12] and Hutchison[13] on lithium chloride. The fractionation factors obtained were respectively 1.07 (upper limit), 1.02, 1.07 and 1.062.

Heavesy and Logstrup[14] have used a nonequilibrium distillation method and achieved some fractionation of the potassium isotopes. Nier and Hustruled[15] also obtained a small fractionation by a similar method during some studies on the radioactivity of potassium. Brewer[16] has found that small changes in the isotope ratio occur in certain natural processes.

Some further experiments on the fractionation factor for the electrolysis of the lithium isotopes have been made by the present authors. A 38 percent lithium chloride solution was electrolyzed in the same cell with the same current density as that used in the electrolysis of lithium hydroxide except that a graphite anode was used and the average temperature was 36°C. The mercury was stirred rapidly and the gases evolved at the anode stirred the solution. After electrolysis of 450 g of lithium chloride to 5 g, the ratio of the isotopes in the solution remaining in the cell was changed to 13.8 which gives a fractionation factor of 1.039. The lower fractionation factor may arise either from the higher current efficiency resulting from stirring the mercury or an effect due to dilution. The other investigators have electrolyzed in several stages taking the first part of the lithium electrolyzed for the next stage such that they were always electrolyzing from a concentrated solution. In the "batch" method as we have used there is a continual decrease in concentration.

In the course of the experiments on the electrolysis of lithium hydroxide, a sample was found in which the ratio of Li^7 to Li^6 was 12.5 instead of the normal ratio 11.6. It appeared that somewhere in the manufacture appreciable fractionation had occurred. Upon tracing the original source of the material, the manufacturer was hesitant in giving any information regarding its manufacture. One of the most general methods used to obtain lithium is to heat the silicate base ore, such as lepidolite, with potassium sulphate.[17] A zeolite type reaction takes place in which potassium is exchanged for lithium. Such replacement may be preferential with respect to one of the isotopes. This led to an investigation of the possibility of fractionation of the isotopes of the alkali elements by the so-called base exchange reactions with zeolites. A preliminary report was made by Taylor and Urey[11] in which small changes in the ratio of the lithium and the nitrogen isotopes from ammonium ion were described. Further experiments were made on potassium isotopes and reported by Taylor, Brewer and Urey.[18] The results have also been

[1] Urey and Greif, J. Am. Chem. Soc. **57**, 321 (1935).
[2] Huffman and Urey, J. Ind. Eng. Chem. **29**, 531 (1937).
[3] Urey, Aten and Keston, J. Chem. Phys. **4**, 622 (1936).
[4] Urey and Aten, Phys. Rev. **50**, 575 (1936).
[5] Urey, Huffman, Thode and Fox, J. Chem. Phys. **5**, 856 (1937).
[6] Thode, Gorham and Urey, J. Chem. Phys. **6**, 296 (1938).
[7] Rumbaugh and Hofstad, Phys. Rev. **50**, 681 (1936).
[8] Smythe and Hemmendinger, Phys. Rev. **51**, 128 (1937).
[9] Lewis and Macdonald, J. Am. Chem. Soc. **58**, 2519 (1936).
[10] Eucken and Bratzler, Zeits. f. physik. Chemie **174**, 269 (1935).
[11] Taylor and Urey, J. Chem. Phys. **5**, 597 (1937).
[12] Holleck, Zeits. f. Elektrochem. **44**, 111 (1938). See also Zeits. f. angew. Chemie **49**, 33 (1936).
[13] Hutchison, Dissertation. Ohio State University (1938).
[14] Heavesy and Logstrup, Zeits. f. anorg. allgem. Chemie **161**, 1 (1928).
[15] Nier and Hustruled. Private communication.
[16] Brewer, J. Am. Chem. Soc. **58**, 365 (1936); ibid. **59**, 869 (1937).
[17] Osborg, *Lithium* (The Electrochemical Society, New York City, 1935).
[18] Taylor, Brewer and Urey, Rochester Meeting Am. Chem. Soc.

confirmed by Brewer and Reed[19] in their laboratory using "greensand." A more complete discussion of the process and the results obtained on lithium and potassium will follow.

Mass Spectrometer for Analysis of Isotopic Abundance Ratios

One of the most difficult problems in studies on the separation of isotopes is the analysis of the results. One may make direct determinations on a mass spectrometer or make determinations of the atomic weight, density, or other properties associated with the mass of the atoms. During the first part of the investigation, methods of analysis were tried which depended upon a difference in the density of compounds containing the isotopic atoms in different ratios. One was an isopiestic density method and another the flotation of lithium fluoride crystals in an organic liquid mixture by variation of temperature and pressure. Both were comparative methods and required a long time for purification so a mass spectrometer of the Dempster type, as modified by Bainbridge[20] and Brewer[16] was built to analyze the samples.

The apparatus was constructed and operated in most respects as described by Brewer.[16] The positive ions produced by heating a platinum disk previously impregnated with the sample were drawn through the slit by applying 270 volts for potassium and 810 volts for lithium. The resolving chamber had a radius of 5 cm and was evacuated to a pressure of 1 to 5×10^{-6} mm Hg. The positive ion current was carried to a DuBridge[21] balanced circuit amplifying system and the amplified current was read from the galvanometer scale. In making a determination the accelerating voltage was held constant and the magnetic field was varied until the galvanometer deflection was a maximum. This current was taken as a measure of the abundance of the isotope. The magnetic field was then changed until another isotope was brought to the collecting plate. If the amplification was linear, the ratio of the galvanometer deflections could be taken as the ratio of the abundances of the isotopes. However, this was not always the case and the potentiometric method described by Cohn and Urey[22] was used to get the exact ratios.

The principal source of error in the determinations was in reading the height of the smaller peak which is from one-tenth to one-fifteenth the larger one (25 to 40 cm). If the zero point drifts owing to unsteadiness of the amplifier, or if the positive ion beam is not constant, as is often the case, then there are further errors involved in reading the deflections. In general the readings are probably not better than 0.3 to 0.4 mm which means a possible error of 2 to 2.5 percent in a single determination. Averages for different filaments of the same material sometimes varied as much as 2 to 3 percent, especially if solid material remained on the disk or a gas was given off during heating. Also if the pressure was not kept sufficiently low the resulting poor resolution decreased the accuracy. However, there was generally no background between the lithium peaks and seldom as much as 0.1 to 0.2 mm between the potassium peaks. In general, it is considered that the ratios reported are not better than ±0.1 to ±0.2 from the true ratio, that is, about 1.5 to 2 percent.

Determinations of the normal ratio of the lithium isotopes made by different investigators[12] show wide variations. While our purpose was not the accurate determination of the abundance ratio, it is of interest to report our results in confirmation of the accepted values. The average value of the ratio $Li^7 : Li^6$ taken from determinations made at different times during the research is 11.71 ± 0.14 which is in fair agreement with those of Bainbridge[20] and Brewer,[23] 11.51 ± 0.31 and 11.60 ± 0.06, respectively. Using 6.940 for the atomic weight of lithium and Bainbridge's[24] values for the masses of the atoms (6.0145 ± 0.0003 and 7.0145 ± 0.0007), a value of 11.84 is obtained calculated on the basis of Aston's[25] new value for He (4.00391). This is in good agreement with the mass-spectrographic measurements.

Potassium is known to have three isotopes with mass numbers 39, 40 and 41. The ratio of K^{39} to

[19] Brewer and Reed, Nitrogen Fixation Laboratory. Private communication.
[20] Bainbridge, J. Frank. Inst. **212**, 3176 (1931).
[21] DuBridge and Brown, Rev. Sci. Inst. **4**, 532 (1933).
[22] Cohn and Urey, J. Am. Chem. Soc. **60**, 679 (1938).
[23] Brewer, Phys. Rev. **47**, 571 (1935).
[24] Bainbridge, Phys. Rev. **46**, 56 (1933).
[25] Aston, Nature **137**, 357 (1936).

K^{41} is given by Nier[26] as 13.96 ± 0.1 and Brewer[27] as 14.25 ± 0.03. Manley[28] using the ratio of the heights of the peaks in nuclear magnetic moment studies obtains the value 13.4 ± 0.5. The isotope of mass 40 is radioactive and is present to the extent of only one part in 8000.[29] The ratio of K^{39} to K^{41} obtained from the average of a number of determinations made during this research is 14.10 ± 0.09 which is in good agreement with the above values.

The Nature of Zeolitic Substances

Many substances, particularly those which tend to form negative ionic micelles in the sense used by McBain,[30] possess the property commonly referred to as base exchange. That is, they exchange a positive ion held by the substance for one in solution. The most widely studied and well-known of these substances are the complex hydrated alumino-silicates known as zeolites. A review of the voluminous literature on these substances is given by Randall and Cann,[31] Wiegner[32] and Jenny.[33] Many naturally occurring zeolites such as the glauconites (greensands), the ultramarines, sodalite, nephilite, and the clays such as bentonite, all show a greater or less tendency to exchange their bases depending on their composition and structure. Artificial zeolites have been made by many different methods as, for example, the reaction of a solution of sodium aluminate and sodium silicate in such proportions that the product has an approximate composition represented by the empirical formula $Na_2O \cdot Al_2O_3 \cdot 6SiO_2 \cdot xH_2O$. The sodium is replaceable by other positive ions, a divalent ion replacing two monovalent ones.

It has been known for a long time that the alkali and alkaline earth elements exhibit specific ion effects and may be arranged in a lyotrope or Hofmeister series according to their effectiveness in replacing one another from the zeolite. For an artificial zeolite, the ions of the alkali and alkaline earth elements give the following series:

$$Mg^{++} < Li^{+} < Ca^{++} < Na^{+} < Ba^{++} < NH_4^{+} < K^{+} < Rb^{+} < Cs^{+} < H^{+}.$$

That is, lithium is the least strongly held while cesium is the most strongly held of the alkali elements. The order of the alkali elements is that of the decreasing size of the hydrated ion.[33] A change in the conditions affecting the solvation of the ions such as a change in the nature of the solvent[32] or temperature[34] will change the behavior of the ions. Such change in behavior with change of conditions may prove important in the fractionation of isotopes.

If one shakes a quantity of zeolite with a solution of another positive ion, equilibrium is established in ten to twenty minutes and partial replacement takes place according to the reaction

$$A^{+} + BZ \leftrightarrows B^{+} + AZ.$$

Addition of more ion A^{+} replaces more B^{+} approximately in conformity to the mass law up to a certain maximum, beyond which only a small increase in exchange will take place with increased addition of A^{+}. The amount of exchange is essentially independent of dilution[35] but depends principally on the ratio of the total quantities of the two electrolytes and zeolites present. Concentrations may be used only when reference is made to a fixed total volume. Specific properties of the zeolites and ions make it impossible to give equations applicable to all systems. Of the many proposed, one of the most generally applicable equations used to express the amount of base exchange at equilibrium is one of the type given by Rothmund and Kornfeld:[36]

$$K \doteq \frac{AZ}{BZ} \cdot \left[\frac{(B^{+})}{(A^{+})}\right]^{\beta}, \tag{1}$$

where K is an equilibrium constant, (A^{+}) and (B^{+}) are the concentrations of the two ions in the given volume of solution, and AZ and BZ are the equivalents of the two forms of zeolites present. The constant β varies from 0.5 to 1.0 for different

26 Nier, Phys. Rev. **48**, 283 (1935).
27 Brewer, Phys. Rev. **48**, 640 (1935).
28 Manley, Phys. Rev. **48**, 921 (1936).
29 Nier, Phys. Rev. **48**, 283 (1935). Brewer, Phys. Rev. **48**, 640 (1935).
30 McBain, Trans. Faraday Soc. **9**, 99 (1913); Kolloid Zeits. **40**, 1 (1926).
31 Randall and Cann, Chem. Rev. **7**, 369 (1930).
32 Wiegner, J. Soc. Chem. Ind. **50**, 65T (1931).
33 Jenny, J. Phys. Chem. **36**, 2217 (1931).
34 Jaeger, Trans. Faraday Soc. **25**, 320 (1939).
35 Wiegner and Muller, Zeits. Pflanz Dung. **14A**, 321 (1929).
36 Rothmund and Kornfeld, Zeits. f. anorg. allgem. Chemie **103**, 129 (1918).

systems and expresses the fact that the ratio of the two forms of zeolite are not necessarily the same as the ratio of the ions in solution.

According to the scheme used by Austerweil[37] the zeolite is considered to be a solid solvent nonmiscible with the solution of the cation which exchanges. The laws of extraction should apply and there should be a distribution of the two cations depending on the "solubility" or concentration ratio of the two zeolites. To extract a total fraction E of a cation from solution there would be required n extractions as given by $(1-\alpha')^n = 1-E$ where α' is the exchange or distribution coefficient.

Starting with the properties and behavior of zeolites as outlined above and the assumption that one isotope is taken up more readily than the other, a method must be devised to multiply the single stage effect many times. Any of the batch or batch-counter current methods of extraction[38] may be followed. A continuous counter current method as used in the separation of the nitrogen isotopes[5] by exchange of NH_3 and NH_4^+ would be more effective but the difficulties of counter current motion are much greater for a solid and liquid than for a liquid and gas. The use of a long column packed with zeolite seems to be the most favorable scheme. A number of different procedures may be followed. A solution of lithium may be passed through a column containing a sodium zeolite. All the way down the column one isotope should be held back more than the other and the first lithium coming through should have a changed isotope ratio. This first sample will be called the "leading sample." After the column is converted to a lithium zeolite a solution of sodium chloride may be passed through the column. One isotope is replaced more readily than the other by the sodium and the last lithium coming through the column should have the isotope ratio changed in the opposite direction to that of the leading sample. This last sample will be called the "tailing sample."

Another procedure is one similar to that used in the separation of closely related organic substances by the recently developed methods of chromatographic analysis.[39, 40] According to this method a tube is filled with a suitable absorbing substance. The mixture to be separated is added at the top of the column where all the substances are absorbed. A solvent is passed through the column and the substances most strongly held remain in a layer near the top while those less strongly absorbed form successive layers down the column. By continued washing, the substances may be made to descend one by one through the bottom of the column. This method has been studied by Schwab and Jockers[41] for the separation of inorganic cations on adsorbents like aluminum oxide. Analogous procedures can be used with zeolites as adsorbents and solutions of ions as the solvent.

A counter current system similar to that of a counter flow column with total reflux at both ends can be simulated as follows. The column can be filled with a zeolite having an easily replaceable cation compared to the isotopic ions. A quantity of a salt containing the isotopic ions is then added at the top of the column. Following this is a solution containing a cation very strongly held to the zeolite such that practically all the isotopic ions are replaced and descend down the column. The net result will be that effectively the quantity of zeolite containing the isotopic ions moves down the column with reflux at the leading and tailing end.

Experiments on the Fractionation of the Lithium Isotopes

Determination of the fractionation factor

In order to determine the possible effectiveness of the base exchange reactions in separation of the isotopes, a determination of the fractionation factor was made. It is given by the ratio $Li^6Z : Li^7Z / Li^{6+} : Li^{7+} = \alpha$, that is, the distribution of the isotopes between the two phases. If α were large enough, a single stage equilibrium could be used for its determination. However, it is small in most isotopic exchange reactions and thus the single stage must be multiplied many times. The method chosen was to extract a large quantity of

[37] Austerweil, Bull. Soc. Chem. France **51**, 732 (1932).
[38] Hunter and Nash, J. Ind. Eng. Chem. **27**, 836 (1935).
[39] Zechmeister and Cholnoky, *The Chromatographic Adsorption Method* (Springer, Austria, 1937).
[40] Lederer, Chemie and Industrie **33**, 1072 (1932).
[41] Schwab and Jockers, Angew. Chemie **50**, 546 (1937).

lithium chloride solution with successive small quantities of sodium zeolite so that α is approximately given by the Rayleigh distillation formula,

$$\left(\frac{1-N_0}{1-N}\right)^{1/(\alpha-1)}\left(\frac{N}{N_0}\right)^{\alpha/(\alpha-1)}=\frac{W_0}{W},$$

where N_0 is the mole fraction of the heavy isotope in the initial amount of material W_0, and N the mole fraction in the final amount of material W. A batch containing 300 g of lithium chloride in a 20 percent solution was extracted with a large number of successive 30 g portions of sodium zeolites. The zeolite was a synthetic one called "Decalso" and kindly furnished by the Permutit Company for this research. It had a calculated exchange capacity of about 0.32 equivalents per 100 grams and was graded by Tyler screens from 40 to 60 mesh. After each addition the mixture was shaken vigorously for twenty minutes before filtration. At various intervals the sodium chloride resulting from the replacement of the sodium ion from the zeolite was separated from the remaining lithium. When the amount of lithium had been reduced by a factor of 70, the ratio of the isotopes in the remaining solution was determined. These experiments were made before our mass spectrometer was constructed and the samples were kindly analyzed by A. Keith Brewer and later checked. The ratio of $Li^7 : Li^6$ had changed from 11.6 to 12.7. Substitution in the Rayleigh formula gives 1.022 for the fractionation factor. It is difficult to estimate the accuracy or exact meaning of this value because of the variable concentration of lithium and sodium ions, and because the Rayleigh formula does not apply very closely towards the end of the extraction where the quantities of zeolite added were comparable to the amount of lithium remaining. However, the value does give us a fair approximation to what one can expect in a column since somewhat similar conditions occur.

The reverse procedure was also tried in which 2 kg of lithium zeolite were extracted with 36 successive liters of 2 percent chloride solution. The lithium extracted by the last batches of solution showed a ratio of the isotopes of 10.7 which gives a fractionation factor of 1.023 calculated as above.

Experiments on zeolite columns

As pointed out previously, one of the most convenient ways of multiplying the single stage effect is the use of a column of zeolite. A number of preliminary experiments were made to determine if possible the best procedure and conditions for a fractionation. It was not known, for example, whether dilute or concentrated solutions would be more advantageous. One form of the zeolite may be more effective than another. Thus, the use of a more stable zeolite like potassium or barium in place of a sodium zeolite may possibly be better since this would afford more competition for the two isotopes. The possible significance of the rate of diffusion or reaction during exchange was not known. The operation of the column would be different if the rate of extraction or adsorption rather than some equilibrium process were the controlling mechanism. The answers to all these questions have not yet been determined, but the following experiments do give us some ideas as to the nature of the fractionation process.

The first column constructed was a three-quarter inch stainless steel pipe thirty-five feet long packed with about 2 kg of sodium zeolite. The second column was made of a one and one-quarter inch hard rubber pipe thirty feet long containing about 4 kg of sodium zeolite. A third column of one and one-quarter inch hard rubber pipe one hundred feet long was constructed in one of the stairwells of the Chandler Chemical Laboratory. This was filled with about 13 kg of sodium zeolite furnished by the Permutit Company. These columns will be referred to as A, B, and C, respectively. In all cases care had to be taken to make sure that no air bubbles or pockets were left in the column. The most convenient way to fill the column was to fill each ten-foot section as it was added. The section was filled with water and the wet zeolite poured in, the water displaced being drawn off by suction. Tees fitted with one-quarter inch outlets were placed between each section so that samples could be taken. The velocity of the liquid flow was regulated by pinch clamps on heavy pressure tubing at the bottom. The solutions were fed by

TABLE I. *Result on fractionation of lithium isotopes.*

Experiment	Column	Form of Zeolite	Solution and Procedure	Sample of Li through Column	Ratio Li^7/Li^6 Normal Ratio 11.7
1	*A*	NaZ	3% LiCl passed through	Leading	13.3±0.3
2	*A*	NaZ	1% LiCl passed through	Leading	12.5
3	*A*	NaZ	1% LiCl passed through	Leading	12.3
4	*B*	NaZ	15 g LiCl washed down with 0.5% NaCl	Leading Tailing	12.6 8.9
5	*C*	NaZ	15 g LiCl washed down with 3.0% NaCl	Leading Tailing	14.1 8.8

inverting a large bottle over the top of the column. When the liquid ran below the level of the outlet tube in the bottle, air entered the bottle and more liquid ran out, thus maintaining a fairly constant level.

The samples used for analysis of the isotope ratio were evaporated just to dryness and extracted with amyl alcohol. A little P_2O_5 was added to the lithium chloride and put on a platinum disk in the form of a paste. After heating to redness on a tungsten filament for ten to fifteen minutes under vacuum, the disk was cleaned and spot welded to the filament of the mass spectrometer and the ratio of the isotopes determined as previously described. The results reported are the average taken from eight to fifteen determinations.

The results for a number of experiments with lithium are given in Table I.

In experiments 1, 2, and 3 the column was initially full of distilled water and sodium zeolite. The lithium chloride was allowed to run in at the top at a rate of about 2 to 3 cc per minute. Tests were made at intervals on a spectroscope to determine when the first lithium came through the bottom of the column. The first lithium through was taken as the leading sample and was analyzed on the mass spectrometer as described above. About one-half the replaceable sodium in the column was exchanged before the first lithium appeared. Experiments 2 and 3 were made to test the effect of concentration, but the very first lithium coming through was missed in both of these and therefore the results do not necessarily show the influence of dilution. A receiver system was used thereafter which allowed the effluent to successively fill 200 cc bottles. Between each experiment the zeolite had to be completely regenerated by passing sodium chloride solution through the column.

Experiments 4 and 5 were carried out according to the methods of chromatographic analysis. The columns were initially filled with sodium zeolite and with the same solution which was to be used for washing down the lithium chloride. The charge of lithium chloride was added at the top of the column in a fairly concentrated solution (10 percent) and the sodium chloride solution was allowed to run in at the rate of about 2–3 cc per minute. The first lithium appearing was taken as the leading sample and the last to come through as the tailing sample. In experiment 4 about 70 g of sodium chloride in a 0.5 percent solution were required before the first lithium came through the column. Experiment 5 was made with the 100 foot column during an experiment on potassium. We had expected much more change than was obtained. The 3 percent sodium chloride washed the lithium through the column rather rapidly. A more dilute solution would probably have given better results. Further experiments are being made to test this point. All of the solution coming through the column was saved and samples were analyzed to determine the isotope ratio and the total amount of lithium coming through up to the time of the sample. The amount of lithium was determined by the spectroscopic method.[42] While this method of analysis does not give results to better than 5 percent to 10 percent, this accuracy is sufficient to show the variation in isotope ratio as the lithium comes through the column. Such a curve will give us an estimate of the quantities of changed material which can be expected by this procedure. A curve for experiment 5 is given in Fig. 1.

In all the experiments on lithium, the leading sample had a lower concentration of Li^6 while the tailing sample had a higher concentration. This means that Li^{6+} is preferentially taken up by the zeolite and is not so readily removed as Li^{7+}.

[42] Skinner and Collins, U. S. Dept. of Agri. Bull. Chem. 153.

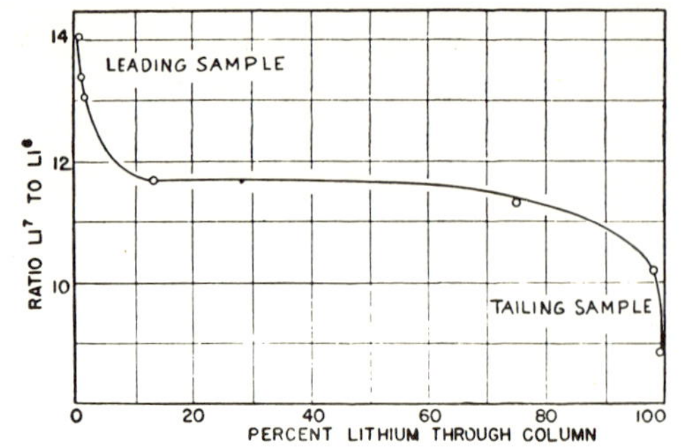

FIG. 1. Distribution of lithium isotopes coming through column in experiment 5.

Experiments on the Fractionation of the Potassium Isotopes

The same general procedures as those used for the lithium isotopes were tried for potassium and for the nitrogen isotopes in NH_4^+. A fractionation of the potassium isotopes with sodium zeolite was found to be more difficult especially in regards to the leading sample. This may possibly be due to the great difference in the binding of the lithium and potassium to the zeolite. Most of the preliminary experiments were made to see whether or not a change could be made on the leading sample.

The solution coming through the column was tested for potassium with sodium cobaltinitrite. When its presence was shown by this test, a quantity of the liquid was evaporated and the potassium precipitated as the cobaltinitrite. This was, decomposed by heating, and after filtering off the cobalt oxide, the sodium and potassium nitrites were converted to the chlorides. A little P_2O_5 was added and the ratio of the isotopes determined as described for lithium. The presence of sodium did not interfere with the analysis.

Table II shows the results of a number of the experiments.

In experiment 1 the 1 percent KCl solution was passed through the column of sodium zeolite at the rate of about 2 to 3 cc per minute. About three-fourths of the calculated replaceable sodium was exchanged before potassium came through in sufficient quantities to give a test with cobaltinitrite. No change was found in the leading sample, but after washing the potassium from the column with about forty liters of 5 percent sodium chloride solution, the ratio of K^{39} to K^{41} was found to be 13.4. For experiment 2, the column was converted to calcium zeolite, filled with water, and a 3 percent potassium chloride solution started through as before. When sufficient potassium had come through to give a test with cobaltinitrite the ratio was 14.2 showing no significant change. However, about five liters of the first solution through the column was saved and later evaporated to recover any potassium. The small amount obtained (about 20 mg of $K_2NaCo(NO_2)_6$) was analyzed and found to have increased in ratio to 14.8.

Experiment 3, using 0.5 percent KCl instead of 3 percent KCl, showed no marked advantage in using more dilute solutions. However, under conditions where an appreciable fractionation is obtained, the more dilute solutions may yet prove to be more effective. The equilibrium constants as given by Eq. (1), or the constants from other equations,[33] are generally about twice as large for replacement of ions from a zeolite by K^+ as for Na^+. It was thought that a mixture containing two equivalents of sodium chloride to one of potassium chloride when passed down the column may give a better fractionation. Accordingly for experiment 4, the column was filled with sodium zeolite and 0.30 N NaCl solution. A mixture of 0.15 N KCl and 0.30 N NaCl was then started through. The leading sample again showed no significant change in isotope ratio. Experiments made in this manner using sodium or calcium zeolites apparently will not give large changes in the ratio of the potassium isotopes in the leading sample.

Experiment 5, carried out according to the methods of chromatographic analysis, gave no significant change in the leading sample, but the tailing sample was changed to a ratio of 12.9. The same experiment was tried on the 100 foot column but the potassium spread throughout the whole column so much during the passage of the 3 percent sodium chloride solution that the experiment was finally stopped when the ratio of the potassium isotopes in a sample taken from the top section of the column was 13.2. The use of larger quantities of potassium chloride and higher concentration of sodium chloride may have been better. Further experiments will be made later. The experiment using a column of sodium zeolite and a solution of barium chloride to follow the potassium down shows some promise since a change was observed in both the leading and tailing samples. The use of ammonium zeolite and ammonium ion which is similar to potassium in replacing power may also prove more effective than the sodium zeolite.

In all those experiments in which a change in the ratio of the potassium isotopes was obtained, K^{41} was decreased in the leading sample and increased in the tailing sample. It seems, therefore, that the heavier isotope is taken up more readily by the zeolite and is more difficult to replace. This is the opposite of the effect observed for lithium.

Experiments on the exchange of ammonium ion with a sodium zeolite and a carbonacious zeolite given to us by the Permutit Company showed a small change in the isotope ratio in the same direction as observed for potassium, i.e. the heavier isotope was more readily taken up by the zeolites. The analyses were made on a mass spectrometer by Marvin Fox in this laboratory.

TABLE II. *Experiments on fractionation of potassium and nitrogen isotopes.*

Experiment	Column	Form of Zeolite	Solution Passed through Column	Sample of K through Column	Analysis $K^{39}:K^{41}$ 14.10
Potassium 1	*A*	NaZ	1% KCl passed through	Leading	14.1 ±0.2
1*a*	*A*	KZ	5% NaCl passed through	Tailing	13.4
2	*A*	CaZ_2	3% KCl passed through	Leading (*a*) Leading (*b*)	14.8 14.2
3	*A*	CaZ_2	0.5% KCl passed through	Leading	14.2
4	*A*	NaZ	0.15 *N* KCl +0.30*N* NaCl	Leading	14.1
5	*B*	NaZ	100 g KCl washed down with 5% NaCl	Leading Tailing	14.3 12.9
6	*C*	NaZ	100 g KCl washed down with 3% then 5% KCl	—	13.20
7	*B*	NaZ	15 g KCl washed down with 1 *M* $BaCl_2$	Leading Tailing	14.6 13.5
Nitrogen 1	(NH^+_4) *A*	NaZ	2% NH_4Cl passed through	Normal Leading	$N^{14}N^{14}:N^{14}N^{15}$ 124.3 136.8
2	8 ft.	HZc	1% NH_4Cl passed through	Normal Leading	127.4 130.8

Discussion

The fractionation of the isotopes by this method depends upon a difference in binding of the two isotopic ions with water and with the zeolite. The differences are due to a difference in the zero point energies and the vibrational and rotational frequencies of the two isotopic combinations in much the same way as those calculated for gaseous systems by Urey and Greif[1] and demonstrated experimentally for the liquid-gas chemical exchange systems previously referred to. Solvation of the ions and the specific nature of the zeolite are, therefore, important considerations. Any change in the system affecting the nature of the solvent medium will affect the binding of the two isotopic ions, and if this is not in the same ratio for the solvation and binding of the ions to the zeolite, there will be a change in the effectiveness of the fractionation. We cannot predict whether it would improve or decrease the desired fractionation, nor can calculations be made for such a complicated system. Thus far we have studied the fractionation with only one or two forms of the same zeolite and have not markedly changed the nature of the solvent medium. Other types of zeolites such as the bentonite clays, the natural "greensands" or the newly developed organic exchange resins are to be studied. Addition of alcohol or other solvent may also influence the fractionation obtained.

If the process by which the isotopes are fractionated is an equilibrium one, we may write the following equations:

$$Li^{6+} + NaZ \leftrightharpoons Li^6Z + Na^+$$

$$Li^{7+} + NaZ \leftrightharpoons Li^7Z + Na^+$$

$$Li^{6+} + Li^7Z \leftrightharpoons Li^{7+} + Li^6Z \quad K = 1.022.$$

Similarly

$$K^{39+} + K^{41}Z \leftrightharpoons K^{41+} + K^{39}Z \qquad K < 1$$

$$N^{14}H_4^+ + N^{15}H_4Z \leftrightharpoons N^{15}H_4^+ + N^{14}H_4Z \quad K < 1.$$

The lighter lithium isotope is preferentially taken up by the zeolite while the reverse is true for the potassium and ammonium ions. The process responsible for the fractionation is essentially an equilibrium one rather than one due to differences in the rate of diffusion or reaction, since otherwise the lighter isotopes would all diffuse faster resulting in changes of the isotope ratio in the same direction. In those experiments in which lithium chloride solutions were passed through the column filled with sodium zeolite, differences in the rate of diffusion of the two isotopic ions would tend to decrease the concentration of Li^6 and therefore increase the fractionation in the leading sample. If diffusion or rate of reaction were the important process, the tailing sample produced by passing sodium chloride solution through a column of lithium zeolite should also give a solution containing a higher concentration of Li^7 since Li^6 should diffuse out of the zeolite faster. This is not the case. In those experiments made according to the procedures of chromatographic analysis, the decrease of Li^6 in the leading sample is approximately the same as the increase of Li^6 in the tailing sample.

The experiments are not yet extensive enough to draw definite conclusions as to which method is the most effective procedure. The procedure according to the methods of chromatographic analysis will give samples with both higher and lower amounts of Li^6. However, if a change is desired in one direction, it seems that the first procedure should be better. That is, to obtain a higher concentration of Li^6, the column should be converted to lithium zeolite and sodium chloride solution slowly passed in at the top so as to displace the lithium from the zeolite. The sodium ion replaces Li^{7+} more readily than the Li^{6+} such that the last lithium replaced from the column has a higher concentration of Li^{6+}. Since there is a distribution of the lithium between the two phases following the laws previously described, one may obtain the number of stages by methods similar to those used in extraction problems. By assuming the number of effective isotopic exchanges is the same as the number of stages thus determined, it is possible to calculate the separation which can be expected once the fractionation factor is known.

Acknowledgments and thanks are due Dr. A. Keith Brewer for his suggestions and advice in construction of the mass spectrometer and for some of the analyses during the first part of the investigation. Thanks are also due Dr. John G. Dean and the Permutit Company for the zeolites used in the research.

Summary

1. The methods used in fractionation of the lithium and potassium isotopes have been discussed and some additional results on the electrolysis of lithium chloride reported.

2. The properties of zeolitic substances significant in studies on the separation of isotopes by chemical exchange are described.

3. The natural abundance ratio of the lithium and potassium isotopes have been determined by analysis on the mass spectrometer and compared to the values reported in the literature. The average value obtained for the ratio $Li^7 : Li^6$ was 11.71 ± 0.14 and for $K^{39} : K^{41}$ was $14.10 \pm .09$.

4. The fractionation factor for the lithium isotopes in the exchange with sodium zeolite was found to be 1.022.

5. A fractionation of the lithium, potassium and nitrogen isotopes using long columns filled with zeolites has been effected. Both an increase and decrease of about 25 percent from the normal abundance ratio of the lithium isotopes has been obtained. A change of about 10 percent has been obtained for the potassium and nitrogen isotopes.

6. Some of the factors involved in the mechanism of the process are discussed.

13

Reprinted from *J. Amer. Chem. Soc.*, 77, 6125–6132 (Dec. 1955)

A Laboratory Method for Separating Nitrogen Isotopes by Ion Exchange[1]

BY F. H. SPEDDING, J. E. POWELL AND H. J. SVEC

RECEIVED JUNE 24, 1955

Various aspects of the separation of N^{15} and N^{14} isotopes from each other by an ion-exchange process have been discussed. A number of experiments have been performed to establish the feasibility of using the process and to check the countercurrent theory of separation which has been developed. The separation factor for the exchange of N^{15} and N^{14} between a solution of dilute ammonium hydroxide and the ammonium form of Dowex 50-X12 was measured and found to be 1.0257 ± 0.0002. It was shown that an adsorbed band of ammonium ion must be eluted about 39 times its length in order to achieve separation of the isotopes. A method of continuous operation using a series of ion-exchange beds has been described and factors which influence the HETP at a steady state have been investigated.

I. Theoretical Considerations

A. Introduction.—The remarkable success achieved in separating adjacent rare earths in high purity and good yields by ion-exchange methods suggested that isotopes might also be separated by a similar procedure. It was found[2,3] that, if a solution of mixed rare earth ions was adsorbed on the top of a column of sulfonated styrene–divinylbenzene copolymer in the hydrogen state, little separation occurred in the adsorption step. However, if the band was eluted with an eluent of citric acid, buffered with ammonia in an appropriate *p*H range, the individual rare earth ions separated to form sharp bands which traveled head to tail down the column. The individual rare earths could then be successively eluted off the resin bed in a pure state, with only a small amount of overlap between bands.

In order to obtain this type of elution,[4] the eluent had to contain one or more anions which complexed the rare earth ions strongly enough to compete with the resin in removing the triply charged rare earth cations from solution. It was found necessary that distinct chemical reactions occur at the front and rear edges of the adsorbed band so that the rare earths would be sharply confined between two boundaries. Under these conditions, the original rare earth band would spread out to a fixed equilibrium length which was determined solely by the composition of the eluent. Once this equilibrium was attained, the length of the band remained constant regardless of how far the band was eluted down a resin bed. As a consequence, all the boundaries of the individual rare earth bands moved at the same rate. It was also ascertained that, if the concentration of any ion in the eluate was plotted *versus* the volume of effluent solution while an individual rare earth band was passing off the column, a flat elution curve was obtained showing that the concentrations of all ions remained constant. When subsequent individual rare earth bands passed off the column the concentrations of all ionic species shifted slightly but remained at characteristic fixed values while each individual rare earth moved from the bed. The automatic adjustment, which arises primarily from the stability constants of the various complexed rare earth species, results in a self-sharpening of the bands.

It was hoped that the slight differences in the exchange constants of various isotopes of an element might also cause banding of the isotopes within a chemically constrained band, if the band was eluted under proper conditions.[5] While the chemical constraints at the ends of a band cause the over-all band to attain equilibrium in a short distance of travel, the much smaller exchange constants of the isotopes would require that the band travel a much greater distance before complete isotopic equilibrium would be approached.

Since the isotopic exchange constants for light elements are considerably larger than for heavier elements, the first experiments were carried out using nitrogen in the form of ammonium hydroxide in an attempt to separate N^{15} from N^{14}.

B. Chemical Conditions Required to Obtain a Sharp Adsorbed Ammonium Band.—In order to develop and maintain a sharp adsorbed ammonium band it is necessary to have a uniformly packed resin bed. Under these conditions, as the solution percolates down the column, a horizontal boundary between two solutions can be maintained. In practice, if sufficient care is taken in preparing the resin beds, tilting and channeling can be kept less than a few millimeters for columns 2 to 6 inches in diameter. When one ion replaces another on the resin bed there is always some change in the volume of the resin. In small diameter tubes, bridging frequently occurs during expansion and contraction of the bed and results in uneven packing of the resin particles. This leads frequently to serious tilting of the band.

To separate nitrogen isotopes the resin beds are

(1) Work was performed in the Ames Laboratory of the Atomic Energy Commission.

(2) (a) F. H. Spedding, E. I. Fulmer, J. E. Powell and T. A. Butler, THIS JOURNAL, **72**, 2354 (1950); (b) F. H. Spedding, E. I. Fulmer, J. E. Powell, T. A. Butler and I. S. Yaffe, *ibid.*, **73**, 4840 (1951).

(3) F. H. Spedding and J. E. Powell, *Chem. Engr. Progr., Symposium Series*, **50**, No. 14, 7 (1954).

(4) F. H. Spedding and J. E. Powell, THIS JOURNAL, **76**, 2545, 2550 (1954).

(5) F. H. Spedding, J. E. Powell and H. J. Svec, *ibid.*, **77**, 1393 (1955).

first converted to the hydrogen cycle by passing a dilute mineral acid through the columns until the beds are saturated. The beds are then rinsed with water and a dilute solution of ammonium hydroxide is added until a band of ammonium ion of the desired length is obtained. When Dowex-50 resin is in the ammonium state it exhibits a somewhat different shade of brown than it does in the hydrogen cycle or the sodium cycle so that the boundaries of the ammonium band can be seen distinctly.

If a boundary is to remain sharp, a chemical reaction which goes virtually to completion must take place at the boundary. If ammonium ion from a salt solution such as ammonium chloride reacts with hydrogen resin the boundary will not be very sharp, because the reaction

$$H^+_R + NH^+_{4_S} \rightleftharpoons NH^+_{4_R} + H^+_S \quad (1)$$

has an apparent equilibrium constant

$$K' = K \frac{\gamma_{(NH^+_{4_S})} \times f_{(H^+_{4_R})}}{\gamma_{(H^+_S)} \times f_{(NH^+_{4_R})}} = \frac{C_{(NH^+_{4_R})} \times C_{(H^+_S)}}{C_{(NH^+_{4_S})} \times C_{(H^+_R)}} \quad (2)$$

which does not differ greatly from one. While the forward motion of the solution sweeps the H^+_S down-stream and tends to shift the equilibrium to the right, this alone is not sufficient to maintain a sharp boundary. On the other hand, when ammonium hydroxide is used the over-all reaction becomes

$$H^+_R + NH_4OH_S \rightleftharpoons NH^+_{4_R} + H_2O \quad (3)$$

For this reaction

$$K' = \frac{C_{(NH^+_{4_R})}}{C_{(H^+_R)} \times C_{(NH_4OH)}} \quad (4)$$

has a value on the order of 10^9. Here the equilibrium is far to the right, resulting in an extremely sharp boundary. It might be mentioned that considerable heat is liberated at the boundary by the above reaction and this limits the concentration of NH_4OH which can be used conveniently.

The adsorbed ammonium band is driven down the column with an eluting solution of sodium hydroxide. Here the reaction is

$$NH^+_{4_R} + Na^+_S + OH^-_S \rightleftharpoons NH_4OH_S + Na^+_R \quad (5)$$

and the apparent equilibrium constant

$$K' = \frac{C_{(NH_4OH_S)} \times C_{(Na^+_R)}}{C_{(NH^+_{4_R})} \times C_{(Na^+_S)} \times C_{(OH^-_S)}} \quad (6)$$

has a value of about 10^5. This again produces a sharp boundary.

The adsorbed ammonium band is, accordingly, sharply confined between these two boundaries and is in equilibrium throughout its length with a solution of ammonium hydroxide. The concentration of ammonium hydroxide solution in contact with the ammonium band is constant throughout and is determined by the concentration of the sodium hydroxide used as the eluent.

The length of the adsorbed ammonium band is primarily determined by the amount of ammonia in the column and the capacity of the resin. The band length will depend to a certain extent upon the concentration of the ammonium hydroxide in the resin pores, which is fixed by the concentration of the eluent. Once the concentration of the sodium hydroxide has been chosen, the length of the ammonium band remains constant as it is eluted down the resin bed.

C. Isotopic Equilibrium.—If the ammonium hydroxide solution in the pores of the bed is maintained for a sufficient time in contact with a particular section of resin bed, the system will attain isotopic equilibrium according to the reaction

$$N^{14}H^+_{4_R} + N^{15}H_4OH_S \rightleftharpoons N^{15}H^+_{4_R} + N^{14}H_4OH_S \quad (7)$$

The equilibrium constant is

$$K = K' = \frac{[N^{15}H^+_{4_R}][N^{14}H_4OH_S]}{[N^{14}H^+_{4_R}][N^{15}H_4OH_S]} = \frac{[N^{15}_R][N^{14}_S]}{[N^{14}_R][N^{15}_S]} = \frac{R_R}{R_S} = 1.0257(8)^6$$

where R_R and R_S are the N^{15}–N^{14} ratios in the resin and aqueous phases, respectively.

If the length of the ammonium band remains constant as the band is eluted down the resin bed, each time an equivalent of sodium ion is deposited at the rear edge of the band it displaces an equivalent of ammonium ion. The displaced ammonium ion comes to chemical equilibrium within the system very rapidly and isotopic equilibrium tends to be approached as the solution passes over the resin. When the ammonia in solution reaches the front edge of the ammonium band it is redeposited as ammonium ion on the resin bed. At all times, the ratio of N^{15} to N^{14} on the resin is greater than it is in the solution contacting it. From equation 7

$$R_R = KR_S = (1 + \epsilon)R_S = 1.0257R_S \quad (9)$$

This presupposes that the solution remains in contact with an increment of resin of a given composition for sufficient time for equilibrium to be attained. In the case of the isotopic exchange for ammonia, the reaction is sufficiently rapid that, within our experimental error under the condition we have operated, equilibrium is attained at the critical points in the system even when the band is moving at the rate of 30 inches per hour down a bed. The band becomes richer in N^{15} at the rear edge and richer in N^{14} at the front edge as the elution proceeds.

Figure 1 shows a plot of how the isotopic ratio of the nitrogen varies across the band with different distances of travel. It can be seen that the maximum isotope enrichment or depletion occurs at the ends of the band. In the plateau region, the ratio of the isotopes in the resin phase and solution phase are not changing, although they differ from each other by the amount ϵR_S, where $\epsilon = (K - 1)$. The N^{15}–N^{14} ratio on the resin, R_R, is equal to KR_S from equation 9; therefore

$$KR_S - R_S = (K - 1)R_S = \epsilon R_S \quad (10)$$

Since the solution is flowing past the resin in the plateau region carrying ammonia richer in N^{14} and leaner in N^{15} than the resin it contacts, there is a net transfer of N^{14} toward the leading edge of the band. If the plateau region is sufficiently long that isotopic equilibrium is attained within this re-

(6) See Part II, Section B.

gion, the net transfer of N^{14} per equivalent of ammonia passed over the plateau and redeposited is the same irrespective of the degree of equilibrium reached in the rest of band. It should be borne in mind that the rate at which ammonia is picked up and redeposited is not exactly the same as the rate at which the sodium ion is introduced into the column in equivalents, since some of the sodium ions must remain in the liquid which fills the pore spaces left behind as the ammonium band advances. The total amount of ammonia in the pore spaces of the ammonium band does not change, as this solution simply moves along at the same rate as the band advances. The percentage of the total ammonia in the pores is small, being on the order of a few per cent. when 0.6 *N* NaOH is used as the eluent and becomes increasingly smaller as more dilute NaOH solutions are used. It is apparent from Fig. 1 that the plateau region moves along with the band, but gets smaller and ultimately vanishes as the band progresses down the column. If a correction is made for the amount of sodium ion left in the pore spaces, then the rate at which ammonia is picked up and laid down can be calculated. Under these conditions, it is simpler to fix a plane of reference perpendicular to the moving band at a point within the plateau region. This plane moves with the band. Under these circumstances, each equivalent of ammonium ion picked up at the rear of the band flows through the reference plane and is redeposited at the front. Simultaneously, one equivalent of ammonium ion on the resin passes through the plane in the opposite direction, resulting in no net transfer of total ammonia across the plane of reference. In the case of the nitrogen isotopes, however, there will be a net transfer of Δn equivalents of N^{14} forward and Δn equivalents of N^{15} in the opposite direction across the reference plane for each equivalent of ammonia displaced from the resin and redeposited. This results from the difference in the isotopic ratios existing in the solution and resin phases which are moving past each other.

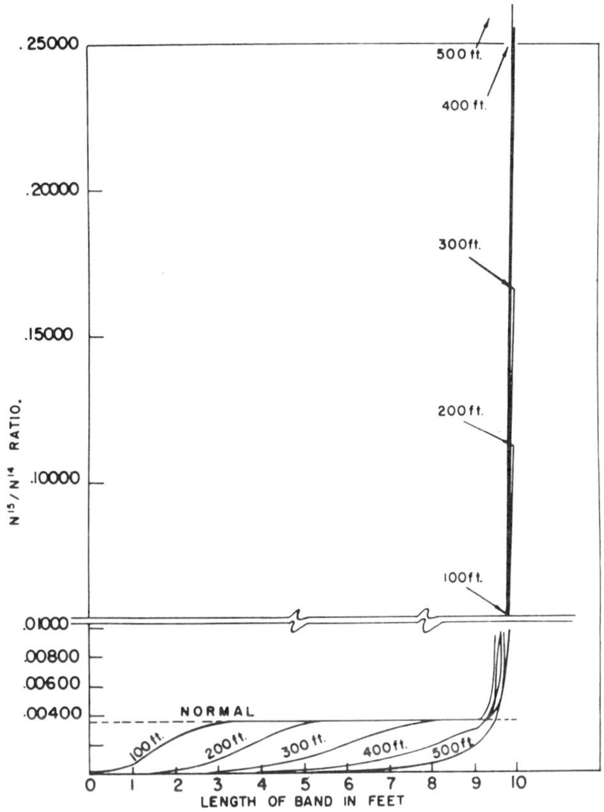

Fig. 1.—Profiles of an adsorbed ammonium band after various distances of elution. The plateau portion is magnified to illustrate the behavior in this part of the band. The depleted area increases at a constant rate up to about 40 times the length of the adsorbed band.

The transfer of N^{14} and N^{15} can be calculated, since Δn for one equivalent of total ammonia transported is equal to $N_R - N_S$ where N_R and N_S are the mole fractions of N^{15} in the resin and solution phases, respectively, in the plateau region. Since $R = N/(1 - N)$ in general, equation 9 becomes

$$\frac{N_R}{1 - N_R} = (1 + \epsilon)\frac{N_S}{1 - N_S} \qquad (11)$$

Solving for N_R and N_S

$$N_R = \frac{(1 + \epsilon)N_S}{1 + \epsilon N_S} \qquad (12)$$

$$N_S = \frac{N_R}{1 + \epsilon(1 - N_R)} \qquad (13)$$

Therefore

$$\Delta n = N_R - N_S = \frac{\epsilon N_S(1 - N_S)}{1 + \epsilon N_S} = \frac{\epsilon N_R(1 - N_R)}{1 + \epsilon(1 - N_R)} \qquad (14)$$

In the plateau region N_S is the mole fraction of N^{15} in the solution originally fed to the column ($N_S = N_0 = 0.00365$). Since there is no net transfer of isotopic species between the resin and solution phases as they pass each other in the plateau region, Δn moles of N^{14} are transported into and distributed throughout the front section of the band, *via* the solution phase, for each mole of total ammonia displaced from the resin bed. Conversely, Δn moles of N^{15} leave the front section, are transported across the plateau, *via* the resin phase, and distributed throughout the rear section. In the front and rear sections, beyond the plateau region, a net transfer of isotopic species between the resin and solution phases will occur as the transported isotopes redistribute.

If it is desired to obtain the maximum rate of separation of isotopic species from a given feed material, one strives for the maximum mass transfer of the isotopes and adjusts the system so that an effectively continuous operation can be carried out.

It is convenient, when operating a fixed bed ion-exchange plant in a continuous manner, to have the resin bed in sections. This permits the injection of feed solution and the withdrawal of products from the band as it moves from section to section by simple valving operations. For this reason, a number of short columns connected in series are used rather than a single long column. This arrangement permits an adsorbed band to be eluted in a cyclic manner using only a few columns, since the spent resin can be regenerated for re-use before the band comes around the next time. Other advantages of such an arrangement are that it is simpler to maintain uniform resin beds in short columns, and the resistance to flow of the eluent and regenerant solutions can be minimized by having the solutions flow only through the desired portion of the bed system at any given time.

Experience has shown that it is only necessary to have a band of ammonia 7 to 10 feet long in order to provide a suitable plateau region and sufficient distance on either side to permit the concentrations of the isotopes to approach 100% N^{14} at the front and 100% N^{15} at the rear. After the ammonium band is adsorbed on the bed, it is eluted for a distance of 200–300 feet. This allows the system to approach the condition where continuous operation can be started. Feed solution can then be injected at the top of each column whenever the plateau region moves past this point. Similarly, products can be removed as the front and rear edges of the ammonium band pass the bottom of a column. In practice, since less valving is required, the feed is injected and the products withdrawn only once each time the band completes a circuit of the beds. Proportionately larger injections and withdrawals are made when the latter procedure is used.

Since the natural abundance of N^{15} is low and the resin bed has an appreciable capacity, it is desirable to permit the N^{15} to accumulate without withdrawal until its concentration has built up to the desired level. Withdrawal of N^{14} can begin at the time feed injections are started. As the ammonium band is eluted from one resin bed with sodium hydroxide eluent, the front section is allowed to proceed down the next column until the plateau region is passing between columns. At this time, a valve between the two columns is closed and a quantity of ammonium hydroxide (say one mole) of the same normality as the sodium hydroxide eluent and containing a mole fraction of N^{15} equal to N_0 is allowed to flow into the top of the leading column. At the same time, or at some prior or subsequent time $(1 - N_0)$ moles of ammonium hydroxide containing virtually pure N^{14} is allowed to flow from the bottom of the column. In practice, the withdrawal can be made slightly before or after the injection so as to occur when the front of the band is moving off a column. This section of the band on the column will now be longer by N_0 moles. Further, since only $(1 - N_0)$ moles of N^{14} were introduced in the mole of feed and $(1 - N_0)$ moles of pure N^{14} were withdrawn, the gain in this section of the band is N_0 moles of N^{15}. However, since it was specified that the reference plane move at the same rate as the front edge of the band, it will have progressed down the column N_0 moles during this operation. Therefore, to restore the front section of the band to its original condition it is necessary to effect a net transport of $(N_0 - N_0N_R)$ moles of N^{15} rearward across the boundary and a corresponding amount of N^{14} in the opposite direction. The N_0N_R moles of N^{15} represent the amount of N^{15} entering the rear section when the reference plane was shifted forward N_0 moles.[7] Since the net number of moles of N^{15} or N^{14} transported per mole of band displacement is Δn, defined by equation 14, the number of moles of displacement the band must undergo, is

$$\nu = \frac{N_0(1 - N_R)}{\Delta n} = \frac{(1 - N_R)(1 + \epsilon N_0)}{\epsilon(1 - N_0)} \quad (15)$$

From equation 12

$$1 - N_R = \frac{1 - N_0}{1 + \epsilon N_0} \quad (16)$$

Substituting for $1 - N_R$ in equation 15

$$\nu = \frac{N_0(1 - N_0)}{\Delta n(1 + \epsilon N_0)} = \frac{1}{\epsilon} = \frac{1}{0.0257} = 39 \quad (17)$$

Therefore, the total advance of the band between subsequent injections of a mole of natural ammonium hydroxide is $(39 + N_0)$ moles. Since a mole of dilute sodium hydroxide solution advances the band about 1.75 inches on a 4-inch diameter resin bed and N_0 is normally small, the band would have to advance approximately 68 inches between one mole additions. It should be noted that a definite number of moles of sodium hydroxide are required to resolve one mole of nitrogen into its isotopes. Once the length of the band required to produce products of the desired purity is determined, increasing the cross-sectional area of the beds would result in a proportionate increase in the yield and would require only a proportionate increase in consumption of sodium hydroxide.

If the band is allowed to progress down a column system about 50 or 60 band lengths without addition or withdrawal of ammonia, it approaches very closely the condition known in countercurrent extraction as steady state in total reflux. Here the plateau range disappears entirely and the exchange between the resin and the solution reaches the steady-state equilibrium condition so that, if one examines the net transport of either chemical or isotopic species across an imaginary plane at any fixed arbitrary distance from the front of the band, the net transport is zero. Under these conditions, the concept of a theoretical plate height becomes unambiguous and the mathematics of a countercurrent system can be utilized, provided the equations used are those which apply to the case in which no product is withdrawn. Since there is no net transport into or out of a plate, the resin stream leaving the plate differs from the resin stream entering it by the factor K and the same is true in reverse for the solution phase. Then if two points on the resin are chosen a distance, L, apart the isotopic ratios at the

(7) This assumes that the ammonia in this section of the column is virtually all adsorbed on the resin bed. If it were all in solution phase the forward shift would be N_0^2 moles. Actually, with dilute solution N_0N_R is the better approximation.

two points will differ by the factor K^n where n is the number of theoretical plates between the points.

$$R_n = K^n R \quad (18)$$

or

$$\log \frac{R_n}{R} = n \log K \quad (19)$$

Under these limiting conditions if the logarithm of the ratio of the isotopic ratios is plotted against n, the number of theoretical plates, a straight line is obtained whose slope is log K. The value of n is not known, in general, but if the reasonable assumption is made that, under the conditions of steady state and total reflux, the height of a theoretical plate is constant across the band, then the plate height can be calculated, since

$$n = \frac{L}{\text{HETP}} \quad (20)$$

giving

$$\log \frac{R_n}{R} = \frac{\log K}{\text{HETP}} L \quad (21)$$

If the plot of log R_n/R *versus* L is linear, the above assumption is justified and the slope is log K divided by HETP. Since K can be determined independently, the height equivalent to a theoretical plate can be found. In our experiments, this has usually ranged from 0.15 to 1.5 mm. depending upon the flow rate, solution concentrations, etc.

When the columns are operated under plateau conditions, as is normally done, care must be exercised in utilizing equations derived for cascade or countercurrent extraction systems. Almost invariably, in deriving these formulas, simplifying assumptions have been made which are not applicable to the resin system as the operations are being carried out. Even in those cases where attempts have been made to adapt the equations to the type systems described here, the mathematics has become so complicated that the equations have not been solved. Frequently, terms have been introduced which not only have not been evaluated, but which are difficult to evaluate independently. Therefore, this theory has not yet been especially helpful in predicting quantitatively the shape of the curve preceding and following the plateau region. Nevertheless, it is clear that certain factors must be considered if the process is to be efficient: (1) the exchange reaction employed should have as large a K as possible; (2) the band should be eluted as fast as practical to obtain the maximum transport per unit time, but not so fast that band boundaries become diffuse or the HETP excessively large; (3) the material should be eluted under conditions which permit maximum utilization of the active points on the resin bed by the ions being separated isotopically; (4) the length of the adsorbed band should be selected so that sufficient plates are included within the band to give the desired products at the steady state, but not so long that excessive resistance to flow is encountered.

II. Experimental

A. Preliminary Investigation of the Approach to a Steady State in the Elution of Ammonium Ion with Sodium Hydroxide (in collaboration with S. A. Harmon and R. Harrington).—A series of ten ion-exchange columns were prepared. Each column consisted of a 5-foot section of 4-inch, i.d., flanged Pyrex pipe closed at the ends by type 316 stainless steel plates bolted to standard cast iron flange sets. The end plates of the columns were center-tapped and fitted with $^1/_8$-inch stainless steel nipples, two inches long. A circular baffle plate, two inches in diameter, supported by three $^1/_2$-inch long rods of stainless steel, was welded to the bottom side of each top plate to prevent the influent solution from disturbing the resin bed. The resin bed was supported by one thickness of Saran filter cloth backed up by a 30-mesh stainless steel screen. The circles of Saran and stainless steel gauze were rubber-cemented into a 5-layer sandwich between three 6-inch, o.d., 4-inch, i.d., by $^1/_8$-inch thick neoprene gaskets. The screen sandwich was supported an inch above the bottom plate by a Pyrex pipe spacer to prevent the screen from blocking the outlet in the bottom plate. Neoprene gaskets were used with both the column plates to obtain a water-tight seal. Individual columns of the series were interconnected by means of flexible $^5/_{16}$-inch Tygon tubing slipped over the $^1/_8$-inch pipe nipples which were screwed into the end plates. By using a binding of Scotch electrical tape over the junctions and $^1/_2$-inch brass tubing clamps, in addition, the system could be used up to an internal pressure of two atmospheres without leakage.

The resin beds of the columns consisted of 100–200 mesh spheres of Dowex 50-X12 and were 58–59 inches long in the H^+ cycle and 53–54 inches long in the Na^+ cycle. The resin beds were backwashed by removing the top plate and adding an auxiliary 5-foot section of Pyrex pipe. All exceptionally fine particles, which either did not settle readily after backwashing or which settled in a thin layer at the top of the resin beds, were removed by means of a siphon so that the finished beds offered a minimum of resistance to liquid flow. Under a 60-foot liquid head a flow rate of about a liter per minute could be obtained through a single unit.

The resin beds were given a preliminary treatment with 2 N NaOH, rinsed with distilled water and restored to the H^+ cycle with 1 N H_2SO_4. The excess H^+ ion was rinsed from the columns by means of distilled water, and five liters of 15 N NH_4OH, diluted 30-fold with water, was passed downflow through three of the columns connected in series. This resulted in an adsorbed NH_4^+ band approximately 10 feet long.

The ammonia band was eluted down the series of columns with 0.6 N NaOH at a flow rate of 120–180 ml. per minute. As the band progressed down the series, the eluent feed tube and the discharge tube were moved along so that only three columns were connected together at any one time. As the band passed, the resin left behind was reconverted from the Na^+ to the H^+ cycle, as before, and the columns were then ready to be reused. In all, the adsorbed band was eluted around the 10-column series ten times. Profiles were taken every twenty columns by withdrawing small samples of the solution periodically as the band passed between columns. Profiles B-20, 40, 60, 80 and 100 are shown in Fig. 2. It can be seen that the normal plateau is gradually removed by the transport of N^{15} from the front to the rear of the band. Up to B-80, which represents about 36 displacements of the adsorbed band, the amount of N^{15} transported was proportional to the distance the band was eluted. Beyond B-80 the plateau disappeared and the efficiency fell off markedly. At B-100 the band was approaching the steady state and very little improvement could be expected from this point on. However, the last 6 inches of the band containing practically all the N^{15} was allowed to pass onto a series of 22 mm., i.d., by 4.5-foot resin beds and was eluted further until a mole fraction of N^{15} of 0.81 was observed at the rear edge of the band. The 22 mm. diameter columns were closed at the bottom with fritted glass plates; the plates were not uniformly porous and, as a consequence, considerable channelling and tilting of the band boundaries were observed. The experiment did, however, show that separation would continue when steady-state conditions were destroyed by decreasing the diameter and increasing the length of the enriched portion of the band.

B. Determination of the Equilibrium Constant for the Exchange of Nitrogen Isotopes between Ammonium Hydroxide Solution and Ammonium Ion Adsorbed on Dowex-50 (in collaboration with D. M. Provow and J. Capellen).—If one passes a dilute solution of NH_4OH containing a mole fraction of N^{15} equal to N_0 or a N^{15}–N^{14} ratio of R_0 down a

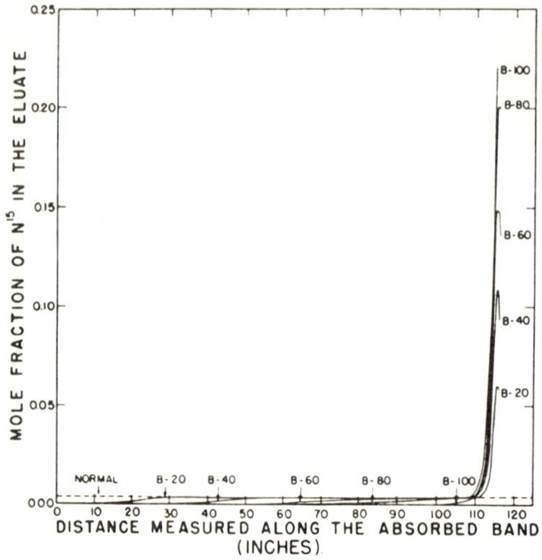

Fig. 2.—Profiles of a 10-foot long adsorbed ammonium band taken after traversing 20, 40, 60, 80 and 100 beds of 100–200 mesh Dowex 50-X12, each 4 inches in diameter and 58–59 inches long.

bed of cation exchanger in the H^+ cycle, the H^+ ion is completely displaced by NH_4^+ due to the large constant for the reaction which takes place at the front of the band.

When all the H^+ ion on the resin has been replaced by NH_4^+ ion, NH_4OH solution will begin to flow from the resin bed. In the first fractions of NH_4OH solution that are collected, the ratio of N^{15} to N^{14}, R, will be less than the ratio R_0, which existed in the feed, due to the exchange reaction given by equation 7. The ratio R will approach R_0 as more effluent solution is collected. When the ratio in the effluent solution becomes equal to R_0, the ratio of N^{15} to N^{14} in the resin phase must differ from R_0 by the factor K due to the relationshp

$$K = \frac{R_R}{R_S} = \frac{R_R}{R_0} \tag{22}$$

Now equation 22 can be written in the form

$$K = \frac{N_R}{1 - N_R} \times \frac{1 - N_0}{N_0} \tag{23}$$

where

$$N_R = \mathrm{n}/Q \tag{24}$$

Q is the total exchange capacity of the resin bed in equivalents and n is the total number of equivalents of $N^{15}H_4^+$ adsorbed on the resin bed.

If K in equation 22 turned out to be exactly 1, then there would be no tendency for either enrichment or depletion of $N^{15}H_4^+$ in the resin phase when it is equilibrated with a solution of ammonia. That is, n_0/Q would equal N_0. However, if K differs from 1, then n can be defined as

$$\mathrm{n} = \mathrm{n}_0 + \Delta\mathrm{n} \tag{25}$$

Therefore

$$N_R = \frac{\mathrm{n}_0 + \Delta\mathrm{n}}{Q} = N_0 + \frac{\Delta\mathrm{n}}{Q} \tag{26}$$

and

$$K = \frac{\left(N_0 + \frac{\Delta\mathrm{n}}{Q}\right)(1 - N_0)}{\left(1 - N_0 - \frac{\Delta\mathrm{n}}{Q}\right)\mathrm{N}_0} = 1 + \frac{\Delta\mathrm{n}}{\mathrm{N}_0 Q\left(1 - N_0 - \frac{\Delta\mathrm{n}}{Q}\right)} \tag{27}$$

and

$$\epsilon = K - 1 = \frac{\Delta\mathrm{n}}{N_0 Q\left(1 - N_0 - \frac{\Delta\mathrm{n}}{Q}\right)} \tag{28}$$

Solving for $\Delta\mathrm{n}/Q$, one obtains the relationship that

$$\frac{\Delta\mathrm{n}}{Q} = \frac{\epsilon N_0(1 - N_0)}{1 + \epsilon N_0} \tag{29}$$

If ϵ is small compared to 1, then $\Delta\mathrm{n}/Q$ is of the order of ϵN_0 $(1 - N_0)$ and is also much less than either 1 or N_0. Consequently, equation 28 can be written

$$\epsilon = \frac{\Delta\mathrm{n}}{QN_0(1 - N_0)} \tag{30}$$

Now $\Delta\mathrm{n} = \Delta\mathrm{n}$ resin $= -\Delta\mathrm{n}$ effluent, so $\Delta\mathrm{n} = \sum_{i=1}^{m} V_i C_i$ $(N_0 - N_i)$, where V_i is the volume; C_i, the ammonia concentration; and N_i, the mole fraction of N^{15} in the ammonia of the ith fraction of the m fractions which were collected before R, the ratio of N^{15} to N^{14} in the effluent, returned to the value R_0, the feed ratio. Equation 30 now becomes

$$\epsilon = \frac{\sum_{i=1}^{m} V_i C_i (N_0 - N_i)}{QN_0(1 - N_0)} \tag{31}$$

Since

$$N_i = \frac{R_i}{1 + R_i} \text{ and } N_0 = \frac{R_0}{1 + R_0} \tag{32}$$

then

$$\epsilon = \sum_{i=1}^{m} \frac{V_i C_i \left(\frac{R_0}{1 + R_0} - \frac{R_i}{1 + R_i}\right)(1 + R_0)^2}{QR_0} \tag{33}$$

Finally, since R_0 for normal nitrogen is 0.00365 $(1 + R_i) \cong (1 + R_0) \cong 1.00$ and

$$\epsilon = \sum_{i=1}^{m} \frac{V_i C_i (R_0 - R_i)}{QR_0} \tag{34}$$

If extremely high precision is desired and experimental accuracy warrants, ϵ can be determined from equation 28 at the expense of more involved computations.

For convenience, NH_4OH solutions are oxidized with NaOBr to give N_2 gas prior to analysis on the mass spectrometer. The most sensitive determination at low concentrations of N^{15} is the ratio of N_2^{29} to N_2^{28}. For normal and below normal values of N^{15}, the ratio of N_2^{29} to N_2^{28} is, to a high precision, equal to twice the N^{15}–N^{14} ratio. Since the ratio appears in both numerator and denominator of equation 34, it makes no difference whether one uses the N_2^{29}–N_2^{28} ratios or the N^{15}–N^{14} ratios for R_0 and R_i in computing the value of ϵ.

The value of $K = 1 + \epsilon$ has been determined in the manner described above and was found to be 1.0257 ± 0.0002.

C. The Investigation of Factors Which Influence the Theoretical Plate Height at the Steady State (in collaboration with J. Capellen).—In Part I it was shown that when an ammoniun band was eluted down a column 40 to 50 band lengths, a steady-state gradient of the isotopes was built up similar to that obtained in a countercurrent extraction system operating at total reflux (no product withdrawn). Under these conditions equation 21, written in the form

$$\log R_n = \frac{\log K}{\mathrm{HETP}} L + \log R \tag{35}$$

states that, if the logarithm of the isotopic ratio in any plate leaving the column is plotted against the length of the band which has progressed off the column, L, a straight line plot should be obtained. The slope of the line is equal to the logarithm of the equilibrium constant for the exchange reaction, K, divided by the height equivalent to a theoretical plate, HETP. Since K is known from Section B, the HETP can be calculated from these data.

It is desirable in determining HETP's for a column that sufficient N^{15} be present in the original mixture to allow a reasonably high mole ratio of N^{15} to be attained in the N^{15}-rich end of the band. A 50-50 mixture of N^{14} and N^{15} would be ideal, but such a quantity of N^{15} was not available. Since normal ammonia contains only one part of N^{15} to 273 parts of N^{14}, enriched ammonia from several previous runs was adsorbed onto a 2-inch diameter resin bed to form a band 45 inches long for the following experiments.

The band was eluted down 60 2-inch by 5-foot columns of Dowex 50-X12, 100-200 mesh resin, at a flow rate of 125 ml. per minute with 0.6 *N* NaOH solution. This was more than sufficient to attain a steady state. Log N^{15}/N^{14} *vs.* *L* is plotted in Fig. 3, curve A, and a straight line was obtained for a considerable portion of the band. It will be noted, for reasons given later, that deviations from linearity occurred at low N^{15}–N^{14} ratios.

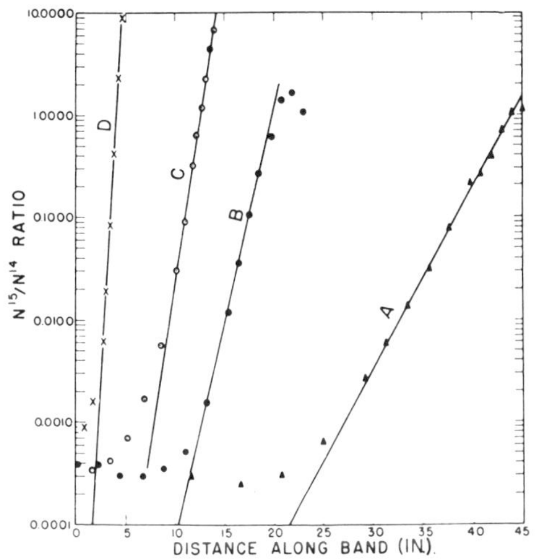

Fig. 3.—Profiles of adsorbed ammonium bands under various steady state conditions on 2-inch by 59-inch beds of 100–200 mesh, Dowex 50-X12 resin: A, 0.6 *N* NaOH at 125 ml. per minute; B, 0.6 *N* NaOH at 25 ml. per minute; C, 0.12*N* NaOH at 125 ml. per minute; D, 0.12 *N* NaOH at 25 ml. per minute. Some abnormal behavior is apparent at the ends of the bands, especially at low concentrations of N^{15}.

The last 23 inches of this band, containing most of the original N^{15} adsorbed on the column, was next eluted with the same NaOH solution down 30 additional columns at one-fifth the original flow rate (25 ml. per minute). Since the length of the band was less, a shorter distance of elution was required to ensure the attainment of steady-state conditions. The results are plotted in Fig. 1, curve B.

The last 14 inches of this band were then eluted down 10 additional columns at the original flow rate of 125 ml. per minute, but with one-fifth the original concentration of NaOH, *i.e.*, 0.12 *N* (see curve C).

Finally, the last 5.5 inches of the band from the previous experiment were eluted down 3 more columns with 0.12 *N* NaOH at 25 ml. per minute (see curve D).

The HETP value for the linear portion of each of the curves in Fig. 1 was calculated and tabulated in Table I.

In general, it can be seen that a 5-fold change in concentration had a greater effect than a 5-fold change in flow rate by comparing experiments B and C, respectively, to experiment A. The apparent discrepancy between experiments A and E which had identical linear flow rates can be explained satisfactorily by the fact that the tubing used to connect consecutive four-inch and two-inch columns was identical and not scaled down proportionately. This would lead to greater re-mixing percentagewise in the case of the two-inch columns as the band passed from column to column during elution. The rates of band progress in B and C were equal and only one-fifth the rate of A or E. Experiment D represents an additional 5-fold decrease in the rate of band travel.

TABLE I

EXPERIMENTALLY DETERMINED VALUES OF THE HEIGHTS EQUIVALENT TO A THEORETICAL PLATE AT VARIOUS FLOW RATES AND CONCENTRATIONS WITH DOWEX 50-X12, 100–200 MESH RESIN

Expt.	Flow rate (ml./min.)	Column diameter (in.)	*N* of base (equiv./l.)	Plate height (HETP in mm.)
A	125	2	0.6	1.55
B	25	2	.6	0.66
C	125	2	.12	0.44
D	25	2	.12	0.17
E[a]	500	4	.6	1.24

[a] Data from a previous larger scale run not described here.

It will be noted that under the best conditions described above the nitrogen isotopic percentages would change from 99% N^{14} to 99% N^{15} in about 2.5 inches of resin bed.

In practice, straight line relationships are obtained in the log N^{15}/N^{14} *vs.* *L* plots when the isotopic ratios lie between 0.002 and 50.0, provided one disregards ratios measured for plates that lie near the band boundaries. Tilted or channelled boundaries give erroneous data.

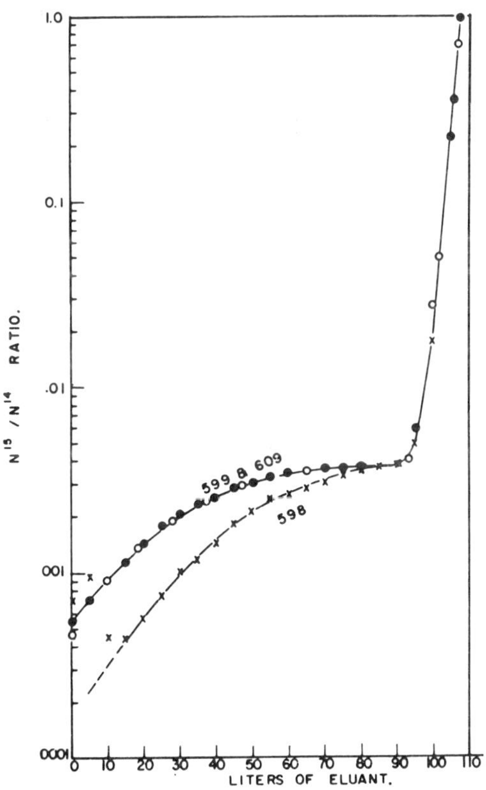

Fig. 4.—Profiles of the adsorbed ammonium band on a series of 4-inch by 59-inch, 100–200 mesh, Dowex 50-X12, resin beds. The adsorbed band was eluted in a semi-continuous manner by adding the equivalent of 550 ml. of 15 *N* ordinary NH_4OH and withdrawing the same amount of NH_4OH depleted in N^{15} every circuit of the 10 column series. Profile 598 was taken just prior to, and profiles 599 and 609 were taken just after, an addition and withdrawal. Profiles 599 and 609 are exactly one cycle apart and are virtually identical.

Deviations will also be observed if any processes occur where part of the enriched N^{15} is picked up at the rear in a form which does not exchange, but which does redeposit at the front edge of the band. For example, if there is carbonate in the sodium hydroxide, this might react with ammonia to give ammonium carbamate. If the carbamate hydrolyzes only slowly as it is swept across the adsorbed ammonium band, it may transport enriched N^{15} from the rear to the front of the band without permitting equilibration. Acid conditions at the front boundary could bring about hydrolysis, redeposition and contamination of the N^{14} at this point.

Deviations may also occur if the driving solution contains any nitrogenous impurities, or if the hydrogen-form resin bed contains nitrogen in a form which can yield ammonia. The latter case apparently occurs in our experiments and sets a lower limit for the N^{15}–N^{14} ratio in the leading plates of the band. Any ammonia in the sulfuric acid or water used to recondition the resin will introduce ammonia into the resin stream and prevent the attainment of low N^{15}–N^{14} ratios. An even more serious possible mechanism for resin contamination is that a small amount of the ammonia might react with the resin (perhaps with the sulfonic acid group) to give a compound which does not exchange rapidly or break down completely in cold caustic or the sulfuric acid regenerant solution. Such a compound might be broken down fairly completely by the heat generated by the reaction of the NH_4OH with hydrogen-form resin at the front edge of the band during the next cycle of elution. Such an effect can be minimized by passing hot caustic over the resin after the band has passed. Analysis of hot caustic strip solution used for this purpose has shown traces of ammonia enriched in N^{15} left behind on the resin. This effect is small and is not observed when the isotopic ratio at the front edge is high. The effect does become important, however, where the N^{15}–N^{14} ratio to be measured is low, particularly, if the compound causing the trouble was formed sometime previously when the N^{15}–N^{14} ratio was high.

Finally, deviations from linearity at very low and very high isotopic ratios may arise from analytical errors in the mass spectrograph as a result of background or air contamination of the gas samples.

D. Continuous Operation with a System of Fixed Bed Ion-exchange Columns (in collaboration with J. Capellen, D. M. Provow and H. Burkholder).—The theory of continuous operation with a system of fixed bed ion-exchange columns was discussed previously. The value of ϵ given in Section II-B indicated that a total of 39 moles of band displacement would be required to resolve each mole of nitrogen (added as ammonium hydroxide) into the individual nitrogen isotopes. Calculation showed that each 4-inch by 5-foot column of Dowex 50-X12, 100–200 mesh resin had an exchange capacity of about 32 equivalents. If an addition of ammonia were made once each cycle of ten columns, a total of 320 equivalents of resin would be traversed between injections. Consequently, 8.2 equivalents of ammonia should be added per cycle.

In order to test the validity of these calculations, a series of ten 4-inch diameter columns five feet long was prepared.

The original adsorbed band consisted of 3.5 liters of 15 *N* ammonium hydroxide (diluted 30-fold for loading). The band was eluted down 30 columns with 0.6 *N* NaOH at a flow rate of 400 ml. per minute before any additional ammonia was added or any N^{14}-rich product was withdrawn. Thereafter, 550 ml. of 15 *N* ammonium hydroxide in 15 liters of solution was added each cycle and a corresponding amount of N^{14} product was withdrawn. No enriched N^{15} product was removed, except when profiles of the adsorbed band were taken to determine its condition. Injections were initially made nine inches ahead of the tail-end of the band, but as the N^{15} concentration in the rear part of the band increased, it became necessary to gradually move the point of injection forward in order to stay on the normal plateau. The injection point was 16 inches from the tail-end of the band by the 500th column.

The depleted product being withdrawn early in the experiment had a N^{15}–N^{14} ratio of about 0.00012, but the ratio tended to increase gradually. When the N^{15}–N^{14} ratio of the depleted product exceeded 0.00050, the length of the band was increased by making an addition but no withdrawal. In all, two such increases were made so that the band was about 100 inches long by column number 550 and contained the equivalent of 4.6 liters of 15 *N* ammonia or 69 moles. Figure 4 is a profile of the band taken as the band passed from the 598th to the 599th column just before an injection and withdrawal were made. Profiles were also taken at the bottoms of columns 599 and 609 just after additions and withdrawals were made. The fact that the data for profiles 599 and 609, exactly one cycle apart, fall on the same curve demonstrates that a more or less steady state had been achieved for this semi-continuous method of operation.

It is planned to operate these semi-continuously fed columns under conditions where 99.99% N^{14} is withdrawn at the front and greater than 99% N^{15} is bled slowly from the rear edge of the band. Before this can be done, however, sufficient N^{15} will have to be accumulated in the band to fill the necessary N^{15}-rich plates. It would, of course, be possible to begin withdrawing highly purified N^{15} from the present band, if it were eluted for a few cycles at a reduced flow rate with a lower concentration of NaOH. This would shorten the HETP and, thereby, decrease the amount of total N^{15} required to fill the number of plates required to attain a mole fraction of N^{15} equal to 0.99. This would not be especially desirable, since the decreased hold-up of N^{15} would hardly justify the decreased rate of production caused by the slower rate of elution.

Ames, Iowa

14

Reprinted from *Analusis*, **2**(1), 37-48 (1973)

THÉORIE DU DÉVELOPPEMENT PAR DÉPLACEMENT EN CHROMATOGRAPHIE D'ÉCHANGE D'IONS

Application à la séparation de corps très voisins (*)†

par Patrick Conrard et Robert Rosset

Laboratoire de Chimie analytique,
École supérieure de Physique et de Chimie de Paris,
10, rue Vauquelin, 75005-Paris

L'équation du front de permutation en régime stationnaire dans le cas du développement par déplacement sur échangeur d'ions d'un mélange binaire d'espèces voisines est établie en utilisant un modèle purement chromatographique. On en déduit une relation approchée donnant la variation d'enrichissement de la bande en fonction du déplacement. Les enrichissements ainsi calculés sont comparés aux enrichissements vrais obtenus par simulation du développement par déplacement sur calculatrice numérique, ce qui justifie l'approximation faite. On en déduit le calcul rigoureux du parcours nécessaire pour atteindre l'état stationnaire avec un écart donné. La relation classique $l = l_0/\varepsilon$ (l : déplacement; l_0 : longueur de la bande : $\varepsilon = K - 1$) est discutée. Une hypothèse simple sur l'enrichissement marginal permet alors d'effectuer l'étude de la production sans avoir recours à une calculatrice numérique.

I. — INTRODUCTION

Les possibilités de la chromatographie d'échange d'ions sont actuellement en pleine évolution avec l'apparition de nouveaux types de résines, de structures bien définies et, surtout, de granulométrie fine et très homogène.

Si le développement par élution est largement utilisé dans le domaine de l'analyse, son intérêt en chimie préparative reste limité, en raison de la faible quantité de matière qu'il met en jeu, ainsi que de la dilution qui accompagne obligatoirement le développement.

La technique du développement par déplacement ne présente pas ce double inconvénient et autorise, de ce fait, de longs parcours; c'est pourquoi elle est tout à fait indiquée pour la séparation préparative d'espèces aux propriétés voisines.

S'il existe un certain nombre d'applications pratiques souvent remarquables dans le cas de séparations difficiles (acides aminés, terres rares, isotopes), les études théoriques sont, en revanche, beaucoup plus limitées, spécialement dans le cas d'espèces, dont les propriétés sont très proches (isotopes, par exemple).

Grâce à la simulation des phénomènes sur calculatrice numérique, J. Persoz [1] a fourni un certain nombre de résultats, dans des cas bien précis.

En utilisant comme elle la théorie des plateaux et en nous limitant au cas de mélanges binaires, nous nous proposons d'étudier l'évolution des phénomènes au cours du développement et d'établir les règles simples qui président aux calculs de production.

Après avoir rappelé les caractéristiques du développement par déplacement, nous étudierons la répartition des espèces à l'état stationnaire, puis nous établirons la loi de variation de l'enrichissement au cours du développement.

Enfin, nous montrerons comment le développement par déplacement peut être appliqué à la chimie préparative et nous étudierons le problème des prélèvements en cours de développement.

II. — ÉQUATION THÉORIQUE DES FRONTS DE PERMUTATION EN RÉGIME STATIONNAIRE

1° Généralités

Rappelons tout d'abord [2] que la technique du développement par déplacement dans le cas d'un mélange de deux espèces A et B consiste à déplacer une bande de mélange fixée sur une colonne de résine échangeuse d'ions initialement sous la forme d'un ion I, par une solution d'un ion D appelé développeur, dans des conditions telles que l'ordre des affinités de chacune des espèces pour la résine soit le suivant :

$$I \ll A < B \ll D.$$

Au cours du déplacement, la bande garde une longueur constante et s'enrichit en A en tête et en B en queue.

(*) Conférence présentée par M. Patrick Conrard à l'Assemblée annuelle de la Société chimique de France, dans le cadre de la Division de Chimie analytique, Lille, 17 mai 1972.

†*Editor's Note:* An English translation follows this article.

A l'intérieur de la bande, on a l'équilibre

$$A_R + B_S \rightleftarrows A_S + B_R$$

auquel correspond la constante d'échange

$$K_A^B = \frac{|A|_S |B|_R}{|A|_R |B|_S} > 1$$

les indices R et S désignant les phases résine et solution respectivement.

Dans le cas où la constante d'échange est assez élevée, on obtient, après un certain parcours un état stationnaire, dans lequel le front de permutation entre A et B (où isochrone) est une courbe bilogarithmique d'équation :

$$\log \frac{A}{B} = 2 \frac{l}{h} \log K_A^B$$

h désignant la hauteur équivalente à un plateau théorique (HEPT), l l'abscisse comptée le long de la colonne en prenant comme origine le point où $|A| = |B|$ et les études entreprises ont toujours supposé que l'on obtenait les corps A et B purs aux extrémités de la bande.

Le problème est tout différent lorsque les affinités des espèces A et B pour l'échangeur sont très voisines ($K_A^B \# 1$). On constate alors que l'on tend encore vers une répartition stationnaire des espèces, mais cette fois sans obtenir les corps A et B purs en tête et en queue de bande.

Dans le cas d'une constante d'échange très peu différente de 1 (1,001), d'une bande courte (50 plateaux) et d'un mélange contenant les deux espèces à des teneurs très différentes (1 % pour l'un des constituants, 99 % pour l'autre), J. Persoz [1] a montré, en effectuant les calculs de répartition dans chaque plateau et à chaque transfert au moyen d'une calculatrice numérique, que la répartition des concentrations le long de la bande était linéaire à l'état stationnaire.

Pour des valeurs plus élevées de la constante d'échange, et pour des mélanges quelconques, diverses séparations publiées dans la littérature montrent des bandes dont le profil n'est ni linéaire, ni bilogarithmique, bien que le déplacement soit assez long.

Nous nous proposons donc d'abord d'établir l'équation générale du front de permutation en régime stationnaire, dans le cas de constantes voisines de 1, et de préciser les conditions qui doivent être remplies pour aboutir à une répartition linéaire.

2° Schéma de déplacement d'une bande

Considérons une colonne de résine échangeuse d'ions sous la forme d'un ion X quelconque, la concentration de X dans la résine est C_R meq/ml de résine. Cette résine est baignée par une solution de X de concentration C_S meq/ml de solution. Dans un volume V de colonne, la solution occupe un volume Vx et la résine un volume V$(1 - x)$; x représente la fraction de volume interstitiel, en général de l'ordre de 35 %.

Nous désignerons par m le rapport des quantités de X dans la résine et dans la solution, pour un volume donné de colonne.

$$m = \frac{V(1-x)C_R}{V_x C_S} = \frac{1-x}{x}\frac{C_R}{C_S}$$

Pour la plupart des résines existantes, la valeur de C_R est de plusieurs milliéquivalents par millilitre de résine. Compte tenu des concentrations des solutions habituellement utilisées (généralement inférieures à 0,5 M), le rapport m est toujours de plusieurs dizaines.

Afin de simplifier les démonstrations, nous choisirons pour m des valeurs entières, ce qui ne restreint en rien la généralité des résultats obtenus.

Supposons maintenant qu'une bande de mélange A + B soit fixée sur la colonne.

On peut montrer que $m + 1$ transferts de la solution de chaque plateau dans le suivant sont nécessaires pour faire progresser la bande d'un plateau (3).

3° Cas d'une bande courte constituée par le mélange de deux éléments très voisins, l'un étant présent à une concentration très faible par rapport à l'autre

Par la suite, nous ne nous préoccuperons plus de l'ion initialement fixé sur la résine, ni de l'ion développeur. Nous supposerons simplement que les déplacements de I par le mélange A + B, et du mélange A + B par D sont quantitatifs.

Soient y_R, x_R, y_S et x_S les fractions molaires de A et de B dans la phase résine et dans la phase solution d'un plateau.

Soit x la fraction molaire de B dans le plateau (résine + solution). Il vient

$$x = \frac{m x_R + x_S}{m+1}. \tag{1}$$

La constante d'échange s'exprime par :

$$\frac{x_R \cdot y_S}{x_S \cdot y_R} = K = 1 + \varepsilon \tag{2}$$

B est supposé avoir plus d'affinité pour la résine que A, mais les deux ions ont un comportement très voisin; ε est donc positif, mais voisin de zéro.

La somme des fractions molaires de A et de B dans une phase est égale à 1. On peut donc écrire (2) sous la forme

$$\frac{x_R}{x_S} = 1 + \varepsilon(1 - x_R) \tag{3}$$

relation qui, dans le cas où la fraction molaire de B est très faible ($x_R \ll 1$), devient :

$$\frac{x_R}{x_S} = 1 + \varepsilon. \tag{4}$$

Dans le cas où A aurait été l'ion le moins concentré, le raisonnement aurait été le même avec cette fois :

$$\frac{y_R}{y_S} = 1 - \varepsilon.$$

L'équation (4) est valable tout le long de la bande, si l'enrichissement reste peu important, soit si $\varepsilon \ll 1$ et si le nombre de plateaux de la bande est faible.

Posons, pour alléger l'écriture

$$\delta = 1 + m + m\varepsilon$$

De (1) et (4), on tire :

$$x_R = \frac{(m + 1)(1 + \varepsilon)x}{\delta} \quad (5)$$

$$x_S = \frac{(m + 1)x}{\delta}. \quad (6)$$

Soit z le facteur par lequel il faut multiplier les fractions molaires d'un plateau pour obtenir celles du plateau suivant.

Si x_{1R} et x_{1S} sont les fractions molaires avant transfert dans le plateau p, celles du plateau $p + 1$ seront $x_{1R}z$ et $x_{1S}z$. Au cours du transfert, la solution venant du plateau p est mise, dans le plateau $p + 1$, au contact de la résine de ce plateau pour réaliser l'équilibre.

Il vient donc, dans le plateau $p + 1$, d'après (1)

$$x_2 = \frac{mx_{1R}z + x_{1S}}{m + 1}.$$

Soit, d'après (5) et (6)

$$x_2 = \frac{\left[m(1 + \varepsilon) + \frac{1}{z}\right]zx_1}{\delta}.$$

Il suffit donc, pour décrire le transfert dans le plateau $p + 1$, de multiplier sa fraction molaire avant transfert zx_1 par le facteur

$$y = \frac{m(1 + \varepsilon) + \frac{1}{z}}{\delta}.$$

Supposons qu'il existe un état stationnaire, caractérisé par une valeur constante de z dans chaque plateau. Pour progresser de un plateau, nous avons vu qu'il faut réaliser $m + 1$ transferts, donc multiplier x par y^{m+1}.

Après ces $m + 1$ transferts, la fraction molaire x_2 de A dans le plateau $p + 1$ doit être égale à celle qui existait dans le plateau p avant les transferts, c'est-à-dire x_1.

Il faut donc que l'on ait

$$y^{m+1}z = 1.$$

Si nous posons :

$$\frac{1}{z} = u^{m+1},$$

cette condition devient :

$$u^{m+1} - (1 + m + m\varepsilon)u + m(1 + \varepsilon) = 0. \quad (7)$$

La solution à retenir doit être voisine de 1; en effet, puisque ε est petit, la variation de fraction molaire d'un plateau au plateau voisin est faible.

La résolution de (7), conduite en posant $u = 1 + V$ et en développant en V en se limitant au premier ordre donne :

$$u = 1 + \frac{2\varepsilon}{m + 1}$$

soit, comme ε est petit

$$z = 1 - 2\varepsilon$$

L'ion B étant le plus fixé, la fraction molaire diminue lorsque l'on progresse de la queue vers la tête de la bande. C'est pourquoi z est inférieur à 1. Si nous numérotons les plateaux dans le sens des x croissants, et en prenant pour origine le plateau pour lequel $x = x_0$, fraction molaire du mélange initial, il vient :

$$x = x_0(1 - 2\varepsilon)^{-p}$$

Si le nombre total de plateaux P est assez faible pour que $P\varepsilon \ll 1$, cette relation devient :

$$x = x_0(1 + 2p\varepsilon) \quad (8)$$

La répartition des espèces A et B est alors linéaire tout le long de la bande.

4° Cas d'une bande de longueur quelconque constituée par le mélange de deux éléments très voisins en proportions quelconques

Dans le cas où B n'est pas présent en faible proportion, la relation (4) reste valable à condition de remplacer ε par

$$\varepsilon' = \varepsilon(1 - x_R).$$

Dans un domaine assez étroit pour que x_R varie peu, il est alors possible d'utiliser la relation (8), en remplaçant ε par ε'.

D'autre part, m étant grand et ε petit, il est possible de confondre x et x_R et d'écrire

$$\varepsilon' = \varepsilon(1 - x)$$

En passant du plateau p au plateau $p + \Delta p$, assez voisin pour que x varie peu, nous aurons

$$\Delta x = 2x\Delta p\varepsilon'$$

à la limite pour une variation dp petite, il vient :

$$dp = \frac{dx}{2\varepsilon x(1 - x)}$$

soit, en intégrant et en prenant l'origine au plateau pour lequel $x = x_0$:

$$p = \frac{1}{2\varepsilon}\left(\text{Ln}\frac{x}{1 - x} - \text{Ln}\frac{x_0}{1 - x_0}\right)$$

soit :

$$x = \frac{1}{1 + \frac{1 - x_0}{x_0}\exp - 2p\varepsilon}. \quad (9)$$

Soit M l'abscisse du milieu de la bande par rapport au plateau $p = 0$. On voit que $M = 0$ si et seulement si $x_0 = 0{,}5$. Dans ce cas, en effet, la bande est parfaitement symétrique et la fraction molaire moyenne est égale à x_0. Dans tous les autres cas, la bande n'est pas symétrique, par rapport au plateau d'origine. P étant le nombre total de plateaux de la bande, celle-ci s'étend de $M - P/2$ à $M + P/2$, et nous pouvons écrire (conservation des espèces) :

$$\int_{M-P/2}^{M+P/2} x(p)dp = Px_0$$

ce qui conduit, après intégration à

$$M = \frac{1}{2\varepsilon} \operatorname{Ln} \frac{1 - x_0}{x_0} \cdot \frac{\exp 2P\varepsilon x_0 - 1}{\exp \varepsilon P - \exp \varepsilon P(2x_0 - 1)}$$

soit

$$M = \frac{1}{2\varepsilon} \operatorname{Ln} \frac{1 - x_0}{x_0} \cdot \frac{\operatorname{sh} \varepsilon P x_0}{\operatorname{sh} \varepsilon P(1 - x_0)}. \quad (10)$$

5° Détermination pratique de la répartition des espèces à l'équilibre

Les conditions expérimentales fixent les valeurs de x_0, ε, P (et de m qui n'intervient pas à l'état stationnaire). Il suffit alors, sur la courbe générale représentative de l'équation

$$x = \frac{1}{1 + \dfrac{1 - x_0}{x_0} \exp - 2p\varepsilon}$$

de placer le plateau pour lequel $x = x_0$, puis, à partir de ce plateau, de placer le milieu de la bande à l'abscisse M, enfin de porter P/2 plateaux de part et d'autre de M.

La *figure 1* représente cette construction dans le cas où $\varepsilon = 0{,}012$; $x_0 = 0{,}2014$; $P = 200$ correspondant au déplacement jusqu'à l'état stationnaire d'une bande de bore naturel de 200 plateaux (dans des conditions dans lesquelles l'isotope ayant le plus d'affinité pour la résine est ^{10}B, dont la fraction molaire initiale est $x_0 = 20{,}14\,\%$. La relation (10) donne alors $M = -21{,}35$.

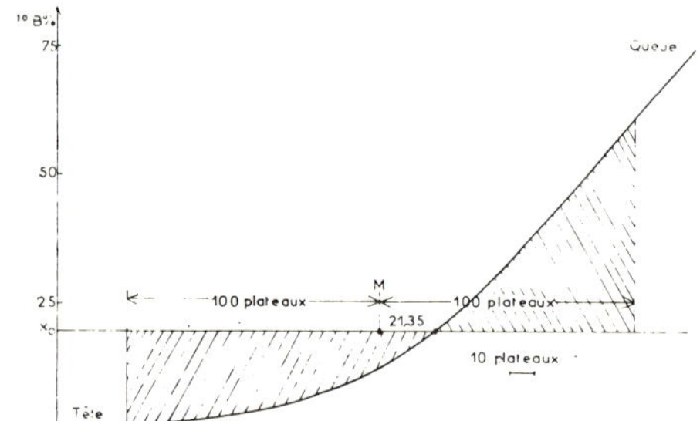

Fig. 1. — Répartition des espèces dans une bande de bore naturel de 200 plateaux à l'état stationnaire ($\varepsilon = 0{,}012$; $x_0 = 0{,}2014$); courbe théorique.

La *figure 2* correspond à une bande de 1 000 plateaux avec $\varepsilon = 0{,}001$ et $x_0 = 0{,}0072$ (cas des isotopes de l'uranium). On trouve alors $M = -79{,}6$.

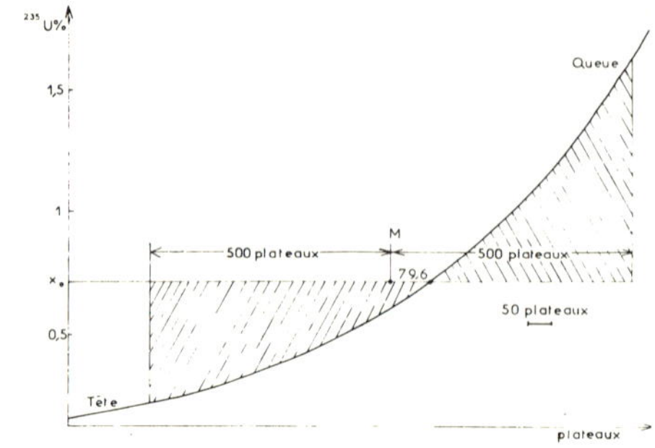

Fig. 2. — Répartition des espèces dans une bande d'uranium naturel de 1 000 plateaux à l'état stationnaire ($\varepsilon = 0{,}001$; $x_0 = 0{,}0072$); courbe théorique.

En conclusion, dans le cas des constantes d'échange voisines de $1 (\varepsilon \ll 1)$, la forme du front peut, à l'état stationnaire, varier selon les conditions expérimentales, de la répartition linéaire ($P\varepsilon \ll 1$) au front bilogarithmique (bande longue, x_0 voisin de 0,5) en passant par des portions de courbes bilogarithmiques dont les *figures 1* et *2* sont des exemples.

III. — ÉTUDE DE LA VARIATION DE L'ENRICHISSEMENT AU COURS DU DÉPLACEMENT

1° Définitions et notations

L'étude des phénomènes avant que l'état stationnaire ne soit atteint, peut se traiter en deux étapes :

a) Variations de l'enrichissement au cours du déplacement, c'est-à-dire comportement global de la partie enrichie en l'une des espèces.

b) Évolution de la forme du front.

Dans de nombreux cas, et, en particulier, comme nous le verrons plus loin, pour l'étude des conditions de production, seule la connaissance de la variation de l'enrichissement est nécessaire.

Soit x la fraction molaire de l'ion ayant le plus d'affinité pour la résine, dans un plateau quelconque, et x_0 sa fraction molaire dans le mélange initial. Nous définirons d'abord un certain nombre de termes relatifs aux variations des fractions molaires au cours du déplacement.

Enrichissement absolu

Il correspond à la différence $x - x_0$ dans le plateau considéré. Changé de signe, il représente également l'enrichissement en l'ion ayant le moins d'affinité pour la résine.

Enrichissement relatif

Il est défini par

$$\frac{x - x_0}{x_0}.$$

Enrichissement relatif réduit

L'enrichissement relatif présente deux inconvénients :

— D'une part, si l'on a étudié complètement une séparation, une autre séparation effectuée dans les mêmes conditions, mais avec une valeur de ε légèrement différente peut s'en déduire de manière assez exacte en remarquant que, dans une zone de variation de ε donnée, les enrichissements sont proportionnels à ε. Il est alors intéressant de faire intervenir la quantité

$$\frac{x - x_0}{x_0\varepsilon}.$$

— D'autre part, cet enrichissement relatif dépend de l'espèce choisie pour le définir. Ainsi, dans le cas d'un mélange tel que $x_0 = 0{,}2$, si l'enrichissement absolu est $x - x_0 = 0{,}1$, l'enrichissement relatif sera, selon l'ion considéré :

$$\frac{x - x_0}{x_0} = 0{,}5 \quad \text{ou} \quad \frac{y - y_0}{y_0} = -\,0{,}125.$$

Or, ces deux valeurs traduisent le même phénomène. C'est pourquoi nous introduisons la notion d'enrichissement relatif réduit, indépendant à la fois de ε et de l'ion considéré, et défini par :

$$E = \frac{x - x_0}{x_0(1 - x_0)\varepsilon}.$$

Dans tout ce qui suit, nous désignerons simplement cette grandeur par le terme enrichissement.

Remarquons que cet enrichissement correspond en fait à l'enrichissement relatif, rapporté non plus à ε, mais à $\varepsilon' = \varepsilon(1 - x_0)$ facteur dont nous avons déjà noté l'importance lors de l'étude de la répartition des espèces à l'état stationnaire.

Enrichissement total d'une zone

Si nous considérons une zone de N plateaux, comprise entre les abscisses a et b, nous appellerons enrichissement total de cette zone la somme des enrichissements de chaque plateau de la zone.

$$E_N = E_a^b = \sum_a^b \frac{x - x_0}{x_0(1 - x_0)\varepsilon}.$$

Dans le cas où une relation mathématique traduit la variation de x en fonction de p (état stationnaire, par exemple), il vient

$$E_a^b = \int_a^b \frac{x - x_0}{x_0(1 - x_0)\varepsilon}\,dp.$$

Enrichissement marginal

L'évolution de l'enrichissement au cours du développement sera caractérisée par l'enrichissement marginal, variation de l'enrichissement de la zone enrichie ($p > 0$) lorsque la bande est déplacée de un plateau.

Si t représente le déplacement de la bande exprimé en nombre de plateaux, l'enrichissement marginal représente la valeur de ΔE_p pour $\Delta t = 1$.

En pratique, l'enrichissement marginal sera confondu avec $\frac{dE}{dt}$.

Transfert

Nous avons vu ci-dessus que $m + 1$ transferts étaient nécessaires pour faire progresser la bande de un plateau. Comme nous exprimerons par la suite le déplacement en nombre de plateaux, c'est la succession de $m + 1$ transferts élémentaires que nous désignerons par le terme de transfert et que nous noterons t.

2° Calcul de l'enrichissement limite

Lorsque l'état stationnaire est atteint, la forme de l'isochrone est définie par les relations (9) et (10). Il est alors possible de calculer l'enrichissement entre les abscisses a et b :

$$E_a^b = \int_a^b \frac{x - x_0}{x_0(1 - x_0)\varepsilon}\,dp.$$

et il vient :

$$E_a^b = \frac{b - a}{x_0\varepsilon} = \frac{1}{2\varepsilon^2 x_0(1 - x_0)}\,\mathrm{Ln}\,\frac{x_{(b)}}{x_{(a)}}. \qquad (11)$$

Pour l'ensemble de la zone enrichie ($p > 0$), en faisant $a = 0$ et $b = P/2 + M$, on trouve, en appelant x_e la fraction molaire à l'extrémité enrichie de la bande :

$$E_l = \frac{P/2 + M}{x_0\varepsilon} - \frac{1}{2\varepsilon^2 x_0(1 - x_0)}\,\mathrm{Ln}\,\frac{x_e}{x_0}. \qquad (12)$$

De même, l'étude de la zone appauvrie ($p < 0$) conduit, si l'on appelle x_a la fraction molaire à l'extrémité appauvrie de la bande, à :

$$E_l' = \frac{P/2 - M}{x_0\varepsilon} + \frac{1}{2\varepsilon^2 x_0(1 - x_0)}\,\mathrm{Ln}\,\frac{x_a}{x_0}. \qquad (13)$$

La conservation des espèces impose

$$E_l = -\,E_l'$$

d'où :

$$\text{Ln}\,\frac{x_e}{x_a} = 2P(1 - x_0)\varepsilon.$$

3° Variation de l'enrichissement

Afin d'étudier la variation de l'enrichissement, nous avons simulé sur calculatrice un certain nombre de développements, en faisant varier les divers paramètres de façon à couvrir de manière satisfaisante la totalité du domaine expérimental intéressant ($0{,}001 < \varepsilon < 0{,}1$; $20 < P < 320$; $0{,}01 < x_0 < 0{,}8$), en tenant compte du fait que les phénomènes sont symétriques pour deux valeurs x_0 et x_0' telles que $x_0 + x_0' = 1$. Connaissant la valeur de E_l dans chacun des cas, nous avons tracé les courbes $E = f(t)$, afin d'étudier leur équation. Leur examen ne permet pas de leur attribuer une équation simple. Par contre, la courbe dérivée, qui traduit l'évolution de l'enrichissement marginal a une forme beaucoup plus simple (*fig. 3*) qui conduit à lui attribuer une équation du type :

$$\frac{dE}{dt} = \frac{\alpha}{1 + \left(\frac{t}{\beta}\right)^2}$$

on constate dans tous les cas que si $t = 0$, l'enrichissement marginal vaut 1 (ce qui justifie *a posteriori* la définition de l'enrichissement relatif réduit que nous avons choisie).

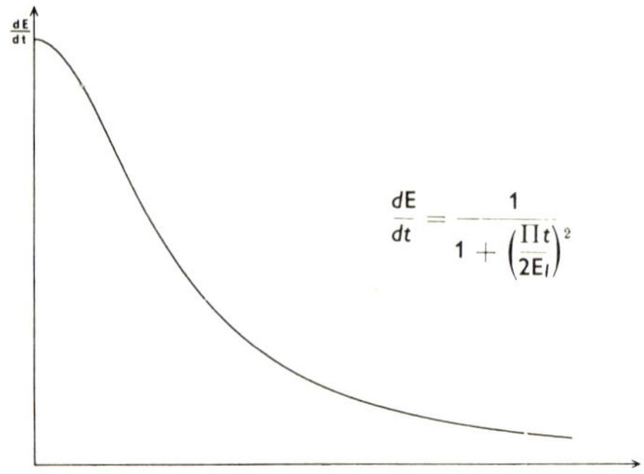

Fig. 3. — Évolution de l'enrichissement marginal $\frac{dE}{dt}$ avec le déplacement.

Il vient donc $\alpha = 1$.

L'intégration de $\frac{dE}{dt}$ donne la loi de variation de E :

$$E = \beta \text{ Arc tg } \frac{t}{\beta}$$

connaissant la valeur limite E_l lorsque $t \to \infty$ (12), on en déduit la valeur de β, ce qui donne :

$$E = \frac{2E_l}{\Pi} \text{ Arc tg } \frac{\Pi t}{2E_l} \qquad (15)$$

$$\frac{dE}{dt} = \frac{1}{1 + \left(\frac{\Pi t}{2E_l}\right)^2}. \qquad (16)$$

La *figure 4* représente la variation de l'enrichissement E en fonction du déplacement *t*.

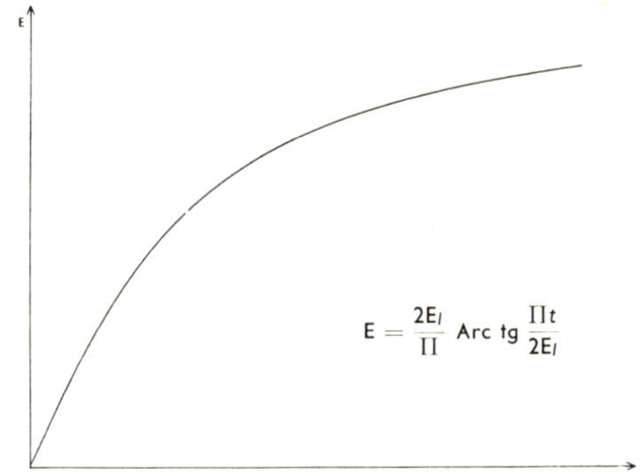

Fig. 4. — Évolution de l'enrichissement E avec le déplacement.

Le *tableau I* rassemble les caractéristiques des diverses simulations réalisées. Pour chacune d'elles, l'écart maximum entre la valeur réelle de E et la valeur calculée grâce à la relation empirique (15) n'est jamais supérieure à l'enrichissement E_l correspondant au plateau extrême de la bande, ce qui est tout à fait satisfaisant.

A titre d'exemple, le *tableau II* compare l'évolution de l'enrichissement calculé à partir de la simulation et de celui déterminé grâce à la relation (15), dans un des cas du *tableau I*.

4° Application : calcul du parcours nécessaire pour atteindre l'état stationnaire

J. Coursier et J. Hure [4], puis B. Trémillon [5] ont proposé l'expression suivante, qui donne le déplacement nécessaire l d'une bande de longueur l_0 pour atteindre l'état stationnaire, dans le cas de deux espèces voisines :

$$l = \frac{l_0}{\varepsilon}.$$

Plus récemment, J. Persoz [1] a montré, par simulation sur calculatrice, que ce déplacement était, en réalité, nettement insuffisant, sans pour autant donner l'expression permettant de le calculer.

En toute rigueur, ce déplacement doit être infini, mais connaissant la variation de l'enrichissement en fonction du nombre de plateaux déplacés, il est possible de calculer le déplacement nécessaire pour atteindre, avec un écart plus ou moins faible, l'état stationnaire.

TABLEAU I

Caractéristiques des simulations étudiées pour déterminer la relation liant l'enrichissement au déplacement

	I	II	III	IV	V
ε	0,01	0,01	0,1	0,001	0,012
P	50	20	50	100	320
x_0	0,2	0,8	0,05	0,01	0,2014
M	− 1,2	0,2	− 12,4	− 0,8	− 48,0
x_e	0,2868	0,8307	0,3935	0,01102	0,7874
E_1	613,88	98,22	340,36	2 528,89	16 891,27
t_{max}	1 000	100	300	1 000	3 200
t_{max}/P	20	5	6	10	10

t_{max} est le nombre maximum de transferts étudiés,
t_{max}/P représente le déplacement maximum (en longueurs de bandes).

TABLEAU II

Comparaison entre l'enrichissement « vrai » et l'enrichissement calculé dans le cas de l'essai I du tableau I

t	E vrai (obtenu par simulation)	E calculé	$\Delta E = E$ vrai − E calculé	$\frac{\Delta E}{E_{vrai}}$ (%)	Enrichissement dans le plateau extrême $p = M + P/2$
100	98,52	97,90	0,62	0,6	14,08
200	187,35	184,86	2,49	1,3	20,78
300	260,31	255,86	4,45	1,7	25,96
400	319,62	311,48	8,14	2,5	30,13
500	367,14	354,61	12,53	3,4	33,51
600	405,27	388,26	17,01	4,2	36,23
700	435,73	414,88	20,85	4,8	38,42
800	460,38	436,29	24,09	5,2	40,18
900	480,05	453,78	26,27	5,4	41,58
1 000	495,82	468,28	27,54	5,5	42,71

Nous dirons que l'état stationnaire est atteint à δ près lorsque l'enrichissement total E sera tel que :

$$\frac{E - E}{E_l} = \delta.$$

Cherchons la relation existant entre δ et t :

$$\delta - 1 - \frac{2}{\Pi} \text{Arc tg} \frac{\Pi t}{2E_l}$$

d'où :

$$\text{Arc tg} \frac{\Pi t}{2E_l} = \frac{\Pi}{2}(1 - \delta)$$

$$\frac{\Pi t}{2E_l} = \text{tg} \frac{\Pi}{2}(1 - \delta) = \frac{1}{\text{tg} \frac{\Pi}{2} \delta}$$

soit

$$\frac{2E_l}{\Pi t} = \text{tg} \frac{\Pi}{2} \delta$$

δ étant petit, on peut confondre la tangente et l'arc, d'où :

$$t = \frac{4E_l}{\Pi^2 \delta}. \qquad (17)$$

En reportant dans l'expression de l'enrichissement marginal (16), il vient :

$$\frac{dE}{dt} = \frac{1}{1 + \frac{4}{\Pi^2\delta^2}}$$

soit, δ étant petit

$$\frac{dE}{dt} \sim \frac{\Pi^2\delta^2}{4}.$$

Exemple

Considérons le déplacement d'une bande ayant les caractéristiques suivantes :

$$x_0 = 0{,}1; \quad P = 150; \quad \varepsilon = 0{,}05.$$

Le calcul donne $E_l = 2\,334$.

Pour $\delta = 0{,}1$ (état stationnaire atteint à 10 % près), on trouve $t = 9\,460$, soit 63 fois environ la longueur de la bande, pour $\delta = 0{,}01$ (état stationnaire atteint à 1 % près), le déplacement doit être 10 fois plus grand, soit 630 fois la longueur de la bande.

En revanche, la relation de Coursier et Hure $l = \frac{l_0}{\varepsilon}$ prévoit que l'état stationnaire sera atteint pour un déplacement de 20 fois la longueur de la bande. L'écart est considérable.

Discussion

D'un point de vue purement expérimental, on considère que l'état stationnaire est atteint lorsque, à la précision des analyses, la forme du front n'évolue plus de manière sensible lorsqu'on continue à déplacer la bande. Cette définition pratique de l'état stationnaire est très différente de la définition rigoureuse fondée sur l'écart à l'enrichissement limite, que nous avons adoptée :

Considérons, en effet, le cas où $\delta = 10\ \%$; on a alors $\frac{dE}{dt} = 0{,}025$.

Si nous reprenons la bande précédente, un déplacement supplémentaire apporte une augmentation de l'enrichissement total égale à 0,025, qui va se répartir sur l'ensemble de la zone enrichie.

Pour connaître cette répartition, il faudrait réaliser la simulation sur calculatrice avec une grande précision. Or, l'étude de l'évolution des fronts montre que l'enrichissement augmente toujours plus vite dans les plateaux extrêmes qu'au milieu de la bande.

Afin de donner un ordre de grandeur, nous supposerons donc que cet enrichissement supplémentaire se répartit d'une manière uniforme dans les 25 plateaux de tête, les autres plateaux de la zone enrichie ne variant pas.

La variation d'enrichissement dans un plateau est alors :

$$\Delta E = \frac{0{,}025}{25} = 10^{-3}.$$

Si la bande est déplacée de une fois sa longueur, chaque plateau de tête verra son enrichissement augmenter de $150\ \Delta E = 0{,}15$, soit une augmentation de fraction molaire

$$\Delta x = 150\ \Delta E x_0(1 - x_0)\varepsilon = 6.10^{-4}$$

ce qui, par rapport à la fraction molaire du mélange initial correspond à une variation relative

$$\frac{\Delta x}{x_0} = 6.10^{-3} = 0{,}6\ \%,$$

valeur à la limite de la précision habituelle des dosages.

Si $\varepsilon = 0{,}01$, la variation relative de la fraction molaire serait 5 fois plus petite, soit 0,12 %.

Autrement dit, une bande peut être assez loin de l'état stationnaire et les concentrations n'évoluer que trop lentement pour que leurs variations après un déplacement de une fois la longueur de la bande puissent être mises en évidence compte tenu de la précision habituelle des analyses.

Il faut donc être très prudent lorsqu'on applique des résultats théoriques valables seulement à l'état stationnaire à un front expérimental qui paraît ne plus évoluer et qui, en réalité, évolue mais trop lentement pour qu'un faible déplacement de la bande (de une fois sa longueur par exemple) fasse varier de façon appréciable la forme du front.

Cherchons à quelle valeur de δ correspond la relation de Coursier et Hure : $l = \frac{l_0}{\varepsilon}$ soit, avec nos notations

$$t = \frac{P}{\varepsilon}$$

en reportant dans (17), il vient :

$$\delta = \frac{4E_l\varepsilon}{\Pi^2 P}$$

soit ici $\delta = 0{,}3$, valeur pour laquelle on ne peut plus, en toute rigueur, confondre l'arc et la tangente. (L'écart encore faible n'a pas d'importance ici où nous cherchons simplement un ordre de grandeur).

Pour cette valeur, la variation relative du titre d'un plateau est de l'ordre de 1 à 2 %, si l'on déplace la bande de une fois sa longueur (avec $\varepsilon = 0{,}05$).

Expérimentalement, on peut considérer qu'elle n'évolue plus et *a fortiori* si ε est plus faible, alors que l'enrichissement n'a atteint que 70 % de sa valeur limite.

L'écart entre état stationnaire vrai et état stationnaire expérimental lorsque le front ne paraît plus évoluer à la précision des analyses est donc considérable.

IV. — ÉTUDE DE LA PRODUCTION

1° Description des opérations

Si, dans une bande, amenée ou non à l'état stationnaire, on prélève une certaine quantité de mélange en tête et en queue (de telle sorte que la composition moyenne x_0 reste inchangée) et que l'on introduise, à l'endroit de la bande où $x = x_0$ une quantité de mélange initial égale à la quantité prélevée, il est possible, en déplaçant à nouveau la bande, de retrouver le profil de départ. C'est ce que montre la simulation sur calculatrice numérique, dont un exemple est donné par la *figure 5*.

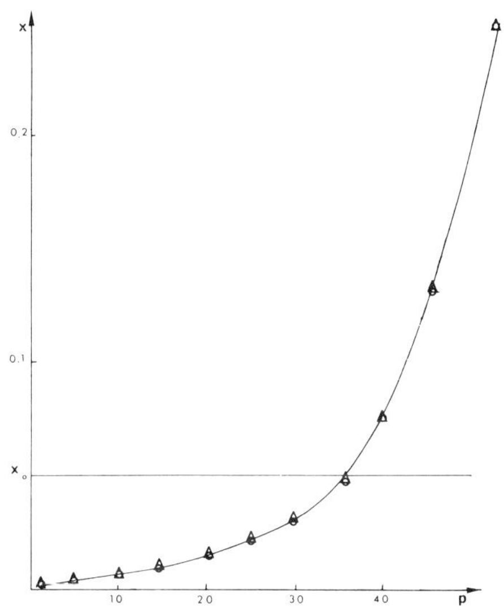

Fig. 5. — Déplacement d'une bande de 50 plateaux $K = 1{,}1$; $x_0 = 0{,}05$.

O Isochrone après un déplacement de 300 plateaux, soit 6 fois la longueur de la bande.

△ Après ce même déplacement, prélèvement de 5 plateaux enrichis et 17 plateaux appauvris, injection de 22 plateaux à la composition initiale et déplacement de 218 plateaux.

En renouvelant le cycle prélèvement, injection, déplacement, on peut produire, de manière semi-continue, des quantités notables de mélange enrichi (et appauvri).

Le problème que nous nous proposons maintenant de résoudre est de déterminer le déplacement nécessaire pour retrouver, après prélèvement et injection, l'enrichissement initial.

En ce qui concerne la forme des fronts, la *figure 5* a montré dans un cas particulier que si l'on part d'une bande déplacée de manière classique, la séquence prélèvement-injection-déplacement ramène à ce même profil. Ce phénomène, vérifié sur de nombreux autres exemples, est tout à fait général, ce qui était prévisible. En effet, la perturbation est d'autant plus forte que l'enrichissement prélevé est important (correspondant à l'injection dans la bande d'un nombre important de plateaux au titre initial x_0). Mais alors, pour retrouver l'enrichissement initial, il faut effectuer un long déplacement, qui permettra aux effets de la perturbation de disparaître.

La simulation complète du phénomène permet naturellement de donner une réponse à ce problème, mais nous allons voir qu'une hypothèse simple permet de la résoudre directement de manière très satisfaisante.

2° Hypothèse

Pour une bande de longueur donnée, nous supposerons que l'enrichissement marginal ne dépend pas de la forme de l'isochrone, mais uniquement de l'enrichissement total.

Ainsi, une bande constituée artificiellement avec des plateaux très enrichis à une extrémité, et très appauvris à l'autre, et comportant une zone importante à la composition x_0, verra son enrichissement évoluer de la même manière qu'une bande de même longueur déplacée de manière que son enrichissement total soit le même que celui de la bande artificielle *(fig. 6)*.

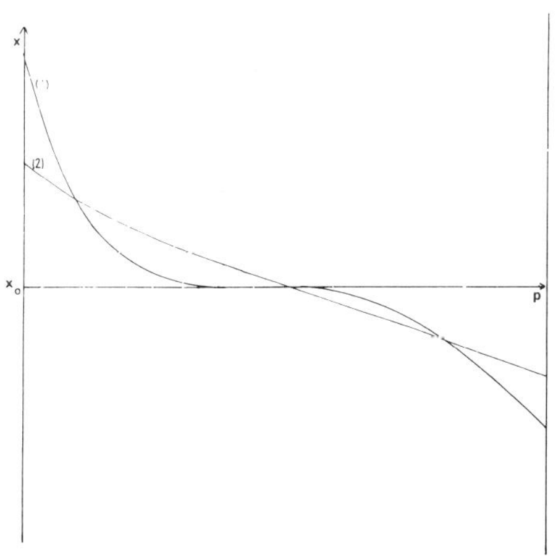

Fig. 6. — Isochrones correspondant à deux bandes de même enrichissement total.

La bande (1) est constituée artificiellement avec du mélange très appauvri, du mélange très enrichi et une zone de composition x_0 importante.

La bande (2) a le même enrichissement que la précédente, mais a été amenée dans cet état par un déplacement convenable.

La *figure* 7 représente l'évolution de l'enrichissement total de la zone enrichie d'une bande quelconque, conformément à l'équation (15), en fonction du déplacement t. A un instant donné, l'état de la bande est représenté par le point $A(t_0, E_0)$. On prélève alors une zone de p_1 plateaux dans l'extrémité enrichie et p_2 plateaux dans l'extrémité appauvrie (la composition moyenne restant égale à x_0).

Les p_1 plateaux prélevés correspondent à un enrichissement E_2. La bande, qui comporte alors $P-(p_1+p_2)$ plateaux, a un enrichissement $E_1 = E_0 - E_2$.

Si, à l'endroit où l'on retrouve la fraction molaire x_0, on injecte $p_1 + p_2$ plateaux de mélange à la composition x_0, l'enrichissement total est inchangé et l'on a constitué artificiellement une bande de P plateaux dont l'enrichissement total est E_1 (correspondant au point B de la *figure* 7).

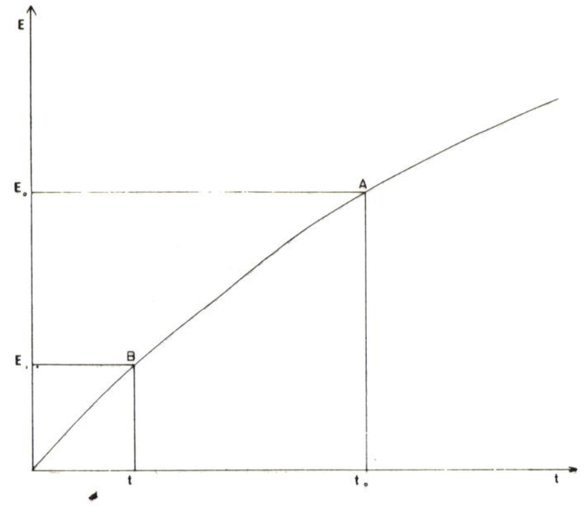

Fig. 7. — Évolution de l'enrichissement en fonction du déplacement dans le cas général. Étude graphique de la production.

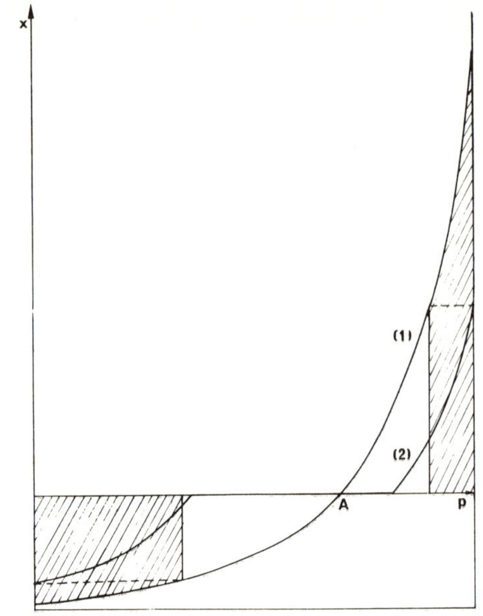

Fig. 8. — Réalisation du cycle déplacement-prélèvement injection-déplacement.

On prélève sur l'isochrone (1) les fractions enrichies et appauvries correspondant aux parties hachurées (dont les aires sont égales), puis on injecte en A la même quantité de mélange à la composition initiale. La bande à déplacer a alors la composition représentée par le profil (2).

L'hypothèse que nous avons faite nous permet de dire que, pour que cette bande retrouve son enrichissement initial E_0, il faut déplacer de $t_2 = t_0 - t_1$ plateaux.

La *figure 8* représente la suite des opérations effectuées au cours d'un cycle.

Nous allons maintenant montrer que l'hypothèse précédente se vérifie d'une manière très satisfaisante par simulation sur calculatrice numérique.

3° Vérification de l'hypothèse par simulation

Dans le *tableau III*, nous avons fait figurer les résultats relatifs à un certain nombre de déplacements réalisés par simulation (permettant de connaître la forme de l'isochrone pour toute valeur de t).

Dans chaque cas, nous donnons l'enrichissement obtenu après t_0 déplacements, avant et après prélèvement, ainsi que le déplacement nécessaire pour retrouver l'enrichissement initial, après injection. Nous donnons également le titre de la tête de bande après le déplacement théorique (correspondant à la zone qui serait prélevée lors d'un cycle ultérieur).

Remarquons que l'obligation de prélever un nombre entier de plateaux ne permet de compenser qu'approximativement par un prélèvement dans la zone appauvrie celui effectué dans la zone enrichie, ce qui introduit de légères fluctuations dans la composition moyenne de la bande, permettant d'expliquer l'écart que l'on observe parfois entre l'enrichissement théorique après $t_0 - t_1$ et celui calculé par simulation complète.

Le *tableau IV* est un tableau simplifié qui, en reprenant les valeurs du *tableau III*, facilite la comparaison entre les résultats fournis par simulation et par calcul à partir de notre hypothèse.

Compte tenu des approximations mentionnées plus haut, nous pouvons considérer que l'hypothèse simple utilisée permet une excellente approche des résultats réels.

Ainsi, à partir d'un profil connu et de conditions de prélèvement déterminées, il est possible de prévoir la longueur du déplacement à effectuer pour retrouver l'état initial, et ceci sans réaliser la simulation. Cette étude réduit donc considérablement l'utilisation de la calculatrice numérique, et permet de résoudre de façon simple tous les problèmes de production d'une bande en développement par déplacement.

V. — CONCLUSION

Nous avons montré comment, à partir d'un petit nombre d'hypothèses très simples, basées sur l'étude de la simulation du processus sur calculatrice numé-

TABLEAU III

Simulation sur calculatrice numérique du développement par déplacement de 6 bandes (numérotées de I à VI) dans des conditions variées

Numéro de la bande	I	II	III	IV	V	VI
Nombre de plateaux de la bande	20	20	50	21	50	100
k	1,01	1,01	1,01	1,1	1,1	1,001
x_0	0,2	0,8	0,2	0,25	0,05	0,01
Déplacement t_0	100	100	550	63	300	998
Enrichissement total E_0	66	66	387	47,5	230,5	900
Enrichissement marginal	0,32	0,34	0,34	0,5	0,5	0,83
Tête — Nombre de plateaux prélevés	2	2	5	4	5	10
Tête — $E_0 - E_1$	25	24	156	34	156	394
Tête — Titre moyen de la partie prélevée	0,22	0,819	0,250	0,408	0,198	0,0104
Queue — Nombre de plateaux prélevés	2	2	6	5	17	10
Queue — Titre moyen de la partie prélevée	0,18	0,791	0,156	0,127	0,001	0,096
Enrichissement total E_1 de la bande après prélèvement	41	42	231	14	75	506
Déplacement correspondant t_1	47	49	258	15	82	523
Reste à déplacer $t_0 - t_1$	53	51	292	48	218	475
Enrichissement après le déplacement $t_0 - t_1$	67	60	395	47,5	230	900
Titre de tête à $t_0 - t_1$	0,22	0,819	0,250	0,409	0,198	0,010
Déplacement effectivement nécessaire pour retrouver E_0	50	47	272	48	219	475

TABLEAU IV

Comparaison entre les principaux paramètres calculés selon l'hypothèse simple et ceux obtenus par simulation sur calculatrice numérique (valeurs reprises du tableau III)

Déplacement théorique $t_0 - t_1$	53	51	292	48	218	475
Déplacement effectivement nécessaire pour retrouver E_0	50	47	272	48	219	475
E_0	66	66	387	47,5	230,5	900
E après le déplacement $t_0 - t_1$	67	68	395	47,5	230	900
Titre de la tête en %	22	81,9	25	40,8	19,8	1,04
Titre de la tête à $t_0 - t_1$ en %	22	81,9	25	40,9	19,8	1,04

rique, il était possible de prévoir l'ensemble des phénomènes, lorsque l'on désire séparer un mélange donné par développement par déplacement.

La décroissance de l'enrichissement marginal avec le déplacement montre qu'il est préférable de travailler loin de l'état stationnaire (pour lequel la forme du front a été précisée), afin qu'un déplacement donné corresponde à l'enrichissement le plus important possible.

En fonction de critères économiques, tels que la quantité de résine immobilisée, la durée des opérations et les volumes de solutions employés d'une part, et des paramètres expérimentaux comme la constante d'échange K et la HEPT h d'autre part, il est alors possible, pour une séparation déterminée, d'optimiser les divers facteurs expérimentaux, tels par exemple que longueur de la bande, section des colonnes, montée en régime initiale, importance des prélèvements.

BIBLIOGRAPHIE

[1] PERSOZ, J. : *Bull. Soc. Chim.*, 848, 1968.
[2] TREMILLON, B. : Les séparations par les résines échangeuses d'ions. Gauthier-Villars, Paris, 249, 1965.
[3] CONRARD, P. et CAUDE, M. : *Ch. et Ind., Gén. Chim.*, **109**, 2431, 1971.
[4] COURSIER, J. et HURE, J. : *Anal. Chim. Acta*, **18**, 272, 1958.
[5] TRÉMILLON, B. : *Bull. Soc. Chim.*, 502, 1958 et 508, 1958.

Manuscrit reçu le 28 septembre 1972.

14

THEORY OF DEVELOPMENT BY DISPLACEMENT IN ION-EXCHANGE CHROMATOGRAPHY: Application to the Separation of Very Similar Substances

Patrick Conrard and Robert Rosset

This condensed translation was prepared expressly for this Benchmark volume by Harold F. Walton, University of Colorado, Boulder Colorado, from "Théorie de développement par déplacement en chromatographie d'echange d'ions: application à la séparation de corps très voisins," Analusis, 2*(1), 37–48 (1973)*

Abstract

The equation for the stationary-state profile of binary mixture of similar species on an ion exchanger is established for the case of development by displacement, using a purely chromatographic model. An approximate relation is derived that gives the variation in enrichment as a function of the displacement. The enrichment calculated in this way is compared with the enrichment found by computer simulation of development by displacement, and the approximation is seen to be justified. The extent of displacement needed to reach the stationary state, within a given margin of error, is calculated. The classical relation $l = l_0 / \epsilon$ (l = displacement, l_0 = length of band, $\epsilon = K - 1$) is discussed. A simple hypothesis for marginal enrichment allows one to make production studies without the need for a computer.

I: INTRODUCTION

At the present time the possibilities of ion-exchange chromatography are evolving rapidly, thanks to the appearance of new resins of well-defined structure and, above all, of fine and highly uniform particle size.

Elution chromatography is widely used for chemical analysis, but its utility in preparative chemistry is limited by the small amounts of material that can be handled, as well as by the dilution that always accompanies the development. The technique of displacement does not suffer from these two drawbacks. It allows one to use long columns and long distances of travel, and is, therefore, appropriate for the preparative separation of species having very similar ion-exchange properties.

There are some notable examples of the use of this technique in difficult separations, such as those of the amino acids, rare earths, and isotopes. Few theoretical studies have been made, however, particularly for the case of very similar species like isotopes. J. Persoz (1) made a limited number of very precise calculations using computer simulation.

Using plate theory, as Persoz did, and limiting ourselves to binary mixtures, we propose to study the progress of displacement development and to establish some simple rules for making calculations. Having reviewed the characteristic features of displacement development, we shall study the distribution of the two species in the stationary state, and then show how the isotopic enrichment progresses in the course of development. Finally, we shall show how displacement development may be applied to preparative chemistry.

II: THEORETICAL EQUATION FOR THE STEADY-STATE PROFILE

1. General

We recall (2) that the technique of development by displacement, in the case of a mixture of two species A and B, consists in placing a band of the mixture on a column of ion-exchange resin that is initially in the ionic form I, then displacing it by an ion D, which is called the developer, under conditions such that the order of affinity of the different species for the resin is the following:

$$\mathrm{I} \ll \mathrm{A} < \mathrm{B} \ll \mathrm{D}$$

As the displacement proceeds, the length of the band remains constant, and the band becomes richer in A at its head and in B at its tail. Within the band there is the equilibrium

$$\mathrm{A}_R + \mathrm{B}_S \rightleftharpoons \mathrm{A}_S + \mathrm{B}_R$$

to which corresponds an exchange constant

$$K_{\mathrm{A}}^{\mathrm{B}} = \frac{[\mathrm{A}]_S\,[\mathrm{B}]_R}{[\mathrm{A}]_R\,[\mathrm{B}]_S} > 1$$

The subscripts R and S indicate the resin and solution phases, respectively.

When the exchange constant is fairly high, one obtains, after a certain distance of travel, a stationary state in which the isochrone, or concentration profile of A and B along the column at a given moment, is a bilogarithmic curve whose equation is

$$\log \frac{\mathrm{A}}{\mathrm{B}} = 2\frac{l}{h} \log K_{\mathrm{A}}^{\mathrm{B}}$$

where h is the height equivalent of the theoretical plate, and l the abscissa reckoned from the point at which [A] = [B].

Previous studies have always assumed that the components A and B were obtained in the pure state at the ends of the band. The situation is quite different when the affinities of A and B for the exchanger are nearly equal ($K_A^B \simeq 1$). Again a stationary-state distribution is approached, but this time without getting pure A and B at the head and tail of the band. If the exchange constant is very close to 1 (1.001), the band is short (50 plates), and the amounts of the two species are very different (1 percent of one species, 99 percent of the other), J. Persoz (1) has shown that the concentration profile is linear at the stationary state.

For higher values of the exchange constant and for mixtures having any desired composition, separations published in the literature show bands whose profile is neither linear nor bilogarithmic, even though the displacement has been long. We propose, therefore, to establish a general equation for the concentration profile in the stationary state for exchange constants close to 1, and to define the conditions that must be met for a linear distribution to be approached.

2. Scheme for Band Displacement

Consider a column of ion-exchange resin carrying exchangeable ions X; let the concentration of X in the resin be C_R mEq/ml of resin. The resin is surrounded by a solution of X whose concentration is C_S mEq/ml of solution. In a volume V of the column, the solution occupies a volume Vx and the resin a volume $V(1-x)$; x is the interstitial volume fraction, usually about 35 percent. Let m be the ratio of the amounts of X in the resin and in the solution, for a given segment of the column; then

$$m = \frac{V(1-x)C_R}{VxC_S} = \frac{1-x}{x} \quad \frac{C_R}{C_s}$$

For most ion-exchange resins, C_R is of the order of several milliequivalents per milliliter of resin. The solutions usually used have concentrations of 0.5 M or less; the ratio m is therefore 10 or more.

To simplify our calculations we shall choose integral numbers for m, but this will in no way lessen the generality of the results. Suppose, now, that a band containing a mixture of A and B is placed on the column. One can show that $(m+1)$ transfers are necessary to advance the band by the width of one plate (3).

[*Editor's Note:* References for this translation will be found in the preceding original article.]

Part V

SEPARATION OF METALS BY ANION EXCHANGE OF CHLORIDE AND FLUORIDE COMPLEXES

Editor's Comments on Papers 15 Through 19

When ion exchange is used today to separate metal ions from one another, or to concentrate trace metals from natural environments, the method of choice is likely to be anion exchange in hydrochloric acid solutions. Most metals form anionic chloride complexes. The stabilities of these complexes vary widely, and so do their affinities for anion-exchange resins. Anion-exchange resins are by nature more selective than cation-exchange resins; witness the large differences in affinities of halide ions and other singly charged ions for a strong-base, anion-exchange resin, compared to the much smaller differences in the affinities of alkali–metal cations, ammonium, and hydrogen ions for a strong-acid, cation-exchange resin. By adjusting the hydrochloric acid concentration, therefore, one can make a large number of separations of metal ions on columns of anion-exchange resins. The only metals that do not bind to such resins at any concentration of aqueous hydrochloric acid are the alkali and alkaline-earth metals, and a few others such as aluminum, lanthanum, thorium, and nickel.

The first ion-exchange separations of metals through their anionic halide complexes were actually made with fluoride ions, not chloride ions. A classic "benchmark" separation was that of zirconium and hafnium described in Paper 15. These elements have almost equal ionic radii; because of this, their chemical properties are very similar and their separation is difficult. Their nuclear properties happen to be very different, and the atomic energy program provided the incentive to separate them.

It was natural to experiment with fluoride complexes of these elements, because their salts are so strongly hydrolyzed that the only practical way to keep them in a stable aqueous solution is as a fluoride complex. The same is true of niobium and tantalum. Paper 16 describes the separation of zirconium and niobium on a strong-base, anion-exchange resin, using mixed hydrochloric acid–hydrofluoric acid eluents. The paper refers to tantalum also. Successful separations of niobium, tantalum, and other elements in hydrochloric acid–hydrofluoric acid mixtures were later made by other workers (e.g., Wilkins, 1955).

Paper 16 is included for another reason. It is the first of an important series, collectively titled "Anion Exchange Studies," in which Kraus, Moore, and Nelson of Oak Ridge National Laboratory showed what could be done with anion-exchange resins in hydrochloric acid. Papers 17 and 18 are also from this series. Distribution ratios were measured over the whole range of hydrochloric acid concentrations, from 0 to 12 M, for most of the elements of the periodic table. The results of 5 years of work were collected and presented by Kraus and Nelson at the First United Nations Conference on Peaceful Uses of Atomic Energy. Their report is reproduced here as Paper 19. A feature of this document is the set of graphs of distribution coefficient against hydrochloric acid concentration presented for each element in the form of the periodic table (see Figure 1, which has been reproduced in many textbooks and reviews). With the help of this "road map" it is possible to plan a very large number of separations, using only a strong-base, anion-exchange resin and different concentrations of hydrochloric acid. And, as the authors point out, complexing anions other than chloride can be used equally well; for example, uranium(VI) is absorbed from solutions of sulfuric and nitric acids over different concentration ranges.

An important advantage of hydrochloric acid for analytical purposes is that the acid excess is easily removed by evaporation. Moreover, hydrochloric acid can be purified by distillation, an essential consideration in trace analysis where reagent blanks are important.

In the years following Kraus and Nelson's paper, systematic studies have been made of distribution ratios between anion- and cation-exchange resins and a number of acids, such as hydrofluoric, nitric, sulfuric, and phosphoric acids, by various authors, including Faris (Faris, 1960; Faris and Buchanan, 1964) and Strelow (Strelow and Bothma,

1967; Strelow and Weinert, 1970; Strelow and Sondorp, 1972; Strelow et al., 1971, 1972). The periodic table arrangement of semilogarithmic graphs has become the standard way to present the results of these studies.

REFERENCES

Faris, J. P. 1960. *Anal. Chem., 32,* 520.
——, and R. F. Buchanan. 1964. *Anal. Chem., 36,* 1158.
Strelow, F. W. E., and C. J. C. Bothma, 1967. *Anal. Chem., 39,* 595.
——, and H. Sondorp. 1972. *Talanta, 19,* 1113.
——, and C. H. S. W. Weinert. 1970. *Talanta, 17,* 1.
——, A. H. Victor, C. R. van Zyl, and C. Egloff. 1971. *Anal. Chem., 43,* 870.
——, C. H. S. W. Weinert, and C. Egloff, 1972. *Anal. Chem., 44,* 2352.
Wilkins, D. H. 1955. *Talanta, 2,* 355.

15

Reprinted from *J. Amer. Chem. Soc.*, **71**, 4147 (Dec. 1949)

THE ANION-EXCHANGE SEPARATION OF ZIRCONIUM AND HAFNIUM

E. H. Huffman and R. C. Lilly

An effective separation of zirconium and hafnium as negative fluo-ions has been obtained by using one of the strongly basic anion exchange resins recently developed, although earlier attempts using the weakly basic resins first available were unsuccessful.

Twenty milligrams of zirconium and 10 mg. of hafnium, as oxides, were added to zirconium and hafnium tracers in a small amount of hydrofluoric and nitric acids, warmed until dissolved, and fumed with 0.5 ml. of sulfuric acid. The residue was dissolved in water and the hydroxides precipitated with ammonium hydroxide. The precipitate was centrifuged, washed with water, dissolved in 5 ml. of 0.64 *M* hydrofluoric acid and diluted to 10 ml. with water. Six hundred milligrams of 200–325 mesh Amberlite IRA-400 resin, in its original chloride form, was added to the sample and the mixture shaken for three hours. The resin was separated from the solution and washed well with 10 ml. of water. Tracer count of the solution and washings indicated that 96% adsorption had taken place. This portion of resin was slurried onto the top of a column of the same resin 30 cm. in length and 0.78 sq. cm. in cross section. Elution with a solution of 0.2 *M* hydrochloric acid and 0.01 *M* hydrofluoric acid at the rate of 6 ml. per hour gave the results shown in Fig. 1. The solid parts of the curve were obtained by counting Zr^{95} and Hf^{181} tracers and the dotted parts by spectrographic analysis. The order of elution of the two elements is the reverse of that obtained by cation-exchange.[2]

Combined fractions of elutriant from 300 ml. to 653 ml., containing 13.8 mg. of zirconium or 69% of the starting material, were found to contain no hafnium detectable by spectrographic analysis. The portion from 300 ml. to 686 ml., containing 17.0 mg. of zirconium or 85% of the starting material, was found to contain 0.04% hafnium. Spectrographic analysis of the 752–1020-ml. portion showed 0.02% zirconium in the 6.9 mg. of hafnium (69% of the starting material). Similarly, 0.03% zirconium was found in the 704–1020-ml. portion containing 8.3 mg. of hafnium (83% of the starting material). The amounts of the major constituents in these portions were determined from the curves.

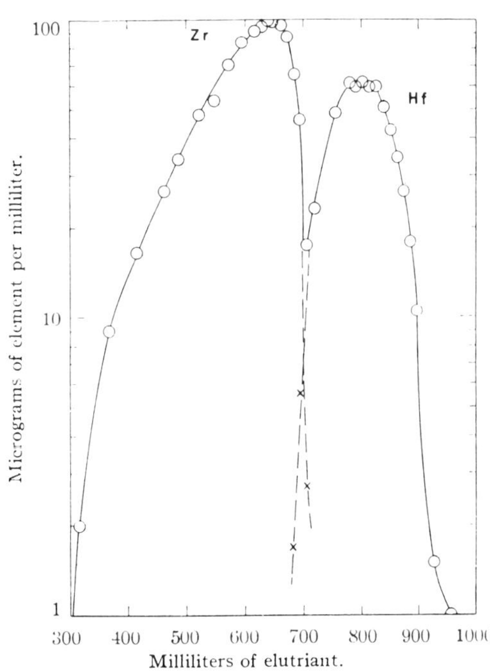

Fig. 1.—Elution of fluozirconate and fluohafniate with 0.2 *M* hydrochloric acid and 0.01 *M* hydrofluoric acid.

This work was done under the auspices of the Atomic Energy Commission.

(1) While official declassification of this paper was being awaited, a communication on a similar separation appeared by Kraus and Moore, This Journal, **71**, 3263 (1949). The separation reported here uses a different resin, a much shorter column, a more dilute eluting solution and macro quantities of zirconium and hafnium instead of micro quantities.

(2) Kenneth Street, Jr., and G. T. Seaborg, This Journal, **70**, 4268 (1948).

16

Reprinted from *J. Amer. Chem. Soc.*, 73, 9–13 (Jan. 1951)

Anion Exchange Studies. I. Separation of Zirconium and Niobium in HCl–HF Mixtures[1]

By Kurt A. Kraus and George E. Moore

The specificity of anion exchange resins for negatively charged ions suggests their use for studies of negatively charged complexes of metal ions and for separations of metals which form such complexes. Anion exchange resins should be particularly useful with metal ions which would precipitate if not complexed (*e. g.*, 4th and 5th group elements), where cation exchange resins could not be used to full advantage. This paper, which deals with the separation of zirconium and niobium, is the first one of a series[2] dealing with these two groups of elements. Although the studies were restricted to HF–HCl mixtures it is expected that similarly effective separations could be attained with other complexing agents.

Experimental

Columns, Column Operation, Materials.—The adsorption columns were constructed of "Transflex" tubing with cross-sectional area *ca.* 0.03 sq. cm., as determined by weighing a column of water of known length. The resin used was Dowex-1, a "strong-base," quaternary amine polymer of mesh size 200–230 as obtained by wet screening. Only the chloride form of the resin was used. The eluent was contained in 50-ml. bakelite burets and the height of the liquid column was occasionally adjusted to maintain approximately the desired flow rate. It was not attempted to maintain constant flow rates. Instead, the volume of eluent which had passed was determined from the buret readings at frequent intervals.

In most cases the zirconium and niobium were added in a small volume to a column which had previously been equilibrated with the desired acid mixture. The added solution was of approximately the same composition as the solution in the column and the eluent. In a few cases the additions were carried out with the column liquid at a composition to cause strong adsorption, and the material was then eluted by changing the composition of the eluent.

The experiments were carried out with tracer zirconium (Zr^{95}, β-emitter $T_{1/2}$ = 65 d.)[3] and tracer niobium (Nb^{95}, β-emitter, $T_{1/2}$ = 35 d.),[3] which were obtained by neutron bombardment of spectroscopically pure zirconium metal[4] in the ORNL reactor.[5] The metal was dissolved in HF–HCl mixtures. Since Nb^{95} is the daughter of Zr^{95}, it was "carrier-free" while by necessity some inactive zirconium was present (*ca.* 0.5 mg. per experiment). The experiments were carried out in a thermostated room at 25 ± 1°. C. P. chemicals were used throughout.

Analytical Procedures. Direct Scanning of Columns.—The data were obtained through continuous monitoring of the activity of the effluent,[6] and through direct scanning of the columns with a detector to locate the activities (bands) and determine their elution rates. After elution identity of the tracers was established through standard radiochemical assays.

The determination of elution rates by scanning offers a number of advantages over the method of continuous monitoring of the effluent. Thus the elution of a previously identified activity can be determined rapidly and with relatively high precision, since the bands in general are quite sharp. Except in the most unfavorable cases a precision of better than 10% is attainable with the band moving through less than 1 cm. of column. Thus considerable time can be saved particularly when the elution rate of a material is unknown and when the optimum length of the column cannot be predicted. It is possible to measure elution rates of several activities at the same time and under identical conditions. Furthermore, the same band (or series of bands) can be studied under a variety of eluting conditions by changing the composition of the

(1) This document is based on work performed for the Atomic Energy Commission at Oak Ridge National Laboratory.

(2) In two earlier communications (K. A. Kraus and G. E. Moore, This Journal, **71**, 3263, 3855 (1949)) the method was shown to be successful for the separation of zirconium and hafnium, and niobium and tantalum. Since then the zirconium-hafnium separation was confirmed by E. H. Huffman and R. C. Lilly (*ibid.*, **71**, 4147 (1949)).

(3) Information from "Table of Isotopes," by G. T. Seaborg and I. Perlman, *Rev. Mod. Phys.*, **20**, 585 (1948).

(4) The only appreciable impurity was hafnium. It yields Hf^{181} (β-emitter, $T_{1/2}$ = 46 d.[3]) on bombardment. In most cases its behavior was very similar to zirconium and, as a first approximation, all results for zirconium also apply for hafnium. Details for their separation will be given in a later paper. Since in most cases discussed here relative enrichment of zirconium and hafnium was not determined, the uncertainty in the reported elution constants for Zr is relatively high.

(5) For some of the earlier work fission product Nb^{95} was used which was kindly furnished us by W. H. Baldwin and co-workers who isolated it from fission product mixtures.

(6) K. A. Kraus and G. E. Moore, This Journal, **72**, 4293 (1950).

eluent as soon as elution rates are established with sufficient precision. Using the scanning technique columns of standard length (*e. g.*, 100 cm.) can be selected which are long enough to permit satisfactory measurement for rapidly moving bands. The method can furnish a promising check of adsorption theories, particularly changes of band characteristics with elution rate or distance.

For scanning a G–M tube was first mounted on a cathetometer stand. Collimation was achieved with a 0.25 inch lead shield containing a narrow slit. The tube was connected to a standard G–M scaler circuit. The shape and position of the bands was obtained from the counting rate as a function of column distance. In view of its usefulness the method was adapted for continuous operation. The automatic scanner is illustrated in Fig. 1. In this instrument the detector (G–M tube or scintillation counter) is driven by a guide screw and its output fed into a scaler with integrating device (count-rate meter) whose output in turn is recorded on a Brown Electronik Strip Chart recorder. The guide screw is driven by a synchronous motor through a gear box which permits scanning rates of 0.5, 1, 2, 4 and 8 in./min. The carriage also holds a reversing switch which is tripped by two adjustable tripping plates. This reversing switch also controls the power input of the recorder, so that it operates only during the downward travel of the carriage. The results of a typical scanning experiment are shown in Fig. 2 which is a reproduction of several scanning intervals taken from a continuous record. In Fig. 3 the band positions (maxima) in each consecutive interval are given together with the buret readings from which the elution constant E (see below) was calculated.

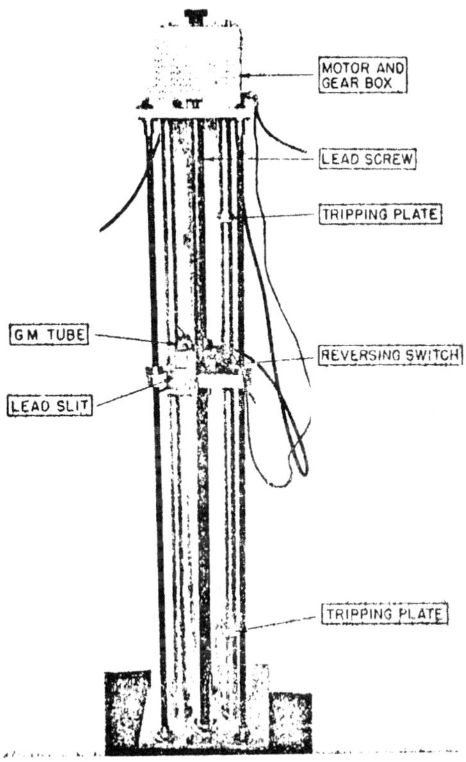

Fig. 1.—Automatic scanner.

Results and Discussion

Elution Constants of Zirconium(IV) and Niobium(V) in HF-HCl Mixtures.—The results of the experiments will be described in terms of the elution constant E, defined by the equation

$$E = dA/V \qquad (1)$$

where d is the distance in cm. a band maximum travels after V cc. of eluent have passed through a column of cross-sectional area A sq. cm.[7] The velocity with which an adsorption band moves through an adsorbent bed under equilibrium conditions is given by Ketelle and Boyd[8,9] as

$$d/t = v/(1 + D/i) \qquad (2)$$

where t is the time, v the linear flow rate of the solution, i the interstitial (fractional void) space and D the bed distribution coefficient (amount per cc. of bed/amount per cc. of solution).[10] Since

$$tv = V/iA \qquad (3)$$

combination of equations (1), (2) and (3) yields

$$E = 1/(i + D) \qquad (4)$$

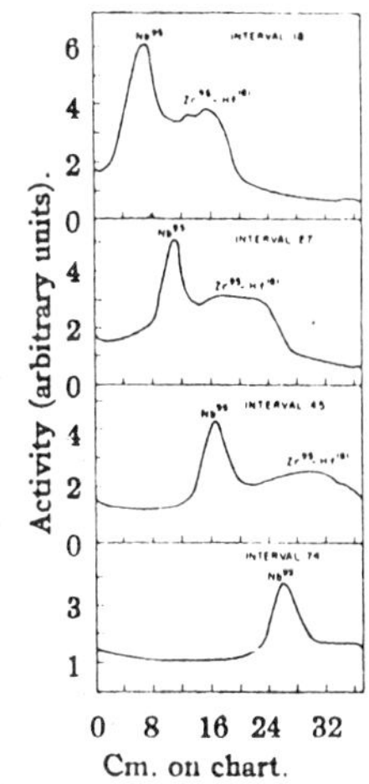

Fig. 2.—Typical intervals from continuous scanning record: Dowex-1 resin, 0.5 M HCl–0.5 M HF, motor speed 1 in./min., 1.00 cm. chart = 2.00 cm. column, flow rate = 0.26 ml. cm.$^{-2}$ min.$^{-1}$.

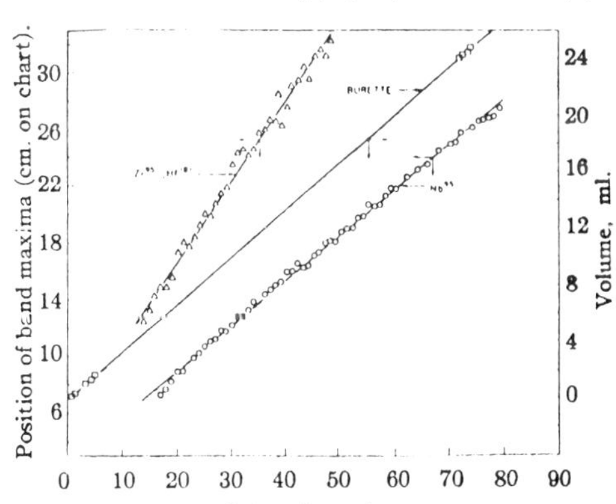

Fig. 3.—Determination of elution constant with automatic scanner: Dowex-1 resin, 0.5 M HCl–0.5 M HF, motor speed 1 in./min., average chart interval 36.86 cm., average column interval 73.69 cm., 1.00 cm. chart = 2.00 cm. column, flow rate 0.26 ml. cm.$^{-2}$ min.$^{-1}$.

(7) E is closely related to the R value frequently used in chromatography which is defined as the ratio of the linear velocities of adsorption band to eluent (A. L. LeRosen, THIS JOURNAL, **64**, 1905 (1942); H. H. Strain, *Anal. Chem.*, **21**, 75 (1949); **22**, 41 (1950)).

(8) B. H. Ketelle and G. E. Boyd, THIS JOURNAL, **69**, 2800 (1947).

(9) This can also be derived from the theory of A. J. P. Martin and R. L. M. Synge, *Biochem. J.*, **35**, 1358 (1941).

(10) For column work D appears to be more useful than K_d, the distribution coefficient in the units amounts per gram oven-dried resin/amounts per cc. of solution. The relationship between these two distribution coefficients is $D = \rho K_d$ where ρ is the bed density (g. oven-dried resin/cc. of bed).

which for $D \gg i$ reduces to

$$E \approx E^* = 1/D \qquad (5)$$

Similarly as pointed out by Ketelle and Boyd,[8] i can be determined by measuring E for a non-adsorbed species (*e. g.*, a positively charged ion for an anion exchanger).

The elution constants of zirconium(IV) and niobium(V) (E_{Zr} and E_{Nb}) were determined for several series of HCl–HF mixtures. The results are summarized in Table I. Zirconium(IV) was found to be very weakly adsorbed under most conditions studied. In 3 and 9 M HCl, at all HF concentrations studied, it has an elution constant (E_{Zr}) which is within a factor of 2 or 3 of the maximum (*ca.* 3, since i = *ca.* $^1/_3$). At lower chloride concentrations zirconium becomes rapidly strongly adsorbed ($E_{Zr} = 8.5 \times 10^{-3}$ in 0.2 M HCl–0.5 M HF). Although the data are fragmentary, it appears that E_{Zr} is independent of HF concentration in the regions studied, suggesting that a single species predominates in these solutions.

TABLE I

ELUTION CONSTANTS OF ZIRCONIUM(IV) AND NIOBIUM(V) IN HF–HCl MIXTURES

M HCl	M HF	Elution constants $E_{Zr} \times 10^3$	$E_{Nb} \times 10^3$	E_{Zr}/E_{Nb}
9	0.004	*ca.* 2000	3.0 ± 0.3[a]	670
9	.01	.....	6.2	...
9	.02	.....	19	...
9	.05	*ca.* 3000	78	38
9	.1	.....	220	...
9	.18	.....	230	...
9	.5	2-3000	72 ± 2	35
9	1.0	.....	27	...
9	6.9	*ca.* 2000	18	110
6	0.5	*ca.* 1000	150 ± 10[a]	7
3	.1	*ca.* 1500	190	8
3	.5	*ca.* 1000	180 ± 10[a]	6
3	2.0	*ca.* 1000	110	9
3	5.0	*ca.* 1000	43	23
3	10.0	.....	21	...
3	17.3	.....	24	...
1	0.5	170	73	2.3
0.5	.05	.....	39	...
.5	.1	65 ± 15[a]	40 ± 1[a]	1.6
.5	.5	65 ± 15[a]	45 ± 5[a]	1.5
.2	.5	8.5	11.5 ± 1.5[a]	0.74

[a] Averages of two or three determinations with maximum deviation given.

The elution constants of niobium (E_{Nb}) vary both with HF and HCl concentrations. Thus in 9 M HCl E_{Nb} is small at low fluoride concentrations (3.0×10^{-3} in 0.004 M HF), rises with increasing HF concentration, passes through a maximum near 0.1 M HF and then rapidly decreases (see Fig. 4). In 3 M HCl data have been obtained only at HF concentration equal to or larger than 0.1 M in which range E_{Nb} decreases up to 17 M HF. E_{Nb} is independent of HF concentration in 0.5 M HCl–0.1 to 0.5 M HF.

On the basis of these data separation of zirconium and niobium should be possible under a variety of conditions. Difficulties would occur only at low HCl concentration where E_{Zr} and

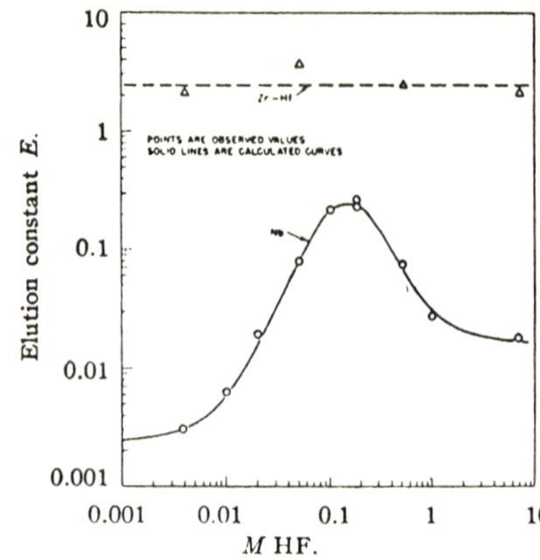

Fig. 4.—Elution constants as a function of HF concentration in 9 M HCl.

E_{Nb} become approximately equal. In fact between 0.2 and 0.5 M HCl in 0.5 M HF there is an inversion of the two elution constants, E_{Zr} being larger in 0.5 M HCl and E_{Nb} in 0.2 M HCl (see Fig. 5). At an intermediate HCl concentration E_{Zr} is equal to E_{Nb} and separation would become impossible. Efficiency of separation of these elements is thus critically dependent on the composition of the eluent. It can be seen from Fig. 5 that in 0.5 M HF, E_{Zr} increases approximately with the square of the HCl concentration, while E_{Nb} increases at first approximately linearly, passes through a maximum and then sharply decreases. The ratio E_{Zr}/E_{Nb}, which determines

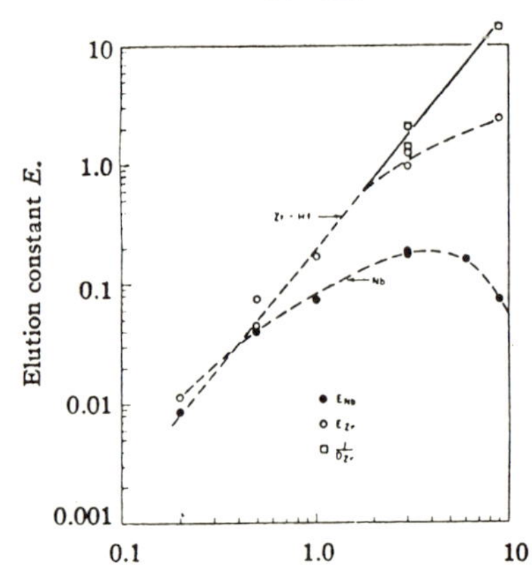

Fig. 5.—Elution constants as a function of HCl concentration in 0.5 M HF

the efficiency of separation, thus increases greatly with increasing HCl concentration. Furthermore until E_{Nb} passes through the maximum, the apparent paradox exists that separation becomes more effective as the elution rate of the slowest component increases. Typical separations are illustrated in Fig. 6 which shows elution curves transcribed from automatic records of the activity of the effluent. It may be noticed that under some conditions the separations are sufficiently good as to permit the use of extremely short columns, make batch separations very effective and permit economical operation of columns under break-through conditions.

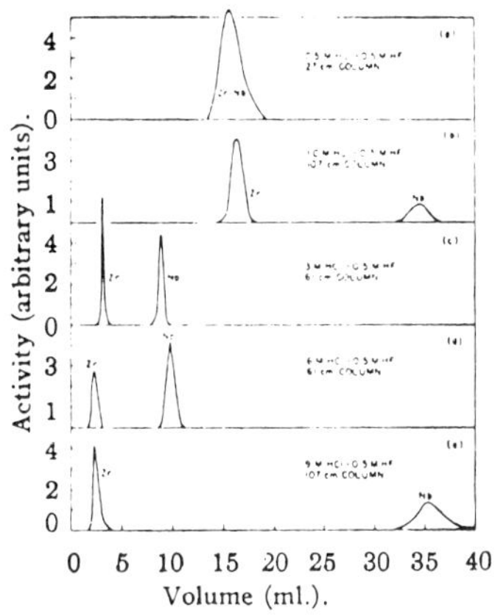

Fig. 6.—Separation of Zr and Nb with Dowex-1 anion exchange columns: average flow rate *ca.* 0.3 ml./sq. cm./min.; cross-section of column: 0.0226 sq. cm.

Ionic Species.—Assuming that the adsorption equilibrium of an ion A^{-n} can be represented by the equation[11]

$$n\text{RCl} + \text{A}^{-n} \rightleftharpoons \text{R}_n\text{A} + n\text{Cl}^- \qquad (6)$$

(where R represents the resin) one obtains the mass action constant

$$K_{\text{A}^{-n}} = \frac{[\text{R}_n\text{A}]}{[\text{RCl}]^n} \frac{(\text{Cl}^-)^n}{(\text{A}^{-n})} \frac{\gamma^n_{\text{Cl}^-}}{\gamma_{\text{A}^{-n}}} \qquad (7)$$

where brackets indicate activities, parentheses concentrations and where γ is the activity coefficient of the ion given as subscript. For tracer solutions the simplifications

$$[\text{R}_n\text{A}] = \alpha(\text{A}^{-n})_r \text{ and}$$
$$[\text{RCl}]^n = \text{constant} \qquad (8)$$

will be assumed where α is a proportionality constant and $(\text{A}^{-n})_r$ the concentration of A^{-n} in the resin. Combining equations (4), (5), (7) and (8) and remembering

$$D_{\text{A}^{-n}} = \frac{(\text{A}^{-n})_r}{(\text{A}^{-n})} \qquad (9)$$

one obtains

$$D_{\text{A}^{-n}} = K^*_{\text{A}^{-n}} \frac{\gamma_{\text{A}^{-n}}}{\gamma^n_{\text{Cl}^-}} \frac{1}{(\text{Cl}^-)^n} = \frac{1 - Ei}{E} = \frac{1}{E^*} \approx \frac{1}{E} \qquad (10)$$

where $K^*_{\text{A}^{-n}} = K_{\text{A}^{-n}}[\text{RCl}]^n/\alpha$ is a proportionality constant. Thus, neglecting the activity coefficient ratio, a plot of log E^* (or log E for small values of E) *vs.* log (Cl^-) for a single (tracer) species has a slope equal to the charge n of the ion. For two ions of different charge two straight lines will result which cross at some hydrochloric acid concentration P. The order of elution will depend on the relative magnitude of the chloride concentration of the eluent and of P. Thus for (HCl) < P the more highly charged ion elutes last (is more strongly adsorbed) and for (HCl) > P it elutes first.

(11) A. P. Vanselow, THIS JOURNAL, **54**, 1307 (1932).

On the basis of the following observations it appears probable that P is considerably smaller than 1 M and possibly smaller than 0.1 M for the type of complexes studied here and that hence in strong hydrochloric acid the more strongly adsorbed ions carry the smaller negative charge and most probably have a charge of minus one. Thus P for the very similar pair of ions HSO_4^-–$\text{SO}_4^=$ was found to be small (*ca.* 0.05).[12] A tantalum complex in HF–HCl solutions for which log E *vs.* log (HCl) was approximately unity, suggesting a charge of minus one, had a very small E (*ca.* 5×10^{-3}) in 9 M HCl.[13] Conversely the zirconium complex studied here for which log E *vs.* log (HCl) was considerably larger than unity, suggesting a charge more negative than minus one, had a large value of E (E^* = *ca.* 10 in 9 M HCl).[14] Addition of hydrogen fluoride to concentrated hydrochloric acid solutions of some ions which are generally considered to be strongly complexed by fluoride ions, causes a marked increase of E.[15] It appears reasonable to assume that increased complexing is usually paralleled by an increase in negative charge.

The complicated hydrogen fluoride dependence of E_{Nb} necessitates assumption of at least three species. Species I would exist (in 9 M HCl) only at very low fluoride concentrations and may not contain fluoride ions. Formation of species II would account for the increase of E with increasing fluoride concentration. The subsequent decrease of E at still higher fluoride concentrations would be due to the formation of species III. On the basis of the above considerations one could tentatively assign a charge of minus one to species I and III and a greater negative charge to species II.[16]

The probable general equation for the formation of species II from species I is

$$\text{NbX}_6^- + n\text{HF} + j\text{H}_2\text{O} \rightleftharpoons \text{H}_{n-j}\text{NbX}_{6-m}\text{F}_n^{-1-j+m} + m\text{X}^- + j\text{H}_3\text{O}^+ \qquad (11)$$

with the (molarity) equilibrium constant

$$K^*_{12} = (2)(\text{X}^-)^m(\text{H}_3\text{O}^+)^j/(1)(\text{HF})^n \qquad (12)$$

where (1) and (2) represent the concentration of species I and II. At constant hydrochloric acid concentration equation (12) reduces to

$$K_{12} = K^*_{12}/(\text{X}^-)^m(\text{H}_3\text{O}^+)^j = (2)/(1)(\text{HF})^n \qquad (13)$$

For the formation of species III from species II the analogous expression for the equilibrium constant is

$$K_{23} = K^*_{23}/(\text{X}^-)^{m'}(\text{H}_3\text{O}^+)^{j'} = (3)/(2)(\text{HF})^{n'} \qquad (14)$$

(12) K. A. Kraus and E. Rona, unpublished results.

(13) K. A. Kraus and G. E. Moore, THIS JOURNAL, **73**, 13 (1951).

(14) log E^* *vs.* log (HCl) has a slope of approximately two for zirconium. The conclusion that the charge of the complex is minus two (*i. e.*, $\text{ZrX}_6^=$ or $\text{HZrX}_7^=$), however, does not appear warranted since the change of the activity coefficient ratio $\gamma_{\text{A}^{-n}}/\gamma^n_{\text{Cl}^-}$ is not known. A charge of minus three appears equally probable (*i. e.*, $\text{ZrX}_7^{\equiv}$).

(15) This is the case with tantalum (ref. 13) and protactinium (K. A. Kraus and G. E. Moore, unpublished results) as well as with niobium at low M HF, but apparently not with niobium at high M HF as will be discussed below.

(16) It should be noted, however, that assumption of specific charges is unnecessary for the evaluation of the fluoride dependence.

Defining

$$D_1 = (1)_r/(1);\ D_2 = (2)_r/(2);\ D_3 = (3)_r/(3) \text{ and } D_0 = T_r/T \quad (15)$$

where D_0 is the observed distribution coefficient, D_1, D_2 and D_3 those of the "pure" species and T_r and T the concentrations of niobium in the resin and solution, respectively, setting (HF) = f and combining equations (13), (14) and (15) one obtains

$$D_0 = (D_1 + D_2K_{12}f^n + D_3K_{12}K_{23}f^{n+n'})/(1 + K_{12}f^n + K_{12}K_{23}f^{n+n'}) \quad (16)$$

In the limits, *i. e.*, when the two reactions do not overlap, equation (16) reduces to

$$D_0 = (D_1 + D_2K_{12}f^n)/(1 + K_{12}f^n)$$

and

$$D_0 = (D_2 + D_3K_{23}f^{n'})/(1 + K_{23}f^{n'}) \quad (17)$$

The 9 M HCl data could not be fitted to equation (16) taking $n = 1$ and $n' = 1$. Satisfactory agreement was obtained using $n = n' = 2$ as shown by the solid line Fig. 4 which represents the calculated values. The constants $D_1 = 415$, $D_2 = 1.0$, $D_3 = 60$, $K_{12} = 1.5 \times 10^4$ and $K_{23} = 1.25$ were evaluated by trial and error. They can only be considered rough approximations in view of the limited number of data available. An analogous calculation for the 3 M HCl data was not attempted since the slight increase in E at very high HF concentration suggests the formation of a fourth species. However, it is apparent that K_{23} must be at least a factor of 10 smaller in 3 M HCl than in 9 M HCl. Since niobium is known to form oxygenated ions (of the type $NbOX_n^{+3-n}$)[17] it appears reasonable to assume that species I and II contain one coördinated oxide ion. This would explain the great difference in anion exchange behavior of niobium and tantalum.[2,13,18] In species III the oxide ion has probably been replaced by fluoride ions. This would make plausible the great decrease in K_{23} with decreasing HCl concentration and the decrease in E as species III is formed since it would give a mechanism for a decrease in negative charge on addition of fluoride ions.

Acknowledgment.—It is a pleasure to acknowledge the valuable technical assistance of Mrs. Hazel Templeton and to thank Messrs. J. East, F. W. Manning, P. P. Williams and the Instrument maintenance group under R. E. Toucey for the assembly and the maintenance of the electronic equipment, Mr. F. King for the construction of the automatic scanner, and Dr. G. E. Boyd for many valuable discussions.

Summary

1. The behavior of zirconium(IV) and niobium(V) on the anion exchange resin Dowex-1 was studied in a number of HF–HCl mixtures.

2. Separation of the two elements is excellent under a number of conditions. Although zirconium is usually eluted first, the elements can be eluted in reverse order under some conditions.

3. Attempts have been made to interpret the data in terms of ionic species and their equilibria. In the case of niobium at least three species were observed in 9 M HCl differing by two fluoride ions each. The first two species, existing at low fluoride concentration, are probably oxygenated.

(17) A. F. Wells, "Structural Inorganic Chemistry," Clarendon Press, Oxford, 1945, p. 424.

(18) In ref. (2) this difference was attributed to a difference in charge between the niobium and tantalum complexes and it was assumed that the more strongly adsorbed tantalum carries a greater negative charge. This was equivalent to the assumption that the cross-over point P would occur at chloride concentrations so high as to be practically unattainable in aqueous solutions. As pointed out above, this assumption now appears very unlikely.

OAK RIDGE, TENN. RECEIVED MAY 15, 1950

17

Reprinted from *J. Amer. Chem. Soc.*, 74, 843 (1952)

Anion Exchange Studies. IV.[1,2] Cobalt and Nickel in Hydrochloric Acid Solutions

By George E. Moore and Kurt A. Kraus

The anion exchange behavior of nickel(II) and cobalt(II) in 0.5 to 12 *M* hydrochloric acid was studied by measuring their elution constants $E = dA/V$, where d is the distance (cm.) an absorption band moves when V (cc.) of eluent have been passed through a column of A ($cm.^2$) cross-sectional area.

Glass columns of the practically colorless resin Dowex-1 (a "strong base" quaternary amine polystyrene–divinylbenzene resin) were used, whose cross-sectional areas were 0.023 $cm.^2$. Since both nickel and cobalt are colored (the latter has a characteristic blue color on the resin) the band positions (and hence E) were determined visually. Approximately 1 mg. of Co and Ni were used per experiment. In addition a few experiments were carried out with tracer cobalt (Co^{60}) and d determined with the automatic scanner previously described.[3]

The results for cobalt are summarized in Fig. 1. E_{Co} rapidly decreases from approximately 2.5 in less than 3 *M* HCl to *ca.* 0.02 in 9 *M* HCl and then increases again to *ca.* 0.055 in 12 *M* HCl. E_{Ni} on the other hand was found to be *ca.* 2.5 throughout the whole acid range. The difference in the adsorption behavior of the two elements is sufficient to permit excellent separation. Since as discussed earlier[3] $E = 1/(i + D)$ where i is the fractional interstitial space and D the distribution coefficient (amount per ml. of resin/amount per ml. of solutions), small values of E indicate good adsorption. Furthermore, E will reach a maximum value ($E_{max} = 1/i$) when D becomes zero. Using sodium and potassium tracers and assuming that their ions are not adsorbed E_{max} = *ca.* 2.5 was found for these columns. Hence there is negligible adsorption of nickel in the range 0.5 to 12 *M* HCl and of cobalt in the range 0.5 to 3 *M*.

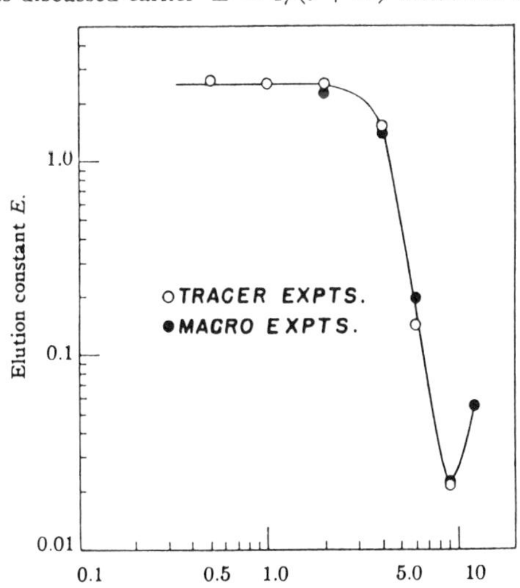

Fig. 1.—Adsorption of cobalt on Dowex-1.

Cobalt is adsorbed as a blue band which is sufficiently intense to permit detection of traces of cobalt (0.5 microgram in 9 *M* HCl was detectable on a 0.023 $cm.^2$ column). Such tests have also been carried out as batch experiments. For example, 1 mg. of Dowex-1 was stirred into 0.5 cc. of 9 *M* HCl containing 4 mg. of reagent grade $NiCl_2 \cdot 6H_2O$ (with 0.13% cobalt). The suspension was slowly dropped onto filter paper. A blue ring of resin (indicating cobalt) formed as the green solution dispersed through the paper.

A comparison of the adsorption data of cobalt with those of iron(III)[4] is of interest. The latter shows very much greater adsorption in concentrated HCl and no maximum as does cobalt near 9 *M* HCl. It had been shown earlier[3] that very strong adsorption in concentrated HCl is probably characteristic of singly negatively charged complex ions and that doubly (or higher) negatively charged complexes in this medium are relatively poorly adsorbed. The adsorbed ion in the case of iron has been identified with $FeCl_4^-$ through resin-capacity measurements,[5] further supporting the earlier conclusions. Additional evidence for the reasonably general applicability of this rule comes from the observation that gallium (probably as $GaCl_4^-$) is adsorbed quantitatively similarly to iron ($FeCl_4^-$).[6] One might thus tentatively assume that $FeCl_4^-$ and $CoCl_3^-$ are also approximately equally strongly adsorbed. On this basis the fraction of cobalt as $CoCl_3^-$ can be estimated to be considerably less than 1% even in 9 *M* HCl, the adsorption maximum.

Since in concentrated HCl a negatively charged complex of cobalt almost certainly exists in considerable concentration[7,8,9] one can conclude that it is not strongly adsorbed. Since this ion probably has a charge of minus two ($CoCl_4^{=}$)[7,8] this relatively poor adsorption in concentrated HCl is in general agreement with the earlier conclusion that in this medium only singly negatively charged complexes are extremely strongly adsorbed.

The non-adsorbability of Ni(II) by the resin suggests very strongly that negatively charged complexes of this element are not formed in appreciable concentration even in concentrated hydrochloric acid.

(1) This document is based on work performed for the Atomic Energy Commission at the Oak Ridge National Laboratory.

(2) Previous paper: K. A. Kraus and G. E. Moore, This Journal, 73, 2900 (1951).

(3) K. A. Kraus and G. E. Moore, *ibid.*, 73, 9 (1951).

(4) G. E. Moore and K. A. Kraus, This Journal, 72, 5792 (1950).

(5) K. A. Kraus and G. E. Moore, unpublished results.

(6) K. A. Kraus, F. Nelson and G. W. Smith, unpublished results.

(7) M. Bobtelsky and K. S. Spiegler, *J. Chem. Soc.*, 143 (1949).

(8) N. V. Sidgwick, "The Chemical Elements and Their Compounds," Vol. II, Clarendon Press, Oxford, 1950.

(9) According to recent paper electromigration work cobalt begins to migrate toward the anode in *ca.* 8 *M* HCl; K. A. Kraus and G. W. Smith, unpublished results; for technique see This Journal, 72, 4329 (1950).

Oak Ridge National Laboratory
Oak Ridge, Tenn.
Received July 18, 1951

18

Reprinted from *J. Amer. Chem. Soc.*, 75, 1460–1462 (Mar. 1953)

Anion Exchange Studies. VI.[1,2] The Divalent Transition Elements Manganese to Zinc in Hydrochloric Acid

By Kurt A. Kraus and George E. Moore

Received September 2, 1952

The anion exchange behavior of the divalent transition elements manganese to zinc was studied in hydrochloric acid solutions. The adsorbabilities of the elements differ sufficiently as a function of hydrochloric acid concentration to permit their separation. The hydrochloric acid concentrations where noticeable adsorption occurs increase in the order Zn, Cu, Co, Fe and Mn with no noticeable adsorption for Ni, even in concentrated hydrochloric acid. It appears that the stability of the negatively charged chloride complexes of these divalent elements decreases in this order.

A survey of the anion exchange behavior of the first row transition elements of oxidation number two from Mn(II) to Zn(II) has been carried out in hydrochloric acid solutions. The results for Co(II) and Ni(II) have been given earlier.[3] In the case of these two elements, it was found that they differed greatly in their adsorbability, Ni(II) not being adsorbed from concentrated hydrochloric acid and Co(II) showing strong adsorption. This difference in anion exchange behavior suggested great differences in the stabilities of the negatively charged complexes of the transition elements, and a systematic study of these elements seemed indicated. In addition, there existed the possibility of establishing good separation procedures for these elements by use of anion exchange techniques.

Experimental

The anion exchange behavior of these elements was studied with the quaternary amine polystyrene–divinylbenzene resin Dowex-1 at room temperature, with a column technique.[4] From it the elution constant $E = dA/V$ is obtained where d is the distance an adsorption maximum travels on passage of V cc. of eluent through a column of cross-sectional area A cm.[2].

The columns, *ca.* 20 cm. long and of cross-sectional area *ca.* 0.03 sq. cm., were made of Pyrex tubing plugged with glass wool and filled with −200 + 230 mesh wet-screened resin in the chloride form. The columns were pretreated with the acid of interest and a small volume (*ca.* 0.025 ml.) of the metal solution containing 1 to 3 mg. of metal in the same acid was placed on the column. After the solutions seeped into the columns, the eluent was added from a buret to permit ready determination of V.

The solutions were prepared from C.P. salts except for the Fe(II) solutions which were prepared by dissolving iron filings in hydrochloric acid. Since Fe(II) and Fe(III)[5] behave very differently on the column and since the two oxidation states can be identified readily analytically, there was no difficulty in the determination of the elution constants of Fe(II). Since copper was found to form a colored band (yellow up to 4 M HCl, brown to 9 M HCl and orange to 12 M HCl) and since the resin is light tan changing to light brown in 12 M HCl, d could be determined visually and E obtained from the slope of a straight line plot of d *vs.* V. The elution constants for the weakly colored and colorless ions Mn(II), Fe(II) and Zn(II) were determined by meas-

(1) This document is based on work performed for the Atomic Energy Commission at the Oak Ridge National Laboratory.

(2) Previous paper, K. A. Kraus and G. E. Moore, This Journal, 75, 1457 (1953).

(3) G. E. Moore and K. A. Kraus, *ibid.*, 74, 843 (1952).

(4) K. A. Kraus and G. E. Moore, *ibid.*, 73, 9 (1951).

(5) G. E. Moore and K. A. Kraus, *ibid.*, 72, 5792 (1950).

uring the volume necessary to elute the adsorption maximum,[6] noting that d is equal to the length of the column. The eluent was collected in drops on a spot plate, the relative concentrations determined by standard spot tests[7] and the volume noted at which the metal appeared in maximum concentration.

Results and Discussion

1. Elution Constants.—Elution constants were determined in the range 0.5 to 12 M HCl. The results of the experiments are summarized in Fig. 1, where for comparison the earlier data on nickel and cobalt have also been included. For all elements E = *ca.* 2.5 was found at low M HCl. Except for Ni(II), E decreases with increasing M HCl, reaches a minimum or approaches such a minimum. E for Ni(II) is *ca.* 2.5 throughout the experimental range.

Since $E = 1/(i + D)$, where i is the fractional interstitial space and D the volume distribution coefficient (amount per ml. bed/amount per ml. solution) a decrease in E implies an increase in D, *i.e.*, increasing adsorption. For negligible adsorption (D = 0) E reaches a maximum value of $1/i$. Apparently E = 2.5 represents this maximum value and corresponds to i = 0.4.

It is of interest that Ni(II) is the only element of this series which does not show adsorption. Mn(II) and Fe(II) show slight adsorption above *ca.* 5 M HCl. Cu(II) shows some adsorption above *ca.* 1 M HCl, is better adsorbed than Mn(II) and Fe(II) at its maximum, but is never as strongly adsorbed as Co(II) for which appreciable adsorption starts above *ca.* 3 M HCl. Zn(II) shows the strongest adsorption of the series. It is considerably adsorbed in 0.1 M HCl and has an adsorption maximum in *ca.* 2 M HCl; *i.e.*, at an HCl concentration where most of these elements show only small adsorption. In addition the adsorbability of Zn(II) greatly exceeds that of the other elements. From these results it appears that the separation of these transition elements from each other can readily be achieved by anion exchange and this separation will be discussed in the next section.

The large differences in the anion exchange behavior of these elements suggest large differences in their complexing properties and considerable irregularity in the extent of complexing as a function of atomic number. If the hydrochloric acid concentration at which noticeable adsorption first takes place (*e.g.*, E = 1) is taken as an index of the strength of complexing, the elements order themselves as shown in Table I, method A.

TABLE I

ORDER OF COMPLEXING OF DIVALENT CHLORIDES

Method	Elements
(A) Anion exchange (E = 1)	Zn > Cu > Co > Fe > Mn > Ni
(B) Anion exchange (position of ads. max.)	Zn > Cu > Co > Mn > Fe > Ni
(C) Solubility of MCl_2 in HCl	Zn > Cu > Co > Fe > Ni
(D) Activity coefficients	Zn > Cu > Mn > Fe > Co > Ni
(E) J. Bjerrum	Zn > Fe > Cu > Co

(6) Some data on the adsorption of Zn using equilibrium experiments with Zn^{65} tracer have been included. These are from unpublished results of F. Nelson.

(7) F. Feigl, "Qualitative Analysis by Spot Tests," Elsevier Publishing Co., New York, N. Y., 1946. Fe(II) was determined with dimethylglyoxine and NH_3, Mn(II) by alkaline oxidation to MnO_4^-, Zn(II) with dithizone and NaOH.

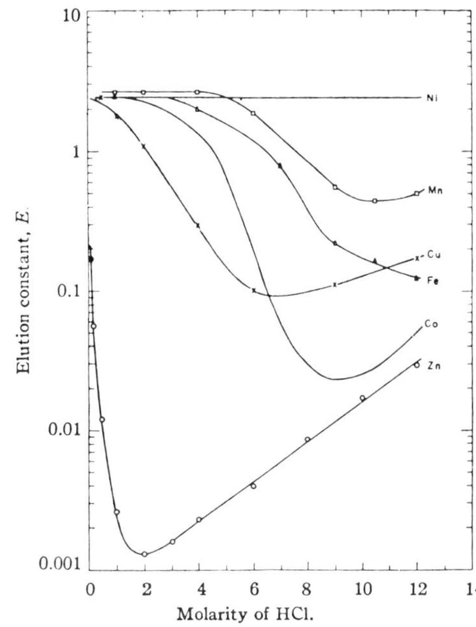

Fig. 1.—Elution constants of some divalent transition elements in hydrochloric acid (data for Zn(II) from unpublished results of F. Nelson).

A very similar order of complexing strength is obtained if the hydrochloric acid concentrations are considered at which adsorption reaches a maximum. This order is listed as method B in Table I. The only differences between the order of method A and method B are the relative positions of Mn(II) and Fe(II), which are somewhat uncertain.

As will be discussed in a later paper, method A probably gives the hydrochloric acid concentrations at which these elements form approximately equal (and small) concentrations of the complex ions MCl_3^-. On the other hand, method B probably gives the hydrochloric acid concentrations at which the complexes $MCl_4^=$ become the principal species in the aqueous phase, while the extent of adsorption at the maximum is related to the concentration of the complexes MCl_3^- under these conditions. Although methods A and B measure two different complexing properties of the elements it is of interest that the two methods give very nearly the same order of complexing strength.

Approximately the same order of complexing strength is also obtained from the solubilities of the metal chlorides MCl_2 in HCl[8] (Table I, method C), if the assumption is made that the undissociated compounds MCl_2 have relatively low solubility and that formation of negative complexes is necessary to yield good solubility at high HCl concentration. Thus the solubility of $NiCl_2$ decreases rapidly with

(8) A. Seidell, "Solubility of Inorganic and Metal Organic Compounds," Third Edition, D. Van Nostrand Co., Inc., New York, N. Y. 1940.

HCl concentration and that of $FeCl_2$ decreases almost as rapidly. The solubility of $CoCl_2$ decreases less rapidly and shows an increase at high HCl concentration. The solubility of $CuCl_2$ is almost identical with that of $CoCl_2$, but shows a slightly greater increase. The solubility of $ZnCl_2$ shows almost no decrease with HCl concentration. However, in this case the conclusions are more tentative since the solubility of $ZnCl_2$ in water is approximately ten times larger than that of the other metal halides, which have approximately the same solubility in water. There appeared to be no data for the solubility of $MnCl_2$ in HCl.

These results may also be compared with conclusions regarding the complexing strength of these elements taken from the values of the activity coefficients ($\gamma_{\pm}$) of the metal chlorides. These have been reported to be[9] 0.682, 0.817, 0.884, 0.938, 0.466 and 0.289 for 2 molal solutions of $MnCl_2$, $FeCl_2$, $CoCl_2$, $NiCl_2$, $CuCl_2$ and $ZnCl_2$, respectively. In such a closely related series of elements, one might expect that the stoichiometric activity coefficients $\gamma_{\pm}$ decrease with increasing complexing. Examination of Table I shows that the order of complexing obtained by this method differs greatly from the order obtained from methods A, B and C. Particularly surprising is the close similarity of the activity coefficients of $NiCl_2^-$ and $CoCl_2$ since they differ greatly with respect to the other methods. It is similarly surprising that $MnCl_2$ and $FeCl_2$ have lower activity coefficients than $CoCl_2$ since negatively charged chloride complexes of these elements seem to appear only at very high hydrochloric acid concentrations. The reason for these differences may be that the activity coefficients of the metal halides become low if the complexes MCl^+ and MCl_2 form, while the anion exchange results and solubilities are closely related to the formation of the complexes MCl_3^- and $MCl_4^{=}$. Comparison of method D with A, B and C, thus indicates that the elements show considerable differences in the relative stability of the various complexes MCl_i^{+2-i}, and that hence the usefulness of the concept of an average stability constant as used by Bjerrum[10] is restricted. The anion exchange results may also be compared with the order of complexing strength suggested by Bjerrum[10] (see Table I, method E). Unfortunately Bjerrum does not include Mn(II) and Ni(II) and a detailed comparison is thus not feasible. However, it should be pointed out that Bjerrum believes that Fe(II) is more strongly complexed than Cu(II). This is in disagreement with method D which we might have expected to yield the same order. Bjerrum's order for Fe(II) is based on the value of the stability constant of $FeCl^+$ determined by Olerup.[11] A check on the value of this constant thus appears desirable.

It is of particular interest that none of these methods yields the simple order of complexing strength (increase to Cu and decrease to Zn) which apparently applies with many chelates.[12]

2. Separation of the Transition Elements Mn to Zn.—The considerable difference in the elution constants of these elements implies that their separation on anion exchange columns of moderate length should be possible. To illustrate the technique, the separation of Mn(II), Fe(III),[13] Co(II), Ni(II), Cu(II) and Zn(II) was carried out, and the results are shown in Fig. 2. Approximately 0.85 ml. of a concentrated hydrochloric acid solution containing *ca.* 6 mg. of each of these elements was placed on a 26 cm. × 0.29 cm.^{2} column containing −200 + 230 mesh Dowex-1. Under these conditions Ni(II) is not adsorbed and immediately appears in the eluent. Mn(II) was removed with 6 *M* HCl, Co(II) with 4 *M* HCl, Cu(II) with 2.5 *M* HCl, Fe(III) with 0.5 *M* HCl and Zn(II) with 0.005 *M* HCl. Mn(II) was determined spectrographically.[14] The concentrations of Fe(III) and Zn(II) were determined by tracer techniques using Fe^{59} and Zn^{65}, respectively, while Co(II), Ni(II) and Cu(II) were determined colorimetrically. As shown in Fig. 2, the separation of these elements is quite satisfactory at the flowrate indicated (0.5 cm./min.) and probably would also be satisfactory at considerably higher flowrates.

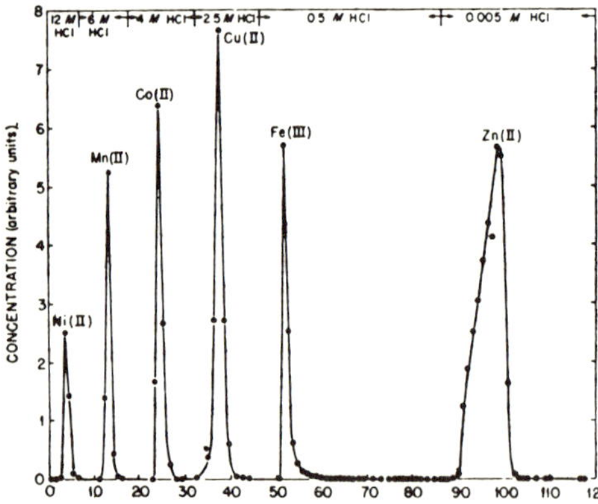

Fig. 2.—Separation of transition elements Mn to Zn (Dowex-1 column; 26 cm. × 0.29 cm.; flowrate = 0.5 cm./min.).

Acknowledgment.—The authors are indebted to Miss Zella L. Bonner for valuable technical assistance.

OAK RIDGE, TENNESSEE

(9) H. S. Harned and B. B. Owen, "Electrolytic Solutions," Second Edition, Reinhold Publishing Corp., New York, N. Y., 1950, pp. 567, 602.

(10) J. Bjerrum, *Chem. Revs.*, **46**, 381 (1950).

(11) H. Olerup, Dissertation, Lund, 1944.

(12) See for example, L. E. Maley and D. P. Mellor (*Australian J. Sci. Res.*, **2**, 92 (1949) (8-hydroxyquinoline, salicylaldehyde, acetylacetone)); Maley and Mellor (*Nature*, **165**, 453 (1950) (amino acids); G. Schwarzenbach and co-workers (*Helv. Chim. Acta*, **33**, 963, 974, 985, 995 (1950) (polyamines)); and M. Calvin and N. C. Melchior (THIS JOURNAL, **70**, 3270 (1948) (5-salicylaldehyde, sulfuric acid)). Calvin and Melchior suggested that the stability of the chelates is related to second ionization potentials of the metals.

(13) In this separation Fe(III) was used rather than Fe(II) since in most practical cases the former would be encountered. Fe(III) is strongly adsorbed by anion exchange resins from concentrated hydrochloric acid solutions (see ref. (5)). It should also be mentioned that the separation of Fe(II) and Fe(III) can readily be carried out in a large range of hydrochloric acid concentrations in view of their large differences in adsorbability.

(14) We are indebted to Mr. C. Feldman and Mr. M. Murray of the Oak Ridge National Laboratory Analytical Division for the spectrographic analyses.

19

Reprinted from *Proc. 1st U.N. Intern. Conf. Peaceful Uses Atomic Energy, Geneva,* Vol. 7, 113–125 (1955)

Anion Exchange Studies of the Fission Products

By K. A. Kraus and F. Nelson,* USA

I. INTRODUCTION

During the last decade the popularity of ion exchange as a tool in analytical, physical and inorganic chemistry has increased tremendously. The field has repeatedly been reviewed in recent years and a number of excellent books and review articles may be cited for general references.[1-9] A glance at the literature shows that the lion's share of attention has gone to cation exchange, while exploitation of anion exchange has not become prominent until a few years ago. Although large scale anion exchange installations now exist, particularly with respect to hydrometallurgy of uranium, exploitation of anion exchange still lags behind that of cation exchange. It is anticipated that, particularly as far as applications in analytical chemistry are concerned, this situation will change in the next few years. Anion exchange, especially of metal complexes, exhibits a number of unique features, and appears to be a new, powerful and versatile tool which should find soon further broad applications. At present, conditions have been found for the adsorption of essentially all metals on anion exchange resins except for the heavy alkali metals. Since the nonmetals of course are more readily handled as negative ions, this implies that anion exchange may be a more general tool than cation exchange.

Anion exchange of metal complexes shares with cation exchange the many advantages which presumably underlie the rapid expansion of this field and make it so attractive for many applications compared with conventional procedures. Ion exchange procedures in general are simple and separations can be achieved with great speed, provided the ions to be separated have sufficient differences in adsorbabilities. The large number of theoretical plates per unit length which ion exchange columns exhibit, particularly with fine mesh resins, is an additional distinct advantage. Further, with conventional ion exchange resins, diffusion rates are usually sufficiently large to permit operation at reasonably rapid flow rates.

Ion exchange separations are characterized by low cross-contamination, probably one of its most attractive features compared with precipitation techniques. Ion exchange techniques are fairly insensitive to the concentration of the materials to be separated. Concentrations ranging from a few atoms per cm³ to essentially molar concentrations may be handled with equal ease and similar efficiency, provided column capacities are properly designed. While loading considerations affect mainly the cross-sectional areas of the columns, their length is determined by relative adsorbabilities of the ions to be separated.

The initial impetus towards detailed study of anion exchange of metal complexes came when it was recognized in our laboratory that certain commercial strong-base anion exchange resins show great, and at times startling, selectivities for certain complex ions, particularly the chloride complexes of a number of metals. It thus became possible to operate effectively at high electrolyte concentrations and even in concentrated acids where the commercial resins show remarkable stability. In contrast, cation exchange often fails in these concentrated electrolyte solutions in no small measure because in these media many metals become complexed and improper targets for cation exchange.

Anion exchange is an ideal tool for the handling of those metal ions which must be complexed to keep them in solution, e.g., to prevent hydrolytic precipitation. Many of these elements in the proper complexing medium apparently form negatively charged complexes in sufficient concentration to permit good adsorption on anion exchange resins.

The suitability of anion exchange resins for adsorption from concentrated electrolyte solutions permits exploitation for analytical purposes of the more unusual complexing properties of the elements. These complexes often are formed in sufficient concentration only at high electrolyte concentrations. Apparently the activity coefficient quotients for the complex (formation) equilibria decrease rapidly with increasing ionic strength and complexes whose formation constants are small may become prominent. In view of the high selectivity of the resins for certain complexes, good adsorption may often be achieved with a small fraction of the metal in the form of negatively charged complexes though, of course, in equilibrium with other species.

With anion exchange techniques, separations of metals can often be achieved which "normally" exhibit similar properties. Such "similar" metals, it appears, may show large differences in the stability constants of the higher complexes, even if they have similar stability constants for the lower complexes.

Since anion exchange techniques thus permit ex-

* Oak Ridge National Laboratory.

ploitation of maximum differences among elements, separations can often be achieved with columns very much shorter than customary in cation exchange work with a resultant decrease in the time required to effect a given separation. In favorable cases good separations may even be achieved in times short compared with a minute. Thus daughter radioactivities with half-lives of the order of seconds and of less than a second have been separated from long-lived parent activities.[10] The speed of separation is greatly aided by the fact that with reversible complex equilibria adsorption may, for example, be carried out at high electrolyte concentration and desorption by a simple change in electrolyte concentration or even with water without change in the overall composition of the resin.

Ion exchange techniques may be used for elucidation of the solution chemistry of the materials studied. In this application, anion exchange in dilute solutions resembles the well-established cation exchange techniques which in recent years have been exploited particularly by Schubert.[11] Since adsorbabilities of many metal complexes on anion exchange resins are very high, extension to concentrated electrolyte solutions becomes very attractive, i.e., to a field where further studies are greatly needed. Although a brief outline of the theoretical considerations leading to identification of species and equilibria will be given later, one might mention here that information gained from these ion exchange studies is essentially thermodynamic (see also reference 12). Ion exchange studies of this type can be reduced to the thermodynamics of two-phase equilibria with the resin considered a special solvent.[13] Anion exchange of metal complexes gains its added strength from the fact that the anion exchanger is intrinsically sensitive to detection and measurement of the negatively charged complexes and thus permits elucidation of complex equilibria starting with the most complexed species rather than with the uncomplexed metal ions, for which many thermodynamic methods are available.

These general remarks regarding the advantages of anion exchange are, of course, directly applicable to fission product chemistry. We may classify the use of anion exchange in fission product work: (1) research with fission products, (2) analytical problems involving fission products, and (3) large scale isolations and separations. At the present time the use of anion exchange unquestionably is most promising for research and analysis. Extensive exploitation at high radiation levels will depend on the solution of a number of serious problems, particularly problems involving the radiation stability of the resins. The resins apparently are stable enough for work in moderate radiation fields. Regarding work at very high radiation levels, where many applications are probably not realizable, one may, however, still often be able to use anion exchange effectively if the processes are designed to adsorb the minor activities and permit passage of the major activities through the columns.

II. ANION EXCHANGE TECHNIQUES

We shall confine ourselves to a brief discussion of the techniques which have been used in this laboratory. These, of course, are similar to and largely based on the cation exchange techniques in general use. Differences principally stem from our rather broad utilization of radiometric methods of analysis. Actually, of the about 65 elements studied in this laboratory, radioactive tracers have been used for approximately 55.

The convenience and speed of the tracer method has often been stressed. In recent years the advent of scintillation counters with high γ-efficiency has further increased the attractiveness of the method. Particularly useful is the re-entrant hole scintillation counter[14] which permits direct measurements of activity of a solution in a small test tube. With this instrument analyses can be completed in a few minutes. Additional advantages of the tracer method may be cited. In many cases adsorbabilities may be studied at very low loading of the resin with respect to the metal of interest. In this way results become directly comparable. In general, adsorbability decreases with increasing loading. This effect is small up to about 1% loading but may amount to a 5 to 20-fold decrease between 1 and 20% loading.

Since general access to tracers is rapidly becoming easier, the main disadvantages of the tracer methods appear to be the strict requirements for purity for many applications and the difficulties engendered when the element of interest may have more than one oxidation state under the conditions studied. We have found the anion exchange technique to be of great value in preparing tracers of very high purity. The solution of the oxidation state problem probably can be achieved by supplementing the tracer techniques with more conventional methods at macro concentrations. However, once the anion exchange behavior of the various oxidation states has been established, the technique may be used for identification of the oxidation states, even at concentrations where conventional methods fail.

It must be stressed, however, that in spite of the enormous advantages which the tracer method offers, a great deal of work in the field of ion exchange and anion exchange may be done rapidly and efficiently without tracers. In this laboratory polarography, spectrophotometry, colorimetry and flame spectrophotometry, as well as standard volumetric techniques, have been used to good advantage. For colored ions visual observation of band motion (chromatography) can of course be used as well as spot testing of the effluent. For the spot testing method it appears advantageous to operate with columns of sufficiently small cross-sectional area so that each drop represents an appreciable fraction of a column volume and can be collected and analyzed without aliquoting.

Most of the separation work which has been carried out has involved short columns (usually less than 5 cm). In general a search is made for conditions where such small columns would be effective and careful chromatographic techniques not necessary. Further, a search was made for eluting conditions where adsorption is sufficiently small to permit effective removal of the ions in a few column volumes or preferably in one column volume. To achieve these objectives it has become standard in this laboratory to measure adsorbabilities over a large range of conditions.

Adsorbabilities have been studied principally by four techniques: (1) column effluent analysis, (2) column scanning, (3) equilibration method, and (4) special methods for measurement of high distributon coefficients.

The column effluent analysis appears to be most suited for conditions where the ion of interest is not strongly adsorbed. It becomes rather cumbersome when more than ca 10 column volumes are necessary to remove the adsorption maximum. The column effluent method directly yields the elution constant, E.[15] From $1/E$, the number of (geometric) column volumes necessary to obtain the eluted ion in maximum concentration in the effluent, the volume distribution coefficient, D_v (amount per liter of bed/amount per liter of solution), may be evaluated by the relationship

$$D_v = 1/E - i \qquad (1)$$

where i is the fractional interstitial space. For more strongly adsorbed ions the scanning method may be used[15] which also yields E.

For strongly adsorbed ions the equilibration technique, however, appears to be more convenient, as pointed out particularly by Schubert.[11] Samples of resin and solution are agitated long enough to establish equilibrium which with the resins used here was usually less than 1 to 2 days. The ratio of resin weight to solution volume is preferably adjusted so that approximately 50% of the material of interest is adsorbed. From the decrease in concentration on equilibration the distribution coefficient D (amount per kg dry resin/amount per liter of solution) may be computed. The values of D may readily be correlated with anticipated column behavior since the number of column volumes necessary to elute the material in maximum concentration is given by $(D_v + i)$ and since D and D_v are related by the bed density (ca 0.45 kg dry resin per liter of bed).

For measurements of extremely high distribution coefficients (e.g., $D > 10^4$ or 10^5) it has been found advantageous first to adsorb the tracer uniformly on the resin and then to measure the concentration of solutions in equilibrium with it. Equilibrium may be achieved either by passing solutions through a bed containing the tracer loaded resin or by agitating resin and solution. Placing the tracer on the resin first permits attainment of equilibrium in each individual experiment in very much shorter times at high values of D than when adsorption is carried out with the tracer originally in the solution.

III. SURVEY OF ANION EXCHANGE STUDIES

For the last six or seven years a systematic survey of the anion exchange behavior of the metals has been carried out in this laboratory. Although the investigations have involved a number of media, the survey in HCl solutions is most nearly complete, and at present includes most of the fission products.

The original intent was to study only those metals which form relatively weak complexes and for which the complexes are in rapid equilibrium with the uncomplexed ions. If these metals show good adsorption at high concentration of complexing agent (e.g., concentrated HCl) adsorption-elution cycles would become particularly simple since negligible adsorption should occur at low concentration of complexing agent, and elution with dilute HCl or even water becomes possible. As this survey progressed, it became desirable, for comparative reasons, to include the more stable complexes, those which are not in rapid equilibrium with the uncomplexed species and those for which no adsorbable complexes were anticipated.

The adsorption data for HCl solutions are summarized in Fig. 1 as a series of graphs of log D_v vs M HCl, where D_v is the volume distribution coefficient and M the molarity. Figure 1 was compiled from previously published and unpublished results obtained at this laboratory. The only exceptions are the data for Nb(V) and Ta(V), taken from the work of Huffman and Lilly, and the data for Te(IV) where the observations of Sasaki were used. Literature references pertinent to Fig. 1 may be found italicized in Table I-A.

In a few cases quantitative information as a function of M HCl was not available, but sufficient experiments had been carried out to indicate that the elements were strongly adsorbed over most of the conditions studied (0.1 M HCl to 12 M HCl). These cases are identified in Fig. 1 by "Str. Ads." Work at this laboratory in hydrochloric acid solutions was carried out with one batch of resin (quaternary amine polystyrene divinyl benzene resin, ca 200 mesh, 10% DVB). Difficulties resulting from differences in selectivities, particularly with resins of different cross-linking, were thus avoided, and the data in Fig. 1 are essentially comparable.

From a glance at Fig. 1 it becomes apparent that the elements can be roughly classified into three groups with respect to their anion exchange behavior:

1. Many elements are not adsorbed from HCl solutions at any concentration. In this class fall the alkali metals, the alkaline earths and largely the elements of Group III-B (Y to Ac). Surprisingly Al(III) is also in this group of non-adsorbable ions, as well as Ni(II) and Th(IV).

2. Many elements show increasing absorption with increasing M HCl and usually good adsorption only

Table 1. Survey of Anion Exchange Studies of Metal Complexes

A. Hydrochloric Acid and Other Chloride Solutions

(Figure 1 based on italicized references)

Atomic no.	and element	References	Atomic no.	and element	References
3	Li	J-1, *K-9*, M-4	45	Rh	B-3, C-3, *K-15*, M-1
4	Be	*K-9*, K-12, M-4	46	Pd	B-3, *K-9*, N-1, M-1
11	Na	J-1, *K-9*, M-4	47	Ag	*K-15*
12	Mg	J-1, *K-9*, M-4	48	Cd	B-5, J-1, K-9, *K-15*
13	Al	B-1, J-1, *K-9*, M-4	49	In	J-1, *K-9*
19	K	J-1, *K-9*, M-4	50	Sn	J-1, *K-15*
20	Ca	J-1, *K-9*, M-4	51	Sb	*K-15*
21	Sc	*K-9*, K-12	52	Te	*S-3*
22	Ti	*K-9*	55	Cs	*K-9*, M-4
23	V	*K-9*	56	Ba	*K-9*, M-4
24	Cr	H-8, *K-15*, M-4, N-1, S-6	57-71	R.E.	*K-9*, M-4
25	Mn	B-1, H-8, J-1, *K-8*, K-12	72	Hf	H-4, H-7, *K-15*
26	Fe	B-1, H-8, J-1, *K-8*(II), K-11	73	Ta	*H-7*, K-6
		K-12, *K-15*(III), M-4, R-2	74	W	*K-13*
27	Co	B1, H-8, J-1, *K-8*, K-12, M-5	75	Re	*K-15*, M-2
28	Ni	B-1, H-8, J-1, K-8, *M-5*	76	Os	*K-15*
29	Cu	J-1, *K-8*(II), *K-15*(I)	77	Ir	C-3, B-3, *K-9*(IV), M-1, *K-15*(III)
30	Zn	B-5, J-1, *K-8*, K-12, M-3	78	Pt	B-3, *K-9*, M-1, N-1
31	Ga	*K-9*, K-12	79	Au	*K-10*, K-12, N-1, S-6
32	Ge	*N-5*, Y-1	80	Hg	*K-15*
33	As	J-1, *N-5*, Y-1	81	Tl	*K-9*
34	Se	S-3, *K-15*	82	Pb	C-1, J-1, *N-2*
37	Rb	*K-9*, M-4	83	Bi	C-1, *N-2*
38	Sr	*K-9*, M-4	84	Po	*S-3*
39	Y	*K-9*	87	Fr	*K-9*
40	Zr	H-4, H-7, *K-15*	88	Ra	*K-9*
41	Nb	*H-7*, K-5	89	Ac	*K-9*
42	Mo	*K-13*, M-2, S-6	90	Th	*K-14*
43	Tc	*K-15*	91	Pa	*K-4*, K-7, K-11
44	Ru	B-3, *K-15*	92	U	*K-13*

B. Anion Exchange Studies of Fission Products in Other Media

Solution	Element	References
Fluoride and HF-HCl Mixtures	Zr	F-2, F-3, F-4, H-6, K-2, K-5, K-7
	Nb	F-4, H-1, K-3, K-5, K-7
	Mo	F-4, H-1, K-13
	Sn	F-4
Bromide and Iodide	Zn	B-5, H-3
	Ga	H-3
	Cd	B-5, F-5, L-1
	Sb	L-2
Hydroxide	Ge	E-1
	As	E-1
	Mo	F-1, H-2, M-2
	Ru	B-3
	Rh	B-3
	Pd	B-3
	$I(IO_3^-)$	K-1
Nitrate	$Br(Br^-)$	A-1, R-1
	$I(I^-)$	A-1
Cyanide	Zn	B-4
	Ag	J-2

Solution	Element	References
Thiocyanate	Tc	A-1, H-2
Sulphate	Cd	L-1
	In	S-5
Acetate	Ce	F-5
Oxalate	Ga	B-2
	Y	C-2
	Zr	W-1
	Nb	G-1, W-1
	Mo	M-2
	Sn	L-2, S-4
	Sb	L-2, S-4
	Te	S-4
Citrate	Sr	N-4, S-1
	Ba	N-4, S-1
	R.E.	H-5
Ethylenediaminetetraacetic acid (EDTA)	Zn	S-7
	Rb	N-3
	Cs	N-3

A-1 Atteberry, R. W. and Boyd, G. E., J. Am. Chem. Soc. *72*: 4805 (1950).

B-1 Blasius, E. and Negwer, M., Naturwissenschaften *39*: 257 (1952).

B-2 Blasius, E. and Negwer, M. Z. anal. Chem. *143*: 257 (1954).

B-3 Blasius, E. and Wachtel, U., Angew. Chem. *66*: 305 (1954).

B-4 Burstad, F. H., Kembler, N. F., Forrest, P. and Weeks, R. A., Ind. Eng. Chem. *45*: 1648 (1953).

B-5 Baggott, E. R. and Willcocks, R. G. W., Analyst *80*: 53 (1955).

C-1 Campbell, E. C. and Nelson, F., Phys. Rev. *91*: 499A (1953).
C-2 Crouthamel, C. E. and Martin, D. S. Jr., J. Am. Chem. Soc. *72*: 1382 (1950).
C-3 Cluett, M. L., Berman, S. S. and McBride, W. A. Analyst *80*: *204* (1955).
E-1 Everest, D. and Salmon, J. J., Chem. Soc. 2438 (1954).
F-1 Fisher, S. A. and Meloche, V. W., Anal. Chem. *24*: 1100 (1952).
F-2 Forsling, W., Arkiv Kemi *5*: 489 (1953).
F-3 Forsling, W., Arkiv Kemi *5*: 503 (1953).
F-4 Freund, H. and Miner, F. J., Anal. Chem. 25: 564 (1953).
F-5 Fronaeus, S., Svensk Kem. Tidskr. *65*: 1 (1953).
G-1 Gillis, J., Eeckhaut, Z., Cornand, P. and Speecke, A., Mededel. Koninkl. Vlaam. Acad. Wetenschap. Belg. *15*: 3 (1953). See also C. A. *48*: 9862[c] (1954).
H-1 Hague, J. L., Brown, E. D. and Bright, H. A., J. Research Nat. Bur. Standards *53*: 261 (1954).
H-2 Hall, N. and Johns, D., J. Am. Chem. Soc. *75*: 5787 (1953).
H-3 Herber, R. H. and Irvine, J. W., J. Am. Chem. Soc. *76*: 987 (1954).
H-4 Huffman, E. H. and Lilly, R. C., J. Am. Chem. Soc. *71*: 4147 (1949).
H-5 Huffman, E. H. and Ostwalt, R. L., J. Am. Chem. Soc. *72*: 3323 (1950).
H-6 Huffman, E. H. and Lilly, R. C., J. Am. Chem. Soc. *73*: 2902 (1951
H-7 Huffman E. H., Lilly, R. C. and Iddings, G. M., J. Am. Chem. Soc. *73*: 4474 (1951).
H-8 Hague, J. L., Maczkowske, E. E. and Bright, H. A., J. Research Nat. Bur. Stds. *53*: 353 (1954).
J-1 Jentzsch, D. and Frotscher, I., Z. anal. Chem. *144*: 17 (1955).
J-2 Jones, L. H. and Penneman, R. A., J. Chem. Phys. *22*: 965 (1954).
K-1 Kikindai, M., Compt. rend. *237*: 250 (1953).
K-2 Kraus, K. A. and Moore, G. E., J. Am. Chem. Soc. *71*: 3263 (1949).
K-3 Kraus, K. A. and Moore, G. E., J. Am. Chem. Soc. *71*: 3855 (1949).
K-4 Kraus, K. A. and Moore, G. E., J. Am. Chem. Soc. *72*: 4293 (1950).
K-5 Kraus, K. A. and Moore, G. E., J. Am. Chem. Soc. *73*: 9 (1951).
K-6 Kraus, K. A. and Moore, G. E., J. Am. Chem. Soc. *73*: 13 (1951).
K-7 Kraus, K. A. and Moore, G. E., J. Am. Chem. Soc. *73*: 2900 (1951).
K-8 Kraus, K. A. and Moore, G. E., J. Am. Chem. Soc. *75*: 1460 (1953).
K-9 Kraus, K. A., Nelson, F. and Smith, G. W., J. Phys. Chem. *58*: 11 (11954).
K-10 Kraus, K. A. and Nelson, F., J. Am. Chem. Soc. *76*: 984 (1954).
K-11 Kraus, K. A. and Moore, G. E., J. Am. Chem. Soc. *77*: 1383 (1955).
K-12 Kraus, K. A., Nelson, F., Clough, F. B. and Carlston, R. C., J. Am. Chem. Soc. *77*: 1391 (1955).
K-13 Kraus, K. A., Nelson, F. and Moore, G. E., J. Am. Chem. Soc. *77*: 3972 (1955).
K-14 Kraus, K. A. and Moore, G. E. (to be published).
K-15 Kraus, K. A. and Nelson, F. (previously unpublished work).
L-1 Leden, I., Svensk Kem. Tidskr. *64*: 145 (1952).
L-2 Luré, Yu. Yu. and Filippova, N. A., Zavodskaya Lab. *14*: 159 (1948).
M-1 MacNevin, W. M. and Crummet, W. B., Anal. Chem. *25*: 1628 (1953).
M-2 Meloche, V. W. and Preuss, A. F., Anal. Chem. *26*: 1911 (1954)
M-3 Miller, C. C. and Hunter, J. A., Analyst *79*: 483 (1954).
M-4 Moore, G. E. and Kraus, K. A., J. Am. Chem. Soc. *72*: 5792 (1950).
M-5 Moore, G. E. and Kraus, K. A., J. Am. Chem. Soc. *74*: 843 (1952).
N-1 Nachod, F., U.S. Patent 2,371,119 (March 6, 1945).
N-2 Nelson, F. and Kraus, K. A., J. Am. Chem. Soc. *76*: 5916 (1954).
N-3 Nelson, F., J. Am. Chem. Soc. *77*: 813 (1955).
N-4 Nelson, F. and Kraus, K. A., J. Am. Chem. Soc. *77*: 801 (1955).
N-5 Nelson, F. and Kraus, K. A., J. Am. Chem. Soc. *77*: 4508 (1955).
R-1 Rieman, W. III. and Lindenbaum, S., Anal. Chem. *24*: 1199 (1952).
R-2 Reents, A. C. and Kohler, F. H., Ind. Eng. Chem. *47*: 75 (1955).
S-1 Samuelson, O., Lunde, L. and Schramm, K., Z. Anal. Chem. *140*: 330 (1953).
S-2 Samuelson, O. and Sjoström, K., Anal. Chem. *26*: 1908 (1954).
S-3 Sasaki, Y., Bull. Chem. Soc. Japan *28*: 89 (1955).
S-4 Smith, G. W. and Reynolds, S., Anal. Chim. Acta *12*: 151 (1955).
S-5 Sunden, N., Svensk Kem. Tidskr. *66*: 173 (1954).
S-6 Sussman, S., Nachod, F. C. and Wood, W., Ind. Eng. Chem. *37*: 618 (1945).
S-7 Samuelson, O., Sjoström, K. and Forsblom, S., Z. Anal. Chem. *144*: 21 (1955).
W-1 Walker, R. E. and Baldwin, W. H., U.S. A.E.C. Report ORNL-637 (1950), Nuclear Sc. Abstr. *4*: 469 (1950). Brit. Abst. C244 (1951).
Y-1 Yoshino, Y., Bull. Chem. Soc. Japan *28*: 382 (1955).

at high *M* HCl. Some of these elements show adsorption maxima.

3. Many elements, located principally in the central part of the periodic table, show only decreasing adsorption with increasing *M* HCl, at least in the region 1 to 12 *M* HCl. In this group fall the second and third row transition elements which are generally recognized to be excellent complex formers with chloride ions.

References to similar work carried out in other laboratories are given in Table I-A. In general, agreement between various laboratories using different resins, though of the same type, has been gratifying. Table I-A is considered reasonably complete except for the large body of work on stable anions, e.g., halides, nitrates, etc., which has been a standard part of the important basic studies on the selectivities and characterization of anion exchange resins (see e.g., references 16 and 17). For more detailed information regarding the ion exchange behavior of these stable anions the many review articles and books on the field may be consulted.

In Table I-A, references to chloride solutions other than hydrochloric acid have been included. Recently (reference K-12 of Table I) a startling difference in the adsorbability of a large number of metal ions in lithium chloride solutions compared with hydrochloric acid solutions has been observed. Further exploitation of these low acidity media appears highly attractive.

A survey of the anion exchange studies of the fission products in other media is given in Table I-B.

As far as our work is concerned, relatively little attention to media other than hydrochloric acid has been given except when special reasons necessitated it. Thus, for example, certain elements particularly of the fourth and fifth groups show highly unfavorable hydrolytic properties in most HCl solutions,

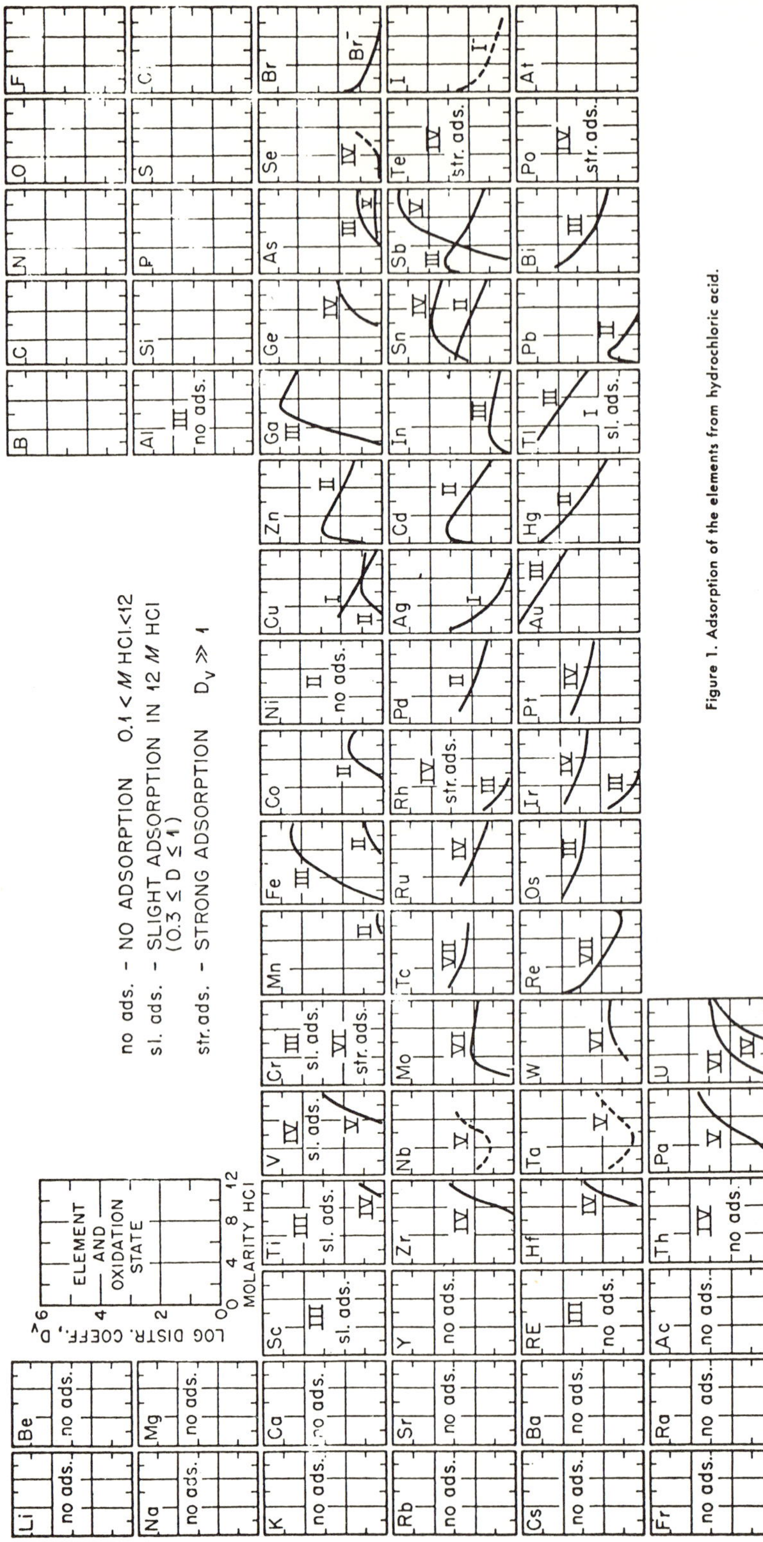

Figure 1. Adsorption of the elements from hydrochloric acid.

which might seriously interfere with separations. In an attempt to avoid these difficulties, studies were carried out in HF-HCl mixtures. Further, for the elements which show essentially negligible adsorption from HCl solutions, other complexing agents are indicated. In many cases citrates or ethylenediaminetetraacetic acid (EDTA) proved satisfactory. Our investigations with the other common mineral acids (nitric acid and sulfuric acid) have so far been limited to uranium and very few other elements. Exploitation of these media appears highly promising.

IV. ANION EXCHANGE STUDIES OF THE FISSION PRODUCTS AND URANIUM

This section will deal principally with the implications and applications of the anion exchange studies at ORNL to separations involving the fission products. A brief discussion of each fission product will be given. Since uranium occupies an obviously unusual position with respect to fission product chemistry, information regarding this element is also included in some detail.

The adsorption data described in Fig. 1 can be used directly to devise a large number of separation schemes. Actually, it may be noticed that most elements differ sufficiently from each other in their adsorbabilities that separations are at least intrinsically possible although for a few of these some reservations must be made.

Anion Exchange Studies of Uranium(VI)

Adsorption behavior of uranium(VI) has been studied in HCl solutions and HF-HCl mixtures (reference K-13, Table I), as well as in nitrate and sulfate solutions (reference K-15). Adsorption of uranium(VI) occurs from all these media although the characteristic adsorption functions (log D vs M HCl) differ widely.

The adsorbability of uranium(VI) rises steeply with increasing HCl concentration from D = ca 1 in 1 M HCl to D = ca 1800 near 9 M HCl. At higher concentrations, D decreases slightly (Fig. 2).

In nitric acid, adsorbability of uranium(VI) becomes significant near 2M HNO_3 where D = ca 1 and then rises to D = ca 20 near 8 M HNO_3 (Fig. 2). Above this concentration the equilibration experiments yielded irreproducible results, probably due to decomposition of the resin by the rather concentrated nitric acid.

Adsorption of uranium(VI) from other nitrate solutions of low acidity has also been studied. At constant nitrate concentration, adsorbability from these media is considerably larger. For example, distribution coefficients of the order of 10^3 have been found for 2.5 M $Al(NO_3)_3$—0.5 M HNO_3 solutions.

The adsorption function of uranium(VI) in sulfuric acid differs markedly from those in HCl or HNO_3 (Fig. 2). Thus, adsorption is very high at low sulfuric acid concentration (D = 35,000 in 0.01 M H_2SO_4) and decreases rapidly with increasing M H_2SO_4 to D = ca 1 in 4 M H_2SO_4. However, at 4 M H_2SO_4, quantitative elution of uranium cannot readily be achieved at room temperature with small volumes of eluent, probably because of slow rates of diffusion. By increasing the elution temperature, satisfactory removal with little tailing can be achieved, even with 1-2 M H_2SO_4.

Adsorption of uranium(VI) from $(NH_4)_2SO_4$ solutions resembles that from H_2SO_4. At low concentrations the distribution coefficients for both media are approximately the same. However, the decrease in D with concentration is considerably less rapid for ammonium sulfate than sulfuric acid. Thus D = 500 in 4 M $(NH_4)_2SO_4$.

Adsorption of uranium(VI) from HF solutions has not been studied. However, data are available on adsorbabilities from HF-HCl mixtures at constant HF concentration (1 M). It was found that presence of 1 M HF affects adsorption of U(VI) little at high M HCl, that at low M HCl adsorption from these mixtures is considerably poorer than from HCl alone, and that at still lower HCl concentration (< 1 M HCl) adsorption of U(VI) rises again rapidly with decreasing M HCl. This behavior is presumably due to formation of positively charged fluoride complexes of U(VI) at high M HCl, which decrease the fraction of U(VI) in the form of an adsorbable chloride complex, and to adsorption of negatively charged fluoride complexes at low M HCl.

The highly varied adsorption properties of uranium(VI) in the various mineral acids should make anion exchange separations of this element from es-

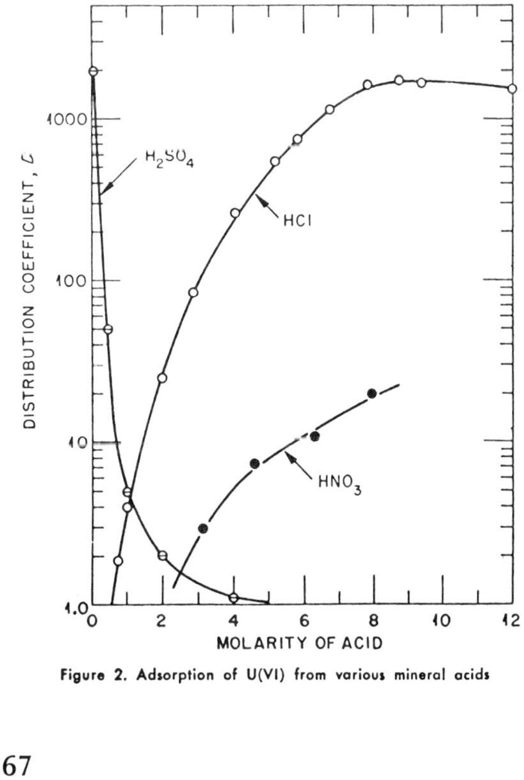

Figure 2. Adsorption of U(VI) from various mineral acids

sentially all other elements relatively simple. Actually, the unusual complexing properties of U(VI) might be expected to produce interesting anion exchange behavior for this ion in many other media.

Rb, Cs, Sr, Ba, Y, Rare Earths

These fission products [except Ce(IV)] show essentially negligible adsorption from HCl solutions and apparently also from LiCl solutions. However, separation of this group of "non-adsorbable" elements from uranium and all other fission products cannot be achieved in a single pass of e.g., a concentrated HCl solution through an anion exchanger. Although under these conditions uranium and most of the fission products will be retained, a number of other elements (e.g., As(V), Rh(III), Ag(I), Br^-) may also appear in the effluent. Of these "interfering" ions, As(V) poses the most serious problem, although it is removable when the solution volume is small compared with the column volume. Most of the other interfering ions may readily be removed by a second pass through the exchanger at low HCl concentration (e.g., 0.5 *M*) i.e., under conditions where these interfering elements show good adsorption.

To achieve separations within this group on HCl-nonadsorbable-elements, other complexing agents are needed. Citrates and ethylenediaminetetraacetic acid (EDTA) have been used to good advantage. Thus the alkali metals show negligible adsorption from citrate solutions while Sr, Ba, Y, rare earths and, presumably, most other elements in the periodic table show good adsorption (see also reference S-1 of Table I). Thus citrate promises to become the basis of a very simple anion exchange method for the specific isolation of the fission products Rb and Cs. Other techniques (e.g., cation exchange) will probably be necessary to separate these two elements from each other. A similar situation appears to occur with EDTA where Rb and Cs may be the only non-adsorbable fission products (see also reference S-7).

Citrate and EDTA media may also be used for more detailed separations within this group of elements. Thus, the separation of Ba from Sr has been reported for citrate solutions. Excellent separation of the rare earths from Ba (and presumably also Sr) can be achieved with either citrates or EDTA (see Fig. 3). Further, partial separation of the rare earths, Pm(III) and Eu(III), has been reported (reference H-5).

Zr and Nb

Both of these elements are strongly adsorbed from concentrated hydrochloric acid solutions. Zr(IV) may be eluted, though with extensive tailing, at low HCl concentrations. At very low metal concentration elution of Nb(V) at low *M* HCl may also be feasible. However, in view of the unfavorable hydrolytic properties of these elements, particularly Nb(V), the use of HCl media for separations does not appear attractive. Instead, operation in stronger complexing solutions is desirable. Successful separations involving Zr and Nb have been achieved in HF-HCl media and details regarding these separations have been described. The insensitivity of uranium(VI) adsorption to presence of HF at high HCl concentrations and its sensitivity at low *M* HCl permits its separation from these elements under a variety of conditions.

Separations involving Zr(IV) and Nb(V) have also been described for oxalate solutions (reference G-1, W-1).

Mo and Tc

Only the higher oxidation states, Mo(VI) and Tc(VII), have been studied. The adsorption function of Mo(VI) in hydrochloric acid is very similar to that of U(VI). This similarity precludes satisfactory separation of these elements in HCl. However, they may readily be separated from each other in HF-HCl mixtures.

Detailed studies on the adsorption behavior of Mo(VI) in oxalate, hydroxide and perchlorate solutions have also been reported (reference F-1, M-2).

Tc(VII) (TcO_4^-) shows good adsorption in the range 0.1 to 12 *M* HCl. Adsorption decreases with increasing *M* HCl in a manner which appears to be characteristic for stable negatively charged ions. Unfortunately, however, adsorption does not become low enough to permit simple removal in 12 *M* HCl. Elution of Tc(VII) from anion exchange columns has

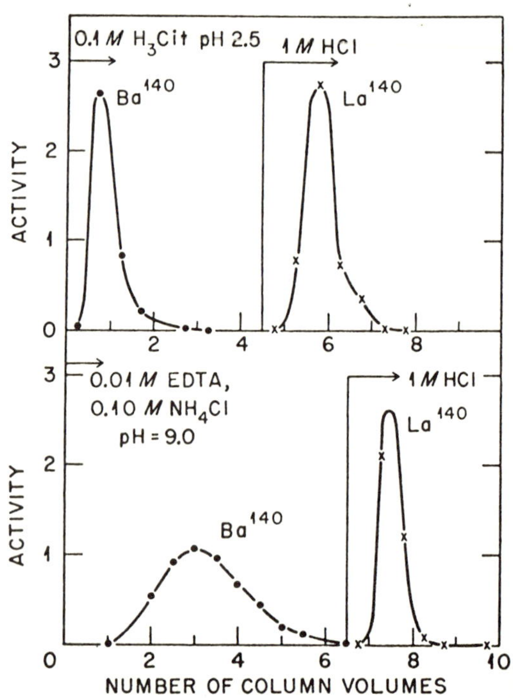

Figure 3. Separation of Ba and La by anion exchange

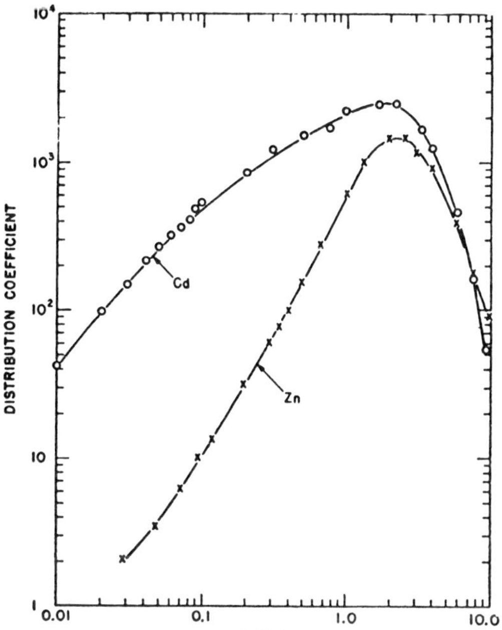

Figure 4. Anion exchange adsorption of Zn(II) and Cd(II)

been reported with SCN^- in connection with the separation of Mo and Tc (reference A-1).

Ru, Rh and Pd

These elements in the various oxidation states studied show decreasing adsorption with increasing HCl concentration of the type characteristic for stable negatively charged complexes. The chloro-complexes of Ru(IV), Rh(IV) and Pd(II) (and presumably also of Pd(IV)) are still strongly adsorbed at high hydrochloric acid concentrations and their removal from the anion exchange columns calls for special techniques. Displacement of the strongly adsorbed Pd(II) complex by other strongly adsorbed chloro-complexes (those of Zn(II) or Cd(II) has proven successful. Rh(III) ($RhCl_6^{---}$) shows only negligible adsorption above 4 *M* HCl and hence it may readily be removed from columns. Further, the low adsorbability of Rh(III) at high *M* HCl may become the basis of anion exchange separations involving this element. The chemistry of Ru has not been studied in sufficient detail to predict its behavior under most conditions apt to be encountered. Unquestionably detailed investigations involving this element in its multiplicity of oxidation states and with its large variety of species (not in equilibrium with each other) must be carried out before it can be handled with confidence.

The anion exchange behavior of the platinum metals in several media including also hydroxides and oxalates has been described (reference B-3).

Ag

This element adsorbs strongly at low *M* HCl and the distribution coefficients decrease with increasing *M* HCl. Adsorbability becomes negligible in concentrated HCl. The anion exchange behavior of Ag(I) is sufficiently unusual to permit rapid isolation of this element, for most purposes in one adsorption-elution cycle, such as adsorption at low *M* HCl and elution at high *M* HCl. The relatively low solubility of Ag in chloride solutions does not appear to present difficulties at tracer concentrations.

Zn and Cd

Adsorption of both of these elements from hydrochloric acid solutions is already high in 0.1 *M* HCl, reaches a maximum near 2 *M* HCl and then decreases with further increasing *M* HCl. Removal of these elements from anion exchange columns can readily be achieved at very low hydrochloric acid concentrations or with water. The differences in the adsorption characteristics of these two elements at low acidities are sufficient to permit their separation from each other (Fig. 4). Thus, Zn may be removed with 0.01 *M* HCl where $D_{Zn} < 1$ and where D_{Cd} is still ca 40. Elution of Cd(II) can then be carried out at *M* HCl $< 10^{-3}$ *M*.

The adsorption functions for these two elements are sufficiently unique to permit isolation of these elements usually in one adsorption-elution cycle.

Ga and In

Adsorption of Ga(III) from HCl solutions becomes appreciable near 2 *M* HCl and rises rapidly to a maximum near 8 *M* HCl. Adsorption of In(III) becomes appreciable near 0.1 *M* HCl, rises slowly to a shallow maximum near 4 *M* HCl ($D = 23$), and then decreases to $D =$ ca 7 in 12 *M* HCl. As described earlier, the separation of Ga from In may readily be achieved and the adsorption functions of these two elements are sufficiently unique to permit ready isolation from most fission products.

Ge and Sn

Adsorption of Ge(IV) becomes appreciable near 4 *M* HCl and increases rapidly with increasing *M* HCl. The adsorption function is sufficiently different from those of the other fission products to make unique isolation of this element feasible. However, the high volatility of Ge(IV) from concentrated HCl solutions and the resulting possibility of loss limits the usable hydrochloric acid concentrations to less than ca 8 *M* if no further precautions are taken.

Both Sn(II) and Sn(IV) adsorb strongly from concentrated hydrochloric acid solutions. The adsorption function of Sn(II) decreases with increasing *M* HCl, while that of Sn(IV) shows a maximum near 7 *M* HCl. Elution of Sn(II) and Sn(IV) with HCl is difficult since these ions hydrolyze at low hydrochloric acid concentrations. However, the high stability of the fluoride complexes, particularly of Sn(IV), suggests that elution may be carried out with HF-HCl mixtures (see also reference F-4 in Table I). Actually, the use of HF-HCl mixtures should prove advantageous for these elements and the elements of group V-A, as they had earlier been shown to be advantageous for separations involving groups IV-B and V-B (see e.g., section on Zr and Nb).

Separations involving Sn(IV) have also been reported for oxalate solutions (reference L-2, S-4).

As and Sb

Adsorption of As(III) becomes appreciable above 4*M* HCl and increases with increasing *M* HCl to $D =$ ca 25 in concentrated HCl. Adsorption of As(V), though appreciable even in 0.1 *M* HCl, never becomes very large, *D* remaining less than 4 even in concentrated HCl. The differences between the two ions are sufficient for separation of the two oxidation states, and they both differ sufficiently from the other fission products and uranium(VI) to make their isolation relatively simple.

Adsorption of Sb(III) is already high near 0.1 *M* HCl. Adsorption of Sb(V) becomes appreciable near 2 *M* HCl. For both ions adsorption first increases with increasing *M* HCl. An adsorption maximum occurs for Sb(III) near 2 *M* HCl. In concentrated HCl both oxidation states are strongly adsorbed. Separations involving antimony are difficult in HCl alone since the rate of desorption of Sb(V) in dilute HCl is slow, possibly because of slow equilibria between various Sb(V) complexes and since for Sb(III) hydrolytic difficulties appear at low *M* HCl, where desorption otherwise should be feasible.

As with Ge and Sn, study of the adsorbabilities of these elements in HF-HCl media should prove profitable for separations.

Work on separations involving Sb(V) in oxalate solutions has been reported (reference L-2, S-4).

Se, Te, Br and I

Se(IV) adsorbs only slightly in solutions of moderate HCl concentration. Distribution coefficients increase rapidly above 6 *M* HCl. Separations involving Se(IV) have been described (S-3) in which Se(IV) is adsorbed from 12 *M* HCl and eluted with 6 *M* HCl. Work in this Laboratory, however, shows (K-15) that if Se(IV) is permitted to age several hours in concentrated HCl, selenium species form (possibly $SeCl_4$ or $SeOCl_2$) which are not in rapid equilibrium with each other and which cannot easily be eluted with HCl solutions. However, rapid elution was achieved with methyl alcohol.

Tellurium has not been studied in this Laboratory. Data on the adsorbability of Te(IV) from HCl solutions (reference S-3) and oxalate solutions (reference S-4) have been reported.

Br^- and I^- adsorb strongly from dilute HCl solutions, as is well known, and adsorbability decreases rapidly with increasing *M* HCl. Since I^- (and to some extent Br^-) tends to be oxidized to the element in strong hydrochloric acid solutions and since the elementary halogens are strongly adsorbed by the resins, separations in these media, particularly at tracer concentrations, become complicated. Separation of the halides in nitrate solutions are described in the literature (reference A-1, R-1).

Separation and Identification of Different Oxidation States of the Same Elements

As one might anticipate, different oxidation states of an element usually show considerable differences in adsorption properties. In many cases these differences may be utilized for separation of the oxidation states from each other. As a corollary, they also permit identification of specific oxidation states. This latter application appears particularly attractive for studies at trace concentrations where the usual tests with more concentrated solutions cannot be used and where precipitation reactions leave a considerable margin of error because of cross-contamination (carrying). Among the elements of interest here, useful differences in adsorption properties are shown in hydrochloric acid solutions as As(III) and As(V), Rh(III) and Rh(IV), and U(IV) and U(VI).

V. SOME CONSIDERATIONS REGARDING SPECIES IN SOLUTION

Detailed discussion of the thermodynamic aspects of anion exchange cannot be undertaken here. For dilute solutions the thermodynamics has repeatedly been discussed for cation exchange and the same considerations apply here. However, since a considerable portion of anion exchange studies is carried out in concentrated electrolyte solutions where available discussions are inadequate, a brief review of the pertinent considerations seems in order. Without significant loss in generality we shall restrict ourselves to those cases where the eluting ion and the complexing ion are the same as occurs, for example, in HCl media.

The mass action expression for the equilibrium

$$aMA_n^{-\nu} + \nu(A^{-a})_r \rightleftharpoons a(MA_n^{-\nu})_r + \nu A^{-a} \quad (2)$$

may be written as

$$K = \frac{(MA_n^{-\nu})^a{}_r\,(A^{-a})^\nu}{(MA_n^{-\nu})^a\,(A^{-a})_r{}^\nu}\,G \quad (3)$$

where

$$G = \frac{(g_{MA_n}{}^{-\nu})^a{}_r\,g_A{}^{-a\nu}}{(g_{MA_n}{}^{-\nu})^a\,(g_A{}^{-a})^\nu{}_r} \quad (4)$$

Parentheses indicate concentration of species and subscript *r* the resin phase. Where no subscript is given, the aqueous phase is implied. *G* is the proper activity coefficient quotient, *g* the activity coefficient of species, and *K* the ion exchange constant, which may be set equal to unity with proper assignment of standard states. Similar equilibria may be written for any species which distribute between the two phases. As pointed out earlier,[13] these mass action expressions imply independent distribution of components, and hence apply for both equivalent and non-equivalent exchange.

The distribution coefficient D may be defined as

$$D = (m_M)_r/m_M \qquad (5)$$

where $(m_M)_r$ is the stoichiometric concentration of M in the resin phase (irrespective of species) and where m_M is the concentration of M in the aqueous phase. In general, both m_M and $(m_M)_r$ represent sums of concentration of a number of species. For example, for the aqueous phase, one may write

$$m_M = (M^{+z}) + (MA^{+z-a}) + \ldots + (MA_n^{-\nu}) = \sum_i (MA_i^{+z-ia}) \qquad (6)$$

We shall assume for simplicity that

$$(m_M)_r = (MA_n^{-\nu})_r \qquad (7)$$

i.e., that only one species is adsorbed by the resin in significant amounts. Although this implies a considerable loss in generality, somewhat more complicated equations involving many species in the resin phase may readily be derived. For present purposes, however, most of the pertinent arguments can be made with equations derived with the restriction of Equation 7. Under these conditions

$$D = \frac{(MA_n^{-\nu})_r}{m_M} \qquad (8)$$

Combination of Equations 3 and 8, after multiplication of the numerator and denominator of Equation 3 by $m_M{}^a$, yields

$$\frac{K}{G} = D^a \frac{m_M{}^a}{(MA_n^{-\nu})^a} \frac{(A^{-a})^\nu}{(A^{-a})_r{}^\nu} \qquad (9)$$

Thus measurement of the distribution coefficients D permits, in principle, evaluation of $(MA_n^{-\nu}/m_M = F^\nu$, the fraction of metal in the form of the complex, through measurement of the concentrations of A^{-a} in the aqueous and resin phases and estimation of K/G.

Before discussing the implications of Equation 9 for measurements in concentrated electrolyte solutions, a few limiting situations will be reviewed.

It has become customary to assume that K/G is constant and independent of medium in dilute electrolyte solutions if the Debye-Hückel term vanishes. If we further restrict ourselves to stable, negatively charged ions (i.e., where $F_\nu = 1$) and to low loading, thus making $(A^{-a})_r =$ constant at low ionic strength, then the charge ν can be evaluated since

$$\frac{d \log D}{d \log m_A{}^{-a}} = -\nu/a \qquad (10)$$

Typical applications of Equation 10 are given in

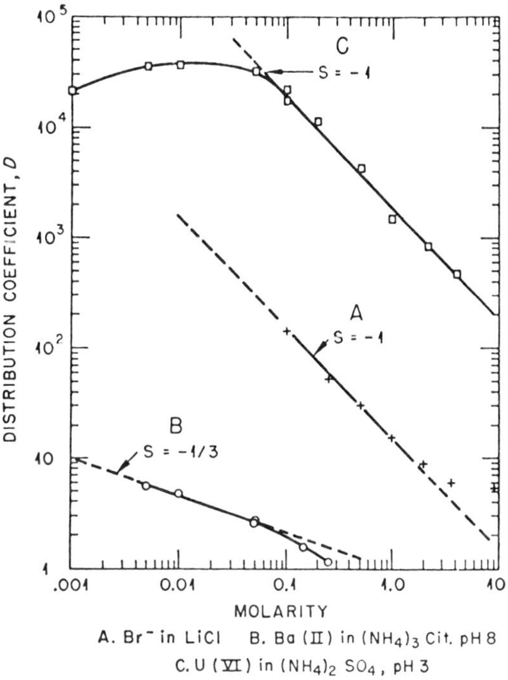

Figure 5. Determination of charge of ions by anion exchange: *(A)* Br^- in LiCl; *(B)* Ba(II) in $(NH_4)_3$ Cit. pH 8; *(C)* U(VI) in $(NH_4)_2SO_4$, pH 3

Fig. 5. For the "known" case of tracer bromide adsorption the slope of a plot of log D_{Br} *vs* log m_{HCl} has a slope of –1 at low m_{HCl}, as anticipated, since here $\nu = 1$ and $a = 1$. Deviations from this straight line, however, become serious at high m_{HCl}, presumably primarily because of non-constancy of G. Curve B (Fig. 5) gives a plot of log D_{Ba} *vs* log $m_{Cit^{---}}$. The slope is approximately –1/3 at low $m_{Cit^{---}}$ Since here $a = 3$, one may conclude that $\nu = 1$, as expected for the complex $BaCit^-$. From measurements of U(VI) in $(NH_4)_2SO_4$ solutions, curve C was constructed and d log $D_{U(VI)}/d$ log $m_{SO_4{}^{--}}$ was found to have a slope of –1 over a rather large range of sulfate concentration. Since here $a = 2$, one may conclude that $\nu = 2$, i.e., that the complex probably has a charge of —2, suggesting the formula $UO_2(SO_4)_2{}^{--}$. The same conclusion can also be obtained from a plot of log $D_{U(VI)}$ *vs* log $m_{H_2SO_4}$ from adsorption studies in sulfuric acid. In this case the slope is approximately —2 in the range 0.1 to 1 M H_2SO_4. Since here the eluting ion is probably HSO_4^-, i.e., $a = 1$, one obtains $\nu = 2$ in agreement with the proposed formula $UO_2(SO_4)_2{}^{--}$.

Most important for applications in concentrated electrolyte solutions is evaluation of variations of G with medium. If we assume that ν and a are known and that $F_\nu = 1$, K/G may be evaluated from Equation 9 through measurement of D, (A^-) and $(A^-)_r$.

Distribution coefficients have been measured for

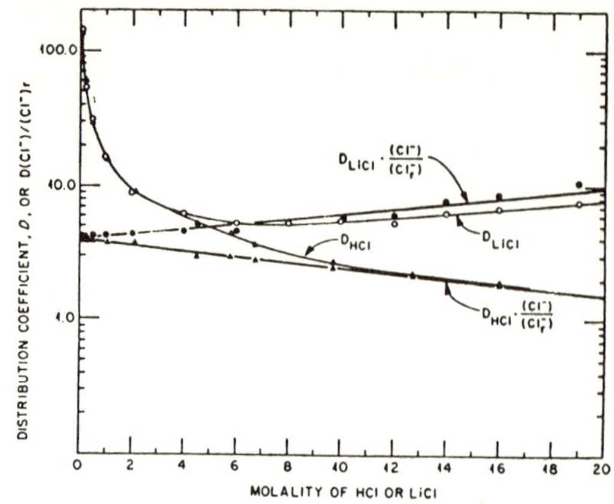

Figure 6. Adsorption of trace bromide from HCl and LiCl solutions

tracer bromide in HCl and LiCl solutions and the concentration of Cl^- in the resin phase determined analytically (see also ref. 18). The values of D and the computed values of K/G are given in Fig. 6 for both LiCl and HCl solutions. Note that in this case $K/G = D(Cl^-)/(Cl^-)_r$. It may be noticed that K/G is not constant even for this relatively simple 1 :1 exchange. Further, K/G is different for LiCl solutions and HCl solutions. For both media log K/G varies linearly with molality. The slopes $(d \log K/G)/dm = -d \log G/dm = b$ have different signs for the two media. Thus, $b_{Br^-}(LiCl) = 0.019$ and $b_{Br^-}(HCl) = -0.021$. These differences probably do not result from differences in the appropriate activity coefficient ratios in the aqueous phase, but probably principally reflect differences in the activity coefficients in the resin phase ($\gamma^r_{HBr}/\gamma^r_{HCl}$ and $\gamma^r_{LiBr}/\gamma^r_{LiCl}$). Considerably greater non-constancy of K/G had earlier been reported[13] for the adsorption of Au(III) ($AuCl_4^-$) ($b_{Au(HCl)} = 0.13$), though, it now appears, for very different reasons.

Rec· ·tly a very large difference was found[19] for the adsorption of a number of metal complexes from HCl and LiCl solutions. Since then a more detailed study has been carried out on the adsorbability of Au(III) and log $D_{Au(III)}$ was found to increase steeply with LiCl concentration which may be compared with the rapid decrease of $D_{Au(III)}$ in HCl solutions. Further, at least in the range $1 < m_{LiCl} < 10$, log D $(Cl^-/Cl^-)_r$ varies essentially linearly with m_{LiCl}. It was suggested[19] that this large difference in adsorbabilities in LiCl and HCl solutions could be caused either by unexpected differences in the activity coefficients in the resin phase or by rather extensive formation of undissociated complex acids at high HCl concentrations. Originally the former explanation was preferred, particularly in view of the great generality of the "lithium chloride effect". The weight of more recent evidence, however, makes the assumption that undissociated complex acids are formed more probable. A study of the activity coefficients of lithium chloride in the resin phase in a manner analogous to the earlier studies[18] of the activity coefficients of HCl revealed no large differences. Further, measurements of the distribution coefficients of Au(III) as a function of acidity at constant total molality ($m = 10$) in LiCl-HCl mixtures have been carried out. The data could be fitted reasonably well by assuming the equilibrium

$$HAuCl_4 \rightleftharpoons H^+ + AuCl_4^- \qquad (11)$$

with concentration quotient k_h^m = ca 0.1 at this ionic strength though this quotient showed some not unexpected drift with composition.

One may thus with reasonable confidence postulate that those chloride complexes which show a pronounced lithium chloride effect form relatively weak complex acids. According to present data this includes the chloro complexes of the fission products Zn(II), Ga(III) and Cd(II), but not Ag(I). Actually, in the case of Ag(I), adsorption from LiCl solutions is only slightly stronger than from HCl solutions, the effect being of about the same order of magnitude as for Br^- adsorption. This implies that Ag(I) does not form undissociated complex acids to any large extent even in concentrated HCl, in agreement with the known acid insensitivity of the solubility of AgCl in chloride media.[20]

Although the differences in the results in LiCl and HCl solutions throw considerable doubt on the earlier qualitative interpretations of the shapes of the adsorption functions present, more detailed work with Ga(III) gives hope that these earlier qualitative interpretations might still be applicable. It had been felt that the steeply rising portions of the distribution functions are due to the formation of the negatively charged complexes, i.e., reflect the rapid rise in F_ν, that the maxima occur (at high ionic strength) at approximately those HCl concentrations where the negatively charged complexes become predominant, and that the region of decreasing D which corresponds to a decrease in D $(Cl^-)^n/(Cl^-)_r^n$ at high M HCl, coincides with the region where the negatively charged complexes or complex acids are essentially completely formed. In the case of Ga(III) this conclusion was confirmed in recent Raman work in HCl and LiCl solutions which was carried out in collaboration with Young and Krawetz (University of Chicago). A Raman line apparently characteristic for $GaCl_4^-$ (or $HGaCl_4$) was found to have essentially constant intensity above 8 M HCl, the position of the adsorption maximum. Further, a rapid decrease in intensity of the line occurred with decrease in M HCl. Thus, the qualitative interpretations suggested appear to apply. One may, therefore, apparently classify the elements according to their tendencies to form negatively charged complexes from the position of the adsorption maxima. The order of stability, on this basis, for uranium and the fission products is as follows:

1. Strong negatively charged complexes, essentially fully formed below 1 M HCl: (Tc(VII)),

Ru(IV), Rh(III), Rh(IV), Pd(II), Ag(I), Sn(II) (?).

2. Negatively charged complexes formed in the range 1 to 12 *M* HCl (in decreasing order of stability): Cd(II) ≈ Sb(III) ≈ Zn(II) > In(III) ≈ Mo(VI) > Sn(IV) > Ga(III) > As(III) ≈ Sb(V) ≈ U(VI) (?).

3. Negatively charged complexes not completely formed in 12 *M* HCl: Ge(IV), Zr(IV), U(IV).

4. No evidence for negatively charged complexes: As(V), Rb(I), Sr(II), Y(III), Cs(I), Ba(II), R.E.(III).

In several instances this order may appear surprising. Most striking is the fact that Cd(II) and Zn(II) appear to form negatively charged complexes at approximately the same HCl concentration, though unquestionably the positively charged complex MCl^+ is much more stable for Cd(II) than Zn(II) (see also Fig. 4). Thus, as had already become apparent with other anion exchange studies, e.g., of the first row transition elements (see ref. K-8, Table I), there appears to be for many cases no simple relation between stability constants within a series of complexes MA_i.

The anion exchange results in general are of course susceptible to considerably more detailed and quantitative interpretation in terms of species formed and the corresponding equilibria, than given here. Most interesting in this respect will be interpretations of the behavior of the metals in concentrated electrolyte solutions. As pointed out, these are intimately tied to evaluation of the activity coefficient term *G*, whose more exact determination is the objective of present research. It is to be hoped that these more quantitative interpretations will become feasible, at least in a few cases, in the not too distant future.

REFERENCES

1. Nachod, F. C., *Ion Exchange: Theory and application*, Academic Press, Inc., New York (1949).
2. Kunin, R. and Myers, R. J., *Ion Exchange Resins*, John Wiley and Sons, Inc., New York (1950).
3. Osborn, G. H., *Applications of Ion Exchange Resins*, Analyst 78: 220 (1952).
4. Lederer, E. and Lederer, M., *Chromatography*, Elsevier Pub. Co., Inc., Houston (1953).
5. Samuelson, O., *Ion Exchangers in Analytical Chemistry*, John Wiley and Sons, Inc., New York (1953).
6. Schubert, J., *Ion Exchange*, Ann. Rev. Phys. Chem. 5: 413 (1954).
7. Kunin, R. and McGarvey, F. X., *Annual Review of Ion Exchange*, Anal. Chem. 26: 104 (1954).
8. Kunin, R. and McGarvey, F. X., *Annual Review of Ion Exchange*, Ind. Eng. Chem. 47: 565 (1955).
9. *Ion Exchange Resins in Medicine and Biological Research*, Ann. N. Y. Acad. Sci. 57: 61 (1953).
10. Campbell, E. C. and Nelson, F., *Chemical Separation of 0.82 sec. Pb^{207m} from Bi^{207}*, Phys. Rev. 91: 499A (1953).
11. Schubert, J., *The Use of Ion Exchangers for the Determination of Physical-Chemical Properties of Substances, Particularly Radiotracers, in Solution. I.*, J. Phys. and Colloid Chem. 52: 340 (1948).
12. Fronaeus, S., *On the Use of Anion Exchangers for the Study of Complex Systems*, Svensk Kem. Tidskr. 65: 1 (1953).
13. Kraus, K. A. and Nelson, F., *Ion Exchange in Concentrated Electrolytes. Gold (III) in Hydrochloric Acid Solutions*, J. Am. Chem. Soc. 76: 984 (1954).
14. Borkowski, C. J., (1949) as reported in ORNL-1160 (1951).
15. Kraus, K. A. and Moore, G. E., *Separation of Zirconium and Niobium in HCl-HF Mixtures*, J. Am. Chem. Soc. 73: 9 (1951).
16. Wheaton, R. M. and Bauman, W. C., *Properties of Strongly Basic Anion Exchange Resins*, Ind. Eng. Chem. 43: 1088 (1951).
17. Kunin, R. and McGarvey, F. X., *Equilibrium and Column Behavior of Exchange Resins*, Ind. Eng. Chem. 41: 1265 (1949).
18. Kraus, K. A. and Moore, G. E., *Adsorption of Hydrochlorid Acid by a Strong Base Anion Exchanger*, J. Am. Chem. Soc. 75: 1457 (1953).
19. Kraus, K. A., Nelson, F., Clough, F. B. and Carlston, R. C., *Adsorption from Lithium Chloride Solutions*, J. Am. Chem. Soc. 77: 1391 (1955).
20. Seidell, A., *Solubilities of Inorganic Compounds*, D. van Nostrand Co., New York (1940).

Mr. KRAUS (USA) presented paper P/837 as follows:

During the last ten to twenty years, as synthetic organic ion exchange resins of high stability became readily available, the popularity of ion exchange as a tool in inorganic and analytical chemistry increased tremendously. Most of these ion exchange studies involve cation exchange resins, resins which can exchange positively charged ions. On the other hand,

Part VI

MIXED SOLVENTS IN ANION AND CATION EXCHANGE

Editor's Comments on Papers 20 Through 22

20 FRITZ and PIETRZYK
Non-aqueous Solvents in Anion-Exchange Separation

21 HAZAN and KORKISCH
Anion-Exchange Separation of Iron, Cobalt and Nickel

22 KOCH and KORKISCH
Anionenaustauschtrennungen der mit Tributylphosphat extrahierbaren Elemente: I. Trennung des Urans vom Eisen in Gegenwart von Tributylphosphat
English translation: *Anion-Exchange Separations of Elements Extractable by Tributyl Phosphate: I. Separation of Uranium from Iron in the Presence of Tributyl Phosphate*

Mixing an organic solvent, like alcohol or acetone, with water lowers the dielectric constant and increases the electrical forces between ions. Complex ions become more stable, and ion pairing is favored. The difference between the resin and solvent environments is magnified, because resins that have ionic groups take up water in preference to less polar organic solvents (see Part XI). Organic solvents, therefore, have profound effects on the distribution of metal ions in both anion and cation exchange.

J. S. Fritz (Fritz and Abbink, 1965) was one of the first to study these effects. He showed that the binding of divalent metal ions to an anion exchanger in hydrochloric acid solutions was strengthened by adding alcohol, and that isopropyl alcohol was more effective than ethyl or methyl alcohol. His 1961 paper written with D. J. Pietrzyk, which is reproduced here, presents a vast amount of data showing the effects of solvent composition and acid concentration on distribution ratios.

A later paper by Fritz and Rettig (1962), which is frequently cited, shows the effect of acetone on cation-exchange distributions in the presence of hydrochloric acid. By stabilizing their complex anions, acetone generally weakens the binding of metals to a cation-exchange resin. It accentuates the differences between different metals, and Fritz and Rettig showed that by progressively increasing the proportions of acetone and hydrochloric acid, good chromatographic separations of several metals could be obtained on a cation-exchange resin column.

Beginning in 1965, J. Korkisch of the University of Vienna has explored the analytical and preparative uses of mixed solvents in anion and cation exchange. He and his colleagues have been publishing at the rate of several papers a year. The book by Korkisch, *Modern Methods for the Separation of Rarer Metal Ions,* is a mine of information on ion-exchange and solvent-extraction techniques (Korkisch, 1969). The author calls the use of ion exchange in mixed solvents "combined ion exchange and solvent extraction," or CIESE (in German, KIALE).

We have chosen two of Korkisch's papers for this collection. Paper 21 is one of his earliest and most dramatic. It shows that the sequence of elution of iron (III), cobalt, and nickel on an anion-exchange resin in hydrochloric acid is changed when acetone is added. The Fe–Co–Ni separation was one of the first to be described by Kraus. Using aqueous HCl, Ni is not absorbed by the resin, Co is eluted by 3 *M,* and Fe(III), which is the most strongly absorbed, is eluted by 1 *M* acid. In 90 percent acetone–10 percent 6 *M* HCl, however, Fe(III) is not retained by the resin. Nickel is weakly retained. Nickel and cobalt are eluted in that order by 70 percent acetone–30 percent 2 *M* HCl.

The reason that iron (III) is not retained is that it probably gives an uncharged, solvated ion pair, $H^+FeCl_4^-$. The iron solution in 90 percent acetone is nearly colorless, in contrast to the bright yellow-brown in aqueous hydrochloric acid.

The practical utility of having iron pass through the resin without being absorbed, while cobalt and nickel are absorbed, is evident. In most natural materials cobalt and nickel are much less abundant than iron. A small resin column now suffices to "filter" these elements from the excess of iron.

Paper 22 is one of Korkisch's most recent papers; it shows how separation by solvent extraction (in this case by tributyl phosphate) can be followed by further separation by ion exchange. Iron, uranium, gold, gallium, and germanium are among the elements that can be recovered in this way from very complex mixtures. This paper is one of a long series that appeared in *Mikrochimica Acta* during 1972 and 1973, many of which had to do with nuclear raw materials. Another important paper, omitted because of space, is that by Korkisch and Steffan (1972) on the concentration and measurement of traces of uranium in geological materials.

Mixed solvents are now used routinely and have added a new dimension to inorganic ion-exchange chromatography. More examples will be seen in Part VII. Strelow (1971) refers to acetone–water mixtures.

REFERENCES

Fritz, J. S., and J. E. Abbink. 1965. *Anal. Chem., 37,* 1274.

——, and T. A. Rettig. 1962. *Anal. Chem., 34,* 1562.

Korkisch, J. 1969. *Modern Methods for the Separation of Rarer Metal Ions,* Elmsford, N.Y., Peragomon Press.

——, and I. Steffan. 1972. *Mikrochim. Acta., 1972,* 837–860.

Strelow, F. W. E., A. H. Victor, C. R. van Zyl, and C. Egloff. 1971. *Anal. Chem. 43,* 870.

20

NON-AQUEOUS SOLVENTS IN ANION-EXCHANGE SEPARATIONS*

JAMES S. FRITZ and DONALD J. PIETRZYK
Institute for Atomic Research and Department of Chemistry
Iowa State University, Ames, Iowa, U.S.A.

(*Received* 18 *August* 1960. *Accepted* 13 *October* 1960)

Summary—Distribution coefficients are measured for the partition of metal ions between anion-exchange resin and organic solvent-water mixtures containing hydrochloric acid. The presence of an organic solvent causes metal ions to be taken up at lower hydrochloric acid concentrations. In many cases, distribution coefficients are significantly higher than in water-hydrochloric acid systems. If other conditions are comparable, the order of distribution coefficients in alcohol-water-hydrochloric acid is: isopropyl > ethyl > methyl alcohol. Column separations of metal ion mixtures can be carried out by eluting with alcohol-water-hydrochloric acid mixtures of different compositions. Successful separations of a number of mixtures are reported.

INTRODUCTION

THE separation of metals as chloro complexes from aqueous hydrochloric acid solutions using anion-exchange columns is now a well-established and valuable analytical procedure.[1-8] In this method, metals that form anionic chloro complexes are taken up from strong hydrochloric acid solutions; other metals pass through the column. Then the metals on the column are eluted one or two at a time using aqueous solutions of hydrochloric acid that are progressively more dilute.

It was observed that metals are taken up strongly and at lower hydrochloric acid concentrations if an appreciable amount of a water-miscible organic solvent is added to the aqueous hydrochloric acid. The purpose of the present work was to study the anion-exchange behaviour of metal chloro complexes in partly non-aqueous media. The primary aim was to find conditions that would improve the convenience and broaden the scope of this type of analytical separation.

Katzin and Gebert[9] studied the absorption of lithium chloride, lithium nitrate, cobalt chloride and nickel nitrate on chloride- and nitrate-form anion-exchange resins from acetone solution. They noted that the whole salt is adsorbed and suggested that the salts might be taken up as complex anions. On the other hand, Davies and Owen[10] postulated that the salts are adsorbed by the resin by a solvent extraction process. Tuck and Welch[11] found that the uptake of plutonium^{IV} by an anion-exchange resin from a nitric acid-diethyleneglycoldibutyl ether solution is the result of anionic complex formation.

In the anion-exchange separation of sulphate, sulphite, thiosulphate, sulphide, selenite and tellurite, Iguchi[12,13] found that the addition of alcohol to the eluting agent increased the adsorbabilities of all but the selenite and tellurite. Burstall *et al.*[14] eluted gold (adsorbed as the cyanide) from an anion-exchange column using an acetone eluent containing 5% hydrochloric acid and 5% water. Korkisch *et al.*[15] separated uranium^{VI} from certain other metals on a chloride-form anion-exchange

* Contribution No. **921.** Work was performed in the Ames Laboratory of the U.S. Atomic Energy Commission.

column using a solution of hydrochloric acid in mixed water and ethyl alcohol. The addition of 10 to 25% methyl alcohol to an aqueous hydrochloric acid eluent has been found to improve the anion-exchange separation of zinc and cadmium.[16] Recently Kojima[17,18] and Yoshino and Kurimura[19] studied the anion-exchange behaviour of several of the transition elements in mixed solvents containing hydrochloric acid. Distribution coefficients of metal ions were measured in various combinations of hydrochloric acid, water and organic solvent, and the results were compared with the distribution coefficients found in aqueous hydrochloric acid solutions by Kraus and Nelson.[1,2] In all cases the incorporation of a non-aqueous solvent in the mixture caused the distribution coefficients to be higher. An elution scheme for several metal ions was devised, and the method was applied to the determination of aluminum and copper in certain zinc alloys.

EXPERIMENTAL

Reagents

Anion-exchange resin: Dowex 1 × 8 "analysed reagent" resin, either 100- to 200-mesh or 200- to 400-mesh. The finer mesh resin gives somewhat sharper separations and was used in most of the column separations. Before use, backwash the resin to remove the very fine particles, then wash a column of resin with 2 to 2·5*M* perchloric acid to remove metallic impurities. Wash with water, then convert the resin to the chloride-form by successive washing with 2 to 2·5*M*, 1*M*, and 0·005*M* hydrochloric acid. Wash with water, then remove the water by suction filtration. Wash the resin with absolute alcohol, then with acetone, and allow it to air-dry. The air-dried resin has a water content of 5 to 10% by weight.

EDTA, (disodium dihydrogen ethylenediaminetetra-acetate dihydrate): Prepare a 0·05*M* stock solution from reagent-grade salt. Standardise by titration of zinc^{II} from primary standard zinc metal or by titration of cadmium^{II} from the primary standard, HCdV.[20] Naphthyl azoxine indicator is recommended for the standardisation.[20,21]

Metal salts: Reagent grade metal chlorides were used when available. Rare-earth, yttrium and scandium chloride solutions were prepared by dissolving the oxides in hydrochloric acid. Thorium^{IV} and uranium^{VI} nitrates, which were converted to chlorides by anion-exchange, were used, and vanadium^{IV} was used as the sulphate. In a few experiments metal perchlorates were employed. The solutions were 0·05*M* to 0·1*M*.

Organic solvent-water-hydrochloric acid mixtures: Mixtures were prepared so that the amount of organic solvent and water was expressed as percentage by volume and the hydrochloric acid concentration as molarity. For example, 1 litre of 1*M* hydrochloric acid in 80% isopropyl alcohol is prepared as follows: Mix together 800 ml of isopropyl alcohol, 83 ml of concentrated hydrochloric acid and 117 ml of water. Disregard any changes in volume due to mixing.

Procedures

Macro amounts of most metal ions were determined by titration with EDTA. Conditions are summarised in Table I. Iron^{III}, chromium^{III} and uranium^{VI} were determined by standard redox methods.

Micro amounts of metal ions were determined by colorimetric methods: copper^{II} with neocuproïne,[28] zinc^{II} with zincon,[29] thorium^{IV} with thorin,[30] uranium^{VI} with arsenazo,[31] bismuth^{III} with thiourea,[32] and iron^{II} with 1:10-phenanthroline.[33]

Measurement of distribution coefficients

Weigh accurately approximately 1 g of air-dried anion-exchange resin into a 125-ml glass-stoppered Erlenmeyer flask. Pipette into the flask 4 ml of metal salt solution and 50 ml of the appropriate organic solvent-water-hydrochloric acid mixture. Stopper the flask and shake for 12 to 18 hr (80% or less of organic solvent) or 22 to 32 hr (more than 80% organic solvent) at 24° ± 1°. Pipette an aliquot from the supernatant liquid, evaporate the acid and organic solvent, and determine the metal ion content by an appropriate analytical method. Determine accurately the water content of the resin and calculate the distribution coefficients on a dry weight basis.

TABLE I.—TITRATION OF METAL IONS WITH EDTA

Ions titrated	pH	Indicator	Reference
Cd^{2+}, Co^{2+}, Cu^{2+}, Ni^{2+}, Zn^{2+}	5.5–6·5	Naphthyl azoxine	21
Mn^{2+}	10	Eriochrome Black T	22
VO^{2+}	5·5–6·5	Naphthyl azoxine (back-titrated with Zn^{2+})	23
Rare earth^{3+}, Y^{3+}	5·5–6·5	Arsenazo	24
Th^{4+}	1·7–3·0	Arsenazo	24
Sc^{3+}	3·0–4·5	Methylthymol blue	22
Bi^{3+}	1·3–2·0	Thiourea	25
Zr^{4+}	1·3–2·0	Thiourea (back titrated with Bi^{3+})	26
Ca^{2+}	10	Thymolphthalexone	27
Mg^{2+}	10	Eriochrome Black T	22

Separation of mixtures

Prepare a sample mixture by pipetting into a beaker known amounts of standard metal salt solutions. Carefully evaporate the sample mixture nearly to dryness. Add 5 ml of 0·3*M* hydrochloric acid in 97% ethyl alcohol. When the residue is dissolved, transfer the sample to an anion-exchange column using 0·3*M* hydrochloric acid in 95% ethyl alcohol as a rinse solution. Use an ion-exchange column having an inside diameter of either 12 mm or 22 mm; the height of Dowex 1 × 8 resin bed for various separations is indicated in Tables II, III, IV. Elute the various sample constituents from the column with the types and quantities of eluent indicated in Tables II, III, IV with a flow rate of 0·25 to 0·33 ml/min. Evaporate the effluent fractions collected, and determine the amount of metal salt present in each effluent fraction by an appropriate titrimetric procedure.

MEASUREMENT OF DISTRIBUTION COEFFICIENTS

Measurement of distribution coefficients of various metal ions over a wide range of conditions is a good way to avoid choosing eluting conditions for column separations by a strictly trial and error method. The batch distribution coefficient, D, is defined:

$$D = \frac{\text{mequiv. metal on resin/g of dry resin}}{\text{mequiv. metal in solution/ml of solution}}$$

Although this distribution coefficient is measured on a batch basis, it can be used to predict elution behaviour for metals eluted from an ion-exchange column. To separate two substances, conditions should be selected such that the distribution coefficient of one is low (preferably 1 or less) so that elution from the column will be rapid. The distribution coefficient of the other substance under the same conditions should be as large as possible so that this substance will be tightly held by the column.

In aqueous hydrochloric acid solution, the distribution coefficient of most metal ions in contact with anion-exchange resin increases as the concentration of hydrochloric acid is increased. After a certain concentration of hydrochloric acid is attained, the distribution coefficient remains about the same or decreases somewhat with further increases in hydrochloric acid concentration. In partly non-aqueous solutions of hydrochloric acid, two factors cause an increase in the value of D; an increase in hydrochloric acid concentration, or an increase in the proportion of non-aqueous solvent in the water-non-aqueous solvent mixture. In solvent mixtures containing an appreciable amount of non-aqueous solvent, a D value sufficiently high to assure quantitative uptake of a substance on an ion-exchange column occurs at a much lower hydrochloric acid concentration than in completely aqueous systems. Also, the maximum distribution coefficient obtainable in partly non-aqueous systems containing hydrochloric acid is frequently much higher than is possible in aqueous hydrochloric acid solutions. The distribution of manganese^{II} between chloride-form anion-exchange resin and hydrochloric acid solution is an outstanding illustration of the latter. In aqueous hydrochloric acid, the distribution coefficient is always less than 10; in alcohol-hydrochloric acid systems containing only a little water a distribution coefficient of several thousand can be realised.

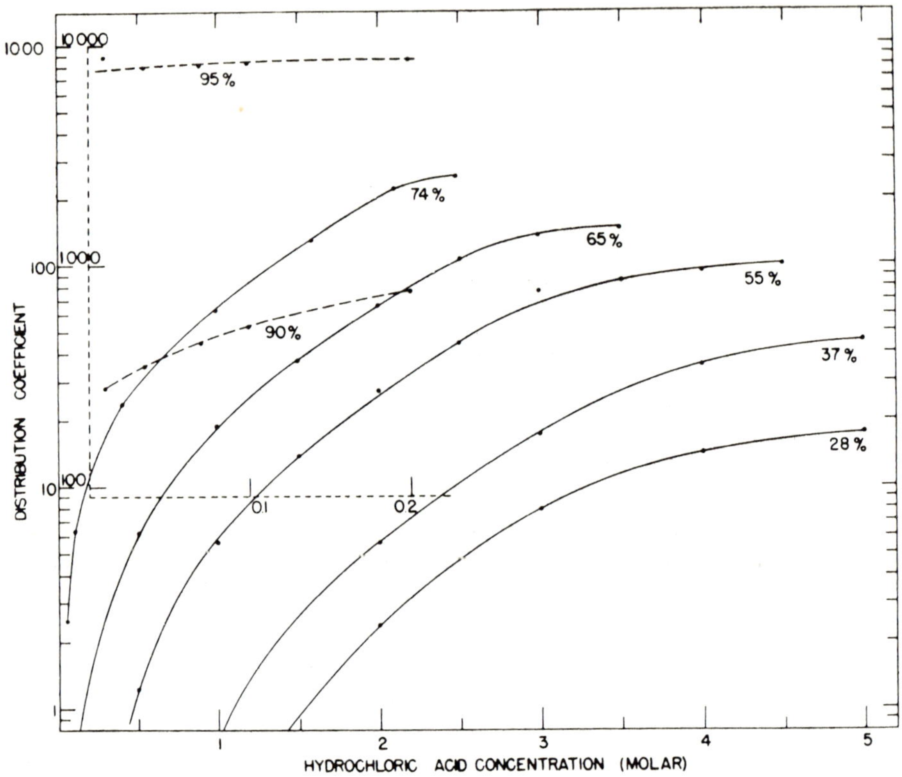

FIG. 1—Distribution coefficients of copper^{II} chloride in various concentrations of isopropyl alcohol *vs.* hydrochloric acid concentrations.

In Fig. 1 the distribution coefficient for copper^{II} is plotted as a function of hydrochloric acid concentration at several isopropyl alcohol-water solvent compositions. It will be noted that the presence of an increasing proportion of isopropyl alcohol has a pronounced effect on the distribution coefficient. In Figs. 2–5, distribution coefficients of other metals are plotted as a function of hydrochloric acid for solvents containing different percentages of isopropyl alcohol.

The effect of different organic solvents on the ion-exchange behaviour of various metal ions was studied. Fig. 6 shows that the various organic solvents studied have about the same effect on the distribution coefficients for copper when the solvent contains an appreciable proportion of water. However, when the solvent contains less than 10 or 15% water, the difference in D in changing from one organic solvent to

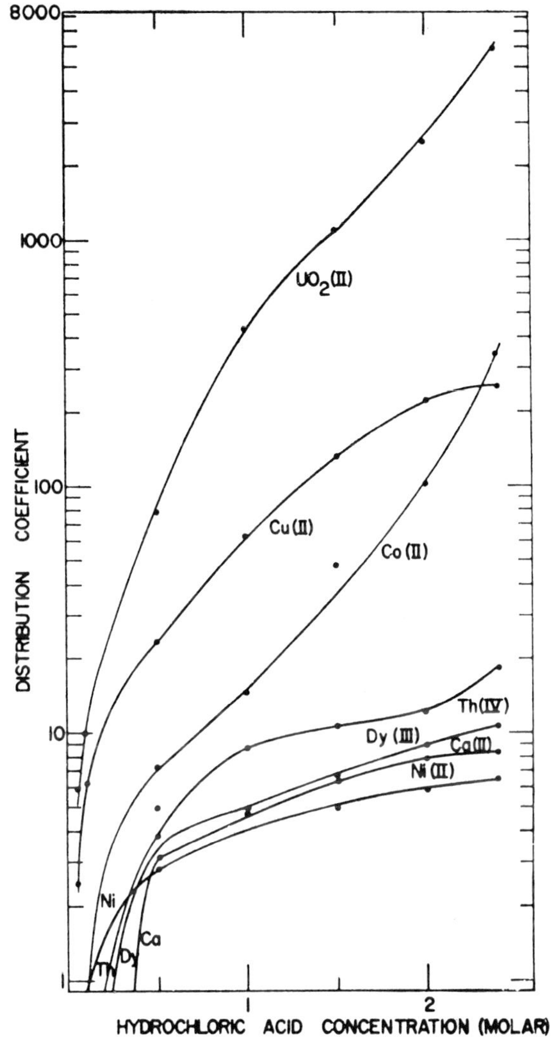

FIG. 2- Distribution coefficients of metal chlorides *vs.* hydrochloric acid concentration in 74% isopropyl alcohol.

another is more pronounced. The distribution coefficients of copper as a function of hydrochloric acid concentration for several solvents are plotted in Fig. 7. Distribution coefficients for several metal ions are plotted in isopropyl alcohol, ethyl alcohol and methyl alcohol in Figs. 8–10. In Figs. 11 and 12, distribution coefficients for various metals are plotted as a function of ethyl alcohol concentration for solutions that are 0·3*M* in hydrochloric acid.

In these figures the concentration of hydrochloric acid is kept low for two reasons. When the proportion of non-aqueous solvent in the mixture is high, only a low concentration of hydrochloric acid is required for metals that form chloro complexes to have high distribution coefficients. Also, the use of much higher concentrations of

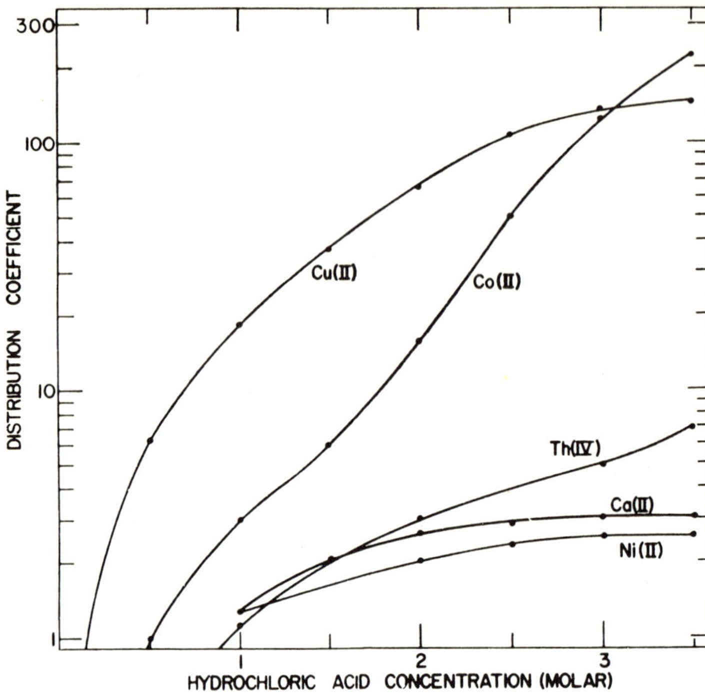

FIG. 3—Distribution coefficients of metal chlorides *vs.* hydrochloric acid concentration in 65% isopropyl alcohol.

hydrochloric acid would make it impossible to maintain the desired concentration of non-aqueous solvent and still use concentrated aqueous hydrochloric acid in preparing the solvent mixture.

Figs. 8–10 show that the distribution coefficients of the metal ions studied are higher in ethyl alcohol than in methyl alcohol, and are still higher in isopropyl alcohol. By mixing the alcohols in varying proportions and keeping the hydrochloric acid concentration constant, a gradual transition of D values from one pure alcohol to another is possible. Figs. 13 and 14 show the distribution coefficients for several metal chlorides in methyl alcohol-isopropyl alcohol and methyl alcohol-ethyl alcohol, respectively.

SEPARATIONS

The batch distribution coefficient, D, can be used to calculate the volume of eluting agent required to elute a metal ion from a given ion-exchange column. The relationships used in such a calculation are as follows:

$$D = D \times \frac{\text{g of resin}}{V}$$

$$U^* = V(D_v + 1)$$

where D is the batch distribution coefficient, D_v is a volume distribution coefficient, V is the interstitial volume (ml) of the resin bed, and U^* is the volume (ml) of eluent required to elute a metal ion to a point where the effluent from the column has the

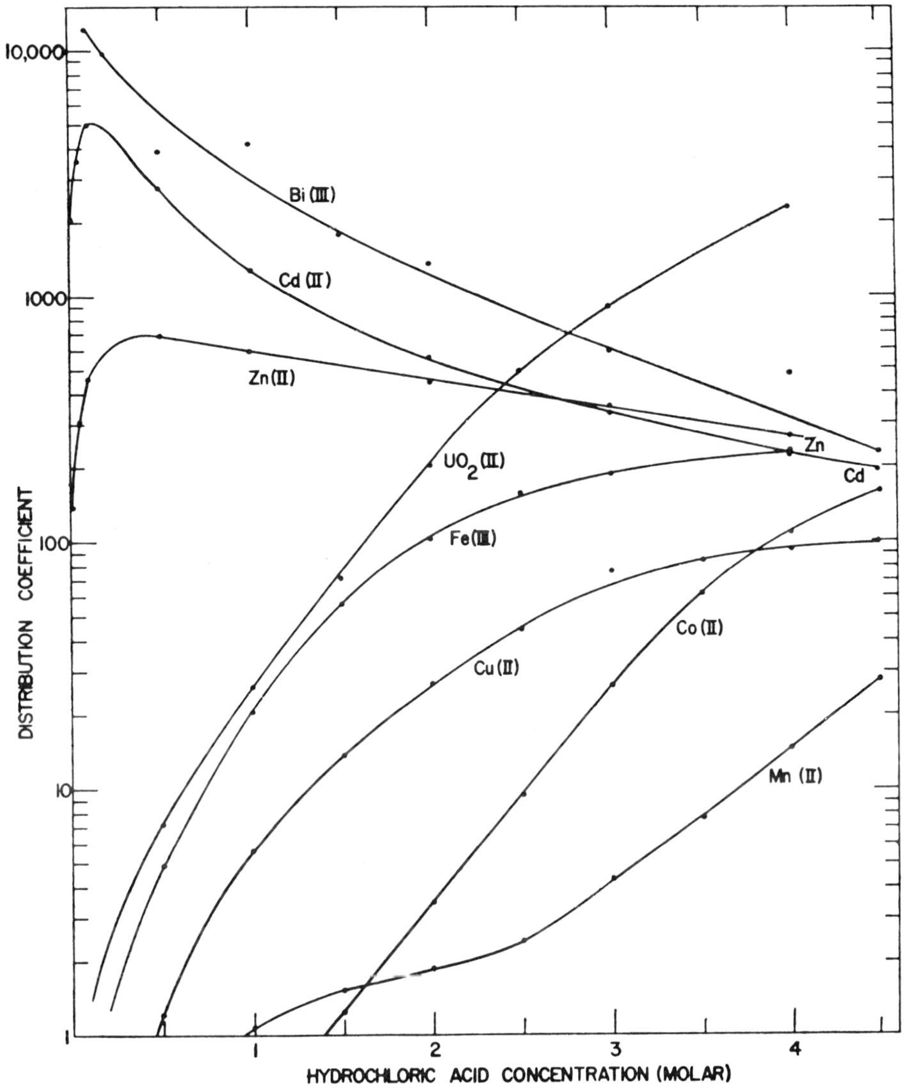

FIG. 4—Distribution coefficients of metal chlorides *vs.* hydrochloric acid concentration in 55% isopropyl alcohol. High adsorption group.

maximum concentration in the metal ion eluted. The volume of eluent required to elute *all* of a metal ion depends on the band width of the metal ion eluted. A good discussion of the selection of column dimensions and operating conditions, when the distribution coefficients of the substances to be separated are known, is given by Cornish.[34]

In the present work, batch distribution coefficients served as a valuable guide in selecting conditions for column separations. However, the volume of solution required to elute various ions from a column was always determined by collecting actual fractions from a column and analysing the fractions. Under the conditions used, this

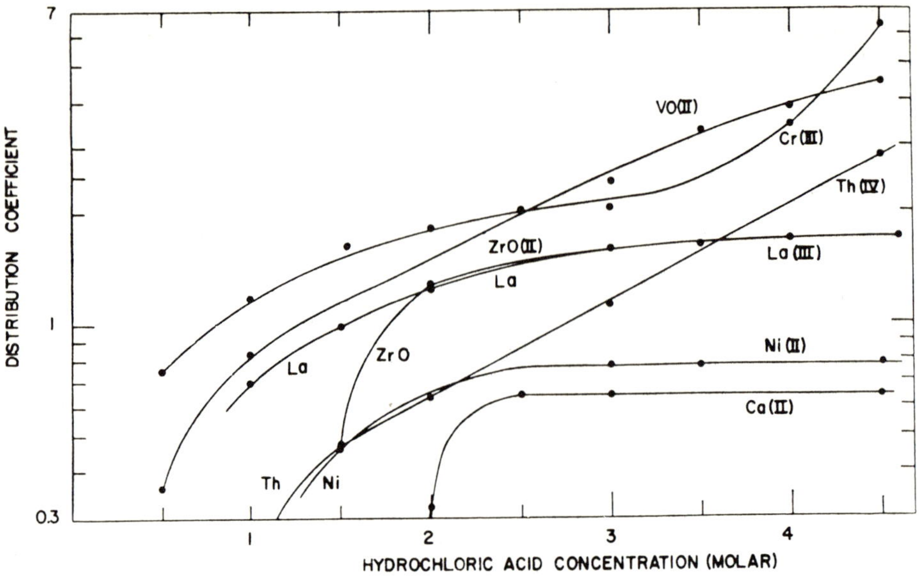

FIG. 5—Distribution coefficients of metal chlorides *vs.* hydrochloric acid concentration in 55% isopropyl alcohol. Low adsorption group.

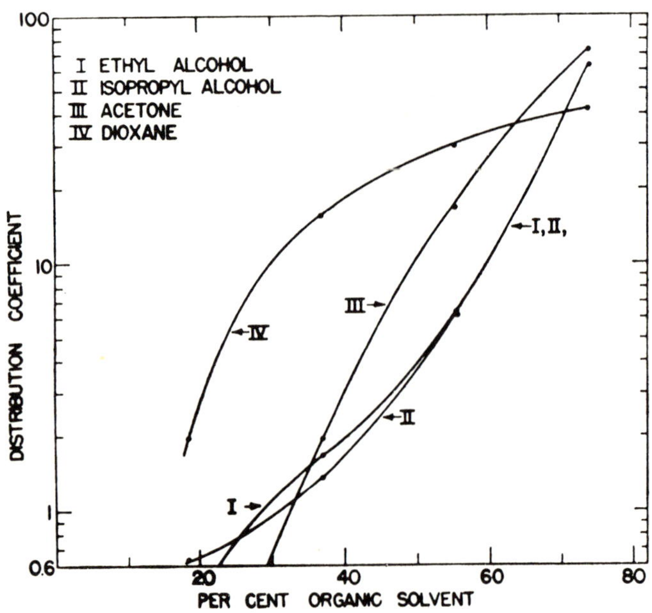

FIG. 6—Distribution coefficients of copper^{II} chloride in $1M$ hydrochloric acid *vs.* varying percentage of organic solvent.

was found to be much more reliable than simply calculating the required volume of eluent from the distribution coefficients.

To separate two metal ions on an anion-exchange column, a solvent composition and hydrochloric acid concentration is chosen such that one metal will be rapidly

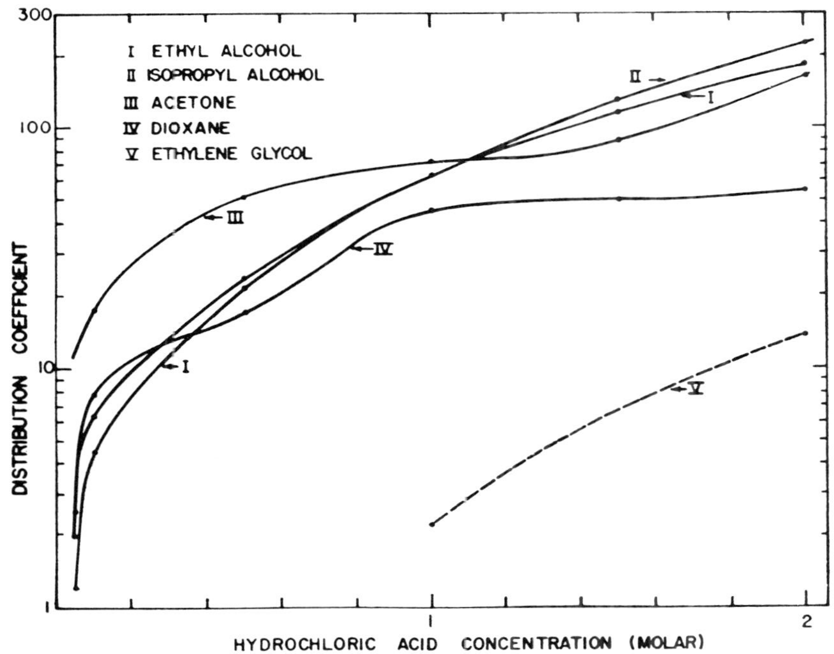

FIG. 7—Distribution coefficients of copper[II] chloride in 74% organic solvent *vs.* hydrochloric acid concentration.

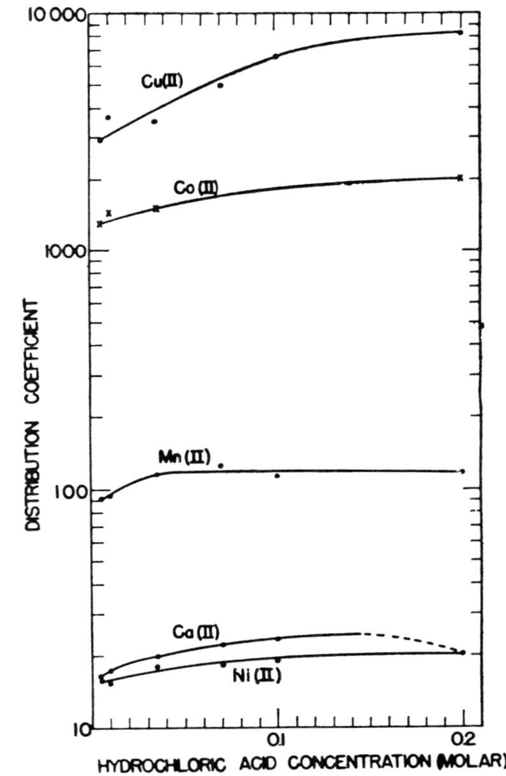

FIG. 8—Distribution coefficients of some metal ions in 96% isopropyl alcohol *vs.* hydrochloric acid concentration.

eluted from the column (D should be low, preferably 1 or below) while the other metal ion has a high D and is tightly held by the column. When several metal ions are to be separated from each other, the scheme is to choose different conditions for successive elutions so that only one metal ion is eluted by each eluting agent. In selecting the composition of eluting agent, we have three variables to work with: the non-aqueous solvent, concentration of non-aqueous solvent in the non-aqueous-water mixture, and concentration of hydrochloric acid.

In alcohol-water mixtures containing a high proportion of alcohol, only a low concentration of hydrochloric acid is required for metal ions to be strongly taken up by an anion-exchange column. As shown in Figs. 7–10 the distribution coefficients of most metals change little with moderate changes in hydrochloric acid concentration. However, inspection of Figs. 11 and 12 reveals that changes in the alcohol content of the solvent affect the D values of metal ions to varying degrees. Thus one elution scheme is based on the use of a series of eluents with a constant hydrochloric acid concentration (0·3*M*), but with decreasing proportions of alcohol. Ethyl alcohol was chosen because the D values of the "non-adsorbable" metal ions such as nickel^{II} and calcium^{II} are too high in isopropyl alcohol (see Fig. 8). Methyl alcohol was not chosen because D values for "adsorbable" metal ions are significantly lower than in ethyl alcohol (Figs. 9 and 10).

The ethyl alcohol elution scheme is as follows:
Sample is added to column in 95% ethyl alcohol-0·3*M* HCl.
Ni^{II} is eluted with 82% to 95% ethyl alcohol-0·3*M* HCl.
Mn^{II} is eluted with 72% to 82% ethyl alcohol-0·3*M* HCl.
Co^{II} is eluted with 72% ethyl alcohol-0·3*M* HCl.
Cu^{II} and Fe^{III} are eluted with 40% ethyl alcohol-0·3*M* HCl.
Zn^{II} is eluted with 0·005*M* aqueous HCl.

Examples of quantitative column separations using this scheme are given in Table II. In some instances the composition of the sample was such that the elution of one of the sample constituents could be accomplished using a lower concentration of ethyl alcohol than stated in the preceding scheme.

Several separations are not possible using this elution scheme. Uranium^{VI}, copper^{II} and iron^{II} can be separated only as a group. Nickel^{II}, calcium^{II}, vanadium^{IV} and chromium^{III} can also be separated only as a group. Manganese^{II} can be separated from this latter group, but the separation is lengthy and only moderately successful. Separation of manganese^{II} from cobalt is not possible.

At higher ethyl alcohol concentrations, tailing of the elution bands is a problem. Tailing of nickel^{II} when eluted with 0·3*M* hydrochloric acid in 95% ethyl alcohol is particularly bad. When nickel is eluted with 0·3*M* hydrochloric acid-82% ethyl alcohol, the tailing is lessened but is still something of a problem. An attempt was made to eliminate tailing by making the 0·3*M* alcoholic solution of hydrochloric acid up to 0·1*M* in perchloric acid. This speeded up all elutions and diminished tailing of the nickel somewhat, but the over-all improvement was not sufficient to recommend the use of perchloric acid in the eluents.

Another elution scheme uses isopropyl alcohol instead of ethyl alcohol (see Figs. 2–5). In this method the isopropyl alcohol concentration is held constant at 55%, and the concentration of hydrochloric acid is varied. The elution scheme is as follows:
Sample is added to the column in 0·3*M* HCl-95% ethyl alcohol.

TABLE II.—SEPARATION AND ANALYSIS OF SYNTHETIC METAL ION MIXTURES ON DOWEX 1 × 8, 200- TO 400-MESH, CHLORIDE-FORM RESIN. METALS APPEAR IN THE ORDER OF ELUTION WITH THE AMOUNT OF WASH SOLUTION INCLUDED WITH THE FIRST ELUTED METAL.

Metal mixture (column dimensions) and eluting agent	Taken, *mg*	Found, *mg*
Ni^{II}-Cu^{II} (3 × 1·1 cm)		
Ni 22 ml 95% EtOH-0·3*M* HCl		
10 ml 74% EtOH-0·3*M* HCl	7·17	7·17
Cu 15 ml 40% EtOH-0·3*M* HCl	7·07	7·01
Ni^{II}-Cu^{II}-Zn^{II} (3 × 1·1 cm)		
Ni 2 ml 95% EtOH-0·3*M* HCl		
9 ml 74% EtOH-0·3*M* HCl	7·17	7·17
Cu 15 ml 40% EtOH-0·3*M* HCl	7·07	7·02
Zn 40 ml 0·005*M* aqueous HCl	7·15	7·16
Mn^{II}-Cu^{II} (3 × 1·1 cm)		
Mn 2 ml 95% EtOH-0·3*M* HCl		
11 ml 74% EtOH-0·3*M* HCl	7·16	7·13
Cu 15 ml 40% EtOH-0·3*M* HCl	7·07	7·02
Ni^{II}-Mn^{II}-Cu^{II} (3 × 1·1 cm)		
Ni 2 ml 95% EtOH-0·3*M* HCl		
41 ml 95% EtOH-0·3*M* HCl	7·17	7·11
Mn 10 ml 72% EtOH-0·3*M* HCl	7·16	7·12
Cu 25 ml 0·005*M* aqueous HCl	7·07	7·02
Ni^{II}-Co^{II} (6 × 1·1 cm)		
Ni 2 ml 95% EtOH-0·3*M* HCl		
13 ml 82% EtOH-0·3*M* HCl	7·17	7·23
Co 15 ml 40% EtOH-0·3*M* HCl	7·64	7·62
Ni^{II}-Co^{II}-Cu^{II}-Zn^{II} (6 × 1·1 cm)		
Ni 2 ml 95% EtOH-0·3*M* HCl		
12 ml 82% EtOH-0·3*M* HCl	7·17	7·20
Co 17 ml 72% EtOH-0·3*M* HCl	7·64	7·62
Cu 15 ml 40% EtOH-0·3*M* HCl	7·07	7·04
Zn 40 ml 0·005*M* aqueous HCl	7·15	7·20
Co^{II}-Fe^{III} (3 × 1·1 cm)		
Co 2 ml 95% EtOH-0·3*M* HCl		
10 ml 72% EtOH-0·3*M* HCl	7·64	7·62
Fe 23 ml 40% EtOH-0·3*M* HCl	6·90	6·92
Ni^{II}-Co^{II}-Fe^{III}-Zn^{II} (6 × 1·1 cm)		
Ni 2 ml 95% EtOH-0·3*M* HCl		
13 ml 82% EtOH-0·3*M* HCl	7·17	7·21
Co 15 ml 70% EtOH-0·3*M* HCl	7·60	7·63
Fe 20 ml 40% EtOH-0·3*M* HCl	6·90	6·90
Zn 40 ml 0·005*M* aqueous HCl	6·91	6·82
Ni^{II}-Fe^{III}-Bi^{III} (3 × 1·1 cm)		
Ni 2 ml 95% EtOH-0·3*M* HCl		
6 ml 72% EtOH-0·3*M* HCl	7·17	7·17
Fe 12 ml 40% EtOH-0·3*M* HCl	6·90	6·88
Bi 25 ml 1*M* aqueous H_2SO_4	6·90	6·82

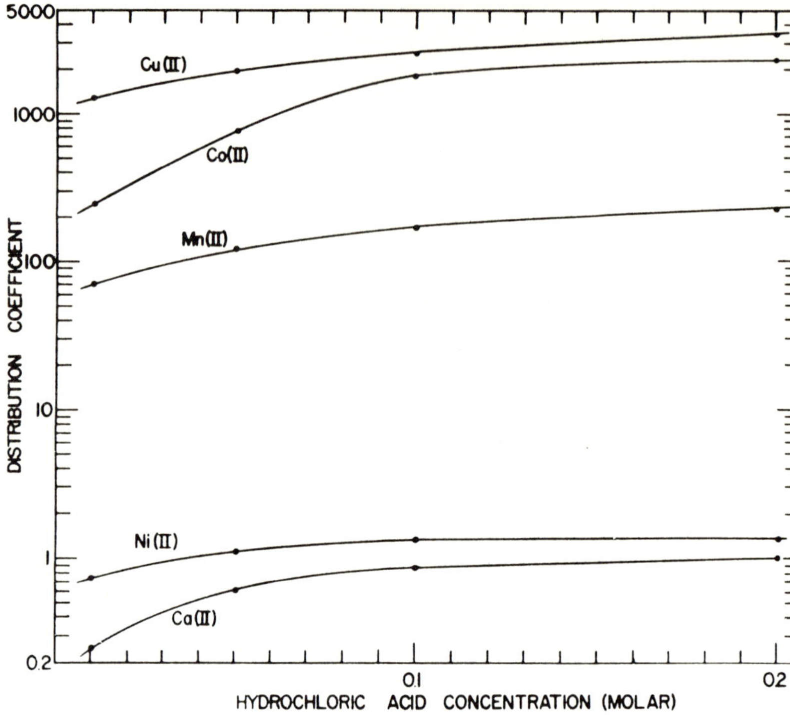

FIG. 9—Distribution coefficients of some metal ions in 96% ethyl alcohol *vs.* hydrochloric acid concentration.

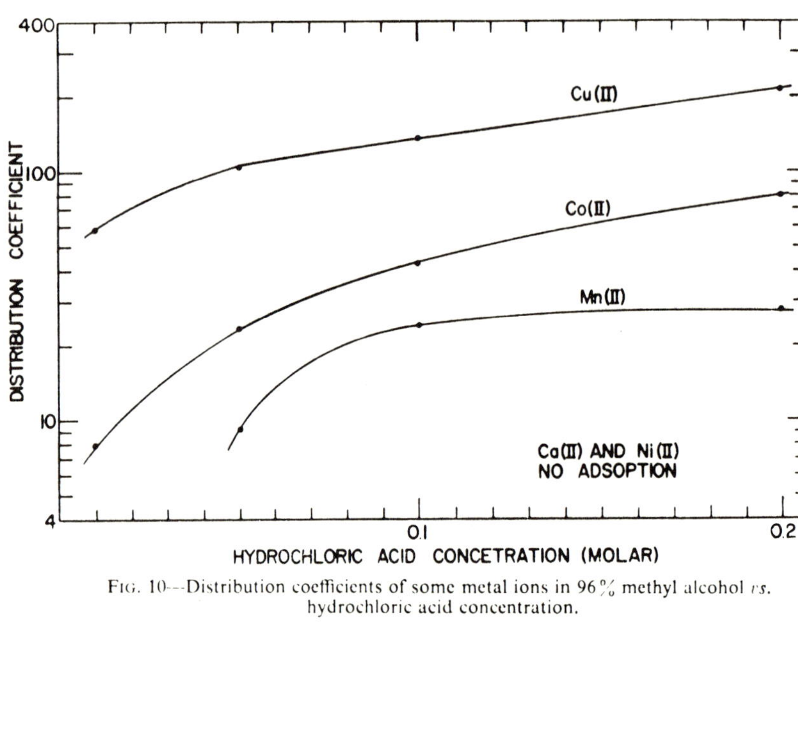

FIG. 10—Distribution coefficients of some metal ions in 96% methyl alcohol *vs.* hydrochloric acid concentration.

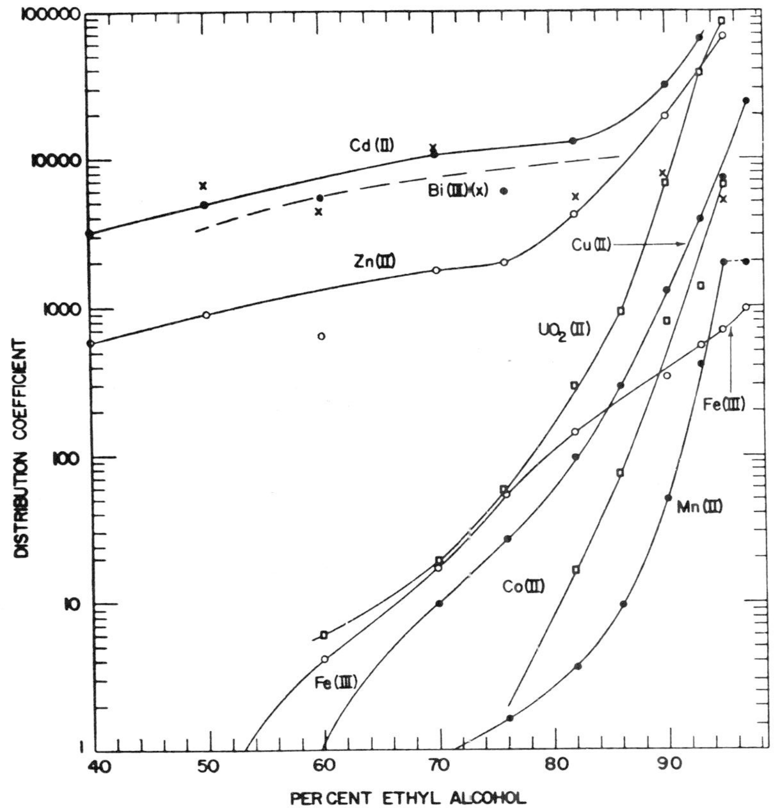

FIG. 11—Distribut on coefficients of metal chlorides in 0·3*M* hydrochloric acid *vs.* percentage of ethyl alcohol. High adsorption group.

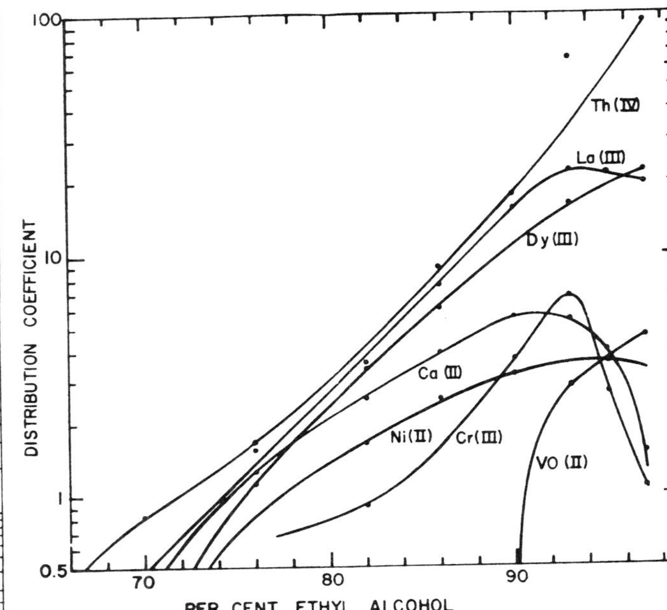

FIG. 12—Distribution coefficients of metal chlorides in 0·3*M* hydrochloric acid *vs.* percentage of ethyl alcohol. Low adsorption group.

Ni^{II}, Mn^{II}, Ca^{II} and Dy^{III} are eluted with 3*M* HCl-55% isopropyl alcohol.
Co^{II} is eluted with 1·3*M* HCl-55% isopropyl alcohol.
Cu^{II} and Fe^{III} are eluted with 0·1*M* HCl-55% isopropyl alcohol.
Zn^{II} is eluted with 0·005*M* aqueous HCl, or Bi^{III} is eluted with 1*M* aqueous sulphuric acid.

TABLE III.—SEPARATION AND ANALYSIS OF SYNTHETIC METAL ION MIXTURES ON DOWEX 1 × 8, 200- TO 400-MESH, CHLORIDE-FORM RESIN. METALS APPEAR IN THE ORDER OF ELUTION WITH THE AMOUNT OF WASH SOLUTION INCLUDED WITH THE FIRST ELUTED METAL.

Metal mixtures (column dimensions) and eluting agent		Taken, *mg*	Found, *mg*
Ni^{II}-Fe^{III}-Bi^{III} (6 × 2·2 cm)			
Ni	4 ml 95% EtOH-0·3*M* HCl		
	50 ml 96% MeOH-0·2*M* HCl	8·96	8·93
Fe	45 ml 0·1*M* HCl-55% 2-PrOH	9·13	9·13
Bi	90 ml 1*M* aqueous H_2SO_4	8·66	8·50
Co^{II}-Fe^{III} (6 × 2·2 cm)			
Co	4 ml 95% EtOH-0·3*M* HCl		
	60 ml 1·3*M* HCl-55% 2-PrOH	9·51	9·58
Fe	45 ml 0·1*M* HCl-55% 2-PrOH	9·13	9·12
Ni^{II}-Mn^{II}-Cu^{II} (6 × 2·2 cm)			
Ni	8 ml 95% EtOH-0·3*M* HCl		
	47 ml 96% MeOH-0·2*M* HCl	8·96	8·94
Mn	80 ml 3*M* HCl-55% 2-PrOH	8·96	9·00
Cu	50 ml 0·1*M* HCl-55% 2-PrOH	8·94	8·89
Ni^{II}-Mn^{II}-Co^{II}-Cu^{II}-Zn^{II} (6 × 2·2 cm)			
Ni	8 ml 95% EtOH-0·3*M* HCl		
	47 ml 96% MeOH-0·2*M* HCl	8·96	8·99
Mn	80 ml 3*M* HCl-55% 2-PrOH	8·96	8·88
Co	70 ml 1·3*M* HCl-55% 2-PrOH	9·51	9·50
Cu	50 ml 0·1*M* HCl-55% 2-PrOH	8·94	8·88
Zn	60 ml 0·005*M* aqueous HCl	8·95	9·06
Ni^{II}-Co^{II}-Cu^{II}-Bi^{III} (6 × 2·2 cm)			
Ni	8 ml 95% EtOH-0·3*M* HCl		
	47 ml 96% MeOH-0·2*M* HCl	8·96	9·06
Co	65 ml 1·3*M* HCl-55% 2-PrOH	9·51	9·51
Cu	40 ml 0·1*M* HCl-55% 2-PrOH	8·95	8·97
Bi	85 ml 1*M* aqueous H_2SO_4	8·66	8·54
Dy^{III}-Mn^{II}-Cu^{II}-Fe^{III}-Bi^{III} (6 × 2·2 cm)			
Dy	8 ml 95% EtOH-0·3*M* HCl		
	47 ml 96% MeOH-0·2*M* HCl	9·22	9·28
Mn	80 ml 3*M* HCl-55% 2-PrOH	8·96	8·99
Cu-Fe[a]	50 ml 0·1*M* HCl-55% 2-PrOH		
	Cu	8·94	9·06
	Fe	9·13	9·06
Bi	85 ml 1*M* aqueous H_2SO_4	8·66	8·66

TABLE III. (*contd.*)

Metal mixtures (column dimensions) and eluting agent		Taken, *mg*	Found, *mg*
Ni^{II}-Th^{IV}-Mn^{II}-Co^{II}-Cu^{II}-Fe^{III} (6 × 2·2 cm)			
Ni	8 ml 95% EtOH-0·3*M* HCl		
	73 ml 48% EtOH-48%*M* OH 0·2*M* HCl	8·96	8·99
Th	85 ml 96% MeOH-0·2*M* HCl	9·06	9·06
Mn	85 ml 3*M* HCl-55% 2-PrOH	8·95	8·96
Co	55 ml 1·3*M* HCl-55% 2-PrOH	9·51	9·56
Cu-Fe[a]	50 ml 0·1*M* HCl-55% 2-PrOH		
	Cu	8·94	8·96
	Fe	9·13	9·03
Ca^{II}-Mn^{II} (5 × 2·2 cm)			
Ca	2 ml 95% EtOH-0·3*M* HCl		
	56 ml 96% MeOH-0·2*M* HCl	8·33	8·40
Mn	80 ml 3*M* HCl-55% 2-PrOH	9·06	9·02
Ni^{II}-Mn^{II} (5 × 2·2 cm)			
Ni	2 ml 95% EtOH-0·3*M* HCl		
	56 ml 96% MeOH-0·2*M* HCl	4·45	4·51
Mn	80 ml 3*M* HCl-55% 2-PrOH	9·05	9·07

[a] Cu-Fe eluted together and analysed by a photometric titration[35] with EDTA as titrant.

This scheme gives sharp separations and virtually eliminates tailing of the bands. In Fig. 15 the elution curves are compared for the separation of $nickel^{II}$ and $cobalt^{II}$ by the isopropyl alcohol and the previous ethyl alcohol elution schemes. In the ethyl alcohol scheme, elution with 0·3*M* hydrochloric acid in 82% ethyl alcohol was continued after elution of the nickel until the cobalt break-through. In the isopropyl alcohol scheme, elution with 3*M* hydrochloric acid in 55% isopropyl alcohol was discontinued shortly after elution of the nickel was complete. At this point the cobalt band was still 4 cm from the bottom of the column, and the estimated cobalt break-through with the 3*M* eluent would be around 250 ml.

The isopropyl alcohol scheme does not offer a convenient separation of $nickel^{II}$ or $calcium^{II}$ from $manganese^{II}$. However, it was found that an excellent separation of $nickel^{II}$ and $manganese^{II}$ can be achieved through the use of 0·2*M* hydrochloric acid in 96% methyl alcohol (see Fig. 10). Although the distribution coefficient for $manganese^{II}$ is only 27 in 0·2*M* hydrochloric acid-96% methanol (compared to a D of around 2000 in 0·3*M* hydrochloric acid-95% ethyl alcohol), column experiments showed that $manganese^{II}$ is still strongly retained by a column. An elution curve for the column separation of $nickel^{II}$ and $manganese^{II}$ is given in Fig. 16. The curve shows sharp bands with very little tailing, and an excellent separation factor.

The recommended separation method is first to separate $nickel^{II}$, $calcium^{II}$ or $dysprosium^{III}$ from $manganese^{II}$ (and other metal ions) using 0·2*M* hydrochloric acid in 96% methyl alcohol. The metal ions remaining on the column are then separated using the 55% isopropyl alcohol scheme outlined above. Fig. 17 shows elution curves for a multi-component mixture separated by this method, and Table III presents results

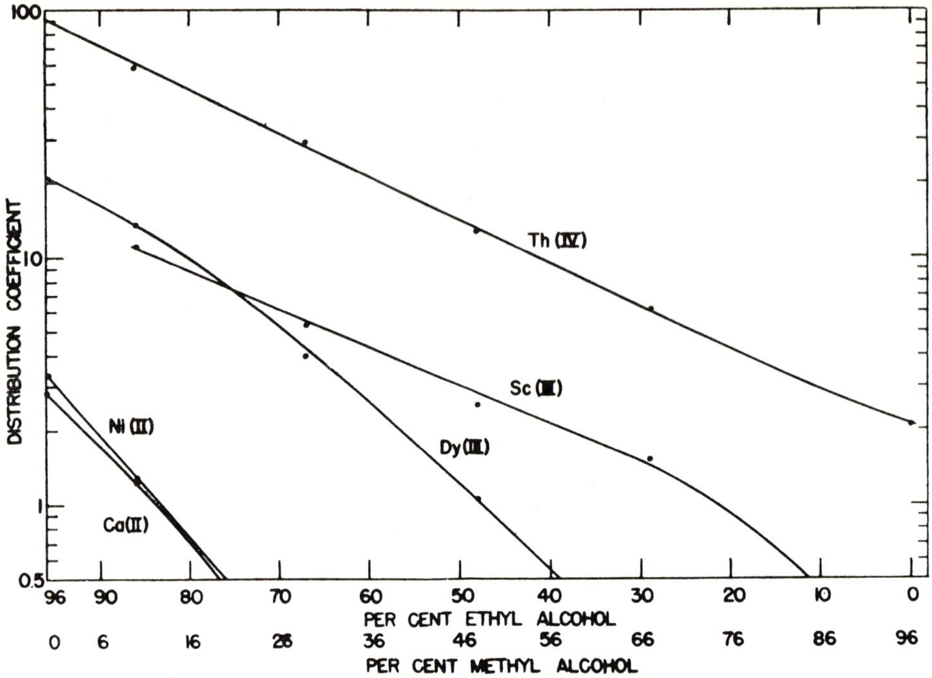

FIG. 13—Distribution coefficients of metal chlorides in 0·2*M* hydrochloric acid *vs.* concentration of methyl alcohol and ethyl alcohol totalling 96%.

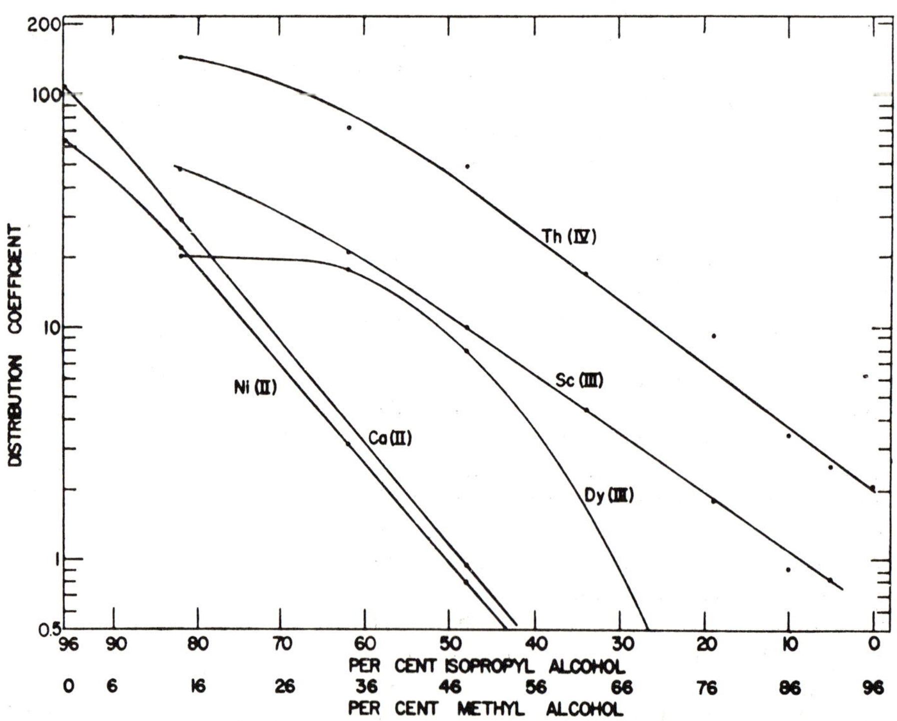

FIG. 14—Distribution coefficients of metal chlorides in 0·2*M* hydrochloric acid *vs.* concentration of methyl alcohol and isopropyl alcohol totalling 96%.

for the quantitative separation of several mixtures. It should be noted that a column of larger diameter was used for these separations than for those using 0·3*M* hydrochloric acid in ethyl alcohol (Table II). The larger column diameter allows greater loading and faster flow rates, but of course requires larger volumes of eluent for separation.

The dis'.ibution coefficient for thorium^{IV} is fairly high in 0·2*M* or 0·3*M* hydrochloric acid in 95% ethyl alcohol or isopropyl alcohol. Under these conditions the distribution coefficients of nickel^{II}, calcium^{II}, *etc.* are a little too high to permit a convenient separation from thorium^{IV}. However, Figs. 13 and 14 show that mixing

TABLE IV.—SEPARATION AND ANALYSIS OF SYNTHETIC METAL ION MIXTURES ON DOWEX 1 × 8, 200- TO 400-MESH, CHLORIDE-FORM RESIN. (5 × 2·2 CM.). QUANTITIES AND TYPE OF ELUTING AGENT ARE SIMILAR TO THOSE IN TABLE III.

Metal mixture	Taken, *mg*	Found, *mg*
Co^{II}-Cu^{II}		
Co	38·46	38·40
Cu	4·46	4·45
Ni^{II}-Co^{II}		
Ni	35·46	35·48
Co	4·82	4·82
Co^{II}-Fe^{III}		
Co	38·46	38·49
Fe	4·63	4·64
Mn^{II}-Co^{II}		
Mn	36·22	36·39
Co	4·82	4·83
Ca^{II}-Mn^{II}		
Ca	33·34	33·36
Mn	4·51	4·60
Ni^{II}-Mn^{II}		
Ni	35·46	35·55
Mn	4·51	4·65

methyl alcohol with isopropyl alcohol or ethyl alcohol provides a way to lower gradually the distribution coefficients of metal ions. In 0·2*M* hydrochloric acid-48% ethyl alcohol-48% methyl alcohol, a quantitative column separation of thorium^{IV} and nickel^{II} is obtained, although some tailing is observed in elution of the nickel. It appears that the use of mixed alcohols may also permit a fair separation of a nickel-dysprosium-thorium mixture.

In the mixtures analysed above, the sample constituents are present in approximately equal molar amounts. In Table IV, results are reported for the successful analysis of mixtures where two components are present in approximately a 10 to 1 ratio. So long as the column is not overloaded, a high ratio of one or more sample constituents to the others does not appear to cause any difficulty.

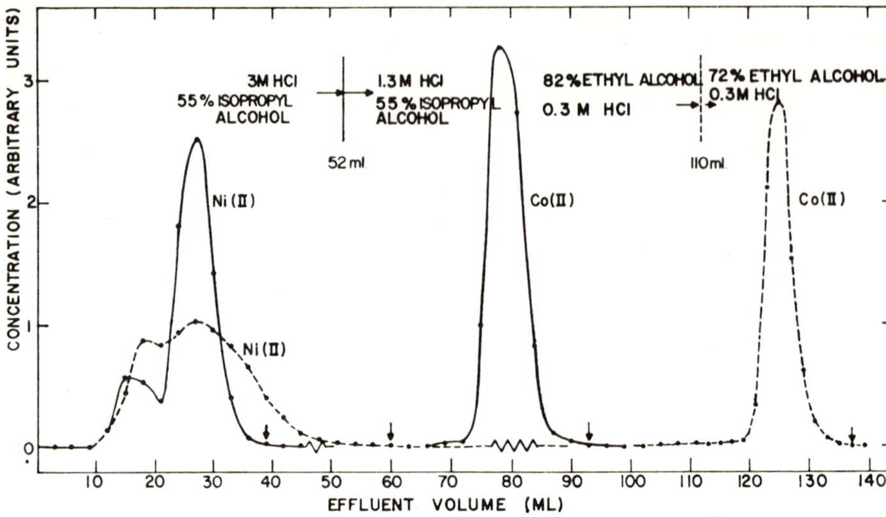

FIG. 15—Elution curves of a Ni^{II}-Co^{II} separation. Solid line represents elution by the varying hydrochloric acid-55% isopropyl alcohol eluting scheme. Dotted line represents elution by the varying percent of ethyl alcohol-0·3*M* hydrochloric acid eluting scheme. Flow rate of 1/4 to 1/3 ml/min and column dimensions of 5 × 2·2 cm (↓ = last detectable trace of metal ion.)

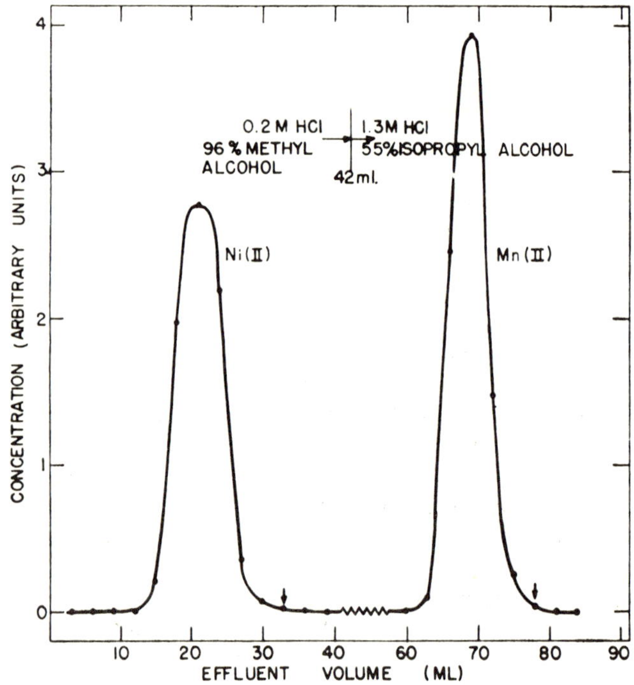

FIG. 16—Elution curve of a Ni^{II}-Mn^{II} separation. Flow rate of 1/4 to 1/3 ml/min and column dimensions of 6 × 2·2 cm. (↓ = last detectable trace of metal ion.)

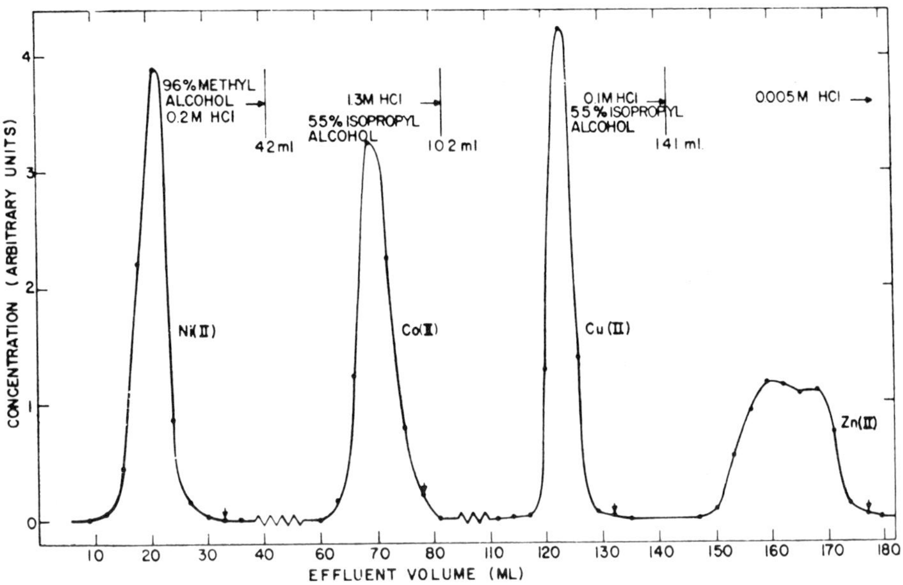

FIG. 17—Elution curve of a Ni^{II}-Co^{II}-Cu^{II}-Zn^{II} separation. Flow rate of 1/4 to 1/3 ml/min and column dimensions of 6 × 2·2 cm. (↓ = last detectable trace of metal ion.)

Zusammenfassung—Die Verteilungskoeffizienten für Metallionen zwischen einem Anionenaustauscher und einer flüssigen Phase (bestehend aus einer Mischung von Wasser und organischem Solvent), die Salzsäure enthält, wurden gemessen. Die Gegenwart von Salzsäure bewirkt, dass die Metallionen bei geringerer Säurekonzentration adsorbiert werden. In vielen Fällen sind die Verteilungskoeffizienten erheblich grösser als in Systemen mit wässriger Salzsäure. Bei sonst gleichen Bedingungen ist die Reihenfolge der Koeffizienten in Alkohol-Wasser-Salzsäure wie folgt: Isopropanol > Äthyl- > Methylalkohol. Säulentrennung von Metallionenmischungen können durchgeführt werden mittels Elution durch Alkohol-Wasser-Salzsäure-Mischungen verschiedener Zusammensetzung. Erfolgreiche Trennung einiger Gemische wird mitgeteilt.

Résumé—Les auteurs ont mesuré des coefficients de partage d'ions métalliques entre une résine échangeuse d'anions et des mélanges eau-solvant organique contenant de l'acide chlorhydrique. La présence d'un solvant organique provoque la fixation des ions métalliques à des concentrations d'acide chlorhydrique plus faibles. Dans de nombreux cas, les coefficients de partage sont beaucoup plus élevés que dans les systèmes eau-acide chlorhydrique. Les autres conditions étant comparables, l'ordre des coefficients de partage dans les mélanges alcool-eau-acide chlorhydrique est le suivant: alcool isopropylique, alcool éthylique, alcool méthylique. Les séparations sur colonne de mélanges d'ions métalliques peuvent être réalisées par élution avec des mélanges alcool-eau-acide chlorhydrique de différentes compositions. Des séparations d un certain nombre de mélanges réalisées avec succès sont mentionnées.

REFERENCES

[1] K. A. Kraus and F. Nelson, *Symposium on Ion Exchange and Chromatography in Analytical Chemistry*. ASTM Special Technical Publication No. 195, American Society for Testing Materials, Philadelphia, 1958, p. 27–57.
[2] *Idem*, Proceedings, Intern. Conference on Peaceful Uses of Atomic Energy, **7**, 113, 131, United Nations, 1956.
[3] D. H. Wilkins and L. E. Hibbs, *Analyt. Chim. Acta*, 1959, **20**, 427.
[4] D. H. Wilkins, *ibid.* 1959, **20**, 271.
[5] *Idem, ibid.*, 1959, **20**, 273.

[6] *Idem, ibid.*, 1958, **18,** 372.
[7] *Idem, ibid.*, 1957, **16,** 449.
[8] D. Jentzsch, *Z. analyt. Chem.*, 1956, **152,** 134.
[9] L. I. Katzin and E. Gebert, *J. Amer. Chem. Soc.*, 1953, **75,** 801.
[10] C. W. Davies and B. D. R. Owen, *J. Chem. Soc.*, 1956, 1676.
[11] D. G. Tuck and G. A. Welch, *J. Inorg. Nuclear Chem.*, 1959, **9,** 302.
[12] A. Iguchi, *Bull. Chem. Soc. Japan*, 1958, **31,** 600; *Chem. Abs.*, **53,** 6906g.
[13] *Idem, ibid.*, 1958, **31,** 748; *Chem. Abs.*, **53,** 11085a.
[14] F. H. Burstall, P. J. Forrest, N. F. Kember and R. A. Wells, *Ind. Eng. Chem.*, 1953, **45,** 1648.
[15] J. Korkisch, P. Antal and F. Hecht, *Z. analyt. Chem.*, 1960, **172,** 400.
[16] E. W. Berg and J. T. Truemper, *Analyt. Chem.*, 1958 **30,** 1827.
[17] M. Kojima, *Japan Analyst*, 1957, **6,** 369.
[18] *Idem, ibid.*, 1958, **7,** 177.
[19] Y. Yoshino and Y. Kurimura, *Bull. Chem. Soc. Japan*, 1957, **30,** 563.
[20] J. E. Powell, J. S. Fritz and D. B. James, *Analyt. Chem.*, 1960, **32,** 954.
[21] J. S. Fritz, W. J. Lane and A. S. Bystroff, *ibid.*, 1957, **29,** 821.
[22] A. J. Barnard Jr., W. C. Broad and H. Flaschka, *The EDTA Titration: Nature and Methods of End-Point Detection.* J. T. Baker Chemical Co., Phillipsburg, New Jersey, 1957.
[23] J. S. Fritz and M. A. Payne, unpublished results.
[24] J. S. Fritz, R. T. Oliver and D. J. Pietrzyk, *Analyt. Chem.*, 1958, **30,** 1111.
[25] J. S. Fritz, *ibid.*, 1954, **26,** 1978.
[26] J. S. Fritz and M. Johnson, *ibid.*, 1955, **27,** 1653.
[27] R. Přibil, *Analyst*, 1958, **83,** 188.
[28] H. Diehl and G. F. Smith, *The Copper Reagents: Cuproine, Neocuproine, Bathocuproine.* The G. Frederick Smith Chemical Co., Columbus, Ohio, 1958.
[29] E. A. Johnson and W. Jablonski, *Zincon.* Hopkin and Williams Ltd., Chadwell Heath, Essex, England, 1959, Monograph No. 37.
[30] E. A. Johnson and E. J. Newman, *Thorin.* Hopkin and Williams Ltd., Chadwell Heath, Essex, England, 1956, Monograph No. 29.
[31] J. S. Fritz and M. Johnson-Richard, *Analyt. Chim. Acta*, 1959, **20,** 164.
[32] F. D. Snell and C. T. Snell, *Colorimetric Methods of Analysis.* D. Van Nostrand Company, Inc., New York, 1949, Vol. II.
[33] H. Diehl and G. F. Smith, *Quantitative Analysis.* John Wiley and Sons, Inc., New York, 1952.
[34] F. W. Cornish, *Analyst*, 1958, **83,** 634.
[35] A. L. Underwood, *Analyt. Chem.*, 1953, **25,** 1910.

21

Reprinted from *Anal. Chim. Acta*, **32**, 46–51 (1965)

ANION-EXCHANGE SEPARATION OF IRON, COBALT AND NICKEL

I. HAZAN AND J. KORKISCH
Analytical Institute, University of Vienna, Vienna (Austria)
(Received June 2nd, 1964)

Earlier investigations[1] concerning the anion-exhange separation of iron, cobalt and nickel in pure aqueous hydrochloric acid solutions have shown that cobalt is the most strongly adsorbed and nickel the least strongly adsorbed of these elements. To separate iron from nickel and cobalt chromatographic elution with varying concentrations of hydrochloric acid must be used, the elements passing into the effluent in the order: nickel, iron and cobalt. This method was found to be highly efficient and can be employed in most cases. If, however, small amounts especially of nickel must be separated from large quantities of iron, as would happen with ferrous alloys or other materials of high iron content, the resin column dimensions would have to be increased considerably in order to adsorb all the iron. In such a case small amounts of cobalt are liable to be replaced in the column by iron ions even during the sorption step, so that some or all of the cobalt might be lost. Ether extraction of iron(III) chloride before passage through the ion exchanger would avoid these adverse effects, but this would introduce another possible source of error which might assume importance in accurate analyses for the nickel and cobalt content of ferrous alloys. We have theretore investigated a large number of hydrochloric acid—organic solvent mixtures in order to find a medium in which cobalt and nickel but not iron (even if present in very large amounts) can be adsorbed on an anion-exchange resin. Since acetone forms a rather strong complex with iron(III), it proved a highly suitable solvent for this purpose.

A remarkable feature of the separation procedure described below is that nickel is adsorbed rather strongly when the acetone medium is used. This is in contrast to its behaviour in pure aqueous hydrochloric acid media, where no adsorption of nickel on strongly basic anion-exchange resins was observed even when concentrated hydrochloric acid was employed[1].

EXPERIMENTAL

Reagents and solutions

Ion-exchange resin. The strongly basic anion-exchange resin Dowex 1-X8 (100–200 mesh, chloride form) was used.

Standard solutions. Stock solutions of iron, cobalt, nickel, and other elements were prepared by dissolving the reagent-grade chlorides of these metal ions in 6 *M* hydrochloric acid. These solutions usually contained 5 mg of metal ion/ml.

Solvent. Acetone of reagent-grade purity.

Wash solutions: (a) 90% (v/v) acetone + 10% 6 *M* hydrochloric acid; (b) 70% (v/v) acetone + 30% 2 *M* hydrochloric acid.

Apparatus

Resin columns of 75 cm length and 1.0 cm in diameter were used.

Determination of the elements

For the determination of iron, cobalt, and nickel as well as of the other investigated elements, titrimetric methods were employed using 0.01 or 0.001 *M* solutions of EDTA (disodium salt) in presence of suitable indicators such as sulfosalicylic acid (for iron), xylenol orange (for cobalt) and murexide, or when manganese was present pyrocatechol violet[2] (for nickel).

Working procedure

Pretreatment of resin bed. The resin suspended in wash solution (a) was transferred to the ion-exchange column and washed with approximately 100 ml of wash solution (a).

Sorption, washing and elution. Dilute 2 ml of 6 *M* hydrochloric acid containing iron, nickel and cobalt to 20 ml with acetone. Pass this solution through the resin column at a flow-rate of 0.3 to 0.4 ml/min. Afterwards wash the resin with wash solution (a) until iron is quantitatively removed from the resin (negative thiocyanate test) (see Fig. 1). Then pass wash solution (b) through the column, effecting the separation of nickel from cobalt. If mg-quantities of nickel and cobalt were originally present in the sorption solution, their green and blue bands respectively can be observed visually on the resin column. After complete elution of nickel, which is the case when 130 ml of wash solution (b) have passed, cobalt appears in the effluent after an eluate fraction of 80 ml (see Fig. 1). Alternatively, cobalt can be eluted with 1 *M* hydrochloric acid; this is advantageous when speed is essential but has the disadvantage that air bubbles are formed in the resin bed and prolong the subsequent regeneration of the resin with wash solution (a) if a second column operation in the same column has to be performed.

Evaporate the respective fractions containing nickel and cobalt to dryness on a water bath and determine the elements as described above.

Analysis of steel samples

Heat 0.5–1.0 g of the sample with concentrated hydrochloric acid and a few ml of 30% hydrogen peroxide until most of the sample has dissolved. Filter off the residue and digest with hydrofluoric acid to remove silica and to dissolve oxides such as titanium oxide. Evaporate the solution to dryness on a water bath, take up the residue in concentrated hydrochloric acid, warm and add excess of boric acid to remove the fluoride ions. Then evaporate the solution to dryness (water bath) twice with a few ml of concentrated hydrochloric acid. Finally, take up the residue in hydrochloric acid and combine the solution with the original filtrate. Filter this solution to remove the insoluble portion of boric acid and evaporate to dryness on a water bath. Then dissolve the residue in 6 *M* hydrochloric acid while transferring to a 10-ml measuring flask. Fill to the mark with 6 *M* hydrochloric acid and dilute an aliquot of 2 ml to 20 ml with acetone. Then pass the mixture through the ion-exchange column as described above.

RESULTS AND DISCUSSION

Based on the results of measurements of distribution coefficients[3] of iron, cobalt, nickel and other metal ions (see Table I) in various mixtures of acetone with hydrochloric acid the most suitable conditions were selected for the quantitative separation of the sum of cobalt and nickel from large quantities of iron.

Table I shows that a medium consisting of 90% acetone and 10% 6 *M* hydrochloric acid could be used effectively. To separate cobalt from the co-adsorbed nickel, a mixture of 70% acetone and 30% 2 *M* hydrochloric acid seemed to be most promising. When this medium was used for the separation, nickel was accompanied into the eluate by manganese and aluminium as well as chromium(III) which was found to behave similarly. On the other hand cobalt was eluted together with copper if present.

Separation experiments with test solutions carried out by the above working

TABLE I

DISTRIBUTION COEFFICIENTS IN ACETONE–HYDROCHLORIC ACID MEDIA

Mixture	*Elements*					
	Fe	*Ni*	*Co*	*Cu*	*Mn*	*Al*
90% acetone + 10% 6 *M* HCl	<1	40	170	70	230	38
80% acetone + 20% 3 *M* HCl	<1	14	87	45	60	25
70% acetone + 30% 2 *M* HCl	<1	2	20	23	3	8

TABLE II

SEPARATION OF IRON, COBALT AND NICKEL IN ACETONE–HYDROCHLORIC ACID MEDIUM

Amounts taken (mg)			*Amounts recovered (mg)*		
Iron	*Cobalt*	*Nickel*	*Iron*	*Cobalt*	*Nickel*
500	1.0	1.0	500.2	0.99	1.01
500	2.5	2.5	500.7	2.5	2.52
500	5.0	5.0	498.6	4.97	5.01
500	5.0	10.0	500.0	5.03	9.97
500	10.0	5.0	500.9	10.06	5.0
500	20.0	20.0	501.2	20.1	20.0
500	50.0	50.0	500.8	49.8	50.1
500	75.0	75.0	498.7	74.9	75.4
500	100.0	100.0	497.9	100.0	99.9
500	1.0	100.0	500.0	0.99	100.3
500	100.0	1.0	501.3	100.4	1.02
100	100.0	100.0	100.0	99.8	99.7

procedure showed that these selected eluants were very effective for the separation. The results recorded in Table II show that in all cases a quantitative separation of nickel and cobalt from each other and large amounts of iron could be accomplished without any loss or cross-contamination of the iron, nickel and cobalt fractions.

To illustrate the effectiveness of this separation method more clearly, a typical chromatogram is shown in Fig. 1; it can be seen that no overlapping of the elution curves would occur even under the most critical conditions, *e.g.* if still larger amounts of iron than those used in this experiment were present.

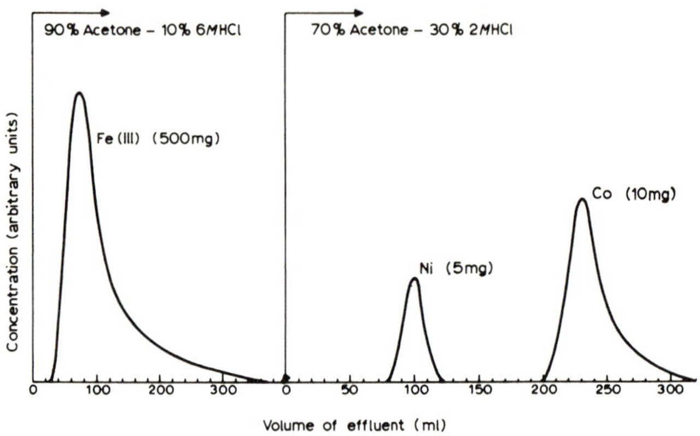

Fig. 1.

The same medium which is used for the elution of iron (wash solution (a)) can also be employed very effectively to separate large amounts of iron from aluminium, the latter element being retained by the resin like nickel, whereas iron appears in the effluent. In pure aqueous hydrochloric acid systems the reverse occurs, *i.e.* iron is strongly adsorbed and aluminium passes into the effluent[1].

Based on all these results the working procedure described in the experimental part was developed and applied to the analysis of 5 steel samples to which in parallel experiments known amounts of nickel and cobalt were added in order to check the applicability of the method to steel analysis. From the results of these investigations (Table III) it can be seen that in all cases a quantitative recovery of the added amounts of nickel and cobalt was achieved.

In conclusion, it is worth mentioning that this method would not only be applicable to the analysis of ferrous alloys but could also be adapted for the analysis of cobalt and nickel in meteorites, soils, ashes of biological materials, crude oils, marine sediments and other natural substances.

Acknowledgement is made to the donors of the Petroleum Research Fund administered by the American Chemical Society, for support of this research (PRF-grant No. 1587-A3).

TABLE III

ANALYSIS OF STEEL SAMPLES

Amount of Ni and Co present in steel sample		*Amount of Ni and Co added to steel sample*		*Recovered amount of Ni and Co*	
Ni(mg)	*Co(mg)*	*Ni(mg)*	*Co(mg)*	*Ni(mg)*	*Co(mg)*
8.2	0.5	5	5	13.3	5.54
		10	10	18.2	10.48
		20	20	28.4	20.62
0.0	0.0	5	5	4.97	5.0
		10	10	9.98	9.89
		20	20	20.0	20.1
0.0	0.0	5	5	5.04	5.02
		10	10	9.97	10.04
		20	20	19.92	20.2
0.15	0.0	5	5	5.2	5.0
		10	10	10.08	10.01
		20	20	20.17	20.04
43.0	0.5	5	5	48.2	5.48
		10	10	52.7	10.53
		20	20	63.0	20.6

SUMMARY

From 90% acetone–10% 6 *M* hydrochloric acid medium, cobalt and nickel are strongly adsorbed on the anion-exchange resin Dowex 1-X8; iron is not adsorbed and can thus be separated from cobalt and nickel. Cobalt and nickel are then separated by elution with 70% acetone–30% 2 *M* hydrochloric acid; nickel is eluted before cobalt. The method can be applied to the determination of nickel and cobalt in materials with high iron content such as steels; compleximetric titrations are used for the final step.

RÉSUMÉ

Le cobalt et le nickel sont fortement adsorbés sur résine Dowex 1-X8 (échangeur d'anions) en milieu acétone/acide chlorhydrique 6 *M* (90/10). Ils peuvent être ensuite séparés par élution avec un mélange acétone/acide chlorhydrique 2 *M* (70/30), le nickel étant élué avant le cobalt. Cette méthode peut s'appliquer au dosage du nickel et du cobalt dans des produits riches en fer, tels que des aciers, le fer n'étant pas adsorbé. On termine par titrages compleximétriques.

ZUSAMMENFASSUNG

Zur Bestimmung von Nickel und Kobalt in Materialien mit hohem Eisengehalt werden Kobalt und Nickel im Gegensatz zum Eisen vom Anionenaustauscher Dowex 1-X8 aus einem Medium von 90% Aceton und 10% 6 *N* Salzsäure stark adsorbiert.

Durch Elution mit einer Lösung aus 70% Aceton und 30% 2 *N* Salzsäure wird Nickel vor dem Kobalt eluiert. Beide Elemente werden komplexometrisch bestimmt.

REFERENCES

1 M. Kojima and A. Iguchi, *Japan Analyst*, 7 (1958) 514.
2 H. Flaschka and R. Püschel, *Z. Anal. Chem.*, 147 (1955) 354.
3 J. Korkisch and G. E. Janauer, *Talanta*, 9 (1962) 957.

22

Reprinted from *Microchim. Acta,* 5, 687–699 (1972)

Anionenaustauschtrennungen der mit Tributylphosphat extrahierbaren Elemente. I*

Trennung des Urans vom Eisen in Gegenwart von Tributylphosphat

Von

W. Koch und J. Korkisch

Mit 4 Abbildungen

(Eingegangen am 20. Januar 1972)

1. Einleitung

Frühere Untersuchungen haben gezeigt, daß eine wesentliche Verbesserung des Adsorptionsverhaltens von Metallionen an stark basischen und sauren Ionenaustauschern durch den Zusatz organischer Lösungsmittel erreicht werden kann[1]. Dadurch wurden Trennungen möglich, die in rein wäßrigen Systemen nicht durchführbar sind.

Eingehendere Untersuchungen haben nun gezeigt[2–4], daß auch komplexe Verbindungen, wie sie sich bei Verwendung von Extraktionsmitteln bilden, in geeigneten organischen Lösungsmitteln gelöst, an Ionenaustauschern ausgetauscht werden. Das Prinzip eines solchen Ionenaustausches beruht auf einem gleichzeitigen Zusammenwirken von Ionenaustausch und Lösungsmittelextraktion, wobei das Harz die stationäre, das organische Lösungsmittel die mobile Phase darstellt. Dabei treten bei der Metalladsorption Effekte auf, die sich wesentlich von denen in rein wäßrigen Systemen und auch von denen in Lösungsmittelgemischen, die keine Extraktionsmittel enthalten, unterscheiden. Einige derartige Systeme, in denen diese allgemein als KIALE-Effekt (Kombinierte Ionen-Austausch-Lösungsmittel-Extraktion) bezeichneten Erscheinungen auftreten, wurden in der Literatur bereits eingehend beschrieben[4–6].

**Editor's Note:* An English translation follows this article.

In der vorliegenden Arbeit wird besonders auf Systeme eingegangen werden, bei denen Tributylphosphat als Extraktionsmittel Verwendung findet. Ähnliche Systeme wurden früher schon zur Abtrennung des Urans aus salpetersauren organischen Lösungsmittelgemischen benützt[7]. Dabei soll am Beispiel der sowohl analytisch wie technisch wichtigen Trennung des Urans vom Eisen gezeigt werden, wie Extraktion und Ionenaustausch in sinnvoller Weise so kombiniert werden können, daß nach Extraktion der genannten Elemente mit Tributylphosphat (TBP) eine Trennung durch Anionenaustausch aus einem TBP-haltigen Lösungsmittelgemisch nach dem vorhin genannten Prinzip erfolgt. Um geeignete Lösungsmittelgemische zu finden, war es zunächst nötig, verschiedene Lösungsmittel auf ihre Brauchbarkeit hin zu untersuchen, ferner die Abhängigkeit der erzielbaren Trennwirkung vom Mischungsverhältnis und der Gesamtacidität der Mischung zu messen. Auf Grund dieser Resultate war es möglich, zwei Lösungsmittelgemische zu finden, die sich zur Abtrennung des Urans vom Eisen besonders gut eignen, und Verfahren zur Trennung und Bestimmung der genannten Elemente zu entwickeln.

2. Experimenteller Teil

Lösungen und Reagenzien

Ionenaustauscher. Es wurde der stark basische Anionenaustauscher Dowex 1-X 8 (100—200 mesh; Chloridform) verwendet. Das Harz wurde durch Waschen mit 1-m Salzsäure solange gereinigt, bis das Filtrat farblos war, danach mit Aceton behandelt und an der Luft getrocknet.

Standardlösungen. Die für die Extraktion verwendeten salzsauren Standardlösungen wurden durch Einwaage der festen Salze und Auflösen dieser in einem bestimmten Volumen 6-m Salzsäure hergestellt. Im Falle des Urans wurde Uranylnitrat durch Behandlung mit Ameisensäure und konz. Salzsäure und anschließendem Eindampfen zur Trockne in Uranylchlorid umgewandelt.

Organische Lösungsmittel. Zur Anwendung gelangten folgende chemisch reine Lösungsmittel: Aceton, Methyl-n-propylketon, Methylisobutylketon (Hexon), Acetylaceton, Cyklohexanon, Tetrahydrofuran, Äthylenglykolmonomethyläther (Methylglykol), Äthylenglykolmonoäthyläther (Äthylglykol), Essigsäure-n-butylester, Methanol, Dimethylsulfoxid (DMSO) und Kerosin (Kp = 40—60° C).

Extraktionsmittel. Chemisch reines Tri-n-butylphosphat (TBP) (Kp = 270° C).

Weitere Reagenzien. Salzsäure ($\simeq$ 12-m; p. a.), Salpetersäure (Dichte 1,40), Sulfosalicylsäure, Hexamethylentetramin, Natrium- und Lithiumfluorid (beide suprapur) und 0,01-m ÄDTA-Lösung.

Apparaturen

Die Bestimmung des Urangehaltes wurde mit einem Galvanek-Morrison Fluorimeter, Mark V der Firma Jarell-Ash durchgeführt. Für die photometrischen Bestimmungen wurden ein Beckman Spektralphotometer, Modell B, und 1-cm-Küvetten verwendet. Die Ionenaustauschtrennungen wurden in Glassäulen von den in Abb. 1 angegebenen Dimensionen ausgeführt.

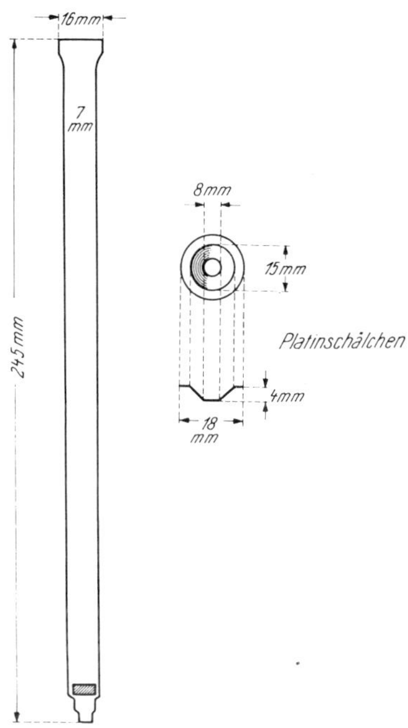

Abb. 1. Ionenaustauschsäule

Bestimmung der Verteilungskoeffizienten und Trennfaktoren

Die Gleichgewichtsverteilungskoeffizienten (Kd-Werte) sind folgendermaßen definiert:

$$Kd = \frac{\mu g \text{ Element} / g \text{ Harz}}{\mu g \text{ Element} / ml \text{ Mischung}} \tag{1}$$

wenn die Batch-Methode (Eintragverfahren) angewendet wird. Wird der Kd-Wert nach der Säulenmethode ermittelt, so gilt die Beziehung

$$Kd = \frac{\overline{V} - 0{,}7}{g} \tag{2}$$

wobei $\bar{V}$ jene Menge Elutionsmittel (in Milliliter) darstellt, die benötigt wird, um das Maximum der Elutionskurve zu erreichen; g = Gramm trockenes Harz und 0,7 (in Milliliter) = Leervolumen (bei Verwendung von 1 g Harz).

Der Trennungsfaktor α ist definitionsgemäß:

$$\alpha = \text{Kd Uran}/\text{Kd Element}$$

Die Bestimmung der Kd-Werte erfolgte beim Uran und teilweise auch beim Eisen nach der Batch-Methode, bei der die Mischungen, welche die genannten Elemente enthielten, zur Gleichgewichtseinstellung 4 Stunden mit 1 g Harz geschüttelt wurden und aus der Konzentrationsabnahme nach (1) der Kd-Wert bestimmt wurde oder nach der Säulenmethode, bei der das Element gelöst in 0,5 ml der jeweiligen Mischung auf das in einer Glassäule (siehe Abb. 1) befindliche, bereits mit der reinen Mischung vorbehandelte Harz aufgebracht wurde und mit weiteren 0,5-ml-Portionen reiner Mischung nachgewaschen wurde. Der Kd-Wert wurde in diesem Fall nach Gl. (2) berechnet. Diese Methode wurde beim Eisen verwendet, wobei zum qualitativen Nachweis eine wäßrige Ammoniumthiocyanatlösung verwendet wurde.

Bei den Batch-Versuchen erfolgte die Bestimmung des Urans fluorimetrisch und die des Eisens durch komplexometrische Titration mit Sulfosalicylsäure als Indikator.

Quantitative Bestimmung der Elemente

Fluorimetrische Bestimmung des Urans. Kleine Urankonzentrationen wurden mit dem Jarrel Ash-Fluorimeter bestimmt. Zu diesem Zweck wurden mittels eines Dosierers etwa 500 mg einer Mischung von 98 Gew.% Natriumfluorid und 2 Gew.% Lithiumfluorid auf ein Platinschälchen von den in Abb. 1 angegebenen Dimensionen aufgebracht und diese Mischung mit einem starken Bunsenbrenner durchgeschmolzen. Auf die erkaltete Schmelze wurden mit einer Lambda-Pipette 20 μl der 1-m salzsauren Probelösung aufgebracht und nochmals geschmolzen. Anschließend wurde der erkaltete Schmelzkuchen in das Fluorimeter gebracht und durch Messung der Intensität des Fluoreszenzlichts und Vergleich mit ebenso bereiteten Standards die Urankonzentration bestimmt.

Photometrische Bestimmung des Urans. Der Urangehalt wurde photometrisch nach der bekannten Thiocyanatmethode[8] bestimmt, die entsprechend modifiziert wurde.

Arbeitsvorschrift. 1 ml des Eluats, 4 ml einer 50 gew.%igen wäßrigen Ammoniumthiocyanatlösung und 1 ml einer 2%igen (2 g Ascorbinsäure in 100 ml Wasser) Ascorbinsäurelösung wurden in einen 10-ml-Meßkolben gegeben und mit 1-m Salzsäure zur Marke aufgefüllt. Nach 30 Min. wurde die Farbintensität des entstandenen Komplexes gegen eine Blindlösung gleicher Zusammensetzung bei 365 nm gemessen. Durch Vergleich mit ebenso bereiteten Standards wurde unter Benutzung einer Eichkurve die Urankonzentration der Probe festgestellt.

Photometrische Bestimmung des Eisens. Zur spektrophotometrischen Bestimmung des Eisens wurde die bekannte Sulfosalicylsäuremethode verwendet, die entsprechend modifiziert wurde.

Arbeitsvorschrift. 1 ml der Probelösung (d. h. des Fe(III)-haltigen Eluats), 2 ml einer Sulfosalicylsäurelösung (5 g der Säure in 100 ml Wasser), 3 ml einer kaltgesättigten Lösung von Hexamethylentetramin (45 g Hexamethylentetramin in 50 ml Wasser) und 6 ml dest.

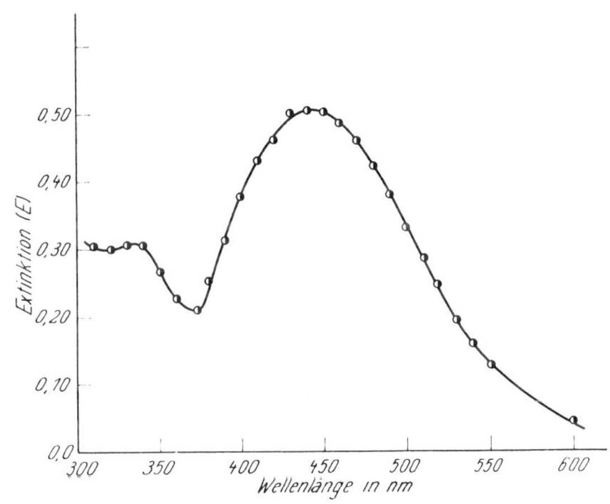

Abb. 2. Abhängigkeit der Adsorption von der Wellenlänge (50 μg Fe(III)/ml Eluat)

Wasser werden durch gründliches Schütteln gut durchgemischt. Nach Phasentrennung wird die wäßrige Phase mit einer Pipette abgetrennt und durch ein trockenes Filter filtriert, um die letzten Spuren der TBP-Kerosinphase zu entfernen.

Wie sich aus Abb 2 ergibt, ist die Wellenlänge von 440 nm für die photometrische Bestimmung am geeignetsten.

Wie aus der in Abb. 3 dargestellten Eichkurve ersichtlich ist, gilt das Beersche Gesetz im Konzentrationsbereich von 0 bis 70 μg Fe pro ml ursprünglichen Eluats.

Komplexometrische Bestimmung des Eisens. Die Bestimmung des Eisen(III)-gehalts kann komplexometrisch mit Sulfosalicylsäure als Indikator erfolgen. Zu diesem Zweck wird die Probe, welche Fe(III) gelöst in TBP-haltigen Mischungen enthält, mit 2 ml einer 5 gew.%-igen Sulfosalicylsäurelösung (5 g in 100 ml Wasser) versetzt und an-

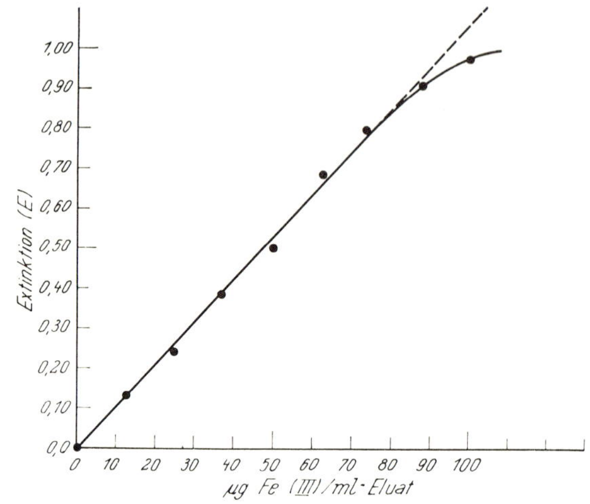

Abb. 3. Eichkurve zur Bestimmung des Fe(III)-Gehaltes

schließend soviel an kaltgesättigter wäßriger Urotropinlösung zugesetzt, bis das pH der wäßrigen Phase etwa 2,5 beträgt, was auch am Maximum der Farbintensität der Lösung ersichtlich ist. Dabei soll mit einem Magnetrührer intensiv gerührt werden. Danach wird mit 0,01-m ÄDTA-Lösung auf farblos titriert. Der Verbrauch entspricht dabei dem theoretisch berechneten, d. h.:

$$1 \text{ ml } 0{,}01\text{-m ÄDTA} = 0{,}5585 \text{ mg Fe(III)}$$

Resultate und Diskussionen

Abhängigkeit des Trenneffekts von der Art des verwendeten Lösungsmittels

Die zur Trennung verwendeten Lösungsmittelgemische müssen außer dem Extraktionsmittel (TBP) noch mindestens ein Lösungsmittel und eine bestimmte Menge wäßriger Säure enthalten, wobei die Säure die Funktion hat, die Bildung des auf dem Ionenaustauscher adsorbierbaren anionischen Metallkomplexes zu ermöglichen, während das Lösungsmittel die Mischbarkeit von TBP und Säure vermitteln soll. Wie sich aber bei den Untersuchungen gezeigt hat, ist die Stärke der Adsorption, die durch den Verteilungskoeffizienten charakterisiert ist, in hohem Maß auch von der Art und Konzentration des Lösungsmittels abhängig, wobei die Solvatisierung der Ionen im Harz eine wesentliche Rolle spielt.

Da sich von den verschiedenen untersuchten Anionformen des Ionenaustauschers Dowex 1-X 8 die Chloridform als besonders geeignet erwies, wurden alle Untersuchungen mit diesem Harz ausgeführt. Daraus ergab sich die Notwendigkeit der Verwendung wäßriger Salzsäure in den Lösungsmittelgemischen. Daher wurden Mischungen aus TBP (mit und ohne Kerosin als inertes Verdünnungsmittel), wäßriger Salzsäure und einer Lösungsmittelkomponente untersucht. Wegen der für die praktische Anwendung wichtigen Voraussetzung genügend hohen Gehaltes an Extraktionsmittel und genügend hoher Acidität wurde die Zusammensetzung des Gemisches zu 30 Vol.% TBP, 60 Vol.% eines organischen Lösungsmittels und 10 Vol% wäßriger Salzsäure gewählt. Die Ergebnisse sind für verschiedene Lösungsmittel in Tab. 1 dargestellt. Sie zeigen, daß sich Mischungen, die als Lösungsmittelkomponente Äthylenglykolmono-

Tabelle 1. Abhängigkeit der Verteilungskoeffizienten und Trennfaktoren von der Lösungsmittelkomponente

Lösungsmittel	Kd Uran(VI)	Kd Fe(III)	Trennfaktor
Aceton	46,1	0,55	83,8
Diäthylketon	29,0	0,55	52,7
Methyl-n-propylketon	37,5	0,80	46,9
Methylisobutylketon	49,3	0,80	61,6
Acetylaceton	11,3	0,30	37,7
Cyklohexanon	22,1	0,55	40,2
Tetrahydrofuran	69,6	0,80	87,0
Methylglykol	1163	0,80	1453,8
Äthylglykol	2230	0,80	2787,5
Essigsäure-n-butylester	380	0,80	475,0

methyläther, Äthylenglykolmonoäthyläther oder Essigsäure-n-butylester enthalten, für Trennungen wegen der damit erzielbaren hohen Trenneffekte besonders gut eignen.

Wegen der geringen Viskosität und der dadurch bedingten höheren Fließgeschwindigkeit wurde aber der Mischung aus 30 Vol.% TBP, 60 Vol.% Methylglykol und 10 Vol.% 12-m Salzsäure der Vorzug gegeben und eine Methode zur Abtrennung des Urans vom Eisen entwickelt (s. Seite 696).

Bei Verwendung von Aceton als Lösungsmittel hat sich der Zusatz von Kerosin bewährt. Besonders günstig erwies sich dabei eine Mischung aus 40 Vol.% TBP (50 vol.%ig an Kerosin) und 60 Vol.% einer Aceton-Salzsäuremischung (95 Vol.% Aceton und 5 Vol.% 12-m Salzsäure). Diese Mischung hat gegenüber der vorhin genannten den Vorteil einer wesentlich höheren Fließgeschwindigkeit. Ein Trennverfahren unter Anwendung dieser Mischung wird auf Seite 696 beschrieben.

Außer den in Tabelle 1 genannten Lösungsmitteln wurden n-Propylketon, Dimethylsulfoxid und Methanol untersucht. Dabei erwiesen sich n-Propylketon und Dimethylsulfoxid wegen Entmischungserscheinungen und Methanol wegen des zu geringen Trennfaktors als ungeeignet.

Abhängigkeit der Trennfaktoren von der Gesamtacidität der zur Trennung verwendeten Mischung

Da, wie bereits erwähnt wurde, für eine Trennung jene Mischungen, die Methylglykol enthalten, besonders gut geeignet sind, wurde in diesen Mischungen die Abhängigkeit der Trennfaktoren von der Gesamtacidität näher untersucht. Dazu wurde das Adsorptionsverhalten von Uran(VI) und Eisen(III) in Mischungen aus 30 Vol.% TBP, 60 Vol.% Methylglykol und 10 Vol.% wäßriger Salzsäure untersucht, deren Molarität 0, 2, 4, 6, 8, 10 bzw. 12 war. Bei der Mischung mit m = 0 war dennoch eine geringe, zur Bildung anionischer Chloridkomplexe ausreichende Menge an Salzsäure vorhanden, die vom Extraktionsschritt (Extraktion von Eisen und/oder Uran mit TBP) herrührt. Die Ergebnisse sind in Tabelle 2 wiedergegeben.

Tabelle 2. Abhängigkeit der Verteilungskoeffizienten und Trennfaktoren von der Gesamtacidität

Gesamtacidität der Mischung (m HCl)	Kd Uran(VI)	Kd Fe(III)	Trennfaktor
0,0	10,64	starke Adsorption	$\ll 1$
0,2	1180	0,80	1475,0
0,4	1410	0,80	1762,5
0,6	1597	0,80	1987,5
0,8	1555	0,80	1937,5
1,0	1540	0,80	1925,0
1,2	1163	0,80	1453,8

Besonders erwähnenswert ist dabei das Adsorptionsverhalten von Fe(III) in der Mischung aus 30 Vol.% TBP, 60 Vol.% Methylglykol und 10 Vol.% reines Wasser: bringt man nämlich bei einem Säulenversuch das Harz in Kontakt mit der klaren Mischung, die Fe(III) enthält, so kommt es infolge Hydrolyse zur Ausfällung von Eisenhydroxid, das die Ausbildung einer deutlich sichtbaren Zone in den obersten Bereichen der Säule bewirkt. Diese Hydrolyse wird offenbar durch die große Oberfläche des Harzes und eine ungleichmäßige Verteilung der Wassermolekel gefördert, die so zustande kommt, daß in der Harzphase infolge der stark polaren funktionellen Gruppen und deren Hydratisierung eine Erhöhung der Wasserkonzentration gegenüber der freien Lösung eintritt. Die Folge ist eine Erniedrigung der Acidität und eine Erhöhung des pH-Wertes, die auch dadurch noch wirksamer wird, daß die Protonen eher in der freien Lösung an TBP als Oxoniumsalz gebunden vorkommen, da vermutlich der TBP-Gehalt in der eher unpolaren freien Lösung höher ist als in der polaren Harzphase. Diese Effekte begünstigen aber eine Hydrolyse des Eisens.

Wie aus Tabelle 2 ersichtlich, reicht schon eine geringe Säurekonzentration aus, um günstige Trennfaktoren zu erzielen. Um die Hydrolyse aber überhaupt zu verhindern, wurden in den zur Trennung verwendeten Mischungen Säurekonzentrationen von 10 Vol.% 12-m Salzsäure verwendet, was einer Gesamtacidität von 1,2-m entspricht. Die Verwendung noch höherer Aciditäten ist auf Grund der begrenzten Mischbarkeit von TBP mit wäßriger Salzsäure und der Tatsache ungünstig, daß dann die beobachteten Trennfaktoren wesentlich niedriger sind.

Abhängigkeit des Trenneffektes von Extraktionsmittel- und Lösungsmittelkonzentration

Zur Untersuchung der Abhängigkeit der Adsorption von Uran und Eisen von der Extraktions- und Lösungsmittelkonzentration

Tabelle 3. Abhängigkeit der Verteilungskoeffizienten und Trennfaktoren von der Extraktionsmittel- und Lösungsmittelkonzentration (0,0 Vol.% 12-m HCl)

Zusammensetzung der Mischungen Vol.% TBP	Vol.% Methylglykol	Kd Fe(III)	Kd UO_2(II)	Trennfaktor
10	90	52,7	121 000	2296
20	80	49,8	45 000	903,6
30	70	46,1	82 700	1794
40	60	22,2	67 360	3034
50	50	17,8	48 100	2702
60	40	11,6	21 150	1823
70	30	6,5	3 817	587
80	20	1,0	1 492	1492
90	10	1,0	223,5	223,5
100	0	1,0	10,97	10,97

wurden Mischungen untersucht, die keine wäßrige Phase enthielten (ausgenommen die geringen Mengen wäßriger Salzsäure, die sich bei der Extraktion in TBP lösen).

Die Ergebnisse sind in Tabelle 3 dargestellt und zeigen eine Abnahme der Adsorbierbarkeit von Uran und Eisen mit zunehmender Extraktionsmittel(TBP)- und entsprechend abnehmender Lösungsmittel(Methylglykol)-Konzentration. Gleichzeitig nehmen auch die Trennfaktoren ab.

Trennung des Eisens vom Uran in einer Mischung aus 30 Vol.% TBP, 60 Vol.% Methylglykol und 10 Vol.% 12-m Salzsäure (Methode I)

Arbeitsvorschrift: Die uran- und eisenhaltige Probe wird in 6-m Salzsäure gelöst und 3mal mit TBP extrahiert. Durch Zugabe von Methylglykol und Salzsäure wird eine Mischung oben angegebener Zusammensetzung hergestellt und diese durch das Harzbett fließen gelassen, das etwa 1 Stunde lang mit der reinen Mischung vorbehandelt worden war. Anschließend wird mit soviel reiner Mischung portionenweise nachgewaschen, bis kein Fe(III) mehr nachweisbar ist; dann wird das Harz mit einer Mischung von 90 Vol.% Methylglykol und 10 Vol.% 12-m Salzsäure zur Entfernung des TBP behandelt, das die fluorimetrische Uranbestimmung stören würde (wobei für 2 g Harz 15 ml ausreichen). Danach wird das Uran mit 1-m Salzsäure eluiert und fluorimetrisch bestimmt. Bei Anwesenheit störender Fremdionen ist es vorteilhaft, nach der Elution des Eisens mit der genannten Mischung das Harz zunächst mit einer ausreichenden Menge 6-m Salzsäure und danach wieder mit einer Mischung von 90 Vol.% Methylglykol und 10 Vol.% 12-m Salzsäure zu waschen, damit die dieses Elutionsverfahren störenden Ionen wie Cu(II) und Co(II) eluiert und so von Fe(III) und U(VI) leichter getrennt weden können.

Mittels des beschriebenen Verfahrens ist es möglich, Uran und Eisen von einer Vielzahl anderer Ionen abzutrennen, die die Bestimmung teilweise stören würden, und zwar von Pt(II, IV), Pd(II, IV), Hg(II), Cu(II), Co(II), Te(IV), Ag(I), Tl(I) und allen anderen Elementen, die von TBP nicht oder nur schlecht extrahiert werden.

Die Bestimmung des Urangehalts erfolgte kolorimetrisch und die Bestimmung des Eisengehalts komplexometrisch (siehe Experimenteller Teil).

Trennung des Eisens vom Uran in einer Mischung aus 40 Vol.% TBP-Kerosin (50 vol.%ig) und 60 Vol.% einer Aceton-Salzsäuremischung (95 Vol.% Aceton und 5 Vol.% 12-m Salzsäure) (Methode II)

Diese Mischung, die eine sehr hohe Fließgeschwindigkeit von 0,82 ml/Minute aufweist, bewährt sich besonders bei salpetersauren Probelösungen.

Arbeitsvorschrift: Die Probe, die 5-m salpetersauer sein soll und die zu bestimmenden Elemente enthält, wird mit einer TBP-Kerosin-Mischung (50 vol.%ig) 3mal extrahiert, dann wird zu den vereinigten TBP-Kerosin-Extrakten soviel 95 vol.%ige Aceton-Salzsäuremischung zugegeben, bis die oben angegebene Zusammensetzung erreicht ist.

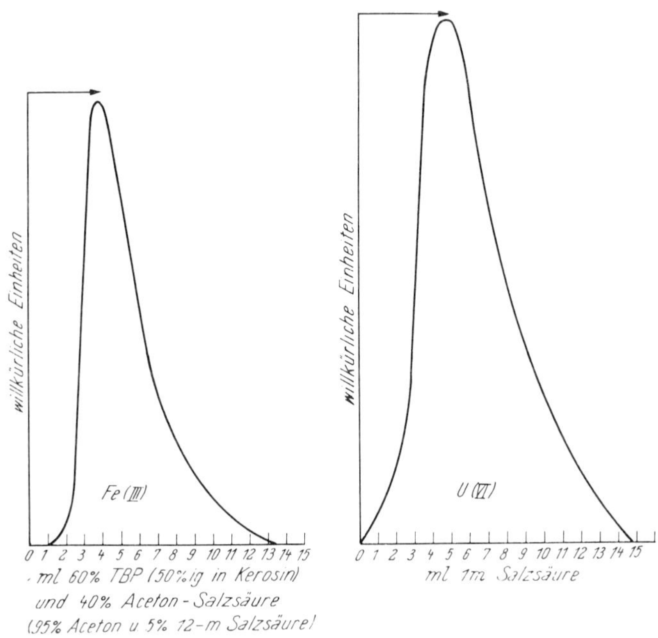

Abb. 4. Elutionskurven von Fe(III) und U(VI)

Diese Lösung wird auf das mit der reinen Mischung vorbehandelte Harz aufgebracht und das Fe(III) mit der reinen Mischung wieder vom Harz eluiert [U(VI) bleibt dabei fest adsorbiert]. Anschließend wird das Harz zur Entfernung von TBP mit einer Mischung von 95 Vol.% Aceton und 5 Vol.% 12-m Salzsäure (15 ml) gewaschen, das Uran mit 1-m Salzsäure eluiert und fluorimetrisch oder spektrophotometrisch bestimmt; die Konzentration von Fe(III) wurde komplexometrisch oder spektrophotometrisch bestimmt.

Die Wirksamkeit der beschriebenen Trennungen wurde an Hand einiger Beispiele untersucht, deren Ergebnisse in den Tabelle 4 und 5 angegeben sind. Sie zeigen die Brauchbarkeit der angegebenen Methoden zur Abtrennung des Urans vom Eisen.

Die gute Trennbarkeit dieser beiden Elemente ist auch aus den in Abb. 4 gezeigten Elutionskurven ersichtlich, für deren Ermittlung Milligrammengen der Elemente verwendet wurden.

Tabelle 4. Eisen-Uran-Trennungen unter Anwendung von Methode I

mg U(VI) eingesetzt	gefunden	Fehler (%)	mg Fe(III) eingesetzt	gefunden	Fehler (%)
1,00	1,02	+2,10	11,00	10,77	−2,13
10,00	9,87	−1,30	11,00	10,77	−2,13
20,00	19,58	−2,10	11,00	10,66	−3,25

Tabelle 5. Eisen-Uran-Trennungen unter Anwendung von Methode II

mg U(VI) eingesetzt	gefunden	Fehler (%)	mg Fe(III) eingesetzt	gefunden	Fehler (%)
1,70	1,76	+3,30	6,67	6,68	+0,15
2,00	2,03	+1,50	10,00	10,25	+2,50
1,80	1,85	+2,78	2,00	1,93	−3,50

Danksagung

Dem Fonds zur Förderung der wissenschaftlichen Forschung wird an dieser Stelle für die Bereitstellung der erforderlichen Mittel bestens gedankt.

Zusammenfassung

Das Verhalten von U(VI) und Fe(III) am stark basischen Anionenaustauscher Dowex 1, X8 (Chloridform) in TBP-haltigen salzsauren Lösungsmittelgemischen wurde systematisch untersucht. Daraus konnten zwei verschiedene Methoden zur Trennung der genannten Elemente auf Ionenaustauschersäulen entwickelt werden, wobei durch Kombination von Extraktion und Ionenaustausch eine Abtrennung des Eisens und des Urans auch von anderen Elementen möglich war.

Die Trennungen erfolgten in Medien folgender Zusammensetzung: 30 Vol.% TBP, 60 Vol.% Methylglykol und 10 Vol.% 12-m Salzsäure sowie 40 Vol.% TBP-Kerosin (50 vol.%ig) und 60 Vol.% einer Aceton-Salzsäuremischung (95 Vol.% Aceton und 5 Vol.% 12-m Salzsäure).

Weiters wurde eine Methode zur spektrophotometrischen Bestimmung von Eisen(III) in einem TBP-haltigen Lösungsmittelgemisch entwickelt.

Summary

Anion Exchange Separations of the Elements that may be Extracted with Tributyl Phosphate. I

A systematic study was made of the behavior of U(VI) and Fe(III) on strongly basic anion exchanger Dowex 1, X8 (chloride form) in hydrochloric acid solvent mixtures containing TBP. The findings led to two different methods for the separation of the named elements on ion exchange columns, whereby by a combination of extraction and ion exchange it was possible to achieve a separation of the iron and the uranium also from other elements.

The separations took place in media of the following composition: 30 volume % TBP, 60 volume % methylglycol and 10 volume % of 12 M hydrochloric acid as well as 40 volume % TBP-kerosene (50 volume %) and 60 volume % of an acetone-hydrochloric acid mixture (95 volume % acetone and 5 volume % 12 M hydrochloric acid). Furthermore a method was developed for the spectrophotometric determination of iron(III) in a solvent mixture containing TBP.

Literatur

[1] *J. Korkisch,* Ion Exchange in Mixed and Non-Aqueous Media, Progress in Nuclear Energy, Analytical Chemistry, Series IX, **6,** 1 (1966), Pergamon Press.

[2] *J. Musich, K. A. Orlandini* und *J. Korkisch,* USAEC Report, ANL-740, **1968.**

[3] *J. Korkisch,* Mikrochim. Acta [Wien] **1967,** 401.

[4] *J. Korkisch,* Österr. Chem.-Ztg., **67,** 309 (1966).

[5] *J. Korkisch,* Separation Science **1,** 159 (1966).

[6] *J. Korkisch,* Modern Methods for the Separation of Rarer Metal Ions. Oxford: Pergamon Press. 1969. S. 34.

[7] *J. Petrement, J. Lopez* und *F. Burriel-Martí,* Anales de la Sociedad Española de Fisica y Quimica **64** (12), 1053 (1968).

[8] *J. Korkisch* und *F. Hecht,* Handbuch der analytischen Chemie, Bd. VI b β: Uran. Berlin—Heidelberg—New York: Springer-Verlag, im Druck.

22

ANION-EXCHANGE SEPARATIONS OF ELEMENTS EXTRACTIBLE BY TRIBUTYL PHOSPHATE: I. Separation of Uranium from Iron in the Presence of Tributyl Phosphate

W. Koch and J. Korkisch

This condensed translation was prepared expressly for this Benchmark volume by Harold F. Walton, University of Colorado, Boulder, Colorado, from "Anionenaustauschtrennungen der mit Tributylphosphat extrahierbaren Elemente: I. Trennung des Urans vom Eisen in Geganwart von Tributyphosphat," Mikrochim Acta, *687–689 (1972)*

INTRODUCTION

Previous studies have shown that the absorption of metal ions on strongly basic and strongly acidic ion exchangers may be significantly enhanced by adding an organic solvent (1). Separations are accomplished in this way that would not be possible in purely aqueous systems.

More careful investigations have shown that even the complex compounds formed between metals and extracting agents in organic solvents can be taken up by ion exchangers. The principle is that of a cooperative effect of ion exchange and solvent extraction in which the resin is the stationary phase and the organic solvent the mobile phase. The absorption process is notably different from absorption in purely aqueous systems and also from absorption in solvent mixtures that contain no extracting agent. Several such systems that show the general phenomenon called CIESE (combined ion exchange and solvent extraction) have been described in the literature (4–6).

This paper will discuss, in particular, systems in which tributyl phosphate is used as the extracting agent. Similar systems have been used to separate uranium from organic solvent mixtures containing nitric acid (7). We shall choose as our example the sepatation of uranium from iron, which is important both analytically and industrially, and illustrates how extraction and ion exchange can be intelligently combined in such a way that, after extracting both elements together by tributyl phosphate, the two elements can be separated from one another by anion exchange in a solvent mixture that contains tributyl phosphate. To find appropriate solvent mixtures, we first had to find good mutual solvents, and then to measure the degree of separation as a function of the mixing ratio and the acidity. On the basis of these results we found two solvent mixtures that were particularly effective in separating uranium from iron, and we were able to develop methods for separating and determining these elements.

RESULTS AND DISCUSSION [p. 692]

Dependence of the Separating Effect on the Nature of the Solvent

The mixture of solvents used for the separation must contain, besides tributyl phosphate (TBP), at least one other solvent and a certain proportion of aqueous acid. The function of the acid is to produce an anionic metal complex that will be absorbed on the resin; that of the solvent is to make the aqueous acid mix with the tributyl phosphate. We found in the course of our work that the strength of adsorption, as expressed by the distribution coefficients, depends very greatly on the nature and concentration of the solvent, because these factors affect the solvation of the ions of the resin.

Of the various ionic forms of the anion exchanger Dowex-1x8 that were studied, the chloride form was the most suitable. We therefore chose this form for our experiments, and, accordingly, we used hydrochloric acid. Mixtures of TBP (with or without kerosene as inert diluent), aqueous hydrochloric acid, and another solvent were studied. For practical purposes we had to maintain a high concentration of extractant and also a high acidity; we chose a proportion, by volume, of 30 percent TBP, 60 percent organic solvent, and 10 percent aqueous hydrochloric acid. The results with different solvents are given in Table I. They show that mixtures containing ethylene glycol monomethyl ether, ethylene glycol monoethyl ether, or *n*-butyl acetate were especially effective.

Because of its low viscosity we favored the mixture containing 30 percent TBP, 60 percent methyl glycol, and 10 percent 12 *M* HCl, and used this to separate uranium from iron [see page 696].

With acetone as a solvent it was best to add kerosene. A good mixture contained 40 percent TBP (50 percent kerosene by volume) and 60 percent acetone–HCl mixture (95 percent acetone, 5 percent 12 *M* HCl by volume).

[*Translator's Note:* A separation procedure using this solvent mixture is described on page 696. The next two sections are entitled "Dependence of the Separation Factors on the Total Acidity" and "Dependence of the Separation Efficiency on the Concentrations of Extractant and Solvent." The authors continue as follows.]

Separation of Iron from Uranium in a Mixture of 30 Percent TBP, 60 Percent Methyl Glycol and 10 Percent 12 *M* Hydrochloric Acid (Method 1)

Procedure. The sample containing uranium and iron is dissolved in 6 *M* HCl and extracted three times with TBP. Methyl glycol and HCl are then added to the TBP extract to produce the composition given above. A bed of Dowex-1x8 resin is pretreated with the same solvent mixture and left to stand for 1 hour; then the solution containing iron and uranium is allowed to flow through it. (Uranium is absorbed, iron passes.) The column is rinsed with the solvent mixture until no more iron is detected in the effluent. Then a mixture of 90 percent methyl glycol and 10 percent 12 *M* HCl by volume is passed to remove TBP, which would otherwise interfere with the determination of uranium. Uranium is now eluted with 1 *M* HCl and determined fluorimetrically. If interfering ions are present, it is best, after eluting the iron, to wash the resin with 6 *M* HCl, followed by 90 percent methyl glycol-10 percent 12 *M* HCl. In this way, ions like Co (II) and Cu(II) are removed and do not contaminate the uranium.

By using this procedure, it is possible to separate uranium and iron from many other ions what would interfere in the determinations, including Pt(II and IV), Pd (II and IV), Hg(II), Cu(II), Co(II), Te(IV), Ag(I), Tl(I), as well as all the elements that were not extracted by TBP.

Separation of Iron from Uranium in a Mixture of 40 Percent TBP-Kerosene (50 Percent by Volume) and 60 Percent Acetone-HCl (95 Percent Acetone, 5 Percent 12 *M* HCl by Volume): (Method II)

This mixture, which allows a very high flow rate, 0.82 ml/min, is especially useful for samples that contain nitric acid.

Procedure. The sample, which must be 5 *M* in nitric acid, is extracted three times with TBP-kerosene mixture (50 percent by volume) and enough 95 percent acetone-5 percent HCl is added to the combined extracts to give the solvent composition shown above. This solution is passed through the resin column, which has been pretreated with the same solvent, and the iron is eluted by passing more of the same solvent [uranium(VI) remains strongly adsorbed]. Then the resin is washed with the acetone-HCl mixture to remove TBP. Uranium is then eluted with 1 *M* HCl and determined fluorimetrically or spectrophotometrically. Iron is determined complexometrically or spectrophotometrically.

The efficacy of the separations was demonstrated by some examples that are shown in Tables 4 and 5. The good separation of uranium and iron from one another is shown in the elution curves of Figure 4, which refer to milligram quantities of the two elements.

[*Editor's Note:* References, tables and figures for this translation will be found in the preceding original article.]

Part VII

CHROMATOGRAPHY OF INORGANIC IONS

Editor's Comments on Papers 23 Through 27

23 **POLLARD** et al.
Separation of Some Lower Oxy-Acids of Phosphorus by Anion-Exchange Chromatography

24 **BROWN** et al.
Production of Bromite and Perbromate Ions in Crystalline $CsBrO_3$ Irradiated with Cobalt-60 γ Rays and the Vibrational Spectrum of the BrO_4^- Ion

25 **STRELOW** et al.
Ion Exchange Selectivity Scales for Cations in Nitric Acid and Sulfuric Acid Media with a Sulfonated Polystyrene Resin

26 **STRELOW** et al.
Distribution Coefficients and the Cation-Exchange Behavior of Elements in Hydrochloric Acid-Ethanol Mixtures

27 **STRELOW** et al.
Accurate Determination of Traces of Sodium and Potassium in Rocks by Ion Exchange Followed by Atomic Absorption Sprectroscopy

For the most part the separations illustrated in this group of papers depend on the intrinsic selectivity of the ion-exchange process itself, rather than on the stabilities of complex ions.

Paper 23 describes the chromatography of a group of closely related anions, those of the oxy-acids of phosphorus, which it would be hard to separate in any other way. The thionic acids and other oxy-acids of sulfur are similar cases; these, too, have been separated by ion-exchange chromatography (Chapman and Beard, 1973).

An important paper that immediately followed Pollard's paper on the phosphorus oxy-acids is from the same research group (Pollard et al., 1963); it deals with the separation of alkaline-earth metals on a cation-exchange column, using a lactate buffer and gradient elution. An automated system was developed and described in a subsequent paper. These papers are fine descriptions of ion-exchange chromatography, but they also illustrate how rapidly an analytical method can become obsolete. The ions appearing in the effluent were detected and measured by flame emission spectrometry. Interelement effects in flame emission are strong, and it was necessary to separate the alkaline-earth ions from one another before flame emission would give accurate results. Atomic absorption spectrometry came into general use just after these papers were published, and in this technique interelement effects are relatively minor. Thus, with the advent of atomic absorption spectrometry, the need for routine ion-exchange separations lessened. Nevertheless, separations are still needed when really accurate and reliable analyses are desired (see the work of Strelow, cited later).

Paper 24 shows the separation of a rare and exotic anionic species, the perbromate ion, BrO_4^-. This paper is not the first to describe a perbromate salt, but it shows the use of ion exchange to separate perbromate from accompanying bromide, bromite, and bromate, and demonstrates the presence of perbromate in certain preparations and its absence from others. Figures 2 and 4 give some insight into anion-exchange selectivities and show that BrO_4^-, like ClO_4^-, is very strongly held by an anion-exchange resin. The ion used to displace it is another four-coordinated, singly charged anion, BF_4^-.

Papers 25, 26, and 27 are chosen from a large number of publications by F. W. E. Strelow and his colleagues at the National Chemical Research Laboratory, Pretoria, South Africa. The work of Strelow's group is "mission-oriented" and is concerned with definitive, accurate analyses of minerals, rocks, and mineral products. Techniques such as emission spectroscopy and atomic absorption are adequate for routine purposes, but when really reliable measurements are needed, particularly in the analysis of trace constituents, there is no substitute for careful chemical separations. Once separated, the elements can be determined by whatever method is most precise, and for major constituents this is usually complexometric titration.

Like many of the papers from Strelow's laboratory, Papers 25 and 26 present extensive tables of distribution ratios, giving the behavior of many elements over a range of solvent compositions. Cation exchange is seen to be as versatile as anion exchange. The tables of Paper 25 show that zirconium and hafnium can be easily separated by cation exchange in sulfuric acid solutions, but there is one caution to be observed that is not adequately pointed out in the text: these and other highly charged ions hydrolyze considerably, even at the acid concentrations used; hydrolysis leads to polynuclear species, however, and for this reason it is slow. The distribution data and the elution curves presented were obtained with solutions that were freshly prepared by diluting solutions in concentrated acid and were not allowed to age.

The chromatographic curves in both papers show beautiful separations. The technique is to use a fairly short and wide column, typically with a 2-cm diameter and 10 to 20 cm long, with resin of commercial size ranges (100–200 or 200–400 mesh). Separations are accomplished with high selectivity by stepwise changes of effluent. Gravity flow is used, and the running times are of the order of several hours. For the purposes of this work, however, slow flow rates and large volumes were not objectionable.

Strelow has experimented with eluents containing complexing organic acids, and one publication (Strelow et al., 1960) describes a comprehensive scheme for the analysis of silicate rocks based on eluents containing oxalic acid. This acid can be easily destroyed by boiling with bromic acid before the effluents are analyzed. It is not suitable for trace determinations, however, because the reagent blank is too big.

Paper 27 shows the power of cation exchange for separating trace constituents, in this case alkali-metal ions. Nitric acid was used as the eluent; it can be purified by distillation and has a very low reagent blank. The chromatography of the five alkali-metal cations shown in Figure 3 is impressive, and so are the data of Table II, which show that this method gives values for the sodium and potassium contents of U.S. Geological Survey standard rocks that are far below the "literature" values, far more reproducible, and in good agreement with the most recent activation analyses.

REFERENCES

Chapman, J. N., and H. R. Beard. 1973. *Anal. Chem., 45,* 2268.

Pollard, F. H., G. Nickless, D. Spincer. 1963. *J. Chromatog., 10,* 215-220.

Strelow, F. W. E., C. J. Liebenberg, and F. von S. Toerien. 1960. *Anal. Chim. Acta, 47,* 251.

23

Reprinted from *J. Chromatog.* .**10**, 212–214 (1963)

SEPARATION OF SOME LOWER OXY-ACIDS OF PHOSPHORUS BY ANION-EXCHANGE CHROMATOGRAPHY

F. H. POLLARD, G. NICKLESS AND M. T. ROTHWELL

Department of Inorganic and Physical Chemistry, The University, Bristol (Great Britain)

(Received July 7th, 1962)

In a previous communication[1], the authors have reported the separation of hypophosphite, phosphite and phosphate using gradient elution anion-exchange chromatography. This scheme has been extended to include the following lower oxy-anions of phosphorus, hypophosphate, diphosphite, pyrophosphite, isohypophosphate, an acid containing three phosphorus atoms and the condensed phosphate, pyrophosphate.

EXPERIMENTAL

Materials

B.D.H. reagent grade sodium hypophosphite, $NaH_2PO_2 \cdot H_2O$ and disodium phosphite, $Na_2HPO_3 \cdot 5H_2O$, were twice recrystallised from aqueous solution. AnalaR grade disodium hydrogen orthophosphate, $Na_2HPO_4 \cdot 12H_2O$ was used without further purification. Disodium hypophosphate, $Na_2H_2P_2O_6$ was prepared by the method of LEININGER AND CHULSKI[2] and three times crystallised from hot water.

Trisodium diphosphite, $Na_3HP_2O_5 \cdot 12H_2O$ was prepared by the hydrolysis of phosphorus trichloride[3,4].

Disodium pyrophosphite $Na_2H_2P_2O_5$ was prepared by heating monosodium phosphite $NaH_2PO_3 \cdot 2.5H_2O$ at 120° under vacuum for 6 h[5]. Iodometric titration of the resultant product showed that a small quantity of phosphite remained[5]. Due to the ease of hydrolysis of pyrophosphite in aqueous solution[6], this solid was used without further purification.

Samples of trisodium isohypophosphate $Na_3HP_2O_6 \cdot 4H_2O$ and tetrasodium pyrophosphate $Na_4P_2O_7 \cdot 10H_2O$ were kindly donated by Albright and Wilson (Mfg.) Ltd. and were used without further purification. The isohypophosphate contained traces of orthophosphate and orthophosphite as impurities and this salt, like pyrophosphite, hydrolyses in aqueous solution[7]

The acid containing three atoms of phosphorus and having the structure[8]:

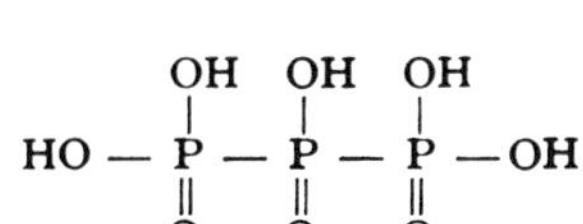

was isolated as $Na_5P_3O_8 \cdot 14H_2O$ from the oxidation products of red phosphorus by sodium chlorite solution. This acid will be referred to as $(\overset{4}{P}-\overset{3}{P}-\overset{4}{P})$ following the nomenclature proposed by BLASER[9].

B.D.H. reagent grade potassium chloride.

Preparation and use of the chromatographic columns

The anion-exchange columns were prepared and used in an identical manner to that described in the previous communication[1]. The conditions are given in Table I. 10 ml fractions of the column effluent were collected, transferred to conical flasks and boiled with 5 ml of bromine water for 30 min and then for a further 30 min after adding 10 ml of concentrated nitric acid. The phosphorus content of each fraction was then determined spectrophotometrically by the phosphovanadomolybdate method[10].

TABLE I

CONDITIONS FOR THE CHROMATOGRAPHIC SEPARATION

Column dimensions (cm cm)	*KCl concentration in mixing vessel (ml/M)*	*KCl concentration in reservoir (M)*	*Temperature (°C)*	*pH*	*Flow rate (ml/h)*
50 × 1.5	750/0.05	0.20	18	6.8	60

Buffer solution: pH 6.8; 25 ml of 2 *M* ammonium acetate solution per litre of potassium chloride. This pH was selected in order to keep the hydrolysis of pyrophosphite and isohypophosphate at a minimum during the chromatographic separation.

RESULTS AND DISCUSSION

An examination of Fig. 1 and Table II shows that good separations were obtained with the exception of that between diphosphite and pyrophosphite. Elution schemes using more dilute potassium chloride solutions were investigated to try and improve this separation, but without success. Neither could the separation be improved by using eluant solutions buffered at a pH of 9, nor by reducing the column temperature to 2°.

TABLE II

RETENTION VOLUMES OF PHOSPHORUS OXY-ANIONS*

Species	*Retention volume (ml)*
Hypophosphite	150
Phosphate	260
Phosphite	330
Hypophosphate	590
Diphosphite	680
Pyrophosphite	720
Isohypophosphate	890
Pyrophosphate	990
$(\overset{4}{P}-\overset{3}{P}-\overset{4}{P})$	1180

* Total percentage of phosphorus recovered = 97%.

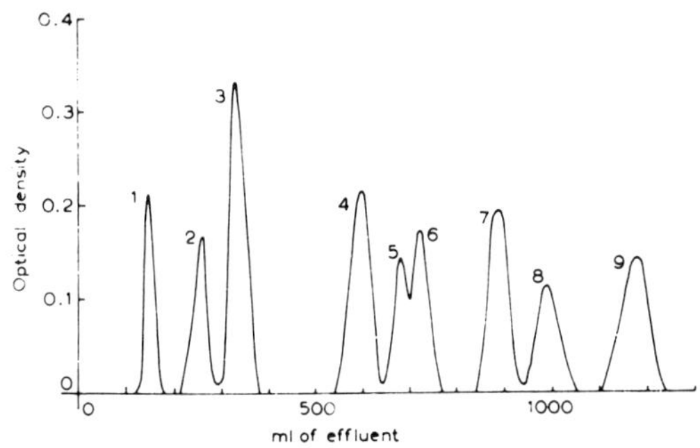

Fig. 1. Elution pattern for the separation of some phosphorus oxy-anions at pH 6.8. (1) Hypophosphite; (2) phosphate; (3) phosphite; (4) hypophosphate; (5) diphosphite; (6) pyrophosphite; (7) isohypophosphate; (8) pyrophosphate; (9) $(\overset{4}{P}-\overset{3}{P}-\overset{4}{P})$.

No appreciable hydrolysis of pyrophosphite or isohypophosphate appeared to occur during the separation. Maximum separations were achieved when the amount of phosphorus per species was less than 2000 μg.

ACKNOWLEDGEMENTS

We wish to thank Mr. D. L. Crone of these laboratories for the sample of the acid $(\overset{4}{P}-\overset{3}{P}-\overset{4}{P})$, and the Department of Scientific and Industrial Research for a Research Studentship to M. T. Rothwell.

SUMMARY

Anion-exchange chromatography employing a gradient elution technique has been used to obtain separations of some lower oxy-anions of phosphorus.

REFERENCES

[1] F. H. Pollard, D. E. Rogers, M. T. Rothwell and G. Nickless, *J. Chromatog.*, 9 (1962) 227.
[2] E. Leininger and T. Chulski, *J. Am. Chem. Soc.*, 71 (1949) 2385.
[3] B. Blaser, *Ber.*, 68 B (1935) 1670.
[4] B. Blaser, *Ber.*, 86 (1953) 563.
[5] S. J. Kiehl and M. F. Moose, *J. Am. Chem. Soc.*, 60 (1938) 47.
[6] S. J. Kiehl and M. F. Moose, *J. Am. Chem. Soc.*, 60 (1938) 257.
[7] B. Blaser and K. H. Worms, *Z. Anorg. Allgem. Chem.*, 301 (1959) 18.
[8] B. Blaser and K. H. Worms, *Z. Anorg. Allgem. Chem.*, 300 (1959) 249.
[9] B. Blaser and K. H. Worms, *Z. Anorg. Allgem. Chem.*, 300 (1959) 225.
[10] C. J. Barton, *Ind. Eng. Chem.* (*Anal. Ed.*), 20 (1948) 1068.

24

Reprinted from *J. Amer. Chem. Soc.*, **91**(9), 2250–2254 (1969)

Production of Bromite and Perbromate Ions in Crystalline $CsBrO_3$ Irradiated with Cobalt-60 γ Rays and the Vibrational Spectrum of the BrO_4^- Ion[1]

L. C. Brown, G. M. Begun, and G. E. Boyd

Contribution from the Oak Ridge National Laboratory, Oak Ridge, Tennessee 37830. *Received December 16, 1968*

Abstract: The formation at *ca.* 30° by ^{60}Co γ rays of perbromate and bromite ions in $CsBrO_3$ was demonstrated by infrared absorption measurements on irradiated crystals and by ion-exchange column chromatographic separations on aqueous solutions of the radiolyzed solid. Vibrational spectra (infrared and laser-excited Raman measurements) were obtained for BrO_4^- ion in solution and in crystalline $RbBrO_4$ and $(C_6H_5)_4AsBrO_4$. The frequency assignments are ν_1 801, ν_2 331, ν_3 878, and ν_4 410 cm^{-1}. Force constants derived from these frequencies suggest that there is a close similarity in bonding between BrO_4^- and IO_4^- and that a significant decrease in the nonbonding interactions occurs in the series ClO_4^-, BrO_4^-, IO_4^- as the size of the central atom increases and the oxygens become less crowded. An unsuccessful search was made for $^{82}BrO_4^-$ ion formed in $^7LiBrO_3$ by thermal neutron capture. The yield of $^{82}BrO_4^-$ ion, if formed, was less than 0.05%.

A continuing objective in our researches on the decomposition of bromate ion in crystalline alkali-metal and alkaline-earth bromates[2] by energetic ionizing radiations has been to relate the species observed in an aqueous solution of the irradiated salt to those trapped in the crystal. It is by no means certain that these are always the same species; in at least one instance[3a] there is evidence that they were quite different. On the other hand, recent studies[3b] with crystalline $KClO_3$ irradiated with ^{60}Co γ rays have shown that within experimental error all the perchlorate ion found in aqueous solution was originally present as ClO_4^- in the crystal lattice. Evidence for stable products in radiolyzed $CsBrO_3$ crystals also has been obtained: cesium bromide in the lattice has been indicated by X-ray diffraction techniques,[2] and reflectance spectra[2] of solid samples show an absorption band in

(1) Research sponsored by the U. S. Atomic Energy Commission under contract with the Union Carbide Corp.

(2) For a review see J. W. Chase and G. E. Boyd, "The Radiolytic Decomposition of Crystalline Alkali-metal and Alkaline-earth Bromates," ASTM Special Publication No. 400, 1966, p 17 ff.

(3) (a) L. J. Sharman and K. J. McCallum, *J. Am. Chem. Soc.*, **77**, 2989 (1955); (b) L. C. Brown and G. E. Boyd, *J. Phys. Chem.*, **73**, 396 (1969).

the ultraviolet (peak at *ca.* 3300 Å) which may be assigned[4] to BrO^-.

Oxygen gas is evolved on dissolving irradiated alkali-metal and alkaline-earth bromates in water, and Br^-, BrO^-, and BrO_2^- ions are found in solution.[2,5,6] The presence of bromite ion in the crystal lattice, however, has not been demonstrated. The amount of oxygen gas was found to increase linearly with the absorbed dose for low and intermediate extents of radiolysis, but for large doses its concentration increased less rapidly than linearly. This result and other considerations led to the suggestion[5,6] that oxygen may be reacting with bromate to give perbromate ion, BrO_4^-, analogous to the formation of ClO_4^- ion in the radiolysis of the alkali-metal chlorates. Perbromate ion until recently, however, has been unknown, despite many attempts to prepare it. The report[7] that it can be prepared by oxidation of bromate ion in aqueous solution by XeF_2 and that crystalline $RbBrO_4$ is stable at room temperature prompted us to renew our search for the species in crystalline $CsBrO_3$ radiolyzed by ^{60}Co γ rays. Rubidium perbromate and tetraphenylarsonium perbromate, $(C_6H_5)_4AsBrO_4$, were prepared, and the vibrational spectrum of BrO_4^- ion has been observed with infrared and laser-excited Raman techniques. The ir "fingerprint" of perbromate ion was used to show that BrO_4^- was formed by γ rays in crystalline $CsBrO_3$. Further demonstration of the production of perbromate ion was obtained with ion-exchange column chromatography.

Experimental Section

Materials. The preparation and analysis of anhydrous $^7LiBrO_3$[8] and $CsBrO_3$[5] have been described. Xenon difluoride (purchased from Peninsular Chemresearch, Inc., Gainesville, Fla. 32601) was used "as received" without further purification. Other chemicals were reagent grade and were used without further purification.

Ir and Raman Spectral Measurements. The infrared spectrum of BrO_4^- ion, to be used for the identification of BrO_4^- in ^{60}Co γ-ray irradiated $CsBrO_3$, was determined with two salts, $RbBrO_4$ and tetraphenylarsonium perbromate, $(C_6H_5)_4AsBrO_4$. Laser-excited Raman spectra of the solids and of aqueous solutions of BrO_4^- ion also were obtained to provide additional structural data on this long-sought species.

Centigram quantities of $RbBrO_4$ were prepared by oxidizing aliquots of aqueous 0.4 *M* $NaBrO_3$ solutions with decigram amounts of XeF_2 as described by Appelman.[7] Crystalline $RbBrO_4$, free of BrO_3^- ion, was obtained by dissolving the initial impure product in 1.5 *M* HBr, sparging the solution with N_2 to remove Br_2 formed by reduction of the BrO_3^-, and finally recrystallizing the slightly soluble compound. The infrared spectrum of the purified salt (Figure 1) did not show any absorptions at *ca.* 365 and 800 cm^{-1} which were observed in the initial salt and are attributable to BrO_3^- ion. Tetraphenylarsonium perbromate was precipitated in centigram quantities by addition of excess tetraphenylarsonium chloride to XeF_2-oxidized $NaBrO_3$ solutions. The salt was filtered and washed with water to remove traces of BrO_3^- ion. The final product was recrystallized from 50% ethyl alcohol-water solution and vacuum dried at room temperature. The infrared spectrum of the final product showed no BrO_3^- ion absorbancies.

Infrared spectra of the solids were obtained with KBr pressed pellet techniques and a Beckman IR-12 grating infrared spectrophotometer operated in the double-beam mode with dry air in the reference beam. Raman spectra were recorded on a modified Cary Model 81 spectrophotometer equipped with a Spectra-Physics 125 helium-neon laser emitting *ca.* 65 mW of radiation at 6328 Å.

(4) C. Ramasastry and Y. V. G. S. Murti, *Indian J. Pure Appl. Phys.*, **2**, 35 (1964).

(5) G. E. Boyd and Q. V. Larson, *J. Phys. Chem.*, **68**, 2627 (1964).

(6) G. E. Boyd and Q. V. Larson, *ibid.*, **69**, 1413 (1965).

(7) E. H. Appelman, *J. Am. Chem. Soc.*, **90**, 1900 (1968).

(8) G. E. Boyd and T. G. Ward, Jr., *J. Phys. Chem.*, **68**, 3809 (1964).

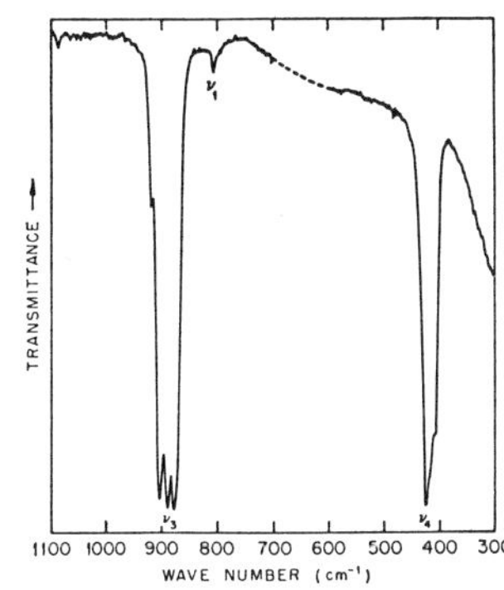

Figure 1. Infrared spectrum of cyrstalline $RbBrO_4$ (KBr pellet; grating change indicated by dashed line).

Polarization measurements were made with a half-wave, retardation-plate polarizer mounted in the laser beam.

Anion-Exchange Column Chromatography. An extension of the anion-exchange chromatographic separation technique employed previously[9] to separate anionic radiobromine species in alkaline solutions was used to separate and identify BrO_4^- ion. It had been shown that BrO_2^-, BrO_3^-, and Br^- ions can be eluted quantitatively from Dowex-1 anion exchanger with aqueous 0.5 *M* HCO_3^-, 1.0 *M* HCO_3^-, and 1.0 *M* NO_3^- solutions, respectively, and that BrO^- ion reacts and remains irreversibly bound at the top of the exchanger bed.

Preliminary experiments to determine the chromatographic behavior of BrO_4^- on Dowex-1 resin were performed with 0.25-ml aliquots of 0.5 *M* solutions of ^{82}Br tagged bromate ion which had been oxidized with 200-mg portions of XeF_2. Bromine-82 labeled BrO_3^- was prepared by neutron irradiation of $^7LiBrO_3$ followed by a waiting period of *ca.* 48 hr to allow for the decay of the short-lived bromine isotopes, *i.e.*, ^{80m}Br ($t_{1/2}$ = 4.38 hr), ^{80}Br ($t_{1/2}$ = 17.6 min), and ^{82m}Br ($t_{1/2}$ = 6.1 min). The XeF_2-oxidized $^{82}BrO_3^-$ solutions were neutralized with $NaHCO_3$, an aliquot was transferred to the top of a cylindrical bed of 200–400 mesh Dowex 1-X10 anion-exchange resin (HCO_3^- form), and the elution sequence described above was followed. After sufficient 1 *M* NO_3^- solution had been passed to reduce the activity of the effluent from the column to the background level, thereby ensuring complete removal of Br^- ion, the eluting agent was changed to freshly prepared 1.0 *M* $NaBF_4$ solution and radioperbromate ion was desorbed. The elution of BrO^- from the top of a resin column by 1.0 *M* $NaBF_4$ was negligible. Chromatograms of XeF_2-oxidized, neutron-irradiated $^7LiBrO_3$ (A) and of irradiated $^7LiBrO_3$ dissolved in 0.1 *M* $NaHCO_3$ (B) are shown in Figure 2.

The ion-exchange column separations were complemented by coprecipitation experiments designed to determine quantitatively the oxidation state of the ^{82}Br species in the XeF_2-oxidized $^7LiBrO_3$ solutions. Bromate ion was separated from solution by precipitation of slightly soluble $AgBrO_3$ at 0°, and the procedure was repeated until an inactive precipitate was obtained. Perbromate ion in the supernatant liquid was determined either by precipitation of $(C_6H_5)_4AsBrO_4$ or by direct counting of an aliquot. Initially ClO_4^- ion was added as a "stand-in" carrier for BrO_4^- before the first $AgBrO_3$ precipitation. After the BrO_3^- ion had been precipitated, $(C_6H_5)_4AsCl$ was added and BrO_4^- ion was coprecipitated with tetraphenylarsonium perchlorate. The direct counting method appeared to be feasible after it had been shown by the chromatographic separations that only BrO_3^- and BrO_4^- ions remained after the XeF_2 oxidation (*cf.* Figure 2). Once bromate ion had been removed by precipitation, an aliquot of the supernatant solution was counted and the amount of perbromate ion was calculated.

Bromine-82 radioactivity was determined by γ counting with a 3 in. × 3 in. NaI (Tl activated) well crystal connected with a single-channel analyzer and scaling system. Empirically determined coincidence corrections were applied when the counting loss exceeded 1%.

(9) G. E. Boyd and Q. V. Larson, *J. Am. Chem. Soc.*, **90**, 254 (1968).

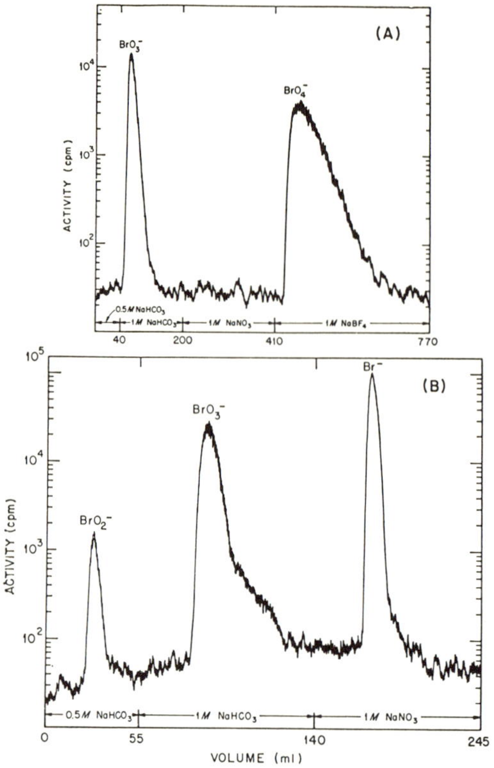

Figure 2. Anion-exchange chromatographic separation of radioperbromate ion in neutron-irradiated $^7LiBrO_3$: (A) irradiated crystals dissolved in H_2O and oxidized with XeF_2; (B) irradiated crysta ls dissolved in 0.1 *M* $KHCO_3$.

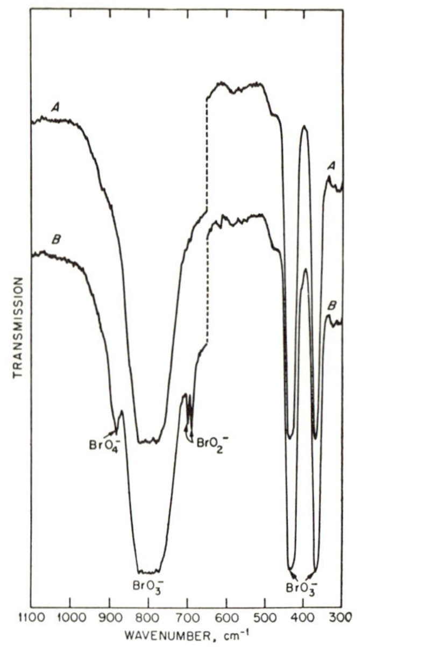

Figure 3. Infrared spectra of crystalline $CsBrO_3$: (A) unirradiated $CsBrO_3$; (B) ^{60}Co γ-ray irradiated $CsBrO_3$, dose = 2.9×10^{24} eV (mole of $CsBrO_3)^{-1}$ (KBr pellets; grating change indicated by dashed line).

Radiolysis of Crystalline $CsBrO_3$. Cesium bromate appeared to be the best compound for detecting radiolytically produced BrO_4^- ion as it is the most easily radiolyzed of all the alkali-metal bromates, and therefore the highest concentration of radiolytic products would be produced. Infrared spectra of ^{60}Co γ-ray irradiated $CsBrO_3$ were obtained with crystals irradiated in a constant geometry ^{60}Co source (equilibrium temperature *ca.* 30°) where the dose rate was 1.89×10^{20} eV min^{-1} (mole of $CsBrO_3)^{-1}$. A sample exposed to a dose of 2.9×10^{24} eV (mole of $CsBrO_3)^{-1}$ was examined. Potassium bromide disks containing *ca.* 2.4% by weight of unirradiated or of irradiated salt were prepared, and infrared spectra were recorded from 300 to 1100 cm^{-1}. A portion of the irradiated salt was heated for 1 hr at 250° before infrared analysis. The spectra of the thermally annealed radiolyzed $CsBrO_3$ and of the unirradiated salt were identical.

The chromatographic detection of BrO_4^- ion in ^{60}Co γ-ray irradiated $CsBrO_3$ was accomplished with ^{82}Br-tagged $CsBrO_3$ samples. Labeled $CsBrO_3$ was prepared from a 25-mg sample of $^7LiBrO_3$ neutron-irradiated as described above. The radioactive $^7LiBrO_3$ was heated at 255° for 1 hr to convert >98% of the bromine-82 to $^{82}BrO_3^-$ ion,[9] and then dissolved in a solution containing 500 mg of inactive $CsBrO_3$. The ^{82}Br-labeled $CsBrO_3$ was recrystallized from hot aqueous solution, air-dried under suction for 30 min, and oven-dried at 140° for 90 min. A 200-mg portion was irradiated in a high-intensity, constant-geometry ^{60}Co γ-ray source (equilibrium temperature *ca.* 30°) where the dose rate was 8.0×10^{20} eV min^{-1} (mole of $CsBrO_3)^{-1}$. A 90-mg portion of the irradiated salt was dissolved in 0.1 *N* arsenous acid–0.1 *M* $NaHCO_3$ solution, and an ion-exchange column separation was performed. A 60-mg portion of the preparation which had not been exposed to ^{60}Co γ rays also was separated, and the oxidation state of the ^{82}Br in it was inferred from the volume at which radioactivity appeared in the eluate from the ion-exchange column.

Search for Perbromate Ion in Neutron-Irradiated $^7LiBrO_3$. Lithium-7 bromate was selected for neutron bombardment in a search for $^{82}BrO_4^-$ ion possibly formed *via* "hot-atom" reactions. Samples (25 mg) of highly irradiated $^7LiBrO_3$ [nvt (thermal) – 1.8 to 5.4×10^{16} neutrons/cm^2] were dissolved in 1-ml aliquots of either 0.1 *N* $NaHCO_3$ or 0.1 *N* arsenous acid–0.1 *N* $NaHCO_3$ and chromatographed as before. Dissolution in the latter medium reduced $^{82}BrO^-$ and $^{82}BrO_2^-$ ions to $^{82}Br^-$ ion and thus allowed the detection of smaller amounts of $^{82}BrO_4^-$ ion. Separate experiments indicated that the reduction of BrO_4^- ion by 0.1 *N* arsenous acid–0.1 *N* $NaHCO_3$ was negligible during the required handling time ($\leq$10 min).

Experimental Results and Discussion

Formation of Perbromate and Bromite Ions in Radiolyzed $CsBrO_3$ Crystals. The infrared spectra of unirradiated and ^{60}Co γ-ray irradiated $CsBrO_3$ [total absorbed dose = 2.9×10^{24} eV (mole of $CsBrO_3)^{-1}$] are shown in Figure 3. The peak observed at 883 cm^{-1} was attributed to the ν_3 vibration of BrO_4^- (see Figure 1 and Table I), whereas the two peaks at 688 and 698 cm^{-1} were assigned to BrO_2^-. The latter frequencies are in approximate agreement with those observed by Evans and Lo[10] for BrO_2^- in aqueous solution (709 and 680 cm^{-1}) by Raman spectroscopy.

The relative order and spacing of the BrO_4^-:BrO_3^-:BrO_2^- bands in Figure 3 are consistent with those for the analogous chlorine oxyanions in ^{60}Co γ-ray irradiated $KClO_3$,[3b] and this lends additional credence to the above band identifications.

Anion-exchange column chromatograms of ^{82}Br-tagged unirradiated and ^{60}Co γ-ray irradiated $CsBrO_3$ [total absorbed dose = 3.2×10^{24} eV (mole of $CsBrO_3)^{-1}$] are shown in Figures 4A and 4B, respectively. The chromatogram for the unirradiated salt illustrates

(10) J. C. Evans and G. Y. S. Lo, *Inorg. Chem.*, **6**, 1483 (1967).

Table I. Vibrational Frequencies of BrO_4^- (cm^{-1})[a]

Compound	$\nu_1(A_1)$	$\nu_2(E)$	$\nu_3(F_2)$	$\nu_4(F_2)$
$RbBrO_4$	802 R	335 R	872, 890 R 879, 890, 904 IR	415 R 410, 420, 426 IR
$RbBrO_4$(aq)	801 R, P	331 R	874 R	411 R
$(C_6H_5)_4AsBrO_4$[b]	799 R	. . .	878 IR	410 IR

[a] R = Raman, IR = infrared, P = polarized, aq = aqueous. [b] The ν_3 and ν_4 bands of $(C_6H_5)_4AsBrO_4$ showed no splitting indicating little or no perturbation of the BrO_4^- ion's T_d symmetry in this compound.

that the initial oxidation state of the ^{82}Br activity was pentavalent, *i.e.*, >99.9% BrO_3^- ion, and that self-radiolysis of the radioactive salt produced no detectable amounts of other bromine oxidation states. The chromatographic separation and identification of BrO_4^- ion in ^{60}Co γ-ray irradiated $CsBrO_3$ are shown in Figure 4B. Analysis of the various chromatographic fractions, Br^-, BrO_3^-, and BrO_4^-, yielded a ^{82}Br activity distribution of 6.4, 92.7, and 0.86%, respectively. The Br^- ion yield, of course, contained contributions from all arsenous acid reducible species, *e.g.*, BrO^- and BrO_2^-. The following "100-eV yields" (*i.e.*, *G* values, molecules/100 eV) were calculated from these data for an exposure of 3.2×10^{24} eV (mole of $CsBrO_3$)$^{-1}$: $G(Br^- + BrO^- + BrO_2^-) = 1.21$; $G(-BrO_3^-) = 1.37$; and $G(BrO_4^-) = 0.16$.

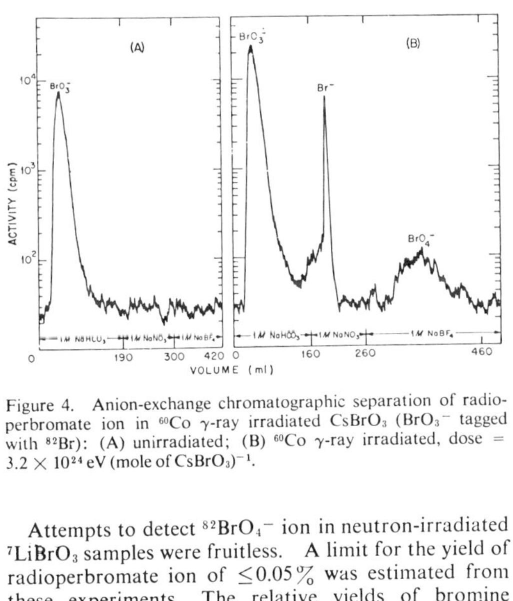

Figure 4. Anion-exchange chromatographic separation of radioperbromate ion in ^{60}Co γ-ray irradiated $CsBrO_3$ (BrO_3^- tagged with ^{82}Br): (A) unirradiated; (B) ^{60}Co γ-ray irradiated, dose = 3.2×10^{24} eV (mole of $CsBrO_3$)$^{-1}$.

Attempts to detect $^{82}BrO_4^-$ ion in neutron-irradiated $^7LiBrO_3$ samples were fruitless. A limit for the yield of radioperbromate ion of ≤0.05% was estimated from these experiments. The relative yields of bromine species for different neutron irradiation times were in good agreement with earlier work.[9]

The Vibrational Spectrum of Perbromate Ion. Raman spectra of crystalline $RbBrO_4$ (B) and of a 0.1 *M* $RbBrO_4$–0.6 *M* HBr aqueous solution (A) are shown in Figure 5. These spectra reveal the characteristic tetrahedral ion pattern analogous to that for ClO_4^- and IO_4^-. Four Raman bands were observed, two of which coincided with two infrared absorptions (*cf.* Figure 1 and Table I). Vibrational frequencies and their assignments, made by analogy with ClO_4^- and IO_4^-, are listed in Table I. The Raman polarization measurements on the aqueous solution showed the ν_1 vibration to be strongly polarized and the other lines to be highly depolarized. Crystalline $RbBrO_4$ exhibits a weak ν_1 absorption in the infrared spectrum and a splitting of the ν_3 and ν_4 degenerate frequencies in both the infrared (Figure 1) and Raman spectra (Figure 5). This splitting is attributed to a lowering of the T_d symmetry of the BrO_4^- ion in the crystal. The shoulder on the high-frequency side of ν_3 in the infrared spectrum (Figure 1) may be caused by a combination band involving a lattice vibration. Similar effects in ClO_4^- ion have been summarized by Hezel and Ross.[11]

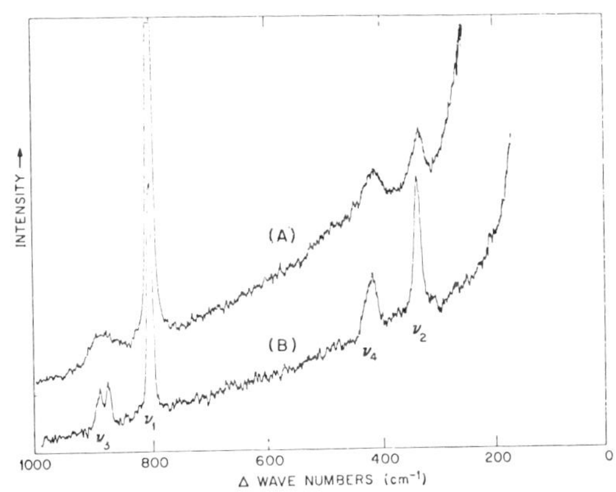

Figure 5. Raman spectra of perbromate ion. (A) 0.1 *M* $RbBrO_4$–0.6 *M* HBr aqueous solution; (B) crystalline $RbBrO_4$.

Venkateswarlu and Sundaram[12] have computed simple force constants for ClO_4^- and IO_4^- ions. Following a similar method which employs the symmetry coordinates of Meister and Cleveland,[13] valence force constants for ClO_4^-, BrO_4^-, and IO_4^- have been calculated with the Wilson *F* and *G* matrix method.[14] The following BrO_4^- ion frequencies were taken from Table I for the calculation: ν_1 801, ν_2 331, ν_3 878, and ν_4 410 cm^{-1}. These values are in good agreement with frequencies obtained recently by Appelman.[15] The Br–O bond distance was assumed to be the same as in $KBrO_3$ (1.68 Å)[15a,16] based on the systematics of the X–O bond-

(11) A. Hezel and S. D. Ross, *Spectrochim. Acta*, **22**, 1949 (1966).
(12) K. Venkateswarlu and S. Sundaram, *J. Chem. Phys.*, **23**, 2365 (1955).
(13) A. G. Meister and F. F. Cleveland, *Am. J. Phys.*, **14**, 13 (1946).
(14) E. B. Wilson, Jr., J. C. Decius, and P. C. Cross, "Molecular Vibrations," McGraw-Hill Book Co., Inc., New York, N. Y., 1955, p 61 ff.
(15) E. H. Appelman, *Inorg. Chem.*, **8**, 223 (1969).
(15a) Note Added in Proof. The force constants reported here have been recalculated and are reported using a value of 1.61 Å for the Br–O bond distance as reported for $KBrO_4$ by S. Siegel, B. Tani, and E. Appelman, private communication; submitted for publication.

length ratios for the halate and perhalate ions, *e.g.*, $ClO_3^-:ClO_4^-$ (1.48 Å:1.48 Å)[16] and $IO_3^-:IO_4^-$ (1.82 Å:1.79 Å).[16] Perchlorate[11,17] and periodate[18] frequencies were selected from the literature and are listed in Table II. It is interesting to note the close similarity between the BrO_4^- ion frequencies and those of the isoelectronic SeO_4^{2-} ion. For SeO_4^{2-} in aqueous solution, the Raman values are ν_1 835, ν_2 342, ν_3 875, and ν_4 415 cm^{-1}.[19]

Table II. Vibrational Frequencies of Perhalate Ions (cm^{-1})

Ion	$\nu_1(A_1)$	$\nu_2(E)$	$\nu_3(F_2)$	$\nu_4(F_2)$
ClO_4^-	935	460	1110	630
BrO_4^-	801	331	878	410
IO_4^-	791	256	853	325

It was possible to fit the frequencies closely by various combinations of valence force constants, and one of the simplest sets is given in Table III. The nonzero force constants used were: f_d, stretching; f_α, bending; $f_{\alpha\alpha'}$, bend–bend interaction between opposite angles; $f_{d\alpha}$ and $f_{d\alpha'}$, stretch–bend interactions. The force constants clearly illustrate the close similarity in bonding between BrO_4^- and IO_4^- ions and the considerable decrease in nonbonding interactions as the central atom increases in size and the oxygens become less crowded.

Table III. Valence Force Constants for ClO_4^-, BrO_4^-, and IO_4^- Ions (mdyn/Å)

Ion	f_d	f_α/d^2	$f_{\alpha\alpha'}/d^2$	$(f_{d\alpha} - f_{d\alpha'})/d$
ClO_4^-	8.24	0.87	−0.21	0.78
BrO_4^-	6.05	0.48	−0.12	0.38
IO_4^-	5.90	0.30	−0.09	0.07

Thermal Stability and Radiolytic Formation of the Perhalates. The similarity in bonding between BrO_4^- and IO_4^- appears to be reflected in some of the properties of these ions when they are present in a crystal lattice. Thus, on heating, $KClO_4$ decomposes at its melting point of 588°[20] to give KCl and $O_2(g)$, whereas $KBrO_4$ decomposes to $KBrO_3$ and $^1/_2O_2(g)$ at 274°[20a] and KIO_4 to KIO_3 and $^1/_2O_2(g)$ at 285°.[21] Crystalline $KBrO_3$ and KIO_3 do not decompose until their fusion temperatures of 396[22] and 560°,[21] respectively, are exceeded; hence, neither $KBrO_4$ nor KIO_4 would be expected as products of halate ion thermolysis. The negative findings in recent thermal decomposition studies on alkali-metal, alkaline-earth, rare earth, and other crystalline bromates[23] appear to confirm this conclusion. In contrast, ClO_4^- ion is formed during the pyrolysis of the alkali-metal chlorates.[24]

Perchlorate ion also is formed in surprisingly good yields in crystalline $NaClO_3$,[25] $KClO_3$,[3b,25] and $Ba(ClO_3)_2$[25] by ^{60}Co γ rays. The amounts of BrO_4^- and IO_4^- produced in $CsBrO_3$ and KIO_3, respectively, by γ rays are relatively much smaller than is the case for ClO_4^-. This difference may result either from the greater intrinsic radiolytic stability of BrO_3^- and IO_3^- ions or because BrO_4^- and IO_4^- are themselves much more easily radiolyzed than ClO_4^- ion. Preliminary measurements by ourselves[21] and others[26] have in fact shown that IO_3^- ion in KIO_3 is exceptionally stable, while IO_4^- ion in KIO_4 is extensively radiolyzed to KIO_3 and oxygen gas.[21] It seems possible that the balance between the rates of radiolytic formation and destruction of the perhalate ions may determine whether or not detectable amounts of them may accumulate in their respective halate crystals.

(16) L. E. Sutton, Ed., "Tables of Interatomic Distances and Configuration in Molecules and Ions," The Chemical Society, London: (a) Special Publication No. 11, 1958; (b) Special Publication No. 18, 1965.

(17) O. Redlich, E. K. Holt, and J. Bigeleisen, *J. Am. Chem. Soc.*, **66**, 13 (1944).

(18) H. Siebert, *Z. Anorg. Allgem. Chem.*, **273**, 21 (1953).

(19) A. S. Ganesan, *Proc. Indian Acad. Sci.*, **A1**, 156 (1935).

(20) M. M. Markowitz, D. A. Boryta, and R. F. Harris, *J. Phys. Chem.*, **65**, 261 (1961).

(20a) NOTE ADDED IN PROOF. DTA and TGA of recrystallized $KBrO_4$ (original $KBrO_4$ solution obtained from E. H. Appelman, Argonne National Laboratory) indicated a gradual exothermic decomposition of BrO_4^- ion over the temperature range 274–308° (heating rate = 0.5°/min). A second, sharp exothermic decomposition, attributable to the decomposition of $KBrO_3$, was observed over the range 391–396°. The observed weight losses for these decompositions agreed well (within ±2%) with the calculated weight losses for the formation of $KBrO_3$ and KBr, respectively.

(21) G. E. Boyd and Q. V. Larson, unpublished research, 1968.

(22) G. E. Boyd, E. W. Graham, and Q. V. Larson, *J. Phys. Chem.*, **66**, 300 (1962).

(23) G. M. Bancroft and H. D. Gesser, *J. Inorg. Nucl. Chem.*, **27**, 1545 (1965).

(24) M. M. Markowitz, D. A. Boryta, and H. Stewart, Jr., *J. Phys. Chem.*, **68**, 2282 (1964).

(25) C. E. Burchill, P. F. Patrick, and K. J. McCallum, *ibid.*, **71**, 4560 (1967).

(26) U. Bertocci, R. B. Jacobi, and G. N. Walton, Symposium on Chemical Effects of Nuclear Transformations, Vol. II, IAEA, 1961, p 337.

25

Reprinted from *Anal. Chem.*, 37(1), 106–111 (1965)

Ion Exchange Selectivity Scales for Cations in Nitric Acid and Sulfuric Acid Media with a Sulfonated Polystyrene Resin

F. W. E. STRELOW, RUTHILD RETHEMEYER, and C. J. C. BOTHMA

National Chemical Research Laboratory, South African Council for Scientific and Industrial Research, Pretoria, South Africa

▶ Cation exchange equilibrium distribution coefficients with BIO-RAD AG 50W-X8, 100- to 200-mesh, a sulfonated polystyrene resin, are presented for 49 cations in nitric acid and 45 cations in sulfuric acid media. The coefficients were determined at a total amount of cation to total resin capacity ratio *q* = 0.4, to ensure comparability with coefficients in hydrochloric acid determined previously. Acid concentrations from 0.1*N* to 4.0*N* were used. The coefficients are arbitrarily arranged into selectivity scales according to their values in 1*N* acid. Included are elution curves for the multicomponent systems V(V)–U(VI)–Sc(III)–Y(III); Na(I)–Be(II)–Ba(II)–Y(III)–Zr(IV); and Hg(II)–Cd(II)–Be(II)–Fe(III)–Ba(II)–Zr(IV) to demonstrate some possibilities inherent in cation exchange separation.

Systematically compiled data on ion exchange equilibrium distribution coefficients are very useful to the analytical chemist for planning separations. This has been generally recognized since Kraus (*6*) and his coworkers presented their pioneer work on the anion exchange of metal chloride complexes. Nevertheless, information of this kind is still fairly limited in the literature. The situation is worse for cation exchange than for anion exchange. Only the hydrochloric acid system (*7*, *9*) seems to have been investigated fairly thoroughly. Nelson (*8*) and coworkers have studied the HCl and $HClO_4$ systems at high acid concentrations. Information of a more limited nature is available for cations in hydrobromic acid (*1*), hydrofluoric acid (*2*), thiosulfate (*5*), and EDTA media (*4*), as well as for hydrochloric acid containing acetone as an organic constituent (*3*).

The senior author (*14*) published

coefficients for 23 cations in nitric acid and nine cations in sulfuric acid. No further compiled data seem to be available for these two systems. This paper therefore presents a fairly comprehensive list of equilibrium distribution coefficients of cations in nitric and sulfuric acid media in the concentration range 0.1*N* to 4.0*N* using BIO-RAD AG 50W-X8 sulfonated polystyrene resin. The coefficients were determined at a ratio of total amount of cation to total resin capacity, $q = 0.4$, to ensure comparability with coefficients in hydrochloric acid determined previously (*9*). Some elution curves for multicomponent systems are presented to demonstrate separation possibilities.

EXPERIMENTAL

Reagents and Apparatus. Analytical reagent grade chemicals were used whenever available. Rare earth oxides and some platinum metals were obtained from L. Light and Co., Ltd., Colnbrook, England, and some platinum metal salts from Fluka A.G., Buchs SG, Switzerland. Hafnium and zirconium oxides were obtained from Johnson, Matthey and Co., Ltd., London.

Solutions of V(V), As(III), and Se(IV) were prepared from the corresponding ammonium or sodium salts by separating the ammonium or sodium ion using cation exchange chromatography. These solutions were prepared in distilled water. Niobium and tantalum solutions were prepared from the respective oxides via fusion with potassium hydroxide. The solutions contained a small excess of potassium hydroxide and were acidified shortly before the equilibrium experiments were started. Solutions of W(VI), Au(III), Pt(IV), and Sb(III) in H_2SO_4 and in HNO_3 were unstable and formed precipitates.

The resin used was the AG 50W-X8 sulfonate polystyrene of 100- to 200-mesh particle size, manufactured by the BIO-RAD Laboratories at Richmond, Calif. Unless stated to the contrary, the resin was used in the hydrogen form.

Borosilicate glass tubes of 50-cm. length and 2-cm. diameter with fused-in glass sinters of No. 2 porosity and a buret tap at the bottom were used as columns. The tops were fitted with ground-glass joints so that dropping funnels of convenient sizes could be attached. Unless stated otherwise, the columns were loaded with 20.0 ± 0.1 grams (dry weight) of the resin.

Equilibrium Distribution Coefficients. An experimental procedure was followed as described for the determination of distribution coefficients in hydrochloric acid (*9*). As an improvement, the resin was dried at 60° C. in a Towers vacuum drying apparatus using P_2O_5 as drying agent. Generally, the determinations were conducted using 2.500 grams of dry resin, 5.00 meq. of the cation, and a total solution volume of 250 ml., but, for some of the rare earths and the platinum metals, the procedure was reduced in scale and 0.500 gram of dry resin, 1.00 meq., and a total solution volume of 50 ml. were used. The coefficients are presented in Tables I and II.

Variation of Distribution Coefficients with Variation of Ratio *q*. Coefficients for a number of ions were determined under similar conditions varying only the ratio *q*. The total solution volume was kept at 250 ml. and the amount of resin was 2.500 grams (dry weight). The amounts of cation taken were 5.00 meq. ($q = 0.4$), 1.25 meq. ($q = 0.1$), 0.125 meq. ($q = 0.01$), and 0.0125 meq. ($q = 0.001$). Procedures were as described before, but the distributions of zinc and cobalt were determined radiometrically using zinc-65 and cobalt-60 as tracers. The radio isotopes were supplied by the Radiochemical Centre, Amersham, Buckinghamshire. The results are presented in Table III.

Table I. *Kd* Values in HNO_3

Cation	0.1*N*	0.2*N*	0.5*N*	1.0*N*	2.0*N*	3.0*N*	4.0*N*
Zr(IV)[a]	$>10^4$	$>10^4$	$>10^4$	6500	652	112	30.7
Hf(IV)	$>10^4$	$>10^4$	$>10^4$	2400	166	61	20.8
Th(IV)	$>10^4$	$>10^4$	$>10^4$	1180	123	43.0	24.8
La(III)[a]	$>10^4$	$>10^4$	1870	267	47.3	17.1	9.1
Ce(III)[a]	$>10^4$	$>10^4$	1840	246	44.2	15.4	8.2
Yb(III)	$>10^4$	$>10^4$	1150	193	41.3	16.0	9.0
Er(III)	$>10^4$	$>10^4$	1100	182	38.2	14.9	8.0
Y(III)[a]	$>10^4$	$>10^4$	1020	174	35.8	13.9	10.0
Sm(III)	$>10^4$	$>10^4$	1000	168	29.8	10.9	7.2
Gd(III)	$>10^4$	$>10^4$	1000	167	29.2	10.8	6.9
In(III)	$>10^4$	$>10^4$	680	118	23.0	10.1	5.8
Sc(III)	$>10^4$	3300	500	116	23.3	11.6	7.6
Cr(III)	5100	1620	418	112	27.8	19.2	10.9
Hg(I)[a]	$>10^4$	7600	640	94	33.5	19.2	13.6
Ga(III)	$>10^4$	4200	445	94	20.0	9.0	5.8
Al(III)[a]	$>10^4$	3900	392	79	16.5	8.0	5.4
Fe(III)[a]	$>10^4$	4100	362	74	14.3	6.2	3.1
Ba(II)[a]	5000	1560	271	68	13.0	6.0	3.6
Sr(II)[a]	3100	775	146	39.2	8.8	6.1	4.7
Pb(II)[a]	$>10^4$	1420	183	35.7	8.5	5.5	4.5
Ca(II)[a]	1450	480	113	35.3	9.7	4.3	1.8
Cd(II)[a]	1500	392	91	32.8	10.8	6.8	3.4
Co(II)[a]	1260	392	91	28.8	10.1	6.1	4.7
Mn(II)[a]	1240	389	89	28.4	11.4	7.1	3.0
Ni(II)[a]	1140	384	91	28.1	10.3	8.6	7.3
Cu(II)[a]	1080	356	84	26.8	8.6	4.8	3.1
Zn(II)[a]	1020	352	83	25.2	7.5	4.6	3.6
Bi(III)	893	305	79	25.0	7.9	3.7	3.0
U(VI)	659	262	69	24.4	10.7	7.4	6.6
Mg(II)[a]	794	295	71	22.9	9.1	5.8	4.1
Tl(I)[a]	173	91	41.0	22.3	9.9	5.8	3.3
Ag(I)[a]	156	86	36.0	18.1	7.9	5.4	4.0
Hg(II)[a]	4700	1090	121	16.9	5.9	3.9	2.8
Cs(I)	148	81	34.8	16.8	7.6	4.7	3.4
Be(II)[a]	553	183	52	14.8	6.6	4.5	3.1
Ti(IV)[a]	1410	461	71	14.6	6.5	4.5	3.4
V(IV)	495	157	35.6	14.0	4.7	3.0	2.5
Rb(I)	118	68	29.1	13.4	6.6	4.1	2.9
K(I)	99	59	26.2	11.4	5.7	3.5	2.6
Te(IV)[b]	40.3	19.7	8.5	5.0	2.4	0.6	0.2
Pd(II)	97	62	23.5	9.1	3.4	2.7	2.5
Rh(III)	78	44.7	19.5	7.8	4.1	2.1	1.0
Na(I)	54	29.4	12.7	6.3	3.4	2.0	1.3
Li(I)	33.1	18.6	8.0	3.9	2.6	1.7	1.1
V(V)	20.0	10.9	4.9	2.0	1.2	0.8	0.5
Mo(VI)	Ppt.	5.2	2.9	1.6	1.0	0.8	0.6
Nb(V)	11.6	6.3	0.9	0.2	0.1	0.1	0.1
Se(IV)	<0.5	<0.5	<0.5	<0.5	<0.5	<0.5	<0.5
As(III)	<0.1	<0.1	<0.1	<0.1	<0.1	<0.1	<0.1

[a] Coefficients published previously but included to provide complete list.
[b] Determined at $q = 0.2$ because of limited solubility at low acid concentrations.

Elution Curves. Three multicomponent elution curves demonstrate the versatility of cation exchange procedures and show how the information presented in the tables can be applied to analytical separation. Up to six cations can be separated in a single column procedure simply by stepwise changing eluents. The eluent concentrations for the following separations were chosen directly from Tables I and II and from a table for coefficients in HCl published previously (*9*). Intermediate coefficients were calculated from plots of the determined coefficients against acid concentration. The relevant coefficients for one of the elution curves are summed up in Table IV.

V(V), U(VI), Sc(III), and Y(III). A solution containing 15.1 mg. of Sc(III), 29.4 mg. of Y(III), 62.3 mg. of V(V), and 128.3 mg. of U(VI) in 25 ml. of 0.5*N* H_2SO_4 plus 2 drops of 30% H_2O_2 was percolated through a

20-gram resin column and the cations were washed onto the resin using a minimum volume of 0.5N H_2SO_4. The cations were then eluted using the following elution sequence: 200 ml. of 0.50N H_2SO_4 for V(V); 350 ml. of 1.00N H_2SO_4 for U(VI); 400 ml. of 2.00N H_2SO_4 for Sc(III); and 500 ml. of 4.00N HCl for yttrium. Before changing eluents, the solution was allowed to drain to the level of the resin bed. The flow rate was kept at 3.0 ± 0.2 ml. per minute. Aliquots of 25 ml. were taken using an automatic Aimer Central fractionator, and the amounts of the cations in the fractions were determined by an appropriate analytical procedure. Because the vanadium is only very weakly absorbed by the resin from 0.5N H_2SO_4, the fractions were taken from the beginning of the adsorption step. The experimental elution curve is presented in Figure 1.

Na(I), Be(II), Ba(II), Y(III), and Zr(IV). A solution containing 23.1 mg. of Na(I), 4.64 mg. of Be(II), 68.9 mg. of Ba(II), and 46.7 mg. of Zr(IV) in 50 ml. of 0.2N HNO_3 was percolated through a 20-gram resin column and the cations were washed onto the resin with 0.1N HNO_3. The following elution sequence was employed: 350 ml. of 0.60N HNO_3 for Na(I); 350 ml. of 1.20N HNO_3 for Be(II), 350 ml. of 2.00N HNO_3 for Ba(II), 300 ml. of 3.00N HNO_3 for Y(III), and 350 ml. of 5.00N HCl for Zr(IV). The yttrium was eluted at a flow rate of 1.3 ± 0.2 ml. per minute. A flow rate of 3.0 ± 0.2 ml. per minute was used for the other cations. The experimental elution curve is presented in Figure 2.

Hg(I), Cd(II), Be(II), Fe(III), Ba(II), and Zr(IV). Another experiment was conducted using a solution containing 99.9 mg. of Hg(II), 57.4 mg. of Cd(II), 4.64 mg. of Be(II), 18.8 mg. of Fe(III), 68.9 mg. of Ba(II), and 46.7 mg. of Zr(IV) in 50 ml. of 0.2N HCl. The elution sequence in this case was: 250 ml. of 0.20N HCl for Hg(II), 250 ml. of 0.50N HCl for Cd(II), 350 ml. of 1.20N HNO_3 for Be(II), 300 ml. of 1.75N HCl for Fe(III), 300 ml. of 2.50N HNO_3 for Ba(II), and 450 ml. of 5.00N HCl for Zr(IV). Fractions were taken from the beginning of the absorption step and the flow rate was 3.0 ± 0.2 ml. per minute throughout. The experimental elution curve is given in Figure 3.

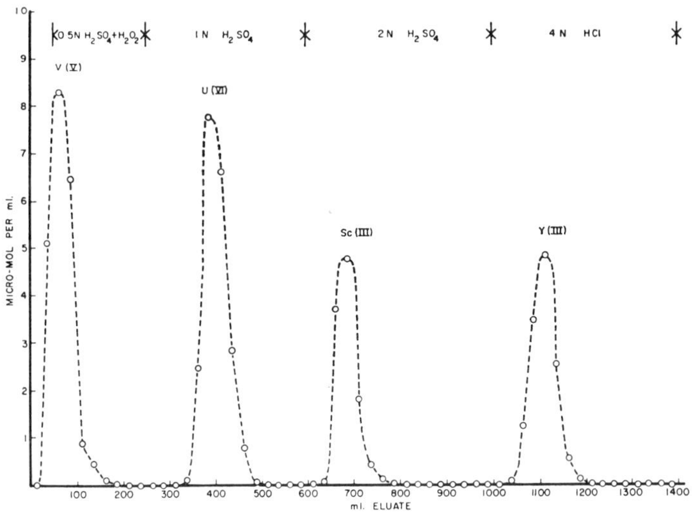

Figure 1. Chromatographic separation of a mixture of V(V), U(VI) Sc(III), and Y(III)

Column of 20.0 grams of dry AG 50W-X8 resin, 100- to 200-mesh. Flow rate 3.0 ± 0.2 ml. per minute

Table II. *Kd* Values in H_2SO_4

Cation	0.1N	0.2N	0.5N	1.0N	2.0N	3.0N	4.0N
La(III)	>10^4	>10^4	1860	329	68	24.3	12.1
Ce(III)	>10^4	>10^4	1800	318	66	23.8	11.8
Sm(III)	>10^4	>10^4	1460	269	56	20.1	10.0
Y(III)	>10^4	>10^4	1380	253	49.9	18.0	9.4
Yb(III)	>10^4	>10^4	1330	249	48.1	17.3	8.8
Gd(III)	>10^4	>10^4	1390	246	46.6	17.9	8.9
Er(III)	>10^4	>10^4	1300	242	48.6	16.7	8.5
Bi(III)[a]	>10^4	>10^4	6800	235	32.3	11.3	6.4
Ga(III)	>10^4	3500	618	137	26.7	10.0	4.9
Al(III)[b]	>10^4	8300	540	126	27.9	10.6	4.7
Hg(II)	7900	1790	321	103	34.7	16.8	12.2
In(III)	>10^4	3190	376	87	17.2	6.5	3.8
Mn(II)[b]	1590	610	165	59	17.4	8.9	5.5
Fe(III)[b]	>10^4	2050	255	58	13.5	4.6	1.8
Cr(III)	198	176	126	55	18.7	0.9	0.2
Th(IV)[b]	>10^4	3900	263	52	9.0	3.0	1.8
Tl(I)	452	236	97	49.7	20.6	11.6	8.7
Tl(III)	6500	1490	205	47.4	12.0	7.2	5.2
V(IV)	1230	490	140	46.6	11.5	2.4	0.4
Ni(II)[b]	1390	590	140	46.0	16.5	6.1	2.8
Fe(II)	1600	560	139	46.0	15.3	9.8	6.6
Cd(II)	1420	540	144	45.6	14.8	6.6	4.3
Zn(II)	1570	550	135	43.2	12.2	4.9	4.0
Co(II)	1170	433	126	42.9	14.2	6.2	5.4
Cu(II)[b]	1310	505	128	41.5	13.2	5.7	3.7
Mg(II)	1300	484	124	41.5	13.0	5.6	3.4
Sc(III)	5600	1050	141	34.9	8.5	4.4	3.4
Be(II)	840	305	79	27.0	8.2	3.9	2.6
Cs(I)	175	108	52	24.7	9.1	4.8	3.5
Rb(I)	148	91	43.8	21.3	8.3	4.4	3.1
K(I)	138	86	41.1	19.4	7.4	3.7	2.9
Rh(III)	80	49.3	28.5	16.2	4.5	2.2	1.3
Pd(II)	109	71	32.5	13.9	6.0	3.8	2.7
Hf(IV)	2690	1240	160	12.1	1.7	1.0	0.7
U(VI)[b]	596	118	29.2	9.6	3.2	2.3	1.8
Ti(IV)[b]	395	225	45.8	9.0	2.5	1.0	0.4
Na(I)	81	47.7	20.1	8.9	3.7	2.6	1.7
Li(I)	48.0	28.2	11.7	5.8	3.0	1.6	1.1
Te(IV)	Ppt.	30.8	9.8	5.2	2.6	0.6	0.3
Zr(IV)[b]	546	474	98	4.6	1.4	1.2	1.0
V(V)	27.1	15.2	6.7	2.8	1.2	0.7	0.4
Nb(V)	14.2	7.4	4.0	1.9	0.7	0.5	0.3
Mo(VI)	Ppt.	5.3	2.8	1.2	0.5	0.3	0.2
Se(IV)	<0.5	<0.5	<0.5	<0.5	<0.5	<0.5	<0.5
As(III)	<0.1	<0.1	<0.1	<0.1	<0.1	<0.1	<0.1

[a] Determined for q = 0.06 because of limited solubility.
[b] Coefficients published previously but included to provide complete list.

DISCUSSION

The experimental distribution coefficients in nitric and sulfuric acid are presented in two tables, in which the cations are arranged arbitrarily into sequences according to the values of the coefficients in 1N acid. Accumulative experimental errors are estimated to be not higher than about 2 to 3% of the results for coefficients between 10 and 500, but are higher for very high or very low coefficients. The main sources of these errors are inhomogeneity of the

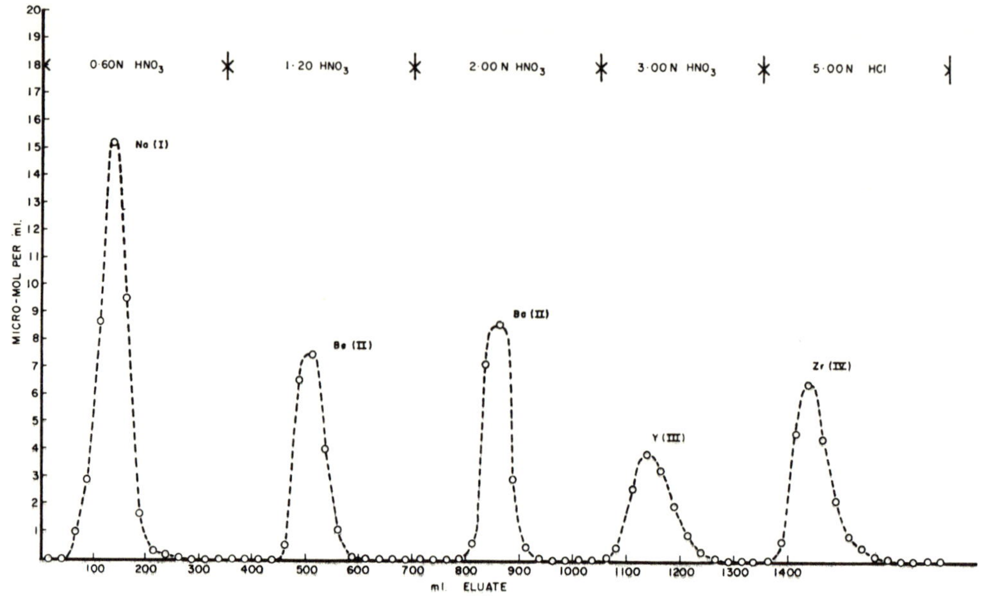

Figure 2. Chromatographic separation of a mixture of Na(I), Be(II), Ba(II), Y(III), and Zr(IV)

Column of 20.0 grams of dry AG 50W-X8 resin, 100- to 200-mesh. Flow rate 1.3 ± 0.2 ml. per minute for Y(III) elution step; 3.0 ± 0.2 ml. per minute for other steps

resin, water content of the resin (the dry resin is a strong desiccant), and analytical errors. Coefficients which show small numerical differences, like Y(III), Yb(III), Gd(III), and Er(III) in 1*N* H_2SO_4 or Co(II), Mn(II), and Ni(II) in 1*N* HNO_3 should be considered as being similar because of the experimental uncertainties.

Generally, distribution coefficients determined at relatively high resin loadings are more dependable than those determined at tracer concentrations, because errors due to secondary effects such as adsorption of the walls of vessels are suppressed. The influence of the factor *q* on the numerical value of the distribution coefficient is shown in Table III. For low coefficients, the values determined at $q = 0.4$ and $q = 0.001$ show negligible differences with regard to the calculation of positions of elution peaks using the equation

$$\bar{v} = Kd \times \text{mass of dry resin in the column} \quad (9)$$

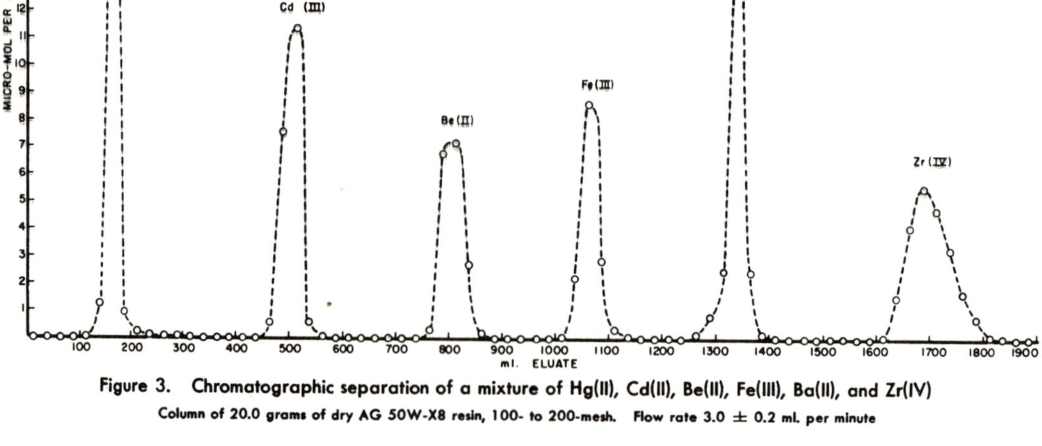

Figure 3. Chromatographic separation of a mixture of Hg(II), Cd(II), Be(II), Fe(III), Ba(II), and Zr(IV)

Column of 20.0 grams of dry AG 50W-X8 resin, 100- to 200-mesh. Flow rate 3.0 ± 0.2 ml. per minute

Table III. Variation in *Kd* with Varying Factor *q*

Cation	Medium	q = 0.4	q = 0.1	q = 0.01	q = 0.001
Cd(II)	0.5*N* HCl	6.5	6.5	6.6	6.6
Al(III)	2.0*N* HCl	12.5	13.4	13.6	13.9
Cu(II)	1.0*N* HCl	17.5	18.9	19.1	19.3
Co(II)	1.0*N* HCl	21.3	23.5	23.8	24
Zn(II)	1.0*N* $HClO_4$	28.8	33.8	35	36
Be(II)	0.5*N* HCl	42.1	51	54	56
Zn(II)	0.5*N* HCl	64	82	88	91
Zn(II)	0.5*N* $HClO_4$	87	135	143	147

Table IV. Distribution Coefficients for Separation of Hg(II)–Cd(II)–Be(II)–Fe(III)–Ba(II)–Zr(IV)

Cation	0.20*N* HCl	0.50*N* HCl	1.20*N* HNO_3	1.75*N* HCl	2.50*N* HNO_3	5.0*N* HCl
Hg(II)	0.9	...	...	...	...	...
Cd(II)	84	6.5	...	...	...	...
Be(II)	117	42	12.7	...	...	...
Fe(III)	3400	225	48	7.1	...	...
Ba(II)	2930	590	44	44	8.4	...
Zr(IV)	>10[4]	>10[4]	4000	930	260	9.5

For higher coefficients, the elution peaks become increasingly flat and wide. In this case, therefore, the coefficient supplies only qualitative information about the appearance of first traces in the eluate and predicted separations have to be confirmed by experimental elution curves.

Distribution Coefficients in HNO_3. Generally, coefficients in HNO_3 are slightly higher than in HCl. Some cations which form relatively stable chloride complexes, such as Hg(II), Bi(III), Cd(II), In(III), Fe(III), and Pd(II), show very substantial differences. Thus, a separation of Fe(III) from the divalent heavy metals Cu(II), Ni(II), Co(II), Zn(II), and Mn(II), which is poor in HCl, becomes feasible in HNO_3 (*11*). Separation of Fe(III) from Ti(IV) or Be(II) is considerably better in HNO_3 than in HCl (*10*). K(I) and Na(I) which accompany Cd(II) and In(III) when these are eluted with 0.5*N* HCl to separate them from Zn(II), Cu(II), Ni(II), Al(III), Fe(III), etc., can be separated from Cd(II) and In(III) in HNO_3 easily.

Ba(II) and, to a lesser degree, Sr(II) and Ca(II) have lower coefficients in HNO_3 than HCl because of ion association. Zirconium(IV), on the other hand, at high acid concentrations is considerably more strongly absorbed from HNO_3 than from HCl. Thus, in HNO_3 it can be separated quantitatively from Ba(II) and Sr(II) as well as from the rare earth elements and Sc(III), a separation which is not satisfactory in HCl. However, zirconium has very slow exchange rates and low flow rates are therefore a necessary condition for a good separation from the rare earths.

The coefficients of Bi(III), Hg(II), and Pb(II) are distinctly lower in HNO_3 than in $HClO_4$ (*16*). This indicates that complex formation or association of these ions occurs in HNO_3, but to a lesser degree than in HCl. A separation of Hg(I) and Hg(II) is feasible in HNO_3.

Cr(III) does not show the normal variation of distribution coefficients with acid concentration, and the experimental values are poorly reproducible. This is probably due to its complex formation characteristics (slow reaction rates for transformation of one complex into another).

While coefficients of the rare earths in HCl decrease steadily in the sequence La(Kd_{1N} = 265), Yb(Kd_{1N} = 153), Y(Kd_{1N} = 115), and Sc(Kd_{1N} = 120), the coefficients in HNO_3 show a minimum at Gd(Kd_{1N} = 167), after which they increase again to Yb(Kd_{1N} = 193). This seems to indicate different patterns of ion association of the rare earths in the two acids.

Distribution Coefficients in H_2SO_4. Coefficients in H_2SO_4 are normally distinctly higher than in either HCl or HNO_3. This is to be expected simply from the mass action law. Sulfuric acid, because of its lower degree of dissociation, will supply a lower concentration of hydrogen ions than the other two acids. On the other hand, a number of cations show marked complex formation or ion association in sulfuric acid, as is indicated by the activity coefficients of their sulfates. This leads to a substantial decrease in the respective distribution coefficients in H_2SO_4, which as a result can even be lower than those in HCl or HNO_3. Cations such as Zr(IV), Th(IV), U(VI), Ti(IV), Sc(III), and to a lesser extent Fe(III), In(III), and Cr(III) show the sulfate complexing effect and therefore take places in the selectivity sequence in H_2SO_4 which are different from those in the other two acids. This opens up possibilities for a number of separations, some of which are specific to this system. Uranium(VI) can be separated from Be(II), Mg(II), Co(II), Cu(II), Mn(II), Fe(II), Fe(III), Al(III), rare earths, Th(IV), and many other elements (*15*), and Ti(IV) can be separated from the same elements, as well as Nb(V), V(V), and Mo(VI) (*12*). Sc(III) can readily be separated from Y(III), La(III), and the other rare earth elements (*13*). V(IV) is most easily separated from V(V) or V(IV) from Mo(VI) and Nb(V) with H_2SO_4 as eluent. Tellurium(IV) forms precipitates in very dilute H_2SO_4 and is distinctly absorbed at slightly higher concentrations. This fact can be used to separate Te(IV) in 0.20*N* H_2SO_4 from Se(IV) and some other cations, which are not absorbed.

The distribution coefficient pattern for Cr(III) is similar to that in nitric acid and the results are poorly reproducible, probably because of the slow reaction rates for the transformation of one complex into another.

Elution Sequences. The information contained in the tables indicates numerous possibilities for separation. It is beyond the scope of this paper to point them out one by one. Intelligent inspection of the relevant *Kd* values will show the analytical chemist whether a separation is feasible and what the most favorable eluent concentration is likely to be. Cation exchange is fully as versatile as anion exchange. This is demonstrated by the multicomponent elution curves presented in Figures 1, 2, and 3. When planning the separation of a multicomponent system, the behavior of each cation in every elution step where it is present on the column has to be considered.

A cation, even if not eluted, may move down the column a considerable distance and with the next eluent may appear in the eluate much earlier than indicated by the coefficient in this eluent alone. By arranging the relevant coefficients for an elution system in a table as has been demonstrated for elution curve 3 in Table IV, a systematic approach is made and fairly accurate predictions about the elution behavior of systems containing a considerable number of different cations become possible.

LITERATURE CITED

(1) Fritz, J. S., Garralda, B. B., Anal. Chem. **34**, 102 (1962).
(2) Fritz, J. S., Garralda, B. B., Karraker, S. K., *Ibid.*, **33**, 882 (1961).
(3) Fritz, J. S., Rettig, T. A., *Ibid.*, **34**, 1562 (1962).
(4) Fritz, J. S., Umbreit, G. R., *Anal. Chim. Acta* **19**, 509 (1958).

(5) Katsura, T., *Bunseki Kagaku* **11,** 769 (1962); *C.A.* **57,** 14415*f* (1962).
(6) Kraus, A., Nelson, F., 1st International Conference on Peaceful Uses of Atomic Energy, Paper 837, 1955.
(7) Mann, C. K., Swanson, C. L., ANAL. CHEM. **33,** 459 (1961).
(8) Nelson, F., Murase, T., Kraus, K. A., *J. Chromatog.* **13,** 503 (1964).
(9) Strelow, F. W. E., ANAL. CHEM. **32,** 1185 (1960).
(10) *Ibid.*, **33,** 542 (1961).
(11) *Ibid.*, p. 994.
(12) *Ibid.*, **35,** 1279 (1963).
(13) *Ibid.*, **36,** 1217 (1964).
(14) Strelow, F. W. E., *J. South African Chem. Inst.* **14,** 51 (1961).
(15) *Ibid.*, **16,** 38 (1963).
(16) Strelow, F. W. E., National Chemical Research Laboratory, Pretoria, unpublished data.

RECEIVED for review July 21, 1964. Accepted September 28, 1964.

26

Reprinted from *Anal. Chim. Acta*, **45**, 81–92 (1969)

DISTRIBUTION COEFFICIENTS AND THE CATION-EXCHANGE BEHAVIOUR OF ELEMENTS IN HYDROCHLORIC ACID–ETHANOL MIXTURES

F. W. E. STRELOW, C. R. VAN ZYL AND C. J. C. BOTHMA

National Chemical Research Laboratory, South African Council for Scientific and Industrial Research, Pretoria (South Africa)

(Received November 4th, 1968)

KEMBER *et al.*[1] probably were the first to employ addition of water-miscible non-aqueous solvents to promote metal–halide complex formation for the selective elution of transition metal ions from cation-exchange resins, while the first systematic survey of cation-exchange distribution coefficients in partly organic solvents has been carried out by FRITZ AND RETTIG[2], who have investigated the exchange behaviour of 14 elements in hydrochloric acid–acetone media. A more comprehensive survey has been undertaken recently by KORKISCH AND AHLUWALIA[3], who determined distribution coefficients for 20 elements in hydrochloric acid containing methanol, ethanol, *n*-propanol, isopropanol, methyl glycol, acetic acid, tetrahydrofuran and acetone as organic solvents. The concentration of hydrochloric acid in these systematic investigations[2,3] has been limited to an upper value of 1.0 and 1.2 M, respectively, and attention has been focussed mainly on elements with differences in their tendencies to form chloride complexes which promised spectacular increases in selectivities. Generally, the effect of increasing amounts of the organic solvent is a reduction of the concentration of water molecules around the cations, a decrease in the forces binding the co-ordinated hydration shell, and a decrease in the size of the outer hydration cloud. One consequence of this is that metal chloride anions are formed at lower concentrations of hydrochloric acid while, generally, the differences in tendencies to chloride-complex formation are enhanced. Although several investigations of the ion-exchange behaviour of small numbers of elements with negligible tendencies to chloride-complex formation have been carried out in partly non-aqueous media[4–11], another consequence of the underlying physicochemical aspects often does not seem to have been fully appreciated. Because the forces binding the hydration cloud depend on the charge density of the cation, they will decrease with an increase in volume for elements in a periodic group. Selective destruction of the hydration cloud will therefore start at a lower organic solvent concentration for the larger cation. As a result, the larger cation (unhydrated size) will enter the resin relatively more easily and differences between distribution coefficients of elements of the same periodic group will be enhanced, provided that other effects are absent. Since the distribution coefficients are measured against H_3O^+, the most strongly hydrated of the cations, they have higher values in partly organic solvents than in aqueous solutions, provided that the effect of chloride-complex formation is negligible. Favourable conditions for separations of elements with negligible tendencies to chloride-complex formation therefore require partly non-aqueous solutions with concentrations of hydrochloric acid which are higher than those at which separations in aqueous solutions are carried out. The range of these concentrations has not been covered in the studies mentioned above[2,3]. Furthermore, only a limited number of elements has been investigated. This paper therefore presents distribution coefficients for 45 elements covering the acid concentration range from 0.1 to 3.0 *M* hydrochloric acid in water–ethanol mixtures, and discusses some of their exchange properties and possibilities for separations.

EXPERIMENTAL

Reagents and apparatus

The resin used was the AG50W-X8 sulphonated polystyrene cation exchanger (100–200 or 200–400 mesh particle size; Bio-Rad Laboratories, Richmond, California). Borosilicate glass tubes of 20 mm i.d. with fused-in glass sinters of No. 2 porosity and a burette tap at the bottom and a B19 ground-glass joint at the top, were used as columns.

Analytical-reagent grade chemicals were used whenever possible. Chlorides of the platinum metals, and of gold, germanium, gallium and indium, and scandium and gadolinium oxides were obtained from Fluka A.G., Buchs, Switzerland. Standard chloride solutions of V(V), Mo(VI), W(VI), As(III) and Se(IV) were prepared by

TABLE I

CATION-EXCHANGE DISTRIBUTION COEFFICIENTS IN 0.10 *M* HCl

Element	*Percentage ethanol*						
	0	*20*	*40*	*60*	*80*	*90*	*95*
Ga(III)	$\sim 10^4$	$>10^4$	$>10^4$	$>10^4$	$>10^4$	6730	153
Sn(IV)	$\sim 10^4$	ppt.	ppt.	ppt.	43.1	3.5	1.1
Fe(III)	9000	$>10^4$	$>10^4$	$>10^4$	$>10^4$	3340	176
U(VI)	758	1300	3200	$>10^4$	$>10^4$	4960	3330
Mn(II)	1360	1580	3020	$>10^4$	$>10^4$	$>10^4$	$>10^4$
Co(II)	1270	1450	2840	6500	$>10^4$	$>10^4$	$>10^4$
Ni(II)	1230	1470	2950	10^4	$>10^4$	$>10^4$	$>10^4$
Fe(II)	1220	1410	2830	6400	$>10^4$	9800	1010
Zn(II)	1030	1200	2150	2250	876	47.5	5.7
Cu(II)	1010	1200	2190	4410	4280	1250	430
Mg(II)	860	990	$>10^3$	$>10^3$	$>10^3$	$>10^4$	$>10^4$
In(III)	806	210	193	153	83	29.2	6.3
Cd(II)	410	367	332	317	123	81	20.8
Mo(VI)	10.9	35.0	43.0	39.8	36.1	38.8	40.9
Rh(III)	4.2	3.1	2.2	1.5	0.9	0.6	<0.5
Tl(III)	2.1	2.4	2.7	1.9	2.2	2.6	2.4
Hg(II)	1.6	1.1	0.8	0.6	<0.5	<0.5	<0.5
Pd(II)	1.6	0.8	1.0	0.8	0.9	0.7	0.5
Pt(IV)	1.4	1.5	1.6	2.1	2.7	3.1	3.4
As(V)	1.4	~ 2	~ 5	~ 3	<1	<1	<1
Ir(IV)	1.4	1.5	1.4	1.3	1.0	1.1	0.9
W(VI)[a]	1.1	1.3	1.7	1.8	8.1	3.0	—
Se(IV)	1.1	1.3	1.0	1.5	1.7	1.0	0.8
Au(III)	0.8	1.1	0.9	1.2	1.3	0.8	1.0
Ge(III)	0.5	0.6	0.4	0.5	1.7	1.8	1.7
Mo(VI)[a]	<0.5	<0.5	<0.5	<0.5	<0.5	<0.5	<0.5

[a] H_2O_2 present.

TABLE II

CATION-EXCHANGE DISTRIBUTION COEFFICIENTS IN 0.20 M HCl

Element	*Percentage ethanol*						
	0	*20*	*40*	*60*	*80*	*90*	*95*
Fe(III)	3400	3600	5100	6000	1410	31.7	15.7
Ga(III)	3040	8680	$>10^4$	$>10^4$	3860	446	6.9
Mn(II)	510	690	1240	2280	5280	5410	4260
Co(II)	460	625	1120	2160	5010	$>10^4$	9330
Ni(II)	450	610	1110	2140	4960	8380	6320
Fe(II)	430	580	1020	2010	4370	1460	249
Cu(II)	380	424	706	1060	1040	308	69
Zn(II)	361	368	489	299	48.3	5.2	1.7
Mg(II)	350	361	369	1510	1990	6400	5100
U(VI)	252	284	469	847	1270	1430	1133
In(III)	110	53	48.8	38.1	17.9	6.0	2.9
Cd(II)	84	118	88	34.8	13.7	2.7	0.6
Sn(IV)	45	3400	1610	18.1	3.6	0.7	0.5
V(V)	7.0	19.7	39.1	84.0	202	ppt.	ppt.
V(V)[a]	6.5	6.6	9.0	57	488	529	336
Mo(VI)	4.5	17.7	25.9	25.7	23.1	12.6	7.4
Mo(VI)[a]	<0.5	<0.5	<0.5	<0.5	<0.5	<0.5	<0.5

[a] H_2O_2 present.

TABLE III

CATION-EXCHANGE DISTRIBUTION COEFFICIENTS IN 0.50 M HCl

Element	*Percentage ethanol*					
	0	*20*	*40*	*60*	*80*	*90*
Ga(III)	260	633	1470	4650	428	3.5
Fe(III)	225	226	304	361	153	6.6
Mn(II)	84	90	162	285	608	730
Mg(II)	74	78	126	234	601	1140
Cr(III)	73	93	118	170	279	421
Co(II)	72	93	159	305	671	1830
Ni(II)	70	96	165	315	689	880
Fe(II)	66	69	131	252	408	136
Cu(II)	65	88	119	176	195	35.9
U(VI)	58	67	111	182	264	220
Zn(II)	64	64	48.3	17.8	4.1	1.5
Cs(I)	44.2	63	108	229	852	—
V(IV)	44.1	53	84	157	286	508
Be(II)	42.3	47.3	69	114	170	—
Ti(IV)	39.1	121	265	634	1700	ppt.
Rb(I)	33.2	42.8	73	165	571	—
K(I)	29.1	47.3	89	201	838	—
Na(I)	13.5	19.1	34.1	79	254	—
Li(I)	8.1	10.8	17.1	28.8	50	—
In(III)	7.6	7.8	8.5	7.2	4.2	2.4
Cd(II)	6.5	6.0	4.6	1.4	<0.5	<0.5
Sn(IV)	6.2	5.9	3.6	2.2	1.3	0.6
V(V)	5.0	19.5	42.6	86	143	142
V(V)[a]	2.1	4.7	23.9	59	129	140
Mo(VI)	<0.5	7.6	11.0	12.8	11.0	7.3
Mo(VI)[a]	<0.5	<0.5	<0.5	<0.5	<0.5	<0.5
Bi(III)	<0.5	<0.5	<0.5	<0.5	<0.5	<0.5

[a] H_2O_2 present.

passing aqueous solutions of the ammonium or sodium salts through a cation-exchange column and adding the required amount of standard hydrochloric acid to the eluate. Solutions of tungsten(VI) thus prepared were stable only in the presence of hydrogen peroxide.

Distribution coefficients

The coefficients were determined using 5 meq (related to exchange) amounts of the elements in 250 ml of solution and 2.500 g of AG50W-X8 resin in the hydrogen form which had been dried at 60° in a vacuum pistol with phosphorus pentoxide as drying agent. In the cases of W(VI), Mo(VI) and V(V), marked with a subscript a in the Tables, the solutions also contained 2.5 ml of 30% hydrogen peroxide. After equilibration for 24 h at 25° in a mechanical shaker, the resin was separated from the aqueous phase by filtration and the amounts of the elements in the resin and in solution were determined by appropriate analytical methods. From the results, weight

TABLE IV

CATION-EXCHANGE DISTRIBUTION COEFFICIENTS IN 1.00 *M* HCl

Element	*Percentage ethanol*				
	0	*20*	*40*	*60*	*80*
Gd(III)	183	259	498	1460	~10^4
Yb(III)	153	178	331	841	4340
Ba(II)	128	193	615	1920	5140
Sc(III)	120	345	925	3680	>10^4
Al(III)	61	74	124	251	502
Sr(II)	60	76	162	543	3490
Ga(III)	42.6	102	204	347	5.8
Ca(II)	41.3	55	106	310	1570
Fe(III)	33.5	31.9	39.4	47.3	6.9
Cr(III)	26.7	35.9	53	83	114
Ni(II)	21.9	19.7	41.8	82	182
Co(II)	21.3	27.6	38.2	78	175
Mn(II)	20.2	28.2	42.4	78	157
Mg(II)	20.1	19.3	33.0	71	166
Fe(II)	19.8	25.8	46.9	81	184
U(VI)	19.2	22.1	34.2	58	70
Cs(I)	19.1	27.0	44.1	96	358
Cu(II)	17.5	17.8	21.9	23.2	24.7
Zn(II)	16.0	9.7	5.3	3.5	2.4
Rb(I)	15.4	21.4	36.8	81	326
K(I)	13.9	20.0	39.3	94	609
Be(II)	13.3	13.5	19.0	31.6	50.5
Ti(IV)	11.9	18.2	57	181	496
V(IV)	7.2	15.4	22.6	38.8	76
Na(I)	6.9	9.7	16.6	36.5	138
Li(I)	3.8	5.1	7.8	13.1	28.8
In(III)	1.8	1.81	1.9	1.3	0.7
Sn(IV)	1.6	1.4	0.8	<0.5	<0.5
Cd(II)	1.6	0.5	0.2	<0.2	<0.2
V(V)	1.1	4.7	10.3	23.1	43.0
Mo(VI)	0.8	4.2	6.9	7.3	5.8
Bi(III)	0.8	0.6	<0.5	<0.5	<0.5
Hg(II)	<0.5	<0.5	<0.5	<0.5	<0.5
Ge(IV)	<0.5	<0.5	<0.5	<0.5	<0.5

equilibrium distribution coefficients

$$D = \frac{\text{amount of element in resin}}{\text{amount of element in solution}} \times \frac{\text{ml of solution}}{\text{g of dry resin}}$$

were calculated. The distribution coefficients in 0.10, 0.20, 0.50, 1.00, 2.00 and 3.00 M hydrochloric acid are presented in Tables I–VI, respectively. Coefficients for solutions containing no ethanol have been included for the sake of completeness. Coefficients for elements like the platinum metals and gold which show only very slight absorption in 0.1 M hydrochloric acid were not determined at all higher concentrations, while coefficients for elements which showed very strong absorption at high acid concentration over the whole range of ethanol concentrations were not determined at the lowest acid concentrations.

Elution curves

Three multicomponent elution curves (Figs. 1, 2 and 3) demonstrate how the

TABLE V

DISTRIBUTION COEFFICIENTS IN 2.0 M HCl

Element	*Percentage ethanol*					
	0	*20*	*40*	*60*	*70*	*80*
Zr(IV)	489	925	2720	6400	$>10^4$	$>10^4$
Th(IV)	239	298	672	3170	$>10^4$	$>10^4$
La(III)	48.1	68	143	474	1200	$\sim 10^4$
Gd(III)	36.2	49.8	116	313	610	1280
Ba(II)	36.0	74	197	951	3260	—
Y(III)	29.7	32.4	72	196	490	1800
Sc(III)	28.8	57	140	470	1140	—
Yb(III)	27.4	36.2	59	151	298	—
Sr(II)	17.8	22.3	49.3	192	542	—
Al(III)	12.5	12.7	16.9	44.8	69	118
Ca(II)	12.2	16.3	31.7	90	202	—
Cs(I)	10.4	10.4	17.1	32.6	60	—
Rb(I)	8.1	10.0	16.4	39.6	77	—
Cr(III)	7.9	7.5	12.3	27.2	33.3	—
Ga(III)	7.8	13.5	20.9	6.3	1.5	<0.5
K(I)	7.4	10.0	21.0	54	118	—
Ni(II)	7.2	7.4	11.7	23.1	31.5	47.3
U(VI)	7.0	7.2	10.4	15.6	16.1	15.4
Co(II)	6.7	8.9	15.4	30.4	44.7	—
Mg(II)	6.2	7.2	12.1	23.8	36.2	55
Mn(II)	6.0	7.7	10.6	18.4	26.2	—
Be(II)	5.2	5.6	7.0	10.8	14.9	—
Fe(III)	5.2	5.0	6.2	5.0	3.1	0.6
V(IV)	5.0	5.8	7.6	13.3	16.4	17.8
Cu(II)	4.2	4.2	4.8	5.1	2.8	0.9
Fe(II)	4.1	4.3	6.9	8.5	6.3	1.8
Na(I)	3.8	5.5	8.9	24.5	57	—
Ti(IV)	3.7	3.8	9.8	41.9	88	123
V(V)	<0.5	2.9	5.2	10.3	13.4	—
In(III), Sn(IV) Bi(III), Hg(II) Ge(IV), Mo(VI)[a]	<0.5	<0.5	<0.5	<0.5	<0.5	<0.5

[a] H_2O_2 present.

information presented in the Tables can be applied to develop analytical separation procedures.

Fe(III), U(VI), Ca(II), La(III). A solution containing about 1 millimole of each element in 25 ml of 0.2 *M* hydrochloric acid was passed through a column (length 10 cm, diameter 19 mm) of 30 ml (10 g) of AG50W-X8 resin (200–400 mesh, hydrogen form), the cations being absorbed in the top resin layer. The elements were then eluted

TABLE VI

DISTRIBUTION COEFFICIENTS IN 3.0 *M* HCl

Element	*Percentage ethanol*				
	0	*20*	*40*	*60*	*70*
Th(IV)	114	142	363	2360	—
Zr(IV)	61	152	343	2800	ppt.
La(III)	18.8	32.6	72	239	948
Ba(II)	18.5	42.6	143	1340	—
Gd(III)	15.3	22.8	51	178	495
Sc(III)	14.9	30.9	76	281	—
Y(III)	13.6	17.7	32.6	128	288
Yb(III)	12.2	15.2	25.2	76	—
Sr(III)	10.0	13.1	29.8	162	—
Ca(II)	7.3	8.4	17.2	69	—
Cs(I)	5.9	5.7	9.3	19.9	—
Rb(I)	5.3	6.5	10.6	28.7	—
K(I)	4.9	7.1	14.3	40.4	—
Cr(III)	4.8	3.9	4.9	11.4	—
Al(III)	4.7	4.3	7.3	13.2	—
Mg(II)	4.3	3.7	5.9	12.3	—
Co(II)	4.2	3.8	6.2	11.8	—
Mn(II)	3.9	3.5	5.5	8.6	—
Fe(III)	3.6	1.8	2.4	2.2	—
V(IV)	3.5	4.3	5.6	7.4	—
U(VI)	3.5	3.5	5.5	6.3	—
Fe(II)	2.9	3.0	4.1	4.3	—
Na(I)	2.7	3.8	6.8	20.5	—
Ti(IV)	2.4	3.1	5.3	22.1 (6.8[a])	—
Ni(II)	2.0	2.1	4.5	6.7	—

[a] H_2O_2 present.

using the following elution sequence: 250 ml of 1.50 *M* hydrochloric acid in 75% ethanol for iron(III), 300 ml of 3.00 *M* hydrochloric acid in 60% ethanol for uranium(VI), 250 ml of 3.00 *M* hydrochloric acid in 30% methanol for calcium(II) and 300 ml of 4.0 *M* hydrochloric acid for lanthanum. Before eluents were changed, the solution was allowed to drain to the level of the resin bed. The maximum flow rate attainable with a solution head of 25–30 cm in the column was maintained throughout. This amounted to about 1–1.2 ml/min for the solutions containing a large percentage of ethanol and hydrochloric acid and about 2.5–3.0 ml/min for pure aqueous solutions. Aliquots of 25 ml were taken with an automatic fractionator, and the amounts of elements in the fractions were determined by appropriate analytical procedures. The experimental elution curve is presented in Fig. 1.

Cd(II), Zn(II), Fe(III), Ca(II), Ba(II). A solution containing 0.5 millimole of each element in 25 ml of 0.1 *M* hydrochloric acid containing 40% ethanol was passed through a pre-equilibrated column of the size and type mentioned above. For the elution of the elements, the following sequence was used: 200 ml of 0.50 *M* hydro-

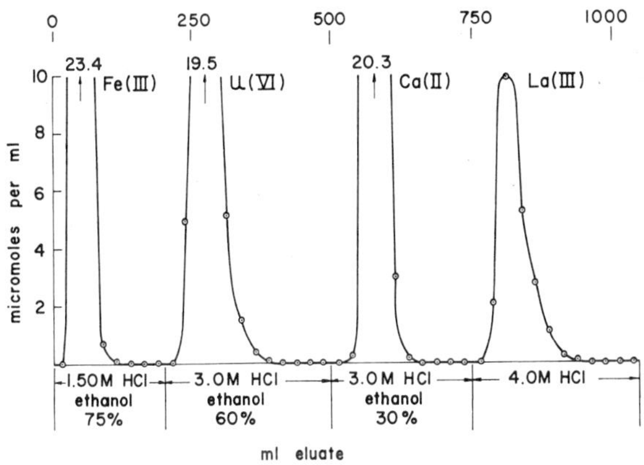

Fig. 1. Elution curve for Fe(III)–U(VI)–Ca(II)–La(III) mixture.

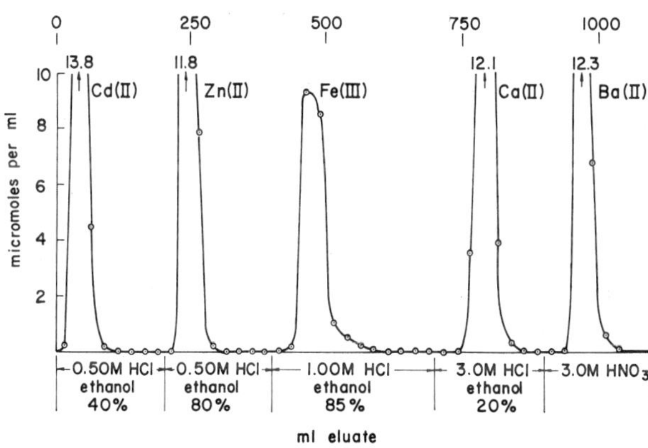

Fig. 2. Elution curve for Cd(II)–Zn(II)–Fe(III)–Ca(II)–Ba(II) mixture.

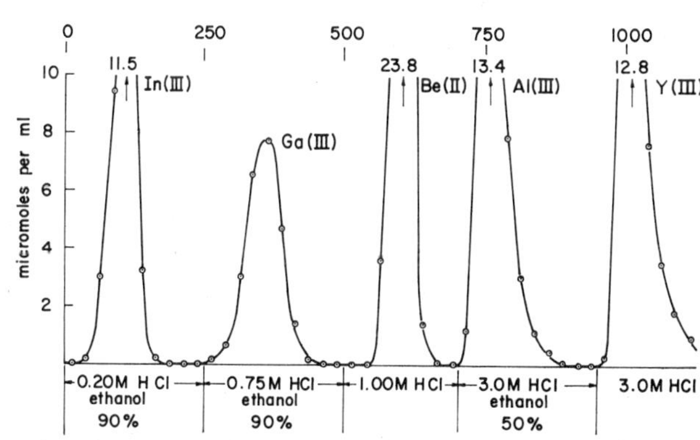

Fig. 3. Elution curve for In(III)–Ga(III)–Be(II)–Al(III)–Y(III) mixture.

chloric acid in 40% ethanol for cadmium(II), 200 ml of 0.5 *M* hydrochloric acid in 80% ethanol for zinc(II); 300 ml of 1.00 *M* hydrochloric acid in 85% ethanol for iron-(III); 200 ml of 3.0 *M* hydrochloric acid in 20% ethanol for calcium(II); and 200 ml of 3.0 *M* nitric acid for barium(II). The flow rates were as described above. The experimental elution curve is presented in Fig. 2.

In(III), Ga(III), Be(II), Al(III), Y(III). A solution containing about 0.7 millimoles of In(III) and Ga(III) and 1 millimole of Be(II), Al(III) and Y(III) in 0.1 *M* hydrochloric acid containing 40% ethanol was passed through a column as described above. The elution sequence was: 250 ml of 0.20 *M* hydrochloric acid in 90% ethanol for indium(III); 250 ml of 0.75 *M* hydrochloric acid in 90% ethanol for gallium(III); 250 ml of 1.00 *M* hydrochloric acid for beryllium(II); 250 ml of 3.0 *M* hydrochloric acid in 50% ethanol for aluminium(III) and 350 ml of 3.0 *M* hydrochloric acid for yttrium-(III). The experimental curve is presented in Fig. 3.

RESULTS AND DISCUSSION

Cation-exchange distribution coefficients in the region from 0.10 to 3.00 *M* hydrochloric acid generally decrease with increasing hydrochloric acid concentration. They increase with increasing concentration of alcohol until the hydration field around the cation is weakened enough so that the water dipoles in the co-ordination shell can be replaced by chloride anions. This causes a sudden decrease in the value of the distribution coefficients. In cases where the tendency to chloride-complex formation is fairly strong, only a slight increase of the coefficients at low alcohol concentration or no increase at all occurs, and the decrease takes place at lower ethanol concentrations. Exchange rates in dilute acid concentrations (< 0.5 *N*) are fast and comparable to those for cation exchange in aqueous solutions. At high acid and alcohol concentrations, the exchange rates seem to be somewhat slower than in aqueous acid and the use of resin of small particle size and slow flow rates often improves chromatographic separations of the strongly absorbed ions, such as Al(III), Ca(II), Sr(II) and Ba(II), considerably.

Alkali metals

Coefficients increase with increasing ethanol concentrations. Those for sodium increase faster than those for lithium and good separations can be obtained by eluting lithium with 1.0 *M* hydrochloric acid in 70% ethanol, while sodium is retained by the column. Separation of the other alkalies from each other is less favourable than in aqueous hydrochloric acid.

Alkaline earths

An excellent separation of calcium from Mg, Al, Fe, Ti and many other elements is possible in 3 *M* hydrochloric acid containing 60% ethanol[12], and barium can be separated from strontium, calcium and many other elements in 3 *M* hydrochloric acid containing 20% ethanol[13]. The separation factors are comparable to those obtained with organic complexing agents. The separation factors of the Mg–Be and Sr–Ca pairs increase only slightly with increase in ethanol concentration and the advantage gained is more than neutralized by the slower exchange rates in solutions containing high concentrations of both hydrochloric acid and ethanol.

Al(III), Ga(III), In(III) and Tl(III)

Tendencies to chloride-complex formation increase with increasing atomic weights and good separations are possible by using the following elution sequence: 0.10 *M* hydrochloric acid containing bromine for thallium(III), 0.20 *M* hydrochloric acid in 90% ethanol for indium(III), 0.75 *M* hydrochloric acid in 90% ethanol for gallium(III) and 3.0 *M* hydrochloric acid for Al(III). Furthermore, aluminium(III) can be separated from Fe(III), Be(II) and U(IV) by eluting these elements with 2 *M* hydrochloric acid in 70% ethanol, and indium(III) can be separated from Cu(II), Co(II), Ni(II), Fe(III) and many other elements by elution with 0.20 *M* hydrochloric acid in 90% ethanol.

Cr(III)

The distribution coefficients are composites of coefficients of different ionic species existing in solution and the composite values depend on the history of the solutions. Different bands, due to the species $[CrCl_3(H_2O)_3]^\circ$ green, $[CrCl_2(H_2O)_4]^+$ green, $[CrCl(H_2O)_5]^{2+}$ dirty blue violet, and $[Cr(H_2O)_6]^{3+}$ red violet, can be observed during column work. The uncharged species passes through the column without absorption, but its presence can be avoided by boiling the solutions in very dilute hydrochloric acid. Coefficients for the other species were estimated from column work to be about 50, 400 and >2000, respectively, in 1.0 *M* hydrochloric acid containing 80% ethanol and 5, 60 and 400 in 2.0 *M* hydrochloric acid containing 60% ethanol. Because the rates of aquo- and chloride-complex formation in ethanolic hydrochloric acid are even slower than in the corresponding aqueous solutions, almost quantitative separation of the different chromium species can be obtained by chromatographic elution, but separation of chromium from most other elements, except Th(IV), Zr(IV), Ba(II) and some rare earths which are strongly absorbed, is prevented by the fact that the different chromium species are eluted under widely different conditions.

V(V), Mo(VI) and W(VI)

Molybdenum(VI) and tungsten(VI) can easily be eluted with 0.1 or 0.2 *M* hydrochloric acid at ethanol concentrations up to 95% and thus be separated from many other elements, provided that hydrogen peroxide also is present. Tungsten(VI) precipitates in the absence of peroxide and molybdenum(VI) shows no agreement between batch distribution coefficients and column behaviour. This apparently is due to the presence of several ionic species with slow conversion reaction rates. Vanadium(V) shows no agreement between distribution coefficients and elution curves in the absence or presence of peroxide. Even with a *D* value of >100 (0.2 *M* hydrochloric acid in 80% ethanol) some vanadium appears in the eluate not far behind the elution front while the bulk remains absorbed at the top of the column and appears in the eluate gradually. This is ascribed to the simultaneous presence of several cationic species such as VO^{3+}, VO_2^+ and their peroxide complexes, and a small amount of neutral species. Effective elution of vanadium(V) can be obtained with 0.01 *M* or more dilute aqueous hydrochloric acid containing a small amount of hydrogen peroxide because only neutral or anionic species are present above pH 2. Alternatively, 0.25 *M* aqueous sulphuric acid or 0.5 *M* aqueous hydrochloric acid containing a small amount of hydrogen peroxide can be used.

Au(III), Pt(IV), Pd(II), Rh(III), Ir(IV), Hg(II) and Ge(IV)

These elements are only very weakly, or not at all, absorbed from all concentrations of hydrochloric acid and ethanol. They can be eluted with aqueous 0.1 *M* hydrochloric acid and thus be separated from most other elements of the periodic table, yet residual traces of rhodium(III) and palladium(II) are retained tenaciously by the resin. Considerably reduced tailing and quantitative elution can be obtained with 0.1 *M* hydrochloric acid in 90 or 95% ethanol as eluent. Rhodium(III) shows two elution peaks on elution with 0.1 *M* hydrochloric acid in 40% ethanol, a large one appearing with the elution front and a small one corresponding to a distribution coefficient of about 10, while only the first elution peak appears in 0.1 *M* hydrochloric acid in 95% ethanol. The two peaks are ascribed to the presence of two different rhodium(III) complexes, probably $[RhCl_3(H_2O)_3]^\circ$, not absorbed, and $[RhCl_2(H_2O)_4]^+$, weakly absorbed, which have slow complex conversion rates.

Zn(II), Cd(II) and Sn(IV)

These elements are only very weakly absorbed from 0.2 *M* hydrochloric acid in 90% ethanol and can easily be separated from Cu(II), Co(II), Ni(II), Mn(II), Fe(III), U(VI) and many others by elution with this reagent.

Fe(III)

Iron(III) can be eluted with 1.0 *M* hydrochloric acid in 85% ethanol while Mg(II), Ca(II), Al(III), Ti(IV), Mn(II), Co(II), Ni(II) and many other elements are retained by the column.

Zr(IV), Th(IV), Sc(III), Y(III), La(II) and the rare earths

These elements are strongly absorbed from 3 *M* hydrochloric acid containing 50 or 60% of ethanol. Most other elements, including Al(III), Fe(III) and Ti(IV), in the presence of water can be eluted and separated quite easily. The separation factor D_{Sc}/D_{Al} in 3 *M* hydrochloric acid containing 60% ethanol is about 21, compared with 3.2 in aqueous 3 *M* hydrochloric acid, while the respective values for D_Y/D_{Al} are 10 and 2.9. The coefficients for scandium(III) increase faster with increasing ethanol concentration than those for yttrium(III). Exchange rates are slower in ethanol solutions than in aqueous hydrochloric acid and resin of small particle size and slow flow rates improve separations of these elements from aluminium(III), etc., considerably.

Cu(II), Co(II), Ni(II), Mn(II) and V(IV)

The last four elements show the normal behaviour of divalent elements with small or negligible tendencies to chloride-complex formation, and the coefficients increase with increasing ethanol concentrations, while those for copper(II) do not increase as strongly and decrease at the highest concentration. Thus copper(II) can be separated from the other elements by elution with 1.50 *M* hydrochloric acid in 80% ethanol. The alkalies, alkaline earths, Al(III), Ti(IV), Zr(IV), Hf(IV), Th(IV), Sc(III), Y(III), La(III) and the rare earths are retained together with Co(II), etc.

General

The system offers many attractive possibilities for separations. Its advantages over the acetone system are the higher boiling point of ethanol and the facts that ethanol does not polymerize in acid solution and that the vapours are less of a health hazard. Its disadvantages are that it promotes chloride-complex formation less strongly and that solutions containing high concentrations of hydrochloric acid in high percentages of ethanol have relatively high viscosities. Exchange rates for elements which require these kind of eluents often are slow in aqueous solutions. In ethanolic solutions they are even slower and resin of a fine particle size leads to considerably improved separations. High viscosities in such cases limit the maximum flow rates attainable without the application of external pressures.

SUMMARY

Cation-exchange equilibrium distribution coefficients with Bio-Rad AG50W-X8, a sulfonated polystyrene resin, are presented for 45 elements in hydrochloric acid–ethanol media. The acid concentration range 0.1–3.0 *M* and the ethanol concentration range 0–95% are covered. The elements are arbitrarily arranged in tables according to their coefficients in aqueous hydrochloric acid. Possibilities for separations inherent in the system are demonstrated by elution curves for the multicomponent systems Fe(III)–U(VI)–Ca(II)–La(III); Cd(II)–Zn(II)–Fe(III)–Ca(II)–Ba(II); and In(III)–Ga(III)–Be(II)–Al(III)–Y(III). Some aspects in the elution behaviour of various elements are discussed.

RÉSUMÉ

Les coefficients de distribution, à l'équilibre, lors d'un échange cationique sur une résine Bio-Rad AG50-X8 au polystyrène sulfoné sont présentés pour 45 éléments dans des milieux acide chlorhydrique–éthanol. Les concentrations de l'acide vont de 0.1 à 3.0 *M*, celles de l'éthanol de 0 à 95%. Les éléments sont classés arbitrairement dans des tables selon leurs coefficients dans les solutions aqueuses d'acide chlorhydrique. Les possibilités de séparation inhérentes au système sont démontrées au moyen des courbes d'élution pour des systèmes à multicomposants Fe(III)–U(VI)–Ca(II)–La(III), Cd(II)–Zn(II)–Fe(III)–Ca(II)–Ba(II); et In(III)–Ga(III)–Be(II)–Al(III)–Y(III). On discute quelques aspects du pouvoir d'élution de divers éléments.

ZUSAMMENFASSUNG

Verteilungskoeffizienten an Kationenaustauschern mit Bio-Rad AG50-X8 und mit sulfoniertem Polystyrolharz werden für 45 Elemente in einem Medium aus Salzsäure und Äthanol angegeben. Die Säurekonzentration lag zwischen 0.1 und 3.0 *M* und die des Äthanols zwischen 0 und 95%. Mögliche Trennungen in diesem System werden demonstriert durch Elutionskurven der Systeme Fe(III)–U(VI)–Ca(II)–La(III); Cd(II)–Zn(II)–Fe(III)–Ca(II)–Ba(II) und In(III)–Ga(III)–Be(II)–Al(III)–Y(III). Einige Aspekte im Elutionsverhalten zahlreicher Elemente werden diskutiert.

REFERENCES

1 N. F. KEMBER, P. J. MACDONALD AND R. A. WELLS, *J. Chem. Soc.*, (1955) 2273.

2 J. S. Fritz and T. A. Rettig, *Anal. Chem.*, 34 (1962) 1562.
3 J. Korkisch and S. S Ahluwalia, *Talanta*, 14 (1967) 155.
4 V. Nevoral, *Z. Anal. Chem.*, 195 (1963) 332.
5 Z. Šulcek, P. Povondra and R Stangl, *Collection Czech. Chem. Commun.*, 30 (1965) 380.
6 M. Tanaka, *Nippon Kagaku Zasshi*, 85 (1964) 117.
7 G. Venturello, C. Gualandini and I. Mazzei, *Ann. Chim. (Rome)*, 49 (1959) 149.
8 H. Okuna, M. Honda and T. Ishimori, *Japan Analyst*, 2 (1958) 428.
9 R. Ratner and Z. Ludmer, *Israel J. Chem.*, 2 (1964) 21.
10 C. W. Davies and B. D. R. Owen, *J. Chem. Soc.*, (1956) 1676, 1681.
11 V. I. Gorshkov, I. A. Kuznetzov and G. N. Panshenkov, *Zh. Analit. Khim.*, 14 (1959) 417.
12 F. W. E. Strelow and C. R. van Zyl, *Anal. Chim. Acta*, 41 (1968) 529.
13 F. W. E. Strelow, *Anal. Chem.*, 40 (1968) 928.

27

Reprinted from *Anal. Chim. Acta*, **50**, 399–405 (1970)

ACCURATE DETERMINATION OF TRACES OF SODIUM AND POTASSIUM IN ROCKS BY ION EXCHANGE FOLLOWED BY ATOMIC ABSORPTION SPECTROSCOPY

F. W. E. STRELOW, F. VON S. TOERIEN AND C. H. S. W. WEINERT
National Chemical Research Laboratory, P.O. Box 395, Pretoria (South Africa)
(Received January 26th, 1970)

The reliable determination of sodium and potassium in silicate rocks still appears to be a problem when these elements are present in trace amounts. This is pertinently demonstrated by recent results for the U.S. geological survey standards PCC-1 and DTS-1[1]. The reason for the wide spread of the results—about an order of magnitude and more—is not quite clear. Inhomogeneous sample composition, interference by the large amount of a complex mixture of other elements, losses in separation procedures, contamination before or during analysis, or a combination of these factors could be responsible. The second and third of these factors can be eliminated by employing an effective method for separation and concentration before the final determination step. In addition this also would increase the sensitivity of the method, a feature which appears to be welcome when looking at the published results[1]. An ion-exchange procedure for the separation of silicate rock elements which has been described recently[2] is not satisfactory because of relatively high reagent blank values. A special procedure therefore was developed for the determination of trace amounts of alkalies. This incorporated the following features. Only specially purified reagents were employed. The eluting agent, 0.50 *M* nitric acid, for separating the alkalies from multivalent elements[2,3] was retained, but only that fraction of the eluate which contained the relatively narrow peak of one particular alkali metal was taken for analysis. Because dunites and peridotites contain between 40 and 50% of magnesium oxide, the peak positions had to be determined in the presence of large and varying amounts of magnesium, which caused a shift to a lower elution volume. Predominantly anionic elements such as vanadium(V), molybdenum(VI) and tungsten(VI) were eluted with 0.01 *M* nitric acid containing hydrogen peroxide[2]. Perchloric or sulphuric acid of the same concentration could be used instead[4]. The alkali metals were retained according to their distribution coefficients[5] and then eluted with 0.5 *M* nitric acid. The separated alkali metals were determined by atomic absorption spectroscopy.

EXPERIMENTAL

Reagents and apparatus

Water was redistilled and all solutions were kept in polyethylene containers. Hydrochloric, hydrofluoric, sulphuric, perchloric and nitric acid were "suprapur" grade (E. Merck A. G., Darmstadt, Germany). Phosphoric acid (A.R. grade) was diluted with 4 volumes of redistilled water and purified by passing through a cation-exchange column. The resin was AG50W-X8 cation exchanger (200–400 mesh, BIO.RAD Laboratories, Richmond, Calif.). Borosilicate glass tubes of 2.5 cm inner diameter fitted with a No. 2 porosity glass sinter and a stopcock at the bottom and a B19 joint at the top were used as columns. All glassware was purified by washing with hot 1:1 nitric acid (A.R.) and resin columns were cleaned by passing through 500 ml of 3 *M* nitric acid (A.R.), followed by 50 ml of 1 *M* nitric acid "suprapur" before use. A Perkin Elmer 303 instrument was used for atomic absorption measurements. Alkali metal determinations were carried out with a propane–butane–air flame and a special flat water-cooled burner[6].

Elution curves

A solution containing 1 mg of each vanadium(V), lithium, sodium and potassium, 2 mg of cesium, 300 mg of magnesium, 55.8 mg of iron(III), 10 ml of 2.0 *M* hydrochloric and 10 ml of 2.5 *M* sulphuric acid and 10 ml of 0.3% (w/v) hydrogen peroxide in a volume of 100 ml, to simulate conditions after decomposition of a rock sample, was passed onto the exchange column and washed onto the resin with 0.01 *M* nitric acid containing 0.03% (w/v) hydrogen peroxide. Vanadium was eluted with 100 ml of the same solution. Then the alkali metals were eluted with 0.50 *M* nitric acid. A flow rate of 3.0 ± 0.5 ml min^{-1} was maintained throughout. Fractions (25 ml) were taken with an automatic fractionator and analysed. Vanadium was determined spectrophotometrically as the 4-(2-pyridylazo)-resorcinol complex[7] and the alkali metals and magnesium were determined by atomic absorption spectrometry using the following lines: 6708 Å for Li, 5890 Å for Na, 7665 Å for K, 8521 Å for Cs, 2852 Å for Mg. A propane–butane–air flame was used for lithium, sodium, potassium and cesium and an acetylene–air flame for magnesium.

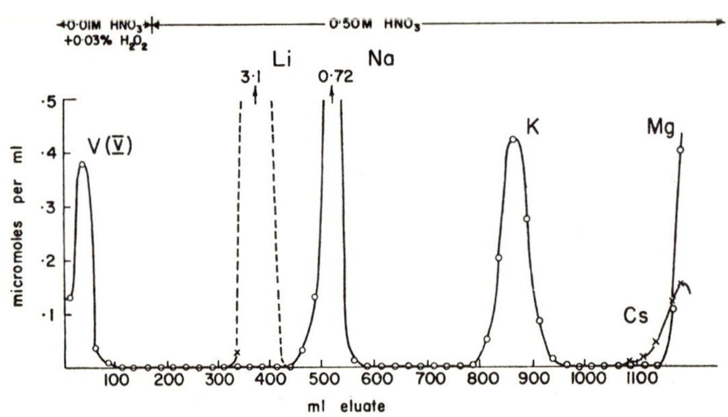

Fig. 1. Elution curve for V(V), Li, Na, K, Cs and Mg with 0.01 *M* HNO_3 and 0.50 *M* HNO_3. 90 ml (30 g) AG50W-X8 resin, 200–400 mesh.

The experimental elution curve for a 90-ml (30-g) resin column (Fig. 1) shows that sodium in microgram amounts will be collected quantitatively in the 250–500 ml fraction of the 0.50 *M* nitric acid eluting agent. The only possible contaminant is lithium, but trace amounts of lithium will not interfere in the determination of sodium. Potassium is collected in the 600–850 ml fraction of the eluting agent. The only possible contaminant, rubidium, which is eluted between the potassium and cesium peaks, is unlikely to be present in observable amounts and will not interfere in the determination unless present in fairly large amounts.

Titanium and beryllium are eluted later than cesium and are separated quantitatively unless present in large amounts (more than 1 mmole). All other di- and multivalent cationic elements such as Ca, Sr, Ba, Cu(II), Co(II), Zn, Cd, Mn(II), Ni(II), Al, Ga, In(III), Fe(III), rare earths(III), Zr(IV) and Th(IV) are more strongly retained than magnesium. As much as 12 mmoles of these elements can be present without interfering.

Figure 2 shows a similar curve for a 75-ml (25-g) resin column to demonstrate the effect of column size. Lithium, rubidium and cesium were not included. The Figure shows that with this column size, separation of 1 mg of potassium from 300 mg of magnesium is too marginal for quantitative work. A good separation can be obtained when only 200 mg of magnesium are present. No magnesium could be detected in the first 1000 ml of 0.5 *M* nitric acid in this case, while the last traces of potassium were eluted with 750 ml.

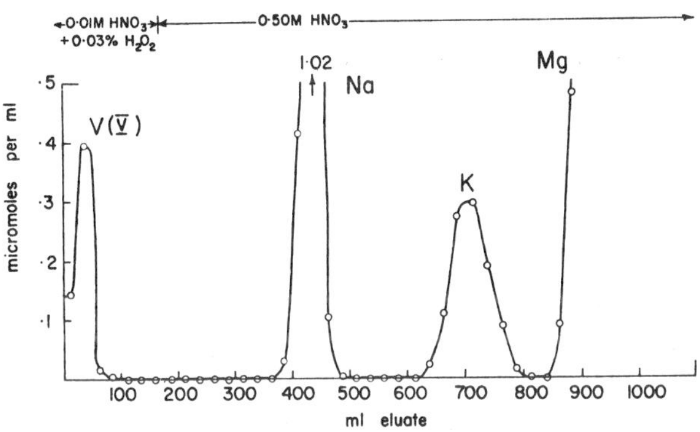

Fig. 2. Elution curve for V(V), Na, K and Mg with 0.01 *M* HNO_3 and 0.50 *M* HNO_3. 75 ml (25 g) AG50W-X8 resin 200–400 mesh.

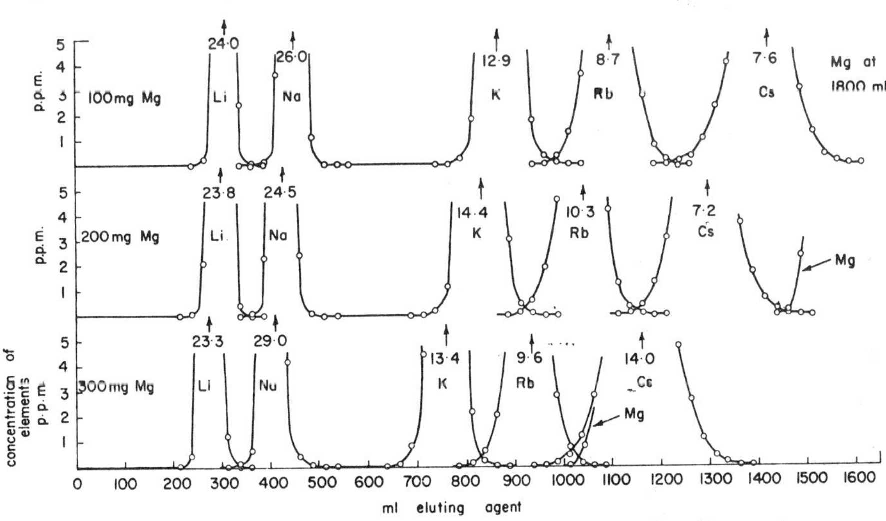

Fig. 3. Elution curve for Li, Na, K, Rb, S and Mg with 0.50 *M* nitric acid at various amounts of Mg. 90 ml (30 g) AG50W-X8 resin, 200–400 mesh.

Effect of column load on alkali peak position

Figure 3 shows the effect of column loading on the position of the sodium and potassium peaks for a 90-ml (30-g) resin column. Only 1 mg of each sodium and potassium and 100, 200 or 300 mg of magnesium, respectively, were present. The cations were absorbed from 0.01 *M* hydrochloric acid in this case. Equivalent loads of iron, calcium and aluminium cause considerably less shifting of the alkali metal peaks. The effect of manganese is similar to that of magnesium. Other divalent heavy metals such as Cu, Co, Zn, Ni and Cd were not investigated, but from their known distribution coefficients[5] it can be expected that their effect will be similar to that of magnesium.

Analysis of synthetic mixtures

Accurate amounts of standard solutions containing known amounts of the alkali metals and magnesium or other elements were measured out, mixed and absorbed on a 90-ml resin column of 2.5 cm diameter and 21 cm length from 0.01 *M* nitric acid or from 0.10 *M* hydrochloric acid when iron(III) or aluminium were present. The cations were washed onto the resin with the same reagent, and the alkalies were then eluted with 0.50 *M* nitric acid. A flow rate of 3.0 ± 0.5 ml min^{-1} was used. The eluate fraction from 300 to 550 ml was taken for the analysis of sodium and that from 650 to 900 ml for the analysis of potassium when 300 mg of magnesium or 700 mg manganese were present (Fig. 3). The same fraction was taken for the determination of sodium when 300 mg of aluminium, 500 mg of calcium or 700 mg of iron were present, but for potassium the fraction from 750 ml to 1000 ml was taken.

The eluates containing the alkali metals were received in acid-cleaned 250-ml volumetric flasks and the concentrations of the alkali metals were measured by atomic absorption spectrophotometry under the conditions described above. The concentrations were read from a calibration curve prepared from standards in 0.50 *M* nitric acid. The results were corrected by subtracting the average of a duplicate blank run and are shown in Table I.

TABLE I

SEPARATION OF SYNTHETIC MIXTURES

Taken		*Found*[a]
μg alkali metal	*mg other element*	*μg alkali metal*
102 Na	300 Mg	101 ± 1
20.4 Na	300 Mg	20.1 ± 0.8
5.1 Na	300 Mg	4.9 ± 0.6
501 Na	300 Mg	499 ± 3
102 Na	500 Ca	100 ± 2
102 Na	700 Mn	101 ± 1
102 Na	300 Al	102 ± 1
102 Na	700 Fe	101 ± 1
104 K	300 Mg	103 ± 2
20.8 K	300 Mg	20.6 ± 1.3
5.2 K	300 Mg	5.0 ± 0.9
502 K	300 Mg	501 ± 3

[a] The results are means of quadruplicate determinations.

Procedure for analysis of dunites and peridotites

About 1-g samples of dunite or peridotite were weighed out and dissolved by heating with mixtures of hydrofluoric, hydrochloric and sulphuric acid as described before[2]. The solution was adjusted to a volume of about 100 ml containing 10 ml of 2 *M* hydrochloric acid and 10 ml of 2.5 *M* sulphuric acid, and passed through a 90-ml (30-g) resin column of 2.5 cm inner diameter. It was washed onto the resin with 0.01 *M* nitric acid containing 0.03% (w/v) hydrogen peroxide. Vanadium(V), molybdenum(VI), tungsten(VI), phosphate and other anionic elements were eluted with another 100 ml of the same reagent. Sodium and potassium then were eluted with 0.50 *M* nitric acid taking the fraction 250–500 ml for the determination of sodium, and the fraction 600–850 ml for the determination of potassium. The fractions were received in 250-ml volumetric flasks and the amounts of the elements were determined by atomic absorption spectrophotometry as described above; scale expansion was used for low concentrations. Specpure potassium (500 p.p.m.) was added to sodium samples and standards. Duplicate runs on reagent and procedure blanks were carried out and the averages subtracted from the results. The results are presented in Table II.

TABLE II

DETERMINATION OF SODIUM AND POTASSIUM IN ROCK SAMPLES

Sample	*p.p.m. found*[a]		*"Average" literature value*[b]	
	Na	*K*	*Na*	*K*
PCC-1	16.2 ± 0.7	30.9 ± 0.8	~ 390	~ 160
DTS-1	36.9 ± 0.9	14.0 ± 1.9	~ 340	~ 190
NIM-D	524 ± 3	226 ± 2	Not available	Not available

[a] The results are means of quadruplicate determinations.
[b] From ref. 1.

DISCUSSION

The described method provides a useful means for the determination of trace amounts of sodium and potassium in silicate rocks. Between 5 and 1000 μg of sodium and potassium can be determined in a 1-g rock sample. In the range 100–1000 μg (0.01–0.1%) results are quite precise (coefficient of variation about ±1 to ±2%), and from the results obtained with synthetic standards (Table I), it is believed that the accuracy is of the same order as the precision. A very small negative bias may be present. At lower amounts, the precision slowly declines to about ±10 to ±20% at the 5-μg level. This is due to low atomic absorption readings and variable blank values. The blank values for sodium were of the order of 2–3 μg, and those for potassium of the order of 0.5–1.0 μg. First attempts to obtain higher accuracy through higher absorption readings by evaporating the sample solutions and reading in a smaller volume failed. Evaporation in glass led to very high and variable blank values (>20 μg sodium) and even evaporation in teflon and platinum gave blank values which were increased by 50 to 100%. Apparently this was due to uptake of alkali metals from dust particles in the air.

The results obtained for the international rock standards PCC-1 and DTS-1 are an order of magnitude lower than the average value (Table II) given by FLANAGAN[1], but FLANAGAN already has pointed out that the "averages" given in his paper for sodium and potassium in PCC-1 and DTS-1 are very much higher than the neutron activation results and probably are much too high. Because the proposed sample decomposition procedure is simple and is included in the blank runs, and because good results were obtained with synthetic mixtures (Table I), it is felt that the results presented are reliable. They also show good agreement with the newest results obtained by neutron activation analysis[7].

The described method is very simple and lends itself to rationalization, because only a small amount of actual work is involved. As much as 20 and even more samples can be done in a working day, provided that an adequate battery of exchange columns is available. The position of the peaks in the ion-exchange separation is critical; apart from resin loading (Fig. 3), it also may vary with long-term variations in the effective cross-linkage of different resin batches. Although no significant variations could be detected when 3 different batches of AG50W-X8 resin were tested, for obtaining reliable results it is advisable to establish the peak positions of sodium and potassium for the kind and amount of sample and the resin batch and column size used before the analysis is started. It may be pointed out that Dowex 50W-X8 cation exchangers will give similar peak positions to AG50W-X8 resin, but Amberlite 120H or other cation exchangers, even with 8% cross-linkage according to their designation, do not.

SUMMARY

Sodium and potassium (5–1000 p.p.m.) in rocks such as peridotites and dunites can be determined accurately by ion-exchange separation followed by atomic absorption spectrophotometry. The samples are decomposed in sulphuric, hydrochloric and hydrofluoric acid mixtures, and after removal of hydrofluoric acid the cations are absorbed on a 90-ml column of AG50W-X8 cation-exchange resin. Vanadium and other anion-forming elements are eluted with 0.01 *M* nitric acid containing hydrogen peroxide. Then the alkali metals are eluted with 0.50 *M* nitric acid, while Mg, Mn, Ca, Ti, Al, Fe and other multivalent elements are retained by the column. The eluate fractions containing the respective alkali metals are measured directly by atomic absorption spectrophotometry against standards in 0.50 *M* nitric acid. Relevant elution curves, results for synthetic mixtures and for three international rock standards are presented and discussed.

RÉSUMÉ

Il est possible de doser avec précision 5 à 1000 p.p.m. de sodium et potassium dans des roches telles que péridotites et dunites, par séparation au moyen d'échangeurs d'ions et spectrophotométrie par absorption atomique. Les échantillons sont décomposés par des mélanges d'acides sulfurique, chlorhydrique et fluorhydrique. Après élimination de l'acide fluorhydrique les cations sont absorbés sur une colonne de résine cationique AG50W-X8. Le vanadium et d'autres éléments pouvant former des anions sont élués avec de l'acide nitrique 0.01 *M* renfermant du peroxyde d'hydrogène. Les métaux alcalins sont ensuite élués avec acide nitrique 0.5 *M*; alors que Mg, Mn, Ca, Ti, Al, Fe et autres éléments multivalents sont retenues sur la colonne.

Les fractions d'éluat contenant les métaux alcalins sont mesurés directement par spectrophotométrie par absorption atomique, en utilisant des étalons en milieu acide nitrique 0.5 *M*. Des résultats sont présentés et discutés.

ZUSAMMENFASSUNG

Natrium und Kalium (5–1000 p.p.m.) in Gesteinen wie Peridotiten und Duniten können durch Ionenaustausch-Trennung und anschliessende Atomabsorptionsspektrophotometrie genau bestimmt werden. Die Proben werden in Mischungen von Schwefel-, Salz- und Flusssäure zersetzt und die Kationen nach Entfernung der Flusssäure in einer 90 ml-Säule vom Kationenaustauscher AG50W-X8 absorbiert. Vanadin und andere anionenbildende Elemente werden mit 0.01 *M* Wasserstoffperoxid enthaltender Salpetersäure eluiert. Dann werden die Alkalimetalle mit 0.50 *M* Salpetersäure eluiert, während Mg, Mn, Ca, Ti, Al, Fe und andere mehrwertige Elemente von der Säule zurückgehalten werden. Die die jeweiligen Alkalimetalle enthaltenden Eluatfraktionen werden direkt durch Atomabsorptionsspektrophotometrie gegen Standards in 0.50 *M* Salpetersäure gemessen. Die Elutionskurven sowie die Ergebnisse von künstlichen Mischungen und von drei internationalen Standardgesteinen werden vorgelegt und diskutiert.

REFERENCES

1 J. F. Flanagan, *Geochim. Cosmochim. Acta*, 33 (1969) 81.
2 F. W. E. Strelow, C. J. Liebenberg and F. Von S. Toerien, *Anal. Chim. Acta*, 47 (1969) 251.
3 F. W. E. Strelow, J. H. C. Coetzee and C. R. Van Zyl, *Anal. Chem.*, 40 (1968) 196.
4 J. S. Fritz and J. E. Abbink, *Anal. Chem.*, 34 (1962) 1080.
5 F. W. E. Strelow, R. Rethemeyer and C. J. C. Bothma, *Anal. Chem.*, 37 (1965) 106.
6 L. R. P. Butler, *J. S. African Inst. Mining Met.*, 62 (1962) 786.
7 E. Steinness, Institutt for Atomenergi, Kjeller, Norway, private communication, July, 1969.

Part VIII

INORGANIC AND SPECIAL EXCHANGERS

Editor's Comments on Papers 28 Through 30

28 KRAUS et al.
Ion Exchange Properties of Hydrous Oxides

29 KOSTER and SCHMUCKLER
Separation of Noble Metals from Base Metals by Means of a New Chelating Resin

30 DAVANKOV et al.
Ligand-Exchange Chromatography of Racemates: Influence of the Degree of Saturation of the Asymmetric Resin by Metal Ions on Ligand Exchange

The first synthetic ion exchanges were aluminosilicates. Outside of water softening, they were of little use, because they disintegrated in acid and alkaline solutions. A new class of inorganic exchangers was discovered in 1955 by K. A. Kraus and his colleagues at Oak Ridge National Laboratory. They were working on the polymeric species formed by hydrolysis of highly charged metal ions, such as zirconium ions, and saw that these polymers were analogous to the resinous, organic, ion-exchanging polymers. They were also aware of the need of the nuclear power industry for ion exchangers that would be resistant to radiation and high temperatures.

Hydrous, precipitated zirconium oxide was found to behave as an anion exchanger in acidic solutions and as a cation exchanger in alkaline solutions (Kraus and Phillips 1956a). Zirconium phosphate acts purely as a cation exchanger (Kraus and Phillips, 1956b) and so does zirconium tungstate (Kraus et al., 1956). These three papers began a series that culminated in our Paper 28.

Amorphous precipitates having ion-exchange properties can be made by combining ions of elements of groups IV, V, and VI of the periodic table, plus other ions of high charge, like Ce(IV) and U(VI), that do not properly fall into these groups. The number of possible combinations is very large. Examples are thorium arsenate, tantalum antimonate, and tin(IV) selenite. Most of these combinations have now been prepared and their ion-exchange properties studied, particularly by M. Qureshi of the Aligarh Moslem University (Qureshi et al., 1973). Another approach has been taken by A. Clearfield of Ohio University (Clearfield,

1973). He has prepared a few of these exchangers in crystalline form, particularly zirconium phosphate, and has studied in detail the physical chemistry of these well-characterized substances.

Other inorganic exchangers are the heteropoly acids, ferrocyanides, and other complex cyanides. Undoubtedly their ion-exchanging powers are related to a porous, honeycomb, microcrystalline structure.

As a group, the inorganic exchangers are distinguished by great selectivity toward the heavier alkali and alkaline-earth ions, particularly cesium. They are used to remove radioactive, long-lived cesium 137 from fission wastes. As stationary phases in ion-exchange chromatography, however, they have been disappointing. Physically, their granules are loose aggregates that break down to powder under the conditions of modern liquid chromatography. A more serious drawback is that their ion-exchange sites are not uniform. The symmetrical elution peaks shown in Reference 1 of Paper 28 are deceptive; they were obtained with tracers, and macro quantities of metal ions tend to give broad and tailing bands. However, these exchangers have analytical potential for the recovery and concentration of strongly absorbed trace constituents.

In the quest for ion-exchange selectivity, much attention has been paid to *chelating resins*, organic exchangers having selective functional groups. The most successful of these is Dowex-1, which has iminodiacetate groups on a crosslinked polystyrene matrix. A selectivity scale for divalent cations on this resin was compiled by Rosset (1966). By analogy with selective organic precipitants, it was hoped to find resins that would have great selectivity for a limited group of metal ions. Many polymers with special functional groups have been prepared, but they have low capacity, are slow to react with metal ions, show large volume changes, and do not have the expected selectivity.

One of the few such polymers that has been successful is described in Paper 29. This resin has isothiouronium (sulfaguanidine) groups on a polystyrene matrix, and it is selective for transition-metal ions having a d^8 electronic configuration, which form square-planar complexes. These include ions of the platinum metals and gold, which can be recovered selectively from dilute solutions by this resin. Ions of mercury (II) and methyl mercury can also be recovered (Law, 1971).

Later papers by Schmuckler et al. (1972, 1974) give more information on the mechanism of coordination of Pd (II) and Pt (II) by this resin. It appears that $PdCl_2$ is bound as a molecule by coordination of the palladium atom to the sulfur atoms of two thiouronium groups, and the guanidine groups are not directly attached. A resin with sulfaguanidine groups, $-SO_2-NH-C(=NH)NH_2$, however, does bind molecules of $PdCl_2$ through its guanidine function.

Paper 30 describes the realization of a long-standing dream, the separation of optical isomers by chromatography on a resin with asymmetric functional groups. It also illustrates the method of ligand-exchange chromatography (see Part XII).

The functional groups are optically active amino acids, in this case proline or hydroxyproline, and they are attached to a polystyrene matrix through the nitrogen atom and methylene bridge. The resin is weakly crosslinked, and the crosslinks are not divinylbenzene units, but the long molecules whose structures are shown on page 361 of the paper. The idea is to make the polymer matrix flexible, so that two proline groups anchored to the polymer can coordinate simultaneously with the same ion of copper (II). When a solution of an amino acid is passed into a column of the copper-loaded resin, one amino acid molecule from the solution can attach itself to a copper ion, displacing a "matrix" amino acid group in so doing. Then the mobile amino acid is displaced from the resin column by passing a solution of ammonia. A resin with fixed *l*-amino acid binds dissolved *l*-amino acids in preference to *d*-acids. If the bound and the mobile amino acids are the same, for example both proline, the steric perference for the *l–l* combination *versus* the *l–d* combination is considerable, amounting to 400 cal·mole^{-1} in the free energy.

An important sequel to Paper 30 is one by Davankov et al. (1974a) on the resolution of amino acids by ligand-exchange chromatography. Different combinations of fixed and mobile amino acid ligands are compared, and nickel is compared with copper as the coordinating metal ion. Another paper (Davankov et al., 1974b) describes the effects of temperature and ammonia concentration on the efficiency of resolution.

While Davankov's group has been the most productive and successful in resolving optical isomers, a very similar approach was taken by Snyder et al. (reference 19, of Paper 30).

REFERENCES

Clearfield, A. 1973. *Ion Exchange and Solvent Extraction.* Vol. 5, Chap. 1. New York, Marcel Dekker.

Davankov, V. A., S. V. Rogozhin, and A. V. Semechkin, 1974a. *J. Chromatog., 91,* 493.

——, S. V. Rogozhin, and A. V. Semechin. 1974b. *J. Chromatog., 93,* 363.

Kraus, K. A., and H. O. Phillips. *J. Amer. Chem. Soc., 78,* 249.

——, and H. O. Phillips. 1956b. *J. Amer. Chem. Soc., 78,* 694.

——, T. A. Carlson, and J. S. Johnson. 1956. *Nature, 177,* 1128.

Law, S. L. 1971. *Science, 174,* 285.

Qureshi, M., N. Zehra, S. A. Nabi, and V. Kumar. 1973. *Talanta, 20,* 609.

Rosset, R. 1966. *Bull. Soc. Chim. France, 1966,* 59.

Schmuckler, G., et al. 1972. *Anal. Chim Acta, 59,* 397.

——, et al. 1974. *Ion Exchange and Membranes, 1,* 225.

28

Reprinted from *Proc. 2nd U.N. Intern. Conf. Peaceful Uses Atomic Energy, Geneva,* Vol. 28, 3-16 (1958)

Ion Exchange Properties of Hydrous Oxides

By K. A. Kraus, H. O. Phillips, T. A. Carlson and J. S. Johnson*

During the last decade ion exchange has become an established separations technique. It has found application not only in water purification, the original major application, but also in analytical chemistry, in separation and isolation of new elements, in hydrometallurgy, and of course in many specialized fields related to the utilization of atomic energy.

The largest fraction of recent ion-exchange literature has dealt with the synthetic organic resins. Their ready availability and stability is probably largely responsible for the present popularity of ion exchange as a separations tool. The inorganic zeolites, with which all early ion-exchange work was carried out, have with the years become less popular for separations presumably because they are unstable, both in strongly acidic and strongly basic media.[2a] While other inorganic materials have also been used either for ion-exchange or adsorption processes, their application has been primarily to chromatography of organic compounds in nonaqueous media and to separations involving nonelectrolytes.

Several years ago the authors began a re-examination of the adsorptive properties of inorganic materials from the point of view of possible ion-exchange utilization.[2-4] Silicates were excluded because of their acid and base lability and because development of new adsorbents was a primary objective.

The time for systematic re-examination of adsorptive properties of inorganic materials seemed overdue because through the availability of radioactive tracers of most elements many cumbersome analytical problems had become greatly simplified. Detailed knowledge of adsorptive properties of inorganic materials had become of considerable importance in many fields such as geochemistry, heterogeneous catalysis and corrosion. Utilization of adsorption processes in the rapidly expanding atomic energy field had created a demand for materials stable in intense radiation fields and at high temperatures. Most inorganic adsorbents are expected to have radiation stability and some of these might also have high temperature stability. In addition, because of the wide variety of materials to choose from, there was a high probability that compounds with unusual selectivities would be found.

The very large class of insoluble hydrous oxides seemed to us an ideal starting point for a general investigation of adsorption and ion exchange with inorganic materials, and the present paper summarizes some of our findings. Adsorptive properties of hydrous oxides are well known[5, 6] and relatively detailed studies have been carried out with one of these, hydrous aluminum oxide. Various chromatographic processes based on aluminum oxide have been described[2e,7-14] and ion-exchange properties have been attributed at times to this[15-19] and other oxides.[20, 21] We shall attempt description of adsorptive properties in terms of ion-exchange terminology. Though in part a semantic device, this is considered useful because it permits direct adaptation of the extensive available information on ion-exchange technology and theory to new adsorbents.

An ion exchanger is usually pictured as a matrix to which charged, or potentially charged, groups are affixed. If the counter-ions which are necessary to give the material overall electrical neutrality are negatively charged, one speaks of an anion exchanger, and if they are positively charged, of a cation exchanger. In a useful exchanger these counter-ions should be readily diffusible. In the organic exchangers the diffusion medium is generally the imbibed water or electrolyte; in many inorganic materials the diffusion medium is probably the solution surrounding the solid particles or crystallites.

Most organic exchangers are considered related to high molecular weight polyelectrolytes, though cross-linked into an insoluble matrix. One is tempted to propose a similar picture for hydrous oxides. During the hydrolysis of many metal ions, positively charged polymers are formed on the acid side of the isoelectric point and negatively charged polymers have been reported (*iso*polyacids) on the alkaline side. If these polymeric aggregates are arranged into an insoluble 3-dimensional network, i.e., by a generalized cross-linking procedure, anion and cation exchangers should result. The insoluble "hydroxides" or "hydrous oxides" might be considered such 3-dimensional networks. Without anionic or cationic impurities they would be reminiscent of weakly acidic cation exchangers in the hydrogen form or weakly basic exchangers in the free base form. The networks would become charged on treatment with acids or bases.

Of course, not all elements form insoluble hydrous oxides (phosphorus, arsenic) and others form oxides which are too dispersible for use in columns (Mo(VI),

---

* Oak Ridge National Laboratory, operated by Union Carbide Corporation for the U.S. Atomic Energy Commission.

W(VI)). The oxides of these elements are very acidic and hence are expected to form cation exchangers if incorporated into an insoluble matrix. This can be done[3b,4] by precipitating certain amorphous insoluble salts, e.g., zirconium phosphate, zirconium tungstate. An excess of the acidic oxide is needed to produce cation-exchange activity. Because of the presence of excess acid in the acid washed precipitates we shall refer to these materials by the generic term "acid salts." They are included in the present paper because of the implied relation to hydrous oxides and because otherwise a comparative discussion of ion-exchange properties of hydrous oxides would be arbitrarily limited by solubility considerations.

Treatment of hydrous oxides with an amount of acid insufficient to dissolve them often yields so-called basic salts, though not necessarily with simple stoichiometric composition. Presumably hydrogen ions of the added acid react with some of the attached hydroxide ions of the solid to yield water molecules which remain attached to the metal. The network acquires a positive charge. The anions of the acid become included in the solid as counter-ions, giving it electrical neutrality. These counter-ions, if not built into the network, should be readily diffusible and hence these basic salts are expected to be anion exchangers.

If the metals on which the hydrous oxides are based are sufficiently acidic, one might expect the networks to react with, or coordinate, hydroxide ions and assume a negative charge. The cations of the base which simultaneously become incorporated in the solid should retain a high degree of diffusibility and a cation exchanger should result.

In some intermediate pH region a given solid might show both anion and cation exchange activity and thus in a sense be a giant zwitterion. At the pH at which both effects are equal (isoelectric or iso-ionic point) salt pickup predominates. In this model adsorption of a salt is directly related to the anion and cation exchange properties of the solid and does not result from an entirely different mechanism.

The model was discussed in terms of crosslinked polymers, which implies amorphous solids. However, crystalline materials should, except for adsorptive capacities, have similar properties, provided of course no phase changes occur during the adsorption process. Reversible reactions would be expected to occur only on the surfaces of these crystals. The large crystals with their relatively small surfaces should have low capacities. The capacity of microcrystalline materials might approach those of amorphous materials, where adsorption would be pictured as occurring on the "surfaces" of chains or sheets.

In some cases reactions of the solids with acids might result in substitution of a hydroxide ion by the anion of the acid, i.e., in formation of first coordination sphere complexes, rather than charged hydrated surfaces. With this mechanism, unique selectivities reflecting unique complexing properties of the metals would be expected. The exchange reactions would be rapid if the complexes formed on the surface are in rapid equilibrium with the ions in solution, which presumably would be the case with most metals.

The seat of exchange activity assumed here for the hydrous oxides is quite different from that usually assumed for zeolites and clay minerals where exchange activity is attributed to substitution in a fixed lattice of ions with lower positive charge for ions with higher charge. In the hydrous oxides, exchange activity would be imparted to the solid through an essentially reversible reaction with the medium.

However, irrespective of possible mechanisms or models, certain basic questions appeared to need experimental resolution before adsorption on hydrous oxides could usefully be described in terms of ion-exchange terminology and used in typical ion-exchange technology.

1. Are reactions of acids and bases with oxides rapidly reversible?
2. Are the adsorbed anions and cations rapidly exchangeable for other ions of the same type? In particular, are rates sufficiently fast for column operation?
3. Are changes in properties (hysteresis) sufficiently slow so that prediction of properties can be made with confidence?
4. Are capacities sufficiently high for applications?
5. How far can idealized ion-exchange relationships be applied to the adsorption reactions? As a corollary, are the adsorptive properties reasonably independent of loading?

Some of these questions had previously been answered affirmatively for hydrous aluminum oxide. As we shall demonstrate with typical results, we have found additional affirmative answers for other hydrous oxides and "acid salts" sufficient to imply that these classes of compounds indeed are suitable for ion exchange applications.

PREPARATION

The adsorbents were prepared by suitable precipitation reactions.[3,4] In the precipitation of metals which are soluble in the common mineral acids (Cr(III), Zr(IV), Sn(IV), Th(IV), etc), acidic solutions were treated with base (usually NH_3) until precipitation was complete. In the case of Nb(V), precipitation was carried out from HF–HNO_3 solutions. For Ta(V), solutions in base were prepared from the products of carbonate fusions (tantalate) and the hydrous oxides precipitated by addition of acid. Similarly, vanadium-(V) was precipitated by addition of acid to a vanadate solution.

Since some of the most acidic oxides are readily dispersible (e.g., Mo(VI), W(VI)), or soluble (e.g., P_2O_5, As_2O_5), their expected ion-exchange characteristics could be utilized only by the more devious technique of precipitating a suitable metal (e.g., Zr(IV)) with an excess of appropriate reagent (Na_2WO_4, H_3PO_4, etc.). The resulting solids shall be referred to as zirconium tungstate, zirconium phosphate, etc. The other solids shall be referred to as hydrous oxides. These designations will be used whether or not the solids have simple stoichiometric composition or special crystal structure.

The usually flocculent or gelatinous precipitates were collected by centrifugation or filtration, dried at room temperature, ground in a mortar, and sized by screening and decantation. For column operation materials of mesh size *ca.* 150 were usually selected. In some cases, portions of the solids were dried or fired overnight at controlled elevated temperatures.

No attempt was made to prepare true gels, although in a few instances (particularly with zirconium tungstate) gels happened to form. It is expected that, in general, dried ground gels (or beaded gels) would have better packing characteristics in columns than the materials obtained from flocculent precipitates, that they would have slower equilibration (diffusion) rates but otherwise have similar adsorptive properties. Indeed Amphlett and coworkers, utilizing principally gels, confirmed our results in substantial detail.[22, 23]

REVERSIBILITY OF ADSORPTION REACTIONS

In the precipitation of hydrous oxides or hydroxides pure compounds are rarely obtained. The precipitates frequently retain some of the anions of the solution or cations of the precipitating base. These "impurities" are usually difficult to remove by washing with water. This is expected from an ion-exchange point of view (electroneutrality) since the concentration of hydrogen and hydroxide ions available for replacing the ionic "impurities" is very small in the wash water. Usually, the impurities may be removed readily by treatment with excess acids or excess base. During this treatment, however, new "impurities", the anions of the acid or the cations of the base, may be introduced.

As a typical example consider hydrous zirconium oxide. The solid precipitated with ammonia from a concentrated solution of zirconium oxychloride, after drying and washing with water, was found to release approximately 0.3 moles of chloride ions per kg of solid on treatment with excess base. If the base used is sodium hydroxide, containing radioactive sodium tracer (e.g., Na^{22}), it can be readily demonstrated that the resulting solid contains substantial amounts of sodium ions. The extent of base pickup, as well as acid pickup (see also ref 24), seems to depend principally on the composition of the solution (pH and salt concentration) as also recently pointed out by Amphlett and coworkers.[22] Through suitable analyses a composition-pH diagram may be constructed. Typical diagrams of this type (titration curves) which are standard for characterization of organic ion exchange resins are illustrated in Fig. 1 for Zr(IV) and Sn(IV). These titration curves were obtained at essentially constant ionic strength (0.1*M*) using NaCl–HCl and NaCl–NaOH mixtures.

For both Zr(IV) and Sn(IV) the chloride content of the solid decreases continuously with increasing pH, becoming essentially negligible in basic solutions. Conversely the sodium content of the solids is negligible in acidic solutions and increases continuously with pH. These titration curves are quite reproducible for a given solid, i.e., they do not differ substantially after several cycles of acid and base treatment. They are also reasonably reproducible for solids prepared in approximately the same way and dried at the same temperature.

Further proof of reversibility may be obtained by studying the equilibrium acidity as a function of supporting electrolyte concentration. If the composition of the solid is kept constant, the activity coefficients in the solid are independent of solution composition and hence, at equilibrium, the activity *a* of the adsorbed electrolyte is also constant. We have attempted to check this relationship by studying the acidity of various NaCl solutions in equilibrium with a hydrous zirconium oxide sample containing *ca.* 0.7 moles of Cl^- per kg of solid and no significant amounts of sodium ions. As expected from the equation

$$a_{HCl} = m_H m_{Cl} \gamma_{\pm}^2 = \text{constant} \qquad (1)$$

the equilibrium acidity m_H was found to decrease with increasing chloride concentration m_{Cl}. Indeed Eq. 1 was verified to better than $\pm 5\%$ in the chloride concentration range 10^{-3} to 1*M*.

A still more stringent test of reversibility is a study of acid and base adsorption as a function of temperature.[25] The studies again were carried out at constant loading of the solid with respect to HCl or NaOH and a typical set of data is illustrated in Fig. 2. From this plot of log m_H (or log m_{OH}) vs. $1/T$, where T is the absolute temperature, the values of ΔH of adsorption in the temperature range studied were found to be -6.9 and -3.8 kcal per mole of HCl and NaOH, respectively. Repetitive runs on the same sample did not exactly reproduce the equilibrium acidities shown in Fig. 2, presumably because of an ageing process accelerated at higher temperatures. Acidities found at a given temperature for the rising and falling branches of a temperature cycle differed at most by a factor of 2; this corresponds to an extremely small change in the composition (loading) of the solid. In any case, the slopes of the lines, particularly those obtained with decreasing temperatures, were quite reproducible.

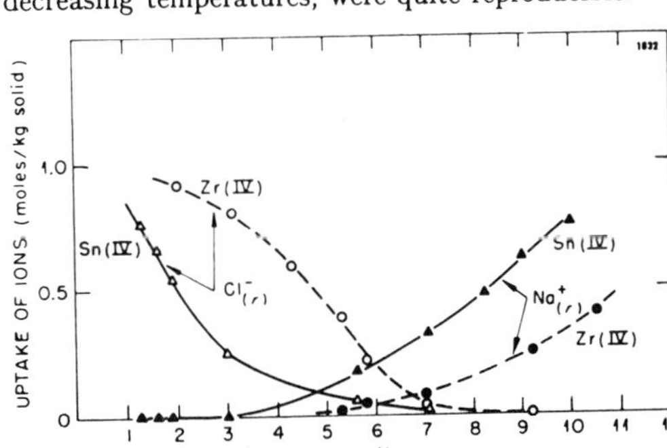

Figure 1. Anion- and cation-exchange properties of hydrous Zr(IV) and Sn(IV) oxides

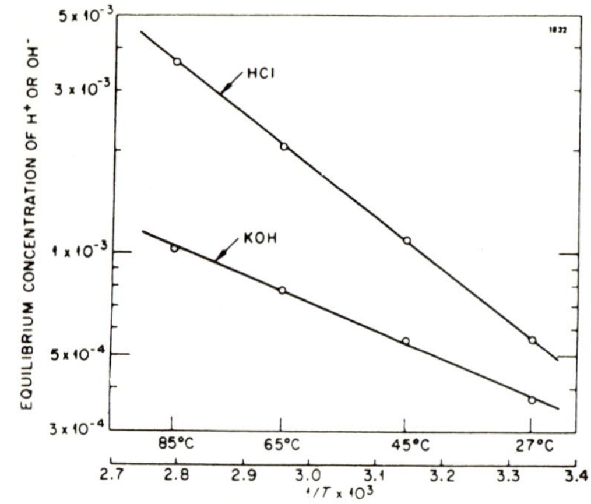

Figure 2. Temperature dependence of HCl and KOH adsorption by hydrous zirconium oxide

In one case, adsorption of HNO_3 by hydrous zirconium oxide, the study was carried to 150°C. After ageing for several hours at 150°C equilibrium acidities were measured with decreasing temperature. A plot of log m_H vs. $1/T$ was a straight line over the whole temperature range. The value of ΔH of adsorption (−8.4 kcal) was in reasonable agreement with that found in the lower temperature studies for HCl.

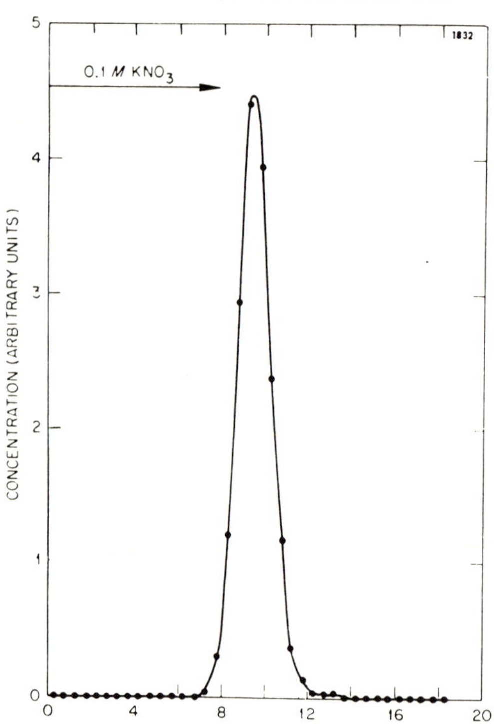

Figure 3. Elution of Br⁻ from a hydrous zirconium oxide column (drying temp. 200°C, 5.0 cm × 0.20 cm² column, flow rate 0.8 cm/min)

We shall give here one additional example illustrating reversibility. Bromide tracer was added to a column (0.2 cm² × 5.0 cm) of hydrous zirconium oxide in the nitrate form (nitric acid washed). The adsorbed tracer was eluted with 0.1M KNO_3 at a flow rate of 0.8 cm/min and the effluent analyzed radiometrically for bromide ions. The resulting elution curve is reproduced in Fig. 3. It has the typical Gaussian shape expected for rapid ion-exchange chromatography, proving that rates of adsorption and desorption are sufficiently rapid even for standard column operation.

APPARENT CAPACITIES

Acid-base titration curves like those of Fig. 1 give directly the apparent capacities of the exchangers (e.g. as moles per kg adsorbent). In a given medium the relative anion exchange (chloride uptake) and cation exchange (sodium uptake) capacities of different oxides may differ substantially. Thus, in acidic solutions Zr(IV) is able to hold more chloride ions than Sn(IV) and, conversely, in basic solutions Sn(IV) holds more sodium ions than Zr(IV). In a study of a large number of hydrous oxides it became apparent that the relative importance of the anion and cation exchange branches may differ widely and in some cases one or the other branch may be absent. Thus no detectable anion-exchange branch was observed for very acidic oxides such as Mo(VI), W(VI) U(VI) and V(V) and only small cation-exchange branches for such basic oxides as Th(IV) and Bi(III).

While the capacities determined by acid-base titration (see Fig. 1) are quite large in terms of moles per kg of solid or moles per liter of bed, considerably higher apparent capacities could have been realized if the solids had been treated with more acidic or more basic solutions or with anions which are more strongly adsorbed. Thus fluoride uptake as high as 4 moles per kg has been realized for hydrous zirconium oxide, which corresponds to almost 1 mole of F^- per mole of Zr(IV).

Similarly, high capacities were observed in acid-base titrations of "acid salts". For example, a zirconium phosphate sample, dried at 25°C was titrated

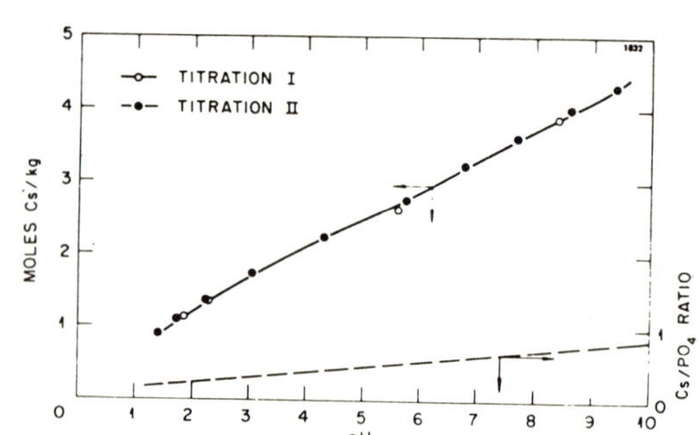

Figure 4. Cesium uptake by "zirconium phosphate" (drying temp 25°C, pretreatment 1M HCl, medium 0.1M CsCl, titrant 0.2M CsOH)

in 0.1*M* CsCl with 0.2*M* CsOH containing Cs^{134} tracer. Apparent capacities, determined radiometrically from the cesium content of the solid, rose from about 1 mole Cs per kg near pH 1.5 to about 4.5 moles per kg near pH 9.5 (see Fig. 4). The higher figure corresponds to almost 1 mole Cs/mole PO_4 (or 2 moles Cs/mole Zr(IV)) which would be one-half of the value expected if the solid had the composition $ZrO(H_2PO_4)_2$, i.e., had four replaceable hydrogen ions per mole of zirconium or two replaceable hydrogen ions per mole of phosphate.

Acid-base titrations of zirconium tungstate or zirconium molybdate are complicated because at sufficiently high pH loss of the anion from the solid becomes serious with a resulting decrease in the capacities computed per kg of original solid (see Fig. 5). However, the maximum ratios of moles Cs per mole molybdenum or tungsten on the solid are also not much less than 1:1.

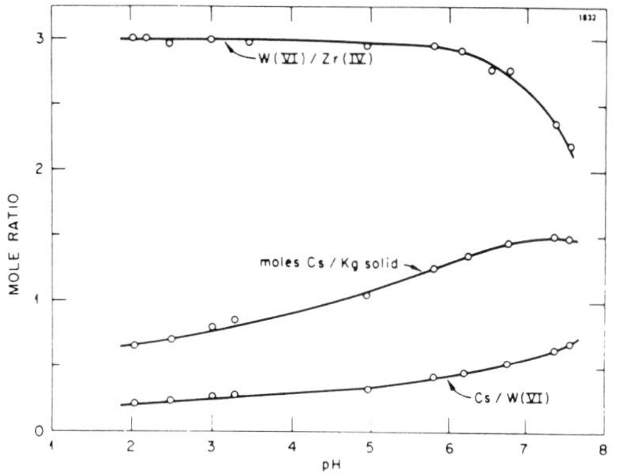

Figure 5. Stability and ion-exchange properties of Zr(IV)–W(VI) precipitates (ammonia titration in 1*M* NH_4Cl, followed by Cs^+–NH_4^+ exchange; drying temp 25°C)

EFFECT OF DRYING TEMPERATURE

Increase in the drying temperature of the solids generally causes a decrease in the apparent capacity. In a few cases, such as zirconium tungstate, drying at temperatures as low as 100°C causes essentially complete destruction of the cation exchange properties. However, for most hydrous oxides the data shown in Table 1, which were obtained for hydrous zirconium oxide, are typical.

Table 1. Effect of Drying Temperature on Cr(VI) Uptake by Hydrous Zirconium Oxide (Column Breakthrough Method with Acidic Cr(VI) Solutions)

Temp. °C	*Moles Cr(VI)* / *kg solid*	*Moles Cr(VI)* / *liter bed*	% H_2O	*Moles Cr(VI)* / *mole Zr(IV)*
32	1.38	0.85	46.4	0.32
105	1.25	1.12	19.4	0.19
202	0.91	0.87	15.4	0.13
320	0.45	0.47	7.0	0.06
500	0.14	0.18	1.1	0.02
800	0.05	0.07	(0) [a]	0.008

[a] Assumed value.

The chromate uptake data were obtained by measuring breakthrough curves of chromate with Cr(VI) solutions containing at least 2 moles of acid (HNO_3 or HCl) per mole of Cr(VI). Exhaustion of the column can be followed visually since the chromate (or dichromate) forms a sharp, strongly colored band. If chromium tracer has been added (Cr^{51}) quantitative breakthrough curves may be established quickly by radiometric methods. This chromate exhaustion technique was found to be a reliable and convenient method for comparing the anion-exchange capacities of oxides.

Increase in drying temperature usually causes a decrease in apparent capacity. In a few cases slight increases in capacity were found after drying at 100 or 200°C, presumably because with loss of excess water the number of moles of metal oxide per kg of solid had been substantially increased. The loss in capacity which occurs on drying at higher temperatures is almost surely caused by a growth in the size of crystallites which causes a decrease in the effective surface area per mole of oxide. This crystallite growth was independently confirmed by the sharpness of the diffraction lines observed in X-ray crystallographic examinations.

ION-EXCHANGE IDEALITY

Ion-exchange equilibria may be written[2b] in the form

$$a\mathrm{B}^{\pm b} + b\mathrm{A}_{(\mathrm{r})}^{\pm a} \rightleftharpoons b\mathrm{A}^{\pm a} + a\mathrm{B}_{(\mathrm{r})}^{\pm b} \qquad (2)$$

with the equilibrium constant

$$K_{\mathrm{AB}} = \frac{m^a_{\mathrm{B(r)}} m^b_{\mathrm{A}}}{m^a_{\mathrm{B}} m^b_{\mathrm{A(r)}}} \cdot \frac{\gamma^a_{\mathrm{B(r)}} \cdot \gamma^b_{\mathrm{A}}}{\gamma^a_{\mathrm{B}} \cdot \gamma^b_{\mathrm{A(r)}}} = K^m_{\mathrm{AB}} \Gamma_{\mathrm{AB}} \qquad (3)$$

where m is the concentration of the ion indicated by subscript, subscript (r) denotes the exchanger phase, no subscript the aqueous phase K_{AB} is the equilibrium constant, which with proper choice of standard states may be set equal to unity, K^m_{AB} is the concentration quotient and Γ_{AB} the appropriate activity coefficient quotient. Equation 3 is general for two-phase equilibria although in some cases the activity coefficients may assume quite unpredictable values. It will be useful as an ion-exchange equation if the activity coefficients in the solid (exchanger) phase do not vary rapidly, at least under some limiting conditions.

With organic exchangers the activity coefficient ratio $\gamma^a_{\mathrm{B(r)}}/\gamma^b_{\mathrm{A(r)}}$ for the exchanger phase is essentially constant, if exchange reactions are carried out in moderately dilute electrolyte solutions, if one of the adsorbed ions ($\mathrm{B}^{\pm b}$) occurs at trace concentrations, and if no complexing or hydrolytic reactions occur. The composition of the exchanger ($m_{\mathrm{A(r)}}$) is then essentially independent of the concentration of the adsorbed tracer ion and Eq. 3 simplifies to

$$D_{\mathrm{B}} m_{\mathrm{A}}^{b/a} \cdot \frac{\gamma_{\mathrm{A}}^{b/a}}{\gamma_{\mathrm{B}}} = \text{constant} \qquad (4)$$

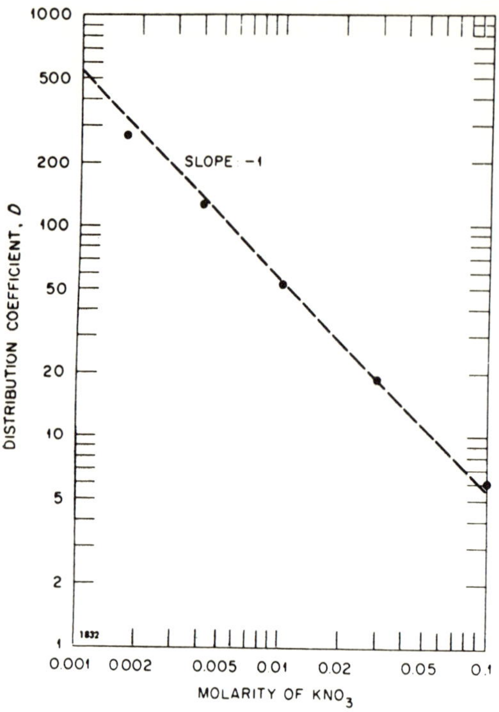

Figure 6. Bromide-nitrate exchange with hydrous zirconium oxide (drying temp 200°C)

where the distribution coefficient, D_B, is defined by

$$D_B = \frac{m_{B(r)}}{m_B} \qquad (5)$$

According to Eq. 4 a plot of log D_B vs. log m_A should, except for activity coefficient corrections in the aqueous phase, be a straight line with slope $-b/a$. This relationship has been verified for several exchange reactions involving tracer ions adsorbed on hydrous oxides as illustrated in Figs. 6 and 7.

For ion-exchange terminology to be generally useful, deviations from ideality should also not be excessive when adsorption experiments are carried out as a function of loading. Equivalently, K_{AB}^{m} should not change rapidly as $m_{B(r)}$ increases and $m_{A(r)}$ decreases, under conditions where the sum $(m_{B(r)} + m_{A(r)})$, the apparent capacity, is approximately constant. This aspect of ideality has been checked in a few cases. Typical is the exchange of Cs^+ and NH_4^+ ions on zirconium tungstate where K^m, though not independent of $m_{Cs(r)}$, varied by less than a factor of two as $m_{Cs(r)}$ increased from zero (trace concentration) to one-half of apparent capacity. There was no evidence for a small number of "active sites" as so often proposed in adsorption studies.

Since adsorbabilities closely follow the ideal dependence on ionic strength and loading, there seems little doubt about the general usefulness of applying ion-exchange terminology to these adsorption reactions. Of course it must be remembered that these ideal relationships can be found only with certain simple ions which do not undergo complexing, polymerization or hydrolytic reactions in the media of interest.

SELECTIVITY AND SEPARATIONS

One of the most attractive features of hydrous oxide adsorbents, compared with standard organic ion-exchange resins, is their unusual and sometimes unique selectivity. The examples which will be discussed have been selected to illustrate this point. For this reason complexing reactions in the aqueous phase were avoided although these are standard devices for enhancing separations with organic exchangers. As long as the inorganic solids are not dissolved by the complexing agents, the improvement in separations should be similar for organic and inorganic exchangers.

Adsorption of Polyvalent Anions

Most hydrous oxides seem to have unusual selectivities for polyvalent anions as compared with monovalent anions. This property was implied in our proposal to use adsorption of chromate as a standard technique for evaluating anion-exchange capacities. Since many heavy metals occur normally in solution as polyvalent anions (e.g., Mo(VI), W(VI), V(V)), the hydrous oxides might find application in problems involving concentration or recovery of these elements. Other ions which are strongly adsorbed include borate, sulfate, phosphate, arsenite and arsenate.

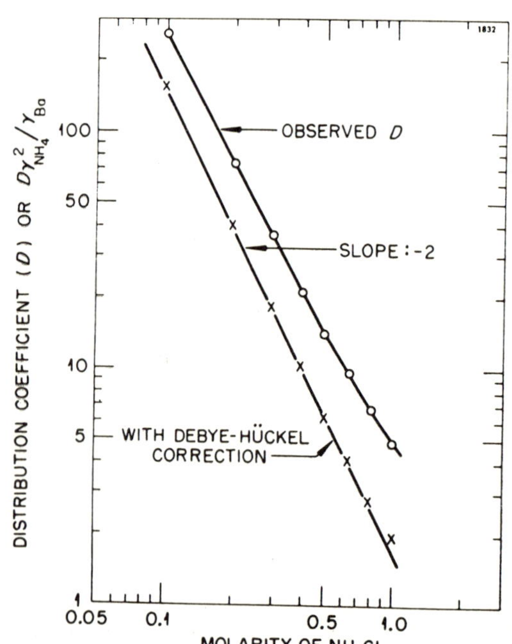

Figure 7. Adsorption of trace Ba^{++} on zirconium tungstate

Since the polyvalent anions are so strongly adsorbed, their removal from the exchangers may at times be difficult. Probably the most effective eluants are strong bases such as NaOH, although weaker bases such as ammonia (and in some cases carbonate, acetate, etc.) may be satisfactory. While displacement of weakly adsorbed anions from the exchangers can usually be achieved essentially quantitatively (see Fig. 3) elution, even by bases, of strongly adsorbed anions may be incomplete. For example, columns loaded with chromate and then eluted either with ammonia or NaOH, frequently retain a yellow cast which can be removed only by prolonged treatment with strong base. The per cent chromate lost in a rapid elution cycle, however, is usually small ($< 5\%$) and seems relatively independent of the amount of chromate originally adsorbed, i.e., approximately the same fraction is retained with tracers as with macro quantities. This small "irreversibility" probably results from changes in the structure of the solids as they are treated with strong bases, and presumably involves covering of exchange sites by the products of slowly reversible hydrolysis reactions.

Special Anion Selectivities

Whenever unusual complexing properties of the metals can be utilized, the adsorbents should have special selectivities. This effect was illustrated by the enormous capacity of zirconium oxide for fluoride ions which is expected from the fact that Zr(IV)is strongly complexed by this anion. Adsorbents prepared from certain transition elements, which are strongly complexed by the higher halides, should have special selectivity for these anions. For this reason studies have been carried out on the adsorption properties of bismuth oxide for higher halide ions and the anticipated special selectivity was confirmed. Since Bi_2O_3 readily undergoes phase changes (oxyhalides, etc.) when treated with acids, the flow characteristics of columns prepared from Bi_2O_3 often deteriorate rapidly. However, this difficulty can be avoided and selectivity sometimes even improved if mixed oxides are prepared by coprecipitating Bi(III) with other metals (e.g., Sn(IV)).

Separation of Negatively Charged Complexes

The remarkable selectivity of exchange resins for certain negatively charged metal complexes has become the basis of many separations processes.[26] Although there is ample reason to presume that this special selectivity is related to properties of the organic network, it was nevertheless of interest to investigate whether negatively charged complexes could also be adsorbed by and separated on hydrous oxides.

Some adsorption of negatively charged complexes on hydrous oxides has been observed. While it is quite modest compared with that found for organic exchangers, it is sufficient to allow some separations. An example, separation of Ag(I) and Au(III) in HCl,

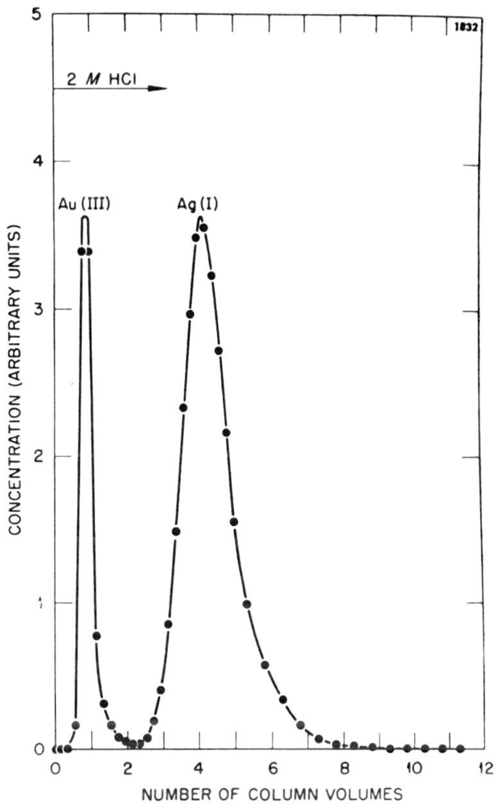

Figure 8. Separation of Au(III) and Ag(I) with hydrous zirconium oxide (drying temp 200°C, 3.6 cm $\times$ 0.28 cm² column, flow rate 0.5 cm/min)

where they are well known to form strong negatively charged complexes, is given in Fig. 8. This separation almost surely involves anion exchange of metal complexes since hydrous zirconium oxide in strongly acidic solution is almost exclusively an anion exchanger, and since adsorbability of both elements increases with decreasing HCl concentration approximately as expected from anion exchange-mass action considerations.

Separation of Alkali Metals

Cation exchangers prepared from acidic oxides (e.g., zirconium phosphate, tungstate or molybdate), when largely in the hydrogen form, were found to have remarkable selectivity for the various alkali metals. Adsorbability increases sharply with atomic number and is extremely high for cesium in acidic solutions. Under these conditions cesium is so strongly adsorbed that it may be isolated essentially uniquely from all other elements in the Periodic Table. It has been removed, for example, from acidic solutions of lithium chloride (13.6M) and of aluminum nitrate (2M). It may also, as a further example, be removed from uranyl sulfate solutions.

Complete separation of the alkali metals from each other has been achieved with relatively small columns as illustrated in Fig. 9. The elements were adsorbed on zirconium tungstate. Elution was carried out with ammonium chloride solutions whose concentrations were increased after each element had been removed to compensate (mass action law) for the high adsorbabilities of the higher alkali metals and to obtain them in narrow bands.

With some adsorbents the elution order of the alkali metals may be reversed, e.g., Cs^+ may be eluted before Na^+. This occurs, for example, with hydrous titanium oxide in base and, interestingly, with uranium-(VI) oxide.

Such reversal of selectivity is also implied, to some extent, by the decrease of selectivity which occurs when the insoluble "acid salts" are neutralized with ammonia (converted from the hydrogen to the ammonium form). Indeed, after complete conversion to the ammonium form, zirconium phosphate, for example, has essentially no selectivity left for the alkali metals and they may readily be eluted as a group from a column (see below).

Separation of Alkaline Earths

The "acid salts" have been found to have excellent selectivity for the alkaline earths. As demonstrated in Fig. 10, the elements from calcium to radium may be completely separated from each other with a relatively small column of zirconium molybdate. Because of the large differences in adsorbability, a stepwise elution technique was chosen, as in the separation of the alkali metals. Particularly interesting is the extremely large difference in the adsorbability of barium and radium, which by this technique can be separated with columns only a few centimeters in length.

Separation of Alkali Metals from Alkaline Earths

The extreme versatility of hydrous oxides and "acid salts" for separations may be illustrated by the large variety of techniques which can be used for separation of two elements, Cs^+ and Ba^{2+}.

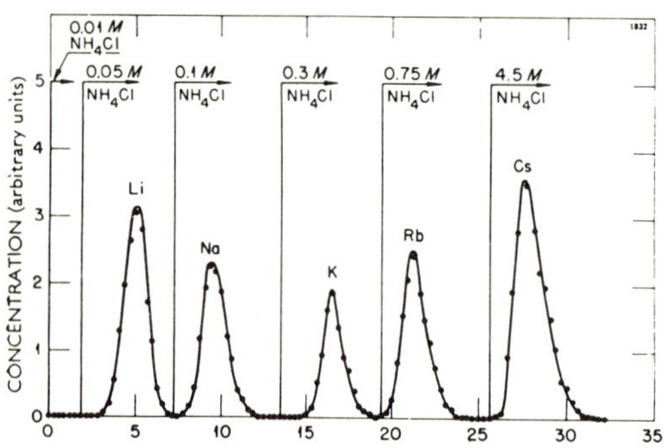

Figure 9. Separation of the alkali metals (zirconium tungstate, 12.3 cm × 0.13 cm² column, flow rate ca. 0.75 cm/min.)

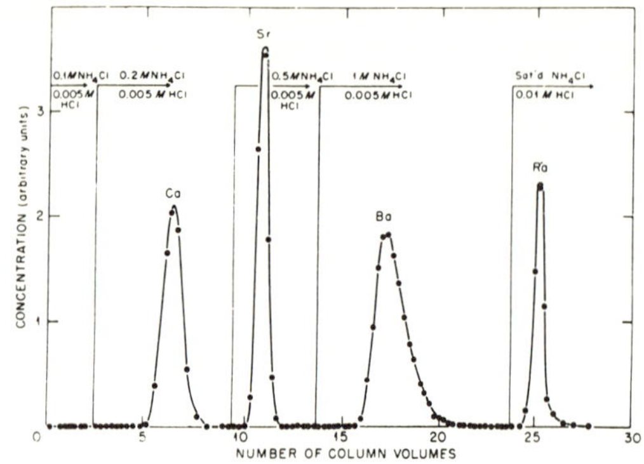

Figure 10. Separation of alkaline earths with zirconium molybdate (drying temp 25°C, 0.19 cm² × 10.0 cm column, flow rate 1.1 cm/min)

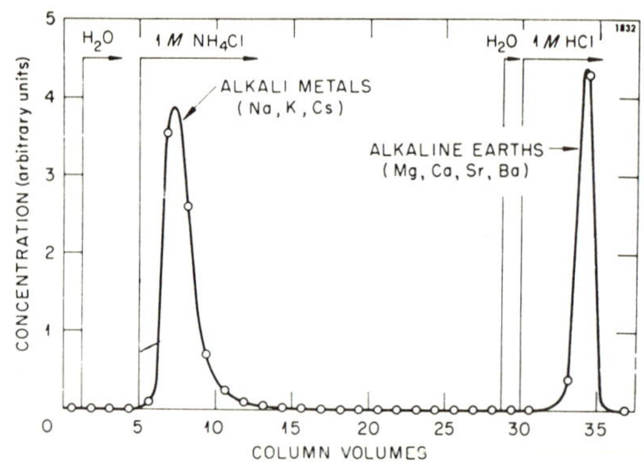

Figure 11. Group separation of alkali metals and alkaline earths with zirconium phosphate (drying temp 25°C, pretreatment 1M NH_3, column 0.2 cm² × 2.0 cm, flow rate 1.1 cm/min, loading 0.052 eq/kg/element)

In strongly acidic solutions, for example, Cs^+ is strongly held by zirconium phosphate. Under these conditions Ba^{2+} is essentially not adsorbed, and separation of these elements can be achieved rapidly with a column less than a centimeter long. The separation is sufficiently rapid to permit complete isolation of the short-lived Ba^{137} from its long-lived parent by a milking procedure involving flushing a column containing Cs^{137} with acid.

The elution order may be inverted and separations carried out equally rapidly and effectively by using a salt form (NH_4^+) of the exchanger and using an ammonium salt as eluant. A group separation involving various alkali metals and alkaline earths which is based on this principle is demonstrated in Fig. 11.

Techniques for separation of alkali metals and alkaline earths can, of course, also be developed with other hydrous oxides. If these separations are carried out in neutral or weakly acidic solutions Ba^{2+} is usually much more strongly adsorbed than Cs^+. The latter can be eluted from short columns with moderately concentrated salt solutions (e.g., NH_4Cl). More

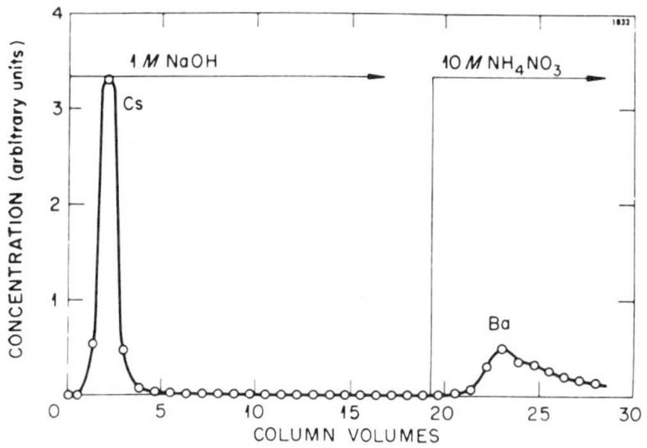

Figure 12. Cation-exchange separation of Cs^+ and Ba^{++} with uranium-(VI) oxide (drying temp 25°C, column 0.2 cm² × 3 cm, flow rate 1 cm/min)

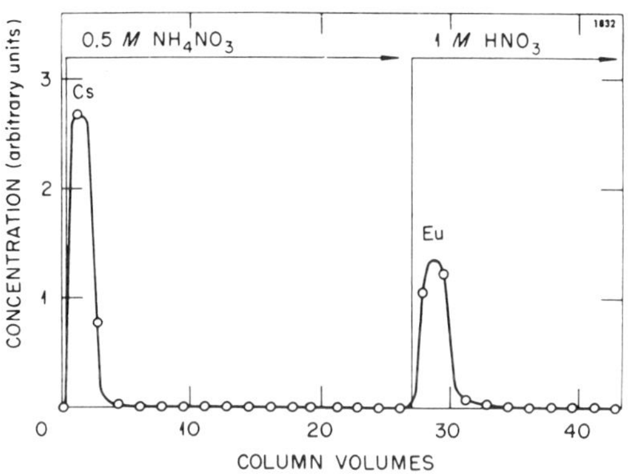

Figure 13. Separation of alkali metals from rare earths with zirconium oxide (drying temp 300°C, pretreatment NaOH, column 0.2 cm² × 1.5 cm, flow rate 1.3 cm/min)

concentrated or acidic solutions are necessary for removal of Ba^{2+}. Separations based on this procedure have been carried out with hydrous oxides such as ZrO_2 and SnO_2. Interestingly, uranium-(VI) oxide may also be used, as illustrated in Fig. 12, though solubility considerations limit the acid range suitable for elution.

Other Separations

In our discussion of cation exchange separations we have so far concentrated on the alkali metals and alkaline earths. This was done because ambiguities regarding adsorption mechanisms could thus best be avoided. These elements do not undergo extensive hydrolytic reactions in the media considered, and questions regarding species in solution normally do not arise. However, separations with hydrous oxides as adsorbents are of course not limited to these elements.

For example, rare earths can be adsorbed on hydrous oxides from basic, neutral and even moderately acidic solutions. With eluants of controlled pH and proper electrolyte concentration they may be separated from alkali metals and alkaline earths. Figure 13, involving separation of Eu^{3+} and Cs^+ with hydrous zirconium oxide is a typical illustration.

Another example, separation of Co(II) and Fe(III), is presented in Fig. 14. Both elements are strongly adsorbed by zirconium tungstate from moderately acidic or neutral solutions. Cobalt can be eluted at some intermediate acidity where iron remains strongly adsorbed. Iron may then be eluted either at very high acidity or by taking advantage of complexing reactions.

Extremely interesting is the strong adsorption of copper by hydrous oxides (e.g., hydrous zirconium oxide) from ammoniacal solutions. The intense blue color of the adsorption band indicates that Cu(II) is held as the characteristic ammonia complex. Effective capacities are high. Elution can be achieved with acid, though it is difficult to remove in this manner the last traces of copper. Nevertheless, many interesting separations processes involving transition elements may be based on this ability of the oxides to adsorb ammonia complexes.

CONCLUSION

We have endeavoured in this brief summary of our work on ion-exchange properties of hydrous oxides to demonstrate the great promise these and other inorganic materials hold for chemical separations. The studies have so far been limited to compounds containing a few types of metals and anions. No doubt, other interesting separations will be found as the studies are extended to other metals and anions, particularly those with strikingly different complexing properties. However, aside from separations, we believe that adsorption studies of the type discussed are particularly intriguing because with their simplicity and high sensitivity they should develop into powerful tools for equilibrium and kinetic studies of surfaces and for comparative studies of amorphous and microcrystalline materials.

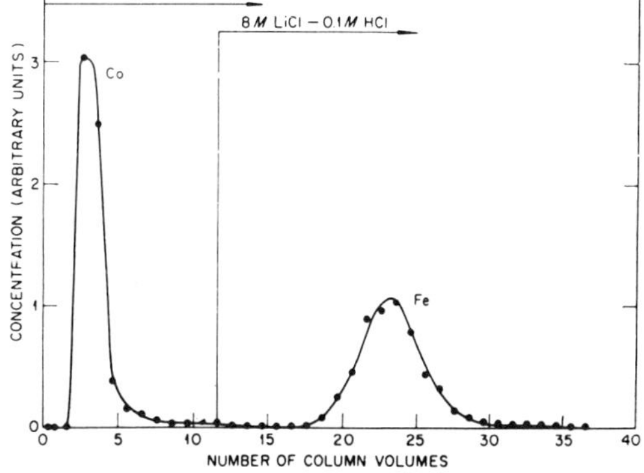

Figure 14. Separation of Co(II) and Fe(III) (zirconium tungstate drying temp 25°C, 5.5 cm × 0.09 cm² column, flow rate 0.5 cm/min)

REFERENCES

1. E. Lederer and M. Lederer, *Chromatography*, (*a*) p. 71, (*b*) chap. 40, 2nd Edition, Elsevier Publishing Co., New York (1957).
2. K. A. Kraus, (*a*) Gordon Research Conference, Ion Exchange, New Hampton, N. H. (June 1955). (*b*) *Trace Analysis* (Symposium, New York Academy of Medicine, November 1955), p. 34, J. H. Yoe and H. J. Koch, Editors, John Wiley and Sons, New York (1957). (*c*) Fall Meeting, American Chemical Society, Atlantic City, N. J. (September 1956), Chem. Eng. News, *34*, 4760 (1956). (*d*) K. A. Kraus, H. O. Phillips, T. A. Carlson, J. S. Johnson and D. J. Coombe, Oak Ridge National Laboratory Reports ORNL-1853 (January 1955), ORNL-1895 (April 1955), ORNL-1943 (July 1955), ORNL-2004 (October 1955), ORNL-2057 (January 1956), ORNL-2386 (June 1957).
3. K. A. Kraus and H. O. Phillips, (*a*) J. Am. Chem. Soc., *78*, 249 (1956), (*b*) J. Am. Chem. Soc., *78*, 694 (1956).
4. K. A. Kraus, T. A. Carlson and J. S. Johnson, Nature, *177*, 1128 (1956).
5. H. B. Weiser, *The Hydrous Oxides*, McGraw-Hill Co., Inc., New York (1926).
6. V. R. Deitz, *Bibliography of Solid Adsorbents*, (*a*) *1900–1942*, Lancaster Press, Inc., Lancaster, Pa. (1944), (*b*) *1943–1953*, National Bureau of Standards, Circular 566 (1956).
7. G. M. Schwab and K. Jockers, Angew. Chem., *50*, 546 (1937).
8. G. M. Schwab and A. N. Ghosh, Angew. Chem., *53*, 39 (1940).
9. O. Erämetsä, Th. G. Sahama and V. Kanula, Ann. Acad. Sci. Fennicae, *57*, Paper No. 3 (1940).
10. H. Kubli, Helv. Chim. Acta, *30*, 453 (1947).
11. R. Lindner, Z. Elektrochem., *54*, 421 (1950).
12. F. Nydahl and L. A. Gustafsson, Acta Chem. Scand., *7*, 143 (1953).
13. S. N. Tewari and S. Ghosh, Kolloid-Z., *130*, 167 (1953).
14. W. Noddack and E. Bankmann, Z. Elektrochem., *58*, 725 (1954).
15. G. M. Schwab and G. Dattler, Angew. Chem., *50*, 691 (1937).
16. G. M. Schwab and A. N. Ghosh, Angew. Chem., *52*, 666 (1939).
17. L. Sacconi, Discussions Faraday Soc., *7*, 173 (1949).
18. J. A. Dean, Anal. Chem., *23*, 202 (1951).
19. G. M. Shuvaeva and E. N. Gapon, Zhur. Anal. Khim., *8*, 50 (1953).
20. M. H. Kurbatov, G. B. Wood and J. D. Kurbatov, J. Phys. Chem., *55*, 1170 (1951).
21. H. Kautsky and H. Wesslau, Z. Naturforsch., *9B*, 569 (1954).
22. C. B. Amphlett, L. A. McDonald and M. J. Redman, Chem. and Ind. (London), 365 (1957).
23. C. B. Amphlett, L. A. McDonald and M. J. Redman, Chem. and Ind. (London), 1314 (1956).
24. E. Wedekind and H. Wilke, Kolloid-Z., *34*, 83, 283; *35*, 23 (1924).
25. K. A. Kraus, D. J. Michelson and R. J. Raridon, Unpublished results.
26. K. A. Kraus and F. Nelson, *Anion Exchange Studies of the Fission Products*, Proceedings of the International Conference on the Peaceful Uses of Atomic Energy, Geneva 1955, P/837, Vol. 7, p. 131, United Nations, New York (1956).

Mr. K. A. Kraus (USA) presented P/1832 above at the Conference and added the following remarks:

With the ready availability of organic ion-exchange resins, ion exchange in recent years has become an established separations technique. Simultaneously—at least for aqueous systems—inorganic exchangers have become less favored, presumably because of the acid and base lability of their most important representatives, the zeolites, glauconites, and similar silicates. However, inorganic exchangers should have a great deal to offer to separations chemistry, particularly in atomic energy applications, provided cheap and stable materials with reasonable capacity and rapid exchange rates could be prepared. In contrast to the organic resins these would probably be stable to radiation and powerful oxidizing agents. Some should be more selective than the organic exchangers or at least exhibit selectivities of a different type. They might also be more stable at high temperatures provided, of course, difficulties related to hydrothermal synthesis of minerals could be avoided or circumvented. For these and other reasons, we initiated several years ago at the Oak Ridge National Laboratory a systematic study of adsorption on inorganic materials from the ion exchange point of view. Since then, a similar study has also been taken up by Amphlett and coworkers at Harwell, who, in general, seem to confirm our results. Some of these results will be presented later in this session. Similar studies were also initiated by Gal and Gal (Yugoslavia) and Huré and coworkers (France) who submitted to this conference Papers 468 and 1155, respectively.

We began with an examination of the properties of "hydrous oxides", particularly of the more insoluble ones. Increased understanding of their adsorptive properties would, of course, have pertinence to fields other than separations chemistry, e.g., to geochemistry, because of the wide abundance of oxides in nature; to corrosion, because corrosion films and corrosion products have that composition; and to catalysis, because many oxides are catalytically active in gas-phase reactions and they might, suspended in aqueous media, catalyze reactions as organic resins do.

We shall use the term "hydrous oxide", or "oxide" for short, in a loose connotation to mean materials which could be precipitated from aqueous media and then dried at various temperatures. They need not have specific water content or crystal structure nor need they be free of other ionic impurities. The preparation of the materials which we have used is simple, consisting essentially of typical precipitation reactions, washing, drying and sizing. Gels, which may also be used, though somewhat more difficult to prepare and often with slower response, have the advantage that they can be obtained in larger, more granular particles than precipitated materials and thus impart to columns better flow characteristics.

We were led to an examination of hydrous oxides from a long-standing interest in the hydrolysis of metal ions. Hydrolytic polymers are an important part of this study. From there it is a short step to visualize "cross-linked" hydrolytic·polymers which in analogy to cross-linked organic polyelectrolytes should have ion-exchange properties. Thus, should the hydrolytic polymers carry a positive charge, as is the case on the acid side of the isoelectric point, and were they connected into a three-dimensional network, anion-exchange properties would be predicted for the solids. The simplest means, conceptually, for establishing cross-links would, of course, be precipitation of hydroxides, hydrous oxides, or basic salts whose study, conversely, is an integral part of any general program of hydrolytic investigations.

Existence of anion-exchange properties can easily be demonstrated for these materials (see Fig. 15). Thus, if an acidic solution of chromate is passed through a column containing a typical hydrous oxide, such as that of Zr(IV), a tightly held, sharp, yellow band of adsorbed chromate forms. If the solution is not too concentrated, many column volumes may be passed through before chromate appears in the effluent. From this chromate retention one may evaluate the apparent capacities of the oxides which are in the general vicinity of 1 to 2 moles per kg of solid, provided the solid was not dried at too high a temperature. Elution may easily be achieved by treatment with base. A small percentage of chromate may remain, which may give a yellow cast to regenerated columns, presumably because hydrolytic reactions during the acid-base cycle may cover or coat some of the surface of the oxide particles.

If a major pH change is avoided, this relatively minor difficulty seems not to occur, as illustrated in Fig. 3 of P/1832. Bromide tracer was adsorbed on acid-washed hydrous zirconium oxide and eluted with nitrate. A symmetrical and narrow band was obtained, implying rapid reversibility, good recovery and, also, a small equivalent height of a theoretical plate.

Table 2. Selected List of Polyvalent Anions Strongly Adsorbed by Hydrous Zirconium Oxide

Borate	Arsenite
Carbonate	Arsenate
Phosphate	Molybdate
Sulfate	Tungstate
Chromate	

From the point of view of selectivities, probably the most striking feature is the ability of hydrous oxides to remove polyvalent anions even from relatively concentrated electrolyte solutions. In a sense this was demonstrated with the chromate adsorption experiment. A selected list of anions for which these exchangers show remarkable selectivity is shown in Table 2. Because of this characteristic feature, these exchangers, no doubt, should find extensive use in metal recovery and industrial waste treatment processes.

Figure 15. Existence of anion-exchange properties in hydrous oxides

Regeneration with base is in most cases relatively simple, as is separation of a few of these anions from each other, as illustrated in Fig. 16, with a separation of chromate from phosphate with sodium hydroxide as eluting agent. Since adsorbabilities of various anions vary widely, many such separations are feasible. Eluting agents, other than alkali hydroxides, which we have used with some success include ammonia, ammonia-ammonium salt mixtures, carbonates, bicarbonates and acetates.

Table 3 lists a number of hydrous oxides which have been studied. Most of these have similar anion selectivities. However, with some metals special selectivities may be achieved. For example, oxides of Bi(III) or materials containing a large amount of

Bi(III) show great selectivity for chloride ions, a reflection of the fact that Bi(III) is strongly complexed by this anion. Most of the other oxides show only moderate selectivities between singly-charged anions, excepting F^-, which may indicate that most of these anions do not penetrate through the hydration layers of the oxide surfaces.

Table 3. Typical Anion-exchange-active Hydrous Oxides

Al(III)	Ti(IV)	
Cr(III)	Zr(IV)	Nb(V)
Fe(III)	Th(IV)	Ta(V)
Bi(III)	Sn(IV)	

As one might expect, the capacity of the oxides to adsorb anions is a function of the acidity of the solutions. From acid-base titrations the extent of adsorption (apparent capacity or uptake) may be established. As shown in Fig. 1 for the oxides of Zr(IV) and Sn(IV), chloride uptake rapidly decreases with increasing pH. Simultaneously, however, sodium uptake increases, which implies that the materials become cation exchangers. Presumably at high pH the polymers or oxide particles become studded with hydroxide ions, giving the matrix a negative charge while the adsorbed cations are replaceable or exchangeable.

These cation-exchange properties of oxides are illustrated in Fig. 17 with a separation of Cs^+ from Na^+ utilizing a hydrous zirconium oxide column and KOH as eluting agent. Interestingly Cs^+ is here less strongly held than Na^+; later we shall show an example where Cs^+ elutes after Na^+, i.e., in the order often considered "normal".

Figure 18 illustrates the separation of an alkali metal, an alkaline earth, and a rare earth with hydrous ZrO_2 as exchanger. Cesium is removed with a basic eluant in which alkaline earths and rare earths remain adsorbed. Concentrated NH_4NO_3 solution is used for elution of Ba^{++}; acidification of the column is necessary to elute the rare earths which, at high pH,

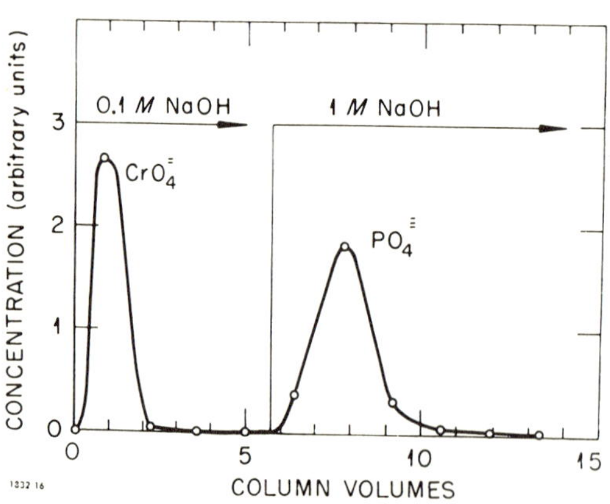

Figure 16. Separation of chromate and phosphate on hydrous titanium oxide (drying temp 105°C, 0.13 cm^2 × 2.8 cm column)

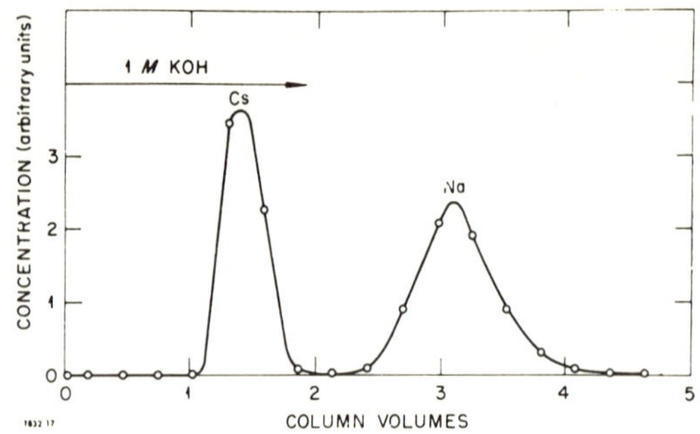

Figure 17. Separation of Cs^+ and Na^+ on hydrous zirconium oxide (drying temp 200°C, 0.2 cm^2 × 3.6 cm column)

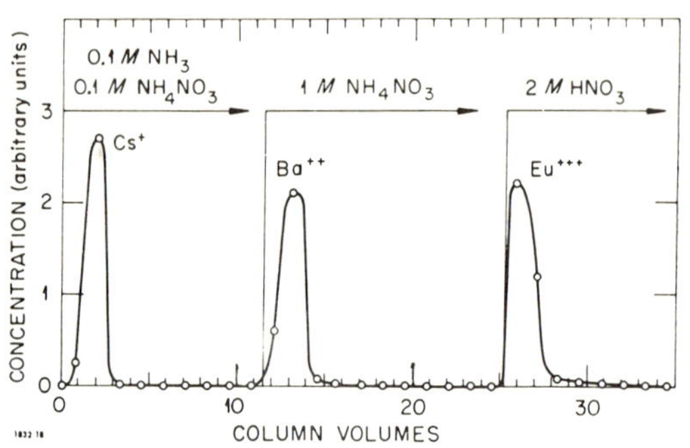

Figure 18. Separation of Cs^+, Ba^{++} and Eu^{+++} on hydrous zirconium oxide (drying temp 500°C, 0.2 cm^2 × 2 cm column)

may be held by the oxide as hydrolyzed species. Although the column is only 2 cm long, one may conclude from the wide separation of the bands that this length is still excessive.

A different and more colorful illustration of cation exchange properties of these oxides, adsorption of the copper ammonia complex by zirconium oxide, is shown in Fig. 19. Removal of the adsorbed copper may be achieved with strong salt solutions or, more effectively, with acid.

The relative importance of cation and anion exchange properties is different for different oxides and seems closely related to the usual concepts of acidity and basicity of oxides. As shown in Fig. 1, Zr(IV) has more anion-exchange capacity in acidic solutions than Sn(IV), and Sn(IV) has more cation-exchange capacity in basic solutions. Some oxides, for example that of Bi(III), exhibit only anion-exchange branches even in basic solutions; others, such as Th(IV), are principally anion exchangers though showing a detectable cation-exchange branch in base; still others, such as U(VI) and V(V), are principally cation exchangers.

The most acidic oxides, which should be cation exchangers even in strongly acidic solutions, tend to disperse too easily for adequate column operation (W(VI), Mo(VI)) and are soluble (H_3PO_4, H_3AsO_4). However, if these acids are held by an insoluble matrix, materials with excellent cation-exchange properties may be prepared. For example, a hydrous oxide, such as that of Zr(IV), may be treated in a column with phosphoric acid. The resulting solid is a cation exchanger with capacity depending on the extent of treatment.

Materials of similar properties though usually with higher capacities may be prepared by precipitation of salts, such as zirconium phosphate, tungstate, and arsenate, with an excess of the acidic component. Other metals such as Ti(IV) or Th(IV), may serve as matrix elements and, according to preliminary work, the matrix element seems to have little effect on the over-all cation-exchange properties. These materials, as a class, could be called mixed hydrous oxides, but we shall refer to them here as "acid salts". If prepared from acidic solutions they contain replaceable hydrogen ions which can be titrated by typical ion-exchange methods.

An example, titration of zirconium phosphate with CsOH in the presence of Cs tracer, is shown in Fig. 4. The apparent capacity increases from *ca.* 1 mole per kg near pH 1.5 to more than 4 moles per kg near pH 9. As with the hydrous oxides these capacities compare favorably with those observed with organic exchangers. Considering that the density of the inorganic materials is higher than that of organic exchangers, the capacities on a volume basis are actually even more attractive.

High capacities for these inorganic materials, if dried at low temperatures, are not surprising. As used by us, the materials are usually amorphous or microcrystalline and thus have a very large effective surface. Note that a cubic aggregate containing 1000 molecules has *ca.* 50% of them in the surface. If the molecular weights are also low, truly remarkable capacities could occur. Thus, assume that hydrous zirconium phosphate has the formula $ZrO(H_2PO_4)_2$, which seems reasonable since a pyrophosphate is obtained on firing. It would then have four replaceable hydrogen ions per mole and a maximum capacity of over nine moles per kg, assuming a 30% water content for the solid. This is about twice the maximum uptake illustrated in Fig. 4.

Acid-base titrations of other acid salts, such as zirconium tungstate or molybdate, are more complicated than the titration of zirconium phosphate, since (see Fig. 5) with these salts substantial removal of the anions from the solids occurs even in essentially neutral solutions. In very basic solutions essentially all the anions are removed, leaving the matrix oxide behind. This might be taken as evidence that these amorphous or microcrystalline acid salts consist of an oxide matrix whose surface is studded with the polyvalent anions.

The formation of the acid salts thus seems intimately related to the anion-exchange properties of the underlying network, and we have the following intricate sequence of reactions (Fig. 20): A hydrous oxide on adsorption of "simple" acids becomes an anion exchanger and its matrix acquires a positive charge. In this form it strongly adsorbs polyvalent anions. When a sufficient amount is adsorbed, a charge inversion takes place and a cation exchanger results.

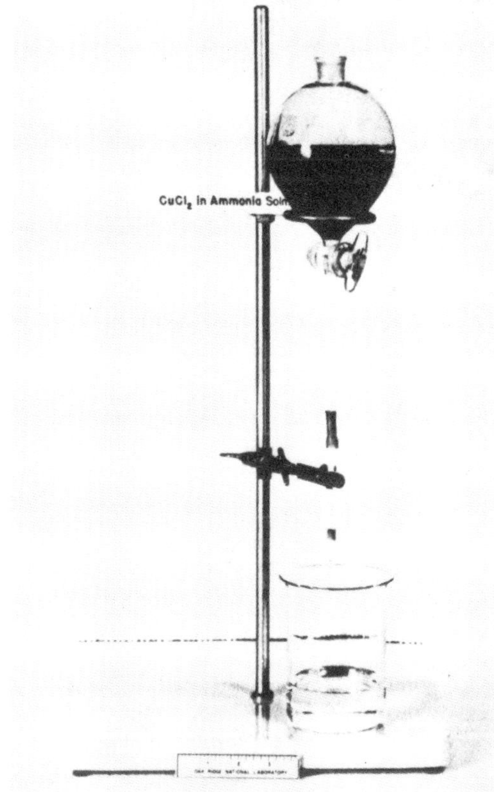

Figure 19. Cation-exchange properties of hydrous zirconium oxide

If now an excess of polyvalent cations is adsorbed, anion-exchange properties will return, and so on. If the cation-exchange-active acid salt or an anion exchanger loaded with polyvalent anions is treated with strong base, the polyvalent anions are removed or exchanged for hydroxide ions. This displacement is

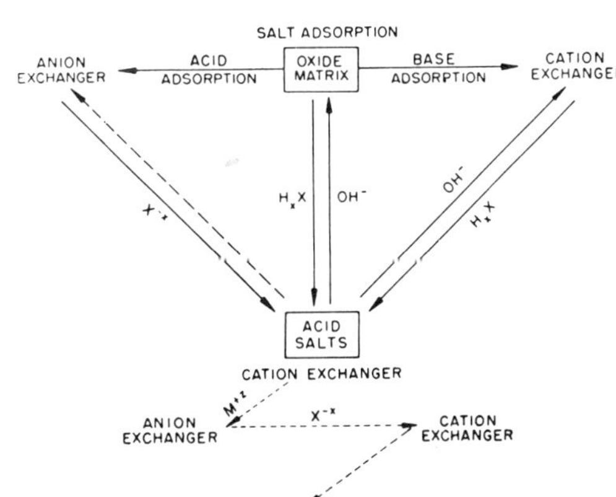

Figure 20. Relationship between oxides and acid salts

aided by possible amphoteric properties of the hydrous oxides because then additional hydroxide ions become attached to the network, giving it a negative charge and making it a cation exchanger.

The acid salts, in some cases, show remarkable selectivities. For example, while the maximum adsorbability ratio observed within the alkali metal group for organic exchangers is usually less than two, it may be more than 1000 for these inorganic exchangers. This makes possible complete separation of the alkali metals from each other with short columns, as shown in Fig. 9. Here Li^+ is difficult to hold on the exchanger except in fairly dilute electrolyte solutions, but Cs^+ can only be removed with fairly concentrated electrolyte solutions. Even better adsorption of Cs^+ may be achieved in more acidic solutions, since for these systems selectivity is strongly dependent on the acidity of the medium, i.e., on the completeness of conversion of the exchanger to the hydrogen form. From a waste disposal point of view this is an important property because it makes possible removal of radiocesium from concentrated electrolyte solutions, even those with large concentrations of polyvalent cations not adsorbed from strongly acidic media.

Similarly excellent separations can be carried out with the alkaline earths, as illustrated in Fig. 10. The selectivity difference between Ba^{++} and Ra^{++} is particularly large so that it is now possible to separate these elements from each other completely with columns only a few centimeters long.

The adsorbability of the alkaline earths decreases rapidly with acidity so that, e.g., in 1*M* acid Ba^{++} is essentially not adsorbed and thus may easily be separated from the strongly adsorbed Cs^+ with columns shorter than a centimeter. This leads to an extremely simple separation of the short-lived Ba^{137} from its long-lived parent Cs^{137} by a milking procedure: Cs^{137} is adsorbed on a small amount of, for example, zirconium phosphate, and Ba^{137} is eluted or milked by flushing the bed with acid.

One final example dramatically illustrates the power of these materials for separations as well as the need for further extensive investigations. We had mentioned a dependence of selectivity on acidity. This change is so great that, for example, zirconium phosphate, which in the hydrogen form can be used for effective separation of alkali metals from each other, has essentially lost this selectivity in the NH_4^+ form and the alkali metals may readily be eluted as a group. In the same medium the alkaline earths are retained. They may be eluted with acid, thus permitting an interesting separation by groups with extremely small columns.

In conclusion, I would like to express the hope that we have demonstrated that hydrous oxides, in a general sense, and presumably also other inorganic materials, have attractive ion-exchange properties and that conversely ion-exchange techniques are relatively simple tools for the study of amorphous and microcrystalline materials which frequently are not considered to be particularly attractive targets of investigations. Considering the enormous number of inorganic compounds from which one may choose to select special properties, the relative simplicity of preparations, the easy availability of starting materials, the high capacities, rapid rates, radiation stability and probable temperature stability, it is probably not too optimistic to anticipate that these materials in the future will occupy an important position in the general field of separations chemistry.

29

Reprinted from *Anal. Chim. Acta*, **38**, 179–184 (1967)

SEPARATION OF NOBLE METALS FROM BASE METALS BY MEANS OF A NEW CHELATING RESIN

G. KOSTER AND G. SCHMUCKLER
Department of Chemistry, Technion—Israel Institute of Technology (Israel)

(Received November 1st, 1966)

The noble metals dealt with in this work—platinum metals and gold—have an important property in common, namely, that in a certain environment they may be stabilized in a d^8 electronic configuration and form square-planar complexes with various ligands. NYHOLM AND TOBE[1] have given many examples of the possibilities of stabilizing the transition metals in various oxidation states by correctly chosen ligands.

In this work an attempt was made to stabilize the oxidation states of these metals (Pt^{2+}, Ir^{+}, Os^{0}, Pd^{2+}, Rh^{+}, Au^{3+}) by means of a chelating ion exchanger in which the active chelating agent contains two resonating amino groups attached to a carbon atom.

BAYER[2] was the first to stress the principle that if a ligand, which in an aqueous solution forms complexes with many different metal ions, is bound to a polymeric matrix, greater selectivity results, because the ligands have to assume a certain geometric arrangement in the resin.

Utilizing this principle, the authors succeeded in synthesizing a number of chelating ion exchangers of the general structure

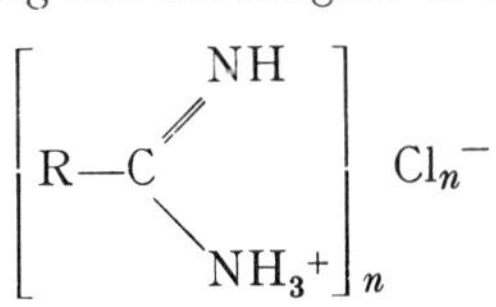

which very strongly bind the noble metals mentioned but do not bind the base metals at all. In order to determine the binding mechanism, some monomeric analogues of the polymeric system were investigated, among them benzylisothiouronium chloride (BTU) and sulphaguanidine (N′-amidinosulfanilamide) (SG). From their properties the processes in the polymer could be deduced.

It was found that the complexes formed by BTU with the noble metals enumerated are dimers with chlorine as the bridging atom, while the complexes formed with SG are monomers with a metal:ligand ratio of 1:2. Both complexes were found to be insoluble in water.

The tendency of these noble metals to form dimers with certain organic reagents is known from the literature, and examples may be found[3,4] in the carbonyl halides of Rh^{+}, in which chlorine serves as the bridging atom, or in the mercaptides[5], in which that function is filled by sulphur.

EXPERIMENTAL AND RESULTS

Reagents

S-Benzylisothiouronium chloride (Fluka AG, puriss) and *sulphaguanidine* (B.D.H., 99.0%) were used.

Chelating ion-exchange resin. The resonating amino groups were introduced into a styrene–divinylbenzene copolymer matrix by a method for which a patent has been applied[6]. The resin was rinsed with 0.1 *N* hydrochloric acid before use.

Preparation of the palladium–S-benzylisothiouronium complex (Pd–BTU)

To a palladium(II) solution (20 mg in 100 ml) at pH 1–2 was added 100 mg of solid S-benzylisothiouronium chloride (BTU) with constant stirring. A brown precipitate soon appeared and collected at the bottom of the vessel. Filtering the precipitate and washing and drying it at 100° produced the pure Pd–BTU. The BTU must be added as the solid and not in solution because of its relatively rapid hydrolysis in water.

Structural analysis of the palladium–BTU complex

Elemental analysis indicated that the complex corresponded to the formula $C_8H_9N_2SPdCl$, *i.e.* palladium(II) combined with one mole of BTU. For ignition of the organic complex at high temperature leaves the pure metal rather than its oxide.

The 1:1 reaction of palladium(II) with BTU was also proved by ordinary gravimetric methods; after the precipitate had been filtered and dried at 100°, its weight was determined, and then the palladium content was found after ignition at 1000°. From the formulae of the complex and of the reagent, it can be seen that two hydrogen atoms are liberated in the complex formation and this was proved by potentiometric measurement of the pH before and after the addition of the ligand. Table I summarizes the experimental data obtained.

TABLE I

pH CHANGES IN THE REACTION BETWEEN PALLADIUM AND BTU

No.	*Pd^{2+} concn. (moles)*	*Initial pH*	*Final pH*	*(H^+) evolved*
1	$3.16 \cdot 10^{-5}$	4.8	4.23	$5.9 \cdot 10^{-5}$
2	10^{-4}	4.71	3.65	$2.04 \cdot 10^{-4}$
3	$3.16 \cdot 10^{-4}$	4.45	3.12	$7.24 \cdot 10^{-4}$
4	10^{-3}	4.24	2.67	$2.08 \cdot 10^{-3}$
5	$3.16 \cdot 10^{-3}$	3.75	2.20	$6.13 \cdot 10^{-3}$
6	10^{-2}	3.79	1.77	$1.84 \cdot 10^{-2}$

The amount of chlorine present in the palladium complex was determined by a semimicro modification of Schöniger's method, the chloride formed being determined by potentiometric titration with 0.05 *M* silver nitrate. The ratio of Pd to Cl was found to be 1:1.

Preparation of the palladium–sulphaguanidine complex

A solution containing 10 mg of palladium was diluted to 50 ml with distilled water, the pH was adjusted to 1.5–2.5 and the solution was heated to 70–80°. Sulphaguanidine (80–100 mg) was dissolved in 50 ml of hot distilled water and this solution was added slowly to the palladium solution with constant stirring. The precipitate did not form immediately; the solution had to be kept on a hot plate for 1½–2 h until the yellow precipitate settled. After the solution had cooled, it was filtered through a fine-pore Gooch crucible, washed with cold distilled water, dried at 100° for 1 h and weighed (gravimetric factor: 5.334).

Properties of the chelating resin

Maximum capacity of the resin. This was determined in 6 columns of 6-mm internal diameter and 15-cm length. Weighed samples of the dried (at 100°) resin were introduced into the columns. The concentration of the metal ions (in the form of their chlorides) flowing through the columns was invariably 1 mg/ml. The solutions were allowed to flow slowly through the columns until breakthrough was reached.

The columns were then flushed with water. The resin was taken out, weighed at 100° (in order to determine the weight of the bound metal), and ignited at 1000° in order to find the weight of the pure metals. To prevent the formation of oxides after the ignition the samples were allowed to cool in an atmosphere of carbon dioxide.

Table II summarizes the maximum capacities of the resin for the six noble metal ions.

TABLE II

MAXIMUM CAPACITY OF RESIN FOR NOBLE METALS

	Os^{8+}	Rh^{3+}	Ir^{4+}	Pd^{2+}	Pt^{4+}	Au^{3+}
Weight of dry polymer (mg)	537	532	537	521	600	524
Weight of polymer with bound metal ions at 110° (mg)	587	730.3	654.7	652.5	1067.1	1072.2
Weight of pure metal after ignition at 1000° (mg)		101.2	92.0	113.6	292.5	577.3
(mg) metal ion/(g) dry resin	93	190	158	218	488	1100
Millimoles metal ion/g dry resin	0.49	1.85	0.82	2.05	2.50	5.58
Volume changes during adsorption of metal ions	0.93	0.8	0.5	0.8	0.75	0.55

Influence of acidity. The noble metals must be adsorbed on the resin in an acidic medium, as in an alkaline environment hydrolysis of the metals as well as of the amine groups of the resin takes place. Increasing acidity, on the other hand, reduces the adsorptive capacity of the resin, since the binding of the metal ions is accompanied by the liberation of hydrogen ions: increasing the hydrogen ion concentration too far reverses the reaction. Figure 1 summarizes the experimental results for the metal ions Au^{3+}, Pt^{4+}, and Pd^{2+}. The experiments were conducted as follows: 100-mg batches of resin were placed in contact with the metal chloride solutions (containing 150 mg of Au^{3+}, 110 mg of Pt^{4+} or 106 mg of Pd^{2+} per 100 ml); to each batch a measured quantity of hydrochloric acid was added, and the solution was intensively stirred

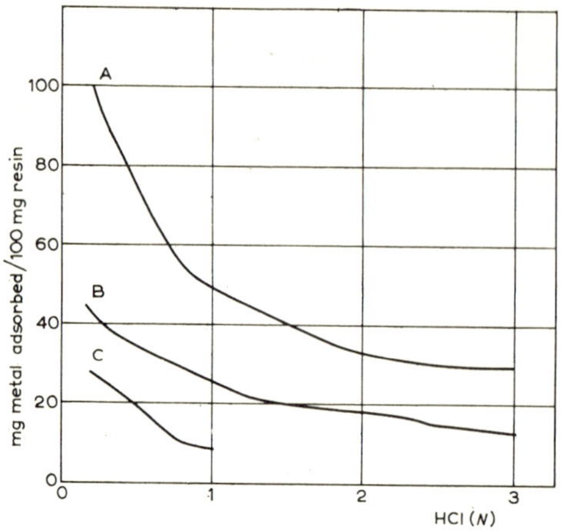

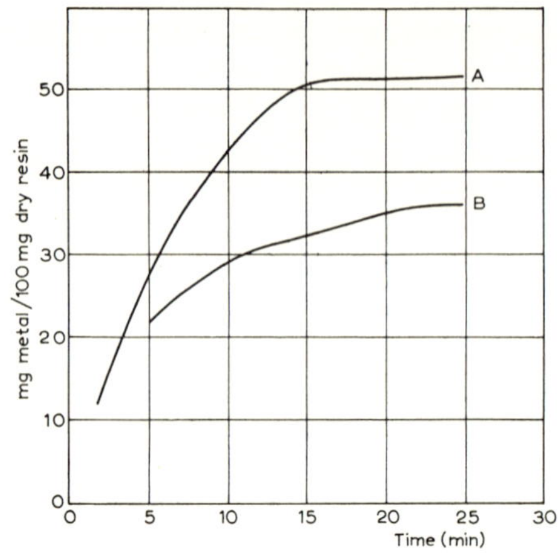

Fig. 1. Influence of acidity on the adsorption. For each batch of 100 mg resin: A, 1.50 g l^{-1} Au^{3+}; B, 1.10 g l^{-1} Pt^{4+}; C, 1.06 g l^{-1} Pd^{2+}.

Fig. 2. Rate of adsorption. A, 1.00 g l^{-1} Au^{3+}; B, 1.00 g l^{-1} Pt^{4+}.

for 24 h. After this the resin and the solution had been separated, and the amount of metal in the resin was determined.

Rate of adsorption. The rate of adsorption was determined for gold(III) and platinum(IV) in the following manner: 100-mg batches of resin were stirred with 100-ml batches of solution each containing 100 mg of the metal ion, and the resin was separated from the solution at fixed intervals. The amount of metal in the resin was found by ignition. Figure 2 summarizes the experimental results; it can be seen that adsorption is relatively rapid, the main portion of the metal ion being adsorbed within the first 15 min.

Adsorption in the presence of other metal ions. To illustrate the selectivity of the chelating resin, mixtures containing small quantities of platinum with large amounts of nickel(II), copper(II) or iron(III) were passed through columns containing 200 mg of resin. The results are summarized in Table III.

TABLE III

SELECTIVE ADSORPTION OF PLATINUM

Pt taken (mg)	*Base metal taken*	*Pt found by ignition of resin (mg)*
9.67	500 mg Ni^{2+}	9.65
9.67	500 mg Cu^{2+}	9.67
9.67	505 mg Fe^{3+}	9.56

DISCUSSION

For the determination of the reaction mechanism between a typical d^8 metal and the chelating reagents, the palladium(II) ion was selected. The reaction in an acidic medium (pH 1–2) between palladium(II) and BTU gives a brown precipitate which, according to the analytical data, should be a dimer:

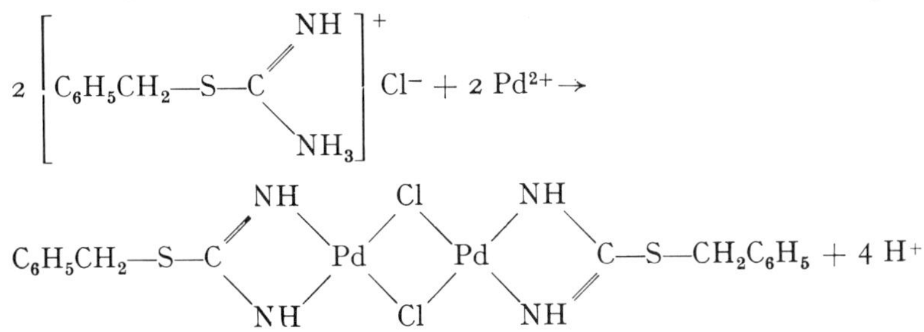

In the resin the reaction is entirely analogous, as regards the brown colour obtained, the amount of hydrogen liberated, and the ratio between palladium and chloride (1:1).

The reaction between palladium and SG produces a yellow precipitate in which the ratio of metal to ligand is 1:2 and coordination occurs through the resonating amino groups:

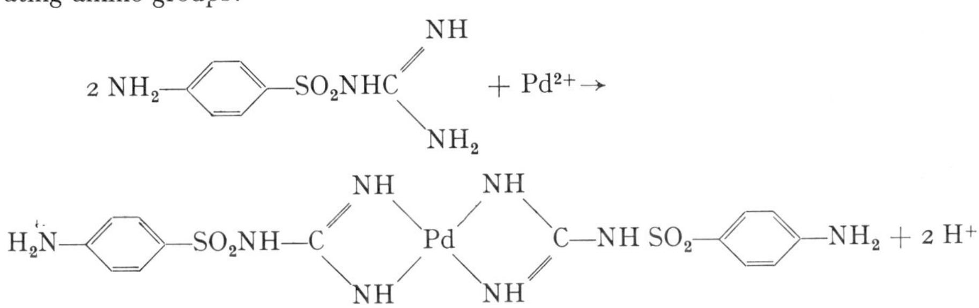

From Table II it can be seen that metal ions of different valencies were introduced into the columns. If the resin does indeed bind only ions with the d^8 electronic configuration forming square planar complexes, then only Pt^{2+} but not Pt^{4+} (in the case of platinum) ought to be adsorbed by the resin. In actual fact, however, the resin contains double bonds, so that it has redox properties. Some of the active groups in the resin are needed to reduce Pt^{4+} to Pt^{2+}, so that platinum(II) can be bound as a square planar complex. This may be proved by the diamagnetism of the compounds in the resin and by the fact that they can be eluted with oxidizing agents. The theoretical capacity of the resin according to the nitrogen analysis is 3.9 mmole/g, but the actual capacity is in most cases, except for gold, much smaller because of the redox reactions. The high capacity for gold is explained by the fact that the resin is able to reduce the gold to metallic gold. The formation of metallic gold on the resin is shown by the appearance of solid yellow spots. The reduction to metallic gold appears to take place in two stages, as suggested by LINGANE[7]; if a small quantity of metallic gold is present the reaction will be:

$$AuCl_4^- + 2\ Au + 2\ Cl^- \rightarrow 3\ AuCl_2^-$$

The univalent gold formed easily undergoes disproportionation to gold(III) and metallic gold, so that it is not surprising that the capacity for gold is very high indeed—5.58 mmole/g.

CONCLUSION

It is possible to stabilize oxidation states of metal ions which tend to form square planar complexes, by chelating in the solid phase in the presence of suitable ligands.

The following ions were examined: Os^{8+}, Rh^{3+}, Ir^{4+}, Pd^{2+}, Pt^{4+}, Au^{3+}.

When bound to the resin they assumed the valencies with which they have a d^8 electronic configuration, *viz.*: Os^0, Rh^+, Ir^+, Pd^{2+}, Pt^{2+}, Au^{3+}.

SUMMARY

A new type of chelating resin, which is highly selective for gold and the platinum metals, is described. Its properties and applications are indicated. The reaction mechanism between this resin and the noble metal ions was investigated with the aid of some monomer analogues.

REFERENCES

1 R. S. NYHOLM AND M. L. TOBE, (The Stabilization of Oxidation States of the Transition Metals), *Advan. Inorg. Chem. Radiochem.*, 5 (1963) 1.
2 E. BAYER, *Angew. Chem.*, 76 (1964) 76.
3 M. A. BENNETT AND P. A. LONGSTAFF, *Chem. Ind. (London)*, (1965) 846.
4 F. A. COTTON AND G. WILKINSON, *Advanced Inorganic Chemistry*, Interscience, New York, 1962, p. 843.
5 L. F. DAHL AND C. H. WEI, *Inorg. Chem.*, 2 (1963) 328.
6 *U.S. Patent*, Application No. 467,995.
7 J. J. LINGANE, *J. Electroanal. Chem.*, 4 (1962) 332.

ADDENDUM BY THE AUTHORS

The correct formula of the dimer with BTU is:

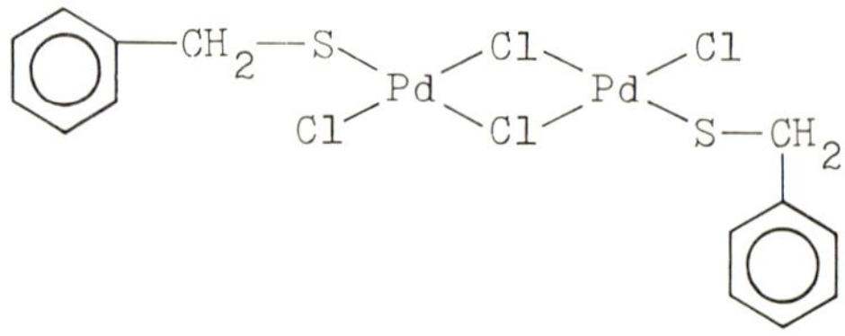

and with SG a monomer is formed the formula of which is described in *Anal. Chim. Acta*, 59, 399 (1972).

30

Reprinted from *J. Chromatog.*, **82**, 359–365 (1973)

LIGAND-EXCHANGE CHROMATOGRAPHY OF RACEMATES

INFLUENCE OF THE DEGREE OF SATURATION OF THE ASYMMETRIC RESIN BY METAL IONS ON LIGAND EXCHANGE

V. A. DAVANKOV, S. V. ROGOZHIN, A. V. SEMECHKIN and T. P. SACHKOVA

Institute of Organo-Element Compounds, Academy of Sciences, Moscow (U.S.S.R.)

(Received March 8th, 1973)

SUMMARY

The application of an asymmetric resin containing optically active L-proline in the Cu^{2+} form for the effective resolution of racemates by means of ligand-exchange chromatography is reported. It was found in the static sorption of proline enantiomers that the difference in its sorption energy reaches 450 cal/mole and depends on the degree of saturation of the resin by copper ions. Chromatographic data such as elution volumes and the concentration of ammonia in the eluent also depend on this factor.

INTRODUCTION

The number of publications that describe the chromatographic resolution of racemic compounds by means of optically active stationary phases[1–3] totals almost 400. In addition to the well known advantages of chromatographic processes, the chromatographic resolution of racemates is superior to traditional enzymatic and crystallization methods in that it is in principle possible[4] to obtain quantitatively both enantiomers at 100% optical purity by using a resolving agent (asymmetric stationary phase) with an optical purity of less than 100%.

Exceptional progress has been made in the gas chromatographic resolution of racemates[3]. However, for resolutions on a larger scale, liquid ligand-exchange chromatography, recently reviewed by Davankov *et al.*[5], is much more promising. The ligand-exchange process differs from adsorption and ion-exchange chromatography in that the interaction between the stationary phase and the sorbates to be separated is due to the formation of coordination bonds inside the coordination sphere of the complex-forming metal ion, which thus ensures that sorbent–sorbate interaction occurs. The complex-forming metal ion itself may be bound to mobile ligands and in this case we deal with the ligand-exchange chromatography of several complexes formed by the metal ion with various ligands. However, the metal ion is more often retained by the fixed ligands of the stationary phase owing to the existence of coordination and/or ionic bonds, and in this case we deal with the chromatography of ligands[5].

The ligand chromatography of racemates on asymmetric complex-forming stationary phases was suggested by Rogozhin and Davankov in 1968[6,7], who described a series of asymmetric resins containing optically active bifunctional[4,8] and trifunctional[9] α-amino acids in a macroreticular isoporous polystyrene framework[10]. The methylene bridges between the aromatic rings of the framework and the NH_2-groups of the amino acids preserve the capability of the amino acids to form complexes with the transition metal ions. Copper ions, with a coordination number of 4, are able[11] to form stationary (fixed) complexes of the form $\bar{R}$–Cu–$\bar{R}$, with two bidentate fixed ligands, $\bar{R}$. The structure of the fixed complex $\bar{R}$–Cu–$\bar{R}$ in a resin containing L-proline (or L-hydroxyproline) as the fixed ligand $\bar{R}$ can be represented by formula I. In fact, the maximum amount of Cu(II) ions absorbed by the resin corresponds to half of their content of L-proline residues. Moreover,

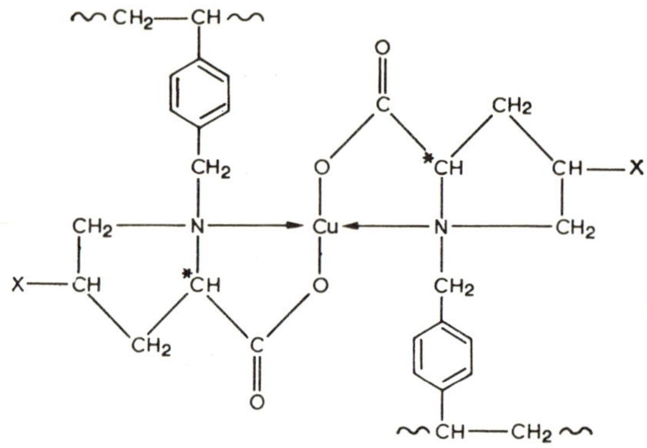

I: X = H
II: X = OH

the analysis of the distribution of Cu(II) ions between the resin and a solution of $N(CH_2COOH)_3$ indicates the formation of fixed complexes containing two fixed ligands per metal ion in the resin phase. It was pointed out earlier by Hering and co-workers[11,12] that the formation of bis-complexes in the resin phase causes a strain in the polymer chains of the framework, which is shown by a decrease in the swelling capacity of the resin when the degree of saturation of the resin by Cu(II) ions increases. The increasing strain in the polymer chains results in the stability of the fixed complexes $\bar{R}$–Cu–$\bar{R}$ formed at low concentrations of Cu(II) ions being higher than that of the complexes formed at higher degrees of saturation of the resin when all of the favourably arranged pairs of fixed ligands have already been complexed.

The sorption of a molecule capable of forming complexes (mobile ligand A) in the ligand-exchange process according to the equilibrium

$$\bar{R}\text{–Cu–}\bar{R} + A \rightleftharpoons \bar{R}\text{–Cu–}A + \bar{R}$$

implies a higher free energy of the mixed sorption complex, $\bar{R}$–Cu–A, in comparison with that of the fixed $\bar{R}$–Cu–$\bar{R}$ complex. As the stability of the latter decreases with the increasing degree of saturation of the resin by Cu(II) ions, the adsorption capacity of the resin with respect to mobile ligands increases as the amount of metal ions in the system increases.

If the mobile ligand A is an amino acid, the mixed sorption complex, $\bar{R}$–Cu–A, formed in the ligand-exchange process can be represented by the formula:

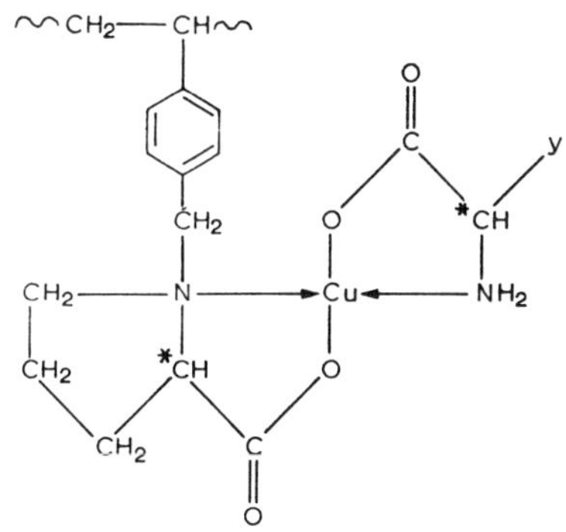

We have found that the stability of mixed sorption complexes depends not only on the nature of the amino acid A, but also on its configuration. These stereoselectivity effects in the formation of mixed complexes $\bar{R}$–Me–A (ref. 13) permit the resolution of optical isomers during the ligand chromatography of the racemate.

EXPERIMENTAL

The asymmetric resin I contained 5 mole% of crosslinking bridges of the structure

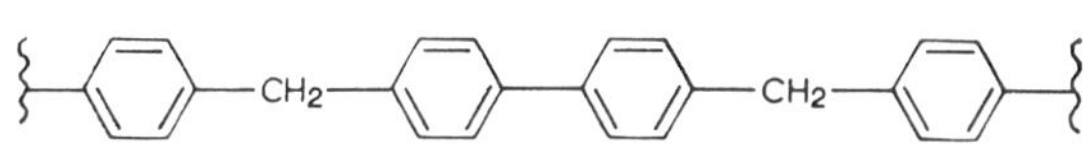

between the polystyrene chains and 1.68 mequiv./g of L-proline residues. The diameter of the particles was 0.2–0.4 mm.

A 1.785-g amount of dry resin (containing 3 mequiv. of active groups) was treated with 25 ml of aqueous solution containing 0.345 g (3 mequiv.) of DL-proline, 0.28–4.19 mequiv. of $Cu(NO_3)_2$ and the appropriate amount of NaOH such that during equilibration of the mixture its pH was 10.3.

In order to prevent crystallization of $Cu(Pro)_2$, the total volumes of the last three of the eleven samples were increased to 30, 35 and 50 ml, respectively, the pH value of the final sample having decreased to 10.0 After shaking the mixtures for 70 h at 25°, the resin was filtered and washed with 5 ml of water. The liquid and resin phases, after extraction with 2 *N* HCl, were analyzed for their total copper content[14], total proline content[14,15] and optical activity (Roussel–Jouan Quick-polarimetre; λ=436 nm; l=7 cm; 1 *N* HCl; for the optically pure proline, $[\alpha]^{25}_{436}$ = 123°).

For the chromatographic experiments, 10-g portions of resin I were treated with solutions of different amounts of $Cu(NO_3)_2$ in 1 *N* ammonia solution, washed with 0.5 *N* CH_3COONa+NaOH solutions of pH 10.3 and loaded on to columns of I.D. 9 mm (thickness of resin layer *ca.* 200 mm). The chromatography of 0.100 g of DL-proline dissolved in 1 ml of 0.5 *N* CH_3COONa+NaOH solution of pH 10.3

was carried out at a flow-rate of 6 ml/h of ammonia solutions prepared by using the above CH_3COONa buffer. Fractions of 5 ml were collected and their optical activities measured at 436 nm in a 2-cm cell.

The asymmetric resin II, containing 5 mole% of crosslinking bridges and 2.96 mequiv./g of L-hydroxyproline residues, had a particle diameter of 0.16–0.20 mm and was used in the Cu(II) form in a column 60 mm long and 9 mm I.D. for the chromatographic resolution of 1.5 mg of DL-proline in a 1.0 *M* pyridine solution at the rate of 15 ml/h and a temperature of 56°.

STATIC SORPTION

The ligand that is mainly sorbed in the case of competitive sorption between two mobile ligands (in our experiments, the L- and D-isomers of proline) is that which produces the stronger mixed complex $\bar{R}$–Cu–A and is able to destroy a greater number of the fixed complexes $\bar{R}$–Cu–$\bar{R}$. From the equilibrium distribution of L-proline (L-Pro) and D-proline (D-Pro) between the solution and the resin (Fig. 1), the difference in the thermodynamic stabilities of the mixed complexes $\bar{R}$–Cu–D-Pro and $\bar{R}$–Cu–L-Pro can be evaluated as follows:

$$\bar{R}\text{–Cu–}\bar{R} + \text{D-Pro} \overset{K_D}{\rightleftarrows} \bar{R}\text{–Cu–D-Pro} + \bar{R}$$

$$\bar{R}\text{–Cu–}\bar{R} + \text{L-Pro} \overset{K_L}{\rightleftarrows} \bar{R}\text{–Cu–L-Pro} + \bar{R}$$

$$\delta \Delta G^0 = \Delta G^0_{\bar{R}\text{–Cu–D-Pro}} - \Delta G^0_{\bar{R}\text{–Cu–L-Pro}}$$

$$= -RT \ln \frac{K_D}{K_L}$$

$$= -RT \ln \frac{[\bar{R}\text{–Cu–D-Pro}]}{[\bar{R}\text{–Cu–L-Pro}]} \cdot \frac{[\text{L-Pro}]}{[\text{D-Pro}]}$$

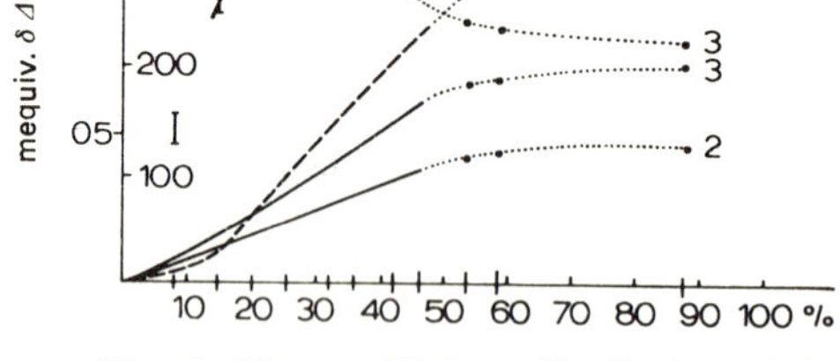

Fig. 1. The equilibrium distribution of Cu(II) ions (1), L-proline (2) and D-proline (3) between the solution phase (– – –) and the resin phase (——) and the difference in the free energies of formation of the two diastereomeric mixed adsorption complexes (4) as a function of the degree of saturation of the L-proline resin I by Cu(II) ions. The conditions of the last three experimental points differ from those of the other eight.

The decrease in the average stability of fixed $\bar{R}$–Cu–$\bar{R}$ complexes with the increase in the total concentration of Cu(II) ions results in a situation when the desorption of metal ions becomes more significant (Fig. 1):

$$\bar{R}\text{–Cu–L-Pro} + \text{L-Pro} \rightleftarrows \bar{R} + \text{Cu(L-Pro)}_2$$

$$\bar{R}\text{–Cu–L-Pro} + \text{D-Pro} \rightleftarrows \bar{R} + \text{Cu(L-Pro)(D-Pro)}$$

$$\bar{R}\text{–Cu–D-Pro} + \text{L-Pro} \rightleftarrows \bar{R} + \text{Cu(L-Pro)(D-Pro)}$$

$$\bar{R}\text{–Cu–D-Pro} + \text{D-Pro} \rightleftarrows \bar{R} + \text{Cu(D-Pro)}_2$$

Some mobile ligands in the solution are coordinated to the desorbed Cu(II) ions, which was taken into account in calculations of the concentration of the free mobile ligands in the solution.

From the results in Fig. 1, it is obvious that with an increase in the degree of saturation of the resin by Cu(II) ions, the amount of L-proline and D-proline adsorbed also increases. A rapid increase in the difference in the amounts of D-proline and L-proline adsorbed is reflected in the resulting tendency of the calculated $\delta\Delta G^0$ values to increase. This tendency, however, may be due to non-selective sorption of proline according to the Donnan distribution, which mainly affects the experimental points of low adsorption values of proline according to the ligand-exchange mechanism.

A large difference between the effective free energies of adsorption of D-proline and L-proline ($\delta\,\Delta G^0 \approx 400$ cal/mole) indicates a high sensitivity of ligand-exchange processes with respect to the steric structures of the isomers to be separated. In a similar experiment, with a cation-exchange resin containing L-proline residues, the difference in the sorption energies of the α-methylbenzylamine enantiomers (according to the ion-exchange mechanism) was found[16] to be *ca.* 5 cal/mole. A very high efficiency of the resolutions of racemates by means of ligand-exchange chromatography[17–19] is in good agreement with this result.

CHROMATOGRAPHY

The increase in the affinity of the resin for the mobile ligands with the increasing degree of saturation of the former by Cu(II) ions is observed in chromatographic experiments through a rapid increase in the retention times of the mobile ligands (Fig. 2). If small amounts of Cu(II) ions are present in the resin, their fixed complexes $\bar{R}$–Cu–$\bar{R}$ are stable so that both enantiomers of proline are readily eluted from the column by water, only a small resolution being achieved. When larger amounts of metal ions are present in the resin, it becomes necessary to use 0.1–0.2 *N* ammonia solution in order to elute L-proline, whereas for the elution of the D-isomer, whose adsorption is higher by 300–400 cal/mole, 1.0–2.0 *N* ammonia solution is required. In these elution processes, the NH_3 molecules compete with the proline molecules for the formation of mixed adsorption complexes:

$$\bar{R}\text{–Cu–Pro} + 2NH_3 \rightleftarrows \bar{R}\text{–Cu–}(NH_3)_2 + \text{Pro}$$

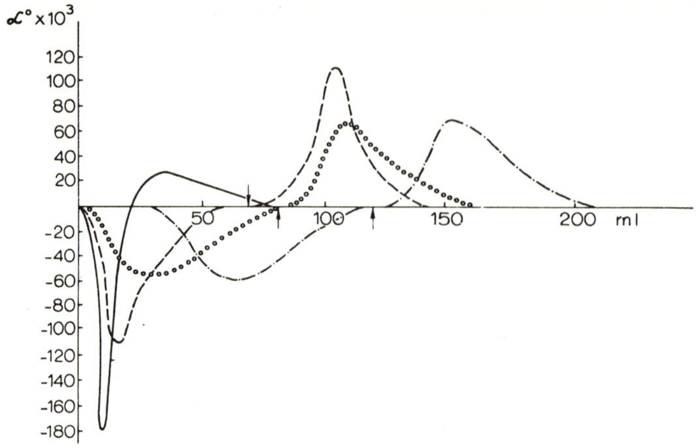

Fig. 2. Optical activity of the eluate during the chromatography of 0.1 g of DL-proline on 10 g of L-proline resin I saturated to 10.1% (——), 21.3% (- - - -), 54.4% (○ ○ ○) and 71.0% (–·–·–·–) of the theoretical amount of Cu(II) ions. The NH_3 contents of starting eluent are 0, 0, 0.1 and 0.2 *N*, respectively. At the points marked with arrows, the starting eluent was changed to 0, 1.0, 1.0 and 1.5 *N* ammonia solution, respectively.

If the degree of saturation of the resin by copper (II) ions does not exceed 50%, desorption of metal from the column by the proline solution passing through is negligible. It is noteworthy that the D-isomer desorbs Cu(II) ions from the resin saturated by the metal to higher degrees than does the L-isomer. This phenomenon is due to the difference in the conditions of elution. A high total concentration of the mobile ligands required for the elution of D-proline facilitates the desorption of copper. Because in the absence of proline a 1 *N* ammonia solution loses its desorption properties (when the saturation of the resin is less than 70%), it may be suggested that copper (II) ions are desorbed in the form of $Cu(D\text{-}Pro)(NH_3)_2$ complexes.

Fig. 3 shows an example of the complete resolution of DL-proline on a small

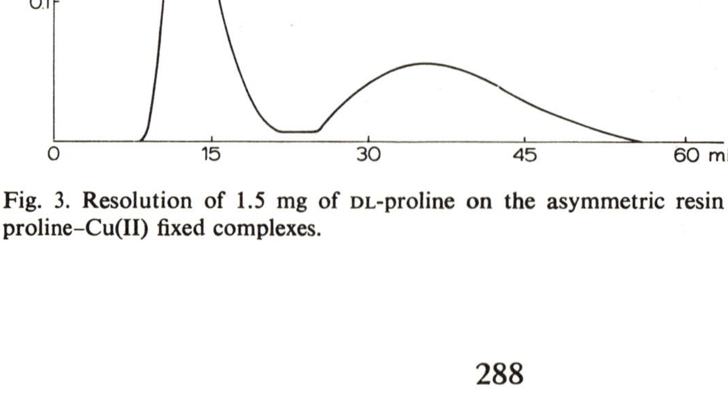

Fig. 3. Resolution of 1.5 mg of DL-proline on the asymmetric resin II containing L-hydroxyproline–Cu(II) fixed complexes.

column of L-hydroxyproline resin (of the above-mentioned formula II) as recorded with a standard amino acid analyzer. The resolving factor of 2.8 far exceeds the results obtained by any other chromatographic technique.

The possibility of greatly changing the sorbent–sorbate affinity by varying the degree of saturation of the resin by metal ions extends significantly the application of the ligand-chromatography technique. Substances with both weak and strong complexing abilities can be successfully separated on the same resin. The amount of metal ions present in the resin for second separations should be relatively small.

As the results of the chromatographic resolution of a racemate depend on the degree of saturation of the resin by metal ions, the optimum conditions for the resolution must be strictly controlled.

REFERENCES

1 R. D. Buss and T. Vermeulen, *Ind. Eng. Chem.*, 60 (1968) 12.
2 S. V. Rogozhin and V. A. Davankov, *Usp. Khim.*, 37 (1968) 1327; *Russ. Chem. Rev.*, 37 (1968) 565.
3 S. V. Rogozhin, V. A. Davankov and I. I. Piesliakas, in A. M. Sladkov (Editor), *Chemistry and Technology of High Molecular Compounds*, Vol. 4, VINITI, Moscow, 1973, p. 45.
4 S. V. Rogozhin, V. A. Davankov, V. V. Korshak, V. Vesa and L. A. Belchich, *Izv. Akad. Nauk SSSR, Ser. Khim.*, (1971) 502.
5 V. A. Davankov, S. V. Rogozhin and A. V. Semechkin, in A. M. Sladkov (Editor), *Chemistry and Technology of High Molecular Compounds*, Vol. 4, VINITI, Moscow, 1973, p. 5.
6 S. V. Rogozhin and V. A. Davankov, *Fr. Pat.* 2,012,102 (1970); *C.A.*, 72 (1970) 90875c; 75 (1971) 6347p.
7 S. V. Rogozhin and V. A. Davankov, *Dokl. Akad. Nauk SSSR*, 192 (1970), 1288; *Chem. Commun.*, (1971) 490.
8 V. A. Davankov, S. V. Rogozhin and I. I. Piesliakas, *Vysokomol. Soedin.*, 14 B (1972) 276.
9 V. A. Davankov, S. V. Rogozhin and I. A. Yamskov, *Izv. Akad. Nauk SSSR, Ser. Khim.*, (1971) 2325 and 2327.
10 V. A. Davankov, S. V. Rogozhin and M. P. Tsjurupa, *Belg. Pat.*, 756,082; *Fr. Pat.*, 2,061,341 (1971).
11 R. Hering, K. Trenne and P. Neske, *J. Prakt. Chem.*, 32 (1966) 291.
12 R. Hering and K. Heilmann, *J. Prakt. Chem.*, 32 (1966) 59.
13 V. A. Davankov and S. V. Rogozhin, *Dokl. Akad. Nauk SSSR*, 193 (1970) 94; *J. Chromatogr.*, 60 (1971) 280.
14 A. Woiwood, *Biochem. J.*, 45 (1949) 412.
15 C. Pope and M. Stevens, *Biochem. J.*, 33 (1939) 1070.
16 G. A. Knudsen, Jr., B. Franzus and J. H. Surridge, *Chem. Commun.*, (1969) 35.
17 V. A. Davankov, S. V. Rogozhin, I. I. Piesliakas, A. V. Semechkin and T. P. Sachkova, *Dokl. Akad. Nauk SSSR*, 201 (1971) 854.
18 K. Bernauer, M.-F. Jeahneret and D. Vonderschmitt, *Helv. Chim. Acta*, 54 (1971) 297.
19 R. V. Snyder, R. J. Angelichi and R. B. Meck, *J. Amer. Chem. Soc.*, 94 (1972) 2660.

Part IX

NUCLEIC ACID DERIVATIVES

Editor's Comments on Papers 31 Through 35

31 **COHN**
The Separation of Purine and Pyrimidine Bases and of Nucleotides by Ion Exchange

32 **COHN**
The Anion-Exchange Separation of Ribonucleotides

33 **HORVATH et al.**
Fast Liquid Chromatography: An Investigation of Operating Parameters and the Separation of Nucleotides on Pellicular Ion Exchangers

34 **HORVATH et al.**
Rapid Analysis of Ribonucleosides and Bases at the Picomole Level Using Pellicular Cation Exchange Resin in Narrow Bore Columns

35 **KIRKLAND**
High Speed Separations of Nucleotides and Nucleic Acid Bases by Column Chromatography Using Controlled Surface Porosity Ion Exchangers

The development of atomic energy raised questions about the effects of radiation on living cells. New analytical methods were needed to study the composition of nucleic acids and their breakdown products. Waldo Cohn, at Oak Ridge National Laboratory, turned to ion exchange, which had been so successful in separating the fission products, and found that it would separate the hydrolysis products of nucleic acid just as well. Complete hydrolysis of nucleic acids yields the purine and pyrimidine bases, adenine, guanine, thymine, uracil, and cytosine. These were separated on short columns of a cation exchanger with hydrochloric acid eluent, or, since they have acidic as well as basic character, by anion exchange with ammonia–ammonium chloride eluent. This work is described in Paper 31. Paper 32 describes the separation of nucleotides, products of partial hydrolysis of nucleic acids that contain base, ribose, and phosphoric acid. Because these compounds are primarily acidic, they are best separated on an anion-exchange resin column.

Elution volumes depend on the net charge of the eluted species, which in turn depends on its ionization constants and on the pH. They also depend on the solubility of the compound. Separations were optimized by regulating the pH and buffer concentration.

The existence of isomers of guanylic and adenylic acids was shown by the elution curves in Paper 32. The separate isomers were isolated in gram quantities by first concentrating the effluent fractions on a short anion-exchange resin column at high pH, then eluting with a small volume of dilute hydrochloric acid; the compounds were stripped from the resin in supersaturated solution and crystallized in the receiver.

The separations took a long time, up to 100 hours. Even the "rapid," partial separation of four nucleotides shown in Figure 2 took nearly 4 hours. Later improvements in technique led to a "60-minute analysis" of mixed ribonucleosides, compounds that contain a base and ribose, but no phosphoric acid (see Uziel et al., 1968).

Papers 33, 34, and 35 report effective separations of bases, nucleosides and nucleotides in less than 10 minutes! While the analysis times were reduced from hours to minutes, the quantities handled were reduced from centigrams to micrograms. The reduction in quantity may be advantageous from the analytical point of view, but it makes peak indentification harder and precludes preparative separations. Moreover, are 10-minute nucleic acid analyses really needed? Nevertheless, these papers are benchmark papers in the fullest sense, for they mark the beginning of modern, high-speed liquid chromatography.

Csaba Horvath, at the Yale University Medical School, introduced the concept of *pellicular packings.* Silica beads 30 μm in diameter were coated with thin, adherent layers of ion-exchange resin. Mass transfer is rapid through resin layers only 2 or 3 μm deep, yet the bead diameter is large enough to permit fluid flow without an impossibly high hydrodynamic resistance. Long columns, to 3 m and more, can be used with high linear flow rates. The columns are made of stainless steel to hold the necessary pressures, and may be quite narrow, a 1- to 2-mm internal diameter being common. The ion-exchange capacity of such columns is of course quite low, and since they can handle only very small samples, a sensitive detector is essential. Dual-path ultraviolet detectors provide the required sensitivity.

Paper 35 describes a somewhat different kind of surface-bonded resin, the popular Zipax resins. An earlier paper (Kirkland, 1969) says more about the packings themselves, but Paper 35 describes their applications to nucleic acid derivatives.

REFERENCES

Kirkland, J. J. 1969. *J. Chromatog. Sci.*, *7*, 361.
Uziel, M., C. K. Koh, and W. E. Cohn. 1968. *Anal. Biochem.*, *25*, 77.

31

Reprinted from *Science*, **109**(2833), 377–378 (1949)

The Separation of Purine and Pyrimidine Bases and of Nucleotides by Ion Exchange

Waldo E. Cohn

Biology Division, Oak Ridge National Laboratory, Oak Ridge, Tennessee

As a prelude to tracer studies on the enzymatic formation and degradation of nucleic acids, we have undertaken to develop ion exchange techniques for the separation and analysis of the various bases, nucleosides, and nucleotides at the milligram level. Although this technique is finding increasing application in biochemical studies for separation of these compounds (*8, 9*) and others (*1, 2, 6, 7*), we were led independently to its use

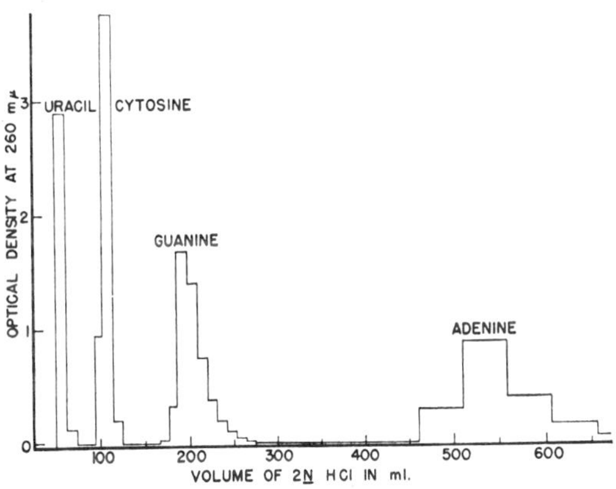

FIG. 1. Separation of purine and pyrimidine bases by cation exchange. *Exchanger*: 8.1 cm × 0.74 cm² Dowex-50, ca. 300 mesh, H⁺ form. *Eluting solution*: 2*N*-HCl at 0.6 ml/min. *Test material*: 0.5 mg uracil, 1.0 mg each of adenine, guanine, and cytosine, in 7.5 ml 2*N*-HCl. *Recoveries*: 98.5–101.5% (based on optical density at 260 mμ).

through our earlier experience with radioisotope separation by ion exchange chromatography (*4, 11*). The present note describes briefly methods we have developed for the separation of each of the purine and pyrimidine bases and of the ribose nucleotides.

Of the five naturally occurring purine and pyrimidine bases, three (cytosine, guanine and adenine) exist as cations in solution of pH $<$ 4 and hence are able to combine with cation exchangers. Since their affinities for the cation exchanger differ, they may be eluted successively with such simple reagents as HCl or NaCl; this is demonstrated in Fig. 1. This procedure does not separate the non-ionized bases, thymine and uracil. Since these will not usually be encountered together, no disadvantage attaches to this fact, but it is easy to separate them, as well as the other three substances, by resorting to anion exchangers[1] and alkaline solutions as shown in Fig. 2.[2] Recoveries by both procedures, based on spectrophotometric absorption in the ultraviolet (260–265 mμ) are essentially quantitative. Identification of components was made by the shape of the ultraviolet absorption curve and by partition chromatography (*3*).[3]

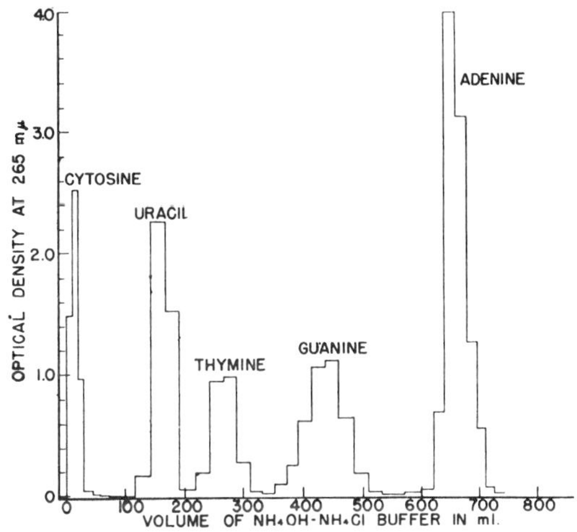

FIG. 2. Separation of purine and pyrimidine bases by anion exchange. *Exchanger*: 8.5 cm × 0.74 cm² Dowex-A1, ca. 300 mesh, Cl⁻ form. *Eluting solution*: 0.2*M*-NH_4OH-NH_4Cl buffer, pH 10.6, 0.025 *M*-Cl⁻ (at 575 ml, pH changed to 10.0, Cl⁻ to 0.1 *M*), at 0.25 ml/min. *Test material*: 1 mg cytosine, 2 mg each of other bases, in 10 ml eluting solution. *Recoveries*: 97.5–99% (based on optical density at 265 mμ).

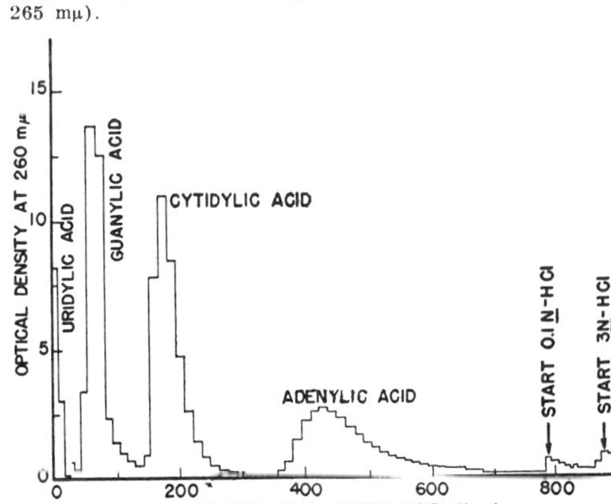

FIG. 3. Separation of mononucleotides of yeast nucleic acid by cation exchange. *Exchanger*: 24 cm × 0.74 cm² Dowex-50, ca. 300 mesh, H⁺ form. *Eluting agent*: 0.1 *M* acetic acid at 0.15 ml/min. *Test material*: 100 mg mixed nucleotides, recovered from their Ba salts following 0.6*N*-$Ba(OH)_2$ hydrolysis of yeast nucleic acid (conversion of nucleic acid to nomonucleotides not quantitative), in 10 ml 0.1 *M* acetic acid. *Recoveries*: 85–100% (based on optical density at 260 mμ).

[1] The Dowex-A1 anion exchanger, a strong base, was kindly furnished us by Dr. L. Matheson of the Dow Chemical Company, Midland, Michigan.

[2] These experiments were carried out by Mr. J. X. Khym.

[3] We are indebted to Dr. C. E. Carter not only for the chromatographic analyses but also for continued encouragement and assistance during the course of these experiments.

The nucleotides of yeast nucleic acid (uridylic, cytidylic, guanylic and adenylic acid) exhibit to some degree the basic properties of the purine or pyrimidine constituent; this is modified by both size and the negative phosphate group with the result that the exchanger-to-ion bond is considerably weaker. Weak acids have been found to be the most practical eluting agents and a separation by means of dilute acetic acid is demonstrated in Fig. 3. Recovery of the starting material, based upon absorption at 260 mμ, varies from 85 to 100%. In the absence of pure starting materials (the commercial preparations showing 5–50% impurity by these methods), it is difficult to do better than this without an extensive series of preparations of pure compounds. It is possible that there is a significant amount of acid hydrolysis of purine nucleotides occurring during their stay in the exchanger bed but we have not been able to prove this conclusively. Regardless of this possibility, the method as it stands will isolate the bulk of each constituent in spectrophotometrically pure form.

Concentration and metathesis of the large volumes collected in the separations are easily achieved (*5, 10*), where evaporation is undesirable, by recycling the effluent through the same column, washing with water to remove residual acid or base, and then eluting with a reagent which removes the adsorbed ion rapidly (e.g., NH_4OH in the acid separations).

Although the separations reported are all at or near the milligram level, we have been able to effect practical separations on the same size columns on up to nearly 100-mg amounts. The lower limit is, of course, set by the limit of detection. At the higher levels, overlapping of bands becomes more prominent because of the relationship between the concentration of an ion and its distribution coefficient (*12*).

The details of these and related separations, with examples of their application to specific problems, will be reported elsewhere.

References

1. Archibald, R. M. *J. biol. Chem.*, 1944, **156**, 121.
2. Cannan, R. K. *J. biol. Chem.*, 1944, **152**, 401.
3. Carter, C. E. (Unpublished data).
4. Cohn, W. E., and Kohn, H. W. *J. Amer. chem. Soc.*, 1948, **70**, 1986.
5. Cohn, W. E., Parker, G. W., and Tompkins, E. R. *Nucleonics*, 1948, **3**, 22.
6. Consden, R., Gordon, A. H., and Martin, A. J. P. *Biochem. J.*, 1948, **42**, 443.
7. Drake, B. *Nature, Lond.*, 1947, **160**, 602.
8. Elmore, D. T. *Nature, Lond.*, 1948, **161**, 931.
9. Harris, R. J. C., and Thomas, J. F. *Nature, Lond.*, 1948, **161**, 931.
10. Kocholaty, W., and Junowicz-Kocholaty, R. *Arch. Biochem.*, 1947, **15**, 55.
11. Tompkins, E. R., Khym, J. X., and Cohn, W. E. *J. Amer. chem. Soc.*, 1947, **69**, 2769.
12. Tompkins, E. R., and Mayer, S. W. *J. Amer. chem. Soc.*, 1948, **69**, 2859.

32

Reprinted from *J. Amer. Chem. Soc.*, **72,** 1471-1478 (Apr. 1950)

The Anion-Exchange Separation of Ribonucleotides

BY WALDO E. COHN

Investigations of nucleic acid chemistry, constitution and metabolism, particularly where tracers are employed, depend to a great degree upon the development of methods for the quantitative separation of the various chemical and enzymatic degradation products of the larger molecules. For greatest usefulness, these methods should be operable at the lowest detectable levels of concentration and amount of material and they should give adequate resolution—complete, if possible—of closely related substances. The classical methods[1a] are based principally upon precipitation which, because of solubility limitations and coprecipitation, do not, in general, meet such quantitative demands as well as do those methods based upon solvent extraction or ion-exchange, which have found such utility in the isolation of carrier-free radioisotopes.[2]

Previous work in the development of ion-exchange procedures for the quantitative separation of fission product ions, including the rare earth elements, from their mixtures at trace concentrations[3,4,5] led to the choice of ion-exchange as a likely mechanism to exploit in order to develop better separation methods for the nucleotides, nucleosides and free bases. The fact that nucleotides and their degradation products contain both acid and basic groups offered the possibility of using both anion and cation exchangers while the differences in pK from compound to compound[1a] indicated that these might prove as conclusive in effecting organic separations as are the complex dissociation constants and base strengths in the rare earth separations.

The experiments and results described below (and described in preliminary form earlier)[6,7] indicate that ion-exchange is a separation tool of the required accuracy and sensitivity and, as with any quantitative and sensitive separation method, it can serve also as a means of analysis, investigation and production.

Preliminary Considerations.—To effect an ion-exchange separation of the members of a family of ions, it is necessary to find a set of conditions under which each exhibits a different degree of affinity (distribution coefficient) with respect to the exchanger. While this affinity is governed by a number of variables,[8,9] it is reasonable to assume that, in a group as closely related as the monoribonucleotides, the net charge per ion will be the most important one. Since these substances possess both acid (phosphate) and basic (amino) groups, the pH of the medium determines the net charge by determining the degree of dissociation of these groups. Hence, a preliminary calculation was made of the degrees of dissociation of these groups as a function of pH, using the pK values quoted by Levene[1a]; these calculations are presented graphically in Fig. 1.

It is immediately apparent from Fig. 1 that (1) the order of increasing net negative charge per molecule is cytidylic, adenylic, guanylic and uridylic acid; (2) cytidylic and adenylic exhibit a net positive charge, and hence are cations, at pH values below about 2.5 and guanylic below 1.5, while uridylic remains anionic down to pH 0, which facts are consistent with the cation-exchange sep-

(1) Operated by Carbide and Carbon Chemicals Corporation under Contract No. W-7405-Eng-26 for the Atomic Energy Commission, Oak Ridge, Tennessee.

(1a) P. A. Levene and L. W. Bass, "Nucleic Acids," Chemical Catalog Co., Inc., New York, N. Y., 1931.

(2) W. E. Cohn, in "Advances in Biological and Medical Physics," Vol. I, Academic Press, New York, N. Y., 1949; also, *Naval Med. Bull. Suppl.*, Mar.-April, 1948, p. 42.

(3) E. R. Tompkins, J. X. Khym and W. E. Cohn, THIS JOURNAL, **69**, 2769 (1947).

(4) W. E. Cohn, G. W. Parker and E. R. Tompkins, *Nucleonics*, **3**, No. 5, 22 (1948).

(5) W. E. Cohn and H. W. Kohn, THIS JOURNAL, **70**, 1986 (1948).

(6) W. E. Cohn, *Science*, **109**, 377 (1949).

(7) W. E. Cohn, THIS JOURNAL, **71**, 2275 (1949).

(8) E. R. Tompkins, *J. Chem. Educ.*, **26**, 32, 92 (1949).

(9) R. Kunin, *Anal. Chem.*, **21**, 87 (1949).

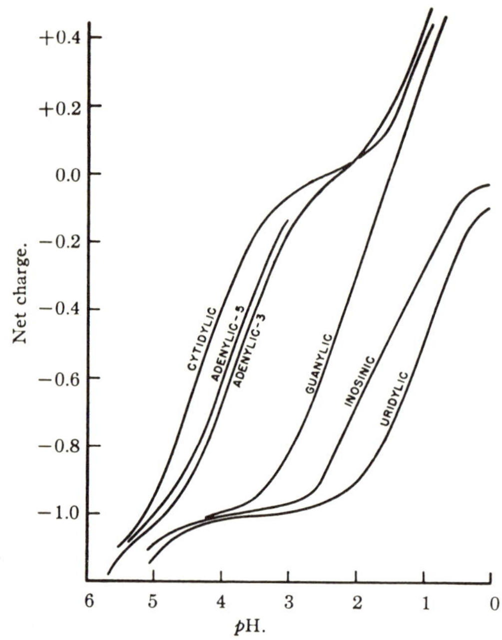

Fig. 1.—Net charge per molecule of ribonucleotide as a function of *p*H (calculated from data of Levene).

arations reported previously[6]; (3) at *p*H values above 5, all are strongly negative. From these observations, it would seem simplest to adsorb the mixture at *p*H values of 6 or greater, and to remove the nucleotides in the order cytidylic, adenylic, guanylic and uridylic acid by lowering the *p*H in a stepwise fashion, increasing the anion concentration with salt near the end if necessary to avoid strongly acid conditions which might hydrolyze the compounds. This was the plan of the experiments reported below.

Experimental

Apparatus.—The type of column is described by Tompkins.[3,8] An automatic sample changer was used to change receivers at fixed time intervals, thus permitting extended runs to be made conveniently. A Beckmann ultraviolet spectrophotometer was employed to assay and identify the samples, taking advantage of the characteristic ultraviolet spectra of the several substances being investigated. The dependency of the shape of these spectra upon *p*H offers a further method of identification. Paper chromatography also served as a means of identification.[10,11]

Ion-exchangers:—The experiments utilized principally 200–400 mesh Dowex-1,[12] a strong-base anion exchanger; Dowex 2[12] and Amberlite IRA-400[13] were also employed with similar results. The exchangers were washed free of fines by decantation, slurried into the columns and washed with acid and base before being converted to the salt form desired for a particular experiment. Columns were used repeatedly without change of resin, often without backwash or resuspension, until accident or change in type of experiment warranted discarding them. One has been used in thirty separations without noticeable loss in combining capacity or efficiency.

Test Substances.—As a source of mixed monoribonucleotides, barium hydroxide hydrolysates of yeast nucleic acid were used.[10,14] Commercial adenylic, guanylic and cytidylic acids were employed without purification as the impurities lent themselves to the investigations in hand. Solutions of uridylic acid were prepared by cation-exchange methods,[6] identified by spectral properties[15] and standardized by phosphorus analysis.[11]

Procedure.—The columns were prepared for use by washing with solutions of stronger eluting power than any which were to be used in the experiment (*e. g.*, 1 *N* hydrochloric acid if replacement by Cl^- was contemplated) until influent and effluent were identical in *p*H and optical density at 240–280 mμ. Following a water wash, the test substance, in amount usually less than 5% of the exchanger capacity, was adsorbed from a solution of low eluting power (*e. g.*, *p*H 8, anion concentration $\leq 0.01M$). The column was again washed with water and elution with the desired reagent was commenced. Zero volume was taken as the point when the effluent attained the same *p*H and ionic strength as the influent solution.

Results

Although many conditions, varying from *p*H 1.8 to 5.6 with various anions and varying anion concentrations, were studied, it is necessary to present only a few typical experiments at the various extremes before summarizing the data obtained. A rapid partial separation, utilizing a stepwise *p*H procedure as outlined above, is shown in Fig. 2. The spectrophotometric ratios (*i. e.*, ratios of optical densities at two wave lengths) of the fractions collected indicate a tendency for guanylic acid to follow uridylic acid, the reverse of the predicted order. The degree of sep-

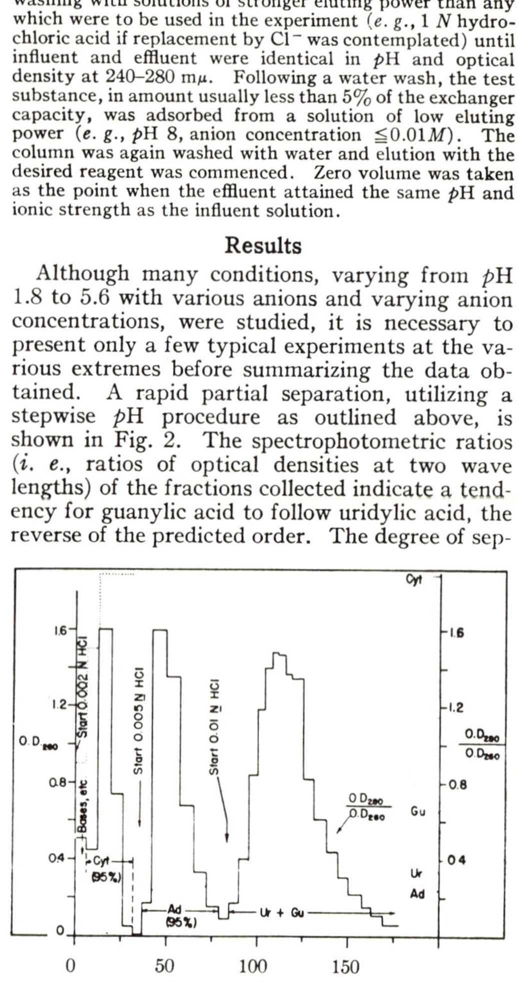

Fig. 2.—Rapid separation of nucleotides: exchanger, Dowex-1, 0.9 sq. cm. × 2 cm.; solution, hydrochloric acid at 0.8 ml./min.; test material, ~4 mg. of mixed nucleotides.

(10) We are indebted to Dr. C. E. Carter for chromatographic and other analyses[11] and also for the nucleic acid hydrolysates and other preparations used in these experiments.

(11) C. E. Carter, THIS JOURNAL, **72**, 1466 (1950).

(12) Obtained through the courtesy of Dr. L. A. Matheson, Dow Chemical Co., Midland, Michigan.

(13) Resinous Products and Chemicals Co., Washington Square, Philadelphia, Pa.

(14) H. S. Loring, P. M. Roll and J. G. Pierce, *J. Biol. Chem.*, **174**, 729 (1948).

(15) J. M. Ploeser and H. S. Loring, *ibid.*, **178**, 431 (1949).

aration of cytidylic and adenylic acids is apparent. Free bases (adenine, guanine, uracil, cytosine) or ribosides exhibit practically no affinity for anion-exchangers at *p*H values below 8 and hence are removed, if adsorbed at all, ahead of cytidylic acid.[6]

As predicted, separations at *p*H values above 4 are less satisfactory and it is necessary to use longer columns in order to achieve even a separation into pairs of nucleotides, as shown in Figs. 3 and 4. In addition, the affinity of the nucleotides for the exchanger here is so much greater than at *p*H 2-3 that the anion concentration must be increased considerably to achieve comparable elution rates, thus hindering readsorption for purposes of concentration (see below). Nevertheless, such a procedure is satisfactory for simple analysis as the spectrophotometric ratios are sufficiently different to permit calculation of the composition of each of the two fractions (pyrimidine nucleotides and purine nucleotides). In addition, satisfactory resolution of the isomers of adenylic and of inosinic acids can be obtained under these same conditions; this, as well as the proof that two kinds of adenylic acid exist in hydrolysates of yeast nucleic acid, is discussed elsewhere.[11,16]

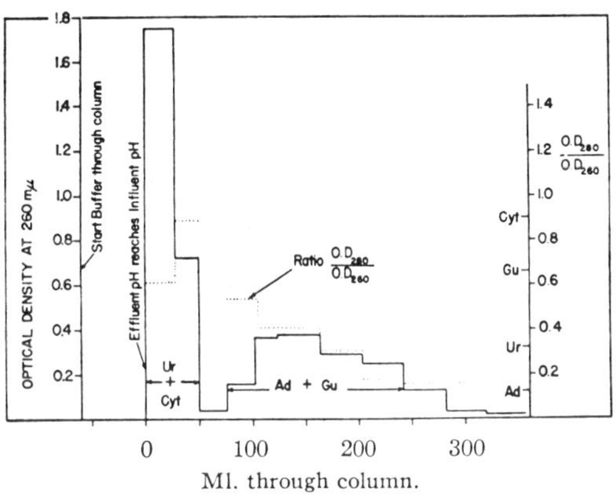

Fig. 3.—Rapid separation of mononucleotides: exchanger, Dowex-1 (Cl^-), 0.74 sq. cm. × 12 cm.; solution, 0.1 *M* sodium chloride in 0.01 *M* acetate buffer, *p*H 5.6, 0.7 ml./min.; test material, ~4 mg. of purified nucleotides.

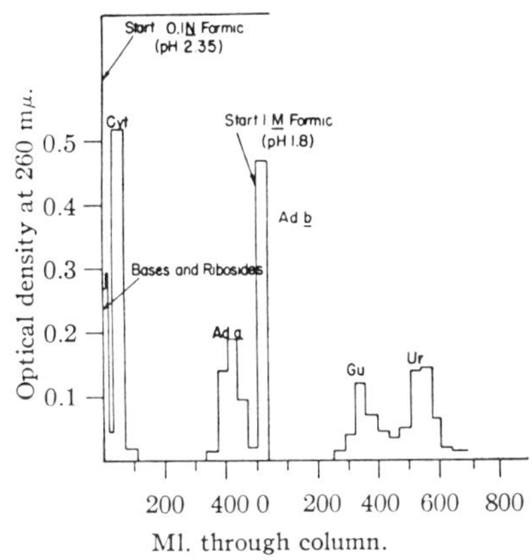

Fig. 5.—Separation of nucleotides: exchanger, Dowex-1 (formate), 0.74 sq. cm. × 13 cm.; solution, formic acid, 0.7 ml./min.; test material, 3 mg. of mixed nucleotides from barium hydroxide hydrolyzate of yeast nucleic acid.

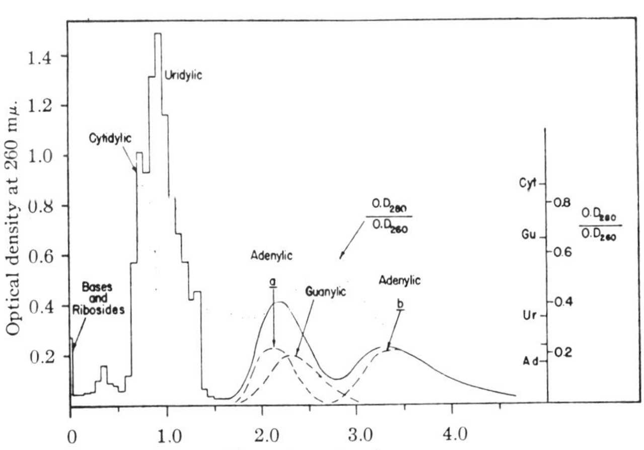

Fig. 4.—Separation of mixed monoribonucleotides: exchanger, Dowex-1 (Cl^-), 10 cm. × 0.94 sq. cm.; solution, 0.02 *M* sodium chloride in 0.01 *M* acetate buffer, *p*H 5.6, 1.4 ml./min.; test material, ~40 mg. of mixed nucleotides from barium hydroxide hydrolyzate of yeast nucleic acid.

Longer columns and higher distribution coefficients give better separations, as demonstrated in Fig. 5. Utilizing formic acid in place of hydrochloric (giving better *p*H control, more available anion and a lower specific replacement power than Cl^- on Dowex-1),[17] adenylic acid is completely separated from the other nucleotides and divided into two parts. (This experiment is the first in which practically complete resolution by ion-exchange of the two adenylic acid isomers derived from yeast nucleic acid was obtained; this separation was verified by paper chromatography and by many subsequent separations and tests.[11,16] It should also be noted that uridylic now follows guanylic, as originally predicted, contrary to the calculations of Fig. 2 (and of Fig. 6, below)).

The best separations of the nucleotides of ribonucleic acid with hydrochloric acid have been obtained at *p*H values

(16) C. E. Carter and W. E. Cohn, to be published; *Federation Proc.*, **8**, 190 (1949).

(17) L. A. Matheson, personal communication.

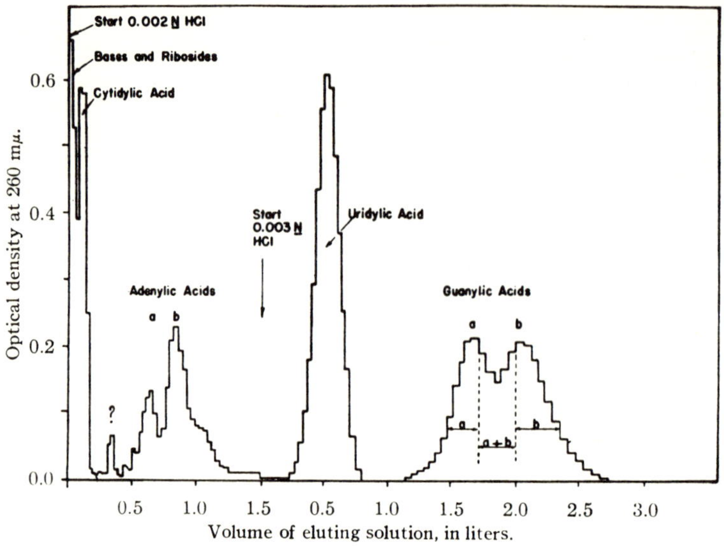

Fig. 6.—Separation of mononucleotides of rat liver ribonucleic acid by anion-exchange: exchanger, Dowex-1, 12.5 cm. × 0.74 sq. cm.; eluent, dilute hydrochloric acid, 0.5 ml./min.; test material, ~14 mg. of mixed nucleotides from barium hydroxide hydrolysis of rat liver nucleic acid in 10 ml. of water.

in the neighborhood of 2.7. A separation of a mixture of nucleotides obtained from a barium hydroxide hydrolysate of ribonucleic acid from rat liver at *p*H 2.7 is shown in Fig. 6. The two adenylic acid peaks are clearly shown while uridylic now precedes guanylic. Of particular interest is the observation of two peaks in the latter. The eluate was therefore divided into the three fractions (a, a + b, b) shown in Fig. 6. When these three fractions were chromatographed,[10,11] the samples a and b migrated to two slightly different positions, while sample a + b occupied both positions. All three fractions were degraded to guanine by acid,[10] which is taken as evidence that both peaks are due to guanylic acid. More evidence is necessary in order to establish the possible existence of two guanylic acids, analogous to the two adenylic acids.[11,16] The nature of the small peak between cytidylic and adenylic acid is not known; in amount, it was too small to be analyzed.

The removal of free base and ribosides ahead of (*i. e.*, at higher *p*H values than) cytidylic acid has been mentioned above in connection with Figs. 2, 4, 5 and 6. In Fig. 7, a separation of a mixture of several free bases and nucleosides from adenylic acid, representing the nucleotides, is shown to prove this point. In this experiment, the concentration of anion (formate) was kept constant throughout, the *p*H being shifted by changing the proportions of ammonium hydroxide and formic acid in the solution. Satisfactory resolution of the pairs shown can be attained by more careful adjustment of conditions; this has already been demonstrated for the five bases[6] and for cytidine and uridine.[18,19] (The probable positions of the latter two substances, obtained from experiments, conducted by Mr. J. X. Khym in this Laboratory, are shown in brackets in Fig. 7,

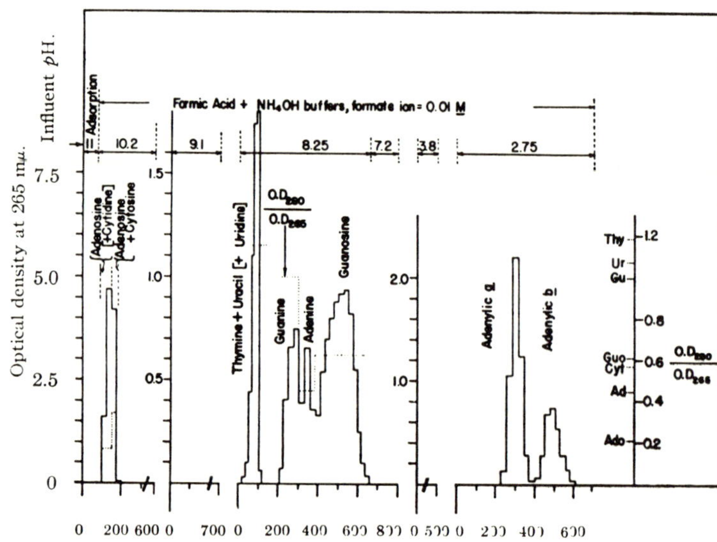

Fig. 7.—Separation of bases, ribosides and adenylic acid: exchanger, Dowex-1, 13 cm. × 0.74 sq. cm.; rate, 0.5 ml./min.

although they were not included in this particular run.) Thus, it is possible to separate most of the free bases, nucleosides and nucleotides of ribonucleic acid from one another in one column run; this procedure can, therefore, be used for the quantitative analysis of acid hydrolysates of nu-

(18) D. T. Elmore, *Nature*, **161**, 931 (1948).
(19) R. J. C. Harris and J. F. Thomas, *ibid.*, **161**, 931 (1948).

cleic acid, which yield adenine, guanine, cytidylic and uridylic acid,[20] in place of the laborious and less accurate precipitation methods, as well as for other types of hydrolysates.

Attempts to obtain complete separations in *p*H regions above 3 have not been successful as certain peaks are too close together. A complete summary of the data, covering the entire range from *p*H 1.8 to 5.6, is given by the points in Fig. 8. The points and the broken lines connecting them denote the positions of the several peaks, these positions being proportional to the distribution coefficients,[21] with respect to that of adenylic acid *a*. When used in conjunction with the absolute data on adenylic acid *a* given in Table I, the curves in Fig. 8 make it possible to estimate the

TABLE I

ML. TO PEAK OF ADENOSINE; PHOSPHATE (ADENYLIC ACID *a*) FOR DOWEX-1 COLUMN, APPROXIMATELY 0.74 SQ. CM. × 13.5 CM. (10 ML.)

Solution	Concentration (*M*)	*p*H	Ml. to peak
Formic acid	1	1.8	25
	0.25	2.2	100
	.1	2.4	340
HCl	.005	2.3	70
	.004	2.4	110
	.002	2.7	600
	.0016	2.8	900
0.01 *M* acetate buffer	.01	3.6	800
plus NaCl	.01	4.6	1800
	.02	4.6	800
	.02	5.6	2100
	.10	5.6	150

positions of the peaks of all ribonucleotides for a given column. The solid curves in Fig. 8 express the calculated net charge of the same nucleotides, again relative to adenylic acid, using the data summarized in Fig. 1; the curves for cytidylic and uridylic acid have, however, been normalized by the arbitrary factor 0.37, the significance of which remains to be determined. While use of this factor gives a good coincidence of experimental and calculated values for uridylic acid, the fit for cytidylic is less satisfactory. This may be due to the fact that cytidylic acid is always eluted first, at low volumes, with a corresponding large uncertainty in zero point (*p*H equilibrium) and hence position of peak.

The data in Fig. 8 indicate clearly the reversal of the relative positions of uridylic and guanylic acid which takes place at *p*H 2.0. Indeed, it is apparent that uridylic can be eluted simultaneously with guanylic acid at *p*H 2, with adenylic acids at *p*H 3.3–3.5, or with cytidylic acid at *p*H values above 4. Also apparent is the fact that the optimum separation of all nucleotides at a single *p*H occurs at about 2.9 (dotted line).

(20) G. Schmidt, R. Cubiles and S. J. Thannhauser, *Cold. Spr. Hbr. Sym. XII*, 161 (1947).

(21) E. R. Tompkins and S. W. Mayer, THIS JOURNAL, **69**, 2859 (1947).

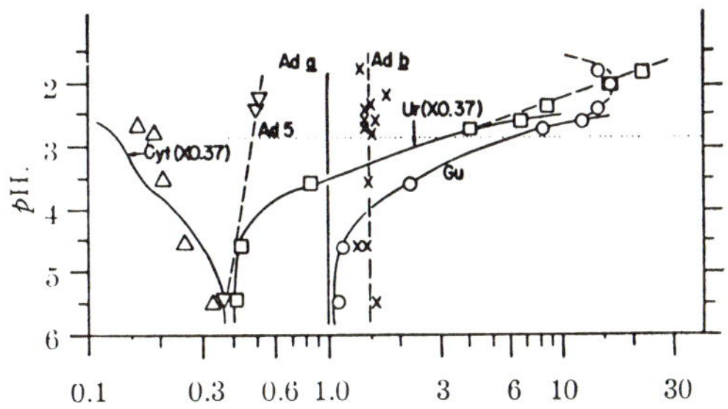

Fig. 8.—Observed relative distribution coefficients (points, broken lines) and calculated relative negative charges (solid lines) of nucleotides as functions of *p*H.

Although the data presented on separations in the region of *p*H 2.9 (*e. g.*, Figs. 6 and 7) indicate that the process may be a slow one and, therefore, somewhat impractical, it should be pointed out that the volume (and hence time) required to remove a given substance from the column at a fixed *p*H is dependent upon the anion concentration of the solution. While this point has not been thoroughly investigated, it appears, as might be expected, that the absolute peak distances (distribution coefficients) are approximately proportional to the (monovalent) anion concentration in the *p*H range of 2 to 5 where the nucleotides are monovalent anions. Therefore, the time of separation of a mixture into its component nucleotides may be reduced several-fold by increasing the anion concentration after the elution of each substance. Thus, each can be obtained in a small volume, comparable to those of cytidylic and uridylic acids in Fig. 6. Alternatively, as has been shown previously[7] and also in Figs. 2, 5 and 6, the *p*H can be changed at appropriate intervals.

The use of formic (or acetic) acids for elution in the *p*H 2–3 region has three advantages over dilute hydrochloric acid: (1) *p*H regulation is more exact; (2) the neutralization of the nucleotides, usually absorbed as divalent anions, is more rapidly achieved, eliminating large dead volumes; (3) simple evaporation may be used to recover the products. However, dilute hydrochloric acid has one very marked advantage in that subsequent concentration by ion-exchange, which requires the conversion of the effluent to a solution of *p*H 8–10 with low (< 0.01 *M*) anion concentration, is possible by the simple addition of a small amount of ammonium hydroxide. Such a solution can readily be stripped of its nucleotide content on 10-fold smaller columns, from which the product can be recovered in concentrated (1–10 mg./ml.) solution by elution with 0.02–0.1 *M* hydrochloric acid. In practice, it has been possible practically to saturate a 2 ml. column with adenylic acid (about 150 mg.) from 2 liters of 0.01 *M* Cl^- solution at *p*H 9; subsequent elution, with 0.1 *M* hydrochloric acid, yields a solution of 5–10 mg./ml., from which the adenylic acid crystallizes on standing. This

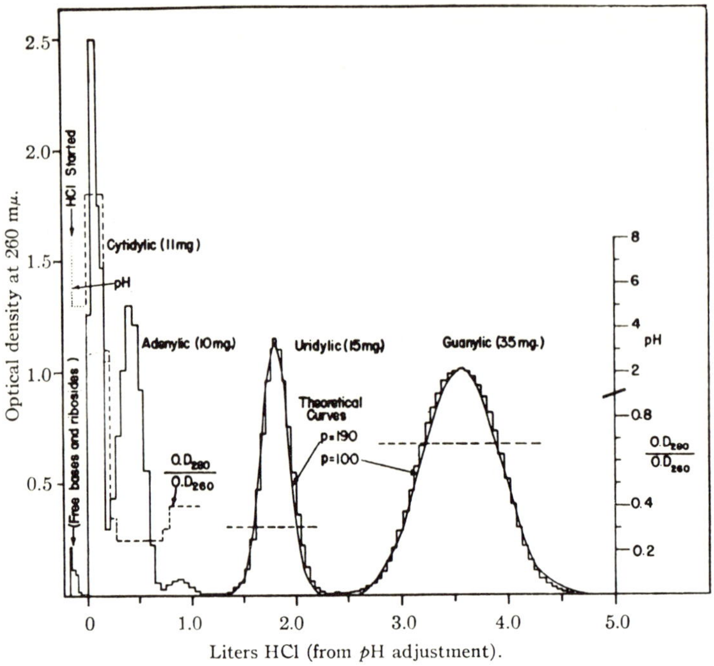

Fig. 9.—Separation of mixed ribonucleotides: exchanger, Dowex-2 (Cl^-), 0.94 sq. cm. × 12.5 cm.; solution, 0.003 *M* hydrochloric acid, 0.8 ml./min.

manufacture of both. Of considerable importance in any of these applications is the fact that the shape of the elution curve, within the range of conditions investigated, is that of the normal curve of error, as predicted by the equilibrium theory of Mayer and Tompkins.[22] It is, therefore, possible to utilize the mathematical relationships of the normal curve of error to predict quantitatively the conditions necessary for a given degree of separation of two substances (length of column, the ratio of the distribution coefficients), the recoveries at given volumes of effluent solution, the concentration at the maximum relative to the distribution coefficient (position of the peak), etc., as has already been demonstrated in the rare earth separations.[22] (A full exposition of these applications is given by Tompkins.[8])

The approach of elution curves for the nucleotides to the theoretical is demonstrated in Fig. 9, in which theoretical curves are superimposed upon the elution curves of uridylic and guanylic acids obtained in a single run at constant *p*H. The theoretical curves were calculated from the formula[22]

$$t = \sqrt{p}\,(F/C - 1) \qquad (1)$$

where F and C are the effluent volumes to a given point and to the maximum of the elution curve, respectively, while the constant p (number of plates) is obtained from a simple modification of the Mayer–Tompkins formula[22]

$$p = 2\pi\,(C_pL_{\text{max.}})^2, \text{ namely}$$

$$p = 2\pi\left(\frac{\text{concentration at peak}}{\text{total solute}} \times \text{ml. to peak}\right)^2 \qquad (2)$$

Since the area under the normal curve of error (amount eluted) corresponding to a given value of t (ordinate of the curve of error) is obtainable from probability tables, only three values are necessary in order to solve, say, for total solute present: concentration at the peak, volume to peak, and p. Although p may vary, on a given column, from compound to compound (*cf.* Fig. 9) and from one eluting condition to another, both it and the volume to the peak are constant for a given set of conditions and a given compound and can be determined from a standardization experiment. Therefore, to solve for the total solute present in an unknown sample, as in a quantitative analysis, it is necessary to determine only the

recycling procedure has been used to prepare gram amounts of the two adenylic acid isomers, in crystalline form, no other operations being necessary. Coupled with a recording flow spectrophotometer and a recording *p*H meter, this procedure, utilizing a 33 sq. cm. column, has afforded a simple method for isolating pure nucleotides on a gram scale; this will be described in detail elsewhere.

No attempt has been made to determine optimum flow rates. Rates up to 1 ml./sq. cm. minute yield elution curves that are apparently symmetrical and practically identical with the normal curve of error shape predicted by the equilibrium theory of Mayer and Tompkins[22] while rates of 2 and 4 do not (*cf.* Fig. 6 *versus* the curve given in[7]). It would seem, then, that flow rates of about 1 ml./sq. cm. min. (at room temperature) permit the attainment of equilibrium between the resin beds employed and the nucleotides.

Discussion

Although these experiments were undertaken for the purposes stated in the introduction, the results reported here and previously[6] have perhaps much greater significance for such basic problems of nucleic acid chemistry as the quantitative assay of known constituents, the detection and isolation of unknown constituents and the

(22) S. W. Mayer and E. R. Tompkins, THIS JOURNAL, **69**, 2866 (1947).

concentration at the peak or to calculate this value, using (1) and the probability tables, from the concentration after a known volume (F) of effluent has passed through the column. It is also possible, once p is known, to predict the fraction eluted when the effluent has arrived at a given fraction of the peak concentration. Thus, when the elution curve has dropped to 50% of its maximum, 88% of the solute has been removed; for a 95% recovery, it is only necessary to proceed to 25% of the peak concentration; etc. Since these relationships hold on the ascending portions of the curves as well, their application to determine the degree of overlap of two substances in a separation procedure is obvious.[8,22]

These useful applications, as well as the use of symmetry as a criterion of a single component, are possible only with elution procedures which result in theoretical curves. Since these cannot be obtained when conditions vary during the course of a separation, it is advisable, in quantitative work or when searching for new substances, to keep flow rate, pH, etc., constant throughout to avoid the asymmetry which results from variations in such conditions.

Of considerable interest, from a theoretical standpoint, is the constant divergence of the observed relative distribution coefficients of the pyrimidine nucleotides from the values predicted on the basis of the ionization constants of the phosphate and amino groups. The data (presented in Fig. 8) indicate that the pyrimidine nucleotides are three times less strongly adsorbed, per unit charge on the molecule, than the purine nucleotides or, conversely, that the latter are three times more strongly adsorbed than would be judged from net charge considerations alone. It would seem, in view of the similarities in molecular weight, constitution, valencies and concentrations of these ions, that degree of ionization would be the sole determining factor.

In searching for an explanation of the observed effect, two family differences between pyrimidine and purine nucleotides come to mind: the much greater solubility of the former[1a] and the presence of the extra imidazole ring in the latter. Evidence favoring the former over the latter as the cause comes from a few (as yet unpublished) observations on inosinic acid, which possesses the high solubility in water characteristic of the pyrimidine nucleotides but also the imidazole ring of the purine nucleotides. The inosinic acids[10] corresponding to the 5, a and b adenylic acids are eluted in that order[16] ahead of the corresponding adenylic acids at pH 5.6 but follow them, again in order, at pH 1.8; at the latter pH, they are scarcely separable from guanylic and uridylic acids. This behavior relates them to uridylic acid (which, like them, possesses no amino group). The factor necessary to normalize the theoretical curve for inosinic acid (based upon dissociation constants only as shown in Fig. 1) to fit the few points determined by elution seems to be about 0.5, which is not too far—considering the number and accuracy of the determinations—from the 0.37 factor noted for the pyrimidine nucleotides.

It should be recalled that the nucleotides were adsorbed from solutions of alkaline pH, where all are very soluble, in the topmost layers of the exchanger bed (as proved by the concentration data mentioned above). The resins employed have apparent combining capacities of about 2–3 meq. per gram of exchanger; thus, the concentration of nucleotide in the exchanger may be as high as one millimole per gram. The elutions were then carried out at lower pH values, between 5.6 and 1.8, and, therefore, amount to a process of dissociating a precipitated acid in a solution in which it is more (pyrimidine nucleotides) or less (purine nucleotides) soluble.

Indeed, it has been found possible, at very low pH and high chloride concentration (*e. g.*, 0.1 N hydrochloric acid), to elute adenylic acid from the exchanger at such a rate that it forms a supersaturated solution, from which it crystallizes in the receiver or, if the flow rate be slow enough, in the exchanger bed itself. In the latter case the crystals may easily be seen dispersed throughout the bed. This indicates that the concentration of nucleotide in the resin bed can exceed greatly its solubility in the aqueous medium, a necessary condition in the hypothesis that solubility is responsible for the differences in distribution coefficient noted.

These methods are being applied in the manner indicated to the analysis of the products of the enzymatic digestion of nucleic acid, to the production of uridylic acid and of the various nucleotide isomers discussed, and to the quantitative assay of the nucleotide content of nucleic acids from various sources. It is intended to apply them in similar fashion to the isolation of desoxyribonucleotides and other nucleotide material of interest in our investigations, including polynucleotides.

Acknowledgment. It is a pleasure to acknowledge the continued advice and assistance of Dr. C. E. Carter in the course of these experiments, in particular for the chromatographic analyses and nucleic acid hydrolysates.

Summary

The nucleotides of ribonucleic acid have been separated from each other and from the corresponding free bases and ribosides by successive elution from strong base anion exchangers with weak acids and buffers of controlled pH and anion concentration. The order of elution at pH 2 is cytidylic, adenylic, guanylic and uridylic acid; at pH 2.5–3, uridylic precedes guanylic acid; at pH 5.5, uridylic and cytidylic are removed together as are adenylic and guanylic acid. In all cases, the bases and ribosides precede cytidylic acid and may be separated from each other as well as from the nucleotides during the procedure. The three isomers of adenylic and of inosinic acids may be separated.

The applications of this technique to analysis, isolation, investigation and manufacture are indicated.

RECEIVED JULY 5, 1949

33

Reprinted from *Anal. Chem.*, 39(12), 1422–1428 (1967)

Fast Liquid Chromatography: An Investigation of Operating Parameters and the Separation of Nucleotides on Pellicular Ion Exchangers

C. G. Horvath, B. A. Preiss, and S. R. Lipsky
Section of Physical Sciences, Yale University School of Medicine, New Haven, Conn.

A liquid chromatographic system featuring high inlet pressures and a sensitive UV detector was designed for fast analysis of nonvolatile organic compounds. To establish optimal operating conditions, band dispersion in the mobile phase was studied by using capillary tubes as well as small bore columns packed with glass beads. Although peak broadening in open tubes was less than predicted by theory, the use of packed columns was more promising for fast separations. Novel pellicular column materials were prepared by coating glass beads with ion exchange resin and other solid phases. Rapid separation of nanomole quantities of ribonucleoside mono-, di-, and triphosphates was achieved by using a pellicular basic ion exchanger and gradient elution with a phosphate buffer. As shown by cellular extract analyses, the stability and efficiency provided by such column materials makes it possible to achieve fast separation of complex mixtures by a liquid chromatographic technique similar in speed, resolution, and quantitative range to gas chromatography.

A TECHNIQUE for the separation of nonvolatile substances which encompasses the speed, efficiency, sensitivity, and versatility of gas chromatography has yet to be devised. The present study was undertaken in an effort to investigate the feasibility of developing a liquid chromatographic system which could approach this goal. The method which evolved was then utilized for the analysis of the nucleoside phosphoric acids.

The time required for the separation of a solute pair is conveniently expressed by the retention time of the second peak, t, as

$$t = NH/u_b \tag{1}$$

where N is the number of plates required for the particular separation, H is the plate height, and u_b is the band velocity. Although H is the basic measure of the peak broadening, H/u_b, the time per plate, is more important if the speed of separation is of concern, as it is directly related to the separation time as shown in Equation 1. Since $u_b = u/(1 + k)$, Equation 1 can be written

$$H/u_b = t/n = H(1 + k)/u \tag{2}$$

where k is the column capacity ratio for the solute, and u the fluid velocity.

The plate height depends on the flow velocity (*1*, *2*) and, as Giddings (*3*) pointed out, if the molecular diffusion is the

(1) J. J. van Deemter, F. J. Zuiderweg, and A. Klinkenberg, *Chem. Eng. Sci.*, **5**, 271 (1956).
(2) M. J. E. Golay, ANAL. CHEM., **29**, 928 (1957).
(3) J. C. Giddings, *Ibid.*, **35**, 2215 (1963).

sole mass transfer mechanism, analogous operating conditions in gas and liquid chromatography would require flow velocities 10^5 times lower in liquid than in gas chromatography because diffusion coefficients are of the order of 10^{-6} cm² second^{-1} in liquids and 10^{-1} cm² second^{-1} in gases: consequently the time of analysis would be increased by the same factor. Obviously, high speed liquid chromatography cannot occur under those analogous conditions.

One can express the plate height (neglecting longitudinal diffusion) as the sum of three terms

$$H = f(H_m/u) + C_{m2}u + C_s u \tag{3}$$

$$H/u = H_m/u + C_{m2} + C_s \tag{4}$$

The first term, H_m, represents the plate height contribution due to peak dispersion in the mobile phase streaming through the column; the dependence of this term on flow velocity is not completely understood. The second term, $C_{m2}u$, stems from the mass transfer resistance by the stagnant mobile phase in the intraparticle pores of the packing. The third term, $C_s u$, is related to nonequilibrium due to the stationary phase. The three terms of Equation 4 should be minimized in order to obtain an increase in speed of analysis.

One conceivable approach is to reduce the particle size in packed columns. With porous granules C_{m2}, and with conventional ion exchange resins, C_{m2} and C_s, decrease with the square of the particle diameter. Thus, extremely fine silica particles were used for fast separation in liquid-solid chromatography (*4*), and significant acceleration of the analysis of amino acids and peptides by ion exchange chromatography (*5*, *6*) has been achieved by using very small uniform spherical resins.

Another approach involves the use of open tubular columns or columns packed with a fluid-impermeable support where $C_{m2} = 0$, and where C_s can be small if the depth of the proper stationary phase is small. Under these conditions the velocity-dependent first term, H_m/u, in Equation 4, which arises from the nonuniform flow pattern, becomes very important.

For laminar flow in straight open tubes according to Taylor (*7*),

$$H_m/u = \frac{d^2}{24\, D_M} \tag{5}$$

where d is the tube diameter and D_M is the diffusion coefficient of the solute in the mobile phase. This term becomes very high for a liquid mobile phase as a consequence of low diffusivity in liquids, and the use of open tubular columns does not appear to be favorable under laminar flow conditions in liquid chromatography.

Pretorius and Smuts (*8*) have proposed the operation of open tubular columns with turbulent flow which is characterized by very low H/u values. In gas chromatography, this effect of turbulence has been proved experimentally (*9*) and an abrupt drop in plate height was found. The exploitation of turbulence in analytical work, however, is limited by the inlet pressure requirements (probably several thousand

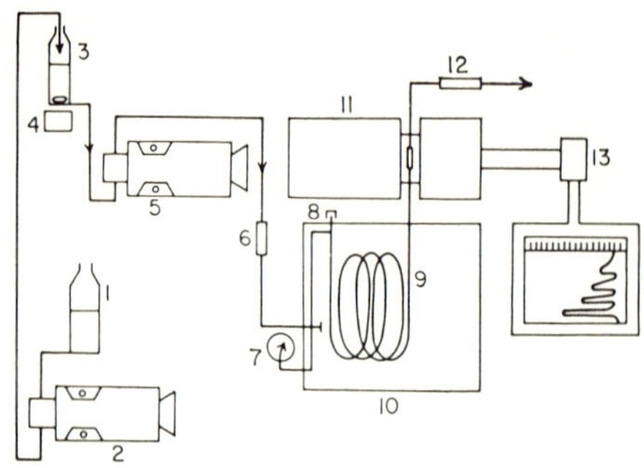

Figure 1. Liquid chromatograph with UV-detector for gradient elution

(1) Reservoir for strong eluent
(2) Pump
(3) Mixing chamber with heating
(4) Magnetic stirrer
(5) Pump
(6) Filter
(7) Pressure gauge
(8) Sample injection port
(9) Column
(10) Air thermostat
(11) Spectrophotometer with micro flow cell
(12) Flow meter
(13) Scale expander
(14) Recorder

atmospheres) for the high liquid velocities necessary to achieve turbulence in small bore columns.

Under laminar and turbulent flow conditions in helical open tubes radial mass transfer is enhanced by secondary flow in the cross sectional plane (*10*). Exploitation of this effect in chromatography has not been attempted as yet.

In packed columns, the fluid flows in the interstitial channels of the packing. At low flow velocities, molecular diffusion is the dominant mass transfer mechanism in the mobile phase, but with increasing flow velocity the contribution of a convective lateral mixing which depends on the packing structure becomes more significant. The result is a flattening of the H *vs.* u curve in the high velocity region. According to Giddings (*11*) the two mechanisms are coupled. He originally proposed for H_m the following expression

$$H_m = (1/2\, \lambda\, d_p + D_M/\omega\, d_p^2\, u)^{-1} \tag{6}$$

where λ and ω are parameters related to packing structure and d_p is the particle size. Giddings has now extended Equation 9 to a summation form which is much more general (*12*).

Knox (*13*) studied this effect in liquid chromatography using columns packed with glass beads and observed a more gradual flattening of the H *vs.* u curve in the high velocity region than predicted by the coupling theory. In liquid chromatography, owing to the low rate of molecular diffusion, coupling becomes significant at relatively low flow velocities and determines H_m over a range of several orders of magnitude in u.

In packed columns there is no sudden onset of turbulence and the decrease of h with increasing fluid velocity is not as pronounced as that in open tubes (*9*).

(4) E. W. Piel, Anal. Chem., **38,** 670 (1966).
(5) P. B. Hamilton, *Ibid.*, **35,** 2055 (1963).
(6) J. V. Benson and J. A. Patterson, *Ibid.*, **37,** 1108 (1965).
(7) G. Taylor, *Proc. Roy. Soc.* (*London*), **219A,** 186 (1953).
(8) V. Pretorius and T. W. Smuts, Anal. Chem., **38,** 274 (1966).
(9) J. C. Giddings, W. A. Manwaring, and M. N. Myers, *Science*, **154,** 146 (1966).
(10) J. A. Koutsky and R. J. Adler, *Can. J. Chem. Eng.*, **43,** 239 (1964).
(11) J. C. Giddings, *Nature*, **184,** 357 (1959).
(12) J. C. Giddings, Anal. Chem., **35,** 2215 (1963).
(13) J. H. Knox, *Ibid.*, **38,** 253 (1966).

In the present study, the effect of various parameters on H_m in helical capillary tubes and in small bore columns packed with glass beads was investigated. The results suggested the use of glass beads coated with a thin skin (pellicle) of stationary phase in a packed column. The rapid separation of nucleoside phosphoric acids was then achieved by utilizing ion exchange resin-coated glass spheres.

EXPERIMENTAL

Figure 1 shows a block diagram of the instrument used. For unsorbed solutes, a differential refractometer (Emmett Watson Associates, Ridgefield, Conn.) having a cell volume of 7 μl was used as detector.

For nucleotides, the detector was a Perkin-Elmer–Hitachi UVIS 139 spectrophotometer modified to accommodate a specially designed micro flow cell having a volume of 8 μl and 4-mm path length. With a scale expander, a full scale signal (1 mV) could be obtained at 0.04 absorbance unit. In this very low absorbance range, at constant flow rate and buffer composition, peak areas were considered to be linearly dependent on the amount of solute present. Analyses were performed at a wavelength of 260 mμ and slit width of 2 mm.

The fluid was forced through the system by a Milton Roy Minipump which provided leak-free operation at pressures up to 4000 psi in the low capacity range (0–20 ml per hour). A 300-ml glass cylinder equipped with an immersion heater and magnetic stirrer served as a solvent reservoir. For gradient elution, a second Milton Roy Minipump was employed to add the stronger eluent to the weaker eluent placed in the reservoir. In certain experiments a Greer Accumulator (Greer Olaer Products, Los Angeles, Calif.) was used to generate nonpulsating flow.

The temperature of the system was controlled by a Statham thermostated chamber, Model SD10-1. A 0–10,000 psi Helicoid gauge with "capillary bleed" was used for monitoring column inlet pressures. In the continuous flow mode, sampling was performed with a modified Loenco sampling valve. In the intermittent flow mode, a 10-μl Hamilton syringe was used to inject the sample onto the column.

The pellicular ion exchange resin used in this study was prepared as follows. Glass beads ($d_p \approx$ 50 microns) were coated with a solution of styrene, divinylbenzene, and benzoyl peroxide in ether and the solvent was evaporated. Polymerization and crosslinking were carried out at 90° C in an aqueous suspension of the coated beads. The product was washed with water, acetone, benzene, and methanol, then converted into the ion exchange resin. The anion-active resin was prepared by chloromethylating the polymer and reacting the product with dimethylbenzylamine. The pellicular ion exchange resin was then sieved and washed with 2*M* NaOH and HCl solutions and dried.

Although the use of this resin only is discussed, other resins were prepared and also found to be useful. For example, cation-active resin was prepared by sulfonating the coating of glass beads ($d_p \approx$ 50 microns) with 98% sulfuric acid at 90° C for 30 minutes. For making crosslinked polyethylene-imine resin, the water from commercial polyethylene-imine (Dow Chemical Co.) was removed by distillation with benzene. Glass beads ($d_p \approx$ 100 microns) were coated with a solution of the dry polyethylene-imine in methylene chloride and the solvent was evaporated. The coated beads were then exposed to methylene bromide vapor at 130° C for 10 minutes in order to crosslink the polymer. The product was washed with acetone, water, diluted hydrochloric acid, then with water, and dried.

The dry column material was packed into straight stainless-steel (No. 316) tubing (1-mm i.d., 0.0625-inch o.d.) by vibration and tapping. The outlet of the column was plugged with a thin disk of porous Teflon. To consolidate the packing, deaerated mobile phase was pumped through the column in perpendicular position for 5–6 hours, at inlet pressures of 3500–4000 psi. This treatment caused the packing to contract a few centimeters. The column was then coiled and connected to the instrument.

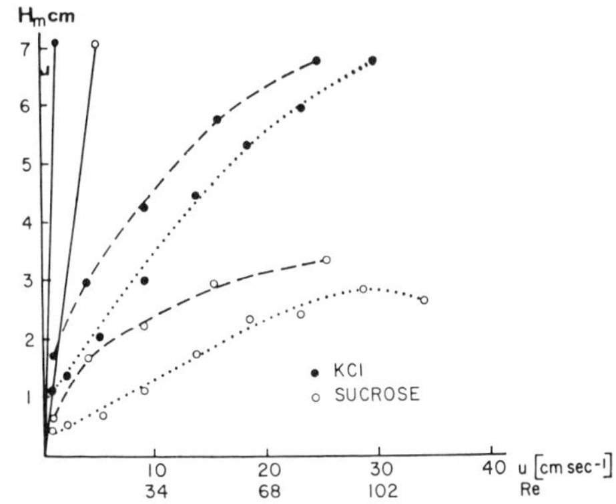

Figure 2. Plot of H_m against flow velocity in helical capillary tubes

Unbroken lines, calculated for straight tube; broken lines, Column I; dotted lines, Column II

After resieving, glass microbeads (Microbeads Inc., Jackson, Miss.) were soaked in chromic sulfuric acid overnight, washed with water and methanol, and dried. Although reagent grade chemicals were used for the preparation of buffers, some were contaminated with ultraviolet absorbing materials which gave rise to a significant base line drift during gradient elution. Reference mixtures were prepared by dissolving the purest available nucleotides in water. Perchloric acid extracts of mouse brain and liver were made in the usual manner and concentrated by lyophilization.

The linear gradient used in these analyses was produced by maintaining the flow rate through the column twice as high as the flow rate of the stronger eluent into the reservor. Samples were injected with a microsyringe at intermittent flow. After closing the sampling ports, both pumps were started at the same time; however, the sample was first eluted with the weak eluent which filled up the volume of the system between the reservoir outlet and the column inlet (approximately 1.7 ml). After each analysis the column was regenerated with approximately 15 ml of liquid. Thus each run was preceded by a 25-minute reconditioning period. The column was used over 5 months with various acidic and neutral salt solutions and occasionally with aqueous ethanol. Nevertheless, retention times remain reproducible on the properly reconditioned column.

RESULTS

Study of Peak Broadening in Columns without Stationary Phase. To assess the significance of secondary flow (*10*) in capillary columns, solute band spreading in coiled uncoated open tubes was investigated. The dimensions of the two helical columns (I and II) made of thin-walled stainless-steel tubing were as follows: Column I was 0.4-mm o.d. by 0.276-mm i.d. with a helix diameter (D_h) of 85 mm. The D_h/i.d. was 308. Column II was 0.4-mm o.d. by 0.287-mm i.d., D_h 1.28 mm, and the D_h/i.d. 4.5. The plots of plate height (H_m) *vs.* flow velocity (u) obtained by injecting KCl and sucrose as tracers in water onto these columns at 38° C are depicted in Figure 2.

Peak dispersion in Column I (broken lines) and Column II (dotted lines) is less than calculated by Equation 5 (unbroken lines) but nevertheless unacceptably large for chromatography.

Instead of the linear increase of H with laminar flow velocity, as predicted by Taylor and Golay, the H *vs.* u curve tends to flatten at high flow velocities. Contrary to the existing belief in chromatography, a decrease in the helix-diameter-to-column-inside-diameter ratio results in a decrease in plate height.

The higher plate heights obtained in both columns for the more slowly diffusing sucrose indicates that molecular diffusion plays a significant role in mass transfer under these conditions.

Although flattening of the H *vs.* u curves in coiled open tubes as shown in Figure 2 was expected as a consequence of secondary flow (*10*), the relatively small effect of coil-to-tube-diameter ratio in comparison to the strong effect of diffusivity suggests that other mechanisms cooperate in reducing H/u. It was thought that the roughness of the inner wall of the tube or the pulsating character of the flow might be responsible for the departure from the parabolic flow pattern. The wall effect is very difficult to assess but microscopic examination of the tube interior revealed no unusual roughness. On the other hand, Aris (*14*) proved theoretically that the effect of pulsation on peak dispersion in open tubes at laminar flow is insignificant unless the mean pressure gradient is smaller than the amplitude of the pressure gradient fluctuations. This is in good agreement with our findings with packed columns under the present experimental conditions because neither the pulsating nor the nonpulsating character of the flow had any demonstrable effect on plate height or retention times.

Although peak dispersion in open capillary tubes continues to be an interesting theoretical problem, the use of open tubular columns under these circumstances does not appear to be a promising approach to high speed liquid chromatography because the H_m/u values calculated from data shown in Figure 2 are much higher than those in gas chromatography.

In order to find optimal conditions for microanalysis, band spreading in two 150-cm long by 1-mm i.d. columns packed with uncoated glass beads was investigated. The small tube diameter was chosen to lessen trans column effects detrimental to column performance (*15*). The contribution of the injection and detector system to peak broadening was negligible in these experiments.

In the tracer experiments (Figure 3) with these columns, solutes of various diffusivities were used in order to observe the effect of particle size, flow velocity, and molecular diffusion on peak spreading. Thus, 10 μg of KCl ($D_m = 2.43 \times 10^{-5}$), sucrose ($D_m = 7.25 \times 10^{-6}$), and albumin ($D_m = 7 \times 10^{-7}$) were injected in the carrier stream of deaerated water. The column temperature was 38° C and flow was laminar. No variation in plate height was found with decreasing the sample size down to 2 μg.

With Column B the flattening of the H_m *vs.* u curves becomes significant at flow velocities of about 0.2 cm second^{-1}. As predicted by the coupling theory, the curves converge at high flow velocities indicating an increasing contribution to lateral mass transfer from convection and decreasing contribution from molecular diffusion. With Column *A* the plate height is significantly greater relative to the particle diameter than expected from data from the chemical engineering literature or from that published by Knox (*13*). The discrepancy is attributed to the small particle and tube sizes used in this study compared to those described in the literature.

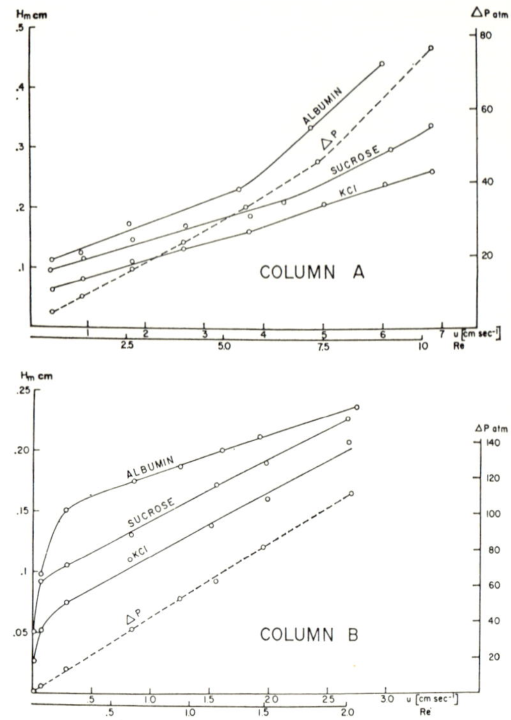

Figure 3. Plots of H_m and pressure drop against flow velocity in Column *A* and Column *B*

Column parameters were: Particle size, 88–105 microns (*A*), 44–53 microns (*B*). Δp per unit length at u = 1 cm second^{-1} (water 38° C) 0.064 atm cm^{-1} (*A*), 0.27 atm cm^{-1} (*B*). Porosity 37.9% (*A*), 39.8% (*B*). Column to particle diameter ratio 10(*A*), 20(*B*)

In chromatography, however, a relatively large surface area of the support material is needed for the fine distribution of the stationary phase. The surface area of glass spheres is inversely proportional to the diameter. Thus, the layer thickness of a given amount of stationary phase in the column decreases with the particle diameter. Because the plate height contribution of the stationary phase is proportional to the square of the layer thickness, a small particle size is attractive to reduce this term. On the other hand, the flow resistance of the column decreases with the square of the particle diameter; therefore, the available inlet pressure limits particle size reduction.

The particle size range selected for these experiments represents a practical compromise for chromatographic application. In both columns *A* and *B*, H_m/u decreased with flow velocity but appeared to approach a minimum value at the highest velocities, although significant reduction was achieved by increasing flow velocity to 1–2 cm second^{-1}. A further increase in u gave only marginal improvement in H_m/u with a large increase in the inlet pressure. Thus, $1 < u < 2$ cm second^{-1} seems a reasonable compromise velocity, comparable in its practical significance to the optimum practical flow velocity defined by Scott (*16*) in gas chromatography.

(14) R. Aris, *Proc. Roy. Soc.* (*London*) **259A**, 370 (1960).
(15) A. B. Littlewood, "Gas Chromatography 1964," A. Goldup, Ed., The Institute of Petroleum, London, 1965, pp. 77–93.

(16) R. P. W. Scott and G. S. F. Hazeldean, "Gas Chromatography 1960," R. P. W. Scott, Ed., Butterworths, Washington D. C., 1960, pp. 144–61.

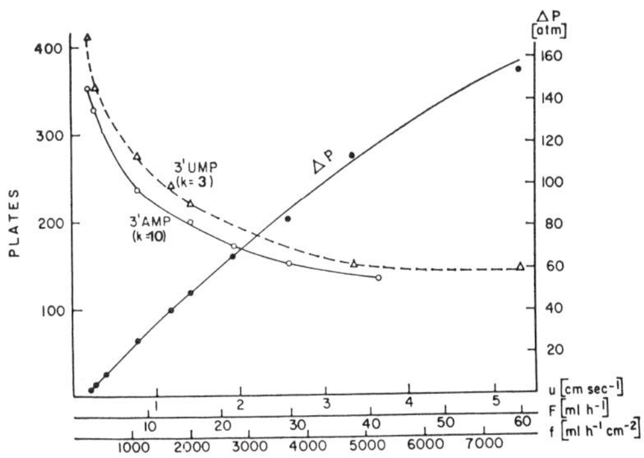

Figure 4. Plate number and pressure drop as a function of flow velocity and flow rate obtained with the pellicular anion exchanger column

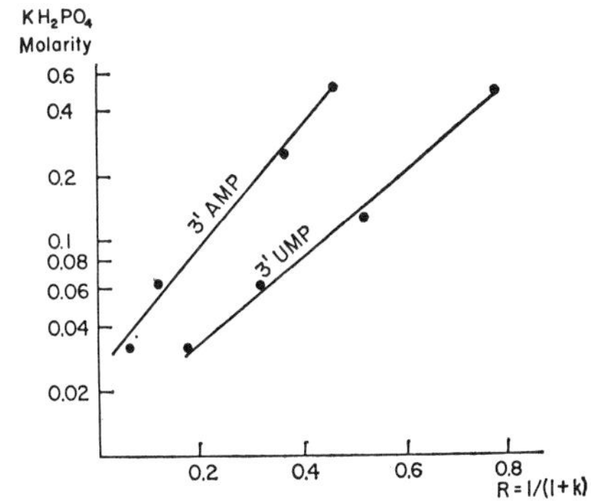

Figure 5. Effect of salt concentration on retention at 60° C

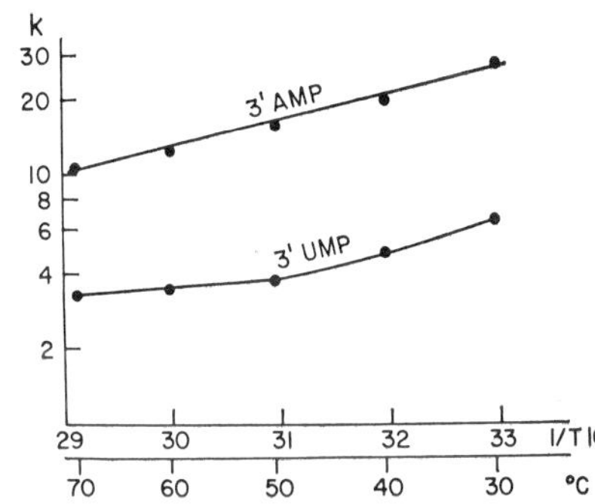

Figure 6. Effect of temperature on retention

Mobile phase: 0.045M KH_2PO_4, 0.002N H_3PO_4

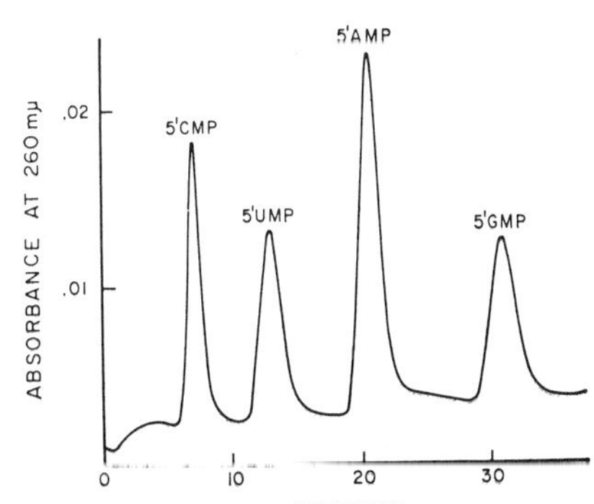

Figure 7. Separation of ribonucleoside-5′-monophosphoric acids

Column: 1-mm i.d., 193-cm length; pellicular strong anion exchange resin, sieve fraction No. 270–325, 8 microequivalent per gram. Eluent: Linear gradient of KH_2PO_4 from 0.025M to 0.325M (starting solution 2.5 × $10^{-3}N$ for H_3PO_4). Temperature: 70° C. Flow rate: 24 ml per hour. Flow velocity: 2.1 cm per minute. Pressure: 75 atm. Sample: 1 μg of 5′CMP, 5′UMP, 5′AMP, 5′GMP

The comparison of open tubular and packed columns indicates that the latter are superior for liquid chromatography. First, more favorable H_m/u values are obtained with packed columns than with open tubes. Second, because of their higher sample capacity and performance, a higher concentration of the emerging solutes is obtained with small-bore packed columns than with open tubes. This was particularly important with respect to the sensitivity of presently available detection systems.

Performance of a Pellicular Ion Exchange Column. For separation of nucleotides [ribonucleoside monophosphoric acids (2′ and 3′ or if not indicated, 5′): CMP (cytidylic acid), UMP (urydylic acid), AMP (adenylic acid), GMP (guanylic acid); ribonucleoside diphosphoric acids: CDP, UDP, ADD, GDP; ribonucleoside triphosphoric acids: CTP, UTP, ATP, GTP], a 193-cm long, 1-mm i.d. column was packed with No. 275-325 sieve fraction of glass beads coated with an anion exchanger containing benzyl-dimethylammonium groups. The capacity of the resin was 8 microequivalents per gram as determined by the method of Helfferich (*17*). Figure 4 shows plots of plate number and pressure drop against the linear flow velocity for 3′UMP and 3′AMP. The column capacity ratios (k) for these solutes were 3 and 10, respectively, large enough for the plate numbers to reflect the resolving power of the column. The flow rate, F, and the flow rate per square centimeter, f, are indicated in Figure 4 to allow comparison of our flow conditions with those in other liquid chromatographic systems. The retention time of an unsorbed solute was measured with cytosine, which was not significantly retarded under the conditions of the experiment. The plate numbers are 50–100 times smaller than those measured with unsorbed solutes because of the stationary phase in the column, and consequently with 3′AMP H/u values were 1–2 orders of magnitude greater. Again, it was not economical to increase the flow velocity above the practical optimum range, as a small additional increase in speed of analysis requires a large increase of inlet pressure. Therefore, for analytical work with this column, the flow velocity was fixed at $u = 1.2$ cm second^{-1}.

Factors Affecting Retention. Because the composition of the mobile phase greatly influences the behavior of solutes in the stationary phase, buffer solutions other than those utilized in this study might provide better column performance

(17) F. Helfferich, "Ion Exchange," McGraw-Hill, New York, 1962, pp. 91–2.

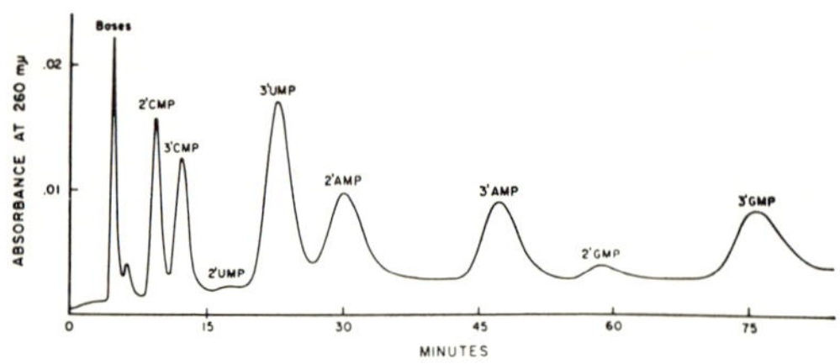

Figure 8. Separation of ribonucleoside-2′ and 3′-monophosphoric acids

Column: 1-mm i.d., 193-cm length; strong basic pellicular anion exchanger, sieve fraction No. 270–325, 8 microequivalent per gram. Eluent: linear gradient of KH_2PO_4 from 0.05*M* to 0.35*M* (starting solution 0.005*N* for H_3PO_4). Temperature: 60° C. Flow rate: 12 ml per hour. Flow velocity: 1 cm second^{-1}. Pressure: 51 atm. Sample: 1 μg of each isomer mixture

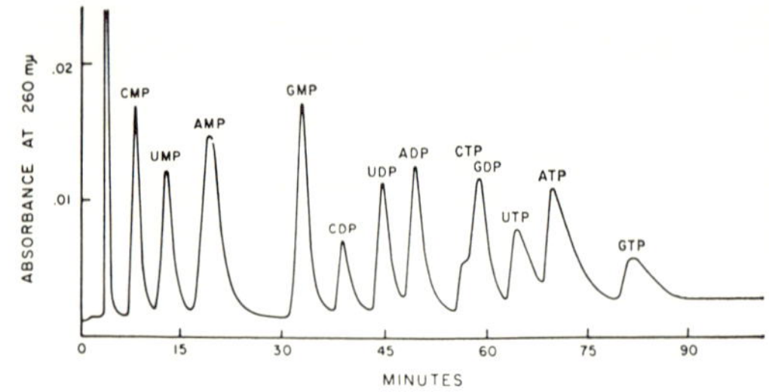

Figure 9. Separation of ribonucleosides mono-, di-, and triphosphoric acids

Column: 1.0-mm i.d., 193-cm length; packed with pellicular anion exchanger (strong basic), sieve fraction No. 270–325; 8 microequivalent per gram. Eluent: linear gradient of ammonium formate buffer, pH 4.35 at 25° C, from 0.04*M* to 1.5*M*. Temperature: 71° C. Flow rate: 12 ml per hour. Flow velocity: 1 cm second^{-1}. Pressure: 51 atm. Sample: 1.5–3.5 nanomoles (0.6–1.5 μg) per component

and peak symmetry, but, at present, no good guide exists for the optimization of the composition of the mobile phase.

The effect of salt concentration *C* on the retardation factor, *R*, is shown in Figure 5 for the elution of 3′AMP and 3′UMP with KH_2PO_4 solutions of different concentrations. The curves correspond to the equation

$$R = a \log C + b \tag{7}$$

where *a* and *b* are constants.

The column performance improved with temperature due to increased solute diffusivity in both phases. At high temperatures, a significant reduction in tailing was also observed. Thus, rapid analysis can be achieved at the highest temperature tolerable for the solutes, the stationary phase, and the mobile phase.

Because increase of temperature reduces *k*, (see Figure 6), this combined with improvements in speed may enable many compounds conventionally analyzed at relatively low temperatures and residing in the column for many hours to be separated at higher temperatures without decomposition. Thus, although nucleotides (particularly nucleoside di- and triphosphates) are relatively labile substances, no decomposition was observed in our work during the chromatographic runs at 70° C.

Columns packed with pellicular ion exchange resins are characterized by a large ratio of mobile phase to stationary phase compared to columns packed with the conventional ion exchange resin. Thus, the eluent strength of the mobile phase can be much lower in the pellicular ion exchange column than in the conventional column in order to obtain the same mass distribution ratio.

Nucleotide Separation. Nucleotides were among the first biochemical substances to be separated by ion exchange chromatography (*18*). A major advance was made by Anderson and his colleagues (*19*) with the development of automated ion exchange chromatography similar to that used for amino acid analysis (*20*). Although rapid separation of relatively simple nucleotides can be achieved by thin layer chromatography (*21*), the analysis of nucleic acid hydrolyzates is most commonly carried out by means of conventional

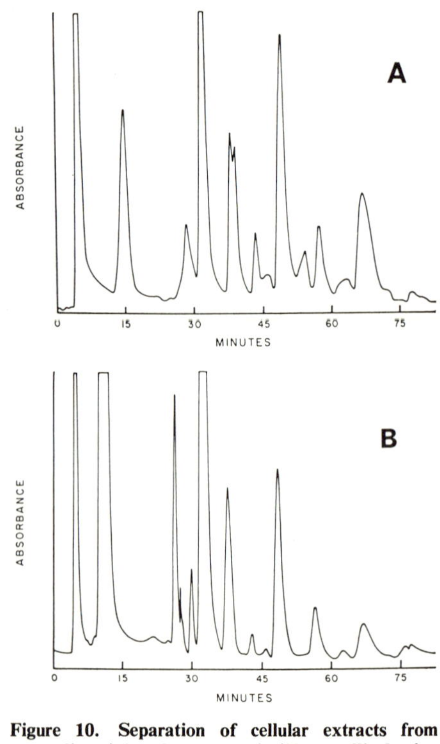

Figure 10. Separation of cellular extracts from mouse liver (*A*) and mouse brain (*B*) on pellicular ion exchanger using linear gradient elution with ammonium formate buffer, pH 4.35

column chromatography on a strong anion exchanger polystyrene resin. This method usually requires at least 20 hours to separate the nucleoside mono-, di-, and triphosphates. Moreover, the column system is not reusable and relatively large samples are necessary.

Recently, automated ion exchange chromatography of amino acid and peptide mixtures has been greatly accelerated by introducing uniform spherical ion exchange resins having small particle size (*6*) and extended for the analysis of nano-

(18) W. E. Cohn, *J. Am. Chem. Soc.*, **72,** 1471 (1950).
(19) N. G. Anderson, J. G. Green, M. L. Barber, and SR. F. C. Ladd, *Anal. Biochem.*, **6,** 153 (1963).
(20) S. Moore and W. H. Stein, *J. Biol. Chem.*, **192,** 663 (1951).
(21) K. Randerath, "Thin Layer Chromatography," Academic Press, New York, 1963, pp. 188–99.

mole quantities of amino acids by using small-bore columns. Recent work of Cohn and Uriel (*22*) shows that the analysis of nucleic acid bases and nucleosides at the nanomole level can be achieved on columns packed with a fine resin at increased detector sensitivity.

Similar modifications must improve the speed and sensitivity of analysis of nucleotides. However, the use of conventional ion exchange resins is beset by some inherent difficulties. Large differences in the equilibrium constants for the mono-, di-, and triphosphate nucleotides require a strong gradient of the eluent: the accompanying shrinking and swelling causes undesirable changes in column permeability and column performance. Ion exchange resins contain low molecular weight contaminants which give rise to column bleeding, which is undesirable when one wishes to collect eluted compounds and which causes base line drift with ultraviolet detectors.

These difficulties are overcome by the use of pellicular ion exchange resins which also permit the fast analysis of minute quantities of nucleoside phosphoric acids. Figures 7–9 show chromatograms of nucleotide mixtures analyzed by this liquid chromatography system using gradient elution with different buffer solutions. The analysis time compares well with that for gas chromatography. Because the order of elution is the same as that with Dowex-1 columns, the retention mechanism for the classical and pellicular ion exchange resin is probably the same.

(22) W. E. Cohn and M. Uriel, Oak Ridge National Laboratory, Oak Ridge, Tenn., personal communication, July 14, 1966.

The efficiency of this complete system is also demonstrated by the chromatograms of cellular extracts of mouse brain and liver in Figure 10. Although the identification of the individual peaks has yet to be completed, the chromatograms indicate more components to be present than are revealed by conventional separation methods.

Of the other pellicular resins mentioned above, pellicular polyethyleneimine resin was useful for the fast separation of nucleic acid bases and nucleosides. Pellicular alumina and carbon black were also effective in the separation of lipids. Column efficiency and optimum operating conditions were similar to those described in this study.

In conclusion, the use of pellicular column materials in small bore columns are widely applicable for liquid chromatography. The stability and efficiency of such materials make it possible to achieve fast separation of complex mixtures by a liquid chromatographic technique similar in speed, resolution, and quantitative range to gas chromatography.

ACKNOWLEDGMENT

We extend our appreciation to R. E. Handshumacher for making available to us the cellular extracts, to E. Canellakis for helpful discussions, and to E. Watson Associates for the loan of the differential refractometer used in this study.

RECEIVED for review May 8, 1967. Accepted August 3, 1967. Work supported by grants from the National Institutes of Health (HE - 03558 - 10 & FR 00356 - 01) and the National Aeronautics and Space Administration (SAR - NGR -07-004-067 & NsG-192-61).

34

Reprinted from *Anal. Chem.*, **41**(10), 1227-1234 (1969)

Rapid Analysis of Ribonucleosides and Bases at the Picomole Level Using Pellicular Cation Exchange Resin in Narrow Bore Columns

Csaba Horvath[1] and S. R. Lipsky

Section of Physical Sciences, Yale University School of Medicine, New Haven, Conn. 06510

The separation of the four major bases and nucleosides of RNA has been investigated under conditions utilizing wide ranges of pH, temperature, and flow rates. Small-bore columns packed with pellicular cation-exchange resin were used in a high-pressure liquid chromatograph equipped with a micro UV detector. Dilute acidic potassium or ammonium phosphate solutions served as eluents. The four bases or nucleosides were analyzed at the subnanomole level in less than 6 minutes using 3-meter colunms and high-flow velocities. With a shorter column operated at relatively low-flow velocities, a few picomoles of ribonucleosides can be separated and determined without resorting to radioactive techniques.

THE DETERMINATION of the base composition of nucleic acids is one of the most important steps in sequence analysis. Further progress in this field largely depends on the development of highly sensitive and rapid techniques for quantitative analysis of nucleic acid constituents. High sensitivity is often required for the minute quantities of nucleic acids or nucleic acid fragments that may be available—*e.g.*, by zonal centrifugation or single cell isolation. On the other hand, the need for a large number of analyses can only be overcome by using a fast and reliable method that is amenable to automation. Thus far, attempts to utilize gas chromatography to analyze nucleic acid hydrolyzates, although promising (*1*, *2*), have not been wholly successful. Thin-layer and paper chromatography (*3*, *4*) are widely used, but quantitative analysis is time consuming and generally inaccurate below the nanomole level (*5*). Using an elaborate radioactive tracer technique, however, the Randeraths (*6*, *7*) recently achieved base composition determination at the picomole level. Higher sensitivity has been obtained only in the quantitation of ATP by the luciferin–luciferase enzyme system (*8*).

Various column chromatographic methods with gels (*9*, *10*), with a ligand exchange resin (*11*), with a liquid ion

[1] Also Department of Engineering & Applied Science, Yale University.

(1) C. W. Gehrke and C. D. Ruyle, *J. Chromatog.*, **38**, 473 (1968).
(2) M. Jacobson, J. F. O'Brien, and C. Hedgcoth, *Anal. Biochem.*, **25**, 363 (1968).
(3) K. Randerath and E. Randerath, in "Methods in Enzymology," Vol. XII, "Nucleic Acids," Part A., L. Grossman and K. Moldave, Eds., Academic Press, New York, 1967, pp 323–50.
(4) G. R. Wyatt, in "The Nucleic Acids Chemistry and Biology," Vol. I, E. Chargaff and J. N. Davidson, Eds., Academic Press, New York, 1955, pp 243–65.
(5) J. M. Gebicki and S. Freed, *Anal. Biochem.*, **14**, 253 (1966).
(6) K. Randerath and E. Randerath, *ibid.*, in press.
(7) E. Randerath, J. W. Ten Broeke, and K. Randerath, *FEBS Letters*, **2**, 10 (1968).
(8) G. E. Lyman and J. P. DeVincenzo, *Anal. Biochem.*, **21**, 435 (1967).
(9) I. Mezzasoma and B. Farina, *Boll. Soc. Ital. Biol. Sper.*, **42**, 1449 (1966).
(10) M. Carrara and G. Bernardi, *Biochim. Biophys. Acta*, **155** (1968).
(11) G. Goldstein, *Anal. Biochem.*, **20**, 477 (1967).

exchanger (*12*), and with crosslinked poly-*N*-vinyl pyrrolidine (*13*), have also been developed for the analysis of nucleic acid hydrolyzates. Yet, ion exchange chromatography, introduced by Cohn (*14*), has retained its leading position in this field due to its versatility and reliability. Both anion and cation exchange resins have been employed successfully to separate various nucleic acid constituents. Utilizing the UV adsorbing properties of these compounds, Anderson *et al* (*15*) devised a continuous column effluent monitoring system which greatly simplified the procedure and gave impetus to a rapid development of the technique. Column efficiency and speed of analysis have been significantly increased by using very fine (particle diameters less than 20 microns) anion or cation exchange resins and high column inlet pressures (*16–18*). Despite these improvements, however, column chromatography had not been employed for the analysis of nucleic acid constituents at the subnanomole level.

It has been shown recently that with small-inside-diameter columns and with a more sensitive detector system, the speed and sensitivity of analysis by ion exchange chromatography (*19*), can be further increased. Conventional ion exchange resins are, however, not suitable for packing long narrow-bore tubes (i.d. 1 mm or less) and no significant improvement in speed and column efficiency can be expected from such materials with respect to the previously described techniques. Therefore, a new type of ion exchange resin called pellicular ion exchange resin has been developed, which consists of thin shells of ion exchange resin bonded to the surface of glass microspheres. This material can not only be packed very easily into long narrow-bore tubes, but has a number of other favorable chromatographic properties.

From chromatographic theory and practice it has been affirmed that high speed and efficiency are achieved at high column inlet pressures in liquid chromatography (*20*). This need arises from using long columns packed with small particles at high eluent velocities. Partly because of the relationship between pressure and performance, the term high pressure liquid chromatography is used to characterize a technique which is comparable to gas chromatography with respect to both the operation (sensitive detection system, temperature and flow control, long life column, convenience of operation) and the performance (high speed, high efficiency, reproducibility).

In this paper we report on the utilization of pellicular cation exchange resins for rapid analysis of subnanomole quantities of the four major bases and nucleosides of RNA with a high pressure liquid chromatographic system. The separation is performed by elution with acidic salt solutions. This elution method for conventional cation exchange resins was developed by Crampton *et al* (*21*) using gradient elution, as well as by Uziel *et al* (*18*), by Katz and Comb (*22*), by Blattner and Erickson (*23*), and by Busch (*24*), using a single eluent. Uziel *et al* (*18*) investigated a wide range of parameters, and their technique made possible the quantitative determination of the four ribonucleosides at the nanomole level within 1 hour.

EXPERIMENTAL

Apparatus. A Picker LCS 1000 nucleic acid analyzer (Picker Nuclear Co., White Plains, N. Y.) was used. The instrument can be operated at inlet pressures up to 3000 psi. The UV detector of this instrument operates at a 254-mμ wavelength and is equipped with a cylindrical flow cell having 1-mm diameter and 10-mm pathlength. The detector output is linear with respect to the solute concentration according to the manufacturer. A Texas Instrument Recorder with a Disc Integrator was used for obtaining the chromatograms. The recorder span was 0 to 10 mV or 0 to 1 mV; thus, full scale absorbancy ranges at the highest detector sensitivity were from 0 to 0.02 or from 0 to 0.002 absorbancy units, respectively.

Columns. Pellicular cation exchanger was prepared by the method described previously (*20*). A sieve fraction U.S.S. No. 270-325 was used. The analysis of the resin showed 0.59% S, and 0.81% C, and 0.12% H, which indicates that the material contains approximately 1% by weight of the dry resin. The ion exchange capacity calculated from the sulfur content is 0.185 meq per gram.

Columns were made of 1-mm i.d. #316 stainless steel tubing. Approximately 150-cm lengths were packed dry under vibration, then the deaerated eluent was pumped through the bed at 200 atm while the column was kept in a perpendicular position. A thin porous Teflon plug was used at the column outlet. Longer columns were assembled with Swagelok unions whose bores were also packed.

Chemicals. Reagent grade KH_2PO_4 or NH_4HPO_4 was used for the preparation of the eluents. The pH of the 0.02*M* solutions was adjusted with 0.1*N* KOH or NH_4OH, respectively, using a Leeds & Northrup pH meter. Samples were prepared with pure nucleosides and bases (Calbiochem), which were dissolved in water or in the eluent. The injected sample volume varied from 2 to 10 μl.

Mode of Operation. All experiments were carried out by elution in the constant eluent strength mode. When the eluent was changed, the system was flushed and the column was conditioned for at least 1 hour at high flow rates before making a chromatographic run. Samples were introduced with a 10-μl Hamilton syringe in the intermittent flow mode as follows. The eluent flow was stopped and the pressure was released through the column outlet. The injection port was then opened and the sample was injected with a syringe onto the column. Thereafter, the port was closed and the pumping was started.

Evaluation of Chromatograms. In the calculation of retention times and plate numbers, the error introduced by the transient flow regime of the chromatographic run was neglected. Adjusted retention volumes, V'_R, (*25*) were calculated from the retention time, t_R, from the liquid hold up time, t_0, and from the steady flow rate, F, by the equation

$$V'_R = (t_R - t_0)\,F \qquad (1)$$

In the case of asymmetrical peaks or doublets, t_R was calculated from the mode of the concentration curve which was obtained by halving the number of counts (Disc Integrator) related to the total peak area. The retention

(12) C. G. Horvath and S. R. Lipsky, *Nature*, **211**, 748 (1966).
(13) J. Lerner, T. M. Dougherty, and A. I. Schepartz, *J. Chromatog.*, **37**, 453 (1968).
(14) W. E. Cohn, *J. Amer. Chem. Soc.*, **72**, 1471 (1950).
(15) N. G. Anderson, J. G. Green, M. L. Barber, and F. C. Ladd, Sr., *Anal. Biochem.*, **6**, 153 (1963).
(16) J. G. Green, C. E. Nunley, and N. G. Anderson, *Nat. Cancer Inst. Monogr.*, **21**, 431 (1966).
(17) C. D. Scott, J. E. Attril, and N. G. Anderson, *Proc. Soc. Exptl. Biol. Med.*, **125**, 181 (1967).
(18) M. Uziel, C. K. Koh, and W. E. Cohn, *Anal. Biochem.*, **25**, 77 (1968).
(19) C. Horvath and S. R. Lipsky, in "Advances in Chromatography 1969," A. Zlatkis, Ed., Preston Technical Abstract Co., Evanston, 1969, pp 268–75.
(20) C. Horvath, B. Preiss, and S. R. Lipsky, ANAL. CHEM., **39**, 1422 (1967).
(21) C. F. Crampton, F. R. Frankel, A. M. Benson, and A. Wade, *Anal. Biochem.*, **1**, 249 (1960).
(22) S. Katz and D. G. Comb, *J. Biol. Chem.*, **238**, 3065 (1963).
(23) F. R. Blattner and H. P. Erickson, *Anal. Biochem.*, **18**, 220 (1967).
(24) E. W. Busch, *J. Chromatog.*, **37**, 518 (1968).
(25) C. Horvath, in "The Practice of Gas Chromatography," L. S. Ettre and A. Zlatkis, Eds., Interscience, New York, 1967, pp 152–61.

time of the small peak, obtained at maximum detector sensitivity by injecting a few microliters of 1*M* KH_2PO_4 into the column, was considered as the liquid hold up time.

RESULTS AND DISCUSSION

The chemical composition of the pellicular resin used in this study was similar to that of the conventional (sulfonated crosslinked polystyrene type) strong cation exchange resins, such as Amberlite IR 120 or Dowex 50. Yet, its unique construction yields properties that are novel in liquid chromatography. In contrast to conventional soft ion-exchange resins, pellicular resins give uniform column packings that are stable at high pressure gradients (*20*) even with small-bore columns. In addition, changing temperature or eluent composition does not give rise to detrimental effects on column permeability or efficiency. Thus, the effect of these variables as well as that of flow velocity over a wide range can be studied with the same column. The lifetime of the column exceeds 6 months even if flow rate, eluent strength, and temperature are frequently varied. Columns of various diameters can be packed dry with pellicular ion exchange resins. This technique is not only very convenient for making long columns, but it is also superior with respect to column efficiency, as shown by Horne *et al* (*26*). Generally, columns packed with pellicular resins have higher column efficiency at high flow velocities then those packed with conventional resins (*20*). Small-bore columns (i.d. ≦ 1 mm) are particularly suitable for the analysis of minute samples because the high column efficiency and low flow rates make possible the elution of the sample components in a small volume of the effluent.

Some of these advantages have been realized in this study involving the rapid determination of the major bases and nucleosides which arise from the hydrolysis of RNA. The bases are uracil (Ura), guanine (Gua), adenine (Ade), and cytosine (Cyt). The nucleosides are uridine (Urd), guanosine (Guo), adenosine (Ado), and cytidine (Cyd).

The separation of nucleosides and bases can be achieved with pellicular cation exchange resins under a variety of conditions using the salt elution mode. The most important variables which determine the equilibrium and mass transfer phenomena are as follows: (a) The resin properties, such as shell thickness, structure of the polymer, degree of sulfonation, particle size, etc. (b) The chemical composition of the salt in the eluent. (c) The concentration of the salt solution. (d) The pH of the eluent. (e) The column temperature. (f) The length and diameter of the column. (g) The flow rate. (h) The elution mode—*e.g.*, constant eluent strength or gradient elution. The effect of a number of these variables on the separation of nucleosides on conventional cation exchange resin column has been investigated recently by Uziel *et al* (*18*).

This study was carried out with resin from one batch process and with acidic potassium and ammonium phosphate solutions, although preliminary experiments have shown that other salts, such as ammonium formate, ammonium acetate, and ammonium citrate, also give good separations. The phosphate solutions were selected for their low UV absorbance and because the stainless steel used for columns and for other wetted parts of the instrument was found to be resistant to them under the conditions used. Acidic ammonium formate appears to interact with #316 stainless steel. The pH of a 0.05*M* ammonium formate solution rose from 5.2 to 7.7 as it was pumped through the system. In addition,

(26) D. S. Horne, J. H. Knox, and L. McLaren, *Separation Sci.*, **1**, 531 (1966).

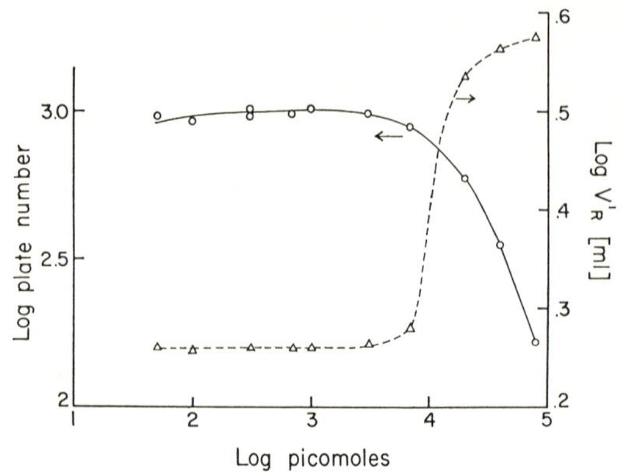

Figure 1. Plots of plate number, *N*, and adjusted retention volume V'_R, against the sample amount

retention values fluctuated with the flow rate and with the flow rate history of the column. No such effects were observed with phosphate solutions.

Sample Capacity. The sample capacity of pellicular ion exchange columns is generally smaller than that of comparable columns containing conventional ion exchange resins because of the relatively small amount of resin present in the column whose volume is largely occupied by inert glass beads. The sample capacity of a column is usually defined by the amount of sample that causes the plate number to drop 10% below the value that would be obtained at zero sample size. Column overloading can also be recognized by the change of the retention values and by the distortion of the peak shapes.

Figure 1 shows logarithmic plots of the plate number and adjusted retention volumes *vs.* the sample amounts. The measurements were made with a 300-cm-long pellicular cation exchange resin column at 69 °C using 0.02*M* $NH_4H_2PO_4$ solution (pH 5.5) as the eluent. Flow rate and inlet pressure were 24 ml/hour and 128 atm, respectively. Cytosine solutions of various concentrations were injected. The "normal" retention time of cytosine was 8 minutes and the capacity factor was 0.7. Under these conditions the loading capacity is somewhat less than 10 nmoles. Significantly higher loading capacity can be obtained with similar columns for peaks having greater capacity factors and at lower flow velocities or if gradient elution is employed. The observed decrease of the retention volume with the sample size and the sharp rear boundary of the peaks at overloading, indicate that the sorption isotherm is of the anti-Langmuir type.

In the analysis of small samples, the sample capacity of the present system is sufficient. Nevertheless, pellicular ion exchange resins can be prepared with high resin content without losing the benefits of this particular structure.

Salt Concentration of the Eluent. The adjusted retention volume, V'_R of a solute is given by the expression

$$V'_R = KV_S \tag{3}$$

where K is the equilibrium constant for solute distribution between the resin and eluent and V_S is small in pellicular resin columns. Therefore, K should be larger in the latter in order to obtain comparable peak retention. When nucleosides and bases are separated by salt elution at fixed pH and temperature, K increases as the salt concentration in the eluent decreases. Consequently they are eluted with less concentrated eluents from a pellicular resin column than from a

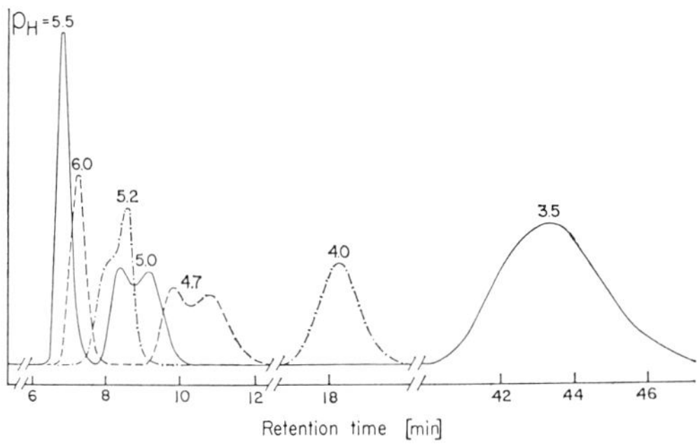

Figure 2. Retention and shape of guanine peaks eluted at different pH

Eluent: $0.02M$ KH_2PO_4; Temperature: 60 °C; Flow rate: 14.8 ml h^{-1}; Column length: 300.7 cm

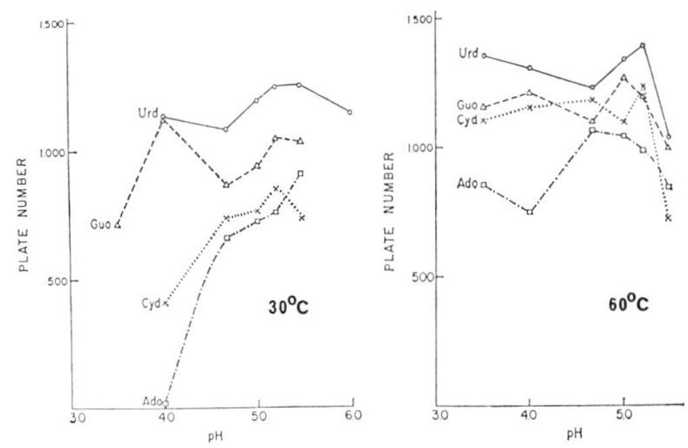

Figure 3. Effect of eluent pH on plate numbers as measured with the four nucleosides at 30 and at 60 °C

Eluent: $0.02M$ KH_2PO_4; Flow rate: 14.8 ml h^{-1}; Column length: 300.7 cm
○ Uridine, △ Guanosine, □ Adenosine, × Cytidine

conventional ion exchange column. On our pellicular resin column, the nucleosides and bases eluted with 0.04 to $0.01M$ salt solutions have retention values similar to those obtained on a conventional ion exchange resin column using 0.4 to $0.2M$ solutions.

The effect of salt concentration on peak retention was investigated with KH_2PO_4 solutions at pH 5.2 and 38 °C. The ratios of the adjusted retention volumes obtained with $0.02M$ eluent to those obtained with $0.01M$ eluent were 1.14 for Urd, 1.23 for Guo, 1.45 for Ado, and 1.6 for Cyd. In further experiments the salt concentration was arbitrarily fixed and $0.02M$ solutions were used for studying the effect of pH, temperature, and flow rate in order to establish optimum conditions for the separation of the four major nucleic acid constituents.

Effect of pH. In the pellicular cation-exchange-resin column the pH of the eluent dramatically affects both the retention and the dispersion of the solute bands. In this investigation salt elution was employed; therefore, the pH range was restricted to pH 3 to 5.6. With strong acidic eluents a different elution pattern is expected, as ionic interactions are more pronounced (*27*). Figure 2 shows the changes in the retention time and in the peak shape of guanine as recorded at various pH values of the $0.02M$ KH_2PO_4 eluent at 60 °C. The double peaks that appear in the pH range from 4.7 to 5.2 suggest that here solute retention is governed by at least two equilibria that give different distribution constants, or by kinetic effects (*28*). Such doublets also have been observed with adenosine. The erratic variations of the plate number with the eluent pH as shown in Figure 3 may be similarly explained. The observed variations could not be related directly to the degree of solute dissociation in the eluent at different pH's.

The variation of the adjusted retention volumes of nucleosides with pH is shown in Figure 4. It is seen that the equilibrium constants decrease with increasing pH both at 30 °C and at 60 °C. The effect of pH on the interaction between the cation exchange resin and these molecules is complicated. In addition to the attraction by the fixed ionic groups, solute retention may be influenced by adsorption on the polymer matrix and by solute–solute interactions, both of which are pH and temperature dependent with respect to their equilibrium and rate constants. In general, this complex nature of ion exchange chromatography, which is invaluable for achieving the high selectivity characteristic for this technique, also makes data interpretation and prediction of solute behavior difficult and optimum separating conditions must often be found by trial and error.

For fast separation of the bases and nucleosides, the pH range from 5.2 to 5.5 was found most adequate. Here the disturbing effects resulting in very broad or asymmetrical peaks are absent, and at 30 °C the relative retentions are adequate to achieve full separation at high flow velocities with the present system.

Effect of Temperature. The temperature affects the separation in two ways. The rate of diffusion and the various kinetic processes increase with the temperature; thus, usually sharper and more symmetrical peaks are obtained. On the other hand, the absolute and relative values of the equilibrium constants vary with the temperature. The absolute values, thus the retention volumes, usually decrease, while the relative retention values often increase with decreasing temperature. The effect of temperature on the separation is demonstrated by the two chromatograms depicted in Figure 5. Comparison of the two runs shows that at low temperature the separation of the last two peaks is poor and the retention volumes are generally large. This is in agreement with earlier observations

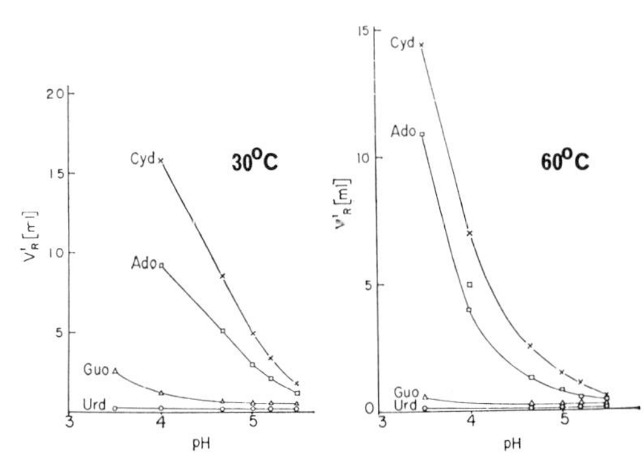

Figure 4. Plot of adjusted retention volume of nucleosides against pH of the $0.02M$ KH_2PO_4 eluent as measured at 30 and 60 °C

(27) W. E. Cohn, *Science*, **109**, 377 (1949).
(28) R. A. Keller and J. C. Giddings, *J. Chromatog.*, **3**, 205 (1960).

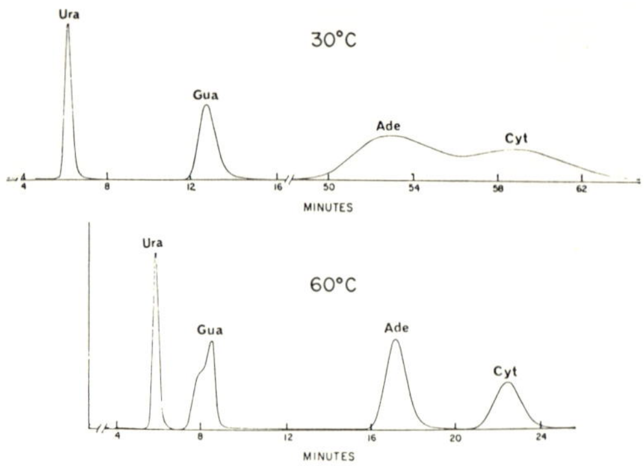

Figure 5. Chromatograms of the bases at 30 and at 60 °C

Eluent: 0.02*M* KH_2PO_4, pH 5.2; Flow rate: 14.8 ml h^{-1}; Column length: 300.7 cm

(*18*) that the effect of temperature on the retention of purines and pyrimidines is different.

These chromatograms were made at a pH of 5.2. At 60 °C guanine appears as a doublet, as also seen in Figure 2. At 30 °C the shape of the guanine peak is normal, probably because of the broadness of this peak having relatively low band velocity.

At pH 5.5 the peaks were symmetrical over a wide temperature range. Figure 6 shows Arrhenius type plots for bases and nucleosides. The adjusted retention volumes were measured at pH 5.5. The straight lines suggest that under these conditions peak retention is controlled by a single mechanism over the full temperature range. On the contrary, changes in the slopes were found with nucleotides on an anion exchange resin column, which were interpreted as changes in the retention-controlling mechanism with the temperature (*20*).

Figure 7 shows the effect of temperature on the adjusted retention volumes over a wide pH range. The ratio of V'_{R30} (adjusted retention volume at 30 °C) to V'_{R60} (adjusted retention volumes at 60 °C) is plotted against the pH. For each solute there is a pH where the effect of the temperature on the retention is the most pronounced and the V'_{R30}/V'_{R60} ratio has a maximum. This shows again that in ion exchange chromatography the combined effect of both the temperature and the pH has to be considered when optimum separating conditions are sought.

It has been shown that both increases in temperature and pH reduce retention values. For obtaining high resolution in the pH range under investigation, high pH appears to be more effective than high temperature alone because of the excessive band broadening and the appearance of spurious peaks observed at low pH even at high temperatures, which

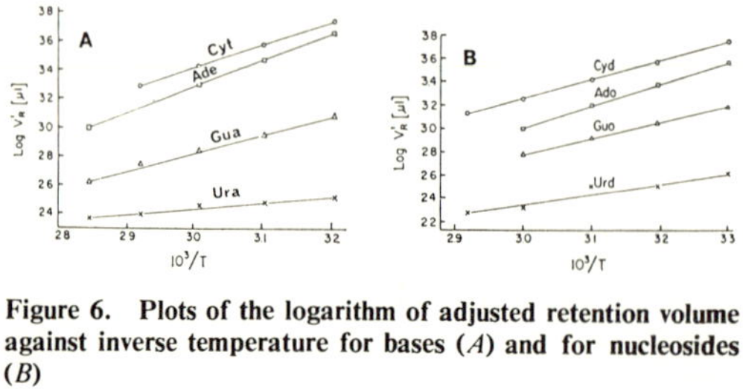

Figure 6. Plots of the logarithm of adjusted retention volume against inverse temperature for bases (*A*) and for nucleosides (*B*)

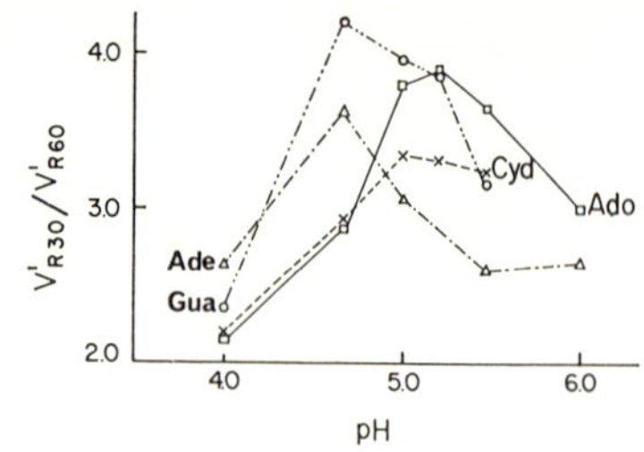

Figure 7. Plots of the ratio of the adjusted retention volumes as measured at 30 and at 60 °C against the pH of the eluent

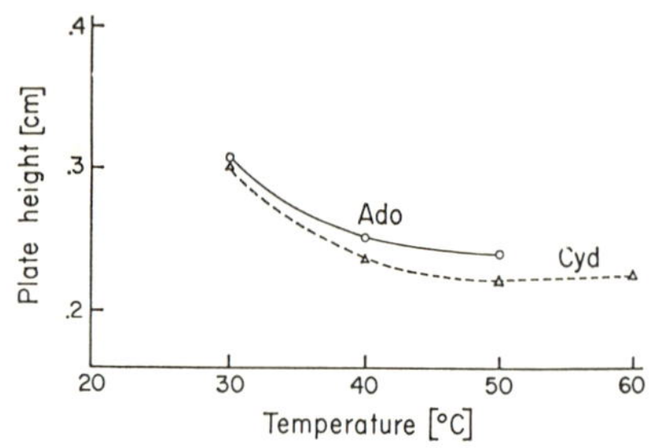

Figure 8. Effect of temperature on the plate height for adenosine and cytidine

Eluent: 0.02*M* KH_2PO_4, pH 5.5

is related to the increasing ionization of the solutes. The optimum temperature is different for bases and nucleosides.

As seen from Figure 8, column efficiency for nucleosides improves only marginally above 40 °C and optimum separating conditions are found at or below this temperature. On the other hand, a high column temperature has been found favorable for the separation of the bases.

Effect of Flow Rate. Plate heights for adenine and cytosine have been measured in the velocity range from 1.4 to 3.4 cm/sec with a 300-cm-long column at pH 5.5 and at 68 °C. The dependence of the plate height on the flow velocity was found to be linear and can be expressed by the following equation

$$h = A + Cu$$

where C encompasses kinetic and diffusional mass transfer resistance, u is the eluent velocity, and A is constant. It indicates that at such high flow velocities, even with pellicular resins, peak spreading is controlled by mass transfer in the stationary phase.

The values of C were: 0.06 sec for adenine and 0.03 sec for cytosine. These compare well with C values obtained in gas chromatography on a glass bead column with tricresyl phosphate as a stationary phase (*29*): 0.03 sec for 2-methyl pentane, 0.02 sec for *n*-heptane and *n*-octane. Since the magnitude of C determines the speed of analysis, this comparison shows that in ion exchange chromatography using pellicular column material, the speed of gas chromatography can be approached at comparable fluid velocities.

(29) A. B. Littlewood, in "Gas Chromatography 1958," D. H. Desty, Ed., Academic Press, New York, 1958, p 23.

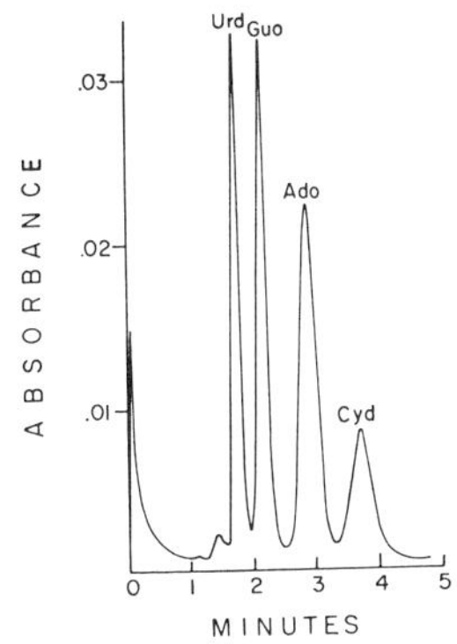

Figure 9. Rapid separation of ribonucleosides

Column: 151.7 cm long; 1-mm i.d.; packed with pellicular cation exchange resin, sieve fraction #270/325
Eluent: 0.02*M* $NH_4H_2PO_4$, pH 5.6. Temperature 39 °C
Flow rate 25.5 ml/h. Inlet pressure 131 atm
Sample size: 300 picomoles of each component
Full scale 0.04 A.U.

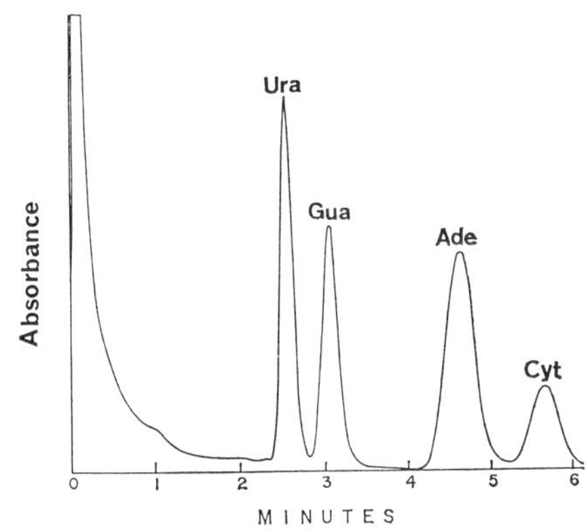

Figure 10. Rapid separation of bases

Column: 300.7 cm long; 1-mm i.d.; packed with pellicular cation exchange resin, sieve fraction #270/325
Eluent: 0.02*M* $NH_4H_2PO_4$, pH 5.5. Temperature 68 °C
Flow rate 33.4 ml/h. Inlet pressure 162 atm
Sample size: 800 picomoles of each component
Full scale 0.02 A.U.

Optimization Procedure. Generally, the present high-pressure liquid chromatographic system which utilizes long columns requires a less rigorous optimization of the operating conditions than that necessary when working with short columns, because a relatively large number of plates can be obtained with the long columns. Yet, in routine analysis the time of separation demands prime consideration and the time of analysis has to be minimized within the technical limits of the instrument system.

From the previously discussed observations, optimization of the operating conditions with pellicular ion exchange resin columns follows similar guide lines which have been applied to conventional ion exchange systems.

First, the optimum pH range is established for a given sample and eluent, because the effect of pH is often the most pronounced. Second, the ionic strength of the eluent and the temperature are selected. The influence of increasing eluent strength on the retention values is usually similar to that of increasing temperature. If it is possible, however, the increase of the temperature should be preferred in order to enhance sorption and diffusion rates. Third, column length and flow rate have to be adjusted to exploit fully the capabilities of the available instrument.

System Sensitivity. The sensitivity of a liquid chromatographic system is determined by the maximum solute concentration in the column effluent and the sensitivity of the detector for the solute in question. Peaks can be eluted from small-bore columns in much smaller volumes than from columns usually employed in liquid chromatography; therefore, the concentration of the effluent peaks is much larger for the same amount injected. The intrinsic sensitivity of UV detectors is high for purine and pyrimidine derivatives, which have large molar absorptivity of the order of 10^4 at 250 mμ. Most commercial UV detectors, however, are not suitable for working with small-bore columns because of their relatively large cell volume, which causes peak distortion and reduction of the concentration maximum. Therefore, a special detector cell was designed and used in previous work with pellicular ion exchange resin columns (*19*). The recently introduced micro UV detector of the Picker instrument has been found to be particularly suitable for analysis with small bore columns. It has a very small volume-to-light-path ratio (about 0.8 mm^2) and low noise level (about 2×10^{-5} absorbancy unit with stagnant liquid), which makes possible the detection of peaks having small volumes and low concentrations. Although this study was not aimed toward exploring the ultimate sensitivity of the system, the analysis of bases and nucleosides as shown by the following chromatograms, could routinely be performed at the subnanomole level. It is possible that the analysis of subpicomole quantities also could be achieved by slight modification of the system, such as a decrease of volume-to-path-length ratio of the detector cell and a decrease of the diameter and length of the column.

Analytical Results. The separation of the four nucleosides within 4 minutes is shown in Figure 9. As discussed previously, optimum pH and temperature for this separation are about 5.5 and 40 °C, respectively. The resolution increases with decreasing flow rate; therefore, flow rates lower than 25 ml/hour are preferred in analytical work. But even at flow rates of 10–12 ml/hour, the analysis of 0.03 to 10 nmoles of nucleosides was performed within 10 minutes.

The chromatogram in Figure 10 shows the separation of the four bases at 68 °C and at a liquid velocity higher than 3 cm/sec. The effect of the temperature and flow rate on this separation is demonstrated in Table I. Plates per second and plates per atmosphere were calculated by dividing the plate number by the retention time and the inlet pressure, respectively. Resolution and resolution per second were calculated as is customary in gas chromatography (*25*). As shown in Table I, the high flow velocity is advantageous for speeding up the separation because a significantly greater number of plates per second and resolution per second can be obtained at high eluent velocities than at low eluent velocities. The pressure drop necessary to obtain a given number of plates increases, however, with the flow rate as seen from

Table I. Separation of Bases

A. Effect of Flow Rate

	Plates/sec		Plates/atm		Resolution per second × 10^3	
Flow rate ml/hour	14.1	33.4	14.1	33.4	14.1	33.4
Uracil	5.1	8.24	26.9	8.05	4.9	8.9
Guanine	3.4	5.45	21.5	6.37	6.4	12.4
Adenine	1.7	2.76	16.5	4.86	2.2	4.7
Cytosine	2.2	3.2	24.4	6.82		

B. Effect of Temperature

	Plate number		Plates/sec		Plates/atm		Resolution		Resolution per second × 10^3	
Temperature °C	39	69	39	69	39	69	39	69	39	69
Uracil	1531	1848	4.0	5.1	14.7	26.9				
Guanine	721	1475	1.2	3.4	6.93	21.5	3.50	1.95	7.2	4.9
Adenine	548	1134	0.38	1.7	5.27	16.5	4.92	3.54	4.8	6.4
Cytosine	894	1738	0.54	2.2	8.60	24.4	0.93	1.61	0 6	2.2

(a) Column length: 300.7 cm; eluent: 0.02*M* $NH_4H_2PO_4$, pH 5.5
(b) Flow rate 14.1 ml/hour
(c) Temperature 69 °C

the comparison of plates per atmosphere at two flow rates. Table I also shows that high temperature has a very favorable effect on the separation of bases in contrast to the effect observed with nucleosides. The improvement in separation efficiency when the temperatures is raised from 39 to 69 °C is mainly due to the large increase in plate numbers. In addition, the resolution of the last two peaks is enhanced by the increase of their relative retention from 1.2 at 39 °C to 1.4 at 69 °C. The increase of the number of plates per atmosphere also reflects the decrease of eluent viscosity with the increasing temperature.

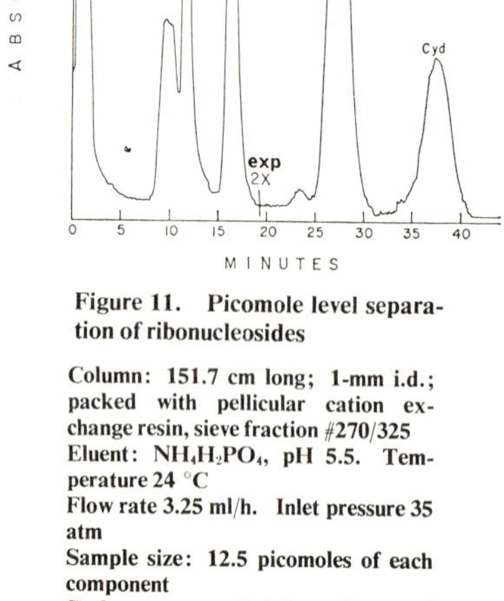

Figure 11. Picomole level separation of ribonucleosides

Column: 151.7 cm long; 1-mm i.d.; packed with pellicular cation exchange resin, sieve fraction #270/325
Eluent: $NH_4H_2PO_4$, pH 5.5. Temperature 24 °C
Flow rate 3.25 ml/h. Inlet pressure 35 atm
Sample size: 12.5 picomoles of each component
Scale was expanded by a factor of two in the middle of chromatogram

Both chromatograms show a large peak at the start. The origin of this phenomenon, which does not appear when the sample is injected into the streaming eluent, has not been established unambiguously.

The start-up peak usually increases with the flow rate as seen by a comparison of Figures 9 and 10. The effect can be partly attributed to the transient flow because the detector is sensitive to drastic changes in flow rate. On the other hand, the highly sensitive microdetector also gives a response to changes in the refractive index of the effluent. Transient changes in the effluent temperature in the start-up flow regime may then explain this phenomenon. Schlieren effects also have been found to give rise to ghost peaks. It has been thought that impurities may be released from the resin or from the column wall into the static eluent during the injection period. Yet, the start-up peak could be reduced neither by extensive purging of the column with strong eluents, nor by using very carefully cleaned tubings made of different materials.

Figure 11 shows a typical chromatogram obtained with 12.5 picomoles of the nucleosides. The increase of sensitivity was achieved by the decrease of the recorder span from 10 mV to 1 mV as well as by the decrease of flow rate and column temperature. The absorbance range was 0 to 0.004 A.U. at the beginning of the chromatogram, but it was expanded by a factor of two in the middle of the run. The sample components were dissolved in tap water whose impurities gave rise to the ghost peak before the uridine peak. Such unresolved peaks at the breakthrough point, however, could not be eliminated with sample solutions in distilled water at this sensitivity. Unfortunately, uracil and uridine are only slightly retarded by the ion exchange resin. Therefore, their quantitation is beset by this spurious peak particularly when RNA hydrolizates are analyzed at high sensitivity. The other three nucleosides, however, can be determined quantitatively under the present conditions if they are present in amounts larger than 3 picomoles. Even smaller amounts give well observable peaks.

Precision and Accuracy of the Analysis. In general, the precision and accuracy of the results are similar to those found in gas chromatography. Preferably, the instrument is calibrated for a particular analysis with known quantities of the sample components. Flow rate and column temperature should be kept constant. With the present system, the flow rate can be kept constant within 2% above 10 ml/hour and the temperature within 1 °C above 35 °C. Under these conditions, the precision of the retention time and peak areas per unit amount of solute was better than 3%. The sample introduction technique of the operator may improve or impair the accuracy of the results. The column life did not affect the precision within a six month period of continuous use at different temperatures.

In practical work, the sample preparation may also greatly influence quantitative results. The optimum sample preparation method should be worked out individually according to the sample origin and history.

ACKNOWLEDGMENT

The technical assistance of Anthony R. DeNoto is gratefully acknowledged.

RECEIVED for review February 24, 1969. Accepted May 21, 1969. This work was supported by grants from the National Institutes of Health (HE 03558) (FR 00356) and the National Aeronautics and Space Administration (SAR-NGR 07-004-067) (SAR-NsG 192-61).

35

Reprinted from *J. Chromatog. Sci.*, **8**, 72–75 (Feb. 1970)

High Speed Separations of Nucleotides and Nucleic Acid Bases by Column Chromatography Using Controlled Surface Porosity Ion Exchangers

by **J. J. Kirkland,** E. I. du Pont de Nemours and Company, Industrial and Biochemicals Department, Experimental Station, Wilmington, Delaware 19898

Abstract

Rapid separations of nucleotides and nucleic acid bases can be carried out by column chromatography using new "Zipax" controlled surface porosity anion and cation exchange packings. For example, the four main ribonucleoside-5′-monophosphoric acids may be completely separated in about six minutes using a strongly basic anion exchange column containing nonporous, spherical support particles uniformly coated with a thin, porous layer of sorbent. The high speed capabilities of the new ion exchangers are primarily due to the greatly improved mass transport effects as compared with conventional ion exchange resins. For analytical studies, the exchangers are normally used in 2 or 3 mm, i.d., columns; consequently, small samples are employed in conjunction with a sensitive, continuous detector. The hard, spherical nature of the new exchangers permit their operation at high column input pressures. Carrier velocities in excess of 3 cm/sec may be employed with good results.

Ion exchange column chromatography has been a preferred method for analyzing nucleic acids and related substances (1-3). Recently, gas-liquid chromatography has been proposed as a means of analyzing the components of nucleic acids (4-6). The GC methods require that volatile derivatives be prepared for analysis, the derivation scheme usually consisting of several reaction steps. In spite of the relative speed of the GC separation, analysis of the nucleotides and nucleic acid bases by ion exchange column chromatography has particular advantages. These compounds may be analyzed directly after their formation from biological material by hydrolysis without having to prepare the volatile derivatives required for GC analysis.

This paper will show that rapid separations of nucleotides and nucleic acid bases can be carried out by column chromatography using "Zipax" (Du Pont's trademark for controlled surface porosity supports) anion and cation exchangers. Fractional microgram mixtures of these compounds may be separated in minutes using linear elution chromatography involving a single carrier.

Experimental

Liquid Chromatographic Apparatus

The equipment used during this study was essentially that previously described (7-9). A Du Pont Model 410 Precision Photometer was used as an ultraviolet detector. This unit is an improved laboratory version of the Model 400 photometric detector described earlier (7).

Chromatographic Columns

The "Zipax" controlled surface porosity cation and anion exchange resins used in this study have previously been described (10). These ion exchange packings consist of a superficially porous crust with an ion exchanger surrounding an impervious, spherical silica core (11). The "Zipax" cation exchange medium is a fluoropolymer containing very strong sulfonic acid groups. This cation exchange material exhibits unusual chemical stability and columns of this packing may be used for months without change in chromatographic characteristics. The cation exchanger used in this study exhibited an exchange capacity of 3.5 microequivalents per gram as measured by its affinity for zinc ion (9). Reproducible columns of both the cation and the anion exchangers may be conveniently prepared using a simple dry-packing technique (9).

The physical characteristics of the strong anion exchange packing used in this study are analogous to the other "Zipax" packings, i.e., a thin porous shell

1. Cohn, W. E., "The Nucleic Acids," E. Chargaff and J. N. Davidson, eds., Academic Press, N.Y., 1965, p. 211.
2. Heftmann, E., "Chromatography," 2nd Ed., Reinhold Publishing Corp., N.Y., 1967, p. 627.
3. Anderson, N. G., Green, J. G., Barber, M. L., and Ladd, F. C., Sr., Anal. Biochem. **6**, 153 (1953).
4. Gehrke, C. W., and Ruyle, C. D., J. Chromatog. **38**, 473 (1968).
5. Jacobson, M., O'Brien, J. F., and Hedgcoth, C., Anal. Biochem. **25**, 363 (1968).
6. Hashizume, T., and Sasaki, Y., Anal. Biochem. **24**, 232 (1968).
7. Kirkland, J. J., Anal. Chem. **40**, 391 (1968).
8. Kirkland, J. J., Anal. Chem. **41**, 218 (1969).
9. Kirkland, J. J., J. Chromatog. Sci. **7**, 7 (1969).
10. Kirkland, J. J., J. Chromatog. Sci. **7**, 361 (1969).
11. Kirkland, J. J., patent pending.

Figure 1. Steroscan Scanning Electron Micrograph of "Zipax" Anion Exchange Particle.

of sorbent surrounding an impervious siliceous core. A steroscanning electron micrograph of a "Zipax" anion exchange particle is pictured in Figure 1. The exchange capacity of the material used in this study was estimated as 12 microequivalents per gram by analyzing for the concentration of nitrogen in the sorbent, the exchange sites consisting of strongly basic tetraalkylammonium groups.

The "Zipax" ion exchangers exhibit excellent mechanical stability. Column input pressures up to 5000 psi have been used without difficulty. Inquiries regarding the commercial availability of ion exchange packings similar to the materials herein described should be directed to the Instrument Products Division, E. I. du Pont de Nemours & Co., Wilmington, Delaware.

The columns used during this study were constructed from precision-bore No. 316 stainless steel, 2.1 mm, i.d., and 6.25 mm, o.d. (Superior Tube Co.).

The nucleic acid bases and nucleotides used were commercially available materials from a variety of sources. All of these materials may be obtained from Mann Research Laboratories, New York, N.Y.

Results and Discussion

Separation of Nucleic Bases By Cation Exchange Chromatography

Mixtures of purines and pyrimidines have traditionally been separated by cation exchange chromatography using gradient elution techniques in which the ionic strength of the carrier is continuously increased during the chromatographic separation (12,13). While excellent separations may be accomplished by this approach, analyses are time consuming (14). The speed of such separations can be significantly increased by using column packings consisting of impervious centers with a thin porous surface of sorbent (8,9,11). "Zipax" ion exchange packings exhibit good efficiency at high carrier velocities and permit rapid separations at moderate column input pressures (10).

Mixtures of the four main nucleic acid bases may be separated with a "Zipax" strong cation exchange column in about 5 minutes as indicated in Figure 2. This separation was carried out with a 1 meter x

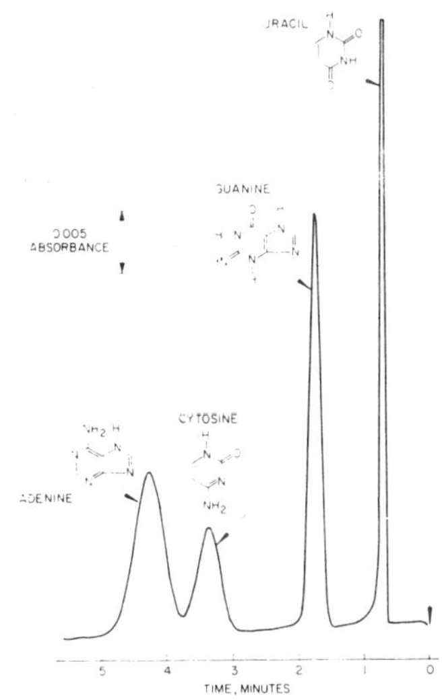

Figure 2. Separation of RNA Nucleic Acid Bases. Column—1 m x 2.1 mm, i.d., "Zipax" strong cation exchanger, 20-37 μ; column temperature—63°C; carrier—0.01 N HNO_3; flow-rate — 2.00 cc/min; column input pressure — 735 psi; sample—1.5 μl of 0.25 mg/ml each in 0.01 N HNO_3.

2.1 mm, i.d., column by linear elution chromatography using a carrier of 0.01 N nitric acid. Since a single eluent was employed, no regeneration of the column was required following the analysis; consequently, the column is in readiness for another separation following the elution of the component peaks. The "Zipax" cation exchange column exhibits approximately 1.4 theoretical plates per second for cytosine with the conditions employed. Similar separations of these components using a 3 meter column of "pellicular" ion exchange resin column have recently been accomplished by others (15,16). These compounds have also been separated by ligand-exchange chromatography in about one hour, the column used exhibiting about 0.33 theoretical plates per second for cytosine (17).

The high permeability of the "Zipax" ion exchangers permits the operation of these materials at high carrier velocities with relatively low column input pres-

12. Crampton, C. F., Frankel, F. R., Benson, F. R., and Wade, A., Anal. Biochem. **1**, 249 (1960).
13. Cohn, W. E., Science, **109**, 377 (1949).
14. Perkin-Elmer Corp., Bulletin on Liquid Chromatography, 1968.
15. Adler, N., Pittsburgh Conference on Analytical Chemistry and Applied Spectroscopy, Cleveland, Ohio, March 5, 1969.
16. Horvath, C., Lipsky, S. R., Anal. Chem. **41**, 1227 (1969).
17. Goldstein, G., Anal. Biochem., **20**, 477 (1967).

sures. Since these ion exchange packings undergo no detectable changes due to swelling effects when buffer systems are varied, column characteristics are not affected by changes in pH, activity, buffer type, etc. As in the case with conventional ion exchangers, column efficiencies can generally be improved by increasing the temperature of the separation. In the cation exchange chromatography of 2-aminobenzimidazole (10), column efficiency is improved by a factor of about two by increasing the temperature from 27° to 63°C. No significant change in chromatographic properties was observed when the column was reused at room temperature in subsequent studies.

Figure 3 shows a chromatogram of a synthetic mixture of the four main nucleic acid bases from RNA plus two of the less common bases, hypoxanthine and 6-methylaminopurine. These six components are separated in about 16 minutes using the same conditions as employed for the separation given in Figure 2. Other mixtures of purines and pyrimidines can be resolved with this same carrier system or with some appropriate modification of pH or ionic strength.

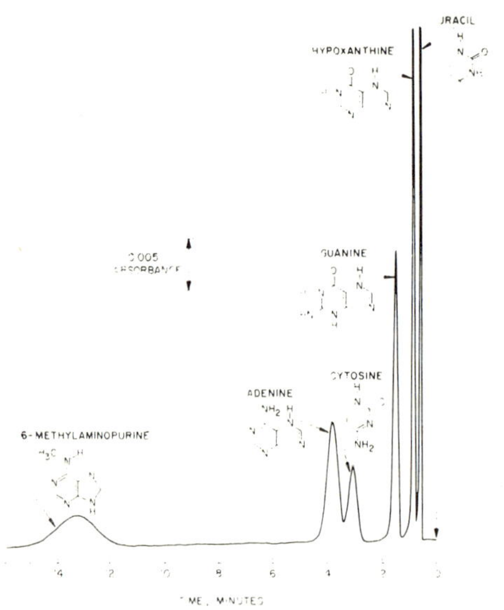

Figure 3. Separation of Some Nucleic Acid Bases. Same conditions as in Figure 2, except sample—2 μl of 0.2 mg/ml guanine, cytosine, adenine and 6-methylaminopurine, 0.1 mg/ml uracil and hypoxanthine in 0.01 N HNO_3.

The lower capacity of the "Zipax" ion exchangers results in less retention of compounds than is obtained with conventional ion exchange resins. With the ion exchangers it is usually necessary to reduce the ionic strength and/or alter the pH of the carrier to resolve mixtures separated by conventional ion exchange resin techniques.

Separation of Nucleotides by Anion Exchange Chromatography

Nucleic acid hydrolysates are often analyzed by conventional column chromatography with a strong anion exchanger in conjunction with gradient-elution techniques (3,18). This approach, while providing excellent separation of the hydrolysate systems, is cumbersome and time consuming, often requiring as much as 24 hours for completion. "Pellicular" anion exchange materials have been used to separate synthetic nucleotide mixtures in 1-2 hours using gradient elution systems (18).

Controlled surface porosity anion exchange columns exhibit relatively high efficiencies when operated at high carrier velocities (10). With these columns relatively complex mixtures of compounds, such as nucleic acid hydrolysates, can also be rapidly separated using linear elution techniques involving a single carrier. Figure 4 shows a synthetic mixture of the ribonucleoside-5′-monophosphoric acids completely separated in slightly over 6 minutes. The amount of sample used in this separation was 1.2 micrograms total. Other column separations of this mixture, some with gradient techniques requiring regeneration steps, have used from ¾-15 hours for the same separation (18-20). The speed of the separation shown in Figure 4 could actually be significantly increased. By reducing the column length to 57 cm and using the same carrier velocity, one should be able to completely separate all peaks (resolution = 1.5) in 3.7 minutes.

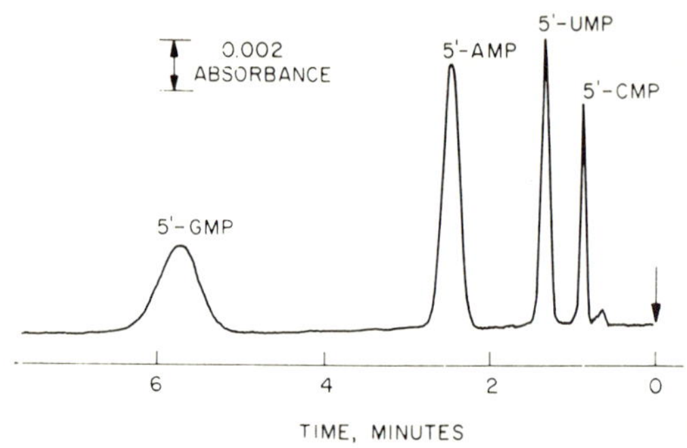

Figure 4.
Separation of Ribonucleoside-5′-monophosphoric Acids. Column—1 m x 2.1 mm, i.d., "Zipax" strong anion exchanger, 20-37 μ; column temperature—60°C; carrier—0.006 M H_3PO_4—0.002 M KH_2PO_4 (PH—3.75); flowrate—1.91 cc/min; column input pressure—900 psi; sample—1.2 μl of 0.25 mg/ml each in carrier.

The same system is capable of resolving mixtures of the 2′(3′)-nucleotides. Figure 5 shows the separation of a synthetic mixture of these compounds. The same conditions were used as in Figure 4, and the sep-

18. Heftmann, E., "Chromatography," 2nd Ed., Reinhold Publishing Corp., New York, N.Y., 1967, p. 636.
19. Horvath, C. G., Preiss, B. A., Lipsky, S. R., Anal. Chem. **39,** 1422 (1967).
20. Blattner, F. R., and Erickson, H. P., Anal. Biochem. **18,** 220 (1967).

aration was complete in about 13 minutes. This separation was carried out on approximately 0.5 μg total sample, using a UV detector sensitivity of 1/4 of maximum. By operating at full detector sensitivity it should be possible to obtain about full-scale recorder deflection for the major components in this mixture on a total of about 100 nanograms of sample. Detector noise at maximum sensitivity is about ±2% of full-scale under these conditions.

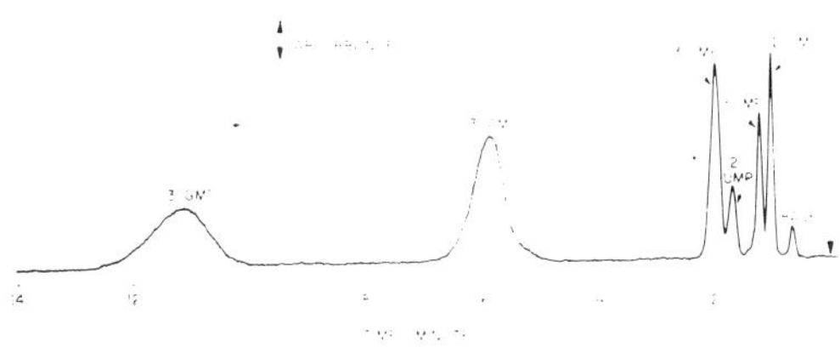

Figure 5.
Separation of Ribonucleoside 2′(3′)-monophosphoric Acids. Same conditions as in Figure 4; sample—2 μl containing approx. 0.5 μg total.

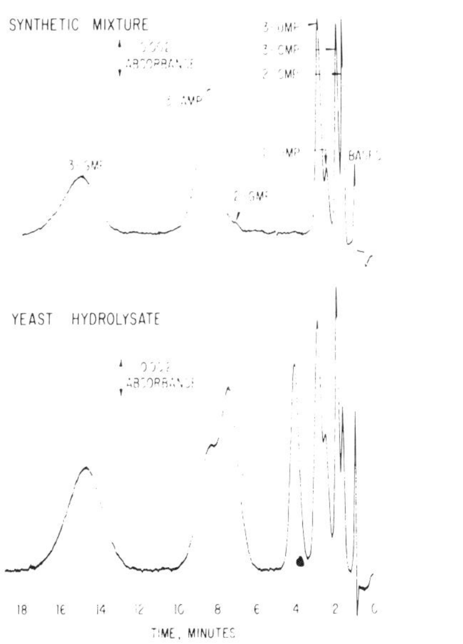

Figure 6.
Ribonucleoside 2′(3′)-monophosphoric Acids from Yeast. Upper—4 μl of synthetic sample containing approx. 1 μg total. Lower—50 μl of yeast hydrolysate; 0.1 mg/ml in carrier. Carrier—0.006 M H_3PO_4; 0.02 M KH_2PO_4; column—similar to that for Figures 4 and 5; flow—1.82 cc/ml; column input pressure—1050 psi; column temperature—60°C.

The components of yeast RNA hydrolysates can be conveniently separated with the anion exchange system described above. The top curve in Figure 6 shows the separation of a synthetic mixture of some 2′(3′)-nucleotides using the same conditions as for the separation in Figure 4, except that a different preparation of "Zipax" anion exchange material was used and the flowrate was slightly lower. A good separation of all components is accomplished in about 18 minutes although baseline resolution is not obtained in all cases. The lower curve in Figure 6 is the chromatogram of a hydrolyzed yeast RNA. To obtain this curve, a commercial sample of yeast RNA (CORE, Mann Research Laboratories) was hydrolyzed for 18 hours at 37°C in 0.15 N KOH, neutralized to pH 4 with 1 M phosphoric acid, and diluted to 10 ml with 0.006 M H_3PO_4-0.02 M KH_2PO_4 buffer. No attempt was made to optimize the separations shown in Figure 6. Baseline separation of all components could be obtained by varying the composition of the carrier or by using a longer column.

The peak eluting just prior to 3′-AMP in Figure 6 is tentatively identified as 2′-GMP. As can be seen in the top curve, a small peak at this same retention time is visible. This peak originated from a 2′(3′)-GMP sample and is presumed to be the 2′-isomer. Interestingly, the relative positions of 3′-AMP and 2′-GMP appear reversed as compared to their elution in other systems (3), indicating the difference in selectivity afforded by the sulfonated fluorocarbon ion exchanger.

Although not attempted, this "Zipax" anion exchange system could probably also be used for separating mono-, di-, and triphosphoric acids. Adjustment of carrier composition, perhaps with gradient elution, might be required to separate a complex mixture of compounds with widely-varying partition coefficients.

Acknowledgment

The assistance of Glenn J. Wallace in obtaining the data in this paper is gratefully acknowledged. The stereoscanning electron microscopy performed by A. N. McKee, Textile Fibers Department, E. I. du Pont de Nemours & Co. is also most appreciated. ■

Manuscript received August 25, 1969

Part X

AMINO ACIDS

Editor's Comments on Papers 36 Through 39

36 **MOORE and STEIN**
Chromatography of Amino Acids on Sulfonated Polystyrene Resins

37 **SPACKMAN et al.**
Excerpts from *Automatic Recording Apparatus for Use in the Chromatography of Amino Acids*

38 **HAMILTON**
Ion Exchange Chromatography of Amino Acids: Effect of Resin Particle Size on Column Performance

39 **STEGEMANN**
Kontinuierliche Aminosäuretrennung an einer Austauschersäule durch Gradientenelution mit einem Puffer
English translation: *Continuous Amino Acid Analysis on an Ion-Exchanging Column by Gradient Elution with a Buffer*

Amino acid analysis is the most familiar use of ion-exchange chromatography today. Hundreds of laboratories in hospitals and medical research organizations throughout the world have automatic amino acid analyzers, and most of them are in constant use. These analyzers use columns of cation-exchange resins with buffers of various pH values as eluents. The effluents are mixed with ninhydrin reagent and retained in heated reaction coils until the red colors have developed; then they are passed into photometers and their absorbances are recorded continuously. In some of the newest instruments provision is made for automatic integration and computer printout.

Paper 36 is the first definitive study of the separation of amino acids by ion exchange. This famous paper has been quoted innumerable times and is beyond all question a benchmark. It helped to win the Nobel Prize for chemistry for 1972, which was awarded to S. Moore, W. H. Stein, and C. Anfinsen for their work on the structure of ribonuclease.

At low pH all amino acids form cations. As cations they are bound by a cation-exchange resin and may be displaced by the cations of the eluting buffer. At higher pH values they are deprotonated, forming uncharged or negatively charged species that no longer are electrically attracted to the resin. The first amino acids to lose their positive charge as the pH is raised are those with two carboxyl groups, like aspartic, glutamic, and hydroxyglutamic acids. Then follow the "neutral" amino acids. The last to lose their positive charges are the "basic" amino acids, those with two amino groups, such as ornithine, lysine, arginine, histidine, and tryptophan. The elution sequence of the amino acids follows roughly this order, but not exactly; for example, serine is eluted before glutamic acid. The presence of hydroxyl groups makes the amino acids more easily eluted, whereas the presence of aromatic groups makes them bind more strongly to the resin. As Moore and Stein point out, "The rate of travel on a column of sulfonated polystyrene resin is a function both of the charge possessed by an amino acid and the nature of its side-chain; it is a result of the affinity of the resin for both the ionic and the non-ionic portions of the molecule."

In their first experiments, Moore and Stein collected fractions and added ninhydrin reagent to each. Then they built equipment with automatic buffer changes, continuous addition of ninhydrin, and continuous recording of the absorbance. Paper 37 describes in full detail the construction of this equipment. There is more detail in the original paper than we need to reproduce here, so only excerpts are shown. A paper immediately preceding this one (Moore et al., 1958) describes the preparation of the columns. The resin was fractionated to give a close range of particle sizes.

Paper 38 describes a detailed study of the effects of particle size and uniformity, and shows very clearly that small and uniform particles are needed to obtain good chromatographic curves. Unlike Moore and Stein, who used a separate column and a separate sample to separate the strongly absorbed basic amino acids, Hamilton used a single column for all the amino acids, including the basic ones, and was able to separate ornithine, lysine, histidine, and arginine very well. In a later paper, Hamilton described a single-column, fully automatic apparatus that used three buffers of successively higher pH and resolved 57 amino acids in submicromole amounts in 20 hours (Hamilton, 1963). The resin particles were spherical, with a diameter of 18 ($\pm$2) μm.

The first successful use of gradient elution to separate all the naturally occurring amino acids on a single column was made by H. Stegemann. His first announcement appeared in the short paper reproduced here along with a complete translation of the text. He has given a more complete account (Stegemann, 1970). Stegemann's work has not had the recognition in the English-language literature that it deserves, possibly because he used gravity flow, simple equipment, and long running times at first to demonstrate that his system would actually work.

More often quoted is a paper by Piez and Morris (1960). These authors describe an apparatus with gradient elution and a single column of highly crosslinked resin (12 percent divinylbenzene) that gives a complete analysis of a protein hydrolyzate in 24 hours. The equipment is simpler than that described in Paper 37.

Both single- and two-column systems have their advantages. The most important factor is the resin, which must be chemically uniform and have small and uniform spherical particles. A recent paper (Benson, 1972) stresses this point.

REFERENCES

Benson, J. V. 1972. *Anal. Biochem., 50,* 477.
Hamilton, P. B. 1963. *Anal. Chem., 35,* 2055.
Moore, S., D. H. Spackman, and W. H. Stein. 1958. *Anal. Chem., 30,* 1185.
Piez, K. A., and L. Morris. 1960. *Anal. Biochem., 1,* 187-201.
Stegemann, H. 1970. *Hoppe-Seyler's Z. Physiol. Chem., 319,* 87-101.

36

Reprinted from *J. Biol. Chem.*, **192**(2), 663–681 (1951)

CHROMATOGRAPHY OF AMINO ACIDS ON SULFONATED POLYSTYRENE RESINS

By STANFORD MOORE AND WILLIAM H. STEIN

(*From the Laboratories of The Rockefeller Institute for Medical Research, New York, New York*)

(Received for publication, May 17, 1951)

The quantitative separation of amino acids by chromatography on columns of starch has been described in previous publications (1–3). In extending these investigations, a systematic study has been made of the separations of amino acids by elution analysis on columns of synthetic ion exchange resins. With a sulfonated polystyrene resin, Dowex-50, it has been possible to develop a chromatographic technique applicable to a wider variety of problems of biochemical interest. Starch columns possess the disadvantage that fluids of high salt content, such as blood plasma or urine, require desalting prior to chromatography (4). Moreover, contamination of the effluent by traces of carbohydrate, coupled with the relatively low capacity of these columns, tends to interfere with the isolation of pure compounds (5). Methods in which columns of Dowex-50 are employed have been found to be free from these difficulties.

Ion exchange resins of various types have been used in the past to separate amino acids into groups, acidic, basic, and neutral, and also to separate some of the individual basic or acidic amino acids (*cf.* (6)). The data of Englis and Fiess (7) indicated that certain of the monoamino acids are adsorbed to different degrees by sulfonic acid type resins, but little information was available as to whether columns of ion exchange resins could yield a degree of resolving power comparable with that obtainable with starch or paper. Recently Partridge and coworkers (8–11) have employed displacement development on ion exchange columns for the separation of amino acids on a preparative scale. Those primary fractions containing more than one amino acid were rechromatographed. Their experiments, together with our results by elution analysis (4), have demonstrated that synthetic resins are capable of separating most of the common amino acids from one another. For analytical experiments, and for isolation work on a small scale, elution analysis would appear to possess higher resolving power and, for this reason, has been employed in the present studies.

In Fig. 1 is shown the result obtained upon passage of a complex mixture of amino acids through a 0.9 × 100 cm. column of Dowex-50. This particular resin (Bauman and Eichorn (12)) is one which is available commercially in a finely divided form well suited for chromatography and has been demonstrated to yield columns of high efficiency for the separation of inorganic cations (*cf.* (13)). In the experiment shown in Fig. 1, the resin is employed in the sodium form and the amino acids are eluted by

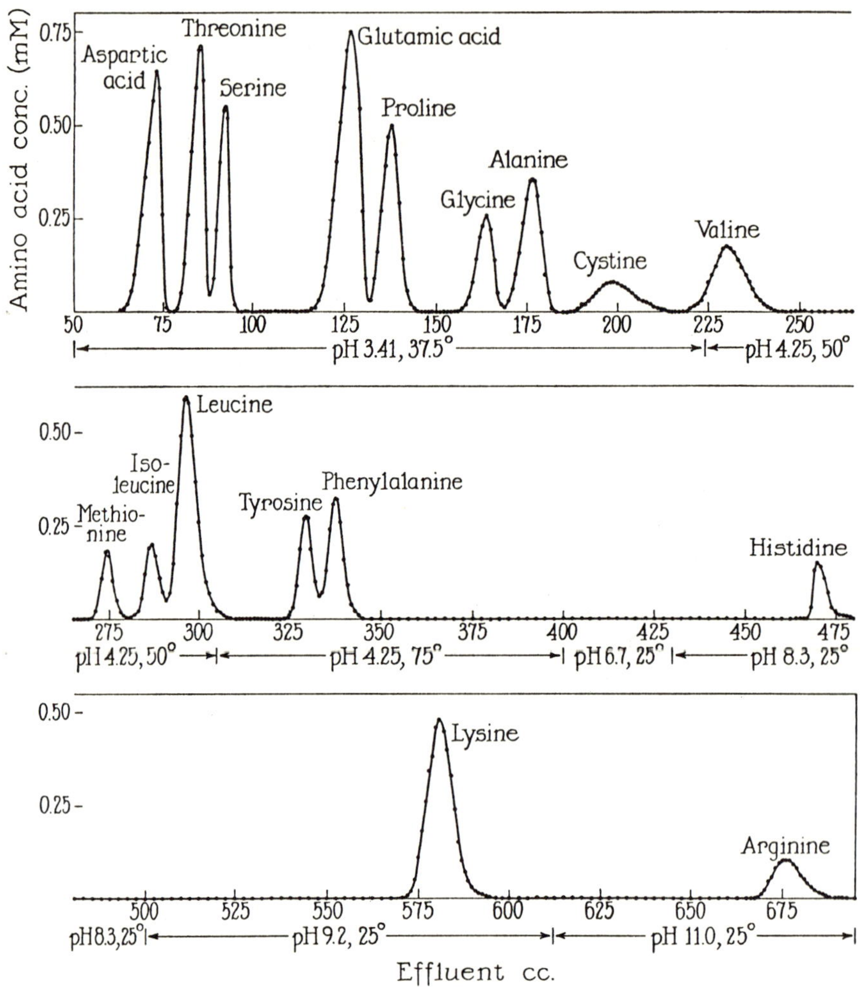

FIG. 1. Separation of amino acids from a synthetic mixture simulating the composition of a protein hydrolysate. The column of Dowex-50, 0.9 × 100 cm., was operated in the sodium form, with buffers of the pH and temperature indicated as eluants. A sample of about 6 mg. of amino acids (see Table II) was chromatographed. The position of an amino acid peak is reproducible, on the average, to better than 5 per cent.

buffers of progressively increasing pH. The results are similar in many ways to those previously reported (5) for columns of Dowex-50 operated in the hydrogen form, from which amino acids were eluted with hydrochloric acid of increasing normality. Although acids as eluants may be useful in isolation experiments, buffers as eluting agents possess several advantages. Many sensitive compounds may be chromatographed with less danger of decomposition when buffers are employed. In addition, a more neutral effluent is easier to analyze by the ninhydrin method.

As may be seen in Fig. 1, it has been possible to arrive at a combination of conditions which, in a single experiment, causes each of the common amino acids to emerge as a discrete peak. In addition, all the amino

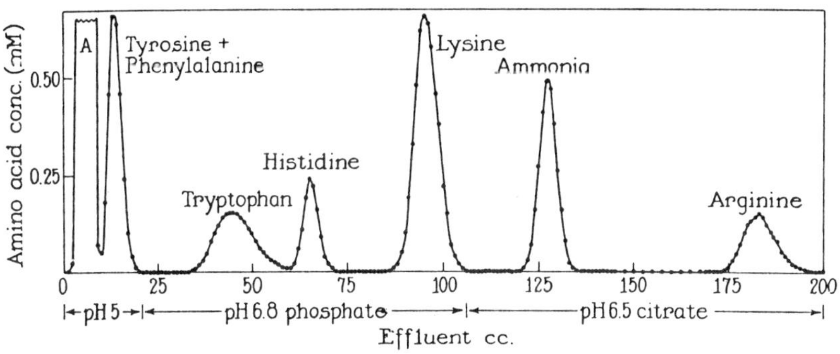

FIG. 2. Separation of the basic amino acids on a column of Dowex-50 0.9 × 15 cm. The column was operated in the sodium form at 25° with the buffers indicated. A sample of about 6 mg. of amino acids (see Fig. 1 and Table II) was employed. The large peak, *A*, comprises all the amino acids emerging before tyrosine in Fig. 1.

acids are recovered quantitatively with the exception of histidine, lysine, and arginine. In order to obtain quantitative recovery of the basic amino acids, it has been found necessary to employ a shorter Dowex-50 column and to elute with buffers below pH 7 (Fig. 2).

It is clear that the separations attained in Fig. 1 cannot be ascribed solely to differences in the ionic natures of the amino acids. The rate of travel on a column of a sulfonated polystyrene resin is a function both of the charge possessed by an amino acid and the nature of its side chain; it is a result of the affinity of the resin for both the ionic and the non-ionic portions of the molecule. Accordingly, columns of Dowex-50 exhibit a resolving power for mixtures of amino acids which compares favorably with that obtained on starch. The procedure employed to obtain curves of the types shown in Figs. 1 and 2 is outlined in the experimental section, followed by a discussion of the influence of variables upon the process and notes on the applicability of the technique to the analysis of protein hydrolysates and of naturally occurring mixtures of amino acids.

Procedure

Preparation of Ion Exchange Columns—The chromatograph tubes used in these experiments were of the Zechmeister-Cholnoky type, with ground joints and coarse sintered glass plates, and had an inner diameter of 0.9 cm.[1] The tubes for columns of resin 100 cm. in height (Fig. 1) are jacketed to permit temperature control by circulation of water from a constant temperature bath.[2] The jacket, similar to that of a condenser, is 108 cm. long and is constructed of 2 cm. (outer diameter) tubing. The inlet should be about 2 cm. from the bottom of the jacket and the outlet about 4 cm. from the top. The jacket is fastened to the chromatograph tube with rubber stoppers. The shorter columns (15 cm. in height, Fig. 2) were operated without a jacket in a room maintained at 25° ± 0.5°.

Dowex-50 (250 to 500 mesh) was purchased from the Microchemical Specialties Company, Berkeley 3, California. 1 pound of the moist resin, as supplied in the hydrogen form, will suffice for the preparation of four 100 cm. columns, with some to spare. The resin (1 pound) is first washed with 4 N HCl on a Büchner funnel with very gentle suction. After washing with 4 to 8 liters of acid, the filtrate, which is initially yellow, should be nearly colorless. After two washes with distilled water (colloidal material may appear in the filtrate), the resin is washed with 2 N NaOH until the filtrate is alkaline. The resulting sodium salt of the resin is suspended in about 3 times its volume of N NaOH and heated over a steam bath for about 3 hours with occasional shaking. The supernatant fluid is decanted or siphoned after about a ½ hour period of settling and replaced with fresh hot N NaOH. This procedure is performed five times in all. The initial supernatant liquids appear very cloudy; the final washes are almost clear. The resin is filtered, washed free of alkali, and passed through a 120 mesh screen with 6 to 8 liters of distilled water. By this procedure a few large particles may be removed. The resin is filtered and stored as the moist sodium salt.

For the preparation of a column, the resin is washed on a filter with a

[1] Purchased from the Scientific Glass Apparatus Company, Bloomfield, New Jersey, catalogue No. J-1663. A length of 115 cm. above the sintered plate must be specified for use with columns 100 cm. in height. The fragility of these long tubes is decreased if heavy walled glass tubing (0.9 cm. inner diameter, 1.5 to 2 mm. wall thickness) is employed in their construction. Tubes 0.9 × 25 cm. are used for the 15 cm. columns. The delivery tip of the chromatograph tube should be beveled.

[2] A constant temperature bath, the temperature of which can be varied readily and which is equipped with a circulating pump (E. H. Sargent and Company, type NS1-33 pump, for example), may be employed.

small amount of 0.1 M sodium citrate buffer, pH 3.4 (not containing the detergent used later), and suspended in about 2 volumes of the buffer. The thickness of the slurry should be such that, when the resin is allowed to settle completely, about 200 cc. of supernatant buffer are present per 100 cc. of settled resin. All bubbles should be removed from the slurry by gentle stirring after it has been allowed to stand for an hour or two. To obtain uniformly efficient 100 cm. columns it has been found desirable to pour the slurry into the chromatograph tube in five portions of approximately equal volume (about 50 cc.). Each portion is poured in through a funnel, the tip of which is bent to direct the stream against the side of the tube. The first portion of resin is allowed to settle under air pressure[3] of 10 to 15 cm. of mercury until no further drop in the height of the surface of the resin occurs. The pressure is maintained until the liquid has fallen to within about 10 cm. of the surface of the resin, whereupon a second portion of the slurry is added. The final additions (or withdrawals) are made to bring the height of the column to 100 cm. The 15 cm. columns are similarly poured in one or two sections. If, at this stage or later, the top surface of the resin is irregular, it should be stirred with a glass rod in the presence of a few cm. of supernatant buffer and allowed to settle again. A separatory funnel (about 300 cc. capacity) containing an appropriate buffer solution is mounted (*cf.* (1)) over the column. The funnel may be attached directly, or, with the aqueous solvents used in the present experiments, a length of plastic tubing may be interposed between the funnel and the column. In the latter case, pressure can be adjusted by moving the funnel as a leveling bulb. After about 100 cc. of buffer have been run through, the column is ready for use.

Preparation of Buffers—The composition of the buffers is given in Table I. Since the reproducibility of results on ion exchange columns is dependent upon pH, the pH values of the buffers have been determined accurately. The measurements have been made at 25° ± 0.5° on a model R pH meter manufactured by the Cambridge Instrument Company, Inc., Ossining, New York. A 0.05 M solution of potassium acid phthalate, pH

[3] Accurate control of air pressure has been obtained by the use of equipment manufactured by The Moore Products Company, H and Lycoming Streets, Philadelphia 24, Pennsylvania. The laboratory compressed air (initially at 65 pounds pressure) is passed through a filter (No. 2306) and a reducing valve (No. 60) set to give 20 pounds pressure on a gage (No. 1169-1). Air of the desired pressure for each chromatogram is withdrawn from this 20 pounds line through a Nullmatic pressure regulator (model 40-15), one being employed for each column. The Nullmatic pressure regulators are equipped with the same 0 to 30 cm. of mercury outlet pressure gages described previously (1). In order to prevent clogging of the pressure-regulating equipment by rust, should the air filter become overloaded, it is advantageous to use brass or copper piping on the clean side of the filter.

TABLE I

Preparation of Buffers

pH	Composition*	HCl or NaOH required to bring 1 cc. buffer to pH 5
3.42 ± 0.01 (Fig. 1)	500 cc. pH 5 sodium citrate† + 110 cc. 1.0 N HCl + 390 cc. water + 0.5 cc. thiodiglycol per 100 cc.	0.1 cc. N NaOH
4.25 ± 0.05 (Fig. 1)	500 cc. pH 5 sodium citrate† + 50 cc. N HCl + 450 cc. water + 0.5 cc. thiodiglycol and 1.0 cc. benzyl alcohol‡ per 100 cc.	0.05 " " "
6.7 ± 0.1 (Fig. 1)	21 gm. citric acid monohydrate + 290 cc. N NaOH to volume of 1 liter	0.05 " 0.5 N HCl
8.3 ± 0.05 (Fig. 1)	250 cc. 0.2 M $NaHCO_3$§, ‖ + 5 cc. 0.1 M Na_2CO_3	0.1 " 2 N HCl
9.2 ± 0.05 (Fig. 1)	500 cc. 0.2 M $NaHCO_3$§ + 90 cc. 0.2 N NaOH	0.1 " 2 " "
11.0 ± 0.05 (Fig. 1)	0.1 M Na_2CO_3	0.1 " 2 " "
5.0 ± 0.1 (Fig. 2)	500 cc. pH 5 sodium citrate† + 500 cc. water + 0.1 gm. disodium versenate¶ and 1.5 cc. benzyl alcohol per 100 cc.	None
6.8 ± 0.03 (Fig. 2)	500 cc. 0.1 M Na_2HPO_4 + 450 cc. 0.1 M NaH_2PO_4 H_2O + 0.1 gm. disodium versenate¶ and 1.5 cc. benzyl alcohol per 100 cc.	0.05 cc. 0.5 N HCl
6.5 ± 0.05 (Fig. 2)	42 gm. citric acid monohydrate + 580 cc. 1.0 N NaOH to volume of 1 liter + 0.1 gm. disodium versenate¶ and 1.5 cc. benzyl alcohol per 100 cc.	0.05 " 2 N HCl

* To all buffers, 1 cc. of BRIJ 35 solution per 100 cc. is added.

† This is the pH 5 (0.2 M) buffer employed for the ninhydrin method (2). A 1:1 dilution of the buffer normally has a pH of 4.95.

‡ The benzyl alcohol serves to improve the resolution of tyrosine and phenylalanine.

§ Freshly prepared before use.

‖ This buffer loses CO_2 on standing and hence becomes more alkaline. Add a 1 cm. layer of mineral oil over the buffer in the separatory funnel to stabilize the pH during the chromatographic experiment.

¶ Analytical grade. We are indebted to the Bersworth Chemical Company, Framingham, Massachusetts, for disodium versenate which contains negligible amounts of ninhydrin-positive material when tested in 0.1 per cent concentration.

4.00, was used as a standard. The various buffers can be prepared in quantity, adjusted if necessary with a few cc. of N HCl or NaOH, and stored in the cold with thymol.

The operation of ion exchange columns with a detergent in the eluant has been found to permit faster flow rates without concomitant broadening of the peaks on the effluent curves. Therefore, for each 100 cc. of buffer taken for a chromatographic experiment, 1 cc. of a detergent solution is added just before use. The detergent solution is prepared by dissolving 50 gm. of BRIJ 35 (Atlas Powder Company, Wilmington, Delaware) in 100 cc. of water. BRIJ 35 is an ether of polyethylene glycol and has the advantage of being a neutral and relatively stable detergent.

For the 100 cm. columns, the buffers all have about the same sodium concentration (0.2 M), in order to minimize the swelling and shrinking of the ion exchange resin which would occur with changes in the ionic strength of the eluant. Before analysis by the ninhydrin method, each effluent fraction must be brought to about pH 5. The amount of acid or alkali required to make this pH adjustment for a 1 cc. effluent fraction is included in Table I for each buffer. The acid or alkali is added from a burette in the manner already described (3). If acid is allowed to remain in a burette for several hours, the tip of the burette must be wiped free of NH_4Cl before use.

It has been observed with ion exchange columns, as with starch chromatograms (1), that losses of methionine occur unless thiodiglycol is added as an antioxidant. Thiodiglycol can be obtained by the redistillation of Kromfax solvent (Carbide and Carbon Chemicals Corporation 30 East 42nd Street, New York) under reduced pressure.

Since the buffer of pH 3.42 ± 0.01 is the only one the performance of which requires checking, it is convenient to prepare it in 10 liter quantities. To check the buffer, a mixture of alanine, cystine, and valine is chromatographed. The pH of the buffer is then adjusted so that the cystine peak is about midway between the neighboring alanine and valine peaks (Fig. 1). The position of the cystine peak is very sensitive to the pH of the buffer. At pH 3.47 cystine emerges almost with alanine, while at pH 3.37 its peak begins to overlap that of valine.

Operation of 100 Cm. Columns—About an hour prior to use, the prepared column is mounted on a fraction collector (1), and water, temperature-controlled at 37.5° ± 0.5°, is started circulating through the jacket. The synthetic mixture of amino acids analyzed in the present work was similar to that employed previously (3). A 0.1 cc. aliquot of the acidic synthetic mixture was added to 1 cc. of the buffer of pH 3.42 and 0.5 cc. of the resulting solution was placed on the column. In the case of more dilute protein hydrolysates, 1 cc. of the hydrolysate can be added to 1 or 2 cc. of buffer, and a 1 or 2 cc. sample placed on the column. A pipette with a bent tip facilitates the addition of the sample to the column without disturbing the surface of the resin. A similar pipette is used to add three 0.3 cc. washes of buffer, as described for starch (1), except that the sample and washings are allowed to drain into the column under gravity. The surface of the resin should not be allowed to remain dry for more than about 5 minutes. A longer time may cause bubbles of air to be trapped in the surface of the column. In adding samples to a buffered column of Dowex-50, it is important that the pH of the sample be the same or *less* than the pH of the column. A pH of 2.5 to 3.0 is optimum. Brief contact with a buffer of higher pH can cause the bands to spread. A convenient load for analytical work is about 5 mg. of total amino acid mixture (0.1 to 0.5 mg. per amino acid). Since the capacity of Dowex-50 columns is high, the total load can be increased several fold, if desired, to increase the accuracy of the determination of components present in small amounts.

When heated columns are employed, it is necessary to use relatively air-free buffers in order to prevent the formation of bubbles in the column as the temperature is raised. At the beginning of the experiment shown in Fig. 1, about 300 cc. of buffer of pH 3.42 were heated to boiling, poured into the separatory funnel, and while still warm, covered with a layer of mineral oil to prevent access of air. The use of boiled buffer is not essential as long as the column is operated at 37.5°. Since, however, it is necessary to have only air-free buffer of pH 3.42 in contact with the column before the temperature is raised to 50°, it is usually convenient to use the boiled solution from the beginning.

The rate of solvent flow through the column should be adjusted to about 4 cc. per hour. A faster rate will cause the peaks to broaden, particularly in the leucine-isoleucine range (Fig. 1). The 100 cm. columns usually run at the appropriate rate under gravity or slight pressure. The effluent is collected in 1 cc. fractions. The photometer tubes employed on the fraction collector do not have to be coated with a silicone film (3) when aqueous eluants are employed. Solvent changes are made in the manner previously described (3). The boiled buffer of pH 4.25 should be run on to the top of the column without letting the surface run dry, that is, while 1 mm. of the previous buffer is still present.

The change of solvent to buffer of pH 4.25 should take place at the beginning of the valine peak (about 225 cc.). During the course of an experiment, the time of change can best be estimated by multiplying the observed position of the proline peak by 1.63. The temperature is also changed at this time to 50°. As the temperature rises, the rate of flow will increase and a lowering of the pressure is required to maintain the desired rate of flow. 75 cc. later (at about 300 cc.) the temperature is increased to 75° to improve the separation of tyrosine and phenylalanine. Frequently the chromatogram is stopped at 375 cc., which is the end of the quantitative part of the experiment. If stopped, the column should be allowed to cool to room temperature before closing the stop-cock of the separatory funnel. If the chromatogram is to be continued for elution of the basic amino acids, subsequent operation is at 25°, with changes of eluant to buffers of pH 6.7, 8.3, 9.2, and 11, as indicated in Fig. 1. The buffer of pH 6.7 is interposed for a 5 to 6 hour period between the buffers of pH 4.25 and 8.3 to avoid the generation of bubbles of CO_2 in the column which might otherwise occur were the shift to be made directly. After the experiment is finished, the column should finally be cleaned with about 100 cc. of 0.2 N NaOH containing BRIJ 35, which may be run through rapidly if necessary. The column is then ready for conversion to pH 3.42 and reuse.

Dowex-50 columns can be used over again indefinitely, so far as we have observed. If not used promptly, the top and bottom of the chromatograph tube should be closed off to prevent drying of the resin. If the surface of the column becomes clogged in the course of use, the top cm. of the adsorbent should be removed and replaced by fresh resin.

Operation of 15 Cm. Columns—The shorter columns employed for the determination of the basic amino acids (Fig. 2) are operated at 25°. The pH of the sample to be analyzed should be about 4. The 15 cm. columns run too rapidly under gravity and require a slight negative pressure, which may be applied by pulling air through a water tower connected to the top of the separatory funnel by rubber tubing, or by lowering the separatory funnel itself. The shifts of buffer are made from pH 5 to 0.1 M phosphate, pH 6.8, and 0.2 M citrate, pH 6.5, as indicated in Fig. 2. If histidine emerges too close to lysine, a phosphate buffer of pH 6.85 or 6.90 may be employed. The column is cleaned with 0.2 N NaOH before reuse.

Analysis of Effluent Fractions—The effluent fractions are brought to about pH 5 with one or two 0.05 cc. drops of the appropriate concentration of HCl or NaOH and analyzed by the photometric ninhydrin method (2).[4] The hold up volume of the 0.9 × 100 cm. columns is about 22 cc. The more acidic buffers, however, have a significant retardation volume. Hence, the amount of N NaOH added to each fraction should be decreased from 0.1 to 0.05 cc. at a point 35 cc. after the buffer of pH 4.25 is added to the top of the column. The subsequent changes in neutralization of the effluent fractions are made at a point 25 cc. after the buffers in question have been added to the column. For the 15 cm. columns (Fig. 2), the corresponding changes in neutralization are made 4 cc. after the change of buffer.

Ammonia emerges from the 100 cm. column (Fig. 1) as a peak very near arginine. Since quantitative recoveries of ammonia are not possible,

[4] Ninhydrin sufficiently pure so that recrystallization is unnecessary is currently obtainable from Dougherty Chemicals, 87-34 134th Street, Richmond Hill, New York.

because of loss by volatilization from the alkaline buffer, it is preferable to remove ammonia completely before analysis. For this purpose, groups of 50 of the tubes containing the buffer of pH 11 are placed in a large vacuum desiccator prior to neutralization, and the desiccator is evacuated continuously by a water pump for 2 hours at room temperature.

In order to obtain accurate analytical values, the 1 cc. effluent fractions require the addition of 2 cc. of ninhydrin reagent and this procedure has been followed in the present work. Nearly the same accuracy can be achieved, however, with 1 cc. of ninhydrin solution, provided a 30 minute heating time is used instead of the usual 20 minutes. With 1 cc. samples and 2 cc. of ninhydrin reagent, the color yields of most of the amino acids have been checked and found to be the same as those obtained before ((2), Table II) with the following changes: aspartic acid 0.93, serine 0.97, alanine 0.99, valine 1.00, and tyrosine and phenylalanine 0.91. The correction factors were calculated according to the equation given previously (2), with the knowledge that no loss in volume because of evaporation occurs during the heating period. For maximum accuracy, it is desirable that the volume of the fractions delivered by the fraction collectors be 1.0 ± 0.1 cc. The volume calculations and the adjustment of the pH are sufficiently sensitive so that a 0.1 cc. variation in the volume of 1.0 cc. effluent fractions can give rise to a 2 per cent error in the final integration.

As has been pointed out previously (1, 3), the choice of blanks against which the amino acid peaks are read is crucial for maximum accuracy. In order to insure a constant blank reading with buffers of pH 3.42 and 4.25, it is essential to take precautions (3) against the uptake of ammonia while the full tubes are on the fraction collector. From Fig. 1 it can be seen that the base-line for the aspartic acid-proline range can be obtained readily from the several valleys in that portion of the curve. The blank after proline can always serve as the point of reference for the glycine and alanine peaks. Depending upon the exact position of cystine, the curve may or may not return to the base-line both before and after the cystine peak. Hence the base-line for cystine and valine can be chosen after valine. The blank reading may change slightly at the point of emergence of the buffer of pH 4.25, just ahead of methionine. The blank for the three peaks in this range is, therefore, best chosen after leucine.

In the determination of the basic amino acids (Fig. 2) the emergence of the buffer of pH 6.5 is accompanied by a rise in the blank of about 0.15 optical density unit. The blank for the ammonia peak should, therefore, be chosen just after the peak rather than before it. The high blank, which results from the mobilization of traces of ammonia adsorbed from the previous buffers, does not prevent accurate integration of the relatively sharp ammonia peak superimposed thereon. The high blank continues as a plateau under the ammonia peak and for a number of fractions thereafter, but falls again to the usual low value before arginine.

Quantitative Analysis of Synthetic Mixtures—The average results obtained by the integration of the curves in chromatograms of the type shown in Fig. 1 are summarized in Table II. Recoveries of 100 ± 2 per cent are obtained for the majority of the amino acids. Low yields are obtained for the basic amino acids, and slightly low results are given by glutamic acid, cystine, and methionine. In the case of glutamic acid, the

TABLE II

Recovery of Amino Acids from Known Mixture

The mixture was chromatographed on Dowex-50 in the manner indicated in Fig. 1. Approximately 6 mg. of the mixture were employed for each chromatogram.

Amino acid	Average recovery	Average deviation from mean	No. of chromatograms
	per cent	*per cent*	
Aspartic acid	99.6	±2.7	12
Threonine	101.9	±3.2	12
Serine	100.4	±2.6	12
Glutamic acid	96.7	±2.1	11
Proline	101.0	±2.4	11
Glycine	99.2	±2.8	12
Alanine	100.7	±2.2	12
Cystine	96.1	±1.9	10
Valine	100.8	±2.9	12
Methionine	95.8	±2.2	11
Isoleucine	101.0	±1.4	7
Leucine	100.6	±2.1	12
Tyrosine	99.8	±2.6	5
Phenylalanine	98.2	±1.4	5
Histidine	70 *Ca.*	±11 *Ca.*	9
Lysine	82 "	±5 "	9
Arginine	71 "	±6 "	8

yield of 96.7 per cent is constant and is independent of the amino acid concentration. Analytical results for glutamic acid can therefore be corrected for the 3 per cent loss, which is probably the result of pyrrolidonecarboxylic acid formation in the feebly acidic solvent. As would be expected on this assumption, the yield is a function of time and of the temperature at which the column is operated. At 25°, essentially quantitative recoveries of glutamic acid are obtained, whereas with columns operated at 60° the recovery drops to below 70 per cent. The slightly low recovery of methionine may be occasioned by failure of the thiodiglycol to protect the sulfur completely from oxidation, or it may be a reflection of slight impurities in the sample of methionine employed.

No satisfactory explanation can be offered at present for the low recoveries of the basic amino acids when alkaline buffers are used as eluants. Experiment has shown that a notable improvement of the recoveries is not obtained if buffers such as phosphate and borate are substituted for the carbonate buffers. The yields are unaffected by the presence of thiodiglycol as an antioxidant or of versene to remove traces of metal ions. Satisfactory recoveries are not obtained if the column is operated at 4° or 37.5°, instead of the usual 25°. The basicity of the amino acid is not a determining factor, since tryptophan, which emerges before lysine (see Fig. 3), is recovered to no greater extent than is arginine. Losses are also encountered with histamine.

It has not been possible to obtain quantitative results for the basic amino acids unless eluting buffers of pH 7 or less are employed. In Fig. 2, the first large peak eluted at pH 5 contains all the amino acids which

Table III

Recovery of Amino Acids from Column of Dowex-50 15 Cm. in Height

The column was operated under the conditions given in Fig. 2.

Constituent	Per cent recovery			
	Chromatogram 908	Chromatogram 909	Chromatogram 910	Chromatogram 911
Tryptophan	100	102	104	108
Histidine	99	98	102	102
Lysine	101	100	102	101
Ammonia	99	96		
Arginine	101	100	102	101

emerge before tyrosine and phenylalanine in Fig. 1. The recoveries for tryptophan and subsequent peaks are 100 ± 3 per cent under these conditions, as may be seen in Table III. To attain quantitative results, the chromatogram must be started with a buffer of pH 5. If it is started with a buffer of pH 6.8, the results are low, particularly for histidine. If versene is not included, low recoveries may be obtained after continued use of the columns. The benzyl alcohol serves to sharpen the tryptophan peak and to move it ahead of histidine.

Chromatographic Analysis of More Complex Mixtures—In many biological media, ninhydrin-positive substances other than the ones shown in Fig. 1 are encountered. The positions of some amino acids and related compounds not discussed thus far are shown in Fig. 3. Thirty-two substances were chromatographed together and thirty peaks can be recognized on the effluent curve, with two complete overlaps. The positions of the individual components in this case, and in earlier instances, have been established by preliminary chromatograms with simpler mixtures. Cysteine, if present, is gradually oxidized and does not yield a peak. The many variations in the nature of the eluting conditions which can be

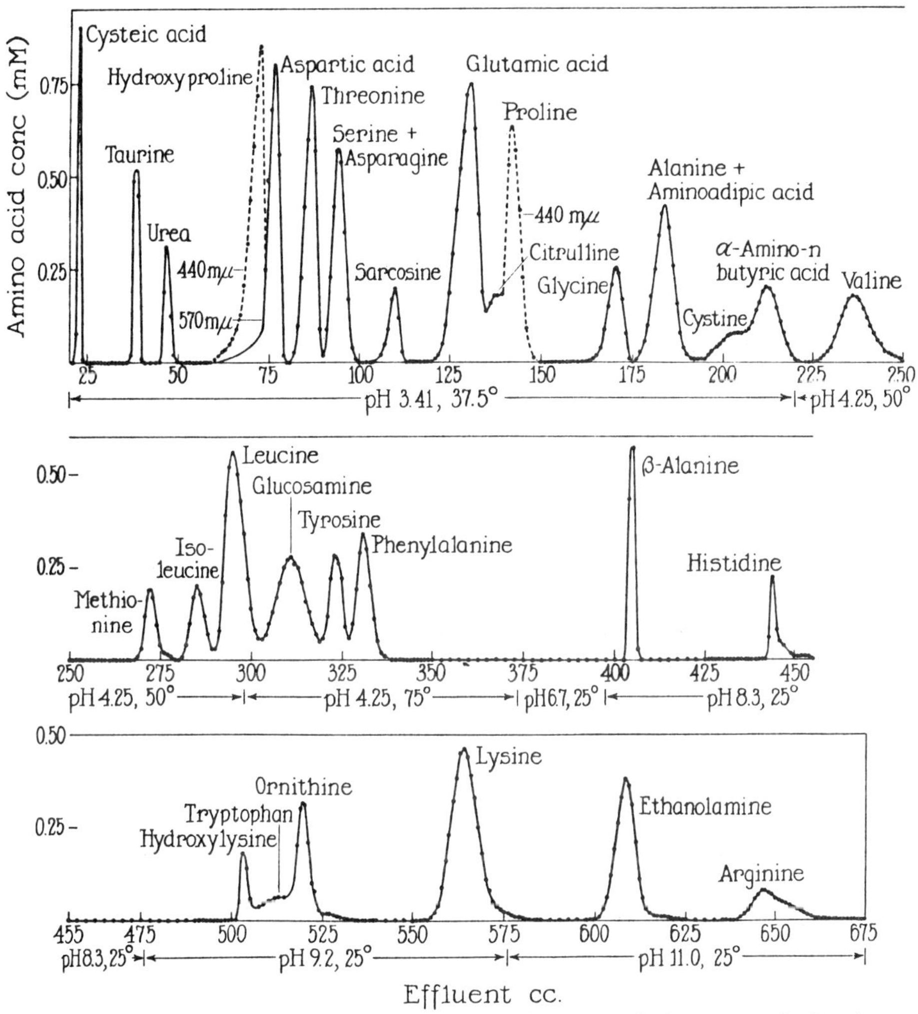

FIG. 3. Separation of amino acids and related compounds from a synthetic mixture containing thirty-two components. The experimental conditions are the same as those given in Fig. 1.

effected with Dowex-50 columns increase the likelihood of resolving substances which are not well separated by the buffer systems employed for the experiment shown in Fig. 3. For example, if a buffer of slightly lower pH, such as 3.2, is employed in place of the buffer of 3.4, hydroxyproline will move farther ahead of aspartic acid. On the other hand, a pH of 3.6 will move cystine to a point well ahead of α-amino-*n*-butyric acid. The importance of applying as many criteria as possible to the identification of the peaks on a complex curve has been stressed for the amino acids in the case of starch (1, 3) and applies with equal emphasis here.

The 15 cm. columns used for the quantitative determination of the basic amino acids do not possess as high a resolving power as do the 100 cm. columns. For example, ornithine emerges with lysine in the experiment shown in Fig. 2 and can only be recognized by the slight asymmetry it confers upon the leading edge of the lysine peak. Hydroxylysine appears well ahead of lysine, but coincides with histidine, while glucosamine yields a sharp peak superimposed upon the center of the broader tryptophan peak. β-Alanine travels very close to tyrosine and phenylalanine. Since the 15 cm. column has limited resolving power, it is recommended that an unknown mixture first be chromatographed in the manner shown in Fig. 1, in order to obtain a qualitative picture of the components in the basic amino acid range.

DISCUSSION

Effect of Variations in Resin—The procedure leading to effluent curves of the type shown in Figs. 1, 2, and 3 has grown out of a study of the characteristics of columns of different preparations of sulfonated polystyrene, operated under varying conditions of pH and with buffers of different anionic and cationic nature, at various flow rates and loads. The nature and constancy of behavior of the resin are, of course, central factors in the problem. Different manufactured lots of the standard Dowex-50 (250 to 500 mesh spheres) have given the same results. A curve similar to the one shown in Fig. 1 was obtained with a "low color" resin prepared by a different process of sulfonation.[5] Major variations in the particle size and the degree of cross-linking of the polystyrene polymers, however, can cause different results. With 100 mesh resin the peaks were about twice as broad as those shown in Fig. 1. The improved performance of finer mesh material has been demonstrated for the separations of the rare earth elements (13, 14). The results of the amino acid studies parallel the observations made with inorganic ions with the exception that, in amino acid chromatography, colloidal agglomerates of Dowex-50 were found to have no advantage over the 250 to 500 mesh beads.

The standard Dowex-50 resin used in this work was prepared from styrene copolymerized with about 8 per cent divinylbenzene. The resulting

[5] The different types of Dowex-50 resins discussed in this section were obtained from Dr. W. C. Bauman of The Dow Chemical Company, Midland, Michigan, whose cooperation we wish to acknowledge with gratitude.

product may be considered to have a molecular structure the holes or pores of which are characteristic of 8 per cent cross-linking. When amino acids were chromatographed on a 16 per cent cross-linked resin (250 to 500 mesh) at a rate of solvent flow similar to that used with the 8 per cent resin, broad, very poorly resolved zones were obtained. This result indicates that the pore size of this resin was too small to permit rapid penetration into the resin particle of substances the size of amino acids. If this supposition is correct, there should also be a molecular weight range above which efficient chromatography on the 8 per cent cross-linked resin will not be feasible. In the present work it has been found that dipeptides such as glycylleucine can be chromatographed well on the standard resin. The tetrapeptide, leucylleucylglycylglycine, gave a somewhat broader peak, however. To obtain sharp peaks with the higher peptides, resins of 4 or 2 per cent cross-linking may be required. Lower cross-linked resins are not so satisfactory for amino acid chromatography, however. It has been found, for example, that, compared to the 8 per cent resin, the 2 per cent resin is somewhat deficient in resolving power. Under conditions similar to those shown in Fig. 1, serine and threonine are not well separated, nor are glycine and alanine. In addition, the greater tendency of the 2 per cent resin to shrink and swell with changes in ionic strength gives rise to columns of poorer mechanical stability. The amino acid peaks obtained from columns of the 2 per cent resin are not sharper than those shown in Fig. 1. Moreover, faster rates of solvent flow cause the peaks to broaden just as much on the 2 per cent resin as was found to be the case for the 8 per cent resin. In general, it appears preferable to use a resin with the highest degree of cross-linking compatible with the separations to be achieved.

Effect of Variations in Eluant—The chromatographic behavior of amino acids on columns of Dowex-50 operated in the sodium cycle can be varied by changes in the pH and the ionic strength of the eluting buffers. In general, low pH values decrease and high pH values increase the rate of travel of the amino acids shown in Fig. 1. Thus, at pH 2.5, the amino acids are retained near the top of the column. At pH 5, on the other hand, all except the basic amino acids move rapidly (see Fig. 2). As might be expected, the positions of aspartic acid, glutamic acid, cystine, and proline are relatively the most sensitive to pH when buffers in the range pH 3 to 5 are employed. If, for example, operation of a column is begun with a buffer of pH 4.0, instead of the buffer of pH 3.4 shown in Fig. 1, aspartic and glutamic acids are accelerated preferentially and emerge still separated but ahead of both threonine and serine. Cystine also moves faster and appears before glycine, while proline is relatively retarded and emerges near alanine. Under these conditions, moreover, the resolution of glycine, alanine, and valine is largely lost, and these three emerge very close together. As has already been noted for the case of hydroxyproline and the basic amino acids, the use of buffers possessing a pH different from any employed in the experiments shown in Figs. 1 and 2 can aid in the resolution of specific overlaps.

The effect of variations in the molarity of the citrate buffers has also been studied. In general, raising the Na^+ concentration produces changes similar to those induced by raising the pH. Thus, 0.2 M sodium citrate at pH 3.1 gives a pattern very similar to that obtained at pH 3.4 with the 0.1 M buffer, except that the aspartic acid-threonine and glutamic acid-proline separations are marginal. With buffers of much lower concentration than 0.1 M, unsatisfactorily broad and asymmetric peaks are obtained. The effect of employing buffers of higher ionic strength has been utilized to advantage in the experiment shown in Fig. 2. A more concentrated buffer (0.2 M citrate), rather than one of higher pH, was employed to accelerate arginine.

The properties of the eluant can also be changed by the addition of an organic solvent. For example, the presence of about 1 per cent benzyl alcohol in the eluant has been found to have a sharpening effect on the peaks of the aromatic amino acids and tends to increase their rates of travel relative to the others. Experiments with citrate buffers containing varying amounts of propyl alcohol have revealed that, in general, organic solvents accelerate preferentially the amino acids possessing the longer non-polar side chains, thus, in effect, tending to reverse the normal order of emergence for the glycine through leucine group of amino acids. With 25 per cent propyl alcohol, alanine is moved ahead of glycine, isoleucine overlaps valine, and threonine is farther in front of serine. For specific separations, particularly in preparative work, alcoholic solvents can frequently be used to advantage.

Effect of Temperature—The separations of amino acids obtained on columns of Dowex-50 are dependent on temperature. Increasing the temperature of the citrate-buffered columns increases the rate of movement of all the amino acids. Part of this effect results simply from the rise in pH which occurs when citrate buffers are warmed. More specific effects also occur, however. For example, methionine and isoleucine appear together when eluted at 25° under conditions otherwise quite similar to those shown in Fig. 1. At 60°, on the other hand, methionine is well ahead of isoleucine.[6] Tyrosine and phenylalanine are also completely resolved at 60°, whereas at 25° and 37° they emerge together. Operation of the column at 60° throughout has the disadvantages that isoleucine

[6] The effect of temperature upon the separability of methionine and leucine has been reported recently by Partridge and Brimley (11).

and leucine are poorly separated, and that the yield of glutamic acid is markedly lowered, as has been mentioned previously. A consideration of all the factors involved suggested the sequence of temperatures shown in Fig. 1. The column is warmed just enough at the beginning to insure the separation of methionine and isoleucine without impairing the separation of the latter from leucine. After the emergence of isoleucine, the temperature is again raised to facilitate the separation of the two aromatic amino acids. It should be mentioned that, as was found with inorganic ions (13, 14), the rate of the exchange of amino acid ions between solution and resin is probably increased at higher temperatures. For this reason somewhat faster rates of solvent flow can be used under these conditions without a concomitant broadening of the amino acid peaks.

Operation of Dowex-50 Columns in Hydrogen and Ammonium Cycles—An effluent curve showing the results which may be obtained with a Dowex-50 column operated in the hydrogen cycle has already been published ((5), Fig. 8). Prior to analysis by the ninhydrin method, the HCl was removed from the effluent fractions by evaporating them to dryness in a vacuum desiccator attached to a water pump and warmed to 40° with an infra-red lamp or a heating mantle. Alternatively, the samples can be neutralized accurately with NaOH and analyzed directly. The increased volume of the neutralized samples requires the use of more ninhydrin solution, however, unless aliquots are pipetted for analysis. For the isolation of many of the amino acids, the HCl system can be useful (*cf.* (15)), although operation in the ammonium cycle with ammonium formate, acetate, and carbonate buffers will probably prove preferable. These can all be removed conveniently by sublimation. A systematic scheme permitting the isolation of amino acids from 2 to 5 gm. of a protein hydrolysate by the use of such buffered columns has been developed by Dr. C. H. W. Hirs and will form the subject of a forthcoming communication.

Operation of Dowex-50 in Lithium and Potassium Cycles—When operated at 25° with lithium citrate buffers, Dowex-50 gives a curve very similar to the one shown in Fig. 1. The resolution of some amino acids is slightly improved, but the separation of glycine and alanine is less complete. It is primarily for this reason that operation in the sodium rather than the lithium cycle has been chosen. With the potassium salt of the resin, very poor resolution and markedly broad peaks were obtained.

Applicability of Method—Compared with the starch method (1, 3, 16), the ion exchange technique is about twice as rapid for the analysis of protein hydrolysates. The 100 cm. column (Fig. 1) requires 4 or 5 days to run through tyrosine and phenylalanine. If the 3 day curve for the basic amino acids (Fig. 2) is obtained concurrently, analysis for all the components of an acid hydrolysate (cysteine excepted) can be completed in 5 days. The determination of tryptophan in alkaline hydrolysates has not been studied by the present method.

The procedure has been applied to blood plasma and urine. The fact that the performance of the column is not altered by the presence of the inorganic salts of the blood was first established with synthetic mixtures containing the appropriate ratio of salt to amino acid. In practice, the dialysate (17) from 5 cc. of blood plasma can be concentrated and chromatographed exactly in the manner described for protein hydrolysates. In the case of urine, only a slight modification in technique is required. The very large amounts of urea and other constituents in a 4 cc. sample of urine of specific gravity 1.020 cause a distortion of the peaks in the aspartic acid-proline range. This distortion can be eliminated if the column is started with a citrate buffer of pH 2.5. After 60 fractions, a buffer of pH 2.92 is introduced for forty additional fractions, followed by the buffer of pH 3.42 and the usual shifts thereafter. The points of emergence of taurine and urea given in Fig. 3 are not noticeably different in the buffer of pH 2.5. The remaining amino acids move so slowly at the lower pH values that the pattern obtained after the shift to the buffer of pH 3.42 is almost identical with that shown in Figs. 1 and 3. There are a number of peaks on urine curves the full interpretation of which requires further investigation.

The authors wish to acknowledge with appreciation the expert technical assistance of Mrs. Gertrude C. Carey.

SUMMARY

A procedure is described for the chromatographic fractionation of mixtures of amino acids by elution analysis on columns of Dowex-50, a sulfonated polystyrene resin. On a 0.9 × 100 cm. column operated in the sodium form, elution of amino acids is effected by the use of buffers of progressively increasing pH, from 3.4 to 11. In experiments with synthetic mixtures of amino acids simulating the composition of protein hydrolysates, a sequence of buffers has been developed which yields in a single chromatogram an effluent curve in which every component emerges as a discrete peak. For analytical work, 3 to 6 mg. of the amino acid mixture are required per chromatogram. Integration of the curves has given quantitative recoveries (100 ± 3 per cent on the average) for all except the basic amino acids. These latter may be determined with the aid of a second column, 0.9 × 15 cm., with buffers in the range pH 5 to 6.8. The position of emergence of some of the less common amino acids and related compounds has also been determined.

The resin columns possess higher resolving power than starch columns and are more convenient to operate. An amino acid analysis of a protein hydrolysate can be carried out in about half the time required with starch. The performance of the ion exchange column is not adversely affected by the presence of inorganic salts in the material chromatographed. Accordingly, blood plasma dialysates and urine may be fractionated directly without preliminary salt removal.

BIBLIOGRAPHY

1. Stein, W. H., and Moore, S., *J. Biol. Chem.*, **176**, 337 (1948).
2. Moore, S., and Stein, W. H., *J. Biol. Chem.*, **176**, 367 (1948).
3. Moore, S., and Stein, W. H., *J. Biol. Chem.*, **178**, 53 (1949).
4. Stein, W. H., and Moore, S., *J. Biol. Chem.*, **190**, 103 (1951).
5. Stein, W. H., and Moore, S., *Cold Spring Harbor Symposia Quant. Biol.*, **14**, 179 (1950).
6. Block, R. J., in Nachod, F. C., Ion exchange, New York, 295 (1949).
7. Englis, D. T., and Fiess, H. A., *Ind. and Eng. Chem.*, **36**, 604 (1944).
8. Partridge, S. M., *Biochem. J.*, **44**, 521 (1949); **45**, 459 (1949).
9. Partridge, S. M., and Westall, R. G., *Biochem. J.*, **44**, 418 (1949).
10. Partridge, S. M., Brimley, R. C., and Pepper, K. W., *Biochem. J.*, **46**, 334 (1950).
11. Partridge, S. M., and Brimley, R. C., *Biochem. J.*, **48**, 313 (1951).
12. Bauman, W. C., and Eichorn, J., *J. Am. Chem. Soc.*, **69**, 2830 (1947).
13. Shubert, J., in Nachod, F. C., Ion exchange, New York, 167 (1949).
14. Mayer, S. W., and Tompkins, E. R., *J. Am. Chem. Soc.*, **69**, 2866 (1947).
15. Ehrensvärd, G., Reio, L., Saluste, E., and Stjernholm, R., *J. Biol. Chem.*, **189**, 93 (1951).
16. Stein, W. H., and Moore, S., *J. Biol. Chem.*, **178**, 79 (1949).
17. Hamilton, P. B., and Archibald, R. M., *Ind. and Eng. Chem., Anal. Ed.*, **16**, 136 (1944).

37

Reprinted from *Anal. Chem.*, **30**, 1190–1191, 1202 (1958)

Automatic Recording Apparatus for Use in the Chromatography of Amino Acids

DARREL H. SPACKMAN, WILLIAM H. STEIN, and STANFORD MOORE
The Rockefeller Institute for Medical Research, New York 21, N. Y.

▶ Quantitative determination of amino acids is made simpler and more rapid by an instrument for automatically recording the ninhydrin color value of the effluent from ion exchange columns. The influent buffer, freed of air, is pumped at a constant rate through a column of sulfonated polystyrene resin. The effluent is met by a capillary stream of ninhydrin reagent delivered by a second pump. The color is developed by passing the mixture of reagent and effluent through a spiral of capillary Teflon tubing immersed in a boiling water bath. The absorbance of the resulting solution is measured continuously at 570 and 440 mμ as it flows through a cylindrical glass cell of 2-mm. bore. The peaks on the recorded curves can be integrated with a precision of 100 ± 3% for loads from 0.1 to 3.0 μmoles of each amino acid. A hydrolyzate of a protein or peptide may be analyzed in less than 24 hours. The more complex mixtures characteristic of blood plasma, urine, and mammalian tissues can be analyzed in 2 days. The instrument is applicable in principle to detection of ninhydrin-positive constituents in the effluent from various types of chromatograph columns.

Quantitative determination of amino acids bears a relationship to the chemistry of proteins and complex peptides similar to that which elementary analysis bears to the chemistry of simpler organic molecules. In recent studies on the structure of ribonuclease, a partial structural formula for the enzyme has been deduced primarily from data obtained by the quantitative amino acid analysis of the peptides formed as a result of the action of trypsin, pepsin, and chymotrypsin (*4*). Determinations of amino acids are also required in investigations of the distribution of amino acids and their derivatives in physiological fluids (*16, 17*) and mammalian tissues (*18*) as well as in the analysis of foods (*10, 14*). To facilitate such studies, there is need for an accurate method of amino acid analysis that is more rapid and less laborious than the techniques employed heretofore. To fill this need, a recording apparatus has been constructed which,

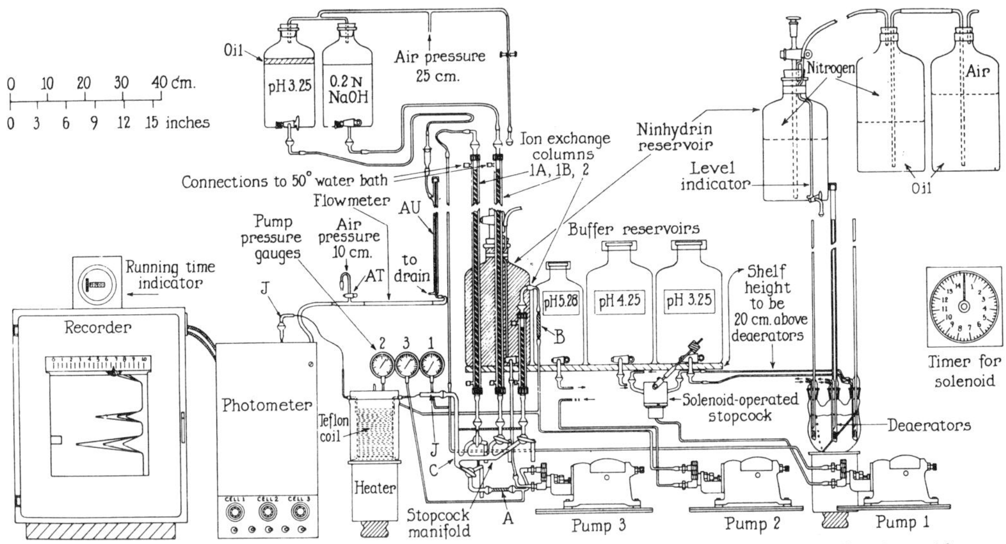

Figure 1. Automatic recording apparatus used in chromatographic analysis of mixtures of amino acids

Influent buffers are drawn from the reservoir bottles through the deaerating equipment and pumped through the ion exchange columns at a constant rate. Ninhydrin reagent is pumped into the effluent stream at C and the effluent-ninhydrin mixture is carried in a Teflon coil through a boiling water bath to develop the colors formed when ninhydrin reacts with amino acids. The absorbance of the colored solution is measured continuously by a photometer assembly and registered on a recorder. (See text for modification in arrangement of four bottles at the top of the figure, and use of a solenoid-operated valve instead of solenoid-operated stopcock.)

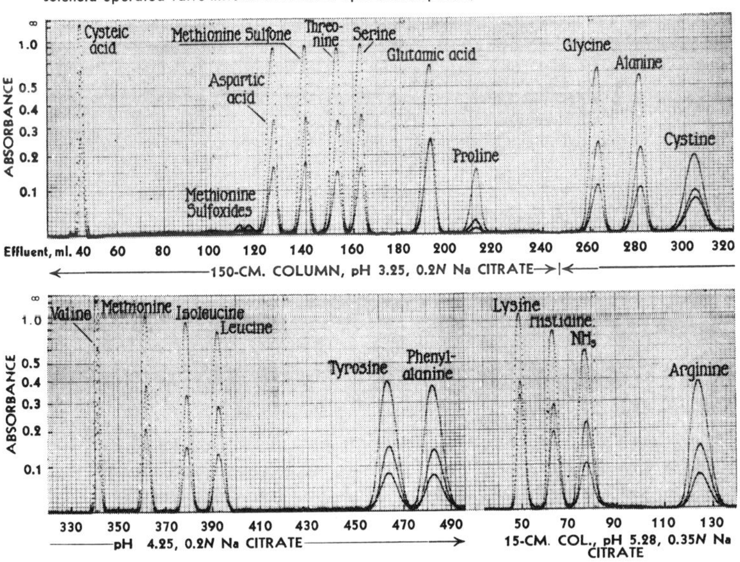

Figure 2. Chromatographic analysis of a mixture of amino acids automatically recorded in 22 hours by equipment shown in Figure 1

The three curves show absorbance at 570 mμ using two different cell depths and at 440 mμ for determination of proline. The chromatographic system is designed for the analysis of protein hydrolyzates (5). The method gives separate peaks for methionine sulfone, the methionine sulfoxides, cysteic acid, and S-carboxymethylcysteine (at 96 ml.). Hydroxyproline can be detected by the presence of a shoulder on the 440-mμ curve at the beginning of the emergence of aspartic acid; full separation of hydroxyproline requires the system illustrated in Figure 9A. Alloisoleucine, frequently formed in trace amounts from isoleucine by racemization during hydrolysis of proteins, appears in a separate peak between methionine and isoleucine (5, Figure 1). Glucosamine and galactosamine follow phenylalanine from the 150-cm. column, and are eluted at 510 and 580 ml., respectively. The change to the pH 4.25 buffer can be delayed 1 hour, if elution of valine as a broader peak with the pH 3.25 buffer is preferred. Any residual tryptophan in an acid hydrolyzate emerges just ahead of lysine from the 15-cm. column; glucosamine, galactosamine, chlorotyrosine, and a ninhydrin-positive degradation product of tryptophan are detectable as constituents which emerge just ahead of tryptophan from the short column. Hydroxylysine and ornithine overlap tryptophan and lysine, respectively; their presence requires use of the system shown in Figure 9B.

in conjunction with improvements in ion exchange chromatography, permits the complete quantitative determination of amino acids in a hydrolyzate of a peptide or a protein every 24 hours with minimum attention.

In the ion exchange methods described previously (*7*, *8*), the effluent from the column was collected as a large number of small fractions of known volume, in each of which the amino acid content was determined individually by a quantitative photometric ninhydrin method (*6*, *9*). In the present automatic technique (*15*), ninhydrin reagent is introduced at a constant rate into the effluent flowing from the column. Color development ensues as the mixture is pumped through a coil of capillary tubing immersed in a boiling water bath, after which the intensity of the blue color (diketohydrindylidene-diketohydrindamine) formed when amino acids are present is determined by continuous photometry at 570 mμ. The yellow colors from proline and hydroxyproline are also monitored at 440 mμ. The results are plotted automatically on a recording microammeter. A schematic drawing of the apparatus is shown in Figure 1, and an effluent curve is shown in Figure 2.

[*Editor's Note:* Material has been deleted at this point.]

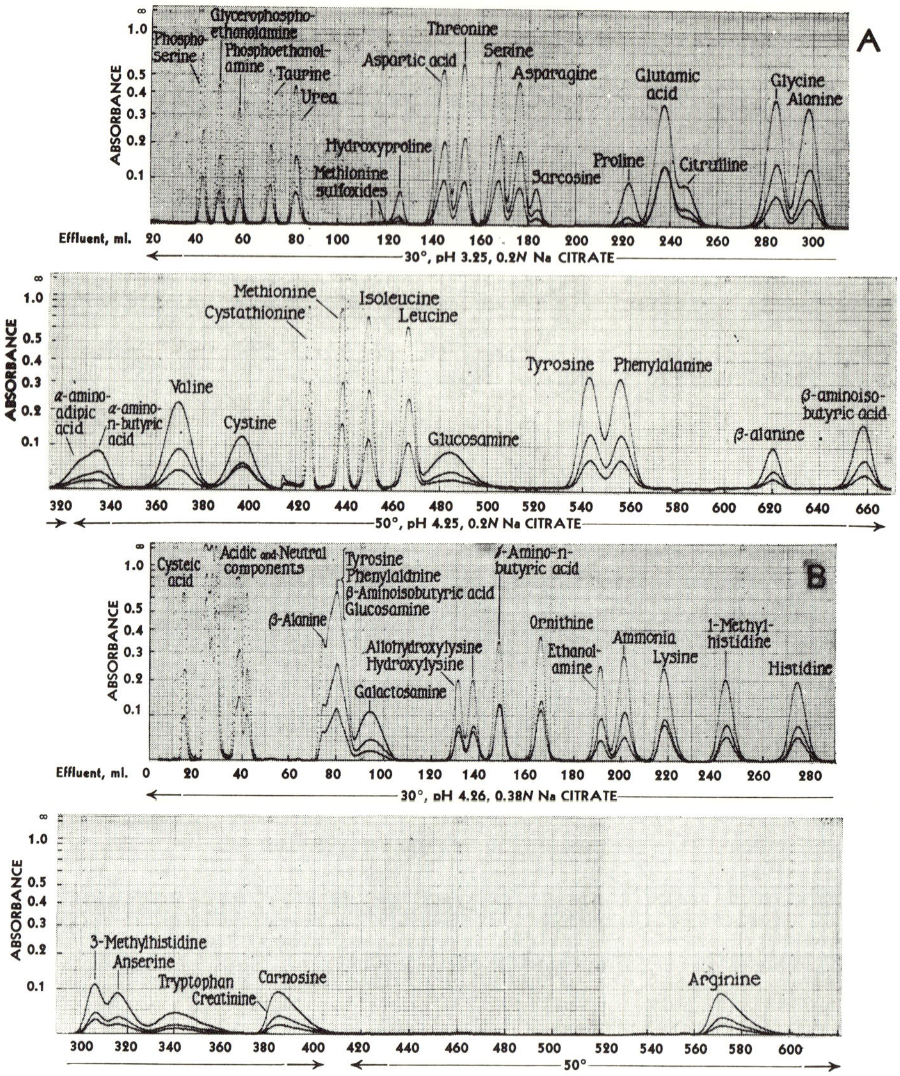

Figure 9. Chromatographic analysis of mixtures of amino acids and related compounds most commonly encountered in physiological fluids

A. 150-cm. column of Amberlite IR-120 at 30° and 50°
B. 50-cm. column at 30° and 50°

Glutamine has the same rate of travel as asparagine; cysteic acid and phosphoserine are not separated; galactosamine, if present, emerges with tyrosine and phenylalanine in *A*; homocystine is eluted as a zone preceding but partially overlapping β-alanine; felinine is slightly more retarded than citrulline, but overlaps it; behavior of glutathione is discussed in the accompanying paper (5). The change in the influent buffer was made at 325 ml., and the break-through of the pH 4.25 buffer is observed at 415 ml.

Reprinted from *Anal. Chem.*, **30**(5), 914–919 (1958)

Ion Exchange Chromatography of Amino Acids

Effect of Resin Particle Size on Column Performance

PAUL B. HAMILTON

Alfred I. du Pont Institute of the Nemours Foundation, Wilmington, Del.

▶A back-washing technique was devised for grading spherical or irregularly shaped particles of sulfonated polystyrene cationic resins, approximately 8% cross-linked, into 20-, 25-, 30-, etc., micron diameter classes, each having 60 to 75% of the particles within ±3 microns of the average diameter of the class, and a diameter range of approximately ±7 microns. Martin's (or accidental) diameter, applicable to spheres and irregular particles alike, is proposed as a designation of particle size in favor of the less precise sieve number. Column performance in the separation of amino acids by ion exchange chromatography at the rapid flow rate of 39.3 ml. per sq. cm. of resin bed per hour was optimum with spherical particles of 25-micron or irregular particles of 30-micron diameter. A range of 20 to 40 microns of pulverized resin was equally serviceable, if size distribution was uniform throughout the range.

THE DEGREE of resolution of amino acids obtained on columns of sulfonated polystyrene resins is a function of many factors, including the pH, ionic strength, and flow rate of the developing buffers, the temperature of operation, and the degree of cross linking and particle size of the resin (*6*, *7*). The sharp peaks, which are an essential attribute of systems exhibiting high resolving power, require rapid attainment of equilibrium between stationary and flowing phase, and it has been recognized since the early studies of Mayer and Tompkins (*3*) that resins of small particle size permit higher rates of flow through the column because equilibrium is attained more rapidly. In attempting to define precisely the optimum conditions for the operation of an ion exchange column, it is advantageous to be able to secure samples of resin of accurately known particle size, possessing a narrow range of particle size distribution. The need for such resins has been rendered more acute by the recent development of automatic recording apparatus for use in the chromatography of amino acids (*9*), and the elaboration in this laboratory of an improved system of analysis with controls for automatically changing temperature and buffer solutions in the conventional fraction collector technique. Both methods permit rapid separations and use rapid flow rates through columns of very finely divided resins.

Details for obtaining resin particles of a serviceable size range by eliminating the fine and coarse material from commercial preparations have been described by Moore and Stein (*6*, *7*). The sodium form of the sulfonated, fully hydrated resin was sieved through a standard 200-mesh screen, with a jet of water, and the particles that passed the 200-mesh screen were reserved. The through-200-mesh resin was then suspended in a large volume of water and after settling, colloidal material and fines were decanted with the supernatant. This procedure is somewhat laborious and lacks precision, as the distribution of particle sizes within the 37- to 74-micron range is not revealed.

The present investigation was initiated by the observation that the resolution of amino acids was better and the peaks were sharper in chromatograms obtained from the first of two consecutively poured columns. This suggested that a higher proportion of finer particles was present in the first column and that particle size might be responsible for this effect. To study this effect in greater detail, a method of classifying the mixed population of the commercial cationic exchange resins in current use in this laboratory was devised, which provided a means of obtaining several classes of particles easily and rapidly. Direct measurement showed that 60 to 75% of the diameters in each class were within ±3 microns of the average diameter of the class. The method of classifying spheres of irregular particles of pulverized resins, and the correlation of particle size with column performance in the chromatographic separation of amino acids, are described in the present paper.

Particle size can be defined in a manner applicable to spheres and irregular particles alike. The accidental or Martin's diameter proposed by Martin, Blythe, and Tongue (*2*) fulfills this condition and entails the simplest measurement. It is defined as the distance between opposite sides of a particle, measured crosswise of the field and on a line which bisects the projected area of the particle. The arithmetical average of these measured distances of a field of particles is the average accidental diameter. Measurement and computation of Martin's diameter can be readily made from enlarged photographs of a microscope field of resin particles taken under medium magnification. According to the thorough discussion by Chamot and Mason (*1*), Martin's diameter gives results in satisfactory agreement with those from the three actual dimensions of the individual particles. Throughout the present study, Martin's diameter, in microns, is used to specify particle size, and the less precise, more cumbersome designation of mesh number is avoided.

RESINS INVESTIGATED

The only resin in bead form investigated was some Dowex 50-X8, hydrogen form, purchased in 1953, which contained a high percentage of spheres with diameters in the 20- to 74-micron range. The Dowex 50-X4, Dowex 50-X8, Dowex 50W-X8, and Dowex 50-X12 (The Dow Chemical Co., Midland, Mich.) from current commercial production contained such small percentages of spheres below 74 microns that they were not studied further. The resin powders investigated were prepared from spheres which were fragmented in a ball mill, shearing plate, or other type of pulverizer. They comprised irregular particles, a high proportion of which were in the 20- to 50-micron range. One of the pulverized resins was Amberlite CG-120, Type III, sodium form (Rohm & Haas Co., Philadelphia, Pa., available from Fisher Scientific Co., Fairlawn, N. J.), which is the resin used for the chromatography of amino acids in the most recent procedures evolved by Moore, Spackman, and Stein (*4*, *8*, *9*), and also by this laboratory. The other resins were Permutit Q-ZP, sodium form (The Permutit Co., Birmingham, N. J.), and Dowex 50W-X8, hydrogen form (Bio-Rad Laboratories, Berkeley, Calif.). All are sulfonated polystyrene polymers, approximately 8 to 8.5% cross-linked, of medium porosity, with an exchange capacity of 5 meq. per gram of dry resin. The wet density of the hydrogen form of Dowex 50-X8 and

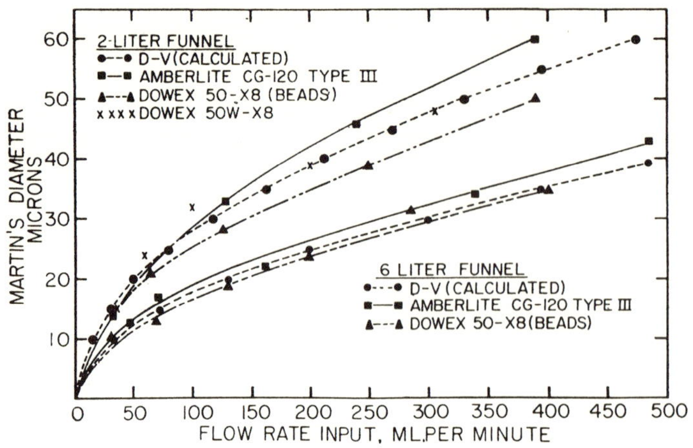

Figure 1. Experimentally determined relationship between flow rate into 2- and 6-liter separatory funnels and diameter of particles removed

Amberlite CG 120 is 1.21; the wet density of the sodium form is 1.27.

PRELIMINARY PREPARATION OF RESINS

All resins were processed in the form in which they were received from the manufacturers—i.e., Dowex resins in the hydrogen form, Amberlite resin in the sodium form. Transfer of the dry resin from shipping container to glass vessel, and the initial hydration, were carried out under adequate ventilation in a fume hood to eliminate inhalation of particles.

Depending on the size of back-washing chamber to be used subsequently, approximately 350 ml. of settled resin volume for a 2-liter vessel, or 1000 ml. for a 6-liter vessel, was suspended in 12 liters of water and stirred violently with a powerful mixer for 20 minutes to break up clumps and to become hydrated. After settling for 4 hours, the supernatant containing much colloidal material and fines was removed with an aspirator pump and discarded. The resin was then resuspended in water, so that 1 liter of slurry contained approximately 175 ml. of settled resin volume. Pulverized resins were violently agitated for 3 hours to break up partially fragmented pieces to minimize later shedding of very fine particles. Excessive colloidal material, which was present in some powders, was further removed, through repetition once or twice more of the suspension, settling, and removal of the supernatant. This was especially necessary with some lots of the Amberlite resin, which generated a voluminous intractable froth when the dry powder was first hydrated.

THEORETICAL CONSIDERATIONS

A sphere falling freely through a liquid reaches a uniform linear downward velocity defined by the relationship:

$$V_{(\text{Stokes})} = \frac{2G(d_1 - d_2)}{9\eta} \times 60\,a^2 \text{ cm./minute (Stokes' law)} \quad (1)$$

where G is the gravitational constant, η the viscosity of water, d_2 the density of water, and d_1 the density and a the radius in centimeters of the sphere. For convenience in present calculations, the computed velocity was converted to centimeters per minute by the factor 60.

As a corollary, a sphere, when placed in a fluid column within which a vertical gradient of linear velocity has been established, maximum at the bottom and minimum at the top, would rise (with diminishing velocity) to the level where $V_{(\text{up})} = V_{(\text{Stokes})}$. Such a gradient of linear velocity would be approximated in a V-shaped vessel, placed vertically, if fluid were to flow at a constant rate up into the vessel. The minimum linear velocity of flow would occur at the widest part and may be computed with sufficient accuracy, as the volume input per minute to the vessel divided by the volume of a layer 1 cm. in depth located at the maximum internal diameter:

$$V_{(\text{up, min})} = \frac{\text{ml./minute input}}{\Pi R^2_{(\text{max})}} = \text{cm./minute} \quad (2)$$

where $R_{(\text{max})}$ is the internal radius in centimeters, at the widest part of the vessel. It follows that those particles with $V_{(\text{Stokes})}$ greater than or equal to $V_{(\text{up, min})}$ will remain below or at the level of the maximum diameter. Particles with $V_{(\text{Stokes})}$ less than $V_{(\text{up, min})}$ would be carried up and out the top of the vessel. Input flow rate may be calculated which, for any particular sized particle located at the widest diameter, $V_{(\text{up})} = V_{(\text{Stokes})}$, by Equation 3:

$$\text{Input flow rate (ml./minute)} = \Pi R^2_{(\text{max})} \times V_{(\text{Stokes})} \quad (3)$$

Linear flow at $R_{(\text{max})}$ is a limiting rate which governs the size of particle liberated from the vessel. Thus, a graded stratification of particles can be achieved in a relatively short V-shaped column; a considerably longer tube would be needed for comparable stratification in a vessel with parallel walls.

For practical purposes these conditions are realized when water flows at a constant rate up into an ordinary separatory funnel of the Squibb type. With resin in the funnel there appears to be sufficient turbulence generated at the point of entry—i.e., immediately above the stopcock—to reduce axial streaming, so that for practical purposes, approximately uniform velocity is obtained throughout the cross-sectional area at any diameter level. To compute $V_{(\text{Stokes})}$ from Equation 1 for particles of diameter 20, 25, 30, etc., microns, the following values were used: G = 980 cm./sec.², η = 0.00894 poise at 25° C., d_2 = 1.00 at 25° C., and d_1 = 1.21, the wet density of the hydrogen form of the resin. Substituting the computed values of $V_{(\text{Stokes})}$ and the appropriate value of $R_{(\text{max})}$—i.e., 7.4 or 11.5 cm.—for the 2- or 6-liter funnel, respectively, Equation 3, input flow rates were calculated. In Figure 1 the dashed curve, *D–V*, illustrates the relationship between particle size and the calculated input flow rates.

APPARATUS FOR FRACTIONATION AND RESULTS

Depending on the amount of resin processed, either a 2-liter (Corning 6400) or 6-liter (Scientific Glass Apparatus Co., J-1875) pear-shaped separatory funnel of the Squibb type was filled with well-mixed slurry (175 ml. of settled resin volume per liter) and clamped vertically. A source of water at constant temperature was joined by rubber tubing through a flowmeter (Emil Greiner and Co., New York, Predictability flowmeter G9145B with calibration chart), to the lower stem of the funnel. In the present instance a 100-gallon reservoir provided a continuous supply of distilled water at an adequate pressure head and at the ambient temperature (25° C.). Cold tap water, warmed to 25° C. by passing through a heat-exchange coil immersed in a warm bath, could probably be used if the water were not too hard; deionized water is preferable. The top of the funnel was closed with a rubber stopper provided with a right-angled glass tube, and a short length of rubber tubing discharged the outflow into suitable collection vessels.

Flow was regulated by the funnel stopcock and the rate determined by means of the flowmeter and its accompanying calibration chart. Alterna-

Table I. Analysis of Dowex 50-X8 Hydrogen Form Spherical Particles with a 2-Liter Separatory Funnel (Squibb Type) at 25° C.

Fraction No.	Flow Rate, Ml./Min.	Settled Volume Ml.	Settled Volume % of total	Martin's Av. Diameter, Microns	Range of Diameter, Microns	No. of Particles within ±3 Microns of Av. Diameter, %
1	30	17.1	4.1	14.5		
2	65	62.4	14.7	21	16–27	73
3	125	100.0	23.6	28	23–36	74
4	240	92.0	21.7	39	33–45	72
5	390	53.4	12.6	50	44–58	64
6	540	22.6	5.3	63	56–71	79
7	Residue	77.2	18.2			

Table II. Analysis of Amberlite CG-120 Type III, Sodium Form, Pulverized Irregular Particles with a 6-Liter Separatory Funnel (Squibb Type) at 25° C.

Fraction No.	Flow Rate, Ml./Min.	Martin's Diameter, Microns	Range of Diameter, Microns	No. of Particles within ± 3 Microns of Av. Diameter, %
1	47	12.5	9–21	86
2[a]	72	17	11–23	77
3[a]	165	22	14–30	69
4[a]	240	30	20–39	57
5[a]	345	34	24–43	60
6[a]	488	42	36–53	53

[a] Settled resin volumes of the fractions were approximately equal to each other; sum of fractions 2 to 6 constituted approximately 55% of total resin volume.

tively, flow rate was determined by timing the collection of a measured volume of outflow; precise adjustment to a predetermined rate was possible but difficult. The outflow was collected in four precipitating jars (13-liter capacity each) connected by siphons, or in a single vessel from which water was removed by suction through a tubular fritted filter (Corning 35000) or on a funnel with fritted disk (Corning 36060) which was connected to an aspirator pump. Fractions of 20 microns or less were collected in jars; the remaining fractions by either of the two latter procedures.

For an experimental determination of the relationship between particle diameter and flow rate, a schedule of values of the latter for the 2- and 6-liter funnels were read from the *D–V* curves corresponding to 15-, 20-, 25-, etc., micron particles. Fractionation was commenced at a rate which washed out colloidal material and particles of less than 15-micron diameter. The outflow from this wash was opaque with fine particles usually for 20 hours or more, clearing after 30 to 36 hours; it was run to waste. The rate was then increased by stepwise increments in accordance with the above schedules and the corresponding fractions were collected. After each increment in rate, the outflow soon became opaque with particles, which reached a maximum and then diminished slowly. The end point of the fraction was assumed to have been reached when the outflow again contained very few particles.

To determine particle size and distribution within the fractions, photographic analysis was carried out as follows.

Each fraction was stirred vigorously and a drop of suspension, on a glass slide under a cover slip, was photographed. Enlarged prints were made to give an over-all magnification of approximately ×700. A substage micrometer was photographed and enlarged to the same degree. The position of the accidental diameter for each particle in the field was determined by inspection and the distance measured with a millimeter rule. The average, the range, and the frequency distribution of the diameters were computed. Measurement of the enlarged micrometer scale, also with the rule, showed that 1 mm. on the photograph was equivalent to 1.4 microns and the measurements (in millimeters) were accordingly converted to microns by multiplying by 1.4.

Typical analyses are shown in Tables I and II. The diameter range of the pulverized resin fractions was a little greater than those obtained for the spherical resin fractions; a higher percentage of the latter was within ±3 microns of the average. The results of analysis of both spherical and pulverized resins using 2- and 6-liter funnels are summarized in Figure 1.

As the computed *D–V* curve is used only as a guide for selecting a schedule of flow rates within the range appropriate for fractionating the resins under investigation, it is immaterial whether the curve is calculated for the wet density of the hydrogen or the sodium form. In the present instance the *D–V* curve was computed for the hydrogen form. Once the diameter–volume input curve for a particular resin is established experimentally, input flow rates are read from the experimental curve to provide a schedule suitable for isolating any particular size of particle of the resin. Figure 1 shows that the flow rate for any specified diameter of spherical particle (Dowex 50-X8, beads) was approximately 20% greater than that indicated by the *D–V* curve for the 2-liter funnel; approximately 15% greater for the 6-liter funnel. This result is to be anticipated where both the computed and experimentally determined curves are for the hydrogen form of the resin spheres, as input flow rate should be higher to provide a linear flow rate at $R_{(max)}$ such that $V_{(up)} - V_{(Stokes)}$ is a positive quantity for any particular sized particle.

The irregular particles of the Amberlite (sodium form) resin were isolated at flow rates numerically less than the *D–V* rate of corresponding diameters. In the absence of further experimental evidence, it is tentatively suggested that the greater surface area and irregular shape of the pulverized resin particles affect the diameter-volume input relationship more profoundly than the change in density from the hydrogen to the sodium form. Study of this observation seemed to add little to the practical aims of the present investigation and it was not pursued further. The particles of pulverized Dowex 50W (hydrogen form) gave a curve similar to the pulverized Amberlite, but the present work offers no worth-while explanation of the differences shown in Figure 1, between these two resins.

A total volume of approximately 40 liters of water was necessary to isolate each fraction from the 2-liter funnel, approximately 60 liters from the 6-liter funnel.

CONDITIONING OF RESINS FOR CHROMATOGRAPHY

Each fraction was conditioned by washing under gravity on a funnel with fritted disk (medium porosity) with 6*N* hydrochloric acid until the effluent was no longer colored yellowish green. About 9 liters of acid per 500 ml. of resin was needed. The colored substances removed were presumably largely copper or iron, which are known to be introduced as contaminants from the machinery during pulverization. Residual acid was washed out with water and the resin converted to the sodium form by 9 liters of 2*N* sodium hydroxide run through under gravity. The alkaline solution contained 0.1% Sequestrene (Geigy Chemical Co., 89 Barclay St., New York 8, N. Y.) to ensure removal of traces of those metals, if any were present, whose chlorides were insoluble in the hydrochloric acid. The first half of the alkaline effluent was slightly colored and somewhat opalescent; the last half became progressively clearer. At this stage, some of the pulverized fractions showed more fine particles, apparently generated as a result of the cycling washes. As it was observed that excessive fines produced higher column operating pressures without contributing to column performance, the above fractions were suspended in 10 liters of 0.2*N* sodium hydroxide, and allowed to settle for 2 hours. The fines in suspension were removed with the supernatant, which was discarded. Occasionally, repetition of this procedure was necessary. When once clean, the fractions, on being suspended in 10 liters of water, exhibited few fines in the supernatant after 60 minutes' settling, and none after 4 hours, nor did they show appreciable production of more fines thereafter.

AMINO ACID CHROMATOGRAPHY

Chromatographic columns, 0.9 × 100 cm., were poured in 20-cm. sections with suspensions of 1 ml. of resin per 2 volumes of 0.2*N* sodium hydroxide, according to the detailed directions of Moore, Spackman, and Stein (*4*).

The bottom of the column was plugged while the initial 60 ml. of slurry was poured in, after which the plug was removed. Each section was settled under air pressure of 7 pounds per square inch until well packed. Residual supernatant solution was removed by suction and the next section was poured directly on the moist surface of the exposed resin bed.

The columns were jacketed and operated at 50° C., and solutions were pumped through at 25 ml. per hour, using a Minipump (225 ml. per hour capacity, Milton Roy Co., Philadelphia, Pa.). This corresponds to a flow rate of 39.3 ml. per sq. cm. of resin bed per hour. The columns were conditioned before analysis with 100 ml. of 0.2*N* sodium hydroxide, followed by 0.1*M* citrate buffer (0.2*N* with respect to sodium), pH 2.90, until the pH (glass electrode) of the effluent was the same. Two cubic centimeters of a test solution of 29 amino acids dissolved in 0.1*N* hydrochloric acid was placed on the column and forced in under air pressure, 10 pounds per square inch, followed by a wash of 1 ml. of 0.1*N* hydrochloric acid. Development of the column was begun with 0.1*M* citrate (0.2*N* sodium), pH 3.10 or 3.00, as noted below. At ml. 400 the buffer was changed to 0.1*M* citrate (0.2*N* sodium), pH 4.15, and again at ml. 600 to 0.4*M* citrate (0.8*N* sodium), pH 5.00. Brij-35 (Atlas Powder Co., Wilmington, Del.) was present in 1% concentration in all conditioning solutions and buffers. The preparation of buffer solutions and general column operation have been described by Moore and Stein (*6*). The pH of buffers, their sequence, and rate of flow were those investigated in this laboratory. Column effluent was collected in 2-ml. fractions and analyzed in accordance with the principles described by Moore and Stein (*5, 6*).

RESULTS OF CHROMATOGRAPHY

In Figure 2, the chromatogram obtained with a column of 55-micron Dowex 50-X8 spherical particles is compared with that obtained from a column of 25-micron particles of the same resin, both columns being operated at a flow rate of 25 ml. per hour. The superior resolving power of the 25-micron column is clear. In this chromatogram the position of methionine was not established. Whether it was degraded on the column or incorporated with the cystine was not determined.

Table III. Conditions for Isolating 20- to 40-Micron Particles from Pulverized Resin Powders

Fraction No.	Flow Rate, Ml./Min.	Martin's Diameter, Microns	Procedure
	2-Liter Separatory Funnel		
1	50	20	Discard
2	250	20–40	Collect
3	. . .	Residue	Discard
	6-Liter Separatory Funnel		
1	120	20	Discard
2	480	20–40	Collect
3	. . .	Residue	Discard

Resolution of a column of 45-micron particles was intermediate between these two. Resolution of a column of 35-micron particles was comparable to that obtained with the 25-micron particles, but the peaks were broader. This sequence of chromatograms provided evidence that as particle size decreased, resolution progressively improved, and that at the relatively high rate of flow of 25 ml. per hour, optimum resolution was obtained with spherical particles 25 microns in diameter. [Spackman, Moore, and Stein (*8*) have found that at lower flow rates of 10 to 15 ml. per hour, comparable resolution may be obtained with coarser particles.] No improvement in either resolution or form of the peaks was observed when 20-micron particles were used, but column operating pressures were increased from 45 to 65 pounds per square inch. A similar series of chromatograms was obtained with columns packed with pulverized Amberlite CG-120 Type III, or with Dowex 50W-X8. Optimum resolution with the pulverized resins was obtained with particles of 30-micron diameter when a flow rate of 25 ml. per hour was used. Cystine and valine were not resolved when column development commenced with pH 3.10 buffer. In later experiments, these two amino acids were resolved with pH 3.00 buffer, as illustrated in Figure 3.

In Figure 3, chromatograms obtained with a column of Amberlite CG-120 Type III, 30-micron particles, before and after washing the resin with 6*N* hydrochloric acid and 2*N* sodium hydroxide are compared. The marked impairment of resolution of the unwashed resin could be attributable to the large amount of metallic ions present, perhaps also to other unidentified substances. The necessity for thorough cleaning of the resins is strikingly apparent.

After the effect of particle size on resolution had been determined with the narrow range fractions for column flow rates of 25 ml. per hour (Figure 2), a 20- and 40-micron fraction of the Amberlite resin, which embraces the widest useful range of particle diameter for this rate of flow, was run for comparison. As indicated in Table II, particle size of this resin was evenly distributed throughout this range. The resolution obtained was similar to that shown in Figure 3,*B*.

Chromatograms obtained with 31- and 39-micron fractions of Dowex 50W-X8 were compared with that of a 20- to 45-micron fraction of the same resin, all operated at flow rates of 25 ml. per hour. All three chromatograms showed equally good resolution of amino acids and the resolution did not differ from that obtained with the Amberlite resin, except in a few minor respects. A single chromatogram using Permutit Q-ZP, 20- to 42-micron range, indicated that this resin was probably serviceable, but it was not studied further.

The results demonstrate that good chromatograms can be obtained from 0.9 × 100 cm. columns at flow rates of 39.3 ml. per sq. cm. per hour with 20- to 45-micron fractions of the pulverized resins, if there is uniform distribution of size throughout the range. If size distribution were shifted markedly toward the upper limit, some loss of resolving power would occur at this high rate of flow. For routine preparative purposes, a more conservative range—e.g., 20 to 40 microns—would ensure columns with good resolving power and sharp peaks, even if a disproportionate distribution toward the upper limit of the range were present, and it would render unnecessary a more detailed preliminary particle analysis. Similar considerations suggest that a 20- to 35-micron range of spherical particles should be most serviceable. These wider diameter ranges allowed flow rates of 25 ml. per hour through 0.9 × 100 cm. columns, at moderate operating pressures of 40 to 50 pounds per square inch. They also constituted reasonable proportions of the particular lots of the unclassified resins investigated: 50% of the Amberlite CG-120 Type III, 75% of the Dowex 50W-X8.

Table III summarizes recommended procedures for isolating 20- to 40-micron particles from pulverized resins, with the 2- and 6-liter separatory funnels. Within the practical limits desired, these conditions are applicable to either the hydrogen or sodium form.

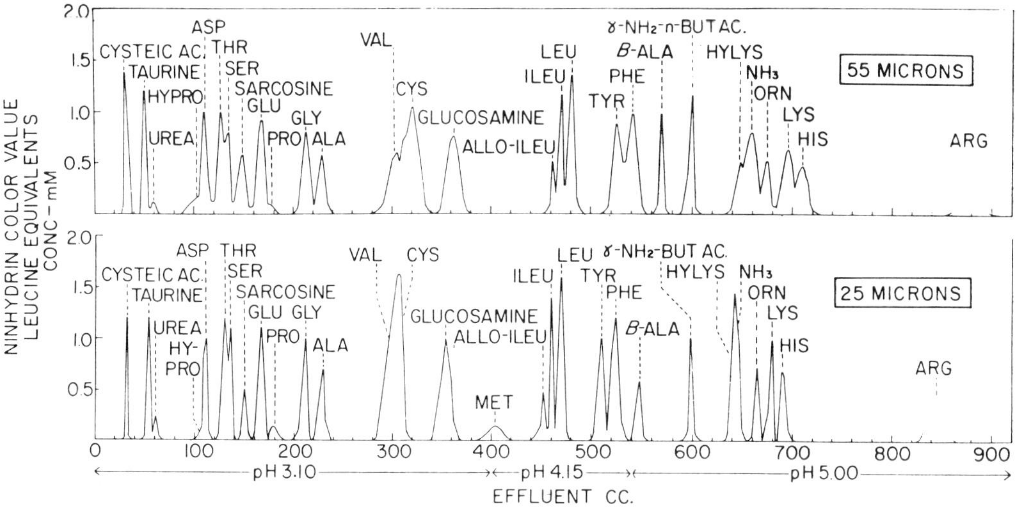

Figure 2. Chromatograms obtained with 0.9 × 100 cm. columns of Dowex 50-X8 spherical particles, 55- and 25-micron diameter

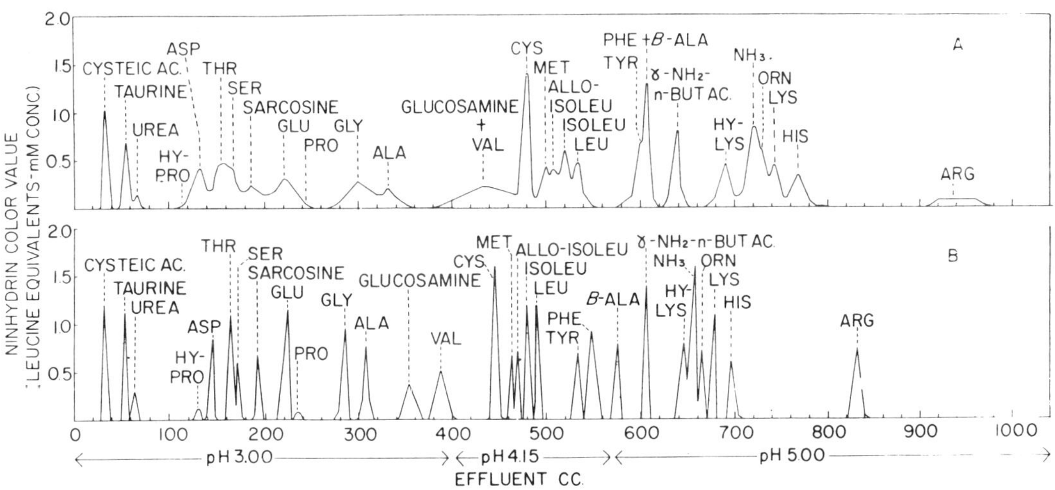

Figure 3. Chromatograms obtained with 0.9 × 100 cm. columns of Amberlite CG-120 Type III, 30-micron particles

A. Before washing *B.* After washing

DISCUSSION

In biochemical laboratories, following industrial custom, the current designation of resin particles has been mesh number, determined by sieving through standard screens. When flow rates are high, optimum performance of ion exchange columns in the separation of amino acids requires greater precision in the control of particle size than can be obtained by sieving. Confusion of terminology has also arisen. In many cases, industrial practice has been to classify the unsulfonated polymer beads (which cannot be hydrated) by screening. After sulfonation, the dry screen (unsulfonated) designation is still applied, even though the polymer will swell rapidly to approximately double its former size on being hydrated. Hence, unless a wet screen classification after sulfonation is specified, particles as much as twice the expected diameter may unwittingly be obtained. For these reasons, it would seem advisable, in the field of analytical chromatography at least, to abandon sieve number and refer to particle diameter (in microns) of the sodium form of the fully hydrated sulfonated resin.

The evidence presented confirms the fact that particle size is an important factor, and shows further that a relatively restricted range is necessary for optimum column performance at high flow rates. There is little else on which to base a choice between current commercial sources of 8% cross-linked sulfonated polystyrene cationic exchange resins for amino acid chromatography, for optimum performance, other than appropriate particle size. Irregular 30-micron particles gave resolution comparable to that obtained with spherical 25-micron particles. A possible explanation may be that equilibrium can take place more rapidly in nonspherical particles than in spheres of the same mass, as the time necessary for diffusion to their centers would be governed by the smallest of the three dimensions. Columns prepared with irregular particles gave higher operating pressures than those filled with spherical particles of the same diameter, presumably because of closer packing of the irregular shapes.

The general relationships among column length, particle size, rate of flow, operating pressure, and column performance merit consideration. If column length is increased—e.g., from 100 to 150 cm.—to obtain higher resolving power, the operating pressure is increased proportionately, and this may be undesirable. Because resistance to flow at constant rate, and therefore pressure, bears an inverse relationship to particle size, the operating pressure can be lowered by employing larger particles. Increase in particle size will, however, partially nullify the gain in resolving power. In turn, this diminution in resolving power may be offset by a reduction in flow rate. Application of these general principles allows resolving power to be manipulated within limits, by the selection of appropriate values for these related parameters. The effects of temperature, pH, and ionic strength of developing solutions on resolving power are beyond the scope of the present discussion.

Some minor differences between resins, probably owing to factors other than particle size, were noted. The resolution of threonine and serine was better on Amberlite CG-120 Type III and Dowex 50-X8 (bead form) than on Dowex 50W-X8 (irregular particles). On the Amberlite columns, the separation of threonine and serine at 50° C. was marginal; at 40° C., complete. They were not separated at either 40° or 50° C. on Dowex 50W-X8. The resolution of tyrosine from phenylalanine on Dowex 50W-X8 was good, but overlapping of β-alanine with tyrosine was sometimes encountered. On the Amberlite resin, the resolution of tyrosine from phenylalanine was not as complete, but β-alanine always emerged well after phenylalanine.

Resolution of the basic amino acids was equally satisfactory on both resins. Ammonia usually overlapped hydroxylysine on Dowex 50W. On the Amberlite resin, ammonia usually emerged with ornithine or with lysine, or in some intermediate position between these two amino acids.

At the commencement of this work, pulverized resins were available only in limited experimental quantities. Now, pulverized resins for analytical chromatography are available from commercial sources.

ACKNOWLEDGMENT

The author wishes to express his appreciation of cooperative exchange of information with W. H. Stein, S. Moore, and D. H. Spackman, Rockefeller Institute, New York, N. Y.

Thanks are due to the Resinous Sales Development Department, Rohm & Haas Co., Philadelphia, Pa., for samples of the Amberlite CG-120 Type III, technical data, and helpful cooperation. Thanks are due to the Organic Sales Department, Dow Chemical Co., Midland, Mich., for samples of Dowex 50-X4, -X8, -X12, Dowex 50W-X8, -X12, and for technical data. The Bio-Rad Laboratories, Berkeley, Calif., supplied pulverized Dowex 50W-X8 and the Permutit Co., Birmingham, N. J., donated the Permutit Q-ZP.

Assistance in the amino acid chromatography and the preparation of illustrations was rendered by Roberta A. Anderson.

LITERATURE CITED

(1) Chamot, E. M., Mason, C. W., "Handbook of Chemical Microscopy," 2nd ed., Vol. I, pp. 412–15, Wiley, New York, 1948.
(2) Martin, G., Blythe, C. E., Tongue, H., *Trans. Ceram. Soc.* (*Eng.*) **23**, 61–118 (1924).
(3) Mayer, S. W., Tompkins, E. R., *J. Am. Chem. Soc.* **69**, 2866 (1947).
(4) Moore, S., Spackman, D. H., Stein, W. H., Anal. Chem., in press.
(5) Moore, S., Stein, W. H., *J. Biol. Chem.* **176**, 367 (1948).
(6) *Ibid.*, **192**, 663 (1951).
(7) *Ibid.*, **211**, 893 (1954).
(8) Spackman, D. H., Moore, S., Stein, W. H., Anal. Chem., in press.
(9) Spackman, D. H., Stein, W. H., Moore, S., *Federation Proc.* **15**, 358 (1956).

Received for review August 5, 1957. Accepted January 27, 1958.

39

Reprinted from *Naturwiss.*, **46**(3), 110, (1959)

KONTINUIERLICHE AMINOSÄURETRENNUNG AN EINER AUSTAUSCHERSÄULE DURCH GRADIENTENELUTION MIT EINEM PUFFER *

Hermann Stegemann

Die Arbeiten von PARTRIDGE[1]) und vor allem von MOORE und STEIN[2a]) haben es ermöglicht, eine Bausteinanalyse von Proteinen durch quantitative Trennung der Aminosäuren an Ionenaustauschern durchzuführen. Die jüngst beschriebene Ausführungsform sieht die Elution der Aminosäuren an einem Sulfogruppen enthaltenden Polystyrolaustauscher vor. An einer Säule werden die sauren und neutralen, an einer anderen die basischen Komponenten getrennt. Zur Elution werden mindestens drei verschiedene Puffer benötigt, die man im geeigneten Moment wechselt. Von einem partiellen Gradientensystem scheinen die Autoren[3]) wieder abgerückt zu sein.

Wir haben seit 2 Jahren die theoretisch Erfolg versprechende totale Gradientenelution untersucht mit dem Ziel, an *einer* Säule bei *einer* Temperatur mit nur *einem* Puffer alle Aminosäuren zu trennen. Der Vorteil liegt nicht allein in der stark vereinfachten Durchführung des Versuches, sondern auch in der automatisch erreichten Reproduzierbarkeit aller Elutionsmaxima der Komponenten. Es ist ohne Extinktionsschreiber sehr schwer, den passenden Zeitpunkt des Pufferwechsels zu bestimmen. Auch im ersteren Falle bedarf es eines großen mechanischen Aufwandes[2b]), die Puffer zu gegebener Zeit umzuschalten. Besonders hier wird die ununterbrochene Elution von der ersten bis zur letzten Fraktion Vorteile bringen.

Als Austauscher benutzen wir genau 100 ml von den feinsten Teilchen befreiten DOWEX 50 × 8, 200 bis 400 mesh, in der mit 0,2 n Natronlauge ins Gleichgewicht gebrachten Form. Die Teilchengröße liegt zwischen 45 und 100 μ, in der Hauptmenge zwischen 50 und 70 μ Durchmesser. Der Austauscher befindet sich in einer auf 40° temperierten, 9 bis 10 mm weiten und 160 cm langen Säule (Liebig-Kühler). Den unteren Abschluß bildet ein kleiner Pfropfen aus Quarzwolle und ein Hahn, der durch eine Polyäthylenkapillare (Innendurchmesser 0,5 mm) mit dem Fraktionssammler verbunden ist. Das obere Ende der Säule trägt eine NS 14-Hülse, in der ein Polyäthylennormalschliff durch eine gleiche Kapillare mit dem Mischgefäß für den Puffer verbunden ist. Als Puffer benötigen wir einen reinen Citratpuffer vom $p_H = 3{,}30 \pm 0{,}02$, der aus 0,2 n Natronlauge und Citronensäure hergestellt wird. Dieser Puffer befindet sich in einer handelsüblichen, 1300 ml fassenden Auslaufflasche mit NS 29-Hals und einem daraufgesetzten Tropftrichter von 680 bis 700 ml Inhalt. Die Lösung wird in

**Editor's Note:* An English translation follows this article.

beiden Gefäßen magnetisch gerührt. Auf den Trichter wird ein zweiter gesetzt, der eine 2,5 m Natriumacetatlösung enthält.

Weiter vereinfachten wir das Verfahren dadurch, daß wir keine Antioxydantien zusetzen. Der Verzicht auf Thiodiglykol bedeutet, auch keine luftfreien Puffer benutzen zu müssen, ist aber mit dem leichter wiegenden Nachteil erkauft, Methionin nur als Methioninsulfoxyd bestimmen zu können. Cystein wird als Cystin erfaßt. Tryptophan ist weder nach unserem noch nach den früher beschriebenen Verfahren quantitativ wiederzugewinnen. Aminosäureamide wurden in den Kreis der Untersuchungen nicht mit einbezogen, da sie ebenso empfindlich und schneller bestimmt werden können[4]). Polyäthylenglykoläther wird nicht zugesetzt, weil sich beim Abtropfen von Polyäthylenkapillaren die Veränderung der Oberflächenspannung des Puffers nicht so stark auf die Tropfengröße auswirkt. Wir verwenden außerdem eine Ninhydrinbestimmung, bei der sich die Veränderung der Tropfengröße nicht wesentlich bemerkbar macht. Schließlich verzichten wir auf die Anwendung von Druck, um den apparativen Aufwand klein zu halten, und nehmen — bedingt durch die gröberen Austauscherpartikeln und langsamere Gleichgewichtseinstellung — eine doppelt so lange Analysendauer (etwa 4 Tage) in Kauf.

Ist das Proteinhydrolysat auf die mit dem Citratpuffer äquilibrierte Austauschersäule gegeben worden, so wird die Säule mit dem Auslauf des Mischgefäßes verbunden. Unter ständigem Einrühren der Acetatlösung eluiert man mit dem Citratpuffer, dessen p_H und Natriumionenkonzentration in charakteristischer Weise ansteigt. Nach 860 ml sind alle Aminosäuren bis auf Arginin (und Tryptophan) eluiert. Arginin erscheint erst in der 1160. bis 1180. Fraktion. Um die Analyse zu verkürzen und die ohnehin notwendige Wäsche mit 0,2 n Natronlauge einzuleiten, wird nach der 840. Fraktion der Zufluß des Puffers beendet und die Säule mit Natronlauge durchspült. Nach weiteren 40 ml tritt Arginin aus.

Sammelt man jeweils 2 ml, so ist nach etwa 450 Fraktionen die Analyse beendet. Zwecks besonders sauberer Trennung bevorzugen wir meist Fraktionen von 1 ml. Die „peaks" der eluierten Komponenten sind scharf, praktisch symmetrisch und gut reproduzierbar, die Halbwertsbreite bei 1 μMol Substanz beträgt im Anfang des Versuches etwa 4 ml, am Schluß etwa 9 ml. Die geringste, noch hinreichend genau bestimmbare Menge liegt bei 0,5 μMol Aminosäure. Die Elutionsmaxima für die Komponenten eines Testgemisches sind in der Tabelle angeführt. Da nach allgemeiner Erfahrung sich DL- und L-Aminosäuren an diesem Austauscher gleich verhalten, werden sie in der Tabelle nicht unterschieden. Die durch die Blindwerte gegebene Basislinie liegt bei 0,07 bis 0,08 Extinktionseinheiten für 16 mm-Röhrchen und ist annähernd konstant über den ganzen Versuch. Detailbeschreibungen der Apparatur wie des Analysenverfahrens, welches mit Natriumdithionit unter Zusatz von Fluorid und hoher Acetatpufferkonzentration arbeitet, folgen an anderer Stelle.

Tabelle. *Elutionsmaxima von ninhydrinpositiven Verbindungen, geordnet nach dem Elutionsvolumen*

Säule mit 100 ml DOWEX 50 × 8, Fraktionen je 1 ml, Extinktion gemessen mit Filter S 57 (Zeiss), Hydroxyprolin und Prolin mit Filter Hg 436.

Cysteinsäure	37	Norvalin	337
Taurin	64	Alloisoleucin	348
Methioninsulfoxyde	101	Isoleucin	368
	106	Leucin	400
Hydroxyprolin	111	Norleucin	415
Asparaginsäure	115	Tyrosin	490**)
Threonin	134	Phenylalanin	500**)
Methioninsulfon	134	β-Alanin	517
Serin	143	β-Aminoisobuttersäure	543
Sarkosin	160	γ-Aminobuttersäure	656
Glutaminsäure	178	δ-Hydroxylysin	671
Prolin	186*)	Allo-δ-hydroxylysin	682
Glycin	219	$NH_4^{(+)}$	737
α-Aminoadipinsäure	225	Lysin	805
Alanin	230	Histidin	848
Cystin	239	Tryptophan	970***)
α-Amino-n-buttersäure	252	Arginin	1167
Valin	271		
α,ε-Diaminopimelinsäure	331		

*) Von zahlreichen europäischen und US-amerikanischen L-Prolinchargen war nur das L-Prolin der Firma SANABO, Wien, einheitlich und enthielt weniger als 0,02% Hydroxyprolin. — **) Mangelhafte Trennung. — ***) Nicht scharf.

Herrn Professor Dr. K. THOMAS danke ich für die stete Förderung dieser Untersuchungen, Fräulein G. BERNHARD für interessierte und unermüdliche Mitarbeit, Herrn H. J. DÜNGER für Hilfe in der letzten Phase der Entwicklung. Der Bergbauberufsgenossenschaft, Bochum, und der Montanunion, Luxemburg, danken wir für finanzielle Unterstützung.

Medizinische Forschungsanstalt der Max-Planck-Gesellschaft, Silikoselabor, Göttingen

Eingegangen am 20. Januar 1959

[1]) PARTRIDGE, S.M.: Biochem. J. **44**, 521 (1949). — [2]) MOORE, S., D. H. SPACKMAN u. W.H. STEIN: Anal. Chem. a) **30**, 1185 (1958) [dort auch frühere Literatur der Autoren]; b) **30**, 1190 (1958). — [3]) MOORE, S., u. W.H. STEIN: J. biol. Chem. **211**, 893 (1954). — [4]) STEGEMANN, H.: Hoppe-Seylers Zeitschr. physiol. Chemie **312**, 255 (1958).

39

CONTINUOUS AMINO ACID ANALYSIS ON AN ION-EXCHANGING COLUMN BY GRADIENT ELUTION WITH A BUFFER

Hermann Stegemann

This condensed translation was prepared expressly for this Benchmark volume by Harold F. Walton, University of Colorado, Boulder, Colorado, from "Kontinuierliche Aminosäuretrennung an einer Austrauschersäule durch Gradient-enelution mit einem Puffer," Naturwiss., **46,** *110 (1959)*

The work of Partridge[1] and above all that of Moore and Stein [2a] has made it possible to analyze the building blocks of proteins by the quantitative separation of amino acids on ion exchangers. The exchangers used today are of the sulfonated polystryrene type. The acidic and neutral components are separated on one column, the basic components on another. At least three different buffers are required, changed at appropriate intervals. This arrangement is apparently preferred to gradient elution.[3]

For the past 2 years we have been investigating a gradient system, believing that it should be possible to separate all the amino acids on *one* column, at *one* temperature and with *one* buffer. The advantage of such a system is not only its simplicity, but the fact that all the elution volumes should be automatically reproducible. Without continuous recording it is very difficult to know the right moment to change the buffer, and even with recording, it takes a complicated mechanism[2b] to change the buffer at the right time. Continuous, uninterrupted elution, from the first component to the last, would have many advantages.

For our exchanger we used exactly 100 ml of Dowex-50x8, 200–400 mesh, freed from fines and equilibrated with 0.2 *N* sodium hydroxide. The particle size was 45 to 100 μm, with most of the material between 50 and 70 μm. The resin was placed in a column 9 to 10 mm in diameter and 160 cm long, maintained at 40° by a condenser jacket. The column was closed at its lower end by a plug of quartz wool and a stopcock from which a polyethylene tube (internal diameter of 0.5 mm) led to a fraction collector. The upper end carried a cap with a polyethylene tube leading to the mixing chamber which contained the buffer. The buffer was prepared from citric acid and 0.2 *N* sodium hydroxide and had pH = 3.30 ± 0.02; it was contained in a 1,300-ml flask, above which was a 700-ml dropping funnel that contained more of the same buffer. Both vessels were magnetically stirred. A second dropping funnel containing 2.5 *M* sodium acetate was placed above the first, and fed into it.

We simplified the procedure by omitting antioxidants. We added no thiodiglycol and our buffers were not air-free. Thus we could only determine methionine as methionine sulfoxide, and cysteine was recovered as cystine. Tryptophan was not recovered quantitatively by the new method, but it was not so recovered by the old methods, either. The amides of amino acids were not included in this study because we already knew how to determine them rapidly at low concentrations.[4]

Polyethylene glycol ether was not added, because the changing surface tension of the buffer did not have much effect on the size of the drops from polyethylene capillaries. Moreover, we used a ninhydrin method that was not significantly affected by variations in drop size. Finally, to keep the apparatus simple we did not apply pressure, and accepted in exchange a twofold increase in the analysis time (to 4 days). This was necessary in any case because of the relatively large particle size of the exchanger.

After equilibrating the column with the citrate buffer, the protein hydrolyzate was placed on the column. Then the column was connected to the outlet of the mixing chamber and was eluted with citrate buffer into which sodium acetate solution was constantly being stirred. The pH and sodium-ion concentration rose in a steady and reproducible manner. After 860 ml, all the amino acids were eluted except arginine and tryptophan. Arginine first appeared in the fractions from 1,160 to 1,180 ml. To shorten the analysis and to start the wash with 0.2 *N* sodium hydroxide, which was necessary in any case, we stopped passing buffer after 840 ml and began passing sodium hydroxide. Arginine appeared after another 40 ml.

If one collected 2 ml at a time, only 450 fractions were needed; however, we usually took 1-ml fractions to get sharper separations. The eluate peaks were sharp, almost symmetrical, and well reproducible. With 1 μmol of each component the half-width of the peaks was 4 ml at the start, and about 9 ml at the end of the run. The smallest quantity that we could measure accurately was 0.5 μmol. The elution maxima for the components of a test mixture are shown in the table. No distinction was observed between *dl*- and *l*-amino acids. The base-line absorbance measured in 16-mm tubes was 0.07 to 0.08 units and was nearly constant throughout the run. A detailed description of the apparatus, and also of an analytical procedure in which we added fluoride, dithionite, and a high concentration of acetate, will be published elsewhere.

[*Editor's Note:* References for this translation will be found in the preceding original article.]

Part XI
CARBOHYDRATES

Editor's Comments on Papers 40 Through 43

40 **KHYM and ZILL**
The Separation of Monosaccharides by Ion Exchange

41 **RÜCKERT and SAMUELSON**
Excerpts from *Die Verteilung von Äthylalkohol und Wasser bei Ionenaustauschern auf Harzbasis*

42 **RÜCKERT and SAMUELSON**
Die Verteilung von Glukose bei Ionenaustauschern auf Harzbasis in Äthylalkohol-Wassergemischen
English translation: *The Distribution of Glucose Between Ion-Exchange Resins and Alcohol–Water Mixtures*

43 **LARSSON and SAMUELSON**
An Automated Procedure for Separation of Monosaccharides on Ion Exchange Resins

Organic compounds that are not ionized themselves may sometimes combine reversibly with simple inorganic ions to form charged complexes, and these charged complexes can be attached to ion-exchange resins. For example, aldehydes combine with bisulfite ions to form anionic bisulfite complexes, which can be bound by an anion-exchange resin. A number of early chromatographic separations of organic compounds depended on this principle.

Sugars and polyhydric alcohols form borate complexes. Therefore, these compounds are absorbed on strong-base, anion-exchange resins in their borate form, and can be eluted selectively by borate buffers. The first chromatographic separations of these compounds were made with ^{14}C -labeled sugars at Oak Ridge National Laboratory by J. X. Khym and L. P. Zill. Their original paper is reproduced here. More detailed studies, to which they refer, were published in the *Journal of the American Chemical Society* (Khym and Zill, 1952).

Soon after this work was published, several investigators noticed that sugars were bound by resins to some extent even in the absence of borate, and that the binding was greatly strengthened by adding alcohol. One of these investigators was Olof Samuelson, the pioneer in the analytical application of ion exchange whose work was cited in our Introduction. He became interested in carbohydrates through his work with the Swedish paper industry, and his work on the ion-exchange chromatography of carbohydrates has continued to this day. Out of a massive quantity of publications we have selected three for reproduction here.

The first paper by Ruckert and Samuelson (1954) reported that sugars were strongly absorbed by anion- and cation-exchangers from solvent mixtures containing 85 to 90 percent ethyl alcohol by weight. Two papers in 1957 (Papers 41 and 42) described a study of the absorption phenomenon. Paper 41, of which we reproduce only three pages, describes the partition of water and alcohol between resins and solution, and is a definitive reference on this subject. Mixed solvents (alcohol–water, acetone–water, and many other combinations) are widely used in ion-exchange chromatography, both for organic and inorganic solutes (see Parts XII and VI), and anyone who works in this field must constantly keep in mind the graphs on pages 310 and 311 of Paper 41. The ordinates are mole fractions of ethyl alcohol in the solution inside the resin; the abscissae are mole fractions in the external solution. The hydrophilic nature of the resin is clearly evident, and so is the effect of the resin counter-ions.

Many of the data of Paper 41 are incorporated into the tables and graphs of Paper 42, which we have translated in part.

Paper 43 describes an automated system of analysis of monosaccharides, using a resin of small particle size, high pressure, and elevated temperature. A specially prepared, macroporous resin was used. Gel-type resins, as Paper 42 shows, shrink as the alcohol concentration is raised, and at high alcohol concentrations the diffusion of sugars becomes very slow. The use of a macroporous resin avoids these difficulties. An anion-exchange resin in the sulfate form gives the best results.

A comprehensive review of the ion-exchange chromatography of sugars and their derivatives has been given by Samuelson (1969).

REFERENCES

Khym, J. X., and L. P. Zill. 1952. *J. Amer. Chem. Soc.*, *74*, 2090.

Ruckert, H., and O. Samuelson. 1954. *Svensk Kem. Tidskr.*, *66*, 337.

Samuelson, O. 1969. In J. A. Marinsky (ed.), *Ion Exchange: A Series of Advances*, Vol. 2, New York, Marcel Dekker.

40

Reprinted from *J. Amer. Chem. Soc.*, **73**, 2399 (1951)

THE SEPARATION OF MONOSACCHARIDES BY ION EXCHANGE[1]

Sir:

The biosynthetic preparation of C^{14}-labeled sugars requires a method for the separation of individual monosaccharides from their mixtures The methods used up to the present time have, in general, been based upon adsorptive chromatography and both columnar and paper partition chromatography.[2] In many cases they are non-quantitative, or difficult, or impossible to apply to preparative isolations, or include undesirable manipulative procedures and chemical conversions.

Since borate ion reacts with sugars to produce negatively charged sugar–borate complexes,[3] it occurred to us that such sugar–borate complexes might be separable by the technique of ion exchange in a fashion similar to that used in separating nucleotides and related compounds.[4]

It was found that fructose, glucose, mannose and galactose, dissolved in weak sodium borate solutions, are quantitatively adsorbed on strong-base anion exchangers. Elution was carried out with dilute sodium borate solutions and the effluent fractions were analyzed for sugar by the quantitative anthrone method of Dreywood[5] as developed by Morris.[6] Identification of the sugars was accomplished by paper chromatography.[7]

(1) Work performed under Contract Number W 7405-Eng-26 for the Atomic Energy Commission.

(2) S. Udenfriend and M. Gibbs, *Science*, **110**, 708 (1949); W. W. Binkley and M. L. Wolfrom, "Chromatography of Sugars and Related Substances, Sugar Research Foundation, Inc., Scientific Report Series No. 10 (1948); B. W. Lew, M. L. Wolfrom and R. M. Goepp, Jr., This Journal, **68**, 1449 (1946); G. R. Noggle and R. A. Bolomey, *Plant Physiol.*, in press.

(3) J. Böeseken, *Advances in Carbohydrate Chem.*, **4**, 189 (1949); H. S. Isbell, J. F. Brewster, N. B. Holt and H. L. Frush, *J. Res. Nat. Bur. Standards*, **40**, 129 (1948); Y. Tuzuki, *Bull. Chem. Soc., Japan*, **16**, 23 (1941).

(4) W. E. Cohn, This Journal, **72**, 1471 (1950); E. Volkin, J. X. Khym and W. E. Cohn, *ibid.*, **73**, 1533 (1951).

(5) R. Dreywood, *Ind. Eng. Chem., Anal. Ed.*, **18**, 499 (1946).

(6) D. L. Morris, *Science*, **107**, 254 (1948).

(7) S. M. Partridge, *Biochem. J.*, **42**, 238 (1948).

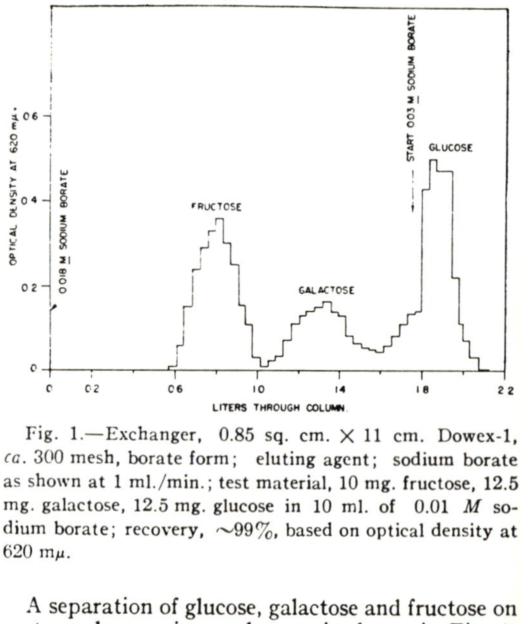

Fig. 1.—Exchanger, 0.85 sq. cm. × 11 cm. Dowex-1, *ca.* 300 mesh, borate form; eluting agent; sodium borate as shown at 1 ml./min.; test material, 10 mg. fructose, 12.5 mg. galactose, 12.5 mg. glucose in 10 ml. of 0.01 *M* sodium borate; recovery, ~99%, based on optical density at 620 mμ.

A separation of glucose, galactose and fructose on a strong-base anion exchanger is shown in Fig. 1. Recoveries of the individual sugars were essentially quantitative.

While a number of variables remain to be explored and the method has not yet been extended to include other hexoses, the pentoses, and the disaccharides, it has promise of wide application to many of the separations and analytical problems involved in investigations of sugars and other compounds which are capable of forming charged borate complexes. Further studies are in progress and will be published later.

Biology Division
Oak Ridge National Laboratory
Oak Ridge, Tennessee

Joseph X. Khym
Leonard P. Zill

Received March 17, 1951

41

Reprinted from *Acta Chem. Scand.*, **11**(2), 303, 310–312 (1957)

Die Verteilung von Äthylalkohol und Wasser bei Ionenaustauschern auf Harzbasis

HEINZ RÜCKERT und OLOF SAMUELSON

Institut für Technische Chemie, Chalmers Technische Hochschule, Göteborg, Schweden

Die Verteilung von Äthylalkohol und Wasser wurde für Kationenaustauscher und Anionenaustauscher in verschiedenen Ionenformen bestimmt. Die Alkoholkonzentration der Harzphase ist im allgemeinen kleiner als in der Aussenlösung (Aussalzen). Je höher die Ionenkonzentration der Austauscherphase ist (kleine Quellung), umso mehr wird der Alkohol ausgesalzt. Unter gewissen Umständen kann man aber auch ein Einsalzen erhalten. Die Quellung der Austauscher nimmt, je nach Beladungszustand und Vernetzungsgrad, entweder mit steigender Alkoholkonzentration kontinuierlich ab oder durchläuft ein Maximum.

Es ist bekannt, dass sich organische Moleküle in wässeriger Lösung ungleichmässig zwischen einem Ionenaustauscher und der ihn umgebenden Lösung (Aussenlösung) verteilen. Das System Äthylalkohol-Wasser ist früher von Gregor und Mitarbeitern [1] untersucht worden und diese Untersuchungen zeigten, dass der Alkoholgehalt der Harzphase (Innenlösung) niedriger ist als in der Aussenlösung.

In einer früheren Arbeit [2] wurde gezeigt, dass Ionenaustauscher zur Adsorption von Zuckerarten aus Mischungen von Wasser und Alkohol verwendet werden können und dass diese Adsorption von der Art der Gegenionen stark abhängig ist. Um die Grundlagen einer Diskussion über die Zuckeradsorption zu schaffen, wird in der vorliegenden Arbeit über die Verteilung von Wasser und Äthylalkohol berichtet und zwar für Austauscher in denselben Ionenformen, wie sie bei den Zuckeruntersuchungen verwendet wurden.

[*Editor's Note:* Material has been omitted at this point.]

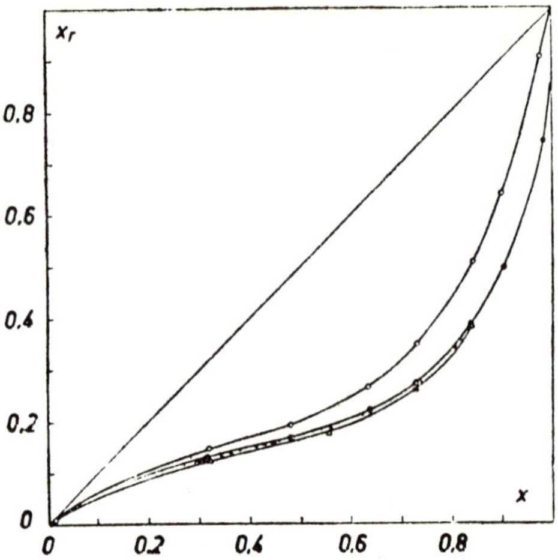

Abb. 4. Äthylalkohol-Wasserverteilungskurven für Dowex 50 in verschiedenen Ionenformen. Molenbruch des Äthylalkohols der Innenlösung x_r als Funktion des Molenbruches für Äthylalkohol der Aussenlösung x.
Ionenform: ○ Li^+ △ Na^+ ● K^+

Die ungleichmässige Verteilung des Alkohols zwischen Harzphase und Aussenlösung kommt am besten zum Ausdruck, wenn man die Versuchsresultate in Form von Verteilungskurven wiedergibt. Abb. 4 zeigt solche Verteilungskurven für den Kationenaustauscher in verschiedenen Ionenformen. Die Zusammensetzung ist durch den Molenbruch Äthylalkohol sowohl für die Aussenlösung (x) als auch für die Innenlösung (x_r) angegeben. Alle Kurven zeigen einen S-förmigen Verlauf und in allen Fällen ist der Alkoholgehalt in der Harzphase kleiner als in der Aussenlösung. Die verschiedenen Gegenionen im Austauscher beeinflussen die Verteilung mehr oder weniger stark. Qualitativ können die erhaltenen Effekte durch einen Aussalzeffekt erklärt werden, d.h. die Ionen des Austauschers wirken auf den Alkohol aussalzend. Beim Kationenaustauscher nimmt der Aussalzeffekt in der Reihenfolge $Li^+ < K^+ < Na^+$ zu. Es sei darauf aufmerksam gemacht, dass sich ein grösserer Aussalzeffekt beim Natriumion bemerkbar macht im Vergleich zum Kaliumion, obwohl der Austauscher in der Na^+-Form mehr quillt als in der K^+-Form, oder mit anderen Worten, dass die Natriumionenkonzentration in der Harzphase kleiner ist. Zu dem geringen Aussalzeffekt bei Lithium trägt wahrscheinlich die Tatsache dazu bei, dass die Quellung hier grösser ist als bei Natrium und Kalium.

Versuche wurden auch ausgeführt mit dem schwach vernetzten Kationenaustauscher (Dowex 50 X1) in Li^+- und K^+-Form. Die Quellung war hier grösser als bei normalem Vernetzungsgrad und es zeigte sich auch eindeutig, dass der Aussalzeffekt kleiner war, was durch die niedrigere Ionenkonzentration in der Innenlösung erklärt werden kann. Da es sich als sehr schwierig erwies bei dem schwach vernetzten Kationenaustauscher reproduzierbare Resultate zu erhalten, wurden diese in den Tabellen nicht wiedergegeben.

Ein eindeutiger Zusammenhang zwischen dem Aussalzeffekt bei den verschiedenen Kationen und der Hydratation liegt in diesem Falle nicht vor. Von anderen Systemen ist es jedoch auch bekannt, dass das Lithiumion eine Ausnahme von den Alkalimetallen macht und dass die Salzeffekte in derselben Reihenfolge zunehmen, wie sie hier beim Kationenaustauscher beobachtet wurden [10].

In Abb. 5 sind die entsprechenden Kurven für den Anionenaustauscher in verschiedenen Ionenformen wiedergegeben. Die Kurven sind symmetrischer als beim Kationenaustauscher und unterscheiden sich in einer wichtigen Beziehung von den Kurven des Kationenaustauschers. Bei niedrigen Alkoholkonzentrationen schneiden nämlich die Kurven die Diagonale. Dies bedeutet, dass man bei niedrigen Alkoholkonzentrationen ein Einsalzen des Alkohols erhält, d.h., der Alkoholgehalt in der Innenlösung ist grösser als in der Aussenlösung. Besonders deutlich tritt diese Erscheinung bei der ClO_4^--Form zu Tage. Die Kurven für die verschiedenen Ionenformen schneiden einander nicht. Der Alkoholgehalt der Innenlösung nimmt im ganzen Konzentrationsbereich nach folgender Reihe ab: $ClO_4^- > Cl^- > SO_4^{2-}$. Innerhalb des Bereiches, wo ein Aussalzen stattfindet, steigt dieses also in der Reihenfolge $ClO_4^- < Cl^- < SO_4^{2-}$ an. In diesem Falle herrscht eine Parallelität zwischen der Alkohol-Wasserverteilung und der Hydratation der Ionen dahingehend, dass eine zunehmende Hydratation zu einer steigenden Wasserabsorption beiträgt (steigender Aussalzeffekt). Dies ist im allgemeinen der Fall, wenn es sich um das Aussalzen von Nichtelektrolyten handelt [10], aber, wie schon oben erwähnt, gibt es auch Ausnahmen von dieser Regel.

[*Editor's Note:* Material has been omitted at this point.]

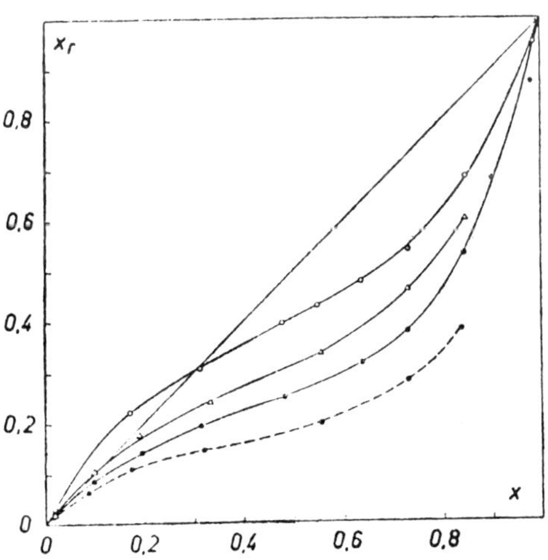

Abb. 5. Äthylalkohol-Wasserverteilungskurven für Dowex 2 (————) und Dowex 2 X1 (– – – –) in verschiedenen Ionenformen. Molenbruch für Äthylalkohol der Innenlösung x_r als Funktion des Molenbruches für Äthylalkohol der Aussenlösung x. Ionenform: ○ ClO_4^- △ Cl^- ● SO_4^{2-}

42

Reprinted from *Acta Chem. Scand.*, **11**(2), 315–323 (1957)

Die Verteilung von Glukose bei Ionenaustauschern auf Harzbasis in Äthylalkohol-Wassergemischen*

HEINZ RÜCKERT und OLOF SAMUELSON

Institut für Technische Chemie, Chalmers Technische Hochschule, Göteborg, Schweden

Die Sorption von Glukose an Ionenaustauscher aus Äthylalkohol-Wasserlösungen wurde für Kationenaustauscher und Anionenaustauscher studiert. Die vom Austauscher aufgenommene Glukosemenge ist sowohl von der Ionenform als auch der Alkoholkonzentration abhängig. Eine gewisse Parallelität zwischen der Verteilung von Äthylalkohol und Wasser und der Glukosesorption wurde festgestellt dahingehend, dass eine hohe Wasserkonzentration in der Austauscherphase eine hohe Glukosesorption begünstigt. Der direkte Salzeffekt der Ionen auf die Glukosemoleküle kann jedoch nicht vernachlässigt werden.

In einer früheren Arbeit [1] wurde gezeigt, dass Ionenaustauscher zur Sorption von Zuckerarten befähigt sind, wenn man als Lösungsmittel Äthylalkohol-Wassergemische verwendet. Die vom Austauscher aufgenommene Zuckermenge kann sehr beträchtlich sein. So werden z.B. von einem Kationenaustauscher in K^+-Form etwa 200 mg Glukose pro Gramm Austauscher aus 73 mol %iger Alkohollösung bei einer Gleichgewichtskonzentration von 10 g Glukose pro Liter aufgenommen. Diese ungleichmässige Verteilung ist in hohem Grade von der Kapazität des Austauschers und der Art der Gegenionen abhängig. Es konnte somit gezeigt werden, dass die Ionen direkt oder indirekt für die Zuckersorption verantwortlich sind. In der vorliegenden Arbeit wurden die Versuche mit Glukose auf andere Alkoholkonzentrationen ausgedehnt, um die Ursachen der Glukoseverteilung näher zu studieren.

EXPERIMENTELLER TEIL

Bezüglich der verwendeten Austauscherharze und des Arbeitsganges für die Bestimmung der Alkoholsorption sei auf eine Arbeit über die Verteilung von Wasser und Äthylalkohol bei Ionenaustauschern auf Harzbasis verwiesen [2]. Hier werden im folgenden nur die fehlenden und geänderten Bestimmungsmethoden, bedingt durch die Gegenwart der Glukose, behandelt. In der früheren Arbeit über die Zuckersorption wurde die vom Austauscher aufgenommene Zuckermenge durch die Konzentrationsänderung der Aussenlösung bestimmt. Bei den hier beschriebenen Versuchen sollte aber neben der Glukose

**Editor's Note:* An English translation follows this article.

auch gleichzeitig der Alkohol in der Innenlösung bestimmt werden, was sich nur dann durchführen lässt, wenn sowohl die Aussenlösung als auch die Innenlösung für sich auf ihre Zusammensetzung analysiert werden. Aus diesem Grunde wurden die beiden Phasen durch Zentrifugieren von einander abgetrennt.

Bestimmung des Äthylalkohols in der Austauscherphase. Da bei der Alkoholbestimmung durch Oxydation mit schwefelsaurer Bichromatlösung auch Glukose oxydiert wird, musste der Alkohol durch Destillation von der Glukose getrennt werden. Zu diesem Zwecke wurde nach der Extraktion des Austauscherharzes mit Wasser 50 ml von dieser Lösung destilliert und das Destillat in einem 100 ml Messkolben aufgefangen. Die Destillationsapparatur wurde im Hinblick auf die geringen Alkoholmengen so klein wie möglich gehalten und bestand aus Normalschliffteilen. Bei allen Versuchen wurden 30 ml Destillat gesammelt, der Messkolben bis zur Marke mit Wasser aufgefüllt und der Alkohol darin bestimmt.

Bestimmung der Glukose in der Austauscherphase. Nach der Extraktion des Austauschers mit Wasser wurde die Glukose mit dem Anthronreagenz [3] colorimetrisch bestimmt.

Bestimmung der Glukose in der Aussenlösung. Eine gewogene Menge der Aussenlösung entsprechend 25—60 mg Glukose wurde auf dem Wasserbad eingedampft und der Rückstand in 20 ml destilliertem Wasser gelöst. Davon gelangten 10 ml zur Glukosebestimmung nach Schoorl [4]. Bei der hier verwendeten modifizierten Methode hat es sich als vorteilhaft erwiesen, zur jodometrischen Titration des Kupferüberschusses eine erheblich grössere Menge an Kaliumjodid zu nehmen, als die CCA-Methode vorschreibt. Der Endpunkt der Titration konnte wesentlich leichter erfasst werden, wenn für jede Titration etwa 10 g Kaliumjodid zugesetzt wurden [5]. Das Eindampfen der alkoholischen Glukoselösung war aus dem Grunde erforderlich, weil der Alkohol den Siedepunkt bei der Reduktion herabsetzte und somit die empirischen Umrechnungsfaktoren ungültig wurden. Ein separater Versuch zeigte, dass sich die Glukose durch das Eindampfen nicht zersetzte.

Einstellung des Gleichgewichtes. Die Einstellung des Gleichgewichtes für die Glukosesorption war sowohl von der Äthylalkoholkonzentration als auch von der Ionenform des Austauschers abhängig. Separate Untersuchungen haben ergeben, dass sich beim Lithiumion das Gleichgewicht bei allen Alkoholkonzentrationen schon innerhalb 50—100 Std. eingestellt hatte und ebenso für den Anionenaustauscher in den untersuchten Ionenformen. Beim Kaliumion hingegen wurde das Gleichgewicht bei 73 Mol % Äthylalkohol nach etwa 200 Std. erreicht. Alle Glukosesorptionsversuche liefen zur Sicherheit 500—1 000 Std.

RESULTATE UND DISKUSSION

Die Resultate der Glukoseverteilung bei verschiedenen Alkoholkonzentrationen sind in Tabelle 1 für den Kationenaustauscher und in Tabelle 2 für den Anionenaustauscher zusammengestellt. Für jede Alkoholkonzentration der Aussenlösung (x) sind die entsprechenden Werte für die Alkoholkonzentrationen der Innenlösung (x_r) angegeben. In die Berechnung des Molenbruches (x_r) wurden nur Alkohol, Wasser und Glukose einbezogen. Die Änderung von x_r mit der Glukosekonzentration lag innerhalb des experimentellen Fehlers. Daher ist für x_r nur jeweils der Mittelwert angegeben. Für jede Alkoholkonzentration wurde die Glukoseverteilung für mindestens zwei verschiedene Glukosekonzentrationen bestimmt. Die Glukosekonzentrationen der Aussen- und Innenlösung sind ebenfalls durch den Molenbruch (x_G bzw. x_{G_r}) ausgedrückt. Da der Molenbruch nur ein relatives Mass darstellt, eignet sich dieser sehr gut dafür, die Konzentration der Innenlösung unabhängig von der Quellung des Austauschers auszudrücken. Vom praktischen Standpunkt aus kann es jedoch von grösserem Interesse sein, ein absolutes Mass für die vom Austauscher aufgenommenen Glukosemengen zu haben. Daher sind auch die Werte für die pro Äquivalent Austauscher absorbierte Anzahl Mol Glukose tabelliert.

Tabelle 1. Glukoseverteilung bei dem Kationenaustauscher Dowex 50.

Austauscher	x	x_r	x_G	x_{G_r}	Mol Glukose / Äq. Austauscher	K^*
Dowex 50 Li+	0,00	0,00	0,00103	0,000323	0,0038	0,31
			0,00254	0,000796	0,0094	
	0,32	0,15	0,00099	0,00091	0,0077	0,89
			0,00243	0,00216	0,0184	
	0,55	0,22	0,00095	0,00254	0,0172	2,67
			0,00237	0,00631	0,0424	
	0,73	0,32	0,00171	0,0115	0,059	7,2
			0,00307	0,0235	0,121	
	0,84	0,51	0,00077	0,0124	0,045	16,1
			0,00191	0,0310	0,110	
	0,90	0,68	0,00072	0,0176	0,052	23,7
			0,00171	0,0398	0,115	
Dowex 50 Na+	0,00	0,00	0,00102	0,00044	0,0047	0,42
			0,00256	0,00107	0,0116	
	0,33	0,12	0,00098	0,00182	0,0133	1,85
			0,00240	0,00440	0,0319	
	0,56	0,18	0,00089	0,0060	0,0323	6,8
			0,00224	0,0152	0,0803	
	0,73	0,24	0,00093	0,0262	0,091	28,2
			0,00248	0,0529	0,189	21,3
	0,84	0,34	0,00050	0,0489	0,117	98
			0,00129	0,0783	0,196	61
Dowex 50 K+	0,00	0,00	0,00102	0,00065	0,0059	0,64
			0,00254	0,00161	0,0147	
	0,32	0,13	0,00097	0,00329	0,0178	3,3
			0,00236	0,00780	0,0427	
	0,55	0,19	0,00083	0,0115	0,045	13,9
			0,00214	0,0260	0,101	12,1
	0,73	0,25	0,00081	0,0434	0,115	54
			0,00226	0,0839	0,223	37
Dowex 50 X1 Li+			0,00059	0,0018	0,033	
	0,73	0,57	0,00118	0,0039	0,069	3,4
			0,00230	0,0079	0,132	
			0,00288	0,0109	0,172	
Dowex 50 X1 K+			0,00064	0,0229	0,094	
	0,73	0,34	0,00119	0,0435	0,146	36
			0,00140	0,0510	0,169	

Tabelle 2. Glukoseverteilung bei dem Anionenaustauscher Dowex 2.

Austauscher	x	x_r	x_G	x_{G_r}	$\frac{\text{Mol Glukose}}{\text{Äq. Austauscher}}$	K^*
Dowex 2 ClO_4^-	0,00	0,00	0,00100	0,00042	0,0024	0,42
			0,00256	0,00110	0,0061	
			0,00127	0,00140	0,0076	
	0,73	0,53	0,00186	0,00202	0,0109	1,1
			0,00306	0,00342	0,0185	
			0,00353	0,00426	0,0228	
Dowex 2 Cl^-	0,00	0,00	0,00101	0,00059	0,0078	0,57
			0,00258	0,00145	0,0198	
	0,73	0,44	0,00112	0,0091	0,056	8,2
			0,00262	0,0219	0,143	
Dowex 2 SO_4^{2-}	0,00	0,00	0,00100	0,00073	0,0110	0,73
			0,00256	0,00188	0,0276	
			0,00086	0,0279	0,181	32,4
	0,73	0,35	0,00131	0,0384	0,252	29,3
			0,00232	0,0571	0,361	24,6
			0,00283	0,0628	0,406	22,2
Dowex 2 X1 ClO_4^-			0,00137	0,00173	0,0052	
	0,73	0,43	0,00195	0,00265	0,0080	1,3
			0,00263	0,00312	0,0110	
			0,00326	0,00426	0,0134	
Dowex 2 X1 SO_4^{2-}			0,00050	0,0270	0,189	54,0
	0,73	0,25	0,00080	0,0397	0,273	49,6
			0,00116	0,0501	0,349	43,2
			0,00150	0,0609	0,425	40,6

In Abb. 1 sind die von dem normal vernetzten Kationenaustauscher Dowex 50 in Li^+-Form aufgenommenen Glukosemengen gegen die Glukosekonzentration der Aussenlösung aufgetragen. Die einzelnen Geraden entsprechen verschiedenen Alkoholkonzentrationen der Aussenlösung. Man sieht, wie die Glukosesorption mit der Alkoholkonzentration kräftig ansteigt. Aus 90 mol %igem Äthylalkohol nimmt dieser Austauscher ungefähr das 20-fache an Glukose auf als aus wässeriger Lösung. Zum Vergleich sind die entsprechenden Kurven für die K^+-Form desselben Austauschers in Abb. 2 wiedergegeben. Auch hier findet man eine mit der Alkoholkonzentration steigende Glukosesorption, die bei allen Alkoholkonzentrationen grösser ist als für Lithium. Bei niedriger Alkoholkonzentration ist die aufgenommene Glukosemenge der Glukosekonzentration der Aussenlösung proportional, bei höheren Alkoholkonzentrationen sind die Kurven gebogen.

Über die Abhängigkeit der Glukosesorption sowohl von der Alkoholkonzentration als auch von der Ionenform des Austauschers gibt am besten der Verteilungskoeffizient K^* Auskunft. Der Verteilungskoeffizient ist durch die

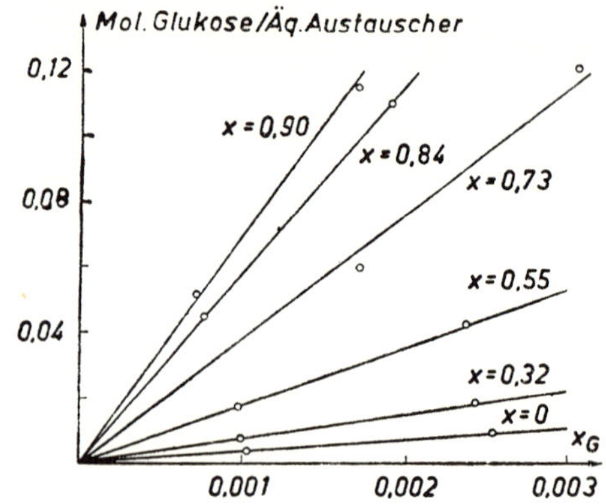

Abb. 1. Aufgenommene Glukosemenge in Mol Glukose/Äq. Austauscher als Funktion des Molenbruches für Glukose der Aussenlösung x_G bei verschiedenen Äthylalkoholkonzentrationen für Dowex 50 Li^+. Die Äthylalkoholkonzentration der Aussenlösung ist durch den Molenbruch x ausgedrückt.

Gleichung $K^* = x_{G_r}/x_G$ definiert. Innerhalb des untersuchten Konzentrationsbereiches ($x_G = 0$ — 0,0025) sind in den Fällen, wo die Sorption niedrig ist, die K^*-Werte von der Glukosekonzentration unabhängig. Bei hoher Sorption nehmen die K^*-Werte mit der Glukosekonzentration ab. Für die Versuchsreihen, bei denen K^* innerhalb der Grenzen des experimentellen Fehlers konstant ist, sind in den Tabellen die Mittelwerte für K^* angegeben, anderenfalls für jede Glukosekonzentration der entsprechende K^*-Wert.

Wie man aus den Tabellen ersieht, steigen die K^*-Werte mit der Alkoholkonzentration. Die Gegenionen lassen sich bezüglich ihres Einflusses auf den Verteilungskoeffizienten bei allen Alkoholkonzentrationen und jedem Austauscher in folgende Reihe für steigendes K^* einordnen: $Li^+ < Na^+ < K^+$ bzw. $ClO_4^- < Cl^- < SO_4^{2-}$. Es wurde versucht, K^* als Funktion der Alkoholkonzentration der Aussenlösung durch die empirische Gleichung

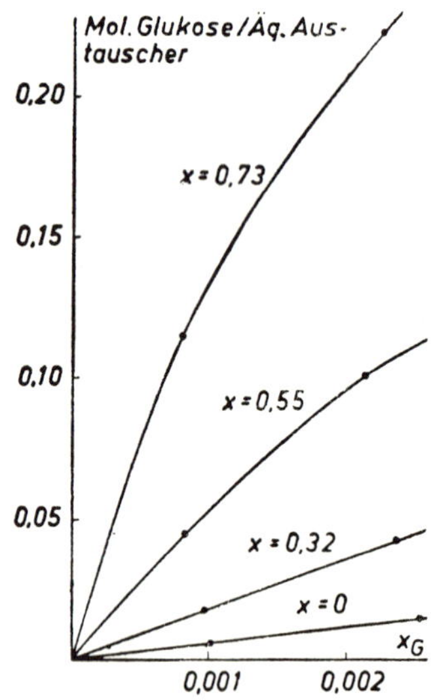

Abb. 2. Aufgenommene Glukosemenge in Mol Glukose/Äq. Austauscher als Funktion des Molenbruches für Glukose der Aussenlösung x_G bei verschiedenen Äthylalkoholkonzentrationen für Dowex 50 K^+. Die Äthylalkoholkonzentration der Aussenlösung ist durch den Molenbruch x ausgedrückt.

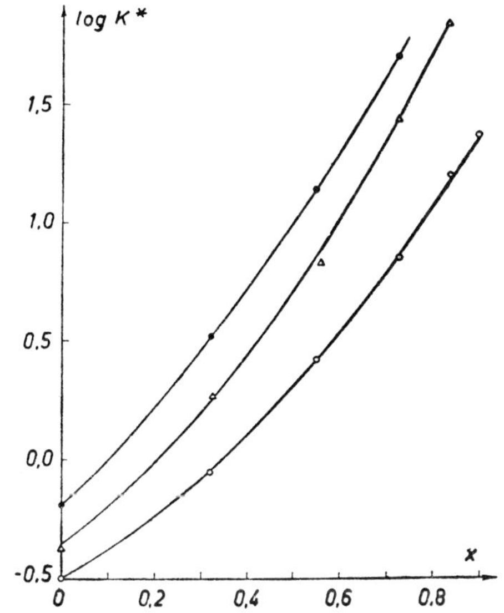

Abb. 3. log K^* als Funktion des Molenbruches für Äthylalkohol der Aussenlösung x für Dowex 50 in verschiedenen Ionenformen.
Ionenform: ○ Li^+, △ Na^+, ● K^+.

$$\log K^* = \log K^*_W + ax + bx^2$$

auszudrücken. In dieser Gleichung sind a und b empirische Konstante und K^*_W bedeutet die Verteilungskonstante in wässeriger Lösung. In Tabelle 3 sind die nach der Methode der kleinsten Quadrate berechneten Konstanten für den normal vernetzten Kationenaustauscher in Li^+-, Na^+- und K^+-Form zusammengestellt. Da, wie schon erwähnt, K^* auch von der Glukosekonzentration abhängig sein kann, wurden die Berechnungen für eine Glukosekonzentration der Aussenlösung von $x_G = 0{,}001$ durchgeführt. Abb. 3 zeigt den empirisch gefundenen Zusammenhang zwischen log K^* und x entsprechend obiger Gleichung mit den Konstanten aus Tabelle 3. Zum Vergleich sind auch in demselben Diagramm die experimentellen Punkte eingezeichnet.

In der Praxis wird man im allgemeinen daran interessiert sein, die pro Austauschereinheit aufgenommenen Glukosemengen berechnen zu können. Da diese auch von der Quellung des Austauschers abhängen, eignet sich obige empirische Gleichung nicht unmittelbar für so eine Berechnung. Die Quellung des Austauschers in Alkohol-Wassergemischen wurde in einer früheren Arbeit [2] behandelt, aus der die Summe der pro Äquivalent Austauscher aufgenommenen Mole Wasser und Alkohol zu entnehmen ist. Wenn zu der Innenlösung auch

Tabelle 3. Konstanten der Gleichung log $K^* = \log K^*_W + ax + bx^2$.

Austauscher	log K^*_W	a	b
Dowex 50 Li^+	−0,501	1,08	1,09
Na^+	−0,356	1,42	1,45
K^+	−0,196	1,91	0,96

die sorbierte Glukose hinzukommt, zeigte es sich, dass diese Summe innerhalb der Grenzen des experimentellen Fehlers unverändert blieb. Man kann also für die Glukosesorption mit hinreichender Genauigkeit die gleichen Quellungswerte wie für das System Äthylalkohol-Wasser verwenden. Mit Hilfe der K^*- und Quellungswerte lassen sich dann auch für jede Alkoholkonzentration die pro Austauschereinheit aufgenommenen Glukosemengen berechnen.

Bei hohen Alkoholkonzentrationen ($x = 0{,}84$ und $x = 0{,}90$) wurde für die K^+-Form das Gleichgewicht noch nicht nach 2 000 Std. erreicht und dasselbe ergab sich für die Na^+-Form bei $x = 0{,}90$. Diese Versuche sind daher auch nicht in den Tabellen verzeichnet. Die lange Dauer für die Einstellung des Gleichgewichtes bei hohen Alkoholkonzentrationen kann in der Schrumpfung des Austauschers seine Erklärung finden. Wenn der Austauscher mit zunehmendem Alkoholgehalt schrumpft, kann bei einer bestimmten Alkoholkonzentration die Maschenweite des Harzskelettes so eng werden, dass die im Verhältnis zu Alkohol und Wasser grossen Glukosemoleküle nur mehr schwer in den Austauscher eindringen können. Dieser beginnende Sperreffekt für Glukosemoleküle tritt für die K^+-Form bei einer Alkoholkonzentration in der Aussenlösung zwischen $x = 0{,}73$ und $x = 0{,}84$ ein. Untersuchungen über die Quellung in Äthylalkohol-Wassergemischen [2] zeigten, dass das Volumen der Na^+-Form bei $x = 0{,}9$ kleiner ist als das der K^+-Form bei $x = 0{,}73$. Der Sperreffekt ist also für die Na^+-Form bei etwa $x = 0{,}9$ zu erwarten, während die Li^+-Form auf Grund der grösseren Quellung bei allen Alkoholkonzentrationen zur ungehinderten Glukosesorption befähigt sein müsste.

Die ungleichmässige Verteilung von Alkohol und Wasser wurde qualitativ durch einen Aussalzeffekt bzw. Einsalzeffekt erklärt [2]. Die Verhältnisse bei der Glukoseverteilung werden dadurch noch komplizierter, weil hier nicht nur die Ionen sondern auch die Alkoholkonzentration die Verteilung beeinflussen. Man könnte sich als Grenzfall denken, dass die Ionen auf die Glukosemoleküle gar nicht einwirken und dass die Glukosesorption eine direkte Folge der ungleichmässigen Verteilung von Wasser und Alkohol zwischen der Aussen- und Innenlösung wäre. Da die Alkoholkonzentration in der Austauscherphase meistens kleiner ist als in der Aussenlösung und die Löslichkeit der Glukose mit steigender Alkoholkonzentration abnimmt, könnte man die Glukosesorption qualitativ als eine Verteilung zwischen zwei Phasen mit verschiedenen Lösungsmitteln erklären, d.h., die Glukose dringt vorzugsweise in die Phase mit dem besseren Lösungsmittel ein, also in die wasserreichere Innenlösung.

Auf Grund von Löslichkeitsbestimmungen für Glukose in Alkohol-Wassergemischen und der Verteilung von Alkohol und Wasser kann eine theoretische Glukoseverteilung berechnet werden. Hierbei wird also angenommen, dass die direkte Einwirkung der Ionen auf die Glukosemoleküle vernachlässigt werden kann. Ein Vergleich dieser berechneten Verteilung mit der experimentell gefundenen zeigte, dass eine solche Approximation unzulässig ist. Bedeutende Abweichungen ergaben sich und zwar in der Richtung, dass man bei niedriger Alkoholkonzentration einen deutlichen Aussalzeffekt für Glukose erhält und bei höherer Alkoholkonzentration statt dessen einen Einsalzeffekt. Der Aussalzeffekt im niedrigen Konzentrationsbereich kann übrigens direkt aus den Tabellen entnommen werden. Die Tatsache, dass in wässeriger Lösung Glukose ausgesalzt wird, ist schon früher bekannt [6].

Tabelle 4. Vergleich des Verteilungskoeffizienten für Glukose K^* mit der Alkoholkonzentration der Innenlösung x_r.

Austauscher	x_r	K^* bei $x = 0,73$ und $x_G = 0,001$
Dowex 50 Li^+	0,32	7,2
Na^+	0,24	27,5
K^+	0,25	50,8
Dowex 50 X1 Li^+	0,57	3,4
K^+	0,34	36,3
Dowex 2 ClO_4^-	0,53	1,1
Cl^-	0,44	8,2
SO_4^{2-}	0,35	31,2
Dowex 2 X1 ClO_4^-	0,43	1,3
SO_4^{2-}	0,25	46,3

Ohne irgend welche anderen theoretischen Berechnungen heranziehen zu müssen, kann man mit dem hier gesammelten experimentellen Material zeigen, dass die Wechselwirkung zwischen den Ionen und Glukosemolekülen nicht vernachlässigt werden kann. In Tabelle 4 sind die K^*-Werte für die untersuchten Austauscher bei einer konstanten Zusammensetzung der Aussenlösung [$x = 0,73$ und $x_G = 0,001$] zusammengestellt. Die Alkoholkonzentration der Innenlösung x_r ist dann für jeden Austauscher verschieden und ebenfalls in der Tabelle verzeichnet. Es soll nun die Alkoholkonzentration der Innenlösung mit K^* verglichen werden. Beim normal vernetzten Anionenaustauscher Dowex 2 findet man eine Parallelität zwischen dem Verteilungskoeffizienten und der Alkoholkonzentration. Wie erwartet steigt die Glukosesorption, wenn die Alkoholkonzentration der Innenlösung abnimmt, d.h. wenn das Lösungsmittel in der Harzphase für Glukose besser wird. Dasselbe ergibt sich für den schwach vernetzten Anionenaustauscher Dowex 2 X1. Vergleicht man diesen bei gleichem Gegenion mit dem normal vernetzten Anionenaustauscher, so nimmt hier auch die Glukosesorption mit fallendem Alkoholgehalt der Innenlösung zu. Für den Kationenaustauscher findet man ebenfalls eine gewisse Parallelität aber auch das Gegenteil. Die Alkoholkonzentration der Na^+-Form des normal vernetzten Kationenaustauschers ist z.B. niedriger als die der K^+-Form. Trotzdem wird von der Na^+-Form viel weniger Glukose aufgenommen als von der K^+-Form. Überhaupt lässt sich eine allgemeine Parallelität zwischen der Alkohol-Wasserverteilung und der Glukosesorption nicht aufstellen. Wenn diese Parallelität vorhanden wäre, müssten alle in Tabelle 4 angeführten Alkoholkonzentrationen die gleiche Reihenfolge für die entsprechenden Verteilungskonstanten geben, was aber nicht zutrifft. Dies zeigt deutlich, dass die Verschiedenheit der Alkoholkonzentration der Innen- und Aussenlösung eine zufriedenstellende Erklärung für die Glukosesorption nicht gibt, sondern dass man ebenfalls auf den Salzeffekt Rücksicht nehmen muss, der auf die Glukosemoleküle ausgeübt wird.

Wir danken dem *Schwedischen Naturwissenschaftlichen Forschungsrat* für die Bereitstellung von Mitteln.

LITERATURVERZEICHNIS

1. Rückert, H. und Samuelson, O. *Svensk Kem. Tidskr.* **66** (1954) 337.
2. Rückert, H. und Samuelson, O. *Acta Chem. Scand.* **11** (1957) 303.
3. Scott, T. A., Jr. und Melvin, E. H. *Anal. Chem.* **25** (1953) 1656.
4. *Svenska Pappers- och Cellulosaingenjörsföreningens tekniska meddelanden,* CCA 11 (1942).
5. Du Rietz, C. *Diss.* Tekniska Högskolan, Stockholm, 1938.
6. Wheaton, R. M. und Bauman, W. C. *Ann. N. Y. Acad. Sci.* **57** (1953) 159.

Eingegangen am 6. November 1956.

42

THE DISTRIBUTION OF GLUCOSE BETWEEN ION-EXCHANGE RESINS AND ALCOHOL—WATER MIXTURES

Heinz Rückert and Olof Samuelson

This condensed translation was prepared expressly for this Benchmark volume by Harold F. Walton, University of Colorado, Boulder, Colorado, from "Die Verteilung von Glukose bei Ionenaustauschern auf Harzbasis in Äthylalkohol-Wassergemischen," Acta Chem. Scand., 11*(2), 315–323 (1957)*

Abstract

The sorption of glucose from ethyl alcohol-water solutions by cation- and anion-exchange resins has been studied. The quantity of glucose absorbed depends on the ionic form of the resin as well as on the alcohol concentration. A correlation was noted between the glucose distribution and the alcohol-water distribution, in the sense that a high concentration of water in the resin favors a high absorption of glucose. The direct salt effect between the ions and the glucose molecules cannot, however, be ignored.

In a previous paper[1] we showed that ion exchangers absorb sugars from alcohol-water mixtures. The quantity of sugar absorbed can be quite considerable. For example, a cation exchanger in the K^+ form absorbed about 200 mg of glucose/g of exchanger from a solution of 10 g of glucose/liter of a solvent containing 73 M percent alcohol. This unequal distribution depends to a large extent on the capacity of the exchanger and the nature of the counter-ions. The ions were responsible, directly or indirectly, for the absorption of the sugar. In the work to be described, the glucose distribution is determined over a range of alcohol concentrations to find out more about the reasons for the glucose absorption.

RESULTS AND DISCUSSION

The glucose distributions at different alcohol concentrations are shown in Table 1 for the cation exchangers and in Table 2 for the anion exchangers. For each alcohol concentration in the external solution the corresponding concentration in the internal solution is given; the mole fractions are x and x_r, respectively. In calculating x_r, only the quantities of alcohol, water, and glucose were considered, not the quantity of resin. The fraction x_r was independent of the gluose concentration, within experimental error; therefore, only its average value is given in the table. The glucose distribution was measured for at least two concentrations of glucose for each alcohol concentration. The glucose concentrations of the external and internal solutions are expressed as mole fractions, x_G and x_{G_r}. Since the mole fraction is only a relative quantity, it expresses the concentration of the internal solution independently of the swelling of the resin. From the practical standpoint, however, it may be more important to know the absolute quantity of glucose taken up by the resin. Therefore, we have tabulated (next-to-last column) the number of moles of glucose absorbed per equivalent of exchanger. (The last column gives the distribution coefficients, $K^* = x_{G_r}/x_G$.)

Figure 1 shows the absorption of glucose plotted against the glucose concentration of the external solution for the Li^+ form of normally crosslinked Dowex-50. The different lines refer to different alcohol concentrations in the external solution. Corresponding curves for the K^+ form are shown in Figure 2.

As is seen from the tables, the distribution coefficients K^* increase with increasing alcoholconcentration, At all alcohol concentrations and for each degree of crosslinking, K^* increases in the following counter-ion sequence: Li^+ -Na^+ -K^+ for cations, ClO_4^--Cl^--SO_4^{2-} for anions. The relation between K^* and the external alcohol concentration is shown in Figure 3 and by the empirical equation given. Least squares values for the empirical coefficients are shown in Table 3.

At high alcohol concentrations ($x = 0.84, 0.90$) the K^+ resin did not reach equilibrium in 2,000 hours; the same was true of the Na^+ resin at $x = 0.90$ . . . . The slow attainment of equilibrium at high alcohol concentrations could be due to the shrinkage of the resin . . . . The Li^+ form of the resin, with its greater swelling, shows no hindrance to glucose absorption, even at the highest alcohol concentrations.

We attributed the unequal alcohol-water distribution to a salting-out or a salting-in effect.[2] The distribution of glucose is more complicated, because it is affected not only by the ions themselves, but also by the alcohol concentration, As a limiting case one could suppose that the ions did not interact with the glucose molecules and that the absorption of glucose was caused solely by the unequal distribution of water and alcohol between the internal and external solutions. Since the alcohol concentration in the exchanger is generally less than that in the external solution, and since the solubility of glucose falls with increasing alcohol concentration, one might explain the glucose absorption qualitatively as a partition between two phases. The glucose moves into the phase that is the better solvent, that is, into the water-rich internal solution.

Knowing the solubility of glucose in alcohol-water mixtures and also the alcohol-water distribution, one can calculate a theoretical distribution of glucose between the resin and the solution, making the assumption, of course, that direct interactions between ions and glucose molecules can be neglected. Comparison of the distribution thus calculated with the distribution found experimentally shows that this assumption is not justified. The deviations are substantial. At low alcohol concentrations there is a distinct salting-out effect on the glucose; at higher alcohol concentrations the reverse is true and there is a salting-in effect.

Without the need to invoke any other theoretical considerations, one can show from the experimental data that the interactions between ions and glucose molecules cannot be neglected. Table 4 shows K* values for all the exchangers studied, at a constant composition of the external solution, The alcohol concentration of the internal solution, x_r, is shown in the table; it is different for different exchangers. Let us now compare the internal alcohol concentrations with K*. With the anion exchanger Dowex-2 of normal crosslinging, K* rises as the internal alcohol concentration drops, just as we would expect. The same relation holds for the weakly crosslinked exchanger, Dowex-2X1. If one compares the weakly crosslinked and the normally crosslinked exchangers having the same counter-ion, again the glucose absorption increases as the internal alcohol concentration decreases. Turning to the cation exchanger, one sees the same relationship, but also its opposite. Thus the alcohol concentration of the Na^+ form of the normally crosslinked exchanger is smaller than that of the K^+ form, yet the Na^+ form absorbs a good deal less glucose. It is not possible to draw a universal parallel between the alcohol-water distribution and the glucose distribution. If such a parallel existed, the distribution coefficients for glucose should follow the same order as the internal alcohol concentrations for all the data of Table 4, which is not the case. Obviously, the difference between the alcohol concentrations inside and outside the exchanger is not in itself sufficient to explain the glucose absorption, and one must take account of the salt effects on the glucose molecules.

[*Editor's Note:* References, tables and figures for this translation will be found in the preceding original article.]

43

Reprinted from *Acta Chem. Scand.,* **19**(6), 1357-1364 (1965)

An Automated Procedure for Separation of Monosaccharides on Ion Exchange Resins

LARS-INGE LARSSON and OLOF SAMUELSON

Department of Chemical Engineering, Chalmers Tekniska Högskola, Göteborg, Sweden

The separation of 16 monosaccharides was studied by partition chromatography on anion exchange resins using an automated procedure. Most separations were quantitative. A sample containing several monosaccharides can be analyzed within a few hours.

In quantitative determinations of the monosaccharides present in wood hydrolyzates partition chromatography on ion exchange resins combined with automatic recording of the eluate concentration has proved to be a more reliable technique than paper chromatography.[1] The aim of this work is to elucidate the possibilities of this technique within other fields of carbohydrate chemistry. The work was started with the monosaccharides known to be present in human gastric mucopolysaccharides [2] (fucose, mannose, galactose, glucose) but has been extended to a number of other monosaccharides. Among these are the monosaccharides usually present in hydrolyzates of soils (rhamnose, fucose, ribose, arabinose, xylose, mannose, galactose, glucose).

EXPERIMENTAL

All experiments were carried out with strongly basic anion exchange resins in the sulfate form. One of the chromatograms reproduced in this paper is taken from experiments with a commercial resin (Dowex 21 K, 1 – 16 μ) obtained from Bio-Rad Laboratories, Richmond, Calif. This resin consisted of irregular crushed particles and a very high pressure drop occurred in the column even at very low flow rates. When this column was started, very broad and irregular elution curves were observed but it was found that improved results were obtained after several runs had been carried out. After a couple of months very sharp elution curves were observed with this column (*cf.* Fig. 1).

All other results reported in this paper were obtained with a spherical macroporous resin specially prepared for our purposes (exchange capacity 3.7 mequiv./g chloride resin, particle size 10 – 35 μ). Even with this resin certain improvements of the elution curves were observed after the column had been used for some time, indicating that the packing of the resin is an important factor.

Jacketed columns were used. These were kept at 75° by means of water circulated from a thermostat. The eluant was fed onto the column by a piston-type pump (*cf.* Ref. 3) and preheated to 75° before entering the column. The eluate was analyzed automatically by means of a Technicon AutoAnalyzer using the orcinol method described in a previous paper.[1]

DISTRIBUTION COEFFICIENTS OF VARIOUS MONOSACCHARIDES

In order to establish the possibilities to separate various monosaccharides from each other the positions of the elution peaks were determined at various ethanol concentrations. From the positions of the elution peaks the distribution coefficients (D_v) were calculated as usual.[3] Small variations in the peak elution volumes were obtained in some duplicate runs, which can be ascribed to slight fluctuations in flow rate, eluant concentration, and elution temperature. The ratios between the distribution coefficients in various runs were, however, unchanged. In order to compensate for such variations, xylose was included as a standard in each run and the separation factor relative to xylose calculated for each individual sugar. As usual [3] the separation factor is defined as the ratio of the distribution coefficient of a given sugar to that of xylose both determined at the same eluant concentration.

Table 1. Distribution factors of xylose at various ethanol concentrations.

86 %	88 %	89 %	90 %	92 %
7.38	9.43	10.68	12.09	18.19

Table 2. Separation factors relative to xylose at various ethanol concentrations (% by weight).

	86 %	88 %	89 %	90 %	92 %
D-Digitoxose	0.10 (0.07)	0.09	0.08	0.07	0.05
2-Deoxy-D-ribose	0.17 (0.13)	0.16	—	0.13	0.11
2-Deoxy-D-galactose	— (0.28)	0.28	0.28	0.27	0.24
2-Deoxy-D-glucose	0.38 —	0.36	0.35	0.34	0.32
L-Rhamnose	0.38 (0.35)	0.36	0.35	0.34	0.32
L-Fucose	0.44 (0.40)	0.43	—	0.41	0.39
D-Ribose	0.55 (0.51)	0.50	0.48	0.48	0.46
D-Lyxose	0.76 (0.70)	0.74	0.73	0.72	0.70
D-Arabinose	0.85 (0.79)	0.83	0.82	0.81	0.80
D-Xylose	1.00 (1.00)	1.00	1.00	1.00	1.00
D-Fructose	1.09 —	1.10	1.11	1.14	1.19
D-Tagatose	1.09 —	1.10	1.11	1.14	1.19
L-Sorbose	1.21 —	1.24	1.25	1.29	1.34
D-Mannose	1.27 (1.30)	1.34	1.39	1.40	1.46
D-Galactose	1.76 (1.85)	1.92	1.97	2.02	2.16
D-Glucose	2.02 (2.21)	2.29	2.41	2.44	2.71

Values within parentheses refer to Dowex 21 K.

The distribution coefficients of xylose at various ethanol concentrations are given in Table 1. The results confirm the earlier observation that the distribution coefficients are extremely sensitive towards changes in eluant concentration, and increase with an increased ethanol concentration.[4]

The separation factors relative to xylose for 15 monosaccharides at various ethanol concentrations are given in Table 2. Using the values given in the two tables the peak elution volume of all sugars can be easily calculated at any column dimension.[3] From Table 2 it is seen that within the range of interest in chromatographic work, the order of elution of all monosaccharides is independent of the ethanol concentration. Similarly, the order of elution is the same with both resins. The selectivity is, however, somewhat better with Dowex 21 K than with the other resin, which is explained by a higher exchange capacity which, as shown previously, results in an increased distribution coefficient.[5] In 86 % ethanol the distribution coefficient of xylose on Dowex 21 K was equal to 14.05 which is almost twice as high as the value observed with the other resin. Hence, these results confirm our earlier observation that an enhanced selectivity is obtained with resins of high exchange capacity.[4]

As a first approximation the sorption of the sugars is explained by the higher solubility in the water-rich resin phase than in the less polar external solution.[6] The distribution coefficient should, therefore, be expected to (*a*) increase with an increased number of hydroxyl groups in the sugar and (*b*) decrease when non-polar groups such as methyl groups are present.

In agreement with the first rule, pentoses exhibit lower distribution coefficients than hexoses. It can also be mentioned that the tetrose, erythrose, is held less firmly than the pentoses. The commercial sample of erythrose used in this investigation was found to contain some impurity and in addition it was found that this sugar was not quite stable during the chromatographic run (75°). The values obtained with erythrose have, therefore, not been included in the table.

In agreement with the second rule the deoxysugars derived from hexoses, 2-deoxy-galactose, 2-deoxy-glucose, rhamnose, and fucose exhibit distribution coefficients which are lower than those of the pentoses. Similarly 2-deoxy-ribose appears ahead of erythrose. As expected from these rules digitoxose (2,6-dideoxy-ribo-hexose) shows the lowest distribution coefficient of all sugars investigated. Fructose is known to be more soluble in ethanol than glucose and it can be anticipated that fructose and other ketohexoses should appear ahead of the aldohexoses. This is confirmed by the results given in Table 2.

The order of elution of the sugars within each group, *e.g.* of the pentoses, cannot be predicted from tnese simple rules. In an earlier paper [6] it was shown that an additional factor which affects the equilibrium uptake of sugars is the interaction between the ions in the resin phase and the sugar. This has been confirmed in a recent work in which a sulfonic acid resin in its potassium form was used instead of an anion exchange resin. With this resin the order of elution of arabinose and xylose was found to be reversed.[7] More experimental results seem to be required if it should be possible to derive any rules about the positions of the individual sugars within each group.

SEPARATION OF VARIOUS MONOSACCHARIDES

From Table 2 it is seen that the separation factors of most sugars differ significantly from each other and that the selectivity increases with an increased ethanol concentration. As shown previously [4] a satisfactory separation of glucose and galactose can be achieved in 88 % ethanol although an improvement is observed at higher ethanol concentration. At 88 % ethanol the separation factor is equal to 2.29/1.92=1.19. With most of the monosaccharides investigated in this work the separation factor is higher, and it can be concluded that most of them can be separated quite easily. An inspection of Table 2 shows, however, that there are some exceptions. Hence, 2-deoxy-glucose and rhamnose exhibit an almost identical behavior and cannot be separated from each other under the conditions used in this work. The same holds true for fructose and tagatose.

The other monosaccharides have been separated completely or, in less favorable systems, sufficiently well for a fairly accurate quantitative determination. Working at a comparatively low ethanol concentration has the advantage that all sugars appear in the eluate within a moderate eluate volume but on the other hand the disadvantage that the separation factors are lower than at high concentration.

A chromatogram from a run with Dowex 21 K in 86 % ethanol is reproduced in Fig. 1. It is seen that despite the low ethanol concentration the

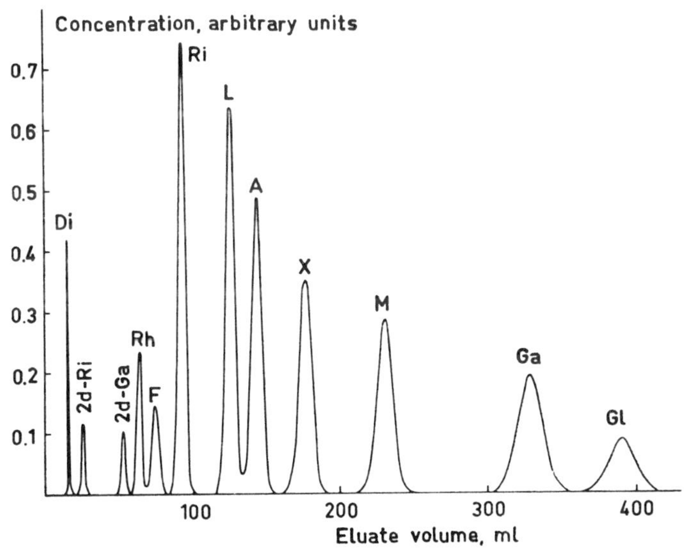

Fig. 1. Separation of various monosaccharides in 86 % ethanol. Resin bed 6 × 440 mm. Dowex 21 K, 1—16 μ. Flow rate 0.9 ml cm^{-2} min^{-1}.
Di = Digitoxose, 0.05 mg; 2d Ri = 2-Deoxy-ribose, 0.10 mg, 2d—Ga = 2-Deoxy-galactose, 0.05 mg; Rh = Rhamnose, 0.05 mg; F = Fucose, 0.05 mg; Ri = Ribose, 0.10 mg; L = Lyxose, 0.10 mg; A = Arabinose, 0.10 mg; X = Xylose, 0.10 mg; M = Mannose, 0.20 mg; Ga = Galactose, 0.30 mg; Gl = Glucose, 0.30 mg.

twelve sugars included in this run can be separated well enough for a quantitative determination of each sugar. The slight overlapping observed between the curves corresponding to rhamnose and fucose (separation factor: 1.14) and between lyxose and arabinose (separation factor: 1.13) does not interfere but can of course be eliminated by using an increased ethanol concentration or an increased column length which, however, results in a greater eluate volume.

Excepting the two pairs of sugars which could not be distinguished from each other the smallest separation factors are those of fructose:xylose (1.11) and mannose:sorbose (1.11). Similarly the separation factor sorbose: fructose (1.13) is unfavorable. The values refer to 89 % ethanol which is the concentration chosen in the chromatogram reproduced in Fig. 2. It is seen that a clear cut separation was achieved with the other sugars included in this run whereas an overlapping of the curves representing these sugars was obtained. The separation is good enough to permit an accurate determination of xylose and fructose in the complicated mixture whereas the determination of sorbose and mannose is less accurate. If a calibration run is made with about the same amounts as present in an unknown sample the error in the determination of the two last-mentioned sugars should not exceed 10 %.

EVALUATION OF CHROMATOGRAMS FOR QUANTITATIVE PURPOSES

In most work reported in this paper, part of the eluate as well the reagents were fed into the analyzing system by means of a peristaltic pump (proportioning pump). This technique has two great advantages. Firstly, part of the eluate can be saved for identification purposes. Secondly, the conditions

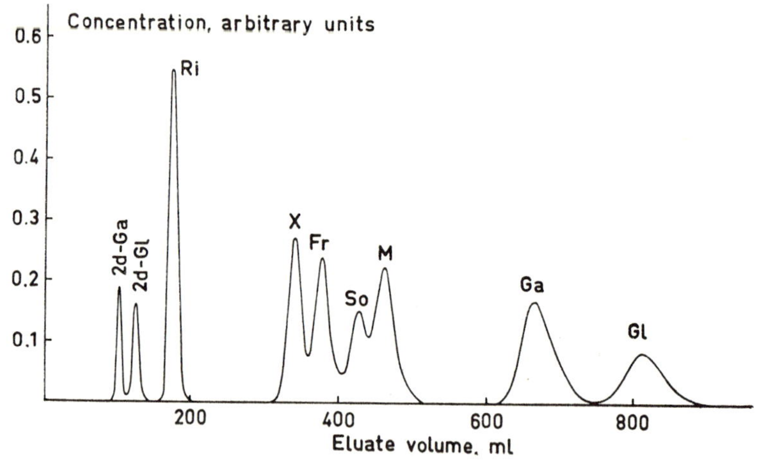

Fig. 2. Separation of various monosaccharides in 89 % ethanol. Resin bed 6 × 760 mm. Flow rate 1.8 ml cm^{-2} min^{-1}. 2d—Ga = 2-Deoxy-galactose, 0.08 mg; 2d—Gl = 2-Deoxy-glucose, 0.08 mg; Ri = Ribose, 0.08 mg; X = Xylose, 0.08 mg; Fr = Fructose, 0.30 mg; So = Sorbose, 0.30 mg; M = Mannose, 0.20 mg; Ga = Galactose, 0.23 mg; Gl = Glucose, 0.23 mg.

Table 3. Comparison between the areas of the elution curves in two runs (I and II).

	Area of elution curve		Area relative to that of ribose		Deviation from mean
	I	II	I	II	%
Ribose (0.10 mg)	13.82	15.00	—	—	—
Fucose (0.10 mg)	6.19	6.53	0.448	0.435	1.5
Mannose (0.20 mg)	11.76	12.55	0.851	0.837	0.8
Galactose (0.30 mg)	14.22	14.91	1.029	0.994	1.7
Glucose (0.30 mg)	6.45	6.89	0.467	0.459	0.9

in the analyzer are not affected when such variables as column diameter and flow rate are altered. One drawback is that the tubings used for feeding the eluate are exhausted very rapidly, which can interfere with the evaluation of the chromatograms for quantitative purposes.

To avoid this difficulty the whole eluate pumped through the column was, in the runs described in this paragraph, mixed with the reagents after the peristaltic pump.

The reproducibility of the method is demonstrated by the results (Table 3) from two experiments (I and II) made under identical conditions. A comparison between the first two columns shows that the areas of the elution curves corresponding to all monosaccharides are larger (5—9 %) in run II than in run I. This is explained by variations in the analyzing system (rate of feed of reagent solutions and other undesired variations). Columns 3 and 4 in the same table show the values recalculated using ribose as an internal standard. It is seen that in this method of calculation the variations are largely compensated for. The deviations from the mean amount to 0.8—1.7 %.

Similar calculations have been carried out in other experiments. In an experiment with the same sugars, mannose was chosen as internal standard. In two runs the area corresponding to the internal standard differed by a factor of 1.162. The maximum deviation from the mean, observed with ribose, was 0.9 %, whereas glucose exhibited a deviation of 0.1 %. These results show that an excellent compensation for variations in the analyzing system can, in favorable systems, be attained by using a suitable internal standard.

With certain mixtures of monosaccharides less favorable results were observed. In some experiments with the same hexoses but with arabinose instead of ribose it was found that the variations in the areas of the arabinose curves differed more than the areas of the other sugars. In an unfavorable series of experiments the areas of the mannose curves differed by a factor of 1.107. This variation can be considered as normal. Using mannose as internal standard the following deviations were observed with the other monosaccharides: fucose 0.6 %; arabinose 5.0 %; galactose 1.4 %, and glucose 0.6 %. A possible explanation is that variations in the relative amount of acid fed into the analyzing system occurred and that these variations have a different effect upon various monosaccharides. The choice of internal standard or internal standards is, therefore, of great importance. No doubt further improvements are possible as far as the quantitative analysis of the eluates is concerned. Such possibilities are being studied at present.

INFLUENCE OF THE FLOW RATE UPON THE SEPARATION

No systematic study of the influence of the flow rate upon the chromatographic separation of monosaccharides has been reported earlier, the chief reason being that certain mechanical troubles, *e.g.* leakage of the pump used for feeding the eluant, occurred when attempts were made to run the elution at high speed. These difficulties have now been overcome.

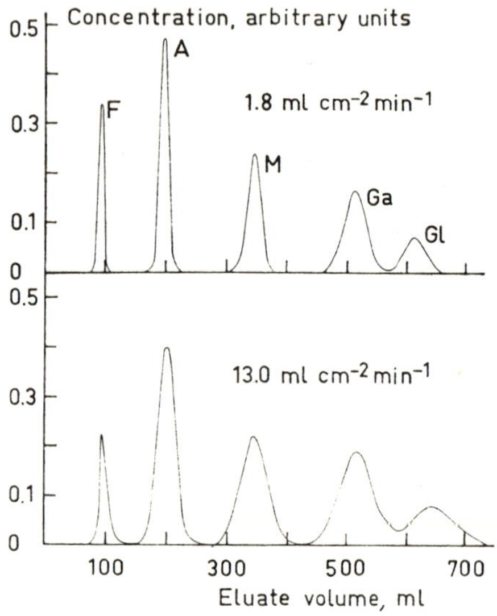

Fig. 3. Influence of the flow rate. Resin bed 6 × 430 mm. Ethanol concentration 92 %.
F = Fucose, 0.10 mg (0.25); A = Arabinose, 0.10 mg (0.25); M = Mannose, 0.20 mg (0.50); Ga = Galactose, 0.30 mg (0.75); Gl = Glucose, 0.30 mg (0.75).
Larger quantities were added in the run represented by the lower chromatogram. The amounts added in this run are given within parantheses.

Some typical results obtained with a sample containing fucose, arabinose, mannose, galactose, and glucose are reproduced in Fig. 3. The upper chromatogram shows that at low flow rate (1.8 ml cm^{-2} min^{-1}) there is under the working conditions chosen, a complete separation of the four first-mentioned sugars and a slight overlapping of the curves corresponding to galactose and glucose. An increased flow rate resulted in a successively increased overlapping of these two elution curves but as can be seen from the lower chromatogram fucose, arabinose, mannose are completely separated from each other even at a flow rate of 13.0 ml cm^{-2} min^{-1}. The separation of galactose and glucose is still good enough to permit a quantitative determination of these sugars by evaluating the chromatograms according to the conventional method used in amino acid analysis.[8] In the run represented by the upper chromatogram the time of elution of the last sugar was 21 h whereas the corresponding time at the increased flow rate was 3.5 h. At the higher flow rate the pressure drop in the column was about 60 atmospheres.

A prerequisite for successful separations at high flow rate is that the eluant be preheated to the elution temperature before it enters the resin bed. For this purpose a jacketed tube with water circulated from a thermostat was inserted before the column.

A few runs have been carried out at even higher flow rates. Successful separations were still observed. Due to the high pressure required to pump the solution through the column mechanical difficulties arose and the results were not reproducible. A better fractionation of the resin particles would no doubt permit reliable results even at higher flow rate. The results given here show that even with non-uniform resin particles reliable chromatograms permitting quantitative determinations of monosaccharides in complicated mixtures can be obtained within reasonable time.

The financial support of the *Swedish Technical Research Council* is gratefully acknowledged. Thanks are also due to Dr. H. W. Holy, Technicon Instruments Co., Chertsey, Great Britain, for gifts of specially prepared ion exchange resins.

REFERENCES

1. Arwidi, B. and Samuelson, O. *Svensk Papperstid.* **68** (1965) 330.
2. Schrager, J. *Nature* **201** (1964) 702, 1220.
3. Samuelson, O. *Ion Exchange Separations in Analytical Chemistry*, Almqvist and Wiksell, Stockholm; Wiley and Sons, New York 1963.
4. Arwidi, B. and Samuelson, O. *Svensk Kem. Tidskr.* **77** (1965) 84.
5. Rückert, H. and Samuelson, O. *Svensk Kem. Tidskr.* **66** (1954) 337.
6. Rückert, H. and Samuelson, O. *Acta Chem. Scand.* **11** (1957) 315.
7. Jonsson, P. and Samuelson, O. *To be published.*
8. Moore, S., Spackman, D. H. and Stein, W. H. *Federation Proc.* **17** (1958) 1107; *Anal. Chem.* **30** (1958) 1190.

Received March 26, 1965.

Part XII

CHROMATOGRAPHY OF NONIONIC COMPOUNDS

Editor's Comments on Papers 44 Through 48

44 SARGENT and RIEMAN
Salting-Out Chromatography: I. Alcohols

45 SHERMA and RIEMAN
Solubilization Chromatography: III. Ethers, Carboxylic Acids, and Hydrocarbons

46 SCOTT
Analysis of Urine for Its Ultraviolet-Absorbing Constituents by High-Pressure Anion-Exchange Chromatography

47 HELFFERICH
"Ligand Exchange": A Novel Separation Technique

48 DE HERNANDEZ and WALTON
Ligand Exchange Chromatography of Amphetamine Drugs

Ion-exchange resins have hydrocarbon skeletons and act like organic solvents. They "dissolve" or absorb organic compounds (and some inorganic compounds, too) even if they have no ionic charge. As Moore and Stein wrote in Paper 36, "The rate of travel on a column of sulfonated polystyrene resin . . . is a result of the affinity of the resin for both the ionic and the non-ionic portions of the molecule."

Where nonionic compounds are concerned, the main chromatographic function of the ions of the resin is to become solvated and thus to cause the resin to swell and become permeable. Their charge sign makes only a minor difference; both cation- and anion-exchange resins are effective absorbents, as we noted with respect to the sugars (Part XI). Cation- and anion-exchange resins behaved very similarly in the systems described in Papers 44 and 45.

Papers 44 and 45 were chosen from several papers by W. Rieman and his students at Rutgers University on the topics "salting-out chromatography" and "solubilization chromatography" (see also Sherma and Rieman, 1958, 1959). In *salting-out chromatography* (Paper 41) the solutes are water-soluble, polar compounds like alcohols, and they are "salted out" of the water and driven into the resin by adding ammonium sulfate. They are eluted from the resin by ammonium sulfate solutions of successively lower concentrations. The higher-molecular-weight alcohols come out last. In *solubilization chromatography* (Paper 45) the solutes are nonpolar, like aromatic hydrocarbons, or of low polarity, like alcohols with long hydrocarbon chains or ethers. They are absorbed by the resins from aqueous acetic acid solutions and eluted first by 6 *M*, then by 8 *M* acetic acid. Aromatic hydrocarbons emerge from the column in order of their molecular weights, the smallest molecules first. The authors note, "Solutes are differentially dissolved from the resin phase into the outside phase."

Solubilization chromatography with mixed aqueous–nonaqueous solvents is common practice today. Absorption is a competitive process between the solvent and the resin, and distribution coefficients are very sensitive to changes in solvent composition. Three factors, at least, affect the absorptive power of the resin: the hydrocarbon matrix, the functional groups, and the counter-ions. The chromatography of a number of aromatic ethers, esters, and amides on several resins, both cation and anion exchanging, having different counter-ions was studied by Larson et al. (1973). A very important factor is the acidic or basic character of the solute, even when the ionization constants are very low. Ionization is accentuated inside the resin. Cation-exchange resins absorb cations of weak bases and promote their binding; they also exclude anions of weak acids, because of the Donnan equilibrium, and this effect becomes noticeable at pH values some three units lower than the pK_a of the acid. Singhal and Cohn (1972) exploited this effect in what they called "ion exclusion chromatography," and used it to separate nucleic acid derivatives.

Paper 46 describes a tour de force of ion-exchange chromatography of nonionized and weakly ionized organic compounds, the resolution of some 140 ultraviolet-absorbing constituents of urine. An anion-exchange resin of very fine and uniform bead size was used, with an aqueous acetic acid–sodium acetate eluent of constant pH whose concentration was increased continuously up to 6 *M*. Besides being a model example of modern high-resolution liquid chromatography, this paper shows clearly the effects of acidic and basic character; bases are eluted first, acids last. Above all, the same aromatic character that makes these solutes absorb ultraviolet radiation favors their retention by the ion-exchange resin, which is of the polystyrene type. If there is one principle that guides the chromatography of organic compounds on ion-exchange resins, it is that aromatic polymer structures attract aromatic solutes.

Paper 47 is the original reference to ligand-exchange chromatography, an example of which we noted in Paper 30. Electron-donating ligands, normally nitrogen containing bases, are bound to a cation exchanger through their complexes with metallic ions such as Cu^{2+} or Ni^{2+}, and are displaced by solutions of ammonia. Helfferich's paper raised high hopes for a powerful and versatile method of chromatography that have only partly been fulfilled. Helfferich's later published experiments (Helfferich, 1962) were particularly promising because the ligand that he concentrated from 10 liters of a 10-ppm solution (paragraph 3 of Paper 47) was 1,3-diamino-2-hydroxypropane, whose hydroxyl group and two amino groups coordinate with the metal ion to form an unusually stable complex. In favorable cases like this, ligand exchange comes into its own. A similar example is the chromatography of amino sugars on a copper-loaded cation-exchange resin (Navratil et al., 1975). In general, however, ligand-exchange chromatography has the following limitations:

1. The metal ions must bind strongly enough to the resin not to be stripped by the ammonia. This requirement limits the choice of metal ions and of exchangers. A small metal-ion leakage can be tolerated, however, and can be compensated by adding metal salt to the eluent.

2. Ligands that form neutral or negatively charged metal complexes, like amino acids, easily strip the metal ions from the resin, although, even here, the stripping may be compensated in some cases by adding metal salt to the eluent.

3. The metal-ligand attraction is superimposed on a matrix-ligand attraction, which gives rise sometimes to broad bands and "tailing." Again the choice of resins may be restricted; an example is seen in Paper 48.

4. A serious limitation is that ligand-exchange chromatography is performed in a strongly alkaline solution, and many organic compounds are hydrolyzed or easily oxidized under these conditions.

Nevertheless, ligand-exchange chromatography has proved to be useful in amino acid analysis, in the separation of amino acids from peptides, and in a number of specialized applications, one of which is illustrated in Paper 48. This paper describes the chromatography of amphetamine, metamphetamine, and related drugs. The resolution was adequate but not good, and the running times were rather long. The acrylic-type resin used in this work is now available in finer and more uniform granules, and there is no question that these separations could now be made in less time with better resolution.

An acrylic resin was used, rather than sulfonated polystyrene, for two reasons. First, functional carboxyl groups hold metal ions more strongly than do sulfonic groups; second, the polystyrene matrix holds aromatic solutes very strongly, giving broad bands and bad tailing. A drawback to the acrylic resin was its softness, which meant that the columns could not be run under high pressures.

The preliminary peaks that appeared at the void volume, labeled "decomposition products" in the paper, were due to small amounts of oxidation products that absorbed very strongly in the ultraviolet.

Recent progress in the chromatography of organic compounds on ion-exchange resins, including ligand-exchange and "matrix-affinity" chromatography, is reviewed by Walton (1974).

REFERENCES

Helfferich, F. 1962. *J. Amer. Chem. Soc., 84,* 3237, 3242.
Larson, P., E. Murgia, Tong-Jung Hsu, and H. F. Walton. 1973. *Anal. Chem., 45,* 2306.
Navratil, J. D., E. Murgia, H. F. Walton. 1975. *Anal. Chem., 47,* 122.
Sherma, J., and W. Rieman. 1958. *Anal. Chim. Acta. 18,* 214.
——, and W. Rieman. 1959. *Anal. Chim. Acta, 19,* 134.
Singhal, R. P., and W. E. Cohn. 1972 *Anal. Biochem., 45,* 585.
Walton, H. F. 1974. *J. Chromatog., 102,* 57.

44

Reprinted from *J. Phys. Chem.*, **61**, 354–358 (Mar. 1957)

SALTING-OUT CHROMATOGRAPHY. I. ALCOHOLS

By Roger Sargent and William Rieman III

Ralph G. Wright Chemical Laboratory, Rutgers University, New Brunswick, New Jersey

Received October 18, 1956

The separation of water-soluble non-electrolytes by elution through a column of ion-exchange resin with an aqueous salt solution as eluent is designated as "salting-out chromatography." Variables such as the functional group, mesh size and cross-linkage of the resin, the sample size and flow rate were studied with several alcohols as solutes. The usefulness of the method is demonstrated by a separation of nine alcohols in a 12-hr. elution.

A process for the chromatographic separation of water-soluble non-electrolytes with ion-exchange resins and aqueous salt solutions as the stationary and mobile phases, respectively, was reported recently.[1] The designation, "salting-out chromatography" is proposed for this method.

In the interpretation of the results, use is made of the equation[2]

$$U^* = CV + V \qquad (1)$$

where U^* is the volume of effluent collected from the addition of the sample to the peak, and C is the distribution ratio of the alcohol (quantity of the alcohol in the resin of any plate divided by the quantity in the interstitial solution of the same plate). The interstitial volume of the column is represented by V.

Experimental

Alcohols.—All alcohols were of the best grade obtainable. No further purification was necessary for this study.

(1) R. Sargent and W. Rieman, *J. Org. Chem.*, **21**, 594 (1956).

(2) J. Beukenkamp, W. Rieman and S. Lindenbaum, *Anal. Chem.*, **26**, 505 (1954).

Resins.—Three types of anion-exchange resins were used. Dowex 1, Dowex 2 and Dowex 3 were converted to the sulfate form (in columns) by the passage of 3.0 M ammonium sulfate until the effluent gave a negative test for chloride. The resins were removed from the columns, slurried with the desired eluent and poured into columns, 30 cm. × 3.90 cm.[2] (Ace Glass Co., drawing T-3422). After 5 min. the excess eluent was drained. An additional 1 l. of eluent was passed through the bed at a maximum flow rate with occasional tapping of the side of the column. This procedure ensures air-free and reasonably uniform packing of the resin. Considerable shrinking or swelling of the resin occurs when the concentration of the eluent is changed markedly. In order to avoid the formation of channels in the column from this cause, the column was repoured before determining distribution ratios with a different concentration of eluent. Two cation exchangers, Dowex 50 and Amberlite IRC-50(H) were similarly converted to the ammonium form and packed.

The interstitial volume, V, was found to be 44.0% of the bed volume for Dowex 1-X8, 200–400 mesh. The values of V for Dowex 50-X8, 100–200 mesh and 200–400 mesh were, respectively, 36.6 and 39.1% of the bed volume. Values of 44.0% for Dowex 1 and 38.0% for Dowex 50 were used to evaluate the elution data.

Elution and Analysis.—The eluent was drained to the top of the bed of resin and the sample of approximately 0.1 mmole of the alcohol in 1 ml. of water was pipetted onto the bed without disturbing the resin. The sample was drained

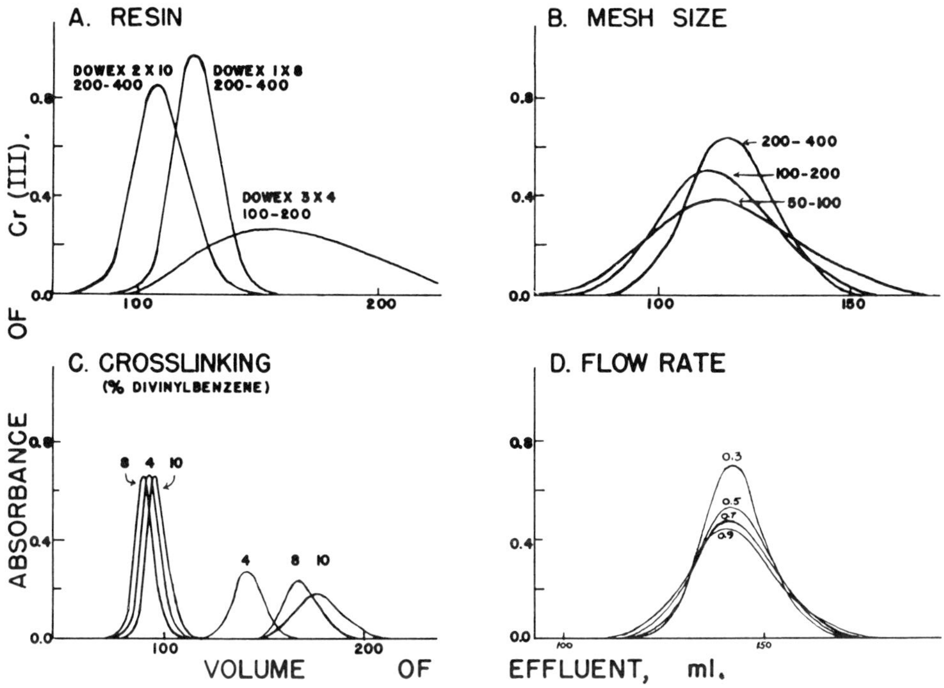

Fig. 1.—Variables in salting-out chromatography.

to the top of the bed, and the sides of the column were washed with two 1-ml. portions of eluent. The alcohol was then eluted with a flow rate of 0.5 cm. per min. unless otherwise specified. Fractions of 6 ml. were collected with the aid of a fraction collector.[2] The fractions were analyzed by mixing them with 0.1 N sodium dichromate dissolved in concentrated sulfuric acid and subsequent measurement of the absorbance of the resulting Cr(III).[3,4]

Solubility Studies.—Salting-out constants for *n*-propyl and *sec*-butyl alcohol were obtained by the determination of relative solubility by shaking an excess quantity of alcohol with 2 to 4 M ammonium sulfate solutions. The relative quantity of alcohol in the aqueous phase was determined by oxidation of an aliquot with dichromate.[3]

Results and Discussion

Resins.—Propylene glycol was eluted with 3.0 M ammonium sulfate from 20-cm. columns of each resin. The elution graphs for Dowex 50 and Amberlite IRC-50 were very similar to the graphs obtained for Dowex 1 and Dowex 3 (Fig. 1a), respectively. Dowex 1 and Dowex 50 were chosen as the best resins for the separation of alcohols.

Mesh Size.—Similar elutions of propylene glycol were run on 20-cm. columns of Dowex 1-X10 of various particle sizes. Figure 1b shows that the smallest mesh, 200–400, produced the sharpest elution graph. No significant shift of U^* was observed for a change in mesh size of resin.

Cross-linking.—The effect of cross-linking was investigated by eluting a mixture of methyl and ethyl alcohols with 3.0 M ammonium sulfate from columns containing 20 cm. of Dowex 1-X10, Dowex 1-X8 and Dowex 1-X4. All resins were 200–400 mesh. Although the selectivity of the 10% divinylbenzene resin (Fig. 1c) was the greatest, Dowex 1-X8 was chosen as the resin that gave good selectivity and the best elution graph. Slight differences in the peaks of methyl alcohol (first peak) with the various resins probably are due to the physical inability to prepare columns whose interstitial volumes are identical, and not to the nature of the resin itself.

Flow Rate.—Figure 1d shows the gradation from narrow to broad elution graphs when ethyl alcohol was eluted from 16.8 cm. of Dowex 1-X8, 200–400 mesh, at flow rates ranging from 0.3 to 0.9 cm. per min. A mixture which is very difficult to separate may require the slower flow rate; conversely, the flow rate for less-difficult separations may be increased. No change in the position of U^* was discernible for a change of flow rate.

Sample Size.—An almost linear decrease in the value of the distribution ratio C was found with increasing the amounts of alcohol eluted. This effect shows up as a shift of the leading edge and the peak of the elution graph toward the left while the trailing edge remains constant. The shift is much greater when the alcohol has a large distribution ratio. The quantity of alcohol that may be taken for an elution so that the value of C remains within 2% of its limiting value, was found to be 0.23, 0.11 and 0.088 mmole for alcohols with C values of 1.02, 5.55 and 12.4, respectively. As a first approximation the sample size in millimoles, S, is governed by the empirical equation $S = 0.022\ A/\log\ (C + 1)$, where A is the cross-sectional area of the column.

Choice of Salt.—The only convenient method known to the authors for the determination of the

(3) R. Sargent and W. Rieman, *Anal. Chim. Acta*, **14**, 381 (1956).
(4) R. Sargent and W. Rieman, *J. Phys. Chem.*, **60**, 1370 (1956).

TABLE I

VALUES OF C FOR VARIOUS ALCOHOLS AND RESINS

Resin	Dowex 1-X8						Dowex 50-X8				
$[(NH_4)_2SO_4]$	0.0	0.5	1.0	2.0	3.0	4.0	0.0	0.5	1.0	2.0	3.0
Glycerol	0.427	0.453	0.473	0.622	0.843	1.16	0.586	0.635	0.688	0.901	1.20
Propylene glycol	.477	.647	.740	1.29	2.21	3.83	.644	.856	1.08	1.77	3.07
Methyl alcohol	.507	.670	.804	1.15	1.68	2.33	.839	.950	1.13	1.59	2.29
Ethyl alcohol	.673	.843	1.12	2.00	3.72	...	.839	1.08	1.46	2.46	4.57
Isopropyl alcohol	.753	1.02	1.53	3.48	7.76	...	.744	1.21	1.61	3.58	8.34
t-Butyl alcohol	.883	1.19	1.96	5.12	13.4	...	.662	1.20	1.90	4.78	12.9
n-Propyl alcohol	1.28	1.71	2.57	5.45	12.3	...	1.03	1.62	2.35	5.01	...
sec-Butyl alcohol	1.56	2.26	3.79	9.30	26.5	...	1.07	1.73	2.88	7.48	...
Isobutyl alcohol	2.47	3.34	5.17	12.3	...	...	1.33	2.20	3.38	8.74	..
n-Butyl alcohol	2.37	3.52	5.67	13.8	...	...	1.28	2.17	3.53	8.44	...
n-Amyl alcohol	6.63	11.0	...	...	...	...	2.69	5.03	8.33	...	...

very small concentrations of alcohol in the effluent fractions is the spectrophotometric determination of chromic ion produced by the reduction of dichromate.[4] Oxidizing anions such as nitrate and reducing ions such as the halides would interfere in this determination. These facts imposed a severe restriction on the salts suitable for use as eluents. Actually, only ammonium, sodium and magnesium sulfates were tried. Ammonium sulfate was best because of its large solubility and great salting-out power.

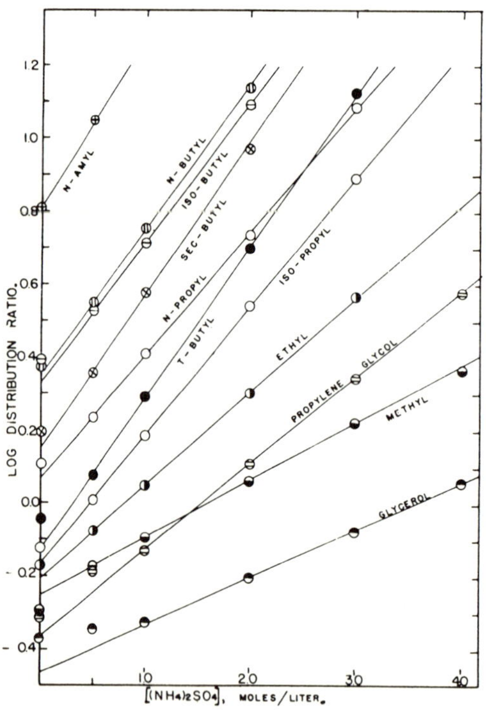

Fig. 2.—Plot of log C for Dowex 1.

Salting-out Constants.—Values of C (Table I) were calculated from the data of the various elutions by eq. 1. The logarithms of these values are plotted against the concentrations of ammonium sulfate for Dowex 1 in Fig. 2. These plots follow fairly closely the equation

$$\log C = \log C_0 + kM \qquad (2)$$

where C_0 is the distribution ratio in pure water, k is the salting-out constant and M is the molarity of ammonium sulfate. The constants of eq. 2 are given in Table III.

Salting-out constants in the absence of resin have been determined[5] by use of eq. 3

$$\log S = \log S_0 - kM \qquad (3)$$

where S is the solubility in ammonium sulfate of molarity M and S_0 is the solubility in water. The agreement of the k values in eq. 2 and 3 is shown by solubility measurements for *n*-propyl and *sec*-butyl alcohol. Table II lists the logarithms of the quantities of alcohol (in arbitrary units of absorbance) that were plotted against the concentration of ammonium sulfate to obtain the slope of the line.

TABLE II

SOLUBILITIES IN SOLUTIONS OF AMMONIUM SULFATE

Arbitrary units

$[(NH_4)_2SO_4]$	2.0	2.5	3.0	3.5	4.0
n-Propyl alcohol	1.240	1.063	0.897	0.781	*a*
sec-Butyl alcohol	1.069	0.867	0.601	0.438	0.233

[a] Ammonium sulfate precipitated.

TABLE III

COMPARISON OF VALUES FOR C_0 AND k

Alcohol	Dowex 1-X8 log C_0	Dowex 1-X8 k	Dowex 50-X8 log C_0	Dowex 50-X8 k	Eq. 3 k
Glycerol	−0.465	0.129	−0.282	0.120	
Propylene glycol	− .362	.210	− .188	.224	
Methyl	− .252	.158	− .095	.151	
Ethyl	− .208	.256	− .083	.243	
Isopropyl	− .170	.353	− .135	.350	
t-Butyl	− .125	.414	− .130	.408	
n-Propyl	.070	.336	.043	.330	0.341
sec-Butyl	.150	.422	.039	.419	.420
Isobutyl	.33	.38	.130	.407	
n-Butyl	.35	.40	.125	.405	
n-Amyl	.82	.45	.44	.50	

Table III reveals the following facts. (1) The C_0 values of the alcohols increase as their hydrophilic nature decreases. (2) Salting-out constants k show a rough tendency to increase as the hydrophilic nature of the alcohol decreases. (3) For any one alcohol, the k values for Dowex 1, Dowex 50 and eq. 3 agree within experimental error.

(5) Setschenow. *Z. physik. Chem.*, **4**, 117 (1889).

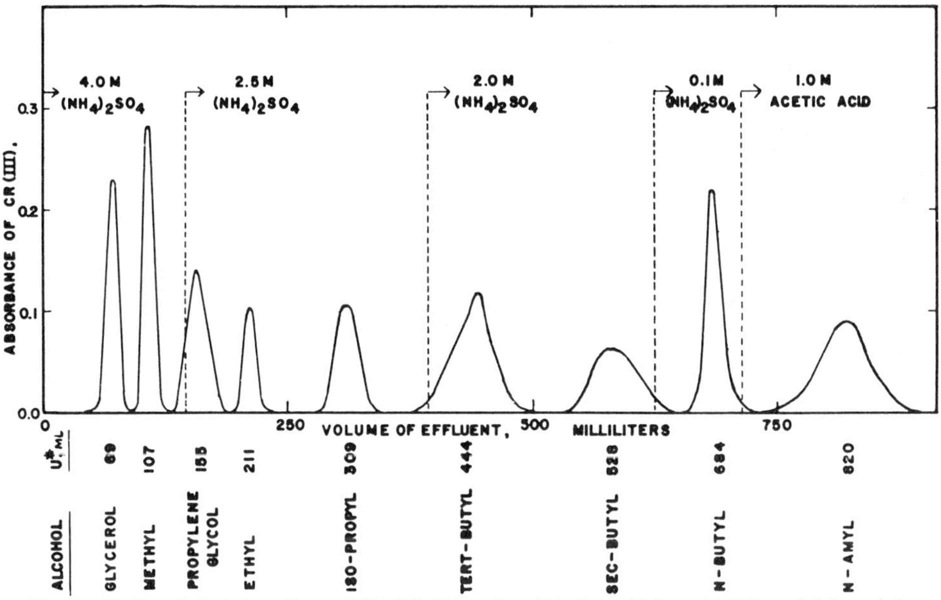

Fig. 3.—Elution of alcohols on Dowex 1X8, 200–400 mesh, sulfate form, 32.0 cm. × 2.28 cm.², 0.6 cm./min.

In salting-out chromatography, a linear increase in the electrolyte concentration of the eluent produces an exponential increase in the distribution ratio for a given alcohol, *i.e.*, the affinity of the resin for the alcohol is greater for high eluent concentrations. This is a major difference from ion-exchange chromatography where an increase in eluent concentration decreases the distribution ratio of the sample constituent.

Separation of Alcohol Mixtures.—Figure 3 shows the separation obtained for a mixture of nine alcohols and the changes of eluent that are required for the separation. Acetic acid was used to "salt-in" the amyl alcohol to the aqueous phase so that it is eluted more rapidly than with water. Acetic acid was chosen since it is one of the few organic compounds which will not reduce dichromate in the analytical procedure. The elution with the specified flow rate required 12 hr. for completion. Normal propyl and isobutyl alcohols, not included in the separation, would be eluted under these conditions with U^* values very close to *t*-butyl and *n*-butyl alcohols, respectively. Straight-chain, primary alcohols above amyl are absorbed by the resins so tenaciously that excessively long elutions would ensue.

The conditions for the separation represented by Fig. 3 were found almost entirely by calculations based on the plate theory. The concentration of ammonium sulfate was determined that would give a ratio of the values of $C \geqq 1.5$ which is necessary to separate the alcohols by elution from a 30-cm. column. For example, 4 M ammonium sulfate is needed to separate the first three alcohols. The eluent concentration is then lowered as much as possible for the rapid elution of the higher alcohols. The positions of peak concentration of the alcohols in the effluent, U^*, were calculated and later compared with the experimental values. Sample calculations for U^* of isopropyl alcohol are given below.

$$\log C = \log C_0 + kM$$
$$\log C = -0.170 + (0.353)(4.0)$$
$$C = 17.5$$
$$U^* = CV + V$$
$$U^* = (17.5)(32.1) + 32.1$$
$$U^* \text{ (4.0 } M \text{ ammonium sulfate)} = 594$$

When the eluent was changed to 2.5 M ammonium sulfate at 150 ml., the isopropyl alcohol had moved down the column

$$\frac{150}{594} \times 32.0 = 8.08 \text{ cm.}$$

At the change of eluent, the interstitial volume above the isopropyl alcohol was 8.1 ml. of 4.0 M ammonium sulfate. Before the 2.5 M ammonium sulfate reaches the isopropyl alcohol, it will have moved an additional 0.4 cm. or 8.5 cm. down the column. The remaining interstitial volume is now 23.6 ml.

$$C \text{ (2.5 } M \text{ ammonium sulfate)} = 5.14$$
$$U^* = 150 + 8.1 + (5.14)(23.6) + 23.6$$
$$U^* = 303$$

The calculated values of U^* for the 9 alcohols were, respectively, 70, 109, 154, 206, 303, 444, 549, 694 and 880. The observed values are listed in Fig. 3. Agreement was excellent for most of the alcohols.

The height of the column necessary for the separation was calculated from the results of a preliminary elution on a 24.8-cm. column of the same resin. This elution indicated that the separation of propylene glycol from methyl alcohol would be the most difficult. Application of the equation[2] for the calculation of column height necessary to separate these alcohols gave a value of 32.0 cm., which was used for the separation in Fig. 3.

Mechanism of Salting-out Chromatography.—It is suggested that the mechanism for the chromatographic process is governed by solvation. Consider a ternary mixture of resin, water and a small amount of alcohol. The ionic groups on the resin are solvated chiefly by water because its solvation power and concentration are larger than those of the alcohol. When a salt is added, such as ammonium sulfate, which is not soluble in (not solvated by) alcohol, the resin shrinks because part of the water is withdrawn to solvate the ions of the salt. The salt, by virtue of its removal of water from the resin, has therefore increased the number of solvation positions around the ionic groups of the resin which can now be occupied by the hydroxyl groups of the alcohols. The hydrocarbon tail of the alcohol would be attracted toward the resin matrix. The process is definitely not simple sorption on the resin matrix since the value of C for the elution of the alcohols from a column of polystyrenedivinylbenzene copolymer with 3.0 M ammonium sulfate is zero.

A study of the salting-out chromatography of the ketones is now being pursued in this Laboratory. Additional studies of other organic compounds and more detailed studies of the nature of the process are planned.

Acknowledgment.—The authors express their gratitude to the Dow Chemical Company for generous financial support of this investigation and for the resins used in the work, also to R. M. Wheaton for reading the manuscript and suggesting several changes.

45

Reprinted from *Anal. Chim. Acta,* **20**(4), 357-365 (1959)

SOLUBILIZATION CHROMATOGRAPHY

III. ETHERS, CARBOXYLIC ACIDS, AND HYDROCARBONS

JOSEPH SHERMA AND WM. RIEMAN III

Ralph G. Wright Chemical Laboratory, Rutgers, The State University, New Brunswick, N.J. (U.S.A.)

In previous papers of this series, solubilization chromatography was applied to the separation of alcohols[1], phenols[1], and ketones[2]; and the general principles of the method were discussed. In the present paper, probably the last of the series, solubilization chromatography is applied to ethers, fatty acids, and hydrocarbons; also some experiments with a special resin of low capacity are described which throw further light on the nature of attraction between the resin and the organic solutes.

EXPERIMENTAL

Materials

All ethers, acids, and hydrocarbons used in this study were of the best grade commercially available. No further purification was required.

The resins used in these studies included: the strong-acid cation exchanger, Dowex-50; the strong-base anion exchanger, Dowex-1; the resinous adsorbent with a phenolic matrix and mixed

References p. 365

polar functional groups, Duolite S-30; the "snake-cage" resin made from Dowex-1 and acrylic acid, Retardion 11A8[3]; polystyrene crosslinked with divinylbenzene; and a Dowex-50 resin of approximately one-half normal capacity.

The ratio of the interstitial volume to the bed volume, V/V_B, was determined for several of the resins used[4], and found to be: 0.333 for Dowex-50-X4, Lot No. 4177-40; 0.402 for Dowex-1-X4, Lot No. 1241; 0.371 for Dowex-50W-X8, Lot No. L·3937-40, capacity (Q) = 1.99 meq/g.

Eluents were prepared by dilution of reagent-grade glacial acetic acid and absolute ethanol.

Elutions

The samples were introduced onto the resin bed as previously reported[5]. The flow rate of eluent was 0.4 to 0.5 cm/min, unless otherwise specified. Fractions of approximately 5.0 ml were automatically collected with the aid of a constant-volume collecting device[6].

Analysis

Fractions containing acetic acid were analyzed by mixing them with 5.0 ml of 0.02M sodium dichromate in concentrated sulfuric acid, diluting with 25 ml of distilled water, and measuring spectrophotometrically the resultant Cr(III)[7]. All the solutes studied in this paper reduce dichromate.

Since the substituted benzenes and naphthalenes studied absorb light in the ultraviolet (225 to 289 mμ) region[8], the ethanolic eluates of these solutes were analyzed by ultraviolet absorption spectroscopy.

RESULTS AND DISCUSSION

Elution variables

Since the effects of such elution variables as choice of resin, particle size of resin, flow rate of eluent, sample size, etc. have been clearly established for both salting-out[5] and solubilization chromatography[1], no further study of these variables was made.

Dowex-50-X4 and Dowex-1-X4 were the resins primarily used. The low percentage of crosslinking was used to permit rapid diffusion of the large molecules encountered in this investigation.

One ml of feed solution was used in each elution. The solution contained 0.20 mmole of solute in an aqueous solution of the organic constituent of the eluent, concentrated enough to cause complete dissolution. Small volumes of feed solutions more concentrated in the organic constituent than the eluent itself were shown to cause insignificant change in the position of the elution maxima[1].

Acetic acid was chosen as the eluent for the ethers and acids because the spectrophotometric determination of the chromic ion produced by the reduction of dichromate is the most convenient method available for the determination of small amounts of these compounds in effluent fractions; and acetic acid is one of the few organic compounds not oxidized by dichromate. Ethanol was employed as an eluent for the hydrocarbons.

With each series of solutes studied, the elution variables just mentioned were kept constant at their optimum values while the most important variable, the concentration of the eluent solution, was varied. Values of the distribution ratio, C, were calculated by the equation[6]

$$U^* = CV + V \quad \ldots\ldots\ldots\ldots\ldots\ldots \quad (1)$$

where U^* is the volume of effluent collected from the point of addition of the sample to the peak of the elution graph and C is the distribution ratio of the solute (the total quantity of compound in the resin of any plate divided by the quantity in the solution of the same plate). The interstitial volume of the column is represented by V.

References p. 365

Ethers

Table I shows the distribution ratios, as calculated by equation (1), for the ethers

TABLE I

ELUTION DATA OF ETHERS ON DOWEX-50-X4, HYDROGEN FORM, 200-400 MESH

[HAc]	*0.0*	*1.0*	*2.0*	*4.0*	*6.0*	*Negative slope of plot of log C vs. M*	*Extent of linearity of plot of log C vs. M*
Diisopropyl ether	3.51	3.11	2.88	2.13	1.30	0.043	0–2*M*
Di-*n*-propyl ether	4.94	4.16	3.37	2.08	1.65	0.083	0–6*M*
Ethyl *n*-butyl ether	5.25	4.19	3.60	2.48	1.69	0.083	0–6*M*
Di-*n*-butyl ether	12.5	9.36	6.81	3.90	1.83	0.127	0–4*M*
Anisole	15.2	11.6	9.09	5.36	3.08	0.116	0–6*M*
Diphenyl ether	—	—	54.6	19.2	6.65	0.226	2–6*M*
Diisoamyl ether	—	—	—	—	2.28	—	—
Di-*n*-amyl ether	—	—	—	—	3.03	—	—

with eluents 0 to 6*M* in acetic acid and the results obtained by plotting log *C vs. M*. These results indicate the following:

1. Because of their greater solubility in aqueous solvents, branched-chain compounds are eluted before (have lower distribution ratios than) their straight-chain isomers.

2. The elution behavior of di-*n*-propyl ether and ethyl *n*-butyl ether is very similar, as is to be expected from their similar solubility characteristics.

3. Contrary to the results of salting-out chromatography[5], but similar to the previously published results of solubilization chromatography[1,2], some of the plots of log *C vs. M* are linear for only a limited region. At large molarities of acetic acid, the graphs tend to curve downward and converge. These results indicate that the change in solubility with molarity of acetic acid for these solutes is probably also nonlinear. This was shown to be true for two alcohols[1].

4. Similar to the earlier results of salting-out chromatography[5] and solubilization chromatography[1,2], the slopes of these plots (solubilization constants) are seen to increase as the ethers become less soluble in aqueous solutions.

5. Values of the distribution ratios of diisoamyl and di-*n*-amyl ethers, for eluents less concentrated than 6*M* acetic acid, could not be determined because of unsatisfactory elution graphs. With 4*M* acetic acid, both ethers gave elution graphs which rose to peaks immediately after the effluent volume equalled the interstitial volume, and then tailed for a large volume of effluent. This is probably due to the fact that most of these very large ether molecules are excluded from the resin by virtue of their size and eluted right after the interstitial volume. The molecules which do diffuse inside the resin are trapped there and cause the considerable tailing. Similar behavior was exhibited by di-*n*-hexyl ether in 6*M* acetic acid.

When introduced onto the resin bed with 6*M*, 4*M* and 2*M* acetic acid eluents, diphenyl ether was precipitated as a white solid on the brown column of Dowex-50.

References p. 365

The precipitate redissolved and the band disappeared as the elution took place. The first two eluents gave completely satisfactory elution graphs; the latter yielded a very flat, severely spread peak. These results are probably due to the rates of redissolving in the various eluents used.

The elution data in Table I were used to plan[6,9] the separation of a synthetic mixture of 0.20 mmole of each of five ethers (Fig. 1). The calculated positions for the ether peaks and the actual U^* values were within experimental error. Di-*n*-propyl ether could have been included in the separation instead of ethyl *n*-butyl ether, but under the given elution conditions, it would have overlapped the peak of diisopropyl ether.

The elution behavior of diisoamyl ether, di-*n*-amyl ether, and di-*n*-hexyl ether was studied on Dowex-50-X2, Duolite S-30, and Retardion 11A8 in an attempt to get elution data which could be used to effect the separation of these compounds. None of these resins gave satisfactory results with eluents of a useful concentration range. It would have been possible to get symmetrical elution graphs for these compounds by use of extremely concentrated solutions of acetic acid. However, with these eluents the resin would show no selectivity among the three solutes. In effecting separations by solubilization chromatography, it is always necessary to find an eluent–resin combination which gives good selectivity and elution peaks which are both symmetrical and narrow. As is seen here, this is not always possible.

Carboxylic acids

Aqueous acetic acid and Dowex-50-X4, hydrogen form, were chosen for the study of the solubilization chromatography of the naturally occurring fatty acids. This system obviated any ion-exchange reactions between the acids and the resin phase.

Fig. 2 shows the separation of approximately 0.20 mmole of each of five acids and the plots of log C *vs.* M from which the conditions for the separation were chosen. Table II contains a summary of the elution data shown in the figure.

It is seen that with the acids, plots of log C *vs.* M of acetic acid are linear over the entire concentration range studied. There seems to be no way of predicting linearity or nonlinearity for a given compound.

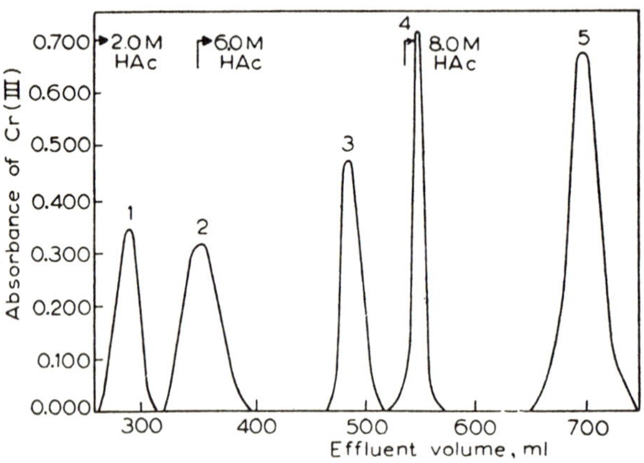

Fig. 1. The separation of a five-component ether mixture. 102.0 cm × 2.28 cm² Dowex-50-X4, 200-400 mesh, hydrogen form, 0.28 cm/min. Aqueous acetic acid eluent changed at 350 ml and 550 ml. 1. diisopropyl ether; 2. ethyl *n*-butyl ether; 3. di-*n*-butyl ether; 4. anisole; 5. diphenyl ether.

References p. 365

TABLE II

ELUTION DATA OF ACIDS ON DOWEX-50-X4, HYDROGEN FORM, 200-400 MESH

[HAc]	C 0.0	C 2.0	C 4.0	C 6.0	Negative slope of plot of log C vs. M	Extent of linearity of plot of log C vs. M
Butyric acid	2.17	1.81	1.45	1.24	0.041	0-6*M*
n-Caproic acid	4.99	3.35	2.01	1.40	0.092	0-6*M*
n-Caprylic acid	18.0	7.74	3.29	1.43	0.184	0-6*M*
n-Capric acid	—[a]	24.4	6.22	1.63	0.294	2-6*M*

[a] Too large to be accurately determined.

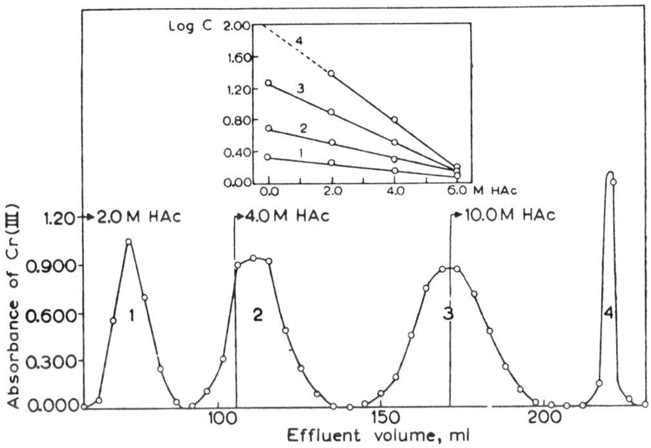

Fig. 2. Plots of log *C* *vs.* *M* for four acids and the separation of a four-component acid mixture planned from these data. 19.0 cm × 3.90 cm² Dowex-50-X4, 200-400 mesh, hydrogen form, 0.28 cm/min. Aqueous acetic acid changed at 105 ml and 170 ml. 1. butyric acid; 2. *n*-caproic acid; 3. *n*-caprylic acid; 4. *n*-capric acid.

From the length of the column used, it is obvious that this series was the easiest to separate of those studied in this and past work. This is because the constituents differed from each other by two carbon atoms, and had correspondingly large differences in their relative solubilities in aqueous solutions of acetic acid.

The twelve-carbon member, lauric acid, was precipitated when introduced onto a bed of resin equilibrated with even 6*M* acid. Some of the compound then appeared in the effluent immediately after the interstitial volume, but most of it was irreversibly sorbed and did not move from the top of the bed even after a large volume of effluent had passed. For this reason the study was stopped with *n*-capric acid.

Hydrocarbons

Because of the partially successful application of reversed-phase partition chromatography to the separation of methyl-substituted benzenes and naphthalenes with aqueous solutions of ethanol as the mobile phase, these eluents were initially tried in an attempt to improve these separations by use of solubilization chromatography.

References p. 365

However, solutions as concentrated as 50% ethanol by volume proved unsuccessful with Dowex-1-X4, Dowex-50-X4, and crosslinked polystyrene as the solid phase. In each case, severe tailing of the elution graphs occurred.

Acetic acid proved a satisfactory eluent for these hydrocarbons in that symmetrical elutions were in general obtained with Dowex-1 and Dowex-50. The use of crosslinked polystyrene again resulted in elution graphs with severe spread. This is due to the fact that polystyrene does not swell in aqueous solutions of acetic acid and adsorption rather than true absorption is the factor controlling equilibrium. Tailing is characteristic of adsorption phenomena. There is no obvious explanation why acetic acid should prove a much superior eluent than ethanol for the hydrocarbons when they were shown[2] to effect ketone elutions in a very similar manner.

A study was made of the effect of concentration of acetic acid on the elution behavior of these various benzene and naphthalene derivatives. Table III shows the results for Dowex-50-X4 and Dowex-1-X4 in terms of values of C. These data show, as was found for the alcohols[1], that Dowex-1 is a much stronger absorbent for organic solutes than is Dowex-50 because of its more hydrocarbon nature.

Toluene was visibly precipitated on the column equilibrated with 1.0M acetic acid and the elution graph spread severely. The same results were found for *m*-xylene with 2.0M acetic acid, mesitylene with 4.0M acetic acid, and durene with 6.0M acetic acid.

TABLE III

ELUTION DATA OF HYDROCARBONS ON DOWEX-50-X4, HYDROGEN FORM, AND DOWEX-1-X4, ACETATE FORM, 200-400 MESH

	Dowex-50-X4 C					*Dowex-1-X4* C	
[*HAc*]	*1.0*	*2.0*	*4.0*	*6.0*	*8.0*	*6.0*	*8.0*
Benzene	6.62	5.69	4.21	2.69		4.55	
Toluene	10.2	8.21	5.36	3.00		5.53	
m-Xylene		14.2	7.76	3.75		7.41	
Mesitylene			13.0	5.12		9.66	
Naphthalene					2.92	15.8	7.48
β-Methylnaphthalene					3.37	22.4	9.40
1,4-Dimethyl-naphthalene					4.21	30.2	11.0

The plots of log C *vs.* M for the benzenes on Dowex-50-X4 begin to curve downward and converge at 4.0M acetic acid.

The elution data of Table III were sufficient to calculate[5,8] the conditions for the separation of 0.20 mmole of each of the seven solutes studied. Fig. 3 shows the conditions used and results obtained for this separation. The calculated positions of the peaks were 305, 359, 462, 542, 774, 860 and 952 ml, as compared with actual U^* values of 294, 354, 464, 549, 770, 869, and 956 ml.

The final two components were not completely separated by use of the given conditions because of the severe spread of the elution peaks. A complete separation of these two peaks could not be accomplished on a convenient-sized column of Dowex-1-X4.

References p. 365

Dowex-1-X4 was chosen in preference to Dowex-50-X4 for this separation because of the greater selectivity exhibited by the former resin for the compounds studied. An elution similar to that shown in Fig. 3 would require a column of Dowex-50-X4 well in excess of 100 cm.

Low-capacity resins

To test the usefulness of low-capacity resins in solubilization chromatography, several alcohols were eluted with acetic acid from a Dowex-50-X8 resin of less than one-half normal exchange capacity. Table IV compares the results of these elutions with those for the same alcohols on a Dowex-50-X8 resin of usual capacity and similar column dimension.

It is seen that although the low-capacity resin exhibits great selectivity among the three solutes studied, the severe spread of the elution peaks more than offsets this beneficial effect. It appears that resins of low capacity are useless for the chromatography of organic compounds of high molecular weight.

Theory of the process

Solubilization chromatography is a partition chromatography on ion-exchange resins. It differs from usual partition chromatography in that the two "phases" between which the solute is partitioned consist of the same solvent, and interactions between the "solid support" and the solute are important. Such interactions include (1) ion-dipole attractions of the functional group of the resin and/or its counter ion for the solute, and (2) London dispersion forces between the solute and the hydrocarbon part of the resin.

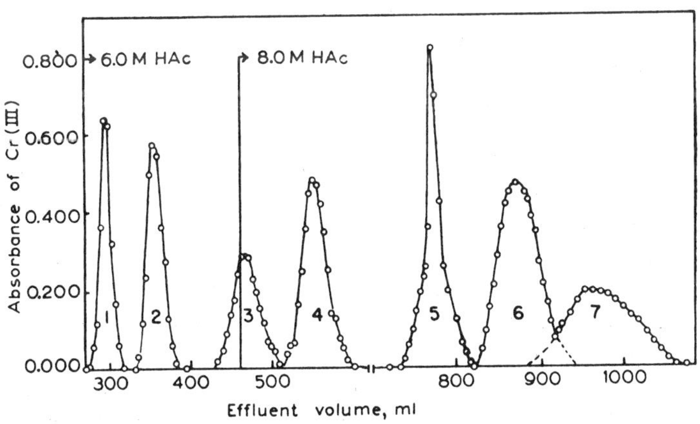

Fig. 3. Separation of a seven-component hydrocarbon mixture. 60 cm × 2.28 cm² Dowex 1 X4, 200–400 mesh, acetate form, 0.28 cm/min. Aqueous acetic acid eluent changed at 460 ml. 1. benzene; 2. toluene; 3. *m*-xylene; 4. mesitylene; 5. naphthalene; 6. β-methylnaphthalene; 7. 1,4-dimethylnaphthalene.

Consider a mixture of Dowex-50-X8, hydrogen form, 2.0*M* acetic acid, and a very small quantity of *n*-octyl alcohol. Both acetic acid and water molecules will be uni-

References p. 365

TABLE IV

EFFECT OF RESIN CAPACITY ON ELUTION BEHAVIOR

Resin	*Dowex-50W-X8* *Q = 1.99 meq/g* *lot No. L3937-40* *bed = 8.3 cm × 3.9 cm²*		*Dowex-50-X8* *Q = 5.2 meq/g* *lot No. 4226-28* *bed = 21.1 cm × 3.9 cm²*	
Eluent	8.0*M* $HC_2H_3O_2$		6.0*M* $HC_2H_3O_2$	
	C	U_T	*C*	U_T
n-Amyl alcohol	4.57	91.6 ml	1.43	53.5 ml
n-Hexyl alcohol	7.66	193.2	1.46	70.4
n-Octyl alcohol	a	a	1.85	101.2

[a] Irreversibly adsorbed on column.
U_T = Total volume of eluent used to elute the solute completely.

formly absorbed throughout the interior of the resin particles (REICHENBERG AND WALL[10] consider the simplest explanation to be that the resin absorbs water which in turn dissolves the acetic acid). The functional groups and counter ions of the resin will no doubt be almost exclusively solvated by water molecules rather than acetic acid molecules because of the greater number and greater polarity of the former. Since this leaves less free water inside the resin in which acetic acid can dissolve, the resinate ions can be said to salt-out acetic acid from the resin phase. Also, the small chain length of acetic acid molecules indicates that little van der Waals interactions will obtain between the benzene rings of the matrix and the acetic acid molecules. Hence the molarity of acetic acid inside the resin will be lower than the molarity outside the resin[10]. Even though this is true, the small amount of octyl alcohol will go almost completely inside the resin due primarily to the strong van der Waals interactions with the resin matrix.

The separations achieved by solubilization chromatography are based on the fact that solutes are differentially dissolved from the resin phase into the outside phase. Hence, the greatest separation among a group of constituents is obtained by use of an eluent containing a small amount of acetic acid since the compounds will have the greatest relative solubility differences in such an eluent. However, the eluent must contain enough acetic acid to give narrow elution graphs.

The results cited above with low-capacity resins indicate that van der Waals forces are much more important than polar interactions between a solute molecule and the resin. The resins of low capacity have a more hydrocarbon nature than the normal resins and hence exert a greater van der Waals attraction for the organic solutes. The fewer functional groups also mean that the resin can less effectively salt the solute out of the resin phase, a factor which would also tend to increase *C* values. If the ion-dipole attractions were important, the *C* values would decrease with diminishing capacity.

Since low-capacity resins contain much less internal water than do normal resins, slower diffusion rates and tailing are to be expected. Crosslinked polystyrene contains no functional groups and cannot swell at all in aqueous solvents. In this case only

References p. 365

surface adsorption, and not true absorption, can occur. Hence, the tailing observed with the hydrocarbons on polystyrene would be characteristic of such an adsorption phenomenon.

ACKNOWLEDGEMENT

The authors express their gratitude to The Dow Chemical Company for financial support of this investigation and for supplying the resins used in the work.

SUMMARY

Synthetic mixtures of aliphatic and aromatic ethers, saturated fatty acids, and substituted benzenes and naphthalenes were separated by elution through columns of ion-exchange resins with aqueous solutions of acetic acid as eluent. An attempt was made to use a resin of low capacity for the separation of such mixtures, and the results of this attempt are used to help elucidate the theory of solubilization chromatography.

RÉUMÉS

Des mélanges d'éthers aliphatiques et aromatiques, d'acides gras saturés, de benzènes et de naphtalènes substitués ont été séparés par élution au moyen d'acide acétique sur des colonnes de résines d'échange d'ions.

ZUSAMMENFASSUNG

Es wird eine Methode beschrieben zur Trennung von Gemischen von aliphatischen und aromatischen Aethern, gesättigten Fettsäuren und substituierten Benzolen und Naphthalinen mit Hilfe von Ionenaustauschern und Eluation mit verd. Essigsäure.

REFERENCES

1 J. SHERMA AND W. RIEMAN, *Anal. Chim. Acta*, 18 (1958) 214.
2 J. SHERMA AND W. RIEMAN, *Anal. Chim. Acta*, 19 (1958) 134.
3 M. J. HATCH, J. A. DILLON AND H. B. SMITH, *Ind. Eng. Chem.*, 49 (1957) 1812.
4 R. SARGENT AND W. RIEMAN, *Anal. Chim. Acta*, 18 (1958) 197.
5 R. SARGENT AND W. RIEMAN, *J. Phys. Chem.*, 61 (1957) 354.
6 J. BEUKENKAMP, W. RIEMAN AND S. LINDENBAUM, *Anal. Chem.*, 26 (1954) 505.
7 R. SARGENT AND W. RIEMAN, *Anal. Chim. Acta*, 14 (1956) 381.
8 P. R. ENSLIN AND D. E. A. RIVETTE, *Chemistry & Industry*, 23 (1956) 23.
9 E. C. FREILING, *J. Am. Chem. Soc.*, 77 (1955) 2067.
10 D. REICHENBERG AND W. F. WALL, *J. Chem. Soc.*, (1956) 3364.

Received July 5th, 1958

46

Reprinted from *Clin. Chem.*, **14**(6), 521–528 (1968)

Analysis of Urine for Its Ultraviolet-Absorbing Constituents by High-Pressure Anion-Exchange Chromatography

Charles D. Scott

An automatic, high-resolution analytic system for the quantitative determination of the ultraviolet-absorbing molecular constituents of urine and other body fluids is being developed. This system is comprised of a heated, high-pressure anion-exchange column for the separation step and a continuous-flow ultraviolet spectrophotometer for the detection step. Separation is achieved by elution with an acetate buffer of gradually increasing concentration. The total analysis time is 40 hr.

Chromatograms showing the absorbance of the column effluent as a function of time are recorded by a strip-chart recorder and by a data acquisition system containing a digital voltmeter and paper tape punch. The output from the latter is then used as the input to a digital computer for analysis of the chromatogram.

More than 140 chromatographic peaks have been resolved from a 2-ml. urine sample. Normal urine samples yield similar chromatograms. A variation in the diurnal cycle has been noted.

HUMAN URINE REPORTEDLY CONTAINS more than 300 molecular constituents (*1–4*), many of which have appreciable ultraviolet (UV) absorbance. Knowledge of the quantities of these constituents represents a wealth of information that can be used to evaluate body functions. This is particularly true of the organic compounds of low molecular weight (less than 1000) since many of them are known to have pathologic significance.

An automatic, high-resolution analytic system is being developed to quantitatively determine many of these UV-absorbing urinary constituents. The primary goal of this effort is to devise a system that will

---

From the Oak Ridge National Laboratory (operated for the U. S. Atomic Energy Commission by Union Carbide Corporation), Oak Ridge, Tenn. 37830.

Supported by the National Institute of General Medical Sciences and the U. S. Atomic Energy Commission.

Presented at the Symposium on Urinary Constituents of Low Molecular Weight at Oak Ridge National Laboratory, Oak Ridge, Tenn., Nov. 30–Dec. 1, 1967.

The fine technical contributions of M. L. Freeman and N. E. Lee of Oak Ridge National Laboratory are gratefully acknowledged.

allow quantitative determination of a large number (greater than 100) of the molecular constituents of human urine with reasonable accuracy. Furthermore, this system will be highly automated so that only a few minutes of operator time is needed per sample. Finally, the resulting data will be automatically reduced to a form that is readily usable by the physician or scientist.

The system is comprised of a high-pressure anion-exchange column and a continuous-flow UV spectrophotometer. The chromatogram is developed with an acetate buffer and is recorded by a strip-chart recorder; the data are punched on paper tape and subsequently evaluated by using a digital computer.

Initial results indicate that eventually up to 140 chromatographic peaks can be resolved from a 2.0-ml. urine sample; to date, over 30 have been tentatively identified. Differences between normal and pathologic samples have been noted; also, differences in the urinary pattern during the diurnal cycle have also been observed.

Experimental System

The present prototype system for the automated analysis of the UV-absorbing constituents of body fluids is composed of a heated, high-pressure anion-exchange column, an acetate buffer elution section, and a recording UV spectrophotometer that has been modified to operate alternately at 2–4 different wavelengths (Fig. 1). There are also provisions for automatic sample injection and automatic data acquisition on punched paper tape.

The high-pressure anion-exchange column is fabricated from standard, Type 316 stainless-steel seamless tubing that is nominal ⅜ in. O.D., 0.62 cm. I.D., and 150–200 cm. long. A 1-in. O.D. stainless-steel water jacket surrounds the column, and a porous stainless-steel filter supports the ion-exchange resin (Fig. 2). Entrance to and exit from the column are provided by standard compression fittings that reduce to ⅛-in. tubing. The filter is included in the bottom fitting by pressing a porous metal disc into an enlarged opening in the fitting. This type of column is relatively inexpensive (approximately $50) and is designed for routine operation at pressures up to 5000 psi.

A strongly basic anion-exchange resin, such as Dowex 1, is used as the separation medium. Since 5 to 10-μ diameter resin is used, an operating pressure of above 2000 psi is required.

The UV detection system is a recording double-beam spectrophotometer that has been modified to operate alternately at 2–4 different wavelengths. Data are accumulated on punched paper tape for computer use (Fig. 3). This data acquisition system uses a multipoint strip-chart recorder as the control for switching the spectrophotometer to different wavelengths (the wavelength changes with each successive recorder

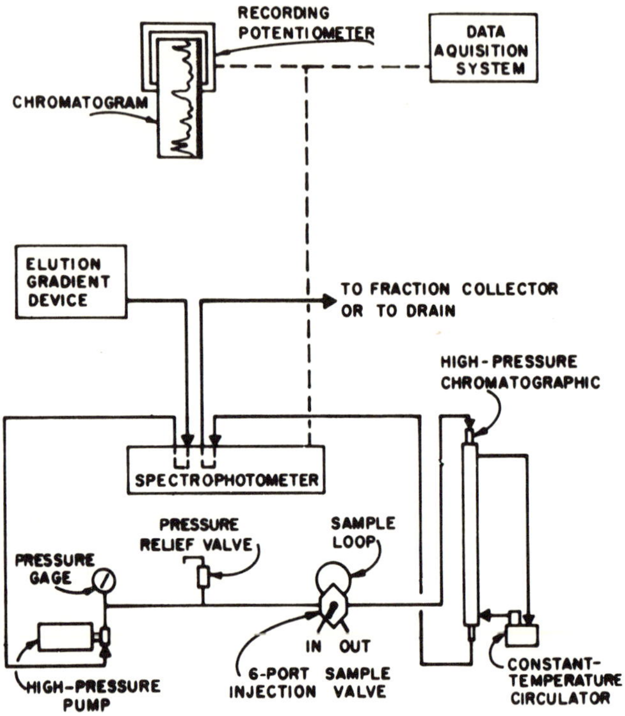

Fig. 1. High-pressure anion-exchange chromatograph for the UV-absorbing constituents of body fluids.

point). A readout control then directs the printed and punched-paper-tape record of the digitized signal from the spectrophotometer as each recorder point is printed. The wavelength is also encoded and punched with each data point.

The chromatogram showing the UV absorbance of the column effluent as a function of time is developed by introducing a urine sample (0.5–2 ml.) into the eluent stream just ahead of the ion-exchange column by means of a six-port injection valve and eluting it from the column by a sodium acetate-acetic acid buffer solution (pH 4.4). The buffer is pumped through the column at a rate of about 100 ml./sq. cm./hr., and the acetate concentration is gradually increased from 0.015 to 6 M. The first 16 hr. of the run is made at a column temperature of 25°; the remainder of the run is made at 60°. This temperature program allows higher resolution at the beginning of the run and a decrease in total run time.

Use of small particles of ion-exchange resin resulted in a system of higher resolution. For example, reducing the resin diameter from an average of 17 μ to 7 μ yielded a 25% increase in the number of chromatographic peaks resolved from the same urine samples. Ion-exchange resin with an average particle size of less than 10 μ is now routinely

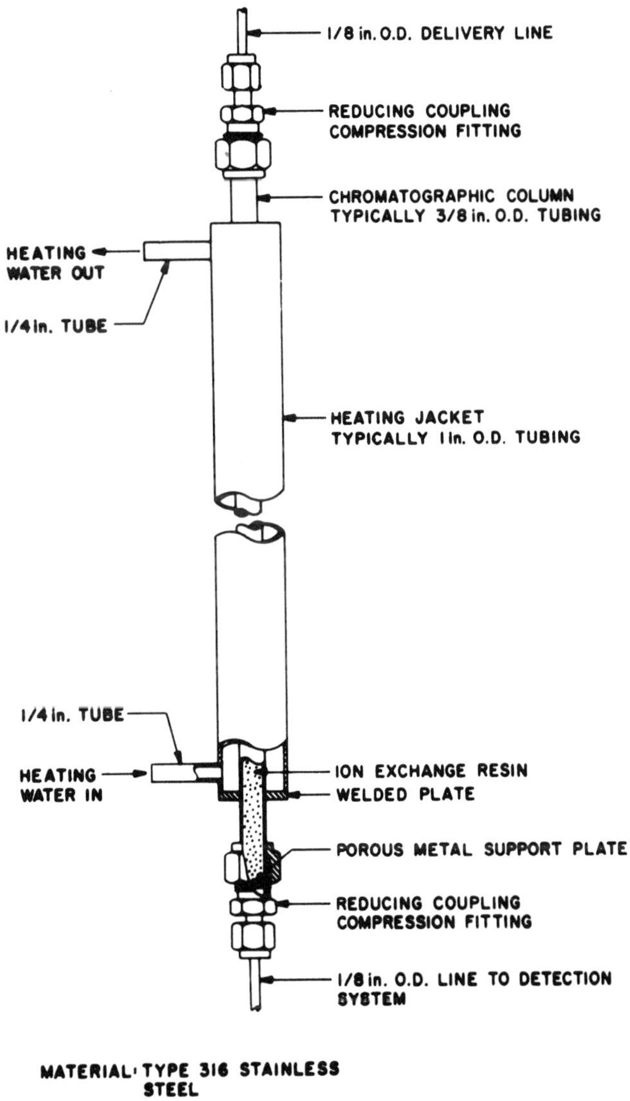

Fig. 2. High-pressure chromatographic column fabricated from stainless-steel tubing.

used in all prototype systems; however, to maintain a normal flow rate and analysis time (40 hr. or less), an operating pressure in the range of 1500–2300 psi is required. In our present systems, pressures of this magnitude do not cause operating difficulties.

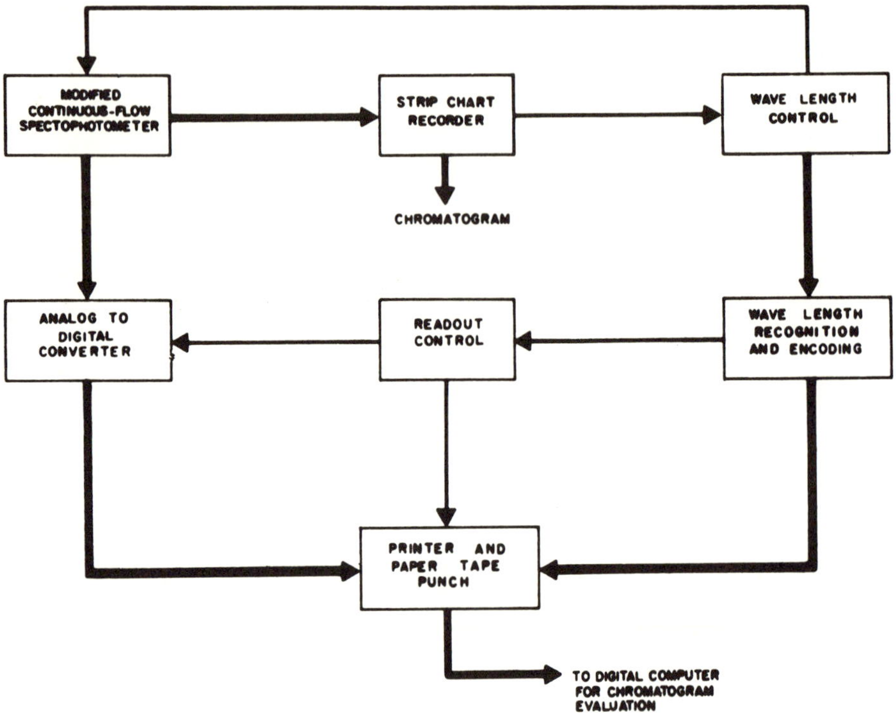

Fig. 3. Data acquisition system for the UV analyzer detection system.

Experimental Results

In the usual urine chromatogram, 80–120 peaks are resolved in 40 hr. Over 30 of these peaks have been tentatively identified by cochromatography, color reactions, and spectrophotometric technics (*5*). Chromatographic patterns of urine from normal subjects were found to be very similar; about 75 specific chromatographic peaks are always seen in the same sequence in such chromatograms (Fig. 4). The sensitivity of this system is in the range of 1–10 μg. for those standards tested.

Since there is variation in the chromatographic pattern during the diurnal cycle (Fig. 4), all urine samples collected for this program have been 24-hr. composites. Refrigeration of the composites during the collection period seems to be necessary. Also, deproteinization of the

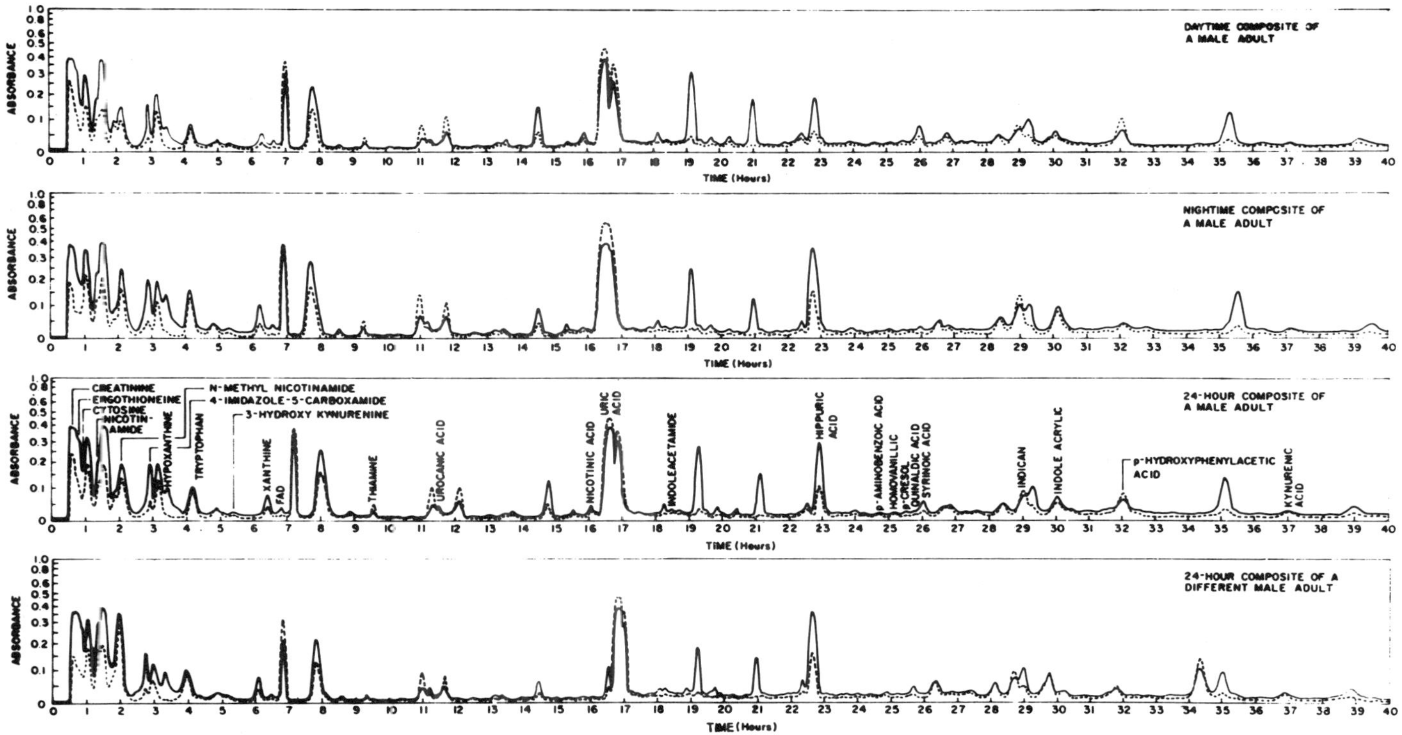

Fig. 4. Typical chromatograms for the UV analyzer showing reproducibility, difference in diurnal cycle, and similarity between different urine samples. Run conditions: column, 0.45 cm. I.D. × 316 stainless steel with 10 μ diameter. Biorad AGI-18; urine sample, 2 ml. each; temperature, 25° increasing to 60° after 15 hr.; pressure 1500–2300 psig; elution, sodium acetate-acetic acid buffer at a pH of 4.4 varying in concentration from 0.015 to 6 M at an average flow rate of 28 ml./hr. Solid lines, at 260 mμ; Broken lines, at 280 mμ.

samples by pressure filtration through dialysis membranes increases the resolution of the peaks in the chromatograms.

Several pathologic urine samples have been run to determine if the system is capable of differentiating between the normal and pathologic states. Significant differences have been observed between normal urine and that from subjects with acute lymphocytic leukemia and schizophrenia (*6*).

Discussion

At the current stage of development, the experimental analyzer is probably suitable for use in some types of biochemical and medical research; however, the time required for analysis is too great to permit extensive, general clinical use. The analysis time has already been reduced by about 16 hr., and a further decrease is anticipated by using a separations system of higher resolution and by optimizing operating parameters.

A significant problem exists in obtaining anion-exchange resin that has high resolution. While some batches of commercially available resin are better than others, there appears to be no reproducibility from batch to batch. We are currently investigating this anomaly in cooperation with resin manufacturers.

Continuous-flow photometers or spectrophotometers that are capable of automatic, multiple wavelength operation with little downtime due to maintenance, are not yet available. We are currently studying several microflow detectors with solid-state electronics to achieve this end.

Ultimately, the processing of data generated by such an instrument must be highly automated for results to be used effectively. At this laboratory, we feel that, in the future, the majority of clinical laboratories either will have, or will have access to, a digital computer; it will be feasible, then, to process all the data from the UV analyzer with such a computer, preferably an on-line type.

Although the instrument described here is presently limited to the analysis of UV-absorbing constituents of body fluids, its capability will ultimately be extended by the incorporation of other types of detection systems.

References

1. Alman, P. L., and Dittmer, D. S., *Blood and Other Body Fluids* (Chapt. 19, *Urine*). Federation of the American Society of Experimental Biology, Washington, D. C., 1961, p. 363.
2. Alman, P. L., and Dittmer, D. S., *Biology Data Book*. Federation of the American Society of Experimental Biology, Washington, D. C., 1964, p. 186.
3. *Handbook of Clinical Laboratory Data*, Damm, H. C., Ed. Chemical Rubber Co., Cleveland, 1965, p. 48.

4. *Biochemists' Handbook* (Chapt. 4, *Chemical Composition of Animal Tissue and Related Data*), Long, C., Ed. Spon, London, 1961, p. 918.
5. Burtis, C. A., and Warren, K., Identification of urinary constituents isolated by anion-exchange chromatography. *Clin. Chem.* **14,** 290 (1968).
6. Scott, C. D., Attrill, J. E., and Anderson, N. G., Automatic, high-resolution analysis of urine for its ultraviolet-absorbing constituents. *Proc. Soc. Exp. Biol. Med.* **125,** 181 (1967).

47

Reprinted from *Nature*, **189**(4769), 1001–1002 (1961)

"LIGAND EXCHANGE": A NOVEL SEPARATION TECHNIQUE

F. Helfferich

A NOVEL technique for analytical and preparative separations of compounds which can form complexes or adducts with metal ions has been worked out in this laboratory. The new technique can appropriately be called 'ligand exchange'. The method combines two fields of chemistry, namely, ion exchange and co-ordination chemistry, in order to accomplish a task that neither could do alone.

An ion exchanger containing a complexing metal ion (Cu^{2+}, Cu^{+}, Ni^{2+}, Ag^{+}, Co^{3+}, etc.) is used as a solid sorbent. The potential ligands are sorbed from solutions or gases and form complexes with the metal in the resin or displace other ligands which have previously complexed the metal. The latter remains in the ion exchanger. In contrast to conventional ion exchange where counter ions are exchanged, ligand exchange is exchange of ligands for others or for solvent molecules in the ionic solvation shells. Ligands which can be isolated or separated include ammonia, organic amines, polyhydric alcohols, olefins, acetylene derivatives, etc., and anions of organic acids and amino-acids. Separations are preferably carried out in a chromatographic column as in ion exchange chromatography or gas-liquid chromatography. Standard chromatographic techniques—elution development, gradient elution, displacement development, frontal analysis, etc.—can be adapted to ligand exchange.

Ligand exchange has several advantages over other, more conventional chromatographic techniques. First, complex formation is a very strong interaction if the proper metal ion is chosen. This provides a very strong 'driving force' for ligand sorption. Thus, ligand exchangers sorb ligands selectively, utilizing almost their full capacity even when the ligand concentration in the external solution or gas phase is extremely low. In this way, ligands can be very efficiently separated from non-ligands. For example, from 10 litres of a 10-p.p.m. solution of most aliphatic amines, essentially all is sorbed by 1 gm. or less of a cation exchanger in Cu^{2+} or Ni^{2+} form. Secondly, complex formation is a highly specific interaction, much more so than ordinary physical adsorption or ion exchange. Complexes of a metal ion with various, even rather similar, ligands differ greatly in their strength. Hence the ligand exchanger can distinguish rather sharply between such ligands, and outstandingly sharp resolutions in separations of ligands from one another can be obtained even in cases in which conventional sorbents fail. A striking illustration of the specificity of complex formation is the separation of the rare earths by ion exchange chromatography ; here, differences in complex-strengths with an eluting or developing agent are used for separating metals. Ligand exchange is a reversal of this technique ; differences in complex-strengths with a metal are used for separating complexing agents. For example, with ammonia and an organic diamine, separation factors well above 100 were easily attained. Thirdly, ligand exchange is very flexible since the metal ion giving the best separation can be chosen from a great number of elements.

A unique and particularly useful feature of ligand exchange is the following. In separations of ligands of different co-ordinative valences the selectivity can be adjusted and even reversed at will, simply by varying the total ligand concentration in the external solution or gas phase. Consider, for example, the 'ligand exchange' of a monodentate ligand A (occupying one co-ordinative valence of the metal) for a bidentate ligand B (occupying two co-ordinative valences) with, say, Ni^{2+} as the complexing metal ion :

$$\mathrm{Ni}A_6 + 3B \rightleftharpoons \mathrm{Ni}B_3 + 6A \qquad (1)$$

The mass action law for this exchange can be written as :

$$\frac{[\mathrm{Ni}B_3][A]^6}{[\mathrm{Ni}A_6][B]^3} = K = \text{const.} \qquad (2)$$

or

$$\frac{[\mathrm{Ni}B_3][A]^3}{[\mathrm{Ni}A_6]} = K\left(\frac{[B]}{[A]}\right)^3 \qquad (3)$$

For a given constant concentration ratio of B to A, the right-hand side of equation (3) remains constant. Thus, high total ligand concentration (high value of $[A]$ on the left-hand side) favours the complex $\mathrm{Ni}A_6$, and low total ligand concentration (low value of $[A]$) favours the complex $\mathrm{Ni}B_3$. By selecting a suitable metal ion one can adjust conditions in such a way that the monodentate ligand is preferentially sorbed from concentrated solutions and the bidentate one from dilute solutions. For example, a bidentate diamine is economically separated from (monodentate) ammonia by selective sorption of the diamine from a dilute mixture by a cation exchanger in metal–

ammonia–complex form. The diamine is then recovered from the column by displacement with concentrated ammonia, which yields the diamine in high concentration. No regeneration is required since, after displacement, the resin is again in metal–ammonia–complex form.

Displacement of the complexing metal ion from the resin by ion exchange with other cations in the external solution interferes with ligand exchange. Such ion exchange can be suppressed by choosing a resin with ionogenic groups which hold the metal firmly by partial complexing. In non-ionic ligand exchange, carboxylic acid groups usually offer sufficient protection. If requirements are more stringent, as is the case in anionic ligand exchange, chelating resins such as 'Dowex *A*-1' (iminodiacetic acid groups) can be used. A possible alternative is the use of an anion exchanger with amino groups. Of course, such protection against ion exchange is achieved at a sacrifice in ligand exchange capacity since the ionogenic groups block co-ordinative valences of the metal ion. For example, with Ni^{2+}, two co-ordinative valences out of six are blocked in resins with carboxylic acid groups, and three in resins with iminodiacetic acid groups.

Ligand sorption by cation exchangers containing complexing metal ions was observed as early as 1954 by Walton[1]. Separations by virtue of differences in complexing strength have occasionally been achieved by other authors[2]. However, it appears that so far the general applicability of the principle, the inherent advantages, and, in particular, the effect of the co-ordinative valences of the ligands have not been recognized.

Separation factors and column performance in ligand exchange operations can be predicted with reasonable accuracy solely on the basis of the tabulated values of the complex-formation constants. Publication of a theoretical treatment and experimental results is anticipated elsewhere.

Shell Development Co.,
Emeryville, California.

[1] Stokes, R. H., and Walton, H. F., *J. Amer. Chem. Soc.*, **76**, 3327 (1954).

[2] Thomas, C. L., U.S. Patent, 2,865,970 (1958). Giessen, J., and Mueller, F., U.S. Patent, 2,916,525 (1959).

48

Reprinted from *Anal. Chem.*, 44(6), 890–894 (1972)

Ligand Exchange Chromatography of Amphetamine Drugs

Cecilia M. de Hernandez[1] **and Harold F. Walton**

University of Colorado, Boulder, Colo. 80302

Amphetamine, phenethylamine, and related compounds have been separated by liquid chromatography on columns of cation-exchange resins loaded with ions of copper, nickel, and cadmium. Polymethacrylate resins are much better than those of the polystyrene type, and Cu(II) was the most effective cation. Eluents were aqueous alcohol solutions containing ammonia, most commonly 0.1*M* ammonia in 33% ethanol. Metamphetamine was eluted first, then ephedrine, amphetamine, norephedrine, and phenethylamine. Detection and quantitative measurement was easily possible at 0.1 mg/ml, using 0.5-ml samples. However, extraneous peaks appeared at one column volume that seemed to be due to decomposition products.

LIGAND EXCHANGE CHROMATOGRAPHY was introduced in 1961 by Helfferich (*1*). He used a column of carboxylic cation-exchange resin loaded with copper(II) ions to absorb a 1,3-diamine from a dilute aqueous solution, and regenerated the column with aqueous ammonia (*2, 3*). Subsequently, other workers adapted the ligand exchange technique to the chromatographic analysis of mixtures (*4, 5*). Applications were made to aliphatic amines and diamines, ethanolamines, substituted hydrazines, and aziridines (*6*). We have extended our study to amphetamine, metamphetamine, and drugs related to phenethylamine.

Amphetamine and related drugs are currently detected and determined by thin-layer (*7*) and gas chromatography (*8, 9*). A method that depends on oxidation by Ce(IV) followed by ultraviolet spectrometry is described (*10*). It is common practice to use ion exchange to separate the drugs from urine (*7, 8*) but ion-exchange chromatography has found little use in separating the drugs from one another, probably because of their very strong absorption on resins having a polystyrene matrix. Amberlite IR-45, a weakly basic anion-exchange resin with a polystyrene matrix, was used to absorb amphetamine, ephedrine, and epinephrine, presumably by nonionic absorption (*11*). Amphetamine, metamphetamine, ephedrine, epinephrine, and related compounds were separated by salting-out chromatography on polystyrene-type cation and anion exchangers (*12*).

[1] Present address Pedagogical Institute, Caracas, Venezuela.

(1) F. G. Helfferich, *Nature*, **189**, 1001 (1961).
(2) F. G. Helfferich, *J. Amer. Chem. Soc.*, **84**, 3237 (1962).
(3) F. G. Helfferich, *ibid.*, p 3242.
(4) A. G. Hill, R. Sedgeley, and H. F. Walton, *Anal. Chim. Acta*, **33**, 84 (1965).
(5) K. Shimomura, L. Dickson, and H. F. Walton, *ibid.*, **37**, 102 (1967).
(6) H. F. Walton, "Ligand Exchange Chromatography," in Advances in Ion Exchange, Vol. 3, J. Marinsky, Ed., Marcel Dekker, New York, N.Y., 1972 (in press).
(7) A. M. Heaton and A. G. Blumberg, *J. Chromatog.*, **41**, 367 (1969).
(8) R. C. Baselt and L. J. Casarett, *ibid.*, **57**, 139 (1971).
(9) E. Anggard, *Int. Symp. Amphetamines Related Cpds., Proc.*, **1969**, 191; *Chem. Abstr.*, **74**, 21643e (1971).
(10) J. E. Wallace, J. D. Biggs, and S. L. Ladd, ANAL. CHEM., **40**, 2207 (1968).
(11) M. G. Vincent, E. Krupski, and L. Fischer, *J. Pharm. Sci.*, **46**, 85 (1957).
(12) J. Halmekoski and A. Vidgren, *Farm. Aikak.*, **75**, 282 (1966); *Chem. Abstr.*, **66**, 22230q (1967).

We made tests with the sulfonated polystyrene cation-exchange resin, Dowex-50 X8, loaded with nickel ions, with ammonia solutions as eluents. Columns of this material are very effective in separating amino compounds (*5, 6*). We found, as expected, that amphetamine and metamphetamine were absorbed strongly. Relatively concentrated ammonia was needed for elution, and the substrates emerged as broad bands with considerable tailing and virtually no separation (Figure 1). We therefore turned our attention to carboxylic cation-exchange resins. These are polymers of acrylic and methacrylic acids that contain no aromatic groups except for cross-links introduced as divinylbenzene. They have a high capacity for holding ions, and bind divalent cations so strongly that they are not washed out by the dilute ammonia solutions used as eluents. The cations used were Cu(II), Ni(II), Cd(II), and Zn(II). Of these, Cu(II) gave the best results.

The principal substrates studied were:

Phenethylamine	$C_6H_5CH_2CH_2NH_2$
Amphetamine	$C_6H_5CH_2 \cdot CHCH_3 \cdot NH_2$
Metamphetamine	$C_6H_5CH_2 \cdot CHCH_3 \cdot NHCH_3$
Norephedrine	$C_6H_5CHOH \cdot CHCH_3 \cdot NH_2$
Ephedrine	$C_6H_5CHOH \cdot CHCH_3 \cdot NHCH_3$

Tests were also made with benzylamine, caffeine, epinephrine, and dopamine. The eluents were solutions of ammonia in mixtures of water with methanol, ethanol, and isopropanol.

EXPERIMENTAL

Materials. The resin was Bio-Rex 70, 200–400 mesh, from the Bio-Rad Corp., Richmond, Calif. It is a methacrylate polymer manufactured by the Diamond Shamrock Corp., Redwood City, Calif., ground and purified for laboratory use. It was converted to the metal forms by stirring with solutions of the metal-ammonia complexes, then transferring to a wide column and washing with more metal-ammonia complex, then water.

In the chromatographic column, the resin tended to pack slowly and unevenly. It was necessary to add more and more resin to the column until it finally stabilized. The packing and flow characteristics were improved by adding 5%–10% by weight of cellulose, Cellex-N (Bio-Rad Corp.), and this was our standard practice. For unknown reasons, the cadmium and zinc forms of the resin were much harder to pack and caused much more flow resistance than the copper and nickel forms. Microscopic examination showed no obvious differences. Under the microscope, the particles of resin had irregular shapes and a fuzzy appearance, indicating a large surface area. This was true of both the copper and the cadmium forms.

The microscopic structure of the grains of Bio-Rex 70 may be significant, for we tested a new product, Amberlite IRC-72, obtained from the Rohm and Haas Co. We ground and screened this resin ourselves. The particles looked smoother and more transparent than those of Bio-Rex 70, and this may be connected with the fact that in the column, IRC-72 gave broad bands with bad tailing.

Amphetamine sulfate, metamphetamine hydrochloride, ephedrine and norephedrine sulfates, dopamine hydrochloride, and epinephrine were obtained from K and K Laboratories, Plainview, N.Y., in their *dl* forms and used without

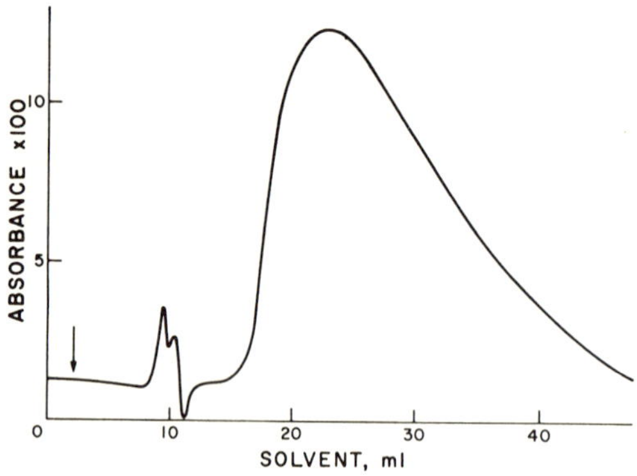

Figure 1. Elution curve from column of nickel–Dowex 50, 6 mm × 51 cm

Metamphetamine and amphetamine, each 2.5 mg/ml: eluent, 2*M* NH_3 in 50% ethanol; 12 ml/hr

purification. Phenethylamine (Eastman Kodak Co.) was distilled before use.

Amphetamine and metamphetamine bases were used in some of the experiments. They were made from the salts by adding a slight excess of sodium hydroxide to the aqueous solutions, then extracting with chloroform and evaporating the chloroform solvent on a steam bath. A solution of amphetamine in 50% alcohol was also prepared by adding to the sulfate solution an equivalent amount of standard barium hydroxide solution, then centrifuging.

Equipment. The columns and fittings, pump, and detector were supplied by Chromatronix, Inc., Berkeley, Calif. Solvents were fed into the columns by a variable-speed pulseless pump, Model CMP 2, and the samples were introduced by a valve, Model SV 8031, with Teflon (Du Pont) slider and 0.5-ml sample loop. As a rule, the materials were injected in the same solvent that was used for elution. Detection was done with a two-cell ultraviolet photometer, Model 200, connected to a Leeds and Northrup Speedomax recorder, which measured the absorbance at 254 nm. The solvent passed from the pump through the reference cell of the photometer, then through the sample injection valve and into the column. From the column outlet, the liquid flowed through the sample cell of the photometer and out into a receiver.

Sometimes samples of effluent were collected and their metal contents measured by atomic absorption spectrometry, using a Varian Techtron spectrometer. Ultraviolet absorption spectra were scanned in some cases, using a Cary Model 17 spectrophotometer with expanded scale.

The columns were 9 mm in internal diameter, except for those used with the nickel-loaded carboxylic and sulfonic resins, which were 6 mm in diameter. The length was 58–60 cm in most experiments. The ionic capacity of the packed columns was not known exactly, but it was at least 1.5 meq of metal ions per milliliter bed volume. (This capacity was measured with columns that had settled under gravity.)

RESULTS AND DISCUSSION

Metal Content of Effluents. The normal metal ion contents were: copper, 0.2–0.5 mg/l; nickel, 0.1 mg/l; cadmium, 0.2–0.5 mg/l. They were no higher when an amine was being eluted than when it was not, except for the few tests made with the catecholamines epinephrine and dopamine. With these two compounds the copper and nickel concentrations rose to 6 mg/l, corresponding to 1 mole of metal to 2–3 moles of base, and the solutions were blue. The absorptivity of the Cu(II)–dopamine complex was over 5000 l. mole^{-1} cm^{-1} at 610 nm. These complexes were eluted at one void column volume and will be the subject of a future report. Except for the catecholamines, one can say that the bases do not remove metal ions from the column.

Table I. Elution from Copper-Loaded Column
(Multiples of bulk column volume)[a]

Alcohol	Ethanol					Isopropanol
Volume concn	33%	33%	50%	50%	66%	33%
Ammonia, *N*	0.06	0.10	0.06	0.11	0.06	0.11
Elution volume of:						
Metamphetamine	1.9	1.75	1.1	...	...	1.05
Amphetamine	2.85	2.15	1.8	1.0	1.3	1.25
Ephedrine	...	2.12	...	...	...	1.25
Norephedrine	...	2.7	...	...	...	1.65
Phenethylamine	5.5	3.9	3.1	...	1.85	2.2
Benzylamine	...	2.9	...	...	...	...
Caffeine	...	0.6	...	...	...	...

[a] Void volume was 0.45 times the bulk volume. Alcohol concentrations are expressed by volume.

Table II. Elution from Cadmium-Loaded Column
(Multiples of bulk column volume)[a]

Alcohol and temp.	Ethanol, 20 °C				Ethanol, 55 °C	Methanol, 60 °C	
Volume concn	33%	33%	50%	75%	33%	33%	50%
Ammonia, *N*	0.09	0.15	0.11	0.09	0.11	0.17	0.17
Elution volume of:							
Metamphetamine	1.0	1.0	...	...	1.15	...	...
Amphetamine	1.65	1.35	1.0	0.95	1.75	1.25	1.05
Ephedrine	1.2	...	...	...	...	...	...
Norephedrine	2.0	...	...	...	...	1.6	...
Phenethylamine	2.7	2.2	1.45	1.15	2.8	2.0	2.0

[a] Void volume was 0.7 times the bulk volume. Alcohol concentrations are expressed by volume.

Conversion of Amine Salts to Bases on the Column. It made no difference to the elution volume whether amphetamine, metamphetamine, and ephedrine were injected as the free bases or as the sulfate or chloride salts. An examination of the spectra of the eluents showed that the spectra were those of the free bases, not the salts. It was also found that sulfate or chloride ions appeared in the effluent just after one void column volume, whereas the amine base appeared later. Thus it seems that the amine cations are converted to bases by the excess of ammonia. Their preferential absorption by the resin, compared to ammonia, probably plays a part, for ammonia is a weaker base than amphetamine by a factor of 5 (*13*).

Elution Volumes; Effects of Metal, Solvent, and Temperature. Elution volumes with Bio-Rex 70 are summarized in Tables I, II, and III. Each compound was run alone before running mixtures. The volumes are quoted as multiples of the bulk column volume to enable comparisons between short (18–20 cm) and long (58–60 cm) columns. Void volumes were estimated from the first rise in ultraviolet absorption caused by decomposition products (see below) or by an added "marker compound," $C_6H_5OC_2H_4OC_2H_5$, which was presumed not to be absorbed.

(13) D. D. Perrin, "Dissociation Constants of Organic Bases in Aqueous Solution," Butterworth, London, 1965.

Table III. Elution from Nickel-Loaded Column, Ethanol–Water

(Multiples of bulk column volume)[a]

Alcohol volume concn	33%	50%
Ammonia, *N*	0.085	0.13
Elution volume of:		
Metamphetamine	1.4	...
Amphetamine	1.55	1.0
Ephedrine	1.55	...
Norephedrine	1.9	...
Phenethylamine	3.1	1.35

[a] Void volume was 0.7 times the bulk volume. Alcohol concentrations are expressed by volume.

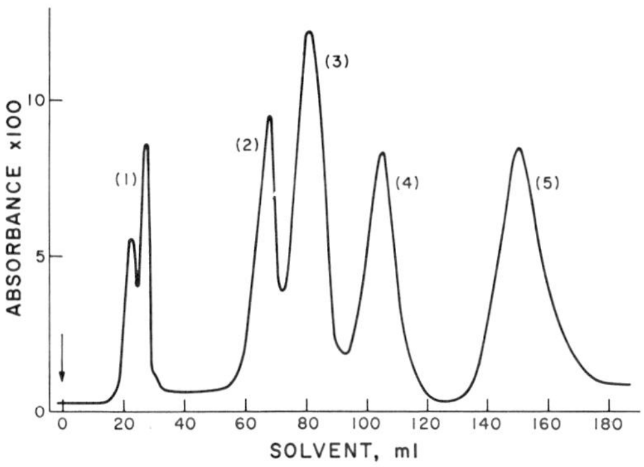

Figure 2. Elution curve from column of copper–Bio-Rex 70, 9 mm × 60 cm

Metamphetamine, amphetamine, and ephedrine, each 1.5 mg/ml; norephedrine and phenethylamine, 2.5 mg/ml; eluent, 0.10*M* NH_3 in 33% ethanol; 12 ml/hr

Peaks; (*1*) decomposition products: (*2*) metamphetamine: (*3*) ephedrine and amphetamine (original shows two distinct peaks separated by less than one standard deviation); (*4*) norephedrine; (*5*) phenethylamine

Figure 1 shows an elution curve for a polystyrene type resin; Figures 2–5 show curves for the elution of mixtures from columns of Bio-Rex 70. No curve is shown for the zinc-loaded resin; this column showed bad tailing and was tested only briefly.

In every case, the elution order was the same. Metamphetamine was eluted first, followed by ephedrine, amphetamine, norephedrine, and phenethylamine. This order is consistent with our earlier experience (*6*) that elution is hastened by placing alkyl groups on the amine nitrogen and on the carbon atom next to the amine nitrogen. The metal–amine complexes in aqueous solution are destabilized in a corresponding way, according to our preliminary (unpublished) tests. Comparing ephedrine with metamphetamine, and norephedrine with amphetamine, we see that the hydroxyl group of ephedrine and norephedrine strengthens the binding to the column, probably by coordinating with the metal ion.

The solvent effect is as expected; the higher the proportion of alcohol the more soluble are the bases, and the more easily they are eluted from the column. Isopropanol is a better solvent than ethanol, and methanol is poorer. Methanol–water solvents behaved in a peculiar way with the copper resin which will be discussed later. Raising the ammonia

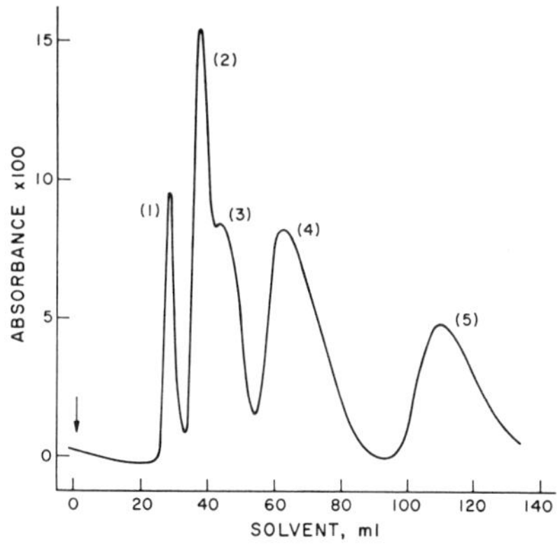

Figure 3. Elution curve from column of cadmium–Bio-Rex 70, 9 mm × 57.5 cm

Each base 2 mg/ml; eluent 0.08*M* NH_3 in 33% ethanol; 12 ml/hr

Peaks: (*1*) metamphetamine, (*2*) ephedrine, (*3*) amphetamine, (*4*) norephedrine, (*5*) phenethylamine

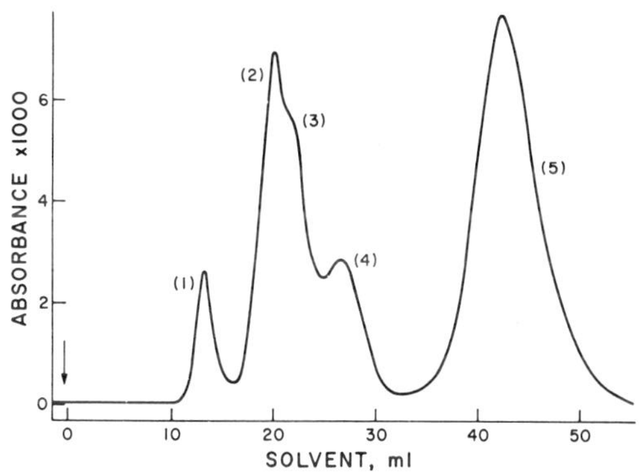

Figure 4. Elution curve from column of nickel–Bio-Rex 70, 6 mm × 49 cm

Phenethylamine 0.6 mg/ml, other bases 0.2 mg/ml; eluent 0.082*M* NH_3 in 33% ethanol; 6 ml/hr

Peaks: (*1*) decomposition products, (*2*) metamphetamine, (*3*) amphetamine, (*4*) norephedrine, (*5*) phenethylamine

Photometer sensitivity was doubled between 16 and 31 ml, and ordinate reading should be divided by two

concentration lowered the elution volume, and the *net* elution volume, after subtracting the void volume, was inversely proportional to the ammonia concentration.

The effect of temperature, as noted with the cadmium resin (Table II), was small. The elution volumes for comparable ammonia concentrations were larger at the higher temperature, and the resolving power was somewhat improved.

Plate Numbers and Resolution. The 58-cm cadmium resin column whose performance is shown in Figure 3 had a plate number of 150, calculated from the sharper (front) side of the phenethylamine peak. This was the best number ever found for this column. The flow rate was 12 ml/hr., temperature 20 °C. When this column was heated to 55 °C, cracks and cavities appeared within the resin bed, yet the

Table IV. Amphetamine and Metamphetamine Peak Areas[a]

Concentration, mg/ml	Flow rate, ml/hr	Area (arbitrary units) Amphet.	Met-amphet.	Ratio to 10 mg/ml area Amphet.	Met-amphet.
				Averages	
10.0	6	219	230		
10.0	12	229	235	10.00 (reference)	
5.0	6	104	107	4.65	4.60
5.0	6	106	109	4.75	4.70
5.0	24	110	121	4.95	5.2
2.0	6	35	40	1.55	1.7
2.0	24	41	45	1.85	1.95
1.0	24	19	20	0.85	0.85

[a] *Note:* the areas on the recorder chart were multiplied by scale factors to allow for flow rate and the photometer sensitivity.

hydrodynamic resistance increased. A short column, 23 cm long, was easier to pack and gave plate numbers of 45 and 75 at 20 and 55 °C for phenethylamine at 24 ml/hr. Both columns were 9 mm in diameter.

Cadmium gave better resolution of ephedrine and amphetamine than other metal ions (see Figure 3) but the packing characteristics of the column were so poor that work with this resin was abandoned in favor of copper- and nickel-loaded resins. The nickel-loaded carboxylic resin was tested in a column of 6 mm internal diameter, 51 cm long, at room temperature only. It separated phenethylamine well from the others, but the separation of metamphetamine, amphetamine, and ephedrine was poor. The plate number for phenethylamine was 90 at 12 ml/hr, 100 at 6 ml/hr.

The copper resin gave the sharpest bands. It packed well and the pressure drop was moderate. Typical plate numbers for phenethylamine in a column 60 cm long and 9-mm internal diameter were 400–500 at 12 ml/hr and 300–400 at 24 ml/hr.

Peculiarities of the Copper-Loaded Column. Two special features of the copper columns can be seen from Figure 5 and less clearly from Figure 2. The first is the shape of the bands; they were unsymmetrical with the sharper side the back, or trailing side. This may be due to a nonlinear absorption isotherm with the distribution ratio increasing as the solute concentration increases. We thought it might be due to a slow chemical decomposition on the column, but the dissymmetry did not change with flow rate, and the absorption spectra of the effluents from the first and last portions of the peak were not significantly different. The second feature is the preliminary wave that appears at one void column volume. The absorption spectrum of the material eluted in this wave is quite different from those of the five solutes investigated. The spectra of these solutes are similar, with a sharp maximum absorbance at 258.5 nm, satellite peaks at 253, 262, 264, and 268 nm, and a marked minimum at 233 nm. The preliminary wave material showed absorbance increasing with decreasing wavelength below 300 nm, with a more rapid increase below 250 nm. In the case of phenethylamine, this material was produced by atmospheric oxidation, for the preliminary wave was greater the longer the solution of phenethylamine base was exposed to air before injection. For the other bases, the preliminary wave appeared whether or not the sample solutions were left exposed to air, and it appeared whether they were injected as salts or as free bases. The preliminary peak was made up of two or more components; see Figures 2 and 5. Their identity is still a mystery.

An even more perplexing phenomenon appeared with the copper column when methanol–water solvents were used.

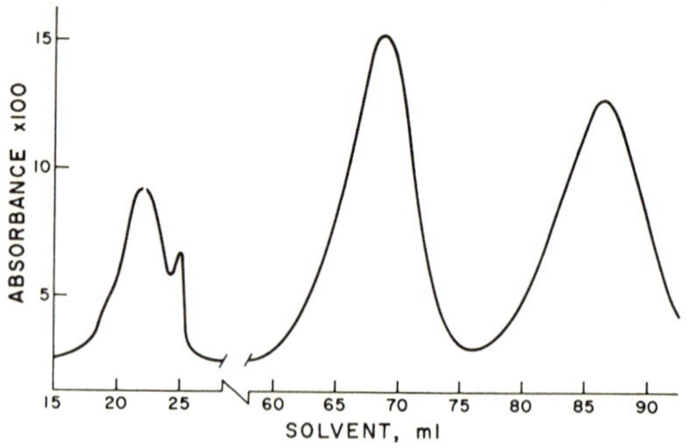

Figure 5. Elution curve from column of copper–Bio-Rex 70, 9 mm × 60 cm

Metamphetamine hydrochloride and amphetamine sulfate, each 5.0 mg/ml. Eluent 0.10*M* NH_3 in 33% ethanol; 6 ml/hr

Peaks are (in order) decomposition products, metamphetamine, amphetamine. Note break in volume scale

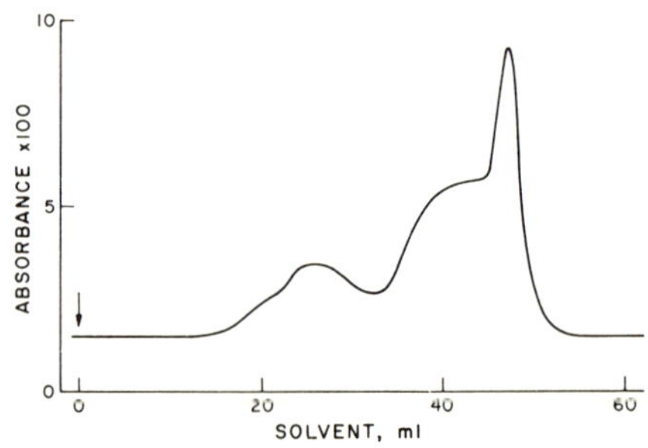

Figure 6. Elution curve from column of copper–Bio-Rex 70, 9 mm × 57 cm

Amphetamine base, 10 mg/ml. Eluent 0.18*M* NH_3 in 50% methanol; 24 ml/hr

Figure 6 shows the kind of waves that appeared. The material that appeared just after one column volume gave the same featureless spectrum that we found with the other preliminary waves. The main solute wave consisted of two parts, a broad diffuse wave followed by a very sharp peak. The absorption spectra of the materials eluted in these two fractions were very similar.

Obviously there is a chemical reaction occurring on the column with methanol, yet it must be a relatively mild reaction that does not affect the ultraviolet-absorbing part of the molecule. We did not investigate it further, since we could get good chromatography with ethanol solvents. However, even with ethanol, the solutes decomposed on the column when the temperature was raised to 60 °C. Therefore we operated the copper-resin column at room temperature.

Quantitative Analysis by Peak Area Measurement. A standard solution containing 10.0 mg each of amphetamine sulfate, and metamphetamine hydrochloride per milliliter was injected into the copper column by means of the 0.5-ml samples loop, and the areas of the metamphetamine and amphetamine peaks were measured. Solutions containing 5.0, 2.0, and 1.0 mg/ml were prepared by dilution and also injected. The peak areas were compared, with the results shown in Table IV.

Roughly the areas were proportional to concentrations, but at lower concentrations they were smaller than they should have been, taking the 10.0 mg/ml solution as reference. The discrepancy exceeds experimental error and may indicate that the samples were decomposing, and that decomposition was more serious at lower concentrations. It may be significant that the preliminary peak area was 24% of the total amphetamine–metamphetamine peak areas at 2 mg/ml but only 13% at 10 mg/ml.

We do not exclude the possibility that air dissolved in the feed solution oxidizes the bases, with copper(II) acting as catalyst. However, boiling the water beforehand and keeping the feed solution under nitrogen did not seem to reduce the preliminary wave.

CONCLUSIONS

Ligand exchange chromatography on metal-loaded carboxylic resin columns gives satisfactory resolution of mixtures of amphetamine bases. A resin loaded with copper(II) ions gives the best resolution and the sharpest bands, but the Cu (II) seems to promote some decomposition of the substrates that we have been unable to characterize or control. Nevertheless it is possible to detect and determine these bases at concentrations down to 0.1 mg/ml, using 0.5-ml samples. Probably this performance could be improved.

From this work and previous work in our laboratory (*5*, *6*), we may make certain generalizations about elution orders in metal ion-loaded columns. Strong bases are more strongly held than weak ones, and primary amines are more strongly held than secondary, tertiary, or heterocyclic amines. The binding to the metal ion-loaded resin is *weakened* by: (a) methyl groups or other alkyl groups on the amine nitrogen; (b) methyl groups one carbon atom removed from the amine nitrogen; (c) hydroxyl groups more than two carbon atoms removed; (d) methoxyl groups (mescaline, for example, is less strongly bound than phenethylamine).

It appears that binding is weakened by steric hindrance around the amine nitrogen and by hydrophilic substituents. The matrix of the ion-exchange resin has a large effect, as we have noted, and it can change elution orders (*5*).

Relatively few of the drugs and biogenic amines that we have studied are strongly held by metal-loaded carboxylic resins, and the selectivity thus shown can be exploited analytically. Obviously, resolution can be improved and the method can be adapted to modern high-speed liquid chromatographic techniques (*14*).

ACKNOWLEDGMENT

We gratefully acknowledge the help of Vernon Shaw, who was sponsored by the National Science Foundation's Summer Research Participation Program for College Teachers, and of Jared Baker. Harold Heim, Dean of the University of Colorado School of Pharmacy, gave valuable advice.

RECEIVED for review November 8, 1971. Accepted January 7, 1972. Work supported by the National Science Foundation, Grant No. GP-25727.

(14) J. J. Kirkland, Ed., "Modern Practice of Liquid Chromatography," Wiley-Interscience, New York, N.Y., 1971.

ADDITIONAL BIBLIOGRAPHY

General Books on Ion Exchange

1. F. G. Helfferich: *Ion Exchange.* New York, McGraw-Hill, 1962.
2. K. Dorfner: *Ion Exchangers, Properties and Applications,* 2nd ed. Ann Harbor (Michigan) Science Publications, 1972.

Books on Ion-Exchange Chromatography

1. J. Inczedy: *Analytical Applications on Ion Exchangers,* translated by A. Pall. Elmsford, N.Y., Pergamon Press, 1962.
2. O. Samuelson: *Ion Exchange Separations in Analytical Chemistry.* New York, Wiley, 1963.
3. J. Korkisch: *Modern Methods for the Separation of Rarer Metal Ions.* Elmsford, N.Y. Pergamon Press, 1969. (Includes ion-exchange and solvent extraction.)
4. W. Rieman and H. F. Walton: *Ion Exchange in Analytical Chemistry.* Elmsford, N.Y., Pergamon Press, 1970.
5. J. X. Khym: *Analytical Ion-Exchange Procedures in Chemistry and Biology.* Englewood Cliffs, N.J., Prentice-Hall, 1974.

Books Containing Chapters on Ion-Exchange Chromatography

1. L. Meites (ed.): *Handbook on Analytical Chemistry.* New York, McGraw-Hill, 1963.
2. J. J. Kirkland (ed.): *Modern Practice of Liquid Chromatography.* New York, Wiley-Interscience, 1971.
3. J. Marinsky and Y. Marcus (eds.): *Ion Exchange and Solvent Extraction: A Series of Advances.* Several volumes, published at intervals. New York, Marcel Dekker.
4. E. Heftmann (ed.): *Chromatography,* 3rd ed., New York, Van Nostrand Reinhold, 1975.

Review Articles

1. Inczedy, in *Reviews of Analytical Chemistry,* Vol. 1, p. 157. CRC Press, Cleveland, Ohio, 1972.
2. P. Jandera and J. Churacek, in *Journal of Chromatography:* Vol. 86, pp. 351–422 (1973)—ion-exchange chromatography of carboxylic acids; Vol. 86, pp. 423–449 (1973)—ion-exchange chromatography of sulfur compounds, phenols, phosphorus compounds, etc.: Vol. 98, pp. 1-54 (1974)—ion-exchange chromatography of nitrogen compounds; Vol. 98, pp. 55–104 (1974)—ion-exchange chromatography of aldehydes, ketones, alcohols, saccharides, etc.
3. H. F. Walton, in *Analytical Chemistry, Fundamental Reviews:* Vol. 46, 398R (1974); Vol. 44, 256R (1972); Vol. 42, 86R (1970); and in previous alternate years.

AUTHOR CITATION INDEX

SUBJECT INDEX